Calculus
Single Variable

Robert A.
Adams

Department of Mathematics
University of British Columbia

FIFTH EDITION

Addison
Wesley
Longman

Toronto

National Library of Canada Cataloguing in Publication Data

Adams, Robert A. (Robert Alexander), 1940-
    Single-variable calculus

5th ed.
Includes index.
ISBN 0-201-79805-0

   1. Calculus. I. Title

QA303.A32 2003   515   C2002-901083-7

ISBN 0-201-79805-0

Vice President, Editorial Director: Michael J. Young
Acquisitions Editor: Leslie Carson
Marketing Manager: Marlene Olsavsky
Developmental Editor: Meaghan Eley
Production Editor: Gillian Scobie
Copy Editor: Betty Robinson
Production Coordinator: Andrea Falkenberg
Permissions/Photo Research: Susan Wallace-Cox
Art Director: Mary Opper
Cover Image: © Art Zone/Photonica
Cover Design: Alex Li

2 3 4 5   07 06 05 04

Printed and bound in Canada.

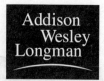

To Anne

Preface

This book consists of the first ten chapters (P–9) and the five appendices of the author's *Calculus: A Complete Course, 5th Edition*, that is, all the material on differentiation and integration of real-valued functions of a single real variable and the material on infinite sequences and series. A companion volume, *Calculus of Several Variables, 5th Edition* covers Chapters 9–16 and repeats the appendices. The main overlap is Chapter 9 on infinite sequences and series.

By the time a textbook reaches its fifth edition, it is to be hoped that the author has got some things right. Little change in the topics presented and their ordering has been made in this edition. The main changes are as follows:

- Much material on calculations with the computer algebra system (CAS) Maple has been added. See Appendix 5 for a list of topics and their locations. (This list covers all the chapters of *Calculus: A Complete Course.*) The additions range in length from brief examples to entire sections. However, it is not intended that this material constitute a lab course on using Maple (for which the author's colleague, Dr. Robert Israel, has written an excellent book, *Calculus: The Maple Way* also published under the Addison-Wesley logo). It is, however, intended to provide the student with some insight into the power of Maple to carry out both symbolic and numeric calculations in the context of calculus, and into the pitfalls that await the unwary user. This material is necessarily specific to the particular CAS Maple. Any attempt to make this material generic so that it would apply as well to Mathematica or Derive or even Matlab (which uses the Maple symbolic engine) would have either added greatly to the length or have rendered it useless.

- *Calculus: A Complete Course* no longer contains a separate chapter on differential equations. That chapter repeated some material from the body of the text and was inadequate for a whole course on the subject. Nevertheless, the close ties, both motivational and historic, between calculus and differential equations demand that some material on the latter be interspersed throughout the text, and this has resulted in a bit more material on differential equations in *Single-Variable Calculus.*

- Numerous small local changes have been made here and there to improve the text. Some material has been rewritten. Some awkward examples have been removed or replaced with more appropriate ones. Some exercises have been removed and others added.

Notwithstanding the changes described above, the main body of this text is actually a couple of pages shorter than the fourth edition. However, this decrease was not achieved at the expense of chopping interesting examples and applications. The process of making a book more reader friendly is always an ongoing one, and relies heavily on comments received from readers.

To the Student

When I took my first course in calculus there was no prescribed textbook but a book called *Calculus Made Easy* was recommended to those of us who felt we needed something beyond the notes we could take during lectures. I bought the book, hoping that it would live up to its title and I would have an easy time learning calculus. It didn't and I didn't.

Is calculus a very difficult subject? No, not really, but it sometimes seems that way to students, especially at the beginning, because of the new ideas and techniques involved, and because success in mastering calculus depends on having a very solid basis in precalculus mathematics (algebra, geometry, and trigonometry) to build upon. You may want to review the background material in Chapter P (Preliminaries) even if your instructor does not refer to it in class. Learning calculus will provide you with many useful tools for analyzing problems in numerous fields of interest, especially those regarded as "scientific." It is worth your while to acquire those tools, but, like any other worthwhile task, this one requires much effort on your part. No book or instructor can remove that requirement.

In writing this book I have tried to organize material in such a way as to make it as easy as possible, but not at the expense of "sweeping the real difficulties under the rug." You may find some of the concepts difficult to understand when they are first introduced. If so, *reread* the material slowly, if necessary several times; *think about it*; formulate questions to ask fellow students, your TA, or your instructor. Don't delay. It is important to resolve your problems as soon as possible. If you don't understand today's topic, you may not understand how it applies to tomorrow's either. Mathematics is a "linear discipline"; it builds from one idea to the next.

Doing exercises is the best way to deepen your understanding of calculus and to convince yourself that you understand it. There are numerous exercises in this text — too many for you to try them all. Some are "drill" exercises to help you develop your skills in calculation. More important, however, are the problems that develop reasoning skills and your ability to apply the techniques you have learned to concrete situations. In some cases you will have to plan your way through a problem that requires several diffferent "steps" before you can get to the answer. Other exercises are designed to extend the theory developed in the text and therefore enhance your understanding of the concepts of calculus.

The exercises vary greatly in difficulty. Usually, the more difficult ones occur towards the end of exercise sets, but these sets are not strictly graded in this way because exercises on a specific topic tend to be grouped together. Some exercises in the regular sets are marked with an asterisk "∗." This symbol indicates that the exercise is *either* somewhat more theoretical, *or* somewhat more difficult than most. The theoretical ones need not be difficult; sometimes they are quite easy. Most of the problems in the *Challenging Problems* section forming part of the *Chapter Review* at the end of most chapters are also on the difficult side though these are not typically marked with a "∗".

Do not be discouraged if you can't do *all* the exercises. Some are very difficult indeed, and only a few very gifted students will be able to do them. However, you should be able to do the vast majority of the exercises. Some will take much more effort than others. When you encounter difficulties with problems proceed as follows:

1. Read and reread the problem until you understand exactly what information it gives you and what you are expected to find or do.

2. If appropriate, draw a diagram illustrating the relationships between the quantities involved in the problem.

3. If necessary, introduce symbols to represent the quantities in the problem. Use appropriate letters (e.g., V for volume, t for time). Don't feel you have to use x or y for everything.

4. Develop a plan of attack. This is usually the hardest part. Look for known relationships; try to recognize patterns; look back at the worked examples in the current or relevant previous sections; try to think of possible connections between the problem at hand and others you have seen before. Can the problem be simplified by making some extra assumptions? If you can solve a simplified version it may help you decide what to do with the given problem. Can the problem be broken down into several cases, each of which is a simpler problem? When reading the examples in the text, be alert for methods that may turn out to be useful in other contexts later.

5. Try to carry out the steps in your plan. If you have trouble with one or more of them you may have to alter the plan.

6. When you arrive at an answer for a problem, always ask yourself whether it is *reasonable*. If it isn't, look back to determine places where you may have made an error.

Answers for most of the odd-numbered exercises are provided at the back of the book. Exceptions are exercises that don't have short answers, for example "Prove that ..." or "Show that ..." problems where the answer is the whole solution. A Student Solutions Manual that contains detailed solutions to even-numbered exercises is available.

Besides ∗ used to mark more difficult or theoretical problems, the following symbols are used to mark exercises of special types:

⬧   exercises pertaining to differential equations or initial-value problems. (It is not used in sections which are wholly concerned with DEs.)

▦   problems requiring the use of a calculator. Often a scientific calculator is needed. Some such problems may require a programmable calculator.

⌁   problems requiring the use of either a graphing calculator or mathematical graphing software on a personal computer.

▭   problems requiring the use of a computer. Typically, these will require either computer algebra software (e.g., Maple, Mathematica, Derive), or a spreadsheet program (e.g., Lotus 123, Microsoft Excel, Quattro Pro).

To the Instructor

As its title suggests, this book is intended to cover all the material usually encountered in the first year of a three- or four-semester real-variable calculus program — all the material involving real-valued functions of a single real variable. Specifically, differential calculus is covered in Chapters 1–4 and integral calculus in Chapters 5–8. A preliminary chapter (Chapter P) reviews those precalculus topics (the real line and Cartesian plane, functions, trigonometric functions and such) that are so essential for a successful initial encounter with calculus. The final chapter (Chapter 9) concerns sequences and series, and its position is rather arbitrary. This book consists of the first half of the author's *Calculus: a Complete Course, 5th Edition.* The second half of that book, dealing with multivariable and vector calculus, also appears as a separate volume, *Calculus of Several Variables, 5th Edition,* which repeats Chapter 9 before going on to the multivariable material.

Many of the chapters have too much material to cover in a normal course. You will have to pick and choose what to include and what to omit based on the background preparation, needs, and interests of your students. This is especially obvious in Chapter 7 which involves a wide variety of applications of integration to different disciplines.

The text is designed for general calculus courses, especially those for science and engineering students. Most of the material requires only a reasonable background in high school algebra and analytic geometry. (See Chapter P.) However, some optional material is more subtle and/or theoretical, and is intended mainly for stronger students. In particular, Appendices II and III explore the theoretical foundations of single-variable calculus, as do some of the Challenging Problems in the Chapter Reviews at the ends of the chapters.

Several supplements are available for use with *Single-Variable Calculus*:

- a **Web Site** www.pearsoned.ca/text/adams_calc
- an **Instructor's Solutions Manual** with detailed solutions to all the exercises, prepared by the author.
- an **Instructor's CRROM** with detailed solutions to all the exercises in .gif and .pdf format and software to enable instructors to publish solutions to specified exercises in HTML documents on their course web sites.
- a **Student Solutions Manual** with detailed solutions to all the even-numbered exercises, prepared by the author.
- a Maple lab manual, **Calculus: The Maple Way**, by Robert Israel (University of British Columbia) with instruction, examples, and problems dealing with the effective use of the computer algebra system Maple as a tool for calculus.

Acknowledgments

The first four editions of this material have been used for classes of general science, engineering, and mathematics majors and honours students at the University of British Columbia. I am grateful to many colleagues and students, at UBC and at many other institutions where these books have been used, for their encouragement and useful comments and suggestions.

In preparing the revision of this text I have had guidance from several dedicated reviewers who provided new insight and direction to my writing. I am especially grateful to the following people:

William J. Anderson	McGill University	Canada
Robert M. Coreless	University of Western Ontario	Canada
Poul G. Hjorth	Technical University of Denmark	Denmark
Jack Macki	University of Alberta	Canada
Mark MacLean	University of British Columbia	Canada
G. R. Nicklason	University College of Cape Breton	Canada
Sixten Nilsson	Linkoping University	Sweden
Viena Stastna	University of Calgary	Canada
Nader Zamani	University of Windsor	Canada

Special thanks go to Olov Johansson (Sweden) who checked the previous edition and the typeset version of this one, and who made many helpful suggestions.

Finally, I wish thank the sales and marketing staff of all Addison-Wesley (now Pearson Canada) divisions around the world for making the previous editions so successful, the editorial and production staff in Toronto, in particular Leslie Carson, Kelly Cochrane, Dave Ward, Pamela Voves, Meaghan Eley, and Gillian Scobie for their assistance and encouragement, and Betty Robinson for her careful copy editing of the typeset manuscript.

I typeset this volume using $\TeX$ and PostScript on a PC running Linux-Mandrake, and also generated all of the figures in PostScript using the mathematical graphics software package **MG** developed by my colleague Robert Israel and myself.

The expunging of errors and obscurities in a text is an ongoing and asymptotic process; hopefully each edition is better than the previous one. Nevertheless, some such imperfections always remain, and I will be grateful to any readers who call them to my attention, or give me any other suggestions for future improvements.

R.A.A.
Vancouver, Canada
April, 2002
adms@math.ubc.ca

Contents

What Is Calculus?

Early in the seventeenth century, the German mathematician Johannes Kepler analyzed a vast number of astronomical observations made by Danish astronomer Tycho Brahe and concluded that the planets must move around the sun in elliptical orbits. He didn't know why. Fifty years later the English mathematician and physicist Isaac Newton answered that question.

Why do the planets move in elliptical orbits around the sun? Why do hurricane winds spiral counterclockwise in the northern hemisphere? How can one predict the effects of interest rate changes on economies and stock markets? When will radioactive material be sufficiently decayed to enable safe handling? How do warm ocean currents in the equatorial Pacific affect the climate of eastern North America? How long will the concentration of a drug in the bloodstream remain at effective levels? How do radio waves propagate through space? Why does an epidemic spread faster and faster and then slow down? How can I be sure the bridge I designed won't be destroyed in a windstorm?

These and many other questions of interest and importance in our world relate directly to our ability to analyze motion and how quantities change with respect to time or each other. Algebra and geometry are useful tools for describing relationships among *static* quantities, but they do not involve concepts appropriate for describing how a quantity *changes*. For this we need new mathematical operations that go beyond the algebraic operations of addition, subtraction, multiplication, division, and the taking of powers and roots. We require operations that measure the way related quantities change.

Calculus provides the tools for describing motion quantitatively. It introduces two new operations called *differentiation* and *integration*, which, like addition and subtraction, are opposites of one another; what differentiation does, integration undoes.

For example, consider the motion of a falling rock. The height (in metres) of the rock t seconds after it is dropped from a height of h_0 m is a function $h(t)$ given by

$$h(t) = h_0 - 4.9t^2.$$

The graph of $y = h(t)$ is shown in the figure below:

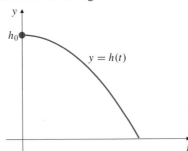

The process of differentiation enables us to find a new function, which we denote $h'(t)$ and call *the derivative* of h with respect to t, that represents the *rate of change* of the height of the rock, that is, its *velocity* in metres/second:

$$h'(t) = -9.8t.$$

Conversely, if we know the velocity of the falling rock as a function of time, integration enables us to find the height function $h(t)$.

Calculus was invented independently and in somewhat different ways by two seventeenth-century mathematicians, Isaac Newton and Gottfried Wilhelm Leibniz. Newton's motivation was a desire to analyze the motion of moving objects. Using his calculus he was able to formulate his laws of motion and gravitation and to *conclude from them* that the planets must move around the sun in elliptical orbits.

Many of the most fundamental and important "laws of nature" are conveniently expressed as equations involving rates of change of quantities. Such equations are called *differential equations*, and techniques for their study and solution are at the heart of calculus. In the falling rock example the appropriate law is Newton's Second Law of Motion:

$$\text{force} = \text{mass} \times \text{acceleration}.$$

The *acceleration*, -9.8 m/s^2, is the rate of change (the *derivative*) of the velocity, which is in turn the rate of change (the *derivative*) of the height function.

Much of mathematics is related indirectly to the study of motion. We regard *lines*, or *curves*, as geometric objects, but the ancient Greeks thought of them as paths traced out by moving points. Nevertheless, the study of curves also involves geometric concepts such as tangency and area. The process of differentiation is closely tied to the geometric problem of finding tangent lines; similarly, integration is related to the geometric problem of finding areas of regions with curved boundaries.

Both differentiation and integration are defined in terms of a new mathematical operation called a **limit**. The concept of the limit of a function will be developed in Chapter 1. That will be the real beginning of our study of calculus. In the chapter called Preliminaries, we will review some of the background from algebra and geometry needed for the development of calculus that follows.

CHAPTER P
Preliminaries

Introduction This preliminary chapter reviews the most important things you should know before beginning calculus. Topics include the real number system, Cartesian coordinates in the plane, equations representing straight lines, circles, and parabolas, functions and their graphs, and, in particular, the trigonometric functions.

Depending on your pre-calculus background, you may or may not be familiar with these topics. If you are, you may want to skim over this material to refresh your understanding of the terms used; if not, you should study this chapter in detail.

P.1 Real Numbers and the Real Line

Calculus depends on properties of the real number system. **Real numbers** are numbers that can be expressed as decimals, for example,

$$5 = 5.00000\ldots$$
$$-\tfrac{3}{4} = -0.750000\ldots$$
$$\tfrac{1}{3} = 0.3333\ldots$$
$$\sqrt{2} = 1.4142\ldots$$
$$\pi = 3.14159\ldots$$

In each case the three dots ... indicate that the sequence of decimal digits goes on forever. For the first three numbers above, the patterns of the digits are obvious; we know what all the subsequent digits are. For $\sqrt{2}$ and π there are no obvious patterns.

The real numbers can be represented geometrically as points on a number line, which we call the **real line**, shown in Figure P.1. The symbol $\mathbb{R}$ is used to denote either the real number system or, equivalently, the real line.

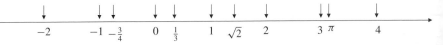

Figure P.1   The real line

The properties of the real number system fall into three categories: algebraic properties, order properties, and completeness. You are already familiar with the *algebraic properties*; roughly speaking, they assert that real numbers can be added, subtracted, multiplied, and divided (except by zero) to produce more real numbers and that the usual rules of arithmetic are valid.

The *order properties* of the real numbers refer to the order in which the numbers appear on the real line. If x lies to the left of y, then we say that "x is less than y" or "y is greater than x." These statements are written symbolically as $x < y$ and $y > x$, respectively. The inequality $x \le y$ means that either $x < y$ or $x = y$. The order properties of the real numbers are summarized in the following *rules for inequalities*:

The symbol $\Longrightarrow$ means "implies."

Rules for inequalities

If a, b, and c are real numbers, then:

1. $a < b \quad\Longrightarrow\quad a + c < b + c$
2. $a < b \quad\Longrightarrow\quad a - c < b - c$
3. $a < b$ and $c > 0 \quad\Longrightarrow\quad ac < bc$
4. $a < b$ and $c < 0 \quad\Longrightarrow\quad ac > bc$; in particular, $-a > -b$
5. $a > 0 \quad\Longrightarrow\quad \dfrac{1}{a} > 0$
6. $0 < a < b \quad\Longrightarrow\quad \dfrac{1}{b} < \dfrac{1}{a}$

Rules 1–4 and 6 (for $a > 0$) also hold if $<$ and $>$ are replaced by $\leq$ and $\geq$.

Note especially the rules for multiplying (or dividing) an inequality by a number. If the number is positive, the inequality is preserved; if the number is negative, the inequality is reversed.

The *completeness* property of the real number system is more subtle and difficult to understand. One way to state it is as follows: if A is any set of real numbers having at least one number in it, and if there exists a real number y with the property that $x \leq y$ for every x in A, then there exists a *smallest* number y with the same property. Roughly speaking, this says that there can be no holes or gaps on the real line—every point corresponds to a real number. We will not need to deal much with completeness in our study of calculus. It is typically used to prove certain important results, in particular, Theorems 8 and 9 in Chapter 1. (These proofs are given in Appendix II but are not usually included in elementary calculus courses; they are studied in more advanced courses in mathematical analysis.) Only when we study infinite sequences and series in Chapter 9 will we encounter any direct use of completeness.

The set of real numbers has some important special subsets:

(i) the **natural numbers** or **positive integers**, namely, the numbers 1, 2, 3, 4, ...

(ii) the **integers**, namely, the numbers 0, ± 1, ± 2, ± 3, ...

(iii) the **rational numbers**, that is, numbers that can be expressed in the form of a fraction m/n, where m and n are integers, and $n \neq 0$.

The rational numbers are precisely those real numbers with decimal expansions that are either:

(a) terminating (that is, ending with an infinite string of zeros), for example, $3/4 = 0.750000\ldots$, or

(b) repeating (that is, ending with a string of digits that repeats over and over), for example, $23/11 = 2.090909\ldots = 2.\overline{09}$. (The bar indicates the pattern of repeating digits.)

Real numbers that are not rational are called *irrational numbers*.

Example 1 Show that each of the numbers (a) $1.323232\cdots = 1.\overline{32}$ and (b) $0.3405405405\ldots = 0.3\overline{405}$ is a rational number by expressing it as a quotient of two integers.

Solution

(a) Let $x = 1.323232\ldots$ Then $x - 1 = 0.323232\ldots$ and

$$100x = 132.323232\ldots = 132 + 0.323232\ldots = 132 + x - 1.$$

Therefore, $99x = 131$ and $x = 131/99$.

(b) Let $y = 0.3405405405\ldots$ Then $10y = 3.405405405\ldots$ and $10y - 3 = 0.405405405\ldots$ Also,

$$10000y = 3405.405405405\ldots = 3405 + 10y - 3.$$

Therefore, $9990y = 3402$ and $y = 3402/9990 = 63/185$.

The set of rational numbers possesses all the algebraic and order properties of the real numbers but not the completeness property. There is, for example, no rational number whose square is 2. Hence, there is a "hole" on the "rational line" where $\sqrt{2}$ should be.[1] Because the real line has no such "holes," it is the appropriate setting for studying limits and therefore calculus.

Intervals

A subset of the real line is called an **interval** if it contains at least two numbers and also contains all real numbers between any two of its elements. For example, the set of real numbers x such that $x > 6$ is an interval, but the set of real numbers y such that $y \neq 0$ is not an interval. (Why?) It consists of two intervals.

If a and b are real numbers and $a < b$, we often refer to

(i) the **open interval** from a to b, denoted by $]a, b[$, consisting of all real numbers x satisfying $a < x < b$.

(ii) the **closed interval** from a to b, denoted by $[a, b]$, consisting of all real numbers x satisfying $a \leq x \leq b$.

(iii) the **half-open interval** $[a, b[$, consisting of all real numbers x satisfying the inequalities $a \leq x < b$.

(iv) the **half-open interval** $]a, b]$, consisting of all real numbers x satisfying the inequalities $a < x \leq b$.

open interval $]a, b[$

closed interval $[a, b]$

half-open interval $[a, b[$

half-open interval $]a, b]$

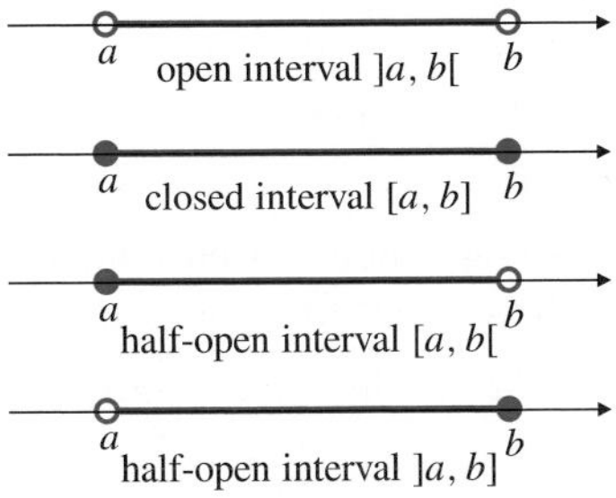

Figure P.2   Finite intervals

These are illustrated in Figure P.2. Note the use of reversed square brackets $]$ and $[$ and hollow dots to indicate endpoints of intervals that are not included in the intervals, and square brackets $[$ and $]$ and solid dots to indicate endpoints that are

---

[1]   How do we know that $\sqrt{2}$ is irrational? Suppose, to the contrary, that $\sqrt{2}$ is rational. Then $\sqrt{2} = m/n$, where m and n are integers and $n \neq 0$. We can assume that the fraction m/n has been "reduced to lowest terms"; any common factors have been cancelled out. Now $m^2/n^2 = 2$, so $m^2 = 2n^2$, which is an even integer. Hence m must also be even. (The square of an odd integer is always odd.) Since m is even, we can write $m = 2k$, where k is an integer. Thus $4k^2 = 2n^2$ and $n^2 = 2k^2$, which is even. Thus n is also even. This contradicts the assumption that $\sqrt{2}$ could be written as a fraction m/n in lowest terms; m and n cannot both be even. Accordingly, there can be no rational number whose square is 2.

included. The endpoints of an interval are also called **boundary points**.

The intervals in Figure P.2 are **finite intervals**; each of them has finite length $b - a$. Intervals can also have infinite length, in which case they are called **infinite intervals**. Figure P.3 shows some examples of infinite intervals. Note that the whole real line $\mathbb{R}$ is an interval, denoted by $]-\infty, \infty[$. The symbol ∞ ("infinity") does *not* denote a real number, so we never include it in an interval.

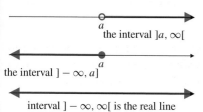

the interval $]a, \infty[$

the interval $]-\infty, a]$

interval $]-\infty, \infty[$ is the real line

Figure P.3   Infinite intervals

Example 2   Solve the following inequalities. Express the solution sets in terms of intervals and graph them.

(a) $2x - 1 > x + 3$
(b) $-\dfrac{x}{3} \geq 2x - 1$
(c) $\dfrac{2}{x - 1} \geq 5$

Solution

(a) $2x - 1 > x + 3$ Add 1 to both sides.

$\qquad 2x > x + 4$ Subtract x from both sides.

$\qquad\quad x > 4$ The solution set is the interval $]4, \infty[$.

(b) $-\dfrac{x}{3} \geq 2x - 1$ Multiply both sides by -3.

$\qquad x \leq -6x + 3$ Add $6x$ to both sides.

$\qquad 7x \leq 3$ Divide both sides by 7.

$\qquad\quad x \leq \dfrac{3}{7}$ The solution set is the interval $]-\infty, 3/7]$.

(c)   We transpose the 5 to the left side and simplify to rewrite the given inequality in an equivalent form:

$$\frac{2}{x-1} - 5 \geq 0 \quad\Longleftrightarrow\quad \frac{2 - 5(x-1)}{x-1} \geq 0 \quad\Longleftrightarrow\quad \frac{7 - 5x}{x-1} \geq 0.$$

The fraction $\dfrac{7 - 5x}{x - 1}$ is undefined at $x = 1$ and is 0 at $x = 7/5$. Between these numbers it is positive if the numerator and denominator have the same sign, and negative if they have opposite sign. It is easiest to organize this sign information in a chart:

The symbol $\Longleftrightarrow$ means "if and only if," or "is equivalent to."

x			1		7/5	
$7 - 5x$	$+$	$+$	$+$		0	$-$
$x - 1$	$-$		0	$+$	$+$	$+$
$(7 - 5x)/(x - 1)$	$-$		undef	$+$	0	$-$

Thus the solution set of the given inequality is the interval $]1, 7/5]$.

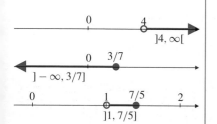

$]4, \infty[$

$]-\infty, 3/7]$

$]1, 7/5]$

Figure P.4   The intervals for Example 2

Sometimes we will need to solve systems of two or more inequalities that must be satisfied simultaneously. We still solve the inequalities individually and look for numbers in the intersection of the solution sets.

Example 3 Solve the systems of inequalities:
(a) $3 \le 2x + 1 \le 5$ (b) $3x - 1 < 5x + 3 \le 2x + 15$.

Solution

(a) Using the technique of Example 2, we can solve the inequality $3 \le 2x + 1$ to get $2 \le 2x$, so $x \ge 1$. Similarly, the inequality $2x + 1 \le 5$ leads to $2x \le 4$, so $x \le 2$. The solution set of system (a) is therefore the closed interval $[1, 2]$.

(b) We solve both inequalities as follows:

$$
\left.
\begin{aligned}
3x - 1 &< 5x + 3 \\
-1 - 3 &< 5x - 3x \\
-4 &< 2x \\
-2 &< x
\end{aligned}
\right\}
\quad \text{and} \quad
\left\{
\begin{aligned}
5x + 3 &\le 2x + 15 \\
5x - 2x &\le 15 - 3 \\
3x &\le 12 \\
x &\le 4
\end{aligned}
\right.
$$

The solution set is the interval $]-2, 4]$.

∎

Solving quadratic inequalities depends on solving the corresponding quadratic equations.

Example 4 **Quadratic inequalities**
Solve: (a) $x^2 - 5x + 6 < 0$ (b) $2x^2 + 1 > 4x$.

Solution

(a) The trinomial $x^2 - 5x + 6$ factors into the product $(x - 2)(x - 3)$, which is negative if and only if exactly one of the factors is negative. Since $x - 3 < x - 2$, this happens when $x - 3 < 0$ and $x - 2 > 0$. Thus we need $x < 3$ and $x > 2$; the solution set is the open interval $]2, 3[$.

(b) The inequality $2x^2 + 1 > 4x$ is equivalent to $2x^2 - 4x + 1 > 0$. The corresponding quadratic equation $2x^2 - 4x + 1 = 0$, which is of the form $Ax^2 + Bx + C = 0$, can be solved by the quadratic formula:

$$
x = \frac{-B \pm \sqrt{B^2 - 4AC}}{2A} = \frac{4 \pm \sqrt{16 - 8}}{4} = 1 \pm \frac{\sqrt{2}}{2},
$$

so the given inequality can be expressed in the form

$$
\left(x - 1 + \tfrac{1}{2}\sqrt{2} \right) \left(x - 1 - \tfrac{1}{2}\sqrt{2} \right) > 0.
$$

This is satisfied if both factors on the left side are positive or if both are negative. Therefore, we require that either $x < 1 - \tfrac{1}{2}\sqrt{2}$ or $x > 1 + \tfrac{1}{2}\sqrt{2}$. The solution set is the *union* of intervals $\left]-\infty, 1 - \tfrac{1}{2}\sqrt{2}\right[\cup \left]1 + \tfrac{1}{2}\sqrt{2}, \infty\right[$.

∎

Note the use of the symbol $\cup$ to denote the **union** of intervals. A real number is in the union of intervals if it is in at least one of the intervals. We will also need to consider the **intersection** of intervals from time to time. A real number belongs to the intersection of intervals if it belongs to *every one* of the intervals. We will use $\cap$ to denote intersection. For example,

$$
[1, 3[\cap [2, 4] = [2, 3[\quad \text{while} \quad [1, 3[\cup [2, 4] = [1, 4].
$$

Example 5   Solve the inequality $\dfrac{3}{x-1} < -\dfrac{2}{x}$ and graph the solution set.

Solution   We would like to multiply by $x(x-1)$ to clear the inequality of fractions, but this would require considering three cases separately. (What are they?) Instead, we will transpose and combine the two fractions into a single one:

$$\frac{3}{x-1} < -\frac{2}{x} \quad \Longleftrightarrow \quad \frac{3}{x-1} + \frac{2}{x} < 0 \quad \Longleftrightarrow \quad \frac{5x-2}{x(x-1)} < 0.$$

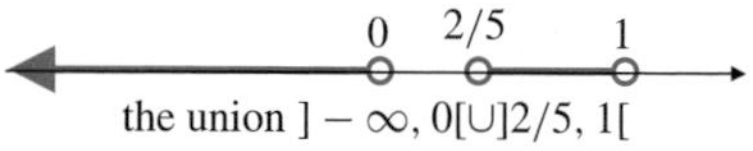

We examine the signs of the three factors in the left fraction to determine where that fraction is negative:

x			0		$2/5$		1	
$5x-2$	$-$	$-$	$-$	0	$+$		$+$	$+$
x	$-$	0	$+$	$+$	$+$		$+$	$+$
$x-1$	$-$	$-$	$-$	$-$	$-$		0	$+$
$\dfrac{5x-2}{x(x-1)}$	$-$	undef	$+$	0	$-$		undef	$+$

The solution set of the given inequality is the union of these two intervals, namely $]-\infty, 0[\cup]2/5, 1[$. See Figure P.5.

Figure P.5   The solution set for Example 5

the union $]-\infty, 0[\cup]2/5, 1[$

The Absolute Value

The **absolute value**, or **magnitude**, of a number x, denoted $|x|$ (read "the absolute value of x"), is defined by the formula

$$|x| = \begin{cases} x, & \text{if } x \geq 0 \\ -x, & \text{if } x < 0 \end{cases}$$

The vertical lines in the symbol $|x|$ are called **absolute value bars**.

Example 6 $|3| = 3, \quad |0| = 0, \quad |-5| = 5.$

Note that $|x| \geq 0$ for every real number x, and $|x| = 0$ only if $x = 0$. People sometimes find it confusing to say that $|x| = -x$ when x is negative, but this is correct since $-x$ is positive in that case. The symbol $\sqrt{a}$ always denotes the *nonnegative* square root of a, so an alternative definition of $|x|$ is $|x| = \sqrt{x^2}$.

Geometrically, $|x|$ represents the (nonnegative) distance from x to 0 on the real line. More generally, $|x - y|$ represents the (nonnegative) distance between the points x and y on the real line, since this distance is the same as that from the point $x - y$ to 0 (see Figure P.6):

It is important to remember that $\sqrt{a^2} = |a|$. Do not write $\sqrt{a^2} = a$ unless you already know that $a \geq 0$.

$$|x - y| = \begin{cases} x - y, & \text{if } x \geq y \\ y - x, & \text{if } x < y. \end{cases}$$

Figure P.6

$|x - y|$ = distance from x to y

The absolute value function has the following properties:

> **Properties of absolute values**
>
> 1. $|-a| = |a|$. A number and its negative have the same absolute value.
> 2. $|ab| = |a||b|$ and $\left|\dfrac{a}{b}\right| = \dfrac{|a|}{|b|}$. The absolute value of a product (or quotient) of two numbers is the product (or quotient) of their absolute values.
> 3. $|a \pm b| \leq |a| + |b|$ **(the triangle inequality)**. The absolute value of a sum of or difference between numbers is less than or equal to the sum of their absolute values.

The first two of these properties can be checked by considering the cases where either of a or b is either positive or negative. The third property follows from the first two because $\pm 2ab \leq |2ab| = 2|a||b|$. Therefore, we have

$$|a \pm b|^2 = (a \pm b)^2 = a^2 \pm 2ab + b^2$$
$$\leq |a|^2 + 2|a||b| + |b|^2 = (|a| + |b|)^2,$$

and taking the (positive) square roots of both sides we obtain $|a \pm b| \leq |a| + |b|$. This result is called the "triangle inequality" because it follows from the geometric fact that the length of any side of a triangle cannot exceed the sum of the lengths of the other two sides. For instance, if we regard the points 0, a, and b on the number line as the vertices of a degenerate "triangle," then the sides of the triangle have lengths $|a|$, $|b|$, and $|a - b|$. The triangle is degenerate since all three of its vertices lie on a straight line.

Equations and Inequalities Involving Absolute Values

The equation $|x| = D$ (where $D > 0$) has two solutions, $x = D$ and $x = -D$: the two points on the real line that lie at distance D from the origin. Equations and inequalities involving absolute values can be solved algebraically by breaking them into cases according to the definition of absolute value, but often they can also be solved geometrically by interpreting absolute values as distances. For example, the inequality $|x - a| < D$ says that the distance from x to a is less than D, so x must lie between $a - D$ and $a + D$. (Or, equivalently, a must lie between $x - D$ and $x + D$.) If D is a positive number, then

$\|x\| = D$	$\Longleftrightarrow$	either $x = -D$ or $x = D$
$\|x\| < D$	$\Longleftrightarrow$	$-D < x < D$
$\|x\| \leq D$	$\Longleftrightarrow$	$-D \leq x \leq D$
$\|x\| > D$	$\Longleftrightarrow$	either $x < -D$ or $x > D$

More generally,

$$
\begin{aligned}
|x - a| = D &\iff \text{either } x = a - D \text{ or } x = a + D \\
|x - a| < D &\iff a - D < x < a + D \\
|x - a| \le D &\iff a - D \le x \le a + D \\
|x - a| > D &\iff \text{either } x < a - D \text{ or } x > a + D
\end{aligned}
$$

Example 7   Solve: (a) $|2x + 5| = 3$ (b) $|3x - 2| \le 1$.

Solution

(a) $|2x + 5| = 3 \iff 2x + 5 = \pm 3$. Thus, either $2x = -3 - 5 = -8$ or $2x = 3 - 5 = -2$. The solutions are $x = -4$ and $x = -1$.

(b) $|3x - 2| \le 1 \iff -1 \le 3x - 2 \le 1$. We solve this pair of inequalities:

$$
\left\{
\begin{aligned}
-1 &\le 3x - 2 \\
-1 + 2 &\le 3x \\
1/3 &\le x
\end{aligned}
\right\}
\quad \text{and} \quad
\left\{
\begin{aligned}
3x - 2 &\le 1 \\
3x &\le 1 + 2 \\
x &\le 1
\end{aligned}
\right\}
$$

Thus the solutions lie in the interval $[1/3, 1]$.

Remark   Here is how part (b) of Example 7 could have been solved geometrically, by interpreting the absolute value as a distance:

$$
|3x - 2| = \left| 3\left(x - \frac{2}{3}\right)\right| = 3\left|x - \frac{2}{3}\right|.
$$

Thus the given inequality says that

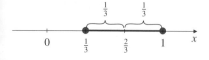

$$
3\left|x - \frac{2}{3}\right| \le 1 \quad \text{or} \quad \left|x - \frac{2}{3}\right| \le \frac{1}{3}.
$$

Figure P.7   The solution set for Example 7(b)

This says that the distance from x to $2/3$ does not exceed $1/3$. The solutions x therefore lie between $1/3$ and 1, including both of these endpoints. (See Figure P.7.)

Example 8   Solve the equation $|x + 1| = |x - 3|$.

Solution   The equation says that x is equidistant from -1 and 3. Therefore x is the point halfway between -1 and 3; $x = (-1 + 3)/2 = 1$. Alternatively, the given equation says that either $x + 1 = x - 3$ or $x + 1 = -(x - 3)$. The first of these equations has no solutions; the second has the solution $x = 1$.

Example 9   What values of x satisfy the inequality $\left| 5 - \dfrac{2}{x} \right| < 3$?

Solution   We have

$$\left|5 - \frac{2}{x}\right| < 3 \quad \Longleftrightarrow \quad -3 < 5 - \frac{2}{x} < 3 \qquad \text{Subtract 5 from each member.}$$

$$-8 < -\frac{2}{x} < -2 \qquad \text{Divide each member by } -2.$$

$$4 > \frac{1}{x} > 1 \qquad \text{Take reciprocals.}$$

$$\frac{1}{4} < x < 1$$

In this calculation we manipulated a system of two inequalities simultaneously, rather than split it up into separate inequalities as we have done in previous examples. Note how the various rules for inequalities were used here. Multiplying an inequality by a negative number reverses the inequality. So does taking reciprocals of an inequality in which both sides are positive. The given inequality holds for all x in the open interval $]1/4, 1[$. ∎

▎Exercises P.1

In Exercises 1–2, express the given rational number as a repeating decimal. Use a bar to indicate the repeating digits.

1. $\dfrac{2}{9}$ **2.** $\dfrac{1}{11}$

In Exercises 3–4, express the given repeating decimal as a quotient of integers in lowest terms.

3. $0.\overline{12}$ **4.** $3.2\overline{7}$

5. Express the rational numbers $1/7$, $2/7$, $3/7$, and $4/7$ as repeating decimals. (Use a calculator to give as many decimal digits as possible.) Do you see a pattern? Guess the decimal expansions of $5/7$ and $6/7$ and check your guesses.

6. Can two different decimals represent the same number? What number is represented by $0.999\ldots = 0.\overline{9}$?

In Exercises 7–12, express the set of all real numbers x satisfying the given conditions as an interval or a union of intervals.

7. $x \geq 0$ and $x \leq 5$ **8.** $x < 2$ and $x \geq -3$

9. $x > -5$ or $x < -6$ **10.** $x \leq -1$

11. $x > -2$ **12.** $x < 4$ or $x \geq 2$

In Exercises 13–26, solve the given inequality, giving the solution set as an interval or union of intervals.

13. $-2x > 4$ **14.** $3x + 5 \leq 8$

15. $5x - 3 \leq 7 - 3x$ **16.** $\dfrac{6 - x}{4} \geq \dfrac{3x - 4}{2}$

17. $3(2 - x) < 2(3 + x)$ **18.** $x^2 < 9$

19. $\dfrac{1}{2 - x} < 3$ **20.** $\dfrac{x + 1}{x} \geq 2$

21. $x^2 - 2x \leq 0$ **22.** $6x^2 - 5x \leq -1$

23. $x^3 > 4x$ **24.** $x^2 - x \leq 2$

25. $\dfrac{x}{2} \geq 1 + \dfrac{4}{x}$ **26.** $\dfrac{3}{x - 1} < \dfrac{2}{x + 1}$

Solve the equations in Exercises 27–32.

27. $|x| = 3$ **28.** $|x - 3| = 7$

29. $|2t + 5| = 4$ **30.** $|1 - t| = 1$

31. $|8 - 3s| = 9$ **32.** $\left|\dfrac{s}{2} - 1\right| = 1$

In Exercises 33–40, write the interval defined by the given inequality.

33. $|x| < 2$ **34.** $|x| \leq 2$

35. $|s - 1| \leq 2$ **36.** $|t + 2| < 1$

37. $|3x - 7| < 2$ **38.** $|2x + 5| < 1$

39. $\left|\dfrac{x}{2} - 1\right| \leq 1$ **40.** $\left|2 - \dfrac{x}{2}\right| < \dfrac{1}{2}$

In Exercises 41–42, solve the given inequality by interpreting it as a statement about distances on the real line.

41. $|x + 1| > |x - 3|$ **42.** $|x - 3| < 2|x|$

43. Do not fall into the trap $|-a| = a$. For what real numbers a is this equation true? For what numbers is it false?

44. Solve the equation $|x - 1| = 1 - x$.

45. Show that the inequality

$$|a - b| \geq \big||a| - |b|\big|$$

holds for all real numbers a and b.

P.2 Cartesian Coordinates in the Plane

The positions of all points in a plane can be measured with respect to two perpendicular real lines in the plane intersecting at the 0-point of each. These lines are called **coordinate axes** in the plane. Usually (but not always) we call one of these axes the x-axis and draw it horizontally with numbers x on it increasing to the right; we call the other the y-axis, and draw it vertically with numbers y on it increasing upward. The point of intersection of the coordinate axes (the point where x and y are both zero) is called **the origin** and is often denoted by the letter O.

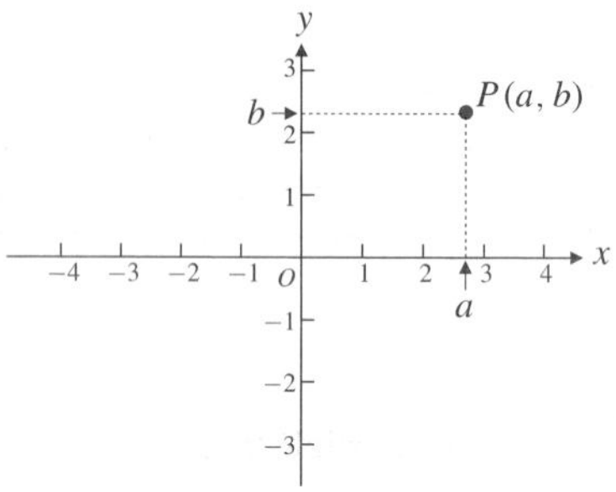

If P is any point in the plane we can draw a line through P perpendicular to the x-axis. If a is the value of x where that line intersects the x-axis we call a the **x-coordinate** of P. Similarly, the **y-coordinate** of P is the value of y where a line through P perpendicular to the y-axis meets the y-axis. The **ordered pair** (a, b) is called the **coordinate pair**, or the **Cartesian coordinates**, of the point P. We refer to the point as $P(a, b)$ to indicate both the name P of the point, and its coordinates (a, b). (See Figure P.8.) Note that the x-coordinate appears first in a coordinate pair. Coordinate pairs are in one-to-one correspondence with points in the plane; each point has a unique coordinate pair, and each coordinate pair determines a unique point. We call such a set of coordinate axes and the coordinate pairs they determine a **Cartesian coordinate system** in the plane, after the seventeenth-century philosopher René Descartes, who created analytic (coordinate) geometry. When equipped with such a coordinate system, a plane is called a **Cartesian plane**.

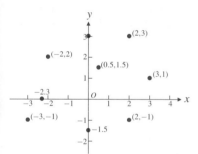

Figure P.8 The coordinate axes and the point P with coordinates (a, b)

Figure P.9 shows the coordinates of some points in the plane. Note that all points on the x-axis have y-coordinate 0. We usually just write the x-coordinates to label such points. Similarly, points on the y-axis have $x = 0$, and we can label such points using their y-coordinates only.

The coordinate axes divide the plane into four regions called **quadrants**. These quadrants are numbered I to IV, as shown in Figure P.10. The **first quadrant** is the upper right one; both coordinates of any point in that quadrant are positive numbers. Both coordinates are negative in quadrant III; only y is positive in quadrant II; only x is positive in quadrant IV.

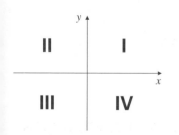

Figure P.9 Some points with their coordinates

Axis Scales

When we plot data in the coordinate plane or graph formulas whose variables have different units of measure, we do not need to use the same scale on the two axes. If, for example, we plot height versus time for a falling rock, there is no reason to place the mark that shows 1 m on the height axis the same distance from the origin as the mark that shows 1 s on the time axis.

When we graph functions whose variables do not represent physical measurements and when we draw figures in the coordinate plane to study their geometry or trigonometry, we make the scales identical. A vertical unit of distance then looks the same as a horizontal unit. As on a surveyor's map or a scale drawing, line segments that are supposed to have the same length will look as if they do, and angles that are supposed to be equal will look equal. Some of the geometric results we obtain later, such as the relationship between the slopes of perpendicular lines, are valid only if equal scales are used on the two axes.

Figure P.10 The four quadrants

Computer and calculator displays are another matter. The vertical and horizontal scales on machine-generated graphs usually differ, with resulting distortions in distances, slopes, and angles. Circles may appear elliptical and squares may appear rectangular or even as parallelograms. Right angles may appear as acute or obtuse. Circumstances like these require us to take extra care in interpreting what we see. High-quality computer software for drawing Cartesian graphs usually allows the user to compensate for such scale problems by adjusting the *aspect ratio* (the ratio of vertical to horizontal scale). Some computer screens also allow adjustment within a narrow range. When using graphing software, try to adjust your particular software/hardware configuration so that the horizontal and vertical diameters of a drawn circle appear to be equal.

Increments and Distances

When a particle moves from one point to another, the net changes in its coordinates are called increments. They are calculated by subtracting the coordinates of the starting point from the coordinates of the ending point. An **increment** in a variable is the net change in the value of the variable. If x changes from x_1 to x_2, then the increment in x is $\Delta x = x_2 - x_1$.

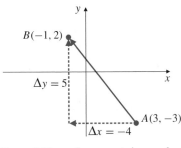

Figure P.11   Increments in x and y

Example 1   Find the increments in the coordinates of a particle that moves from $A(3, -3)$ to $B(-1, 2)$.

Solution   The increments (see Figure P.11) are:

$$\Delta x = -1 - 3 = -4 \quad \text{and} \quad \Delta y = 2 - (-3) = 5.$$

If $P(x_1, y_1)$ and $Q(x_2, y_2)$ are two points in the plane, the straight line segment PQ is the hypotenuse of a right triangle PCQ, as shown in Figure P.12. The legs PC and CQ of the triangle have lengths

$$|\Delta x| = |x_2 - x_1| \quad \text{and} \quad |\Delta y| = |y_2 - y_1|.$$

These are the *horizontal distance* and *vertical distance* between P and Q. By the Pythagorean theorem, the length of PQ is the square root of the sum of the squares of these leg lengths.

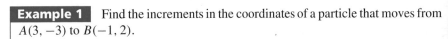

Distance formula for points in the plane

The distance D between $P(x_1, y_1)$ and $Q(x_2, y_2)$ is

$$D = \sqrt{(\Delta x)^2 + (\Delta y)^2} = \sqrt{(x_2 - x_1)^2 + (y_2 - y_1)^2}.$$

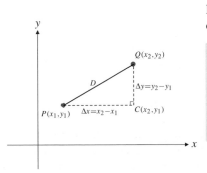

Figure P.12   The distance from P to Q is $D = \sqrt{(x_2 - x_1)^2 + (y_2 - y_1)^2}$

Example 2   The distance between $A(3, -3)$ and $B(-1, 2)$ in Figure P.11 is

$$\sqrt{(-1 - 3)^2 + (2 - (-3))^2} = \sqrt{(-4)^2 + 5^2} = \sqrt{41} \text{ units.}$$

Example 3 The distance from the origin $O(0, 0)$ to a point $P(x, y)$ is

$$\sqrt{(x - 0)^2 + (y - 0)^2} = \sqrt{x^2 + y^2}.$$

Graphs

The **graph** of an equation (or inequality) involving the variables x and y is the set of all points $P(x, y)$ whose coordinates satisfy the equation (or inequality).

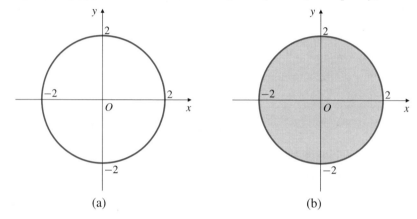

Figure P.13

(a) The circle $x^2 + y^2 = 4$

(b) The disk $x^2 + y^2 \le 4$

(a)   (b)

Example 4 The equation $x^2 + y^2 = 4$ represents all points $P(x, y)$ whose distance from the origin is $\sqrt{x^2 + y^2} = \sqrt{4} = 2$. These points lie on the **circle** of radius 2 centred at the origin. This circle is the graph of the equation $x^2 + y^2 = 4$. (See Figure P.13(a).)

Example 5 Points (x, y) whose coordinates satisfy the inequality $x^2 + y^2 \le 4$ all have distance ≤ 2 from the origin. The graph of the inequality is therefore the disk of radius 2 centred at the origin. (See Figure P.13(b).)

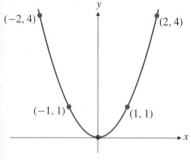

Figure P.14 The parabola $y = x^2$

Example 6 Consider the equation $y = x^2$. Some points whose coordinates satisfy this equation are $(0, 0)$, $(1, 1)$, $(-1, 1)$, $(2, 4)$, and $(-2, 4)$. These points (and all others satisfying the equation) lie on a smooth curve called a **parabola**. (See Figure P.14.)

Straight Lines

Given two points $P_1(x_1, y_1)$ and $P_2(x_2, y_2)$ in the plane, we call the increments $\Delta x = x_2 - x_1$ and $\Delta y = y_2 - y_1$ respectively the **run** and the **rise** between P_1 and P_2. Two such points always determine a unique **straight line** (usually called simply a **line**) passing through them both. We call the line $P_1 P_2$.

Any nonvertical line in the plane has the property that the ratio

$$m = \frac{\text{rise}}{\text{run}} = \frac{\Delta y}{\Delta x} = \frac{y_2 - y_1}{x_2 - x_1}$$

has the *same value* for every choice of two distinct points $P_1(x_1, y_1)$ and $P_2(x_2, y_2)$ on the line. (See Figure P.15.) The constant $m = \Delta y/\Delta x$ is called the **slope** of the nonvertical line.

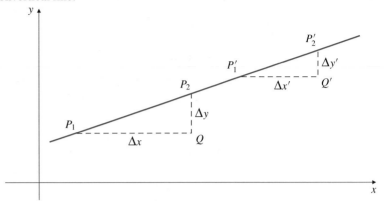

Figure P.15 $\Delta y/\Delta x = \Delta y'/\Delta x'$ because triangles $P_1 Q P_2$ and $P_1' Q' P_2'$ are similar

---

Example 7   The slope of the line joining $A\,(3, -3)$ and $B\,(-1, 2)$ is

$$m = \frac{\Delta y}{\Delta x} = \frac{2 - (-3)}{-1 - 3} = \frac{5}{-4} = -\frac{5}{4}.$$

■

The slope tells us the direction and steepness of a line. A line with positive slope rises uphill to the right; one with negative slope falls downhill to the right. The greater the absolute value of the slope, the steeper the rise or fall. Since the run Δx is zero for a vertical line, we cannot form the ratio m; the slope of a vertical line is *undefined*.

The direction of a line can also be measured by an angle. The **inclination** of a line is the smallest counterclockwise angle from the positive direction of the x-axis to the line. In Figure P.16 the angle ϕ (the Greek letter "phi") is the inclination of the line L. The inclination ϕ of any line satisfies $0° \leq \phi < 180°$. The inclination of a horizontal line is $0°$ and that of a vertical line is $90°$.

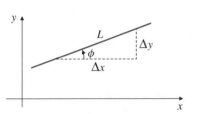

Figure P.16   Line L has inclination ϕ

Provided equal scales are used on the coordinate axes, the relationship between the slope m of a nonvertical line and its inclination ϕ is shown in Figure P.16:

$$m = \frac{\Delta y}{\Delta x} = \tan \phi.$$

(The trigonometric function tan is defined in Section P.6 below.)

Parallel lines have the same inclination. If they are not vertical, they must therefore have the same slope. Conversely, lines with equal slopes have the same inclination and so are parallel.

If two nonvertical lines, L_1 and L_2, are perpendicular, their slopes m_1 and m_2 satisfy $m_1 m_2 = -1$, so each slope is the *negative reciprocal* of the other:

$$m_1 = -\frac{1}{m_2} \qquad \text{and} \qquad m_2 = -\frac{1}{m_1}.$$

(This result also assumes equal scales on the two coordinate axes.) To see this, observe in Figure P.17 that

$$m_1 = \frac{AD}{BD} \qquad \text{and} \qquad m_2 = -\frac{AD}{DC}.$$

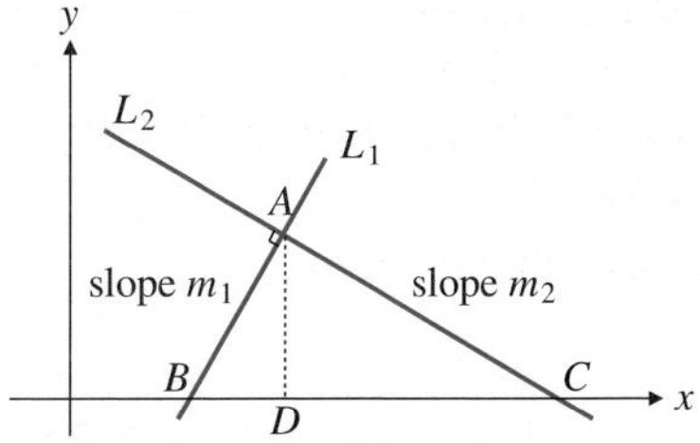

Figure P.17 $\triangle ABD$ is similar to

Since $\triangle ABD$ is similar to $\triangle CAD$, we have $\dfrac{AD}{BD} = \dfrac{DC}{AD}$, and so

$$m_1 m_2 = \left(\dfrac{DC}{AD}\right)\left(-\dfrac{AD}{DC}\right) = -1.$$

Equations of Lines

Straight lines are particularly simple graphs, and their corresponding equations are also simple. All points on the vertical line through the point a on the x-axis have their x-coordinates equal to a. Thus $x = a$ is the equation of the line. Similarly, $y = b$ is the equation of the horizontal line meeting the y-axis at b.

Example 8   The horizontal and vertical lines passing through the point $(3, 1)$ (Figure P.18) have equations $y = 1$ and $x = 3$, respectively.

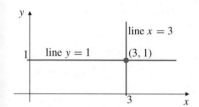

Figure P.18   The lines $y = 1$ and $x = 3$

To write an equation for a nonvertical straight line L, it is enough to know its slope m and the coordinates of one point $P_1(x_1, y_1)$ on it. If $P(x, y)$ is any other point on L, then

$$\frac{y - y_1}{x - x_1} = m,$$

so that

$$y - y_1 = m(x - x_1) \qquad \text{or} \qquad y = m(x - x_1) + y_1.$$

The equation

$$y = m(x - x_1) + y_1$$

is the **point-slope equation** of the line that passes through the point (x_1, y_1) and has slope m.

Example 9   Find an equation of the line of slope -2 through the point $(1, 4)$.

Solution   We substitute $x_1 = 1$, $y_1 = 4$, and $m = -2$ into the point-slope form of the equation and obtain

$$y = -2(x - 1) + 4 \qquad \text{or} \qquad y = -2x + 6.$$

Example 10   Find an equation of the line through the points $(1, -1)$ and $(3, 5)$.

Solution   The slope of the line is $m = \dfrac{5 - (-1)}{3 - 1} = 3$. We can use this slope with either of the two points to write an equation of the line. If we use $(1, -1)$ we get

$$y = 3(x - 1) - 1, \qquad \text{which simplifies to} \quad y = 3x - 4.$$

If we use $(3, 5)$ we get

$$y = 3(x - 3) + 5, \qquad \text{which also simplifies to} \quad y = 3x - 4.$$

Either way, $y = 3x - 4$ is an equation of the line. ∎

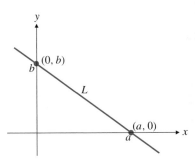

Figure P.19 Line L has x-intercept a and y-intercept b

The y-coordinate of the point where a nonvertical line intersects the y-axis is called the **y-intercept** of the line. Similarly, the **x-intercept** of a nonhorizontal line is the x-coordinate of the point where it crosses the x-axis. A line with slope m and y-intercept b passes through the point $(0, b)$, so its equation is

$$y = m(x - 0) + b \qquad \text{or, more simply,} \quad y = mx + b.$$

A line with slope m and x-intercept a passes through $(a, 0)$, and so its equation is

$$y = m(x - a).$$

> The equation $y = mx + b$ is called the **slope–y-intercept equation** of the line with slope m and y-intercept b.
>
> The equation $y = m(x - a)$ is called the **slope–x-intercept equation** of the line with slope m and x-intercept a.

Example 11 Find the slope and the two intercepts of the line with equation $8x + 5y = 20$.

Solution Solving the equation for y we get

$$y = \frac{20 - 8x}{5} = -\frac{8}{5}x + 4.$$

Comparing this with the general form $y = mx + b$ of the slope–y-intercept equation, we see that the slope of the line is $m = -8/5$, and the y-intercept is $b = 4$.

To find the x-intercept put $y = 0$ and solve for x, obtaining $8x = 20$, or $x = 5/2$. The x-intercept is $a = 5/2$. ∎

The equation $Ax + By = C$ (where A and B are not both zero) is called the **general linear equation** in x and y because its graph always represents a straight line, and every line has an equation in this form.

Many important quantities are related by linear equations. Once we know that a relationship between two variables is linear, we can find it from any two pairs of corresponding values, just as we find the equation of a line from the coordinates of two points.

Example 12 The relationship between Fahrenheit temperature (F) and Celsius temperature (C) is given by a linear equation of the form $F = mC + b$. The freezing point of water is $F = 32°$ or $C = 0°$, while the boiling point is $F = 212°$ or $C = 100°$. Thus

$$32 = 0m + b \qquad \text{and} \qquad 212 = 100m + b,$$

so $b = 32$ and $m = (212 - 32)/100 = 9/5$. The relationship is given by the linear equation

$$F = \frac{9}{5}C + 32 \quad \text{or} \quad C = \frac{5}{9}(F - 32).$$

Exercises P.2

In Exercises 1–4, a particle moves from A to B. Find the net increments Δx and Δy in the particle's coordinates. Also find the distance from A to B.

1. $A(0, 3)$, $\quad B(4, 0)$ **2.** $A(-1, 2)$, $\quad B(4, -10)$

3. $A(3, 2)$, $\quad B(-1, -2)$ **4.** $A(0.5, 3)$, $\quad B(2, 3)$

5. A particle starts at $A(-2, 3)$ and its coordinates change by $\Delta x = 4$ and $\Delta y = -7$. Find its new position.

6. A particle arrives at the point $(-2, -2)$ after its coordinates experience increments $\Delta x = -5$ and $\Delta y = 1$. From where did it start?

Describe the graphs of the equations and inequalities in Exercises 7–12.

7. $x^2 + y^2 = 1$ **8.** $x^2 + y^2 = 2$

9. $x^2 + y^2 \le 1$ **10.** $x^2 + y^2 = 0$

11. $y \ge x^2$ **12.** $y < x^2$

In Exercises 13–14, find an equation for (a) the vertical line and (b) the horizontal line through the given point.

13. $(-2, 5/3)$ **14.** $(\sqrt{2}, -1.3)$

In Exercises 15–18, write an equation for the line through P with slope m.

15. $P(-1, 1)$, $\quad m = 1$ **16.** $P(-2, 2)$, $\quad m = 1/2$

17. $P(0, b)$, $\quad m = 2$ **18.** $P(a, 0)$, $\quad m = -2$

In Exercises 19–20, does the given point P lie on, above, or below the given line?

19. $P(2, 1)$, $\quad 2x + 3y = 6$ **20.** $P(3, -1)$, $\quad x - 4y = 7$

In Exercises 21–24, write an equation for the line through the two points.

21. $(0, 0)$, $\quad (2, 3)$ **22.** $(-2, 1)$, $\quad (2, -2)$

23. $(4, 1)$, $\quad (-2, 3)$ **24.** $(-2, 0)$, $\quad (0, 2)$

In Exercises 25–26, write an equation for the line with slope m and y-intercept b.

25. $m = -2$, $\quad b = \sqrt{2}$ **26.** $m = -1/2$, $\quad b = -3$

In Exercises 27–30, determine the x- and y-intercepts and the slope of the given lines, and sketch their graphs.

27. $3x + 4y = 12$ **28.** $x + 2y = -4$

29. $\sqrt{2}x - \sqrt{3}y = 2$ **30.** $1.5x - 2y = -3$

In Exercises 31–32, find equations for the lines through P that are (a) parallel to and (b) perpendicular to the given line.

31. $P(2, 1)$, $\quad y = x + 2$ **32.** $P(-2, 2)$, $\quad 2x + y = 4$

33. Find the point of intersection of the lines $3x + 4y = -6$ and $2x - 3y = 13$.

34. Find the point of intersection of the lines $2x + y = 8$ and $5x - 7y = 1$.

35. (**Two-intercept equations**) If a line is neither horizontal nor vertical and does not pass through the origin, show that its equation can be written in the form $\frac{x}{a} + \frac{y}{b} = 1$, where a is its x-intercept and b is its y-intercept.

36. Determine the intercepts and sketch the graph of the line $\frac{x}{2} - \frac{y}{3} = 1$.

37. Find the y-intercept of the line through the points $(2, 1)$ and $(3, -1)$.

38. A line passes through $(-2, 5)$ and $(k, 1)$ and has x-intercept 3. Find k.

39. The cost of printing x copies of a pamphlet is $\$C$, where $C = Ax + B$ for certain constants A and B. If it costs $\$5,000$ to print 10,000 copies and $\$6,000$ to print 15,000 copies, how much will it cost to print 100,000 copies?

40. (**Fahrenheit versus Celsius**) In the FC-plane, sketch the graph of the equation $C = \frac{5}{9}(F - 32)$ linking Fahrenheit and Celsius temperatures found in Example 12. On the same graph sketch the line with equation $C = F$. Is there a temperature at which a Celsius thermometer gives the same numerical reading as a Fahrenheit thermometer? If so, find it.

Geometry

41. By calculating the lengths of its three sides, show that the triangle with vertices at the points $A(2, 1)$, $B(6, 4)$, and $C(5, -3)$ is isosceles.

42. Show that the triangle with vertices $A(0, 0)$, $B(1, \sqrt{3})$, and $C(2, 0)$ is equilateral.

43. Show that the points $A(2, -1)$, $B(1, 3)$, and $C(-3, 2)$ are three vertices of a square and find the fourth vertex.

44. Find the coordinates of the midpoint on the line segment P_1P_2 joining the points $P_1(x_1, y_1)$ and $P_2(x_2, y_2)$.

45. Find the coordinates of the point of the line segment joining the points $P_1(x_1, y_1)$ and $P_2(x_2, y_2)$ that is two-thirds of the way from P_1 to P_2.

46. The point P lies on the x-axis and the point Q lies on the line $y = -2x$. The point $(2, 1)$ is the midpoint of PQ. Find the coordinates of P.

In Exercises 47–48, interpret the equation as a statement about

distances, and hence determine the graph of the equation.

47. $\sqrt{(x-2)^2 + y^2} = 4$

48. $\sqrt{(x-2)^2 + y^2} = \sqrt{x^2 + (y-2)^2}$

49. For what value of k is the line $2x + ky = 3$ perpendicular to the line $4x + y = 1$? For what value of k are the lines parallel?

50. Find the line that passes through the point $(1, 2)$ and through the point of intersection of the two lines $x + 2y = 3$ and $2x - 3y = -1$.

P.3 Graphs of Quadratic Equations

This section reviews circles, parabolas, ellipses, and hyperbolas, the graphs that are represented by quadratic equations in two variables.

Circles and Disks

The **circle** having **centre** C and **radius** a is the set of all points in the plane that are at distance a from the point C.

The distance from $P(x, y)$ to the point $C(h, k)$ is $\sqrt{(x-h)^2 + (y-k)^2}$, so that the equation of the circle of radius $a > 0$ with centre at $C(h, k)$ is

$$\sqrt{(x-h)^2 + (y-k)^2} = a.$$

A simpler form of this equation is obtained by squaring both sides.

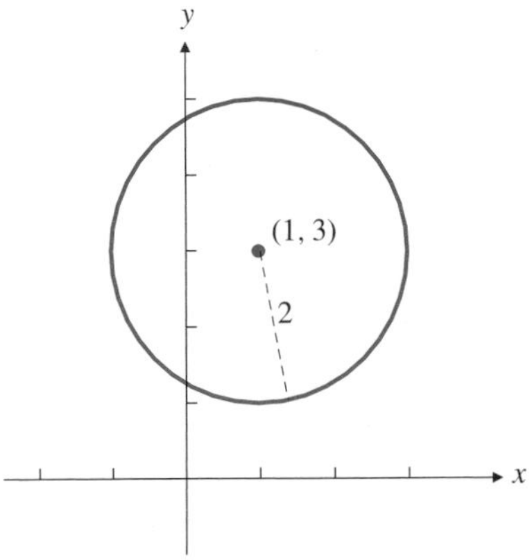

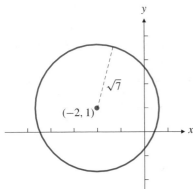

Figure P.20　Circle
$(x-1)^2 + (y-3)^2 = 4$

> **Standard equation of a circle**
>
> The circle with centre (h, k) and radius $a > 0$ has equation
>
> $$(x-h)^2 + (y-k)^2 = a^2.$$
>
> In particular, the circle with centre at the origin $(0, 0)$ and radius a has equation
>
> $$x^2 + y^2 = a^2.$$

Example 1　The circle with radius 2 and centre $(1, 3)$ (Figure P.20) has equation

$$(x-1)^2 + (y-3)^2 = 4.$$

Example 2　The circle having equation $(x+2)^2 + (y-1)^2 = 7$ has centre at the point $(-2, 1)$ and radius $\sqrt{7}$. (See Figure P.21.)

Figure P.21　Circle
$(x+2)^2 + (y-1)^2 = 7$

If the squares in the standard equation $(x-h)^2 + (y-k)^2 = a^2$ are multiplied out, and all constant terms collected on the right-hand side, the equation becomes

$$x^2 - 2hx + y^2 - 2ky = a^2 - h^2 - k^2.$$

A quadratic equation of the form

$$x^2 + y^2 + 2ax + 2by = c$$

must represent a circle, a single point, or no points at all. To identify the graph, we complete the squares on the left side of the equation. Since $x^2 + 2ax$ are the first two terms of the square $(x + a)^2 = x^2 + 2ax + a^2$, we add a^2 to both sides to complete the square of the x terms. (Note that a^2 is *the square of half the coefficient of x*.) Similarly, add b^2 to both sides to complete the square of the y terms. The equation then becomes

$$(x + a)^2 + (y + b)^2 = c + a^2 + b^2.$$

If $c + a^2 + b^2 > 0$, the graph is a circle with centre $(-a, -b)$ and radius $\sqrt{c + a^2 + b^2}$. If $c + a^2 + b^2 = 0$, the graph consists of the single point $(-a, -b)$. If $c + a^2 + b^2 < 0$, no points lie on the graph.

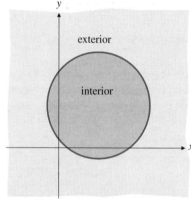

Figure P.22   The interior of a circle (darkly shaded) and the exterior (lightly shaded)

Example 3   Find the centre and radius of the circle $x^2 + y^2 - 4x + 6y = 3$.

Solution   Observe that $x^2 - 4x$ are the first two terms of the binomial square $(x - 2)^2 = x^2 - 4x + 4$, and $y^2 + 6y$ are the first two terms of the square $(y + 3)^2 = y^2 + 6y + 9$. Hence we add $4 + 9$ to both sides of the given equation and obtain

$$x^2 - 4x + 4 + y^2 + 6y + 9 = 3 + 4 + 9 \quad \text{or} \quad (x - 2)^2 + (y + 3)^2 = 16.$$

This is the equation of a circle with centre $(2, -3)$ and radius 4.

The set of all points *inside* a circle is called the **interior** of the circle; it is also called an **open disk**. The set of all points *outside* the circle is called the **exterior** of the circle. (See Figure P.22.) The interior of a circle together with the circle itself is called a **closed disk**, or simply a **disk**. The inequality

$$(x - h)^2 + (y - k)^2 \le a^2$$

represents the disk of radius $|a|$ centred at (h, k).

Example 4   Identify the graphs of:
(a) $x^2 + 2x + y^2 \le 8$ (b) $x^2 + 2x + y^2 < 8$ (c) $x^2 + 2x + y^2 > 8$.

Solution   We can complete the square in the equation $x^2 + y^2 + 2x = 8$ as follows:

$$x^2 + 2x + 1 + y^2 = 8 + 1$$
$$(x + 1)^2 + y^2 = 9.$$

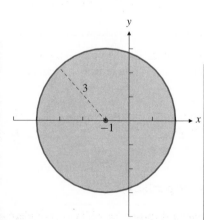

Figure P.23   The disk $x^2 + y^2 + 2x \le 8$

Thus the equation represents the circle of radius 3 with centre at $(-1, 0)$. Inequality (a) represents the (closed) disk with the same radius and centre. (See Figure P.23.) Inequality (b) represents the interior of the circle (or the open disk). Inequality (c) represents the exterior of the circle.

Equations of Parabolas

A **parabola** is a plane curve whose points are equidistant from a fixed point F and a fixed straight line L that does not pass through F. The point F is the **focus** of the parabola; the line L is the parabola's **directrix**. The line through F perpendicular to L is the parabola's **axis**. The point V where the axis meets the parabola is the parabola's **vertex**.

Observe that the vertex V of a parabola is halfway between the focus F and the point on the directrix L that is closest to F. If the directrix is either horizontal or vertical, and the vertex is at the origin, then the parabola will have a particularly simple equation.

Example 5 Find an equation of the parabola having the point $F(0, p)$ as focus and the line L with equation $y = -p$ as directrix.

Solution If $P(x, y)$ is any point on the parabola, then (see Figure P.24) the distances from P to F and L are given by

$$PF = \sqrt{(x-0)^2 + (y-p)^2} = \sqrt{x^2 + y^2 - 2py + p^2}$$
$$PQ = \sqrt{(x-x)^2 + (y-(-p))^2} = \sqrt{y^2 + 2py + p^2}.$$

For P on the parabola we have $PF = PQ$, so their squares are also equal:

$$x^2 + y^2 - 2py + p^2 = y^2 + 2py + p^2,$$

or, after simplifying,

$$x^2 = 4py \qquad \text{or} \qquad y = \frac{x^2}{4p} \qquad \text{(called **standard forms**)}.$$

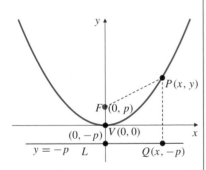

Figure P.24   The parabola $4py = x^2$ with focus $F(0, p)$ and directrix $y = -p$

Figure P.24 shows the situation for $p > 0$; the parabola opens upward and is symmetric about its axis, the y-axis. If $p < 0$, the focus $(0, p)$ will lie below the origin and the directrix $y = -p$ will lie above the origin. In this case the parabola will open downward instead of upward.

Figure P.25 shows several parabolas with equations of the form $y = ax^2$ for positive and negative values of a.

Example 6 An equation for the parabola with focus $(0, 1)$ and directrix $y = -1$ is $y = x^2/4$, or $x^2 = 4y$. (We took $p = 1$ in the standard equation.)

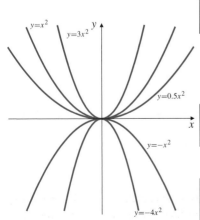

Figure P.25   Some parabolas $y = ax^2$

Example 7 Find the focus and directrix of the parabola $y = -x^2$.

Solution The given equation matches the standard form $y = x^2/(4p)$ provided $4p = -1$. Thus $p = -1/4$. The focus is $(0, -1/4)$ and the directrix is the line $y = 1/4$.

Interchanging the roles of x and y in the derivation of the standard equation above shows that the equation

$$y^2 = 4px \qquad \text{or} \qquad x = \frac{y^2}{4p} \qquad \text{(standard equation)}$$

represents a parabola with focus at $(p, 0)$ and vertical directrix $x = -p$. The axis is the x-axis.

Reflective Properties of Parabolas

One of the chief applications of parabolas involves their use as reflectors of light and radio waves. Rays originating from the focus of a parabola will be reflected in a beam parallel to the axis, as shown in Figure P.26. Similarly, all the rays in a beam striking a parabola parallel to its axis will reflect through the focus. This property is the reason why telescopes and spotlights use parabolic mirrors and radio telescopes and microwave antennas are parabolic in shape. We will examine this property of parabolas more carefully in Section 8.1.

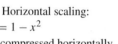

Figure P.26 Reflection by a parabola

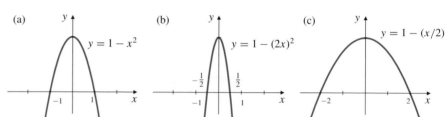

(a) $y = 1 - x^2$
(b) $y = 1 - (2x)^2$
(c) $y = 1 - (x/2)^2$

Figure P.27 Horizontal scaling:
(a) the graph $y = 1 - x^2$
(b) graph of (a) compressed horizontally
(c) graph of (a) expanded horizontally

Scaling a Graph

The graph of an equation can be compressed or expanded horizontally by replacing x with a multiple of x. If a is a positive number, replacing x with ax in an equation multiplies horizontal distances in the graph of the equation by a factor $1/a$. (See Figure P.27.) Replacing y with ay will multiply vertical distances in a similar way.

You may find it surprising that, like circles, all parabolas are *similar* geometric figures; they may have different sizes, but they all have the same shape. We can change the *size* while preserving the shape of a curve represented by an equation in x and y by scaling both the coordinates by the same amount. If we scale the equation $4py = x^2$ by replacing x and y with $4px$ and $4py$, respectively, we get $4p(4py) = (4px)^2$, or $y = x^2$. Thus the general parabola $4py = x^2$ has the same shape as the specific parabola $y = x^2$, as shown in Figure P.28.

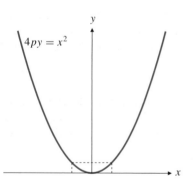

Shifting a Graph

The graph of an equation (or inequality) can be shifted c units horizontally by replacing x with $x - c$ or vertically by replacing y with $y - c$.

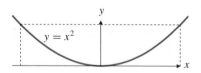

Figure P.28 The two parabolas are similar. Compare the parts inside the rectangles.

> **Shifts**
>
> To shift a graph c units to the right, replace x in its equation or inequality with $x - c$. (If $c < 0$, the shift will be to the left.)
>
> To shift a graph c units upward, replace y in its equation or inequality with $y - c$. (If $c < 0$, the shift will be downward.)

Example 8   The graph of $y = (x - 3)^2$ is the parabola $y = x^2$ shifted 3 units to the right. The graph of $y = (x + 1)^2$ is the parabola $y = x^2$ shifted 1 unit to the left. (See Figure P.29(a).)

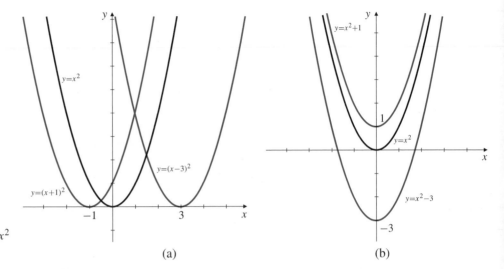

Figure P.29

(a) Horizontal shifts of $y = x^2$

(b) Vertical shifts of $y = x^2$

(a)

(b)

Example 9   The graph of $y = x^2 + 1$ (that is, $y - 1 = x^2$) is the parabola $y = x^2$ shifted upward 1 unit. The graph of $y = x^2 - 3$ (that is, $y - (-3) = x^2$), is the parabola $y = x^2$ shifted downward 3 units. (See Figure P.29(b).)

Example 10   The circle with equation $(x - h)^2 + (y - k)^2 = a^2$ having centre (h, k) and radius a can be obtained by shifting the circle $x^2 + y^2 = a^2$ of radius a centred at the origin h units to the right and k units upward. These shifts correspond to replacing x with $x - h$ and y with $y - k$.

The graph of $y = ax^2 + bx + c$ is a parabola whose axis is parallel to the y-axis. The parabola opens upward if $a > 0$ and downward if $a < 0$. We can complete the square and write the equation in the form $y = a(x - h)^2 + k$ to find the vertex (h, k).

Example 11   Describe the graph of $y = x^2 - 4x + 3$.

Solution   The equation $y = x^2 - 4x + 3$ represents a parabola, opening upward. To find its vertex and axis we can complete the square:

$$y = x^2 - 4x + 4 - 1 = (x - 2)^2 - 1, \qquad \text{so} \quad y - (-1) = (x - 2)^2.$$

This curve is the parabola $y = x^2$ shifted to the right 2 units and down 1 unit. Therefore its vertex is $(2, -1)$ and its axis is the line $x = 2$. Since $y = x^2$ has focus $(0, 1/4)$, the focus of this parabola is $(0 + 2, (1/4) - 1)$, or $(2, -3/4)$. (See Figure P.30.)

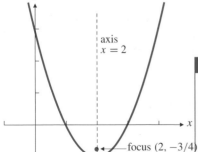

Figure P.30   The parabola $y = x^2 - 4x + 3$

Ellipses and Hyperbolas

If a and b are positive numbers, the equation

$$\frac{x^2}{a^2} + \frac{y^2}{b^2} = 1$$

represents a curve called an **ellipse** that lies wholly within the rectangle $-a \le x \le a$, $-b \le y \le b$. (Why?) If $a = b$, the ellipse is just the circle of radius a centred at the origin. If $a \ne b$, the ellipse is a circle that has been squashed by scaling it by different amounts in the two coordinate directions.

The ellipse has centre at the origin, and it passes through the four points $(a, 0)$, $(0, b)$, $(-a, 0)$, and $(0, -b)$. (See Figure P.31.) The line segments from $(-a, 0)$ to $(a, 0)$ and from $(0, -b)$ to $(0, b)$ are called the **principal axes** of the ellipse; the longer of the two is the **major axis** and the shorter is the **minor axis**.

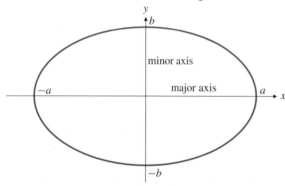

Figure P.31   The ellipse $\dfrac{x^2}{a^2} + \dfrac{y^2}{b^2} = 1$

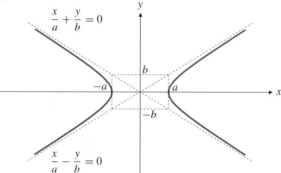

Figure P.32   The hyperbola $\dfrac{x^2}{a^2} - \dfrac{y^2}{b^2} = 1$ and its asymptotes

Example 12   The equation $\dfrac{x^2}{9} + \dfrac{y^2}{4} = 1$ represents an ellipse with major axis from $(-3, 0)$ to $(3, 0)$, and minor axis from $(0, -2)$ to $(0, 2)$. ∎

The equation

$$\frac{x^2}{a^2} - \frac{y^2}{b^2} = 1$$

represents a curve called a **hyperbola** that has centre at the origin and passes through the points $(-a, 0)$ and $(a, 0)$. (See Figure P.32.) The curve is in two parts (called **branches**). Each branch approaches two straight lines (called **asymptotes**) as it recedes far away from the origin. The asymptotes have equations

$$\frac{x}{a} - \frac{y}{b} = 0 \qquad \text{and} \qquad \frac{x}{a} + \frac{y}{b} = 0.$$

The equation $xy = 1$ also represents a hyperbola. This one passes through the points $(-1, -1)$ and $(1, 1)$ and has the coordinate axes as its asymptotes. It is, in fact, the hyperbola $x^2 - y^2 = 2$ rotated $45°$ counterclockwise about the origin. (See Figure P.33.) These hyperbolas are called **rectangular hyperbolas** since their asymptotes intersect at right angles.

We will study ellipses and hyperbolas in more detail in Chapter 8.

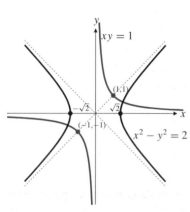

Figure P.33   Two rectangular hyperbolas

Exercises P.3

In Exercises 1–4, write an equation for the circle with centre C and radius r.

1. $C(0, 0)$, $r = 4$ **2.** $C(0, 2)$, $r = 2$

3. $C(-2, 0)$, $r = 3$ **4.** $C(3, -4)$, $r = 5$

In Exercises 5–8, find the centre and radius of the circle having the given equation.

5. $x^2 + y^2 - 2x = 3$ **6.** $x^2 + y^2 + 4y = 0$

7. $x^2 + y^2 - 2x + 4y = 4$ **8.** $x^2 + y^2 - 2x - y + 1 = 0$

Describe the regions defined by the inequalities and pairs of inequalities in Exercises 9–16.

9. $x^2 + y^2 > 1$ **10.** $x^2 + y^2 < 4$

11. $(x + 1)^2 + y^2 \le 4$ **12.** $x^2 + (y - 2)^2 \le 4$

13. $x^2 + y^2 > 1$, $x^2 + y^2 < 4$

14. $x^2 + y^2 \le 4$, $(x + 2)^2 + y^2 \le 4$

15. $x^2 + y^2 < 2x$, $x^2 + y^2 < 2y$

16. $x^2 + y^2 - 4x + 2y > 4$, $x + y > 1$

17. Write an inequality that describes the interior of the circle with centre $(-1, 2)$ and radius $\sqrt{6}$.

18. Write an inequality that describes the exterior of the circle with centre $(2, -3)$ and radius 4.

19. Write a pair of inequalities that describe that part of the interior of the circle with centre $(0, 0)$ and radius $\sqrt{2}$ lying on or to the right of the vertical line through $(1, 0)$.

20. Write a pair of inequalities that describe the points that lie outside the circle with centre $(0, 0)$ and radius 2, and inside the circle with centre $(1, 3)$ that passes through the origin.

In Exercises 21–24, write an equation of the parabola having the given focus and directrix.

21. Focus: $(0, 4)$ Directrix: $y = -4$
22. Focus: $(0, -1/2)$ Directrix: $y = 1/2$
23. Focus: $(2, 0)$ Directrix: $x = -2$
24. Focus: $(-1, 0)$ Directrix: $x = 1$

In Exercises 25–28, find the parabola's focus and directrix, and make a sketch showing the parabola, focus, and directrix.

25. $y = x^2/2$ **26.** $y = -x^2$

27. $x = -y^2/4$ **28.** $x = y^2/16$

29. Figure P.34 shows the graph $y = x^2$ and four shifted versions of it. Write equations for the shifted versions.

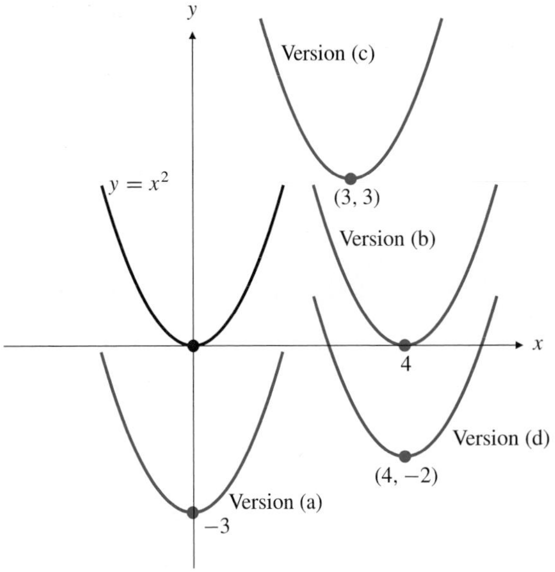

Figure P.34

30. What equations result from shifting the line $y = mx$
   (a) horizontally to make it pass through the point (a, b)
   (b) vertically to make it pass through (a, b)?

In Exercises 31–34, the graph of $y = \sqrt{x + 1}$ is to be scaled in the indicated way. Give the equation of the graph that results from the scaling.

31. horizontal distances multiplied by 3

32. vertical distances divided by 4

33. horizontal distances multiplied by 2/3

34. horizontal distances divided by 4 and vertical distances multiplied by 2

In Exercises 35–38, write an equation for the graph obtained by shifting the graph of the given equation as indicated.

35. $y = 1 - x^2$ down 1, left 1
36. $x^2 + y^2 = 5$ up 2, left 4
37. $y = (x - 1)^2 - 1$ down 1, right 1
38. $y = \sqrt{x}$ down 2, left 4

Find the points of intersection of the pairs of curves in Exercises 39–42.

39. $y = x^2 + 3$, $y = 3x + 1$
40. $y = x^2 - 6$, $y = 4x - x^2$
41. $x^2 + y^2 = 25$, $3x + 4y = 0$
42. $2x^2 + 2y^2 = 5$, $xy = 1$

In Exercises 43–50, identify and sketch the curve represented by the given equation.

43. $\dfrac{x^2}{4} + y^2 = 1$ **44.** $9x^2 + 16y^2 = 144$

45. $\dfrac{(x-3)^2}{9} + \dfrac{(y+2)^2}{4} = 1$ **46.** $(x-1)^2 + \dfrac{(y+1)^2}{4} = 4$

47. $\dfrac{x^2}{4} - y^2 = 1$ **48.** $x^2 - y^2 = -1$

49. $xy = -4$ **50.** $(x-1)(y+2) = 1$

51. What is the effect on the graph of an equation in x and y of
  (a) replacing x with $-x$?
  (b) replacing y with $-y$?

52. What is the effect on the graph of an equation in x and y of replacing x with $-x$ and y with $-y$ simultaneously?

53. Sketch the graph of $|x| + |y| = 1$.

P.4 Functions and Their Graphs

The area of a circle depends on its radius. The temperature at which water boils depends on the altitude above sea level. The interest paid on a cash investment depends on the length of time for which the investment is made.

Whenever one quantity depends on another quantity, we say that the former quantity is a function of the latter. For instance, the area A of a circle depends on the radius r according to the formula

$$A = \pi r^2,$$

so we say that the area is a function of the radius. The formula is a *rule* that tells us how to calculate a *unique* (single) output value of the area A for each possible input value of the radius r.

The set of all possible input values for the radius is called the **domain** of the function. The set of all output values of the area is the **range** of the function. Since circles cannot have negative radii or areas, the domain and range of the circular area function are both the interval $[0, \infty[$ consisting of all nonnegative real numbers.

The domain and range of a mathematical function can be any sets of objects; they do not have to consist of numbers. Throughout much of this book, however, the domains and ranges of functions we consider will be sets of real numbers.

In calculus we often want to refer to a generic function without having any particular formula in mind. To denote that y is a function of x we write

$$y = f(x),$$

which we read as "y equals f of x." In this notation, due to eighteenth-century mathematician Leonhard Euler, the function is represented by the symbol f. Also, x, called the **independent variable**, represents an input value from the domain of f, and y, the **dependent variable**, represents the corresponding output value $f(x)$ in the range of f.

DEFINITION 1

> A **function** f on a set D into a set S is a rule that assigns a *unique* element $f(x)$ in S to each element x in D.

In this definition $D = \mathcal{D}(f)$ (read "D of f") is the domain of the function f. The range $\mathcal{R}(f)$ of f is the subset of S consisting of all *values* $f(x)$ of the function. Think of a function f as a kind of machine (Figure P.35) that produces an output value $f(x)$ in its range whenever we feed it an input value x from its domain.

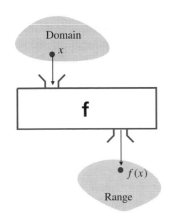

Figure P.35 A function machine

There are several ways to represent a function symbolically. The squaring function that converts any input real number x into its square x^2 can be denoted:

(a) by a formula such as $y = x^2$, which uses a dependent variable y to denote the value of the function;

(b) by a formula such as $f(x) = x^2$, which defines a function symbol f to name the function; or

(c) by a mapping rule such as $x \longrightarrow x^2$. (Read this as "x goes to x^2.")

In this book we will usually use either (a) or (b) to define functions. Strictly speaking, we should call a function f and not $f(x)$, since the latter denotes the value of the function at the point x. However, as is common usage, we will often refer to the function as $f(x)$ in order to name the variable on which f depends. Sometimes it is convenient to use the same letter to denote both a dependent variable and as a function symbol; the circular area function can be written $A = f(r) = \pi r^2$ or as $A = A(r) = \pi r^2$. In the latter case we are using A to denote both the dependent variable and the name of the function.

Example 1 The volume of a ball of radius r is given by the function

$$V(r) = \frac{4}{3} \pi r^3$$

for $r \geq 0$. Thus the volume of a ball of radius 3 ft is

$$V(3) = \frac{4}{3} \pi (3)^3 = 36\pi \text{ ft}^3.$$

Note how the variable r is replaced by the special value 3 in the formula defining the function to obtain the value of the function at $r = 3$.

Example 2 If a function F is defined for all real numbers t by

$$F(t) = 2t + 3,$$

find the output values of F that correspond to the input values $0, 2, x+2$, and $F(2)$.

Solution In each case we substitute the given input for t in the definition of F:

$$F(0) = 2(0) + 3 = 0 + 3 = 3$$
$$F(2) = 2(2) + 3 = 4 + 3 = 7$$
$$F(x + 2) = 2(x + 2) + 3 = 2x + 7$$
$$F(F(2)) = F(7) = 2(7) + 3 = 17.$$

The Domain Convention

A function is not properly defined until its domain is specified. For instance, the function $f(x) = x^2$ defined for all real numbers $x \geq 0$ is different from the function $g(x) = x^2$ defined for all real x because they have different domains, even though they have the same values at every point where both are defined. In Chapters 1–9 we will be dealing with real functions (functions whose input and output values are real numbers). When the domain of such a function is not specified explicitly, we will assume that the domain is the largest set of real numbers to which the function assigns real values. Thus, if we talk about the function x^2 without specifying a domain, we mean the function $g(x)$ above.

> **The domain convention**
>
> When a function f is defined without specifying its domain, we assume that the domain consists of all real numbers x for which the value $f(x)$ of the function is a real number.

In practice, it is often easy to determine the domain of a function $f(x)$ given by an explicit formula. We just have to exclude those values of x that would result in dividing by 0 or taking even roots of negative numbers.

Example 3 **The square root function.** The domain of $f(x) = \sqrt{x}$ is the interval $[0, \infty[$, since negative numbers do not have real square roots. We have $f(0) = 0$, $f(4) = 2$, $f(10) \approx 3.16228$. Note that, although there are *two* numbers whose square is 4, namely, -2 and 2, only *one* of these numbers, 2, is the square root of 4. (Remember that a function assigns a *unique* value to each element in its domain; it cannot assign two different values to the same input.) The **square root function** $\sqrt{x}$ always denotes the *nonnegative* square root of x. The two solutions of the equation $x^2 = 4$ are $x = \sqrt{4} = 2$ and $x = -\sqrt{4} = -2$. ∎

Example 4 The domain of the function $h(x) = \dfrac{x}{x^2 - 4}$ consists of all real numbers except $x = -2$ and $x = 2$. Expressed in terms of intervals,

$$\mathcal{D}(f) =]-\infty, -2[\cup]-2, 2[\cup]2, \infty[.$$

Most of the functions we encounter will have domains that are either intervals or unions of intervals. ∎

Example 5 The domain of $S(t) = \sqrt{1 - t^2}$ consists of all real numbers t for which $1 - t^2 \geq 0$. Thus we require that $t^2 \leq 1$, or $-1 \leq t \leq 1$. The domain is the closed interval $[-1, 1]$. ∎

Graphs of Functions

An old maxim states that "a picture is worth a thousand words." This is certainly true in mathematics; the behaviour of a function is best described by drawing its graph.

The **graph of a function** f is just the graph of the *equation* $y = f(x)$. It consists of those points in the Cartesian plane whose coordinates (x, y) are pairs of input-output values for f. Thus (x, y) lies on the graph of f provided x is in the domain of f and $y = f(x)$.

Drawing the graph of a function f sometimes involves making a table of coordinate pairs $(x, f(x))$ for various values of x in the domain of f, then plotting these points and connecting them with a "smooth curve."

Table 1.

x	$y = f(x)$
-2	4
-1	1
0	0
1	1
2	4

Example 6   Graph the function $f(x) = x^2$.

Solution   Make a table of (x, y) pairs that satisfy $y = x^2$. (See Table 1.) Now plot the points and join them with a smooth curve. (See Figure P.36(a).) ∎

Figure P.36

(a) Correct graph of $f(x) = x^2$

(b) Incorrect graph of $f(x) = x^2$

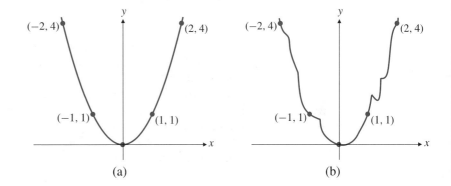

(a)   (b)

How do we know the graph is smooth and doesn't do weird things between the points we have calculated (for example, as shown in Figure P.36(b))? We could, of course, plot more points, spaced more closely together, but how do we know how the graph behaves between the points we have plotted? In Chapter 4, calculus will provide useful tools for answering these questions.

Some functions occur often enough in applications that you should be familiar with their graphs. Some of these are shown in Figures P.37–P.46. Study them for

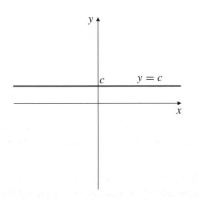

Figure P.37   The graph of a constant function $f(x) = c$

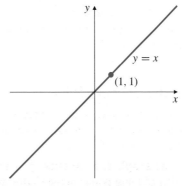

Figure P.38   The graph of $f(x) = x$

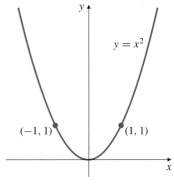

Figure P.39   The graph of $f(x) = x^2$

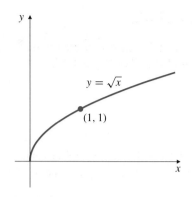

Figure P.40 The graph of $f(x) = \sqrt{x}$

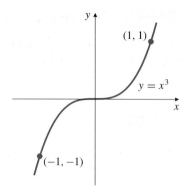

Figure P.41 The graph of $f(x) = x^3$

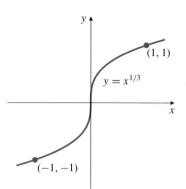

Figure P.42 The graph of $f(x) = x^{1/3}$

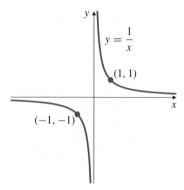

Figure P.43 The graph of $f(x) = 1/x$

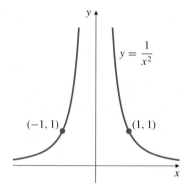

Figure P.44 The graph of $f(x) = 1/x^2$

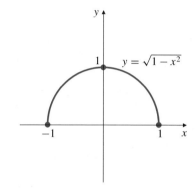

Figure P.45 The graph of $f(x) = \sqrt{1 - x^2}$

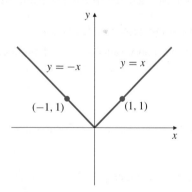

Figure P.46 The graph of $f(x) = |x|$

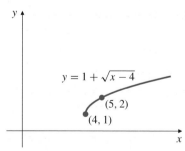

Figure P.47 The graph of $y = \sqrt{x}$ shifted right 4 units and up 1 unit

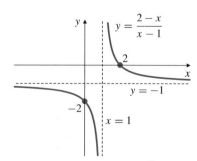

Figure P.48 The graph of $\dfrac{2 - x}{x - 1}$

a while; they are worth remembering. Note, in particular, the graph of the **absolute value function**, $f(x) = |x|$, shown in Figure P.46. It is made up of the two half-lines $y = -x$ for $x < 0$ and $y = x$ for $x \geq 0$.

If you know the effects of vertical and horizontal shifts on the equations representing graphs (see Section P.3), you can easily sketch some graphs that are shifted versions of ones in Figures P.37–P.46.

Example 7 Sketch the graph of $y = 1 + \sqrt{x-4}$.

Solution This is just the graph of $y = \sqrt{x}$ in Figure P.40 shifted to the right 4 units (because x is replaced by $x - 4$) and up 1 unit. See Figure P.47.

Example 8 Sketch the graph of the function $f(x) = \dfrac{2-x}{x-1}$.

Solution It is not immediately obvious that this graph is a shifted version of a known graph. To see that it is, we can divide $x - 1$ into $2 - x$ to get a quotient of -1 and a remainder of 1:

$$\frac{2-x}{x-1} = \frac{-x+1+1}{x-1} = \frac{-(x-1)+1}{x-1} = -1 + \frac{1}{x-1}.$$

Thus, the graph is that of $1/x$ from Figure P.43 shifted to the right 1 unit and down 1 unit. See Figure P.48.

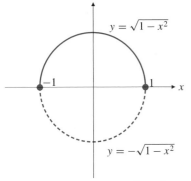

Figure P.49 The circle $x^2 + y^2 = 1$ is not the graph of a function

Not every curve you can draw is the graph of a function. A function f can have only one value $f(x)$ for each x in its domain, so no *vertical line* can intersect the graph of a function at more than one point. If a is in the domain of function f, then the vertical line $x = a$ will intersect the graph of f at the single point $(a, f(a))$. The circle $x^2 + y^2 = 1$ in Figure P.49 cannot be the graph of a function since some vertical lines intersect it twice. It is, however, the union of the graphs of two functions, namely,

$$y = \sqrt{1 - x^2} \qquad \text{and} \qquad y = -\sqrt{1 - x^2},$$

which are, respectively, the upper and lower halves (semicircles) of the given circle.

Even and Odd Functions; Symmetry and Reflections

It often happens that the graph of a function will have certain kinds of symmetry. The simplest kinds of symmetry relate the values of a function at x and $-x$.

DEFINITION 2

Even and odd functions

Suppose that $-x$ belongs to the domain of f whenever x does. We say that f is an **even function** if

$$f(-x) = f(x) \qquad \text{for every } x \text{ in the domain of } f.$$

We say that f is an **odd function** if

$$f(-x) = -f(x) \qquad \text{for every } x \text{ in the domain of } f.$$

The names *even* and *odd* come from the fact that even powers such as $x^0 = 1$, x^2, x^4, ..., x^{-2}, x^{-4}, ... are even functions, and odd powers such as $x^1 = x$, x^3, ..., x^{-1}, x^{-3}, ... are odd functions. Observe, for example, that $(-x)^4 = x^4$ and $(-x)^{-3} = -x^{-3}$.

Since $(-x)^2 = x^2$, any function that depends only on x^2 is even. For instance, the absolute value function $y = |x| = \sqrt{x^2}$ is even.

The graph of an even function is *symmetric about the y-axis*. A horizontal straight line drawn from a point on the graph to the y-axis will, if continued an equal distance on the other side of the y-axis, come to another point on the graph. (See Figure P.50(a).)

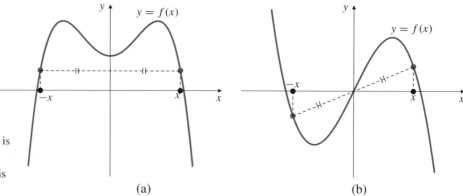

Figure P.50

(a) The graph of an even function is symmetric about the y-axis

(b) The graph of an odd function is symmetric about the origin

(a)

(b)

The graph of an odd function is *symmetric about the origin*. A straight line drawn from a point on the graph to the origin will, if continued an equal distance on the other side of the origin, come to another point on the graph. If an odd function f is defined at $x = 0$, then its value must be zero there: $f(0) = 0$. (See Figure P.50(b).)

If $f(x)$ is even (or odd), then so is any constant multiple of $f(x)$ such as $2f(x)$ or $-5f(x)$. Sums (and differences) of even functions are even; sums (and differences) of odd functions are odd. For example, $f(x) = 3x^4 - 5x^2 - 1$ is even, since it is the sum of three even functions: $3x^4$, $-5x^2$, and $-1 = -x^0$. Similarly, $4x^3 - (2/x)$ is an odd function. The function $g(x) = x^2 - 2x$ is the sum of an even function and an odd function, and is itself neither even nor odd.

Other kinds of symmetry are also possible. For example, the function $g(x) = x^2 - 2x$ can be written in the form $g(x) = (x - 1)^2 - 1$. This shows that the values of $g(1 \pm u)$ are equal, so the graph (Figure P.51(a)) is symmetric about the vertical line $x = 1$; it is the parabola $y = x^2$ shifted 1 unit to the right and 1 unit down. Similarly, the graph of $h(x) = x^3 + 1$ is symmetric about the point $(0, 1)$ (Figure P.51(b)).

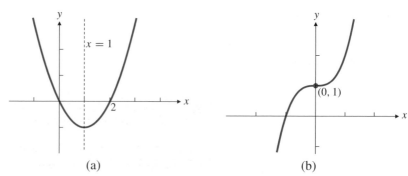

Figure P.51

(a) The graph of $g(x) = x^2 - 2x$ is symmetric about $x = 1$.

(b) The graph of $y = h(x) = x^3 + 1$ is symmetric about $(0, 1)$.

(a)

(b)

Reflections in Straight Lines

The image of an object reflected in a plane mirror appears to be as far behind the mirror as the object is in front of it. Thus, the mirror bisects the line from the object to the image at right angles. Given a line L and a point P not on L, we call a point Q the **reflection**, or the **mirror image**, of P in L if L is the right bisector of the line segment PQ. The reflection of any graph G in L is the graph consisting of the reflections of all the points of G.

Certain reflections of graphs are easily described in terms of the equations of the graphs:

Reflections in special lines

1. Substituting $-x$ in place of x in an equation in x and y corresponds to reflecting the graph of the equation in the y-axis.
2. Substituting $-y$ in place of y in an equation in x and y corresponds to reflecting the graph of the equation in the x-axis.
3. Substituting $a - x$ in place of x in an equation in x and y corresponds to reflecting the graph of the equation in the line $x = a/2$.
4. Substituting $b - y$ in place of y in an equation in x and y corresponds to reflecting the graph of the equation in the line $y = b/2$.
5. Interchanging x and y in an equation in x and y corresponds to reflecting the graph of the equation in the line $y = x$.

Example 9 Describe and sketch the graph of $y = \sqrt{2 - x} - 3$.

Solution The graph of $y = \sqrt{2 - x}$ is the reflection of the graph of $y = \sqrt{x}$ (Figure P.40) in the vertical line $x = 1$. The graph of $y = \sqrt{2 - x} - 3$ is the result of lowering this reflection by 3 units. See Figure P.52(a). ∎

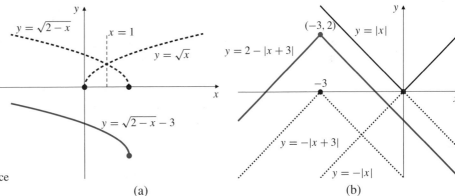

Figure P.52

(a) Constructing the graph of $y = \sqrt{2 - x} - 3$

(b) Transforming $y = |x|$ to produce the coloured graph

(a) (b)

Example 10 Express the equation of the coloured graph in Figure P.52(b) in terms of the absolute value function $|x|$.

Solution We can get the coloured graph by first reflecting the graph of $|x|$ (Figure P.46) in the x-axis and then shifting the reflection left 3 units and up 2 units. The reflection of $y = |x|$ in the x-axis has equation $-y = |x|$, or $y = -|x|$. Shifting this left 3 units gives $y = -|x + 3|$. Finally, shifting up 2 units gives $y = 2 - |x + 3|$, which is the desired equation. ∎

Defining and Graphing Functions with Maple

Many of the calculations and graphs encountered in studying calculus can be produced using a computer algebra system such as Maple or Mathematica. Here and there, throughout this book, we will include examples illustrating how to get Maple to perform such tasks. (The examples were done with Maple 6, but most of them will work with earlier versions of Maple as well.)

We begin with an example showing how to define a function in Maple and then plot its graph. We show in colour the input you type into Maple and in black Maple's response. Let us define the function $f(x) = x^3 - 2x^2 - 12x + 1$.

```
>   f := x -> x^3-2*x^2-12*x+1; <enter>
```

$$f := x \longrightarrow x^3 - 2x^2 - 12x + 1$$

Note the use of := to indicate the symbol to the left is being defined and the use of -> to indicate the rule for the construction of $f(x)$ from x. Also note that Maple uses the asterisk * to indicate multiplication and the caret ^ to indicate an exponent. A Maple instruction must end with a semicolon ; before the Enter key is pressed. Hereafter we will not show the <enter> in our input.

We can now use f as an ordinary function:

```
>   f(t)+f(1);
```

$$t^3 - 2t^2 - 12t - 11$$

Let us plot the graph of f on the interval $[-4, 5]$.

```
>   plot(f(x), x=-4..5);
```

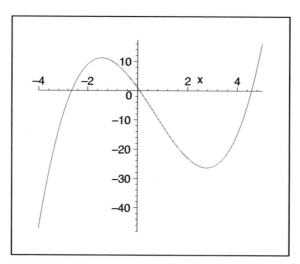

We could have specified the expression x^3-2*x^2-12*x+1 directly in the plot command instead of first defining the function $f(x)$. Note the use of two dots . . to separate the left and right endpoints of the plot interval. Other options can be included in the plot command; all such options are separated with commas. You can specify the range of values of y in addition to that for x (which is required), and you can specify scaling=CONSTRAINED if you want equal unit distances on both axes. (This would be a bad idea for the graph of our $f(x)$. Why?) The modified command looks like this. (The output graph is omitted.)

```
>   plot(f(x), x=-4..5, y=-40..30, scaling=CONSTRAINED);
```

Exercises P.4

In Exercises 1–6, find the domain and range of each function.

1. $f(x) = 1 + x^2$

2. $f(x) = 1 - \sqrt{x}$

3. $G(x) = \sqrt{8 - 2x}$

4. $F(x) = 1/(x - 1)$

5. $h(t) = \dfrac{t}{\sqrt{2 - t}}$

6. $g(x) = \dfrac{1}{1 - \sqrt{x - 2}}$

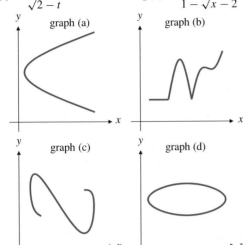

Figure P.53

7. Which of the graphs in Figure P.53 are graphs of functions $y = f(x)$? Why?

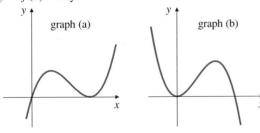

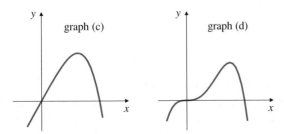

Figure P.54

8. Figure P.54 shows the graphs of the functions: (i) $x - x^4$, (ii) $x^3 - x^4$, (iii) $x(1 - x)^2$, (iv) $x^2 - x^3$. Which graph corresponds to each function?

In Exercises 9–10, sketch the graph of the function f by first making a table of values of $f(x)$ at $x = 0$, $x = \pm 1/2$, $x = \pm 1$, $x = \pm 3/2$, and $x = \pm 2$.

9. $f(x) = x^4$

10. $f(x) = x^{2/3}$

In Exercises 11–22, what (if any) symmetry does the graph of f possess? In particular, is f either even or odd?

11. $f(x) = x^2 + 1$

12. $f(x) = x^3 + x$

13. $f(x) = \dfrac{x}{x^2 - 1}$

14. $f(x) = \dfrac{1}{x^2 - 1}$

15. $f(x) = \dfrac{1}{x - 2}$

16. $f(x) = \dfrac{1}{x + 4}$

17. $f(x) = x^2 - 6x$

18. $f(x) = x^3 - 2$

19. $f(x) = |x^3|$

20. $f(x) = |x + 1|$

21. $f(x) = \sqrt{2x}$

22. $f(x) = \sqrt{(x - 1)^2}$

Sketch the graphs of the functions in Exercises 23–38.

23. $f(x) = -x^2$

24. $f(x) = 1 - x^2$

25. $f(x) = (x - 1)^2$

26. $f(x) = (x - 1)^2 + 1$

27. $f(x) = 1 - x^3$

28. $f(x) = (x + 2)^3$

29. $f(x) = \sqrt{x + 1}$

30. $f(x) = \sqrt{x + 1}$

31. $f(x) = -|x|$

32. $f(x) = |x| - 1$

33. $f(x) = |x - 2|$

34. $f(x) = 1 + |x - 2|$

35. $f(x) = \dfrac{2}{x + 2}$

36. $f(x) = \dfrac{1}{2 - x}$

37. $f(x) = \dfrac{x}{x + 1}$

38. $f(x) = \dfrac{x}{1 - x}$

In Exercises 39–46, f refers to the function with domain $[0, 2]$ and range $[0, 1]$, whose graph is shown in Figure P.55. Sketch the graphs of the indicated functions and specify their domains and ranges.

39. $f(x) + 2$

40. $f(x) - 1$

41. $f(x + 2)$

42. $f(x - 1)$

43. $-f(x)$

44. $f(-x)$

45. $f(4 - x)$

46. $1 - f(1 - x)$

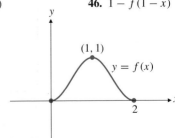

Figure P.55

It is often quite difficult to determine the range of a function exactly. In Exercises 47–48, use a graphing utility (calculator or computer) to graph the function f, and by zooming in on the graph determine the range of f with accuracy of 2 decimal places.

47. $f(x) = \dfrac{x+2}{x^2+2x+3}$ **48.** $f(x) = \dfrac{x-1}{x^2+x}$ **50.** $f(x) = \dfrac{3-2x+x^2}{2-2x+x^2}$

In Exercises 49–52, use a graphing utility to plot the graph of the given function. Examine the graph (zooming in or out as necessary) for symmetries. About what lines and/or points are the graphs symmetric? Try to verify your conclusions algebraically.

49. $f(x) = x^4 - 6x^3 + 9x^2 - 1$

51. $f(x) = \dfrac{x-1}{x-2}$ **52.** $f(x) = \dfrac{2x^2+3x}{x^2+4x+5}$

53. What function $f(x)$, defined on the real line $\mathbb{R}$, is both even and odd?

P.5 Combining Functions to Make New Functions

Functions can be combined in a variety of ways to make new functions. In this section we examine combinations obtained

(a) algebraically, by adding, subtracting, multiplying, and dividing functions;

(b) by composition, that is, taking functions of functions; and

(c) by piecing together functions defined on separate domains.

Sums, Differences, Products, Quotients, and Multiples

Like numbers, functions can be added, subtracted, multiplied, and divided (except where the denominator is zero) to produce new functions.

DEFINITION 3

Combining functions

If f and g are functions, then for every x that belongs to the domains of both f and g we define functions $f+g$, $f-g$, fg, and f/g by the formulas:

$$(f+g)(x) = f(x) + g(x)$$
$$(f-g)(x) = f(x) - g(x)$$
$$(fg)(x) = f(x)g(x)$$
$$\left(\frac{f}{g}\right)(x) = \frac{f(x)}{g(x)}, \qquad \text{where } g(x) \neq 0.$$

A special case of the rule for multiplying functions shows how functions can be multiplied by constants. If c is a real number, then the function cf is defined for all x in the domain of f by

$$(cf)(x) = c\,f(x).$$

Example 1   Figure P.56(a) shows the graphs of $f(x) = x^2$, $g(x) = x - 1$, and their sum $(f+g)(x) = x^2 + x - 1$. Observe that the height of the graph of $f+g$ at any point x is the sum of the heights of the graphs of f and g at that point.

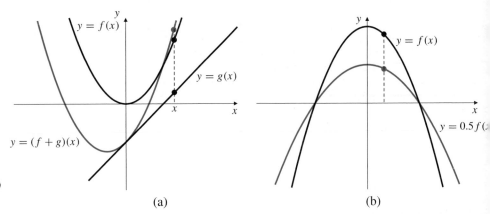

Figure P.56

(a) $(f + g)(x) = f(x) + g(x)$

(b) $g(x) = (0.5)f(x)$

Example 2   Figure P.56(b) shows the graphs of $f(x) = 2 - x^2$ and the multiple $g(x) = (0.5)f(x)$. Note how the height of the graph of g at any point x is half the height of the graph of f there.

Example 3   The functions f and g are defined by the formulas

$$f(x) = \sqrt{x} \quad \text{and} \quad g(x) = \sqrt{1 - x}.$$

Find formulas for the values of $3f, f + g, f - g, fg, f/g,$ and g/f at x, and specify the domains of each of these functions.

Solution   The information is collected in Table 2:

Table 2.   Combinations of f and g and their domains

Function	Formula	Domain
f	$f(x) = \sqrt{x}$	$[0, \infty[$
g	$g(x) = \sqrt{1 - x}$	$]-\infty, 1]$
$3f$	$(3f)(x) = 3\sqrt{x}$	$[0, \infty[$
$f + g$	$(f + g)(x) = f(x) + g(x) = \sqrt{x} + \sqrt{1 - x}$	$[0, 1]$
$f - g$	$(f - g)(x) = f(x) - g(x) = \sqrt{x} - \sqrt{1 - x}$	$[0, 1]$
fg	$(fg)(x) = f(x)g(x) = \sqrt{x(1 - x)}$	$[0, 1]$
f/g	$\dfrac{f}{g}(x) = \dfrac{f(x)}{g(x)} = \sqrt{\dfrac{x}{1 - x}}$	$[0, 1[$
g/f	$\dfrac{g}{f}(x) = \dfrac{g(x)}{f(x)} = \sqrt{\dfrac{1 - x}{x}}$	$]0, 1]$

Note that most of the combinations of f and g have domains

$$[0, \infty[\ \cap \ ]-\infty, 1] = [0, 1],$$

the intersection of the domains of f and g. However, the domains of the two quotients f/g and g/f had to be restricted further to remove points where the denominator was zero.

Composite Functions

There is another method, called **composition**, by which two functions can be combined to form a new function.

> **Composite functions**
>
> If f and g are two functions, the **composite** function $f \circ g$ is defined by
>
> $$f \circ g(x) = f(g(x)).$$
>
> The domain of $f \circ g$ consists of those numbers x in the domain of g for which $g(x)$ is in the domain of f. In particular, if the range of g is contained in the domain of f, then the domain of $f \circ g$ is just the domain of g.

As shown in Figure P.57, forming $f \circ g$ is equivalent to arranging the "function machines" g and f in an "assembly line" so that the output of g becomes the input of f.

In calculating $f \circ g(x) = f(g(x))$ we first calculate $g(x)$ and then calculate f of the result. We call g the *inner* function and f the *outer* function of the composition. We can, of course, also calculate the composition $g \circ f(x) = g(f(x))$, where f is the inner function, the one that gets calculated first, and g is the outer function which gets calculated last. The functions $f \circ g$ and $g \circ f$ are usually quite different, as the following example shows.

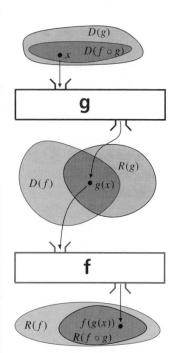

Figure P.57 $f \circ g(x) = f(g(x))$

Example 4   Given $f(x) = \sqrt{x}$ and $g(x) = x + 1$, calculate the four composite functions $f \circ g(x)$, $g \circ f(x)$, $f \circ f(x)$, and $g \circ g(x)$, and specify the domain of each.

Solution   Again, we collect the results in a table

Table 3.   Composites of f and g and their domains

Function	Formula	Domain
f	$f(x) = \sqrt{x}$	$[0, \infty[$
g	$g(x) = x + 1$	$\mathbb{R}$
$f \circ g$	$f \circ g(x) = f(g(x)) = f(x+1) = \sqrt{x+1}$	$[-1, \infty[$
$g \circ f$	$g \circ f(x) = g(f(x)) = g(\sqrt{x}) = \sqrt{x} + 1$	$[0, \infty[$
$f \circ f$	$f \circ f(x) = f(f(x)) = f(\sqrt{x}) = \sqrt{\sqrt{x}} = x^{1/4}$	$[0, \infty[$
$g \circ g$	$g \circ g(x) = g(g(x)) = g(x+1) = (x+1) + 1 = x + 2$	$\mathbb{R}$

To see why, for example, the domain of $f \circ g$ is $[-1, \infty[$, observe that $g(x) = x + 1$ is defined for all real x but belongs to the domain of f only if $x + 1 \geq 0$, that is, if $x \geq -1$. ∎

Example 5   If $G(x) = \dfrac{1 - x}{1 + x}$, calculate $G \circ G(x)$ and specify its domain.

Solution   We calculate

$$G \circ G(x) = G(G(x)) = G\left(\frac{1-x}{1+x}\right) = \frac{1 - \dfrac{1-x}{1+x}}{1 + \dfrac{1-x}{1+x}} = \frac{1+x-1+x}{1+x+1-x} = x.$$

Because the resulting function, x, is defined for all real x, we might be tempted to say that the domain of $G \circ G$ is $\mathbb{R}$. This is wrong! To belong to the domain of $G \circ G$, x must satisfy two conditions:

(i) x must belong to the domain of G, and

(ii) $G(x)$ must belong to the domain of G.

The domain of G consists of all real numbers *except* $x = -1$. If we exclude $x = -1$ from the domain of $G \circ G$, condition (i) will be satisfied. Now observe that the equation $G(x) = -1$ has no solution x, since it is equivalent to $1 - x = -(1 + x)$ or $1 = -1$. Therefore, all numbers $G(x)$ belong to the domain of G, and condition (ii) is satisfied with no further restrictions on x. The domain of $G \circ G$ is $]-\infty, -1[\,\cup\,]-1, \infty[$, that is, all real numbers except -1. ∎

Piecewise Defined Functions

Sometimes it is necessary to define a function by using different formulas on different parts of its domain. One example is the absolute value function

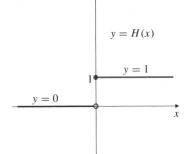

$$|x| = \begin{cases} x & \text{if } x \geq 0 \\ -x & \text{if } x < 0. \end{cases}$$

Here are some other examples:

Figure P.58   The Heaviside function

Example 6   **The Heaviside function.** The Heaviside function (Figure P.58) (or unit step function) is defined by

$$H(x) = \begin{cases} 1 & \text{if } x \geq 0 \\ 0 & \text{if } x < 0. \end{cases}$$

The function $H(t)$ can be used, for example, to model the voltage applied to an electric circuit by a one volt battery if a switch in the circuit is closed at time $t = 0$. ∎

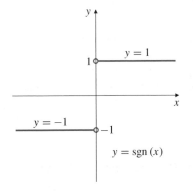

Example 7   **The signum function.** The signum function (Figure P.59) is defined by

$$\operatorname{sgn}(x) = \frac{x}{|x|} = \begin{cases} 1 & \text{if } x > 0, \\ -1 & \text{if } x < 0, \\ \text{undefined} & \text{if } x = 0. \end{cases}$$

The name *signum* is the Latin word meaning "sign." The value of the $\operatorname{sgn}(x)$ tells whether x is positive or negative. Since 0 is neither positive nor negative, $\operatorname{sgn}(0)$ is not defined. The signum function is an odd function.

Figure P.59   The signum function

Example 8   The function

$$f(x) = \begin{cases} (x+1)^2 & \text{if } x < -1, \\ -x & \text{if } -1 \le x < 1, \\ \sqrt{x-1} & \text{if } x \ge 1, \end{cases}$$

is defined on the whole real line but has values given by three different formulas depending on the position of x. Its graph is shown in Figure P.60(a). Note how solid and hollow dots are used to indicate, respectively, which endpoints do or do not lie on various parts of the graph.

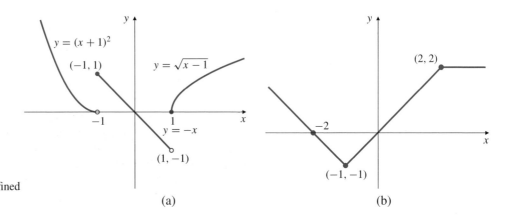

Figure P.60   Piecewise defined functions

(a)   (b)

Example 9   Find a formula for function $g(x)$ graphed in Figure P.60(b).

Solution   The graph consists of parts of three lines. For the part $x < -1$, the line has slope -1 and x-intercept -2, so its equation is $y = -(x+2)$. The middle section is the line $y = x$ for $-1 \le x \le 2$. The right section is $y = 2$ for $x > 2$. Combining these formulas, we write

$$g(x) = \begin{cases} -(x+2) & \text{if } x < -1 \\ -x & \text{if } -1 \le x \le 2 \\ 2 & \text{if } x > 2. \end{cases}$$

Unlike the previous example, it does not matter here which of the two possible formulas we use to define $g(-1)$, since both give the same value. The same is true for $g(2)$.

The following two functions could be defined by different formulas on every interval between consecutive integers, but we will use an easier way to define them.

Example 10   **The greatest integer function.** The function whose value at any number x is the *greatest integer less than or equal to x* is called the **greatest integer function**, or the **integer floor function**. It is denoted $\lfloor x \rfloor$, or, in some books, $[x]$ or $[[x]]$. The graph of $y = \lfloor x \rfloor$ is given in Figure P.61(a). Observe that

$$\lfloor 2.4 \rfloor = 2, \qquad \lfloor 1.9 \rfloor = 1, \qquad \lfloor 0 \rfloor = 0, \qquad \lfloor -1.2 \rfloor = -2,$$
$$\lfloor 2 \rfloor = 2, \qquad \lfloor 0.2 \rfloor = 0, \qquad \lfloor -0.3 \rfloor = -1, \qquad \lfloor -2 \rfloor = -2.$$

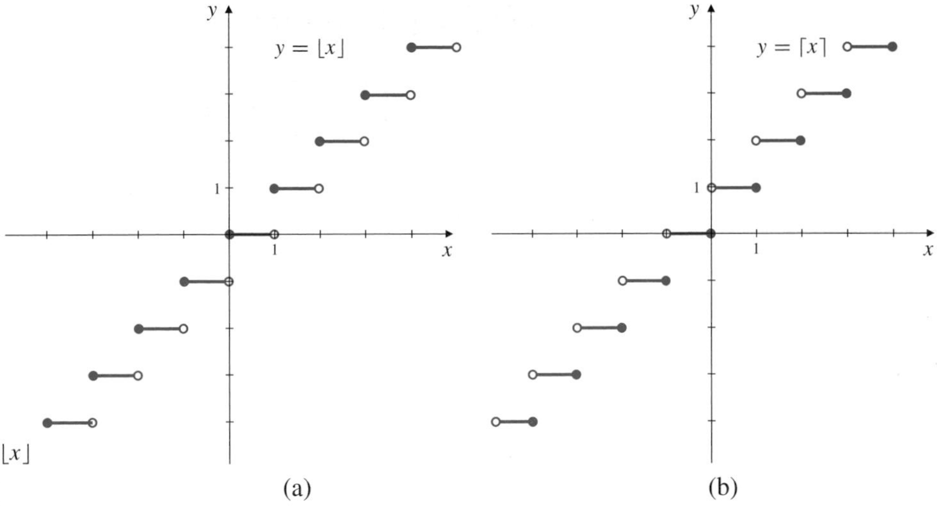

Figure P.61

(a) The greatest integer function $\lfloor x \rfloor$

(b) The least integer function $\lceil x \rceil$

(a)

(b)

Example 11   **The least integer function.** The function whose value at any number x is the *smallest integer greater than or equal to x* is called the **least integer function**, or the **integer ceiling function**. It is denoted $\lceil x \rceil$. Its graph is given in Figure P.61(b). For positive values of x, this function might represent, for example, the cost of parking x hours in a parking lot that charges $1 for each hour or part of an hour.   ∎

Exercises P.5

In Exercises 1–2, find the domains of the functions $f + g$, $f - g$, fg, f/g, and g/f, and give formulas for their values.

1. $f(x) = x$, $g(x) = \sqrt{x - 1}$

2. $f(x) = \sqrt{1 - x}$, $g(x) = \sqrt{1 + x}$

Sketch the graphs of the functions in Exercises 3–6 by combining the graphs of simpler functions from which they are built up.

3. $x - x^2$

4. $x^3 - x$

5. $x + |x|$

6. $|x| + |x - 2|$

7. If $f(x) = x + 5$ and $g(x) = x^2 - 3$, find the following:

(a) $f \circ g(0)$ (b) $g(f(0))$
(c) $f(g(x))$ (d) $g \circ f(x)$
(e) $f \circ f(-5)$ (f) $g(g(2))$
(g) $f(f(x))$ (h) $g \circ g(x)$

In Exercises 8–10, construct the following composite functions and specify the domain of each.

(a) $f \circ f(x)$ (b) $f \circ g(x)$
(c) $g \circ f(x)$ (d) $g \circ g(x)$

8. $f(x) = 2/x$, $g(x) = x/(1 - x)$

9. $f(x) = 1/(1 - x)$, $g(x) = \sqrt{x - 1}$

10. $f(x) = (x + 1)/(x - 1)$, $g(x) = \text{sgn}\,(x)$

Find the missing entries in Table 4 (Exercises 11–16).

Table 4.

	$f(x)$	$g(x)$	$f \circ g(x)$
11.	x^2	$x + 1$	
12.		$x + 4$	x
13.	$\sqrt{x}$		$|x|$
14.		$x^{1/3}$	$2x + 3$
15.	$(x + 1)/x$		x
16.		$x - 1$	$1/x^2$

17. Use a graphing utility to examine in order the graphs of the functions

$$y = \sqrt{x}, \qquad y = 2 + \sqrt{x},$$
$$y = 2 + \sqrt{3 + x}, \qquad y = 1/(2 + \sqrt{3 + x}).$$

Describe the effect on the graph of the change made in the function at each stage.

18. Repeat the previous exercise for the functions

$$y = 2x, \qquad y = 2x - 1, \qquad y = 1 - 2x,$$
$$y = \sqrt{1 - 2x}, \qquad y = \frac{1}{\sqrt{1 - 2x}}, \qquad y = \frac{1}{\sqrt{1 - 2x}} - 1.$$

In Exercises 19–24, f refers to the function with domain $[0, 2]$ and range $[0, 1]$, whose graph is shown in Figure P.62. Sketch the graphs of the indicated functions and specify their domains and ranges.

19. $2f(x)$

20. $-(1/2)f(x)$

21. $f(2x)$

22. $f(x/3)$

23. $1 + f(-x/2)$

24. $2f((x - 1)/2)$

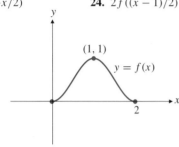

(1, 1)

$y = f(x)$

2

Figure P.62

In Exercises 25–26, sketch the graphs of the given functions.

25. $f(x) = \begin{cases} x & \text{if } 0 \leq x \leq 1 \\ 2 - x & \text{if } 1 < x \leq 2 \end{cases}$

26. $g(x) = \begin{cases} \sqrt{x} & \text{if } 0 \leq x \leq 1 \\ 2 - x & \text{if } 1 < x \leq 2 \end{cases}$

27. Find all real values of the constants A and B for which the function $F(x) = Ax + B$ satisfies:

(a) $F \circ F(x) = F(x)$ for all x.

(b) $F \circ F(x) = x$ for all x.

Greatest and least integer functions

28. For what values of x is (a) $\lfloor x \rfloor = 0$? (b) $\lceil x \rceil = 0$?

29. What real numbers x satisfy the equation $\lfloor x \rfloor = \lceil x \rceil$?

30. True or false: $\lceil -x \rceil = -\lfloor x \rfloor$ for all real x?

31. Sketch the graph of $y = x - \lfloor x \rfloor$.

32. Sketch the graph of the function

$$f(x) = \begin{cases} \lfloor x \rfloor & \text{if } x \geq 0 \\ \lceil x \rceil & \text{if } x < 0. \end{cases}$$

Why is $f(x)$ called *the integer part of x*?

Even and odd functions

33. Assume that f is an even function, g is an odd function, and both f and g are defined on the whole real line $\mathbb{R}$. Is each of the following functions even, odd, or neither?

$$f + g, \quad fg, \quad f/g, \quad g/f, \quad f^2 = ff, \quad g^2 = gg$$

$$f \circ g, \quad g \circ f, \quad f \circ f, \quad g \circ g$$

34. If f is both an even and an odd function, show that $f(x) = 0$ at every point of its domain.

35. Let f be a function whose domain is symmetric about the origin, that is, $-x$ belongs to the domain whenever x does.

(a) Show that f is the sum of an even function and an odd function:

$$f(x) = E(x) + O(x),$$

where E is an even function and O is an odd function. *Hint:* let $E(x) = (f(x) + f(-x))/2$. Show that $E(-x) = E(x)$, so that E is even. Then show that $O(x) = f(x) - E(x)$ is odd.

(b) Show that there is only one way to write f as the sum of an even and an odd function. *Hint:* one way is given in part (a). If also $f(x) = E_1(x) + O_1(x)$, where E_1 is even and O_1 is odd, show that $E - E_1 = O_1 - O$ and then use Exercise 34 to show that $E = E_1$ and $O = O_1$.

P.6 The Trigonometric Functions

Most people first encounter the quantities $\cos t$ and $\sin t$ as ratios of sides in a right-angled triangle having t as one of the acute angles. If the sides of the triangle are labelled "hyp" for hypotenuse, "adj" for the side adjacent to angle t, and "opp" for the side opposite angle t (see Figure P.63), then

$$\cos t = \frac{\text{adj}}{\text{hyp}} \quad \text{and} \quad \sin t = \frac{\text{opp}}{\text{hyp}}.$$

These ratios depend only on the angle t, not on the particular triangle, since all right-angled triangles having an acute angle t are similar.

In calculus we need more general definitions of $\cos t$ and $\sin t$ as functions defined for *all real numbers* t, not just acute angles. Such definitions are phrased in terms of a circle rather than a triangle.

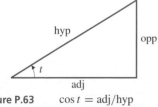

Figure P.63

hyp

opp

t

adj

$\cos t = \text{adj/hyp}$

$\sin t = \text{opp/hyp}$

Let C be the circle with centre at the origin O and radius 1; its equation is $x^2 + y^2 = 1$. Let A be the point $(1, 0)$ on C. For any real number t, let P_t be the point on C at distance $|t|$ from A, measured along C in the counterclockwise direction if $t > 0$, and the clockwise direction if $t < 0$. For example, since C has circumference 2π, the point $P_{\pi/2}$ is one-quarter of the way counterclockwise around C from A; it is the point $(0, 1)$.

We will use the arc length t as a measure of the size of the angle AOP_t. See Figure P.64.

DEFINITION 5

The **radian measure** of angle AOP_t is t radians:

$$\angle AOP_t = t \text{ radians.}$$

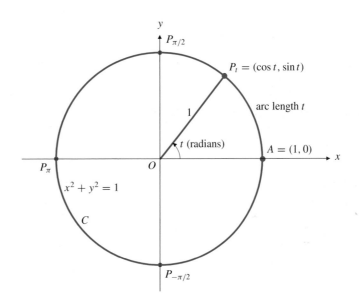

Figure P.64 If the length of arc AP_t is t units, then angle $AOP_t = t$ radians

We are more used to measuring angles in **degrees**. Since P_π is the point $(-1, 0)$, halfway (π units of distance) around C from A, we have

$$\pi \text{ radians} = 180°.$$

To convert degrees to radians, multiply by $\pi/180$; to convert radians to degrees, multiply by $180/\pi$.

Angle convention

In calculus it is assumed that all angles are measured in radians unless degrees or other units are stated explicitly. When we talk about the angle $\pi/3$, we mean $\pi/3$ radians (which is $60°$), not $\pi/3$ degrees.

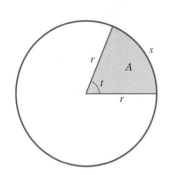

Figure P.65

Arc length $s = rt$

Sector area $A = r^2 t / 2$

Example 1 **Arc length and sector area.** An arc of a circle of radius r subtends an angle t at the centre of the circle. Find the length s of the arc and the area A of the sector lying between the arc and the centre of the circle.

Solution The length s of the arc is the same fraction of the circumference $2\pi r$ of the circle that the angle t is of a complete revolution 2π radians (or $360°$). Thus

$$s = \frac{t}{2\pi} (2\pi r) = rt \text{ units.}$$

Similarly, the area A of the circular sector (Figure P.65) is the same fraction of the area πr^2 of the whole circle:

$$A = \frac{t}{2\pi} (\pi r^2) = \frac{r^2 t}{2} \text{ units}^2.$$

(We will show that the area of a circle of radius r is πr^2 in Section 1.1.)

Using the procedure described above we can find the point P_t corresponding to any real number t, positive or negative. We define $\cos t$ and $\sin t$ to be the coordinates of P_t. (See Figure P.66.)

DEFINITION **6**

Cosine and sine

For any real t, the **cosine** of t (abbreviated $\cos t$) and the **sine** of t (abbreviated $\sin t$) are the x- and y-coordinates of the point P_t.

$$\cos t = \text{the } x\text{-coordinate of } P_t$$
$$\sin t = \text{the } y\text{-coordinate of } P_t$$

Because they are defined this way, cosine and sine are often called **the circular functions**.

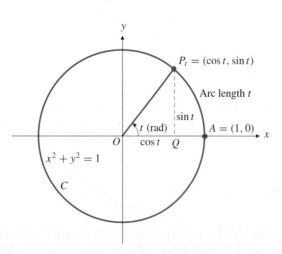

Figure P.66 The coordinates of P_t are $(\cos t, \sin t)$

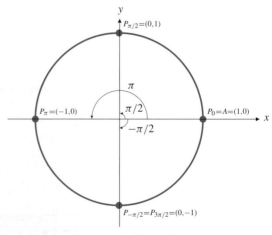

Figure P.67 Some special angles

Example 2 Examining the coordinates of $P_0 = A$, $P_{\pi/2}$, P_π, and $P_{-\pi/2} = P_{3\pi/2}$ in Figure P.67, we obtain the following values:

$$\cos 0 = 1 \quad \cos\frac{\pi}{2} = 0 \quad \cos\pi = -1 \quad \cos\left(-\frac{\pi}{2}\right) = \cos\frac{3\pi}{2} = 0$$

$$\sin 0 = 0 \quad \sin\frac{\pi}{2} = 1 \quad \sin\pi = 0 \quad \sin\left(-\frac{\pi}{2}\right) = \sin\frac{3\pi}{2} = -1$$

Some Useful Identities

Many important properties of $\cos t$ and $\sin t$ follow from the fact that they are coordinates of the point P_t on the circle C with equation $x^2 + y^2 = 1$.

The range of cosine and sine. For every real number t,

$$-1 \le \cos t \le 1 \quad \text{and} \quad -1 \le \sin t \le 1.$$

The Pythagorean identity. The coordinates $x = \cos t$ and $y = \sin t$ of P_t must satisfy the equation of the circle. Therefore, for every real number t,

$$\cos^2 t + \sin^2 t = 1.$$

(Note that $\cos^2 t$ means $(\cos t)^2$, not $\cos(\cos t)$. This is an unfortunate notation, but it is used everywhere in technical literature, so you have to get used to it!)

Periodicity. Since C has circumference 2π, adding 2π to t causes the point P_t to go one extra complete revolution around C and end up in the same place: $P_{t+2\pi} = P_t$. Thus, for every t,

$$\cos(t + 2\pi) = \cos t \quad \text{and} \quad \sin(t + 2\pi) = \sin t.$$

This says that cosine and sine are **periodic** with period 2π.

Cosine is an even function. Sine is an odd function. Since the circle $x^2 + y^2 = 1$ is symmetric about the x-axis, the points P_{-t} and P_t have the same x-coordinates and opposite y-coordinates (Figure P.68).

$$\cos(-t) = \cos t \quad \text{and} \quad \sin(-t) = -\sin t.$$

Complementary angle identities. Two angles are complementary if their sum is $\pi/2$ (or $90°$). The points $P_{(\pi/2)-t}$ and P_t are reflections of each other in the line $y = x$ (Figure P.69), so the x-coordinate of one is the y-coordinate of the other and vice versa. Thus,

$$\cos\left(\frac{\pi}{2} - t\right) = \sin t \quad \text{and} \quad \sin\left(\frac{\pi}{2} - t\right) = \cos t.$$

Supplementary angle identities. Two angles are supplementary if their sum is π (or $180°$). Since the circle is symmetric about the y-axis, $P_{\pi-t}$ and P_t have the same y-coordinates and opposite x-coordinates. (See Figure P.70.) Thus,

$$\cos(\pi - t) = -\cos t \quad \text{and} \quad \sin(\pi - t) = \sin t.$$

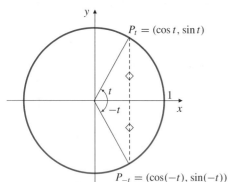

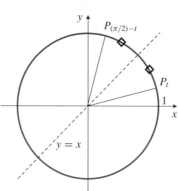

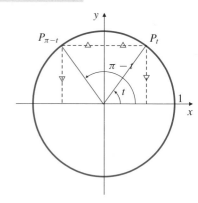

Figure P.70

Figure P.68 $\cos(-t) = \cos t$

$\sin(-t) = -\sin t$

Figure P.69 $\cos((\pi/2) - t) = \sin t$ $\cos(\pi - t) = -\cos t$

$\sin((\pi/2) - t) = \cos t$ $\sin(\pi - t) = \sin t$

Some Special Angles

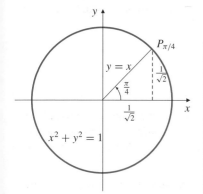

Example 3   Find the sine and cosine of $\pi/4$ (that is $45°$).

Solution   The point $P_{\pi/4}$ lies in the first quadrant on the line $x = y$. To find its coordinates, substitute $y = x$ into the equation $x^2 + y^2 = 1$ of the circle, obtaining $2x^2 = 1$. Thus $x = y = 1/\sqrt{2}$, (see Figure P.71) and

$$\cos(45°) = \cos\frac{\pi}{4} = \frac{1}{\sqrt{2}}, \qquad \sin(45°) = \sin\frac{\pi}{4} = \frac{1}{\sqrt{2}}.$$

Figure P.71 $\sin\dfrac{\pi}{4} = \cos\dfrac{\pi}{4} = \dfrac{1}{\sqrt{2}}$

Example 4   Find the values of sine and cosine of the angles $\pi/3$ (or $60°$) and $\pi/6$ (or $30°$).

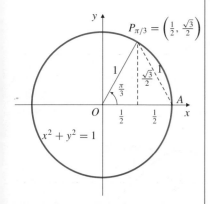

Solution   The point $P_{\pi/3}$ and the points $O(0, 0)$ and $A(1, 0)$ are the vertices of an equilateral triangle with edge length 1 (see Figure P.72). Thus $P_{\pi/3}$ has x-coordinate $1/2$ and y-coordinate $\sqrt{1 - (1/2)^2} = \sqrt{3}/2$, and

$$\cos(60°) = \cos\frac{\pi}{3} = \frac{1}{2}, \qquad \sin(60°) = \sin\frac{\pi}{3} = \frac{\sqrt{3}}{2}.$$

Since $\dfrac{\pi}{6} = \dfrac{\pi}{2} - \dfrac{\pi}{3}$, the complementary angle identities now tell us that

$$\cos(30°) = \cos\frac{\pi}{6} = \sin\frac{\pi}{3} = \frac{\sqrt{3}}{2}, \qquad \sin(30°) = \sin\frac{\pi}{6} = \cos\frac{\pi}{3} = \frac{1}{2}.$$

Figure P.72 $\cos\pi/3 = 1/2$

$\sin\pi/3 = \sqrt{3}/2$

Table 5 summarizes the values of cosine and sine at multiples of $30°$ and $45°$

between $0°$ and $180°$. The values for $120°$, $135°$, and $150°$ were determined by using the supplementary angle identities; for example,

$$\cos(120°) = \cos\left(\frac{2\pi}{3}\right) = \cos\left(\pi - \frac{\pi}{3}\right) = -\cos\left(\frac{\pi}{3}\right) = -\cos(60°) = -\frac{1}{2}.$$

Table 5. Cosines and sines of special angles

Degrees	$0°$	$30°$	$45°$	$60°$	$90°$	$120°$	$135°$	$150°$	$180°$
Radians	0	$\dfrac{\pi}{6}$	$\dfrac{\pi}{4}$	$\dfrac{\pi}{3}$	$\dfrac{\pi}{2}$	$\dfrac{2\pi}{3}$	$\dfrac{3\pi}{4}$	$\dfrac{5\pi}{6}$	π
Cosine	1	$\dfrac{\sqrt{3}}{2}$	$\dfrac{1}{\sqrt{2}}$	$\dfrac{1}{2}$	0	$-\dfrac{1}{2}$	$-\dfrac{1}{\sqrt{2}}$	$-\dfrac{\sqrt{3}}{2}$	-1
Sine	0	$\dfrac{1}{2}$	$\dfrac{1}{\sqrt{2}}$	$\dfrac{\sqrt{3}}{2}$	1	$\dfrac{\sqrt{3}}{2}$	$\dfrac{1}{\sqrt{2}}$	$\dfrac{1}{2}$	0

Example 5 Find: (a) $\sin(3\pi/4)$ and (b) $\cos(4\pi/3)$.

Solution We can draw appropriate triangles in the quadrants where the angles lie to determine the required values. See Figure P.73.

(a) $\sin(3\pi/4) = \sin(\pi - (\pi/4)) = 1/\sqrt{2}$.

(b) $\cos(4\pi/3) = \cos(\pi + (\pi/3)) = -\dfrac{1}{2}$.

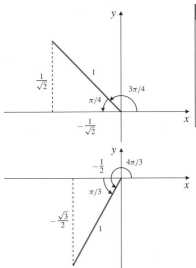

Figure P.73 Using suitably placed triangles to find trigonometric functions of special angles

While decimal approximations to the values of sine and cosine can be found using a scientific calculator or mathematical tables, it is useful to remember the exact values in the table for angles 0, $\pi/6$, $\pi/4$, $\pi/3$, and $\pi/2$. They occur frequently in applications.

When we treat sine and cosine as functions we can call the variable they depend on x (as we do with other functions), rather than t. The graphs of $\cos x$ and $\sin x$ are shown in Figures P.74 and P.75. In both graphs the pattern between $x = 0$ and $x = 2\pi$ repeats over and over to the left and right. Observe that the graph of $\sin x$ is the graph of $\cos x$ shifted to the right a distance $\pi/2$.

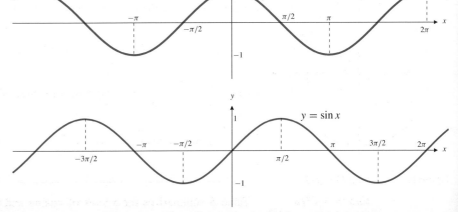

Figure P.74 The graph of $\cos x$

Figure P.75 The graph of $\sin x$

> **Remember this!**
>
> When using a scientific calculator to calculate any trigonometric functions, be sure you have selected the proper angular mode: degrees or radians.

The Addition Formulas

The following formulas enable us to determine the cosine and sine of a sum or difference of two angles in terms of the cosines and sines of those angles.

THEOREM 1

Addition Formulas for Cosine and Sine

$$\cos(s + t) = \cos s \cos t - \sin s \sin t$$
$$\sin(s + t) = \sin s \cos t + \cos s \sin t$$
$$\cos(s - t) = \cos s \cos t + \sin s \sin t$$
$$\sin(s - t) = \sin s \cos t - \cos s \sin t$$

PROOF We prove the third of these formulas as follows: Let s and t be real numbers and consider the points

$$P_t = (\cos t, \sin t) \qquad P_{s-t} = (\cos(s - t), \sin(s - t))$$
$$P_s = (\cos s, \sin s) \qquad A = (1, 0),$$

as shown in Figure P.76.

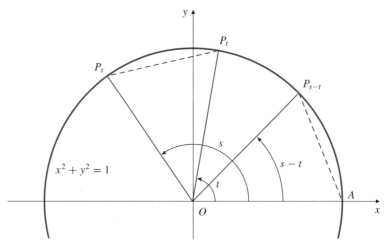

Figure P.76 $P_s P_t = P_{s-t} A$

The angle $P_t O P_s = s - t$ radians = angle $A O P_{s-t}$, so the distance $P_s P_t$ is equal to the distance $P_{s-t} A$. Therefore, $(P_s P_t)^2 = (P_{s-t} A)^2$. We express these squared distances in terms of coordinates and expand the resulting squares of binomials:

$$(\cos s - \cos t)^2 + (\sin s - \sin t)^2 = (\cos(s - t) - 1)^2 + \sin^2(s - t),$$

$$\cos^2 s - 2 \cos s \cos t + \cos^2 t + \sin^2 s - 2 \sin s \sin t + \sin^2 t$$
$$= \cos^2(s - t) - 2 \cos(s - t) + 1 + \sin^2(s - t).$$

Since $\cos^2 x + \sin^2 x = 1$ for every x, this reduces to

$$\cos(s - t) = \cos s \cos t + \sin s \sin t.$$

Replacing t with $-t$ in the formula above, and recalling that $\cos(-t) = \cos t$ and $\sin(-t) = -\sin t$, we have

$$\cos(s + t) = \cos s \cos t - \sin s \sin t.$$

The complementary angle formulas can be used to obtain either of the addition formulas for sine:

$$\begin{aligned}
\sin(s + t) &= \cos\left(\frac{\pi}{2} - (s + t)\right) \\
&= \cos\left(\left(\frac{\pi}{2} - s\right) - t\right) \\
&= \cos\left(\frac{\pi}{2} - s\right)\cos t + \sin\left(\frac{\pi}{2} - s\right)\sin t \\
&= \sin s \cos t + \cos s \sin t,
\end{aligned}$$

and the other formula again follows if we replace t with $-t$.

Example 6   Find the value of $\cos(\pi/12) = \cos 15°$.

Solution

$$\begin{aligned}
\cos\frac{\pi}{12} &= \cos\left(\frac{\pi}{3} - \frac{\pi}{4}\right) = \cos\frac{\pi}{3}\cos\frac{\pi}{4} + \sin\frac{\pi}{3}\sin\frac{\pi}{4} \\
&= \left(\frac{1}{2}\right)\left(\frac{1}{\sqrt{2}}\right) + \left(\frac{\sqrt{3}}{2}\right)\left(\frac{1}{\sqrt{2}}\right) = \frac{1 + \sqrt{3}}{2\sqrt{2}}
\end{aligned}$$

From the addition formulas, we obtain as special cases certain useful formulas called **double-angle formulas**. Put $s = t$ in the addition formulas for $\sin(s + t)$ and $\cos(s + t)$ to get

$$\begin{aligned}
\sin 2t &= 2 \sin t \cos t \qquad \text{and} \\
\cos 2t &= \cos^2 t - \sin^2 t \\
&= 2 \cos^2 t - 1 \qquad \text{(using } \sin^2 t + \cos^2 t = 1) \\
&= 1 - 2 \sin^2 t
\end{aligned}$$

Solving the last two formulas for $\cos^2 t$ and $\sin^2 t$, we obtain

$$\cos^2 t = \frac{1 + \cos 2t}{2} \qquad \text{and} \qquad \sin^2 t = \frac{1 - \cos 2t}{2},$$

which are sometimes called **half-angle formulas** because they are used to express trigonometric functions of half of the angle $2t$. Later we will find these formulas useful when we have to integrate powers of $\cos x$ and $\sin x$.

Other Trigonometric Functions

There are four other trigonometric functions—tangent (tan), cotangent (cot), secant (sec), and cosecant (csc)—each defined in terms of cosine and sine. Their graphs are shown in Figures P.77–P.80.

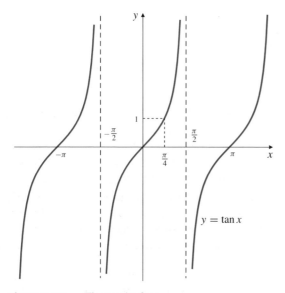

Figure P.77   The graph of tan x

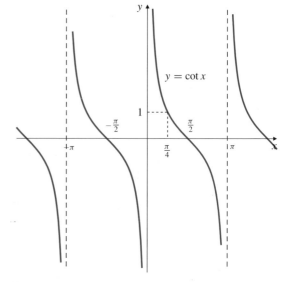

Figure P.78   The graph of cot x

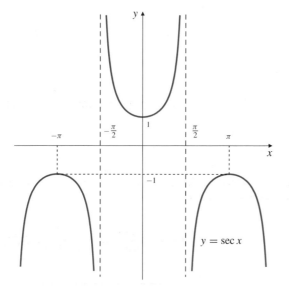

Figure P.79   The graph of sec x

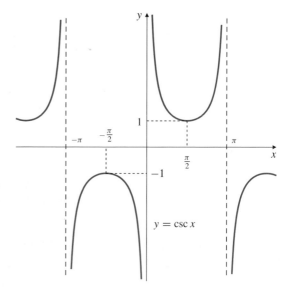

Figure P.80   The graph of csc x

DEFINITION 7

Tangent, cotangent, secant, and cosecant

$$\tan t = \frac{\sin t}{\cos t} \qquad\qquad \sec t = \frac{1}{\cos t}$$

$$\cot t = \frac{\cos t}{\sin t} = \frac{1}{\tan t} \qquad \csc t = \frac{1}{\sin t}$$

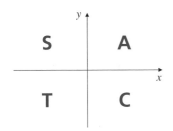

Figure P.81   The CAST rule

Observe that each of these functions is undefined (and its graph approaches vertical lines) at points where the function in the denominator of its defining fraction has value 0. Observe also that tangent, cotangent, and cosecant are odd functions and secant is an even function. Since $|\sin x| \le 1$ and $|\cos x| \le 1$ for all x, $|\csc x| \ge 1$ and $|\sec x| \ge 1$ for all x where they are defined.

The three functions sine, cosine, and tangent are called the **primary trigonometric functions** while their reciprocals cosecant, secant, and cotangent are called the **secondary trigonometric functions**. Scientific calculators usually just implement the primary functions; you can use the reciprocal key to find values of the corresponding secondary functions. Figure P.81 shows a useful pattern called the "CAST rule" to help you remember where the primary functions are positive. All three are positive in the first quadrant, marked A. Of the three, only sine is positive in the second quadrant S, only tangent in the third quadrant T, and only cosine in the fourth quadrant C.

Example 7   Find the sine and tangent of the angle θ in $\left[\pi, \dfrac{3\pi}{2}\right]$ for which $\cos\theta = -\dfrac{1}{3}$.

Solution   From the Pythagorean identity $\sin^2\theta + \cos^2\theta = 1$ we get

$$\sin^2\theta = 1 - \frac{1}{9} = \frac{8}{9}, \qquad \text{so} \quad \sin\theta = \pm\sqrt{\frac{8}{9}} = \pm\frac{2\sqrt{2}}{3}.$$

The requirement that θ should lie in $[\pi, 3\pi/2]$ makes θ a third quadrant angle. Its sine is therefore negative. We have

$$\sin\theta = -\frac{2\sqrt{2}}{3} \qquad \text{and} \qquad \tan\theta = \frac{\sin\theta}{\cos\theta} = \frac{-2\sqrt{2}/3}{-1/3} = 2\sqrt{2}.$$

Like their reciprocals cosine and sine, the functions secant and cosecant are periodic with period 2π. Tangent and cotangent, however, have period π because

$$\tan(x + \pi) = \frac{\sin(x + \pi)}{\cos(x + \pi)} = \frac{\sin x \cos\pi + \cos x \sin\pi}{\cos x \cos\pi - \sin x \sin\pi} = \frac{-\sin x}{-\cos x} = \tan x.$$

Dividing the Pythagorean identity $\sin^2 x + \cos^2 x = 1$ by $\cos^2 x$ and $\sin^2 x$, respectively, leads to two useful alternative versions of that identity:

$$1 + \tan^2 x = \sec^2 x \qquad \text{and} \qquad 1 + \cot^2 x = \csc^2 x.$$

Addition formulas for tangent and cotangent can be obtained from those for sine and cosine. For example,

$$\tan(s+t) = \frac{\sin(s+t)}{\cos(s+t)} = \frac{\sin s \cos t + \cos s \sin t}{\cos s \cos t - \sin s \sin t}.$$

Now divide the numerator and denominator of the fraction on the right by $\cos s \cos t$ to get

$$\tan(s+t) = \frac{\tan s + \tan t}{1 - \tan s \tan t}.$$

Replacing t by $-t$ leads to

$$\tan(s-t) = \frac{\tan s - \tan t}{1 + \tan s \tan t}.$$

Maple Calculations

Maple knows all six trigonometric functions and can calculate their values and manipulate them in other ways. It assumes the arguments of the trigonometric functions are in radians.

```
>   evalf(sin(30)); evalf(sin(Pi/6));
```

$$-.9880316241$$

$$.5000000000$$

Note that the constant Pi (with an upper case "P") is known to Maple. The `evalf()` function converts its argument to a number expressed as a floating point decimal with 10 significant digits. (This precision can be changed by defining a new value for the variable `Digits`.) Without it the sine of 30 radians would have been left unexpanded because it is not an integer.

```
>   Digits := 20; evalf(100*Pi); sin(30);
```

$$Digits := 20$$

$$314.15926535897932385$$

$$sin(30)$$

It is often useful to expand trigonometric functions of multiple angles to powers of sine and cosine, and vice versa.

```
>   expand(sin(5*x));
```

$$16 \sin(x) \cos(x)^4 - 12 \sin(x) \cos(x)^2 + \sin(x)$$

```
>   combine((cos(x))^5, trig);
```

$$\frac{1}{16} \cos(5x) + \frac{5}{16} \cos(3x) + \frac{5}{8} \cos(x)$$

Other trigonometric functions can be converted to expressions involving sine and cosine.

```
>   convert(tan(4*x)*(sec(4*x))^2, sincos); combine(%,trig);
```

$$\frac{\sin(4x)}{\cos(4x)^3}$$

$$4\,\frac{\sin(4x)}{\cos(12x) + 3\cos(4x)}$$

The % in the last command referred to the result of the previous calculation.

Trigonometry Review

The trigonometric functions are so called because they are often used to express the relationships between the sides and angles of a triangle. As we observed at the beginning of this section, if θ is one of the acute angles in a right-angled triangle, we can refer to the three sides of the triangle as adj (side adjacent θ), opp (side opposite θ), and hyp (hypotenuse). The trigonometric functions of θ can then be expressed as ratios of these sides, in particular:

$$\sin\theta = \frac{\text{opp}}{\text{hyp}}, \qquad \cos\theta = \frac{\text{adj}}{\text{hyp}}, \qquad \tan\theta = \frac{\text{opp}}{\text{adj}}.$$

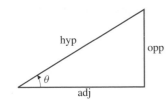

Figure P.82

Example 8 Find the unknown sides x and y of the triangle in Figure P.83.

Solution Here x is the side opposite and y is the side adjacent the $30°$ angle. The hypotenuse of the triangle is 5 units. Thus

$$\frac{x}{5} = \sin 30° = \frac{1}{2} \qquad \text{and} \qquad \frac{y}{5} = \cos 30° = \frac{\sqrt{3}}{2},$$

so $x = \dfrac{5}{2}$ units and $y = \dfrac{5\sqrt{3}}{2}$ units.

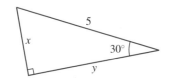

Figure P.83

Example 9 For the triangle in Figure P.84, express sides x and y in terms of side a and angle θ.

Solution The side x is opposite the angle θ and y is the hypotenuse. The side adjacent θ is a. Thus

$$\frac{x}{a} = \tan\theta \qquad \text{and} \qquad \frac{a}{y} = \cos\theta.$$

Hence, $x = a\tan\theta$ and $y = \dfrac{a}{\cos\theta} = a\sec\theta.$

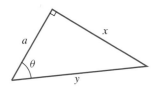

Figure P.84

When dealing with general (not necessarily right-angled) triangles, it is often convenient to label the vertices with capital letters, which also denote the angles at those vertices, and refer to the sides opposite those vertices by the corresponding lower-case letters. See Figure P.85. Relationships among the sides a, b, and c and opposite angles A, B, and C of an arbitrary triangle ABC are given by the following formulas, called the **Sine Law** and the **Cosine Law**.

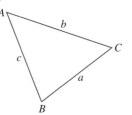

Figure P.85 In this triangle the sides are named to correspond to the opposite angles.

THEOREM 2

Sine Law: $\dfrac{\sin A}{a} = \dfrac{\sin B}{b} = \dfrac{\sin C}{c}$

Cosine Law: $a^2 = b^2 + c^2 - 2bc\cos A$

$b^2 = a^2 + c^2 - 2ac\cos B$

$c^2 = a^2 + b^2 - 2ab\cos C$

PROOF   See Figure P.86. Let h be the length of the perpendicular from A to the side BC. From right-angled triangles (and using $\sin(\pi - t) = \sin t$ if required), we get $c \sin B = h = b \sin C$. Thus $(\sin B)/b = (\sin C)/c$. By the symmetry of the formulas (or by dropping a perpendicular to another side), both fractions must be equal to $(\sin A)/a$, so the Sine Law is proved. For the Cosine Law, observe that

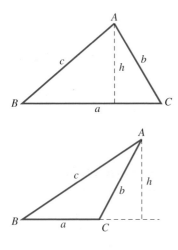

$$c^2 = \begin{cases} h^2 + (a - b\cos C)^2 & \text{if } C \le \dfrac{\pi}{2} \\[2ex] h^2 + (a + b\cos(\pi - C))^2 & \text{if } C > \dfrac{\pi}{2} \end{cases}$$

$$= h^2 + (a - b\cos C)^2 \qquad (\text{since } \cos(\pi - C) = -\cos C)$$
$$= b^2 \sin^2 C + a^2 - 2ab\cos C + b^2 \cos^2 C$$
$$= a^2 + b^2 - 2ab\cos C.$$

Figure P.86

The other versions of the Cosine Law can be proved in a similar way.

Example 10   A triangle has sides $a = 2$ and $b = 3$ and angle $C = 40°$. Find side c and the sine of angle B.

Solution   From the third version of the Cosine Law:

$$c^2 = a^2 + b^2 - 2ab\cos C = 4 + 9 - 12\cos 40° \approx 13 - 12 \times 0.766 = 3.808.$$

Side c is about $\sqrt{3.808} = 1.951$ units in length. Now using Sine Law we get

$$\sin B = b\,\frac{\sin C}{c} \approx 3 \times \frac{\sin 40°}{1.951} \approx \frac{3 \times 0.6428}{1.951} \approx 0.988.$$

A triangle is uniquely determined by any one of the the following sets of data (which correspond to the known cases of congruency of triangles in classical geometry):

1. two sides and the angle contained between them (e.g., Example 10);
2. three sides, no one of which exceeds the sum of the other two in length;
3. two angles and one side; or
4. the hypotenuse and one other side of a right-angled triangle.

In such cases you can always find the unknown sides and angles by using the Pythagorean theorem or the Sine and Cosine Laws, and the fact that the sum of the three angles of a triangle is $180°$ (or π radians).

A triangle is not determined uniquely by two sides and a non-contained angle; there may exist no triangle, one right-angled triangle, or two triangles having such data.

Example 11   In triangle ABC, angle $B = 30°$, $b = 2$, and $c = 3$. Find a.

Solution   This is one of the ambiguous cases. By the Cosine Law

$$b^2 = a^2 + c^2 - 2ac \cos B$$
$$4 = a^2 + 9 - 6a(\sqrt{3}/2).$$

Therefore, a must satisfy the equation $a^2 - 3\sqrt{3}a + 5 = 0$. Solving this equation using the quadratic formula, we obtain

$$a = \frac{3\sqrt{3} \pm \sqrt{27 - 20}}{2}$$
$$\approx 1.275 \quad \text{or} \quad 3.921$$

There are two triangles with the given data, as shown in Figure P.87.

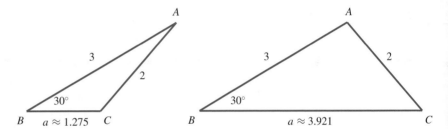

Figure P.87   Two triangles with $b = 2$, $c = 3$, $B = 30°$.

Exercises P.6

Find the values of the quantities in Exercises 1–6 using various formulas presented in this section. Do not use tables or a calculator.

1. $\cos \dfrac{3\pi}{4}$ **2.** $\tan -\dfrac{3\pi}{4}$ **3.** $\sin \dfrac{2\pi}{3}$

4. $\sin \dfrac{7\pi}{12}$ **5.** $\cos \dfrac{5\pi}{12}$ **6.** $\sin \dfrac{11\pi}{12}$

In Exercises 7–12, express the given quantity in terms of $\sin x$ and $\cos x$.

7. $\cos(\pi + x)$ **8.** $\sin(2\pi - x)$ **9.** $\sin\left(\dfrac{3\pi}{2} - x\right)$

10. $\cos\left(\dfrac{3\pi}{2} + x\right)$ **11.** $\tan x + \cot x$ **12.** $\dfrac{\tan x - \cot x}{\tan x + \cot x}$

In Exercises 13–16, prove the given identities.

13. $\cos^4 x - \sin^4 x = \cos(2x)$

14. $\dfrac{1 - \cos x}{\sin x} = \dfrac{\sin x}{1 + \cos x} = \tan \dfrac{x}{2}$

15. $\dfrac{1 - \cos x}{1 + \cos x} = \tan^2 \dfrac{x}{2}$

16. $\dfrac{\cos x - \sin x}{\cos x + \sin x} = \sec 2x - \tan 2x$

17. Express $\sin 3x$ in terms of $\sin x$ and $\cos x$.

18. Express $\cos 3x$ in terms of $\sin x$ and $\cos x$.

In Exercises 19–22, sketch the graph of the given function. What is the period of the function?

19. $f(x) = \cos 2x$ **20.** $f(x) = \sin \dfrac{x}{2}$

21. $f(x) = \sin \pi x$ **22.** $f(x) = \cos \dfrac{\pi x}{2}$

23. Sketch the graph of $y = 2 \cos\left(x - \dfrac{\pi}{3}\right)$.

24. Sketch the graph of $y = 1 + \sin\left(x + \dfrac{\pi}{4}\right)$.

In Exercises 25–30, one of $\sin \theta$, $\cos \theta$, and $\tan \theta$ is given. Find the other two if θ lies in the specified interval.

25. $\sin \theta = \dfrac{3}{5}$, θ in $\left[\dfrac{\pi}{2}, \pi\right]$

26. $\tan \theta = 2$, θ in $\left[0, \dfrac{\pi}{2}\right]$

27. $\cos \theta = \dfrac{1}{3}$, θ in $\left[-\dfrac{\pi}{2}, 0\right]$

28. $\cos \theta = -\dfrac{5}{13}$, θ in $\left[\dfrac{\pi}{2}, \pi\right]$

29. $\sin\theta = \dfrac{-1}{2}$, θ in $\left[\pi, \dfrac{3\pi}{2}\right]$

30. $\tan\theta = \dfrac{1}{2}$, θ in $\left[\pi, \dfrac{3\pi}{2}\right]$

Trigonometry Review

In Exercises 31–42, ABC is a triangle with a right angle at C. The sides opposite angles A, B, and C are a, b, and c, respectively. (See Figure P.88.)

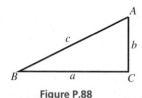

Figure P.88

31. Find a and b if $c = 2$, $B = \dfrac{\pi}{3}$.

32. Find a and c if $b = 2$, $B = \dfrac{\pi}{3}$.

33. Find b and c if $a = 5$, $B = \dfrac{\pi}{6}$.

34. Express a in terms of A and c.

35. Express a in terms of A and b.

36. Express a in terms of B and c.

37. Express a in terms of B and b.

38. Express c in terms of A and a.

39. Express c in terms of A and b.

40. Express $\sin A$ in terms of a and c.

41. Express $\sin A$ in terms of b and c.

42. Express $\sin A$ in terms of a and b.

In Exercises 43–52, ABC is an arbitrary triangle with sides a, b, and c, opposite to angles A, B, and C, respectively. (See Figure P.89.) Find the indicated quantities. Use tables or a scientific calculator if necessary.

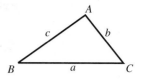

Figure P.89

43. Find $\sin B$ if $a = 4$, $b = 3$, $A = \dfrac{\pi}{4}$.

44. Find $\cos A$ if $a = 2$, $b = 2$, $c = 3$.

45. Find $\sin B$ if $a = 2$, $b = 3$, $c = 4$.

46. Find c if $a = 2$, $b = 3$, $C = \dfrac{\pi}{4}$.

47. Find a if $c = 3$, $A = \dfrac{\pi}{4}$, $B = \dfrac{\pi}{3}$.

48. Find c if $a = 2$, $b = 3$, $C = 35°$.

49. Find b if $a = 4$, $B = 40°$, $C = 70°$.

50. Find c if $a = 1$, $b = \sqrt{2}$, $A = 30°$. (There are two possible answers.)

51. Two guy wires stretch from the top T of a vertical pole to points B and C on the ground, where C is 10 m closer to the base of the pole than is B. If wire BT makes an angle of $35°$ with the horizontal, and wire CT makes an angle of $50°$ with the horizontal, how high is the pole?

52. Observers at positions A and B 2 km apart simultaneously measure the angle of elevation of a weather balloon to be $40°$ and $70°$, respectively. If the balloon is directly above a point on the line segment between A and B, find the height of the balloon.

53. Show that the area of triangle ABC is given by $(1/2)ab\sin C = (1/2)bc\sin A = (1/2)ca\sin B$.

* **54.** Show that the area of triangle ABC is given by $\sqrt{s(s-a)(s-b)(s-c)}$, where $s = (a+b+c)/2$ is the semi-perimeter of the triangle.

* This symbol is used throughout the book to indicate an exercise that is somewhat more difficult and/or theoretical than most exercises.

CHAPTER 1

Limits and Continuity

Introduction Calculus was created to describe how quantities change. It has two basic procedures that are opposites of one another:

- *differentiation,* for finding the rate of change of a given function, and
- *integration,* for finding a function having a given rate of change.

Both of these procedures are based on the fundamental concept of the *limit* of a function. It is this idea of limit that distinguishes calculus from algebra, geometry, and trigonometry, which are useful for describing static situations.

In this chapter we will introduce the limit concept and develop some of its properties. We begin by considering how limits arise in some basic problems.

1.1 Examples of Velocity, Growth Rate, and Area

In this section we consider some examples of phenomena where limits arise in a natural way.

Average Velocity and Instantaneous Velocity

The position of a moving object is a function of time. The average velocity of the object over a time interval is found by dividing the change in the object's position by the length of the time interval.

Example 1 (**The average velocity of a falling rock**) Physical experiments show that if a rock is dropped from rest near the surface of the earth, in the first t s it will fall a distance

$$y = 4.9t^2 \text{ m.}$$

(a) What is the average velocity of the falling rock during the first 2 s?

(b) What is its average velocity from $t = 1$ to $t = 2$?

Solution The *average velocity* of the falling rock over any time interval $[t_1, t_2]$ is the change Δy in the distance fallen divided by the length Δt of the time interval:

$$\text{average velocity over } [t_1, t_2] = \frac{\Delta y}{\Delta t} = \frac{4.9t_2^2 - 4.9t_1^2}{t_2 - t_1}.$$

(a) In the first 2 s (time interval $[0, 2]$), the average velocity is

$$\frac{\Delta y}{\Delta t} = \frac{4.9(2^2) - 4.9(0^2)}{2 - 0} = 9.8 \text{ m/s.}$$

(b) In the time interval [1, 2], the average velocity is

$$\frac{\Delta y}{\Delta t} = \frac{4.9(2^2) - 4.9(1^2)}{2 - 1} = 14.7 \text{ m/s.}$$

Example 2 How fast is the rock in Example 1 falling (a) at time $t = 1$? (b) at time $t = 2$?

Solution We can calculate the average velocity over any time interval, but this question asks for the *instantaneous velocity* at a given time. If the falling rock had a speedometer, what would it show at time $t = 1$? To answer this, we first write the average velocity over the time interval $[1, 1 + h]$ starting at $t = 1$ and having length h s:

$$\text{Average velocity over } [1, 1 + h] = \frac{\Delta y}{\Delta t} = \frac{4.9(1 + h)^2 - 4.9(1^2)}{h}.$$

We can't calculate the instantaneous velocity at $t = 1$ by substituting $h = 0$ in this expression, because we can't divide by zero. But we can calculate the average velocities over shorter and shorter time intervals and see whether they seem to get close to a particular number. Table 1 shows the values of $\Delta y/\Delta t$ for some values of h approaching zero. Indeed, it appears that these average velocities get closer and closer to 9.8 m/s as the length of the time interval gets closer and closer to zero. This suggests that the rock is falling at a rate of 9.8 m/s one second after it is dropped.

Similarly, Table 2 shows values of the average velocities over shorter and shorter time intervals [2, 2 + h] starting at $t = 2$. The values suggest that the rock is falling at 19.6 m/s two seconds after it is dropped.

Table 1. Average velocity over [1, 1 + h]

h	$\Delta y/\Delta t$
1	14.7000
0.1	10.2900
0.01	9.8490
0.001	9.8049
0.0001	9.8005

Table 2. Average velocity over [2, 2 + h]

h	$\Delta y/\Delta t$
1	24.5000
0.1	20.0900
0.01	19.6490
0.001	19.6049
0.0001	19.6005

In Example 2 the average velocity of the falling rock over the time interval $[t, t + h]$ is

$$\frac{\Delta y}{\Delta t} = \frac{4.9(t + h)^2 - 4.9t^2}{h}.$$

To find the instantaneous velocity (usually just called *the velocity*) at the instants $t = 1$ and $t = 2$, we examined the values of this average velocity for time intervals whose lengths h became smaller and smaller. We were, in fact, finding the *limit of the average velocity as h approaches zero.* This is expressed symbolically in the form

$$\text{velocity at time } t = \lim_{h \to 0} \frac{4.9(t + h)^2 - 4.9t^2}{h}.$$

Read "$\lim_{h \to 0} \ldots$" as "the limit as h approaches zero of $\ldots$" We can't find the limit of the fraction by just substituting $h = 0$ because that would involve dividing by zero. However, we can calculate the limit by first performing some algebraic simplifications on the expression for the average velocity.

Example 3 Expand $(t + h)^2$ in the numerator of the average velocity of the rock over $[t, t + h]$ and simplify the result. Hence find the velocity of the falling rock at time t directly, without making a table of values.

Solution The average velocity over $[t, t + h]$ is

$$\frac{4.9(t + h)^2 - 4.9t^2}{h} = \frac{4.9(t^2 + 2th + h^2 - t^2)}{h}$$
$$= \frac{4.9(2th + h^2)}{h}$$
$$= 9.8t + 4.9h.$$

The final form of the expression no longer involves division by h. It approaches $9.8t + 4.9(0) = 9.8t$ as h approaches 0. Thus the velocity of the falling rock is $9.8t$ m/s t s after it is dropped. In particular, at $t = 1$ and $t = 2$ the velocities are 9.8 m/s and 19.6 m/s, respectively.

The Growth of an Algal Culture

In a laboratory experiment, the biomass of an algal culture was measured over a 74 day period by measuring the area in square millimetres occupied by the culture on a microscope slide. These measurements m were plotted against the time t in days and the points joined by a smooth curve $m = f(t)$, as shown in Figure 1.1.

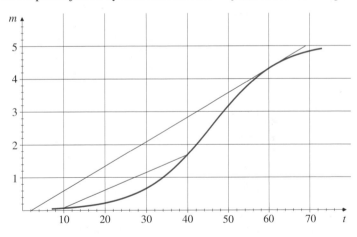

Figure 1.1 The biomass m of an algal culture after t days

Observe that the biomass was about 0.1 on day 10 and had grown to about 1.7 on day 40, an increase of $1.7 - 0.1 = 1.6$ mm^2 in a time interval of $40 - 10 = 30$ days. The average rate of growth over the time interval from day 10 to day 40 was therefore

$$\frac{1.7 - 0.1}{40 - 10} = \frac{1.6}{30} \approx 0.053 \text{ mm}^2/\text{d}.$$

This average rate is just the slope of the line joining the points on the graph of $m = f(t)$ corresponding to $t = 10$ and $t = 40$. Similarly, the average rate of growth of the algal biomass over any time interval can be determined by measuring the slope of the line joining the points on the curve corresponding to that time interval. Such lines are called **secant lines** to the curve.

Example 4 How fast is the biomass growing on day 60?

Solution To answer this question we could measure the average rates of change over shorter and shorter times around day 60. The corresponding secant lines become shorter and shorter, but their slopes approach a *limit*, namely, the slope of the **tangent line** to the graph of $m = f(t)$ at the point where $t = 60$. This tangent line is sketched in Figure 1.1; it seems to go through the points $(2, 0)$ and $(69, 5)$, so that its slope is

$$\frac{5 - 0}{69 - 2} \approx 0.0746 \text{ mm}^2/\text{d}.$$

This is the rate at which the biomass was growing on day 60.

The Area of a Circle

All circles are similar geometric figures; they all have the same shape and differ only in size. The ratio of the circumference C to the diameter $2r$ (twice the radius) has the same value for all circles. The number π is defined to be this common ratio:

$$\frac{C}{2r} = \pi \quad \text{or} \quad C = 2\pi r.$$

In school we are taught that the area A of a circle is this same number π times the square of the radius:

$$A = \pi r^2.$$

How can we deduce this area formula from the formula for the circumference that is the definition of π?

The answer to this question lies in regarding the circle as a "limit" of regular polygons, which are in turn made up of triangles, figures about whose geometry we know a great deal.

Suppose a regular polygon having n sides is inscribed in a circle of radius r. (See Figure 1.2.) The perimeter P_n and the area A_n of the polygon are, respectively, less than the circumference C and the area A of the circle, but if n is large, P_n is *close to* C and A_n is *close to* A. (In fact, the "circle" in Figure 1.2 was drawn by a computer as a regular polygon having 180 sides, each subtending a $2°$ angle at the centre of the circle. It is very hard to distinguish this 180-sided polygon from a real circle.) We would expect P_n to approach the limit C and A_n to approach the limit A as n grows larger and larger and approaches infinity.

A regular polygon of n sides is the union of n nonoverlapping, congruent, isosceles triangles having a common vertex at O, the centre of the polygon. One of these triangles, $\triangle OAB$, is shown in Figure 1.2. Since the total angle around the point O is 2π radians (we are assuming that a circle of radius 1 has circumference 2π), the angle AOB is $2\pi/n$ radians. If M is the midpoint of AB, then OM bisects angle AOB. Using elementary trigonometry, we can write the length of AB and the area of triangle OAB in terms of the radius r of the circle:

$$|AB| = 2|AM| = 2r \sin \frac{\pi}{n}$$

$$\text{area } OAB = \frac{1}{2}|AB||OM| = \frac{1}{2}\left(2r \sin \frac{\pi}{n}\right)\left(r \cos \frac{\pi}{n}\right)$$

$$= r^2 \sin \frac{\pi}{n} \cos \frac{\pi}{n}.$$

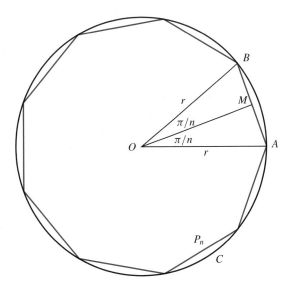

Figure 1.2   A regular polygon of n sides inscribed in a circle

The perimeter P_n and area A_n of the polygon are n times these expressions:

$$P_n = 2rn \sin \frac{\pi}{n}$$

$$A_n = r^2 n \sin \frac{\pi}{n} \cos \frac{\pi}{n}.$$

Solving the first equation for $rn \sin(\pi/n) = P_n/2$ and substituting into the second equation, we get

$$A_n = \left(\frac{P_n}{2}\right) r \cos \frac{\pi}{n}.$$

Now the angle $AOM = \pi/n$ approaches 0 as n grows large, so its cosine, $\cos(\pi/n) = |OM|/|OA|$, approaches 1. Since P_n approaches $C = 2\pi r$ as n grows large, the expression for A_n approaches $(2\pi r/2)r(1) = \pi r^2$, which must therefore be the area of the circle.

Exercises 1.1

Exercises 1–4 refer to an object moving along the x-axis in such a way that its position at time t s is $x = t^2$ m to the right of the origin.

1. Find the average velocity of the object over the time interval $[t, t + h]$.

2. Make a table giving the average velocities of the object over time intervals $[2, 2 + h]$, for $h = 1, 0.1, 0.01, 0.001$, and 0.0001 s.

3. Use the results from Exercise 2 to guess the instantaneous velocity of the object at $t = 2$ s.

4. Confirm your guess in Exercise 3 by calculating the limit of the average velocity over $[2, 2 + h]$ as h approaches zero, using the method of Example 3.

Exercises 5–8 refer to the motion of a particle moving along the x-axis so that at time t s it is at position $x = 3t^2 - 12t + 1$ m.

5. Find the average velocity of the particle over the time intervals $[1, 2], [2, 3]$, and $[1, 3]$.

6. Use the method of Example 3 to find the velocity of the particle at $t = 1, t = 2$, and $t = 3$.

7. In what direction is the particle moving at $t = 1$? $t = 2$? $t = 3$?

8. Show that for any positive number k, the average velocity of the particle over the time interval $[t - k, t + k]$ is equal to its velocity at time t.

In Exercises 9–11, a weight that is suspended by a spring bobs up and down so that its height above the floor at time t s is y ft, where

$$y = 2 + \frac{1}{\pi} \sin(\pi t).$$

9. Sketch the graph of y as a function of t. How high is the weight at $t = 1$ s? In what direction is it moving at that time?

10. What is the average velocity of the weight over the time intervals $[1, 2], [1, 1.1], [1, 1.01]$, and $[1, 1.001]$?

11. Using the results of Exercise 10, estimate the velocity of the weight at time $t = 1$. What is the significance of the sign of your answer.

Exercises 12–13 refer to the algal biomass graphed in Figure 1.1.

12. Approximately how fast is the biomass growing on day 20?

13. On about what day is the biomass growing fastest?

14. The profits of a small company for each of the first five years of its operation are given in Table 3:

Table 3.

Year	Profit ($1,000s)
1990	6
1991	27
1992	62
1993	111
1994	174

(a) Plot points representing the profit as a function of year on graph paper, and join them by a smooth curve.

(b) What is the average rate of increase in the profits between 1992 and 1994?

(c) Use your graph to estimate the rate of increase in the profits in 1992.

1.2   Limits of Functions

In order to speak meaningfully about rates of change, tangent lines, and areas bounded by curves, we have to investigate the process of finding limits. Indeed, the concept of *limit* is the cornerstone on which the development of calculus rests. Before we try to give a definition of a limit, let us look at more examples.

Example 1   Describe the behaviour of the function $f(x) = \dfrac{x^2 - 1}{x - 1}$ near $x = 1$.

Solution   Note that $f(x)$ is defined for all real numbers x except $x = 1$. (We can't divide by zero.) For any $x \neq 1$ we can simplify the expression for $f(x)$ by factoring the numerator and cancelling common factors:

$$f(x) = \frac{(x - 1)(x + 1)}{x - 1} = x + 1 \qquad \text{for} \quad x \neq 1.$$

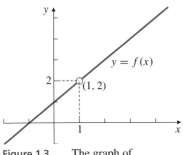

Figure 1.3   The graph of $f(x) = \dfrac{x^2 - 1}{x - 1}$

The graph of f is the line $y = x+1$ with one point removed, namely the point $(1, 2)$. This removed point is shown as a "hole" in the graph in Figure 1.3. Even though $f(1)$ is not defined, it is clear that we can make the value of $f(x)$ *as close as we want* to 2 by choosing x *close enough* to 1. Therefore, we say that $f(x)$ approaches arbitrarily close to 2 as x approaches 1, or, more simply, $f(x)$ approaches *the limit* 2 as x approaches 1. We write this as

$$\lim_{x \to 1} f(x) = 2 \qquad \text{or} \qquad \lim_{x \to 1} \frac{x^2 - 1}{x - 1} = 2.$$

Table 4.

x	$g(x)$
± 1.0	2.0000 00000
± 0.1	2.7048 13829
± 0.01	2.7181 45927
± 0.001	2.7182 80469
± 0.0001	2.7182 81815
± 0.00001	1.0000 00000

Example 2   What happens to the function $g(x) = (1 + x^2)^{1/x^2}$ as x approaches zero?

Solution   The function $g(x)$ is not defined at $x = 0$. In fact, for the moment it does not appear to be defined for any x whose square x^2 is not a rational number. (Recall that if $r = m/n$, where m and n are integers and $n > 0$, then x^r means the nth root of x^m.) Let us ignore for now the problem of deciding what $g(x)$ means if x^2 is irrational and consider only rational values of x. There is no obvious way to simplify the expression for $g(x)$ as we did in Example 1. However, we can use a scientific calculator to obtain approximate values of $g(x)$ for some rational values of x approaching 0. (The values in Table 4 were obtained with a TI-85.)

Except for the last value in the table, the values of $g(x)$ seem to be approaching a certain number $2.71828\ldots$ as x approaches closer and closer to 0. We will show in Section 3.4 that

$$\lim_{x \to 0} g(x) = \lim_{x \to 0} (1 + x^2)^{1/x^2} = e = 2.7\ 1828\ 1828\ 45\ 90\ 45\ldots.$$

The number e turns out to be very important in mathematics.

Observe that the last entry in the table appears to be wrong. This is because the calculator was unable to distinguish $1 + (0.00001)^2 = 1.0000000001$ from 1, and it therefore calculated $1^{10,000,000,000} = 1$. When you use a calculator or computer to evaluate expressions, always be aware of the effects of round-off errors. In cases like this such effects can be catastrophic.

The examples above and those in Section 1.1 suggest the following *informal* definition of limit.

DEFINITION   **1**

An informal definition of limit

If $f(x)$ is defined for all x near a, except possibly at a itself, and if we can ensure that $f(x)$ is as close as we want to L by taking x close enough to a, we say that the function f approaches the **limit** L as x approaches a, and we write

$$\lim_{x \to a} f(x) = L.$$

This definition is *informal* because phrases such as *close as we want* and *close enough* are imprecise; their meaning depends on the context. To a machinist manufacturing a piston *close enough* may mean *within a few thousandths of an inch*. To an astronomer studying distant galaxies *close enough* may mean *within a few thousand light-years*. The definition should be clear enough, however, to enable us to recognize and evaluate limits of specific functions. A more precise "formal" definition, given in Section 1.5, is needed if we want to *prove* theorems about limits like Theorems 2–4, stated later in this section.

Example 3   Find   (a) $\lim\limits_{x \to a} x$,   and   (b) $\lim\limits_{x \to a} c$ (where c is a constant).

Solution   In words, part (a) asks: "What does x approach as x approaches a?" The answer is surely a.

$$\lim_{x \to a} x = a.$$

Similarly, part (b) asks: "What does c approach as x approaches a?" The answer here is that c approaches c; you can't get any closer to c than by *being c*.

$$\lim_{x \to a} c = c.$$

Example 3 shows that $\lim_{x \to a} f(x)$ can *sometimes* be evaluated by just calculating $f(a)$. This will be the case if $f(x)$ is defined in an open interval containing $x = a$ and the graph of f passes unbroken through the point $(a, f(a))$. The next example shows various ways algebraic manipulations can be used to evaluate $\lim_{x \to a} f(x)$ in situations where $f(a)$ is undefined. This usually happens when $f(x)$ is a fraction with denominator equal to 0 at $x = a$.

Example 4   Evaluate:

(a) $\lim\limits_{x \to -2} \dfrac{x^2 + x - 2}{x^2 + 5x + 6}$,   (b) $\lim\limits_{x \to a} \dfrac{\dfrac{1}{x} - \dfrac{1}{a}}{x - a}$,   and   (c) $\lim\limits_{x \to 4} \dfrac{\sqrt{x} - 2}{x^2 - 16}$.

Solution   Each of these limits involves a fraction whose numerator and denominator are both 0 at the point where the limit is taken.

(a) $\lim\limits_{x \to -2} \dfrac{x^2 + x - 2}{x^2 + 5x + 6}$ fraction undefined at $x = -2$

    Factor numerator and denominator.

$\qquad = \lim\limits_{x \to -2} \dfrac{(x + 2)(x - 1)}{(x + 2)(x + 3)}$ Cancel common factors.

$\qquad = \lim\limits_{x \to -2} \dfrac{x - 1}{x + 3}$ Evaluate this limit by substituting $x = -2$.

$\qquad = \dfrac{-2 - 1}{-2 + 3} = -3.$

(b) $\quad \lim\limits_{x \to a} \dfrac{\dfrac{1}{x} - \dfrac{1}{a}}{x - a}$

fraction undefined at $x = a$

Simplify the numerator.

$$= \lim\limits_{x \to a} \dfrac{\dfrac{a - x}{ax}}{x - a}$$

$$= \lim\limits_{x \to a} \dfrac{-(x - a)}{ax(x - a)}$$

Cancel the common factor.

$$= \lim\limits_{x \to a} \dfrac{-1}{ax} = -\dfrac{1}{a^2}.$$

(c) $\quad \lim\limits_{x \to 4} \dfrac{\sqrt{x} - 2}{x^2 - 16}$

fraction undefined at $x = 4$

Multiply numerator and denominator

by the conjugate of the expression

in the numerator.

$$= \lim\limits_{x \to 4} \dfrac{(\sqrt{x} - 2)(\sqrt{x} + 2)}{(x^2 - 16)(\sqrt{x} + 2)}$$

$$= \lim\limits_{x \to 4} \dfrac{x - 4}{(x - 4)(x + 4)(\sqrt{x} + 2)}$$

$$= \lim\limits_{x \to 4} \dfrac{1}{(x + 4)(\sqrt{x} + 2)} = \dfrac{1}{(4 + 4)(2 + 2)} = \dfrac{1}{32}.$$

A function f may be defined on both sides of $x = a$ but still not have a limit at $x = a$. For example, the function $f(x) = 1/x$ has no limit as x approaches 0. As can be seen in Figure 1.4(a), the values $1/x$ grow ever larger in absolute value as x approaches 0; there is no single number L that they approach.

Figure 1.4

(a) $\lim\limits_{x \to 0} \dfrac{1}{x}$ does not exist.

(b) $\lim\limits_{x \to 2} g(x) = 2$, but $g(2) = 1$.

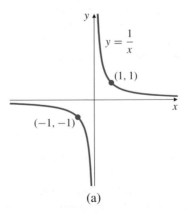

(a)

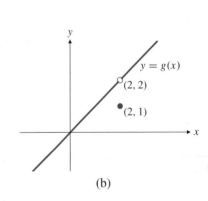

(b)

The following example shows that even if $f(x)$ is defined at $x = a$, the limit of $f(x)$ as x approaches a may not be equal to $f(a)$.

Always be aware that the existence of $\lim\limits_{x \to a} f(x)$ does not require that $f(a)$ exist and does not depend on $f(a)$ even if $f(a)$ does exist. It depends only on the values of $f(x)$ for x *near* but not equal to a.

Example 5 Let $g(x) = \begin{cases} x & \text{if } x \neq 2 \\ 1 & \text{if } x = 2. \end{cases}$ (See Figure 1.4(b).) Then

$$\lim\limits_{x \to 2} g(x) = \lim\limits_{x \to 2} x = 2, \qquad \text{although} \quad g(2) = 1.$$

One-Sided Limits

Limits are *unique*; if $\lim_{x \to a} f(x) = L$ and $\lim_{x \to a} f(x) = M$, then $L = M$. (See Exercise 31 in Section 1.5.) Although a function f can only have one limit at any particular point, it is, nevertheless, useful to be able to describe the behaviour of functions that approach different numbers as x approaches a from one side or the other. (See Figure 1.5.)

DEFINITION 2

Left and right limits

If $f(x)$ is defined on some interval $]b, a[$ extending to the left of $x = a$, and if we can ensure that $f(x)$ is as close as we want to L by taking x to the left of a and close enough to a, then we say $f(x)$ has **left limit** L at $x = a$, and we write

$$\lim_{x \to a-} f(x) = L.$$

If $f(x)$ is defined on some interval $]a, b[$ extending to the right of $x = a$, and if we can ensure that $f(x)$ is as close as we want to L by taking x to the right of a and close enough to a, then we say $f(x)$ has **right limit** L at $x = a$, and we write

$$\lim_{x \to a+} f(x) = L.$$

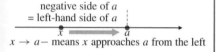

negative side of a
= left-hand side of a

$x \to a-$ means x approaches a from the left

positive side of a
= right-hand side of a

$x \to a+$ means x approaches a from the right

Figure 1.5   One-sided approach

Note the use of the suffix $+$ to denote approach from the right (the *positive* side) and the suffix $-$ to denote approach from the left (the *negative* side).

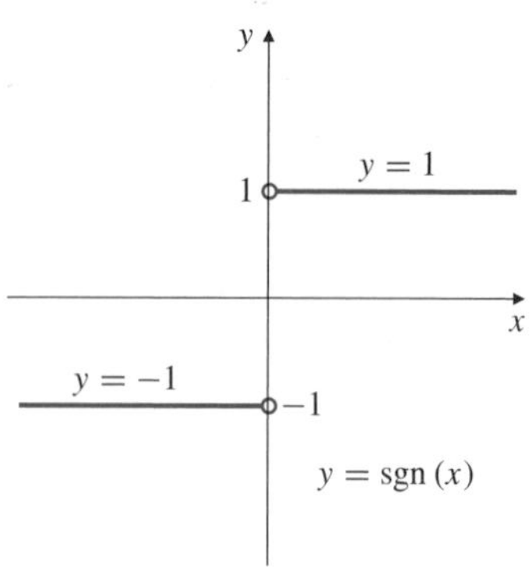

Example 6   The signum function $\operatorname{sgn}(x) = x/|x|$ (see Figure 1.6) has left limit -1 and right limit 1 at $x = 0$:

$$\lim_{x \to 0-} \operatorname{sgn}(x) = -1 \qquad \text{and} \qquad \lim_{x \to 0+} \operatorname{sgn}(x) = 1$$

because the values of $\operatorname{sgn}(x)$ approach -1 (they *are* -1) if x is negative and approaches 0, and they approach 1 if x is positive and approaches 0. Since these left and right limits are not equal, $\lim_{x \to 0} \operatorname{sgn}(x)$ *does not exist.*

Figure 1.6

$\lim_{x \to 0} \operatorname{sgn}(x)$ does not exist, because

$\lim_{x \to 0-} \operatorname{sgn}(x) = -1$, $\lim_{x \to 0+} \operatorname{sgn}(x) = 1$

As suggested in Example 6, the relationship between ordinary (two-sided) limits and one-sided limits can be stated as follows:

THEOREM 1

Relationship between one-sided and two-sided limits

A function $f(x)$ has limit L at $x = a$ if and only if it has both left and right limits there and these one-sided limits are both equal to L:

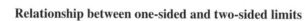

$$\lim_{x \to a} f(x) = L \quad \Longleftrightarrow \quad \lim_{x \to a-} f(x) = \lim_{x \to a+} f(x) = L.$$

Example 7   If $f(x) = \dfrac{|x-2|}{x^2+x-6}$, find: $\lim\limits_{x\to 2+} f(x)$, $\lim\limits_{x\to 2-} f(x)$, and $\lim\limits_{x\to 2} f(x)$.

Solution   Observe that $|x-2| = x-2$ if $x > 2$, and $|x-2| = -(x-2)$ if $x < 2$. Therefore,

$$
\begin{aligned}
\lim_{x\to 2+} f(x) &= \lim_{x\to 2+} \frac{x-2}{x^2+x-6} & \lim_{x\to 2-} f(x) &= \lim_{x\to 2-} \frac{-(x-2)}{x^2+x-6} \\
&= \lim_{x\to 2+} \frac{x-2}{(x-2)(x+3)} & &= \lim_{x\to 2-} \frac{-(x-2)}{(x-2)(x+3)} \\
&= \lim_{x\to 2+} \frac{1}{x+3} = \frac{1}{5}, & &= \lim_{x\to 2-} \frac{-1}{x+3} = -\frac{1}{5}.
\end{aligned}
$$

Since $\lim_{x\to 2-} f(x) \neq \lim_{x\to 2+} f(x)$, the limit $\lim_{x\to 2} f(x)$ does not exist. ∎

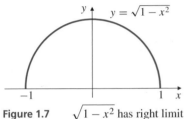

Figure 1.7 $\sqrt{1-x^2}$ has right limit 0 at -1 and left limit 0 at 1

Example 8   What limits does $g(x) = \sqrt{1-x^2}$ have at $x = -1$ and $x = 1$?

Solution   The domain of g is $[-1, 1]$, so $g(x)$ is defined only to the right of $x = -1$ and only to the left of $x = 1$. As can be seen in Figure 1.7,

$$
\lim_{x\to -1+} g(x) = 0 \qquad \text{and} \qquad \lim_{x\to 1-} g(x) = 0.
$$

$g(x)$ has no left limit or limit at $x = -1$ and no right limit or limit at $x = 1$. ∎

Rules for Calculating Limits

The following theorems make it easy to calculate limits and one-sided limits of many kinds of functions when we know some elementary limits. We will not prove the theorems here. (See Section 1.5.)

THEOREM 2

Limit Rules

If $\lim_{x\to a} f(x) = L$, $\lim_{x\to a} g(x) = M$, and k is a constant, then

1. **Limit of a sum:** $\lim\limits_{x\to a} [f(x) + g(x)] = L + M$
2. **Limit of a difference:** $\lim\limits_{x\to a} [f(x) - g(x)] = L - M$
3. **Limit of a product:** $\lim\limits_{x\to a} f(x)g(x) = LM$
4. **Limit of a multiple:** $\lim\limits_{x\to a} kf(x) = kL$
5. **Limit of a quotient:** $\lim\limits_{x\to a} \dfrac{f(x)}{g(x)} = \dfrac{L}{M}$, if $M \neq 0$.

If m is an integer and n is a positive integer, then

6. **Limit of a power:** $\lim\limits_{x\to a} [f(x)]^{m/n} = L^{m/n}$, provided $L > 0$ if n is even, and $L \neq 0$ if $m < 0$.

If $f(x) \leq g(x)$ on an interval containing a in its interior, then

7. **Order is preserved:** $L \leq M$

Rules 1–6 are also valid for right limits and left limits. So is Rule 7, under t⁺ assumption that $f(x) \leq g(x)$ on an open interval extending in the appror direction from a.

In words, part 1 of Theorem 2 says that the limit of a sum of functions is the sum of their limits. Similarly, part 5 says that the limit of a quotient of two functions is the quotient of their limits, provided that the limit of the denominator is not zero. Try to state the other parts in words.

We can make use of the limits (a) $\lim_{x \to a} c = c$ (where c is a constant) and (b) $\lim_{x \to a} x = a$, from Example 3, together with parts of Theorem 2 to calculate limits of many combinations of functions.

Example 9   Find: (a) $\lim_{x \to a} \dfrac{x^2 + x + 4}{x^3 - 2x^2 + 7}$ and   (b) $\lim_{x \to 2} \sqrt{2x + 1}$.

Solution

(a) The expression $\dfrac{x^2 + x + 4}{x^3 - 2x^2 + 7}$ is formed by combining the basic functions x and c (constant) using addition, subtraction, multiplication, and division. Theorem 2 assures us that the limit of such a combination is the same combination of the limits a and c of the basic functions, provided the denominator does not have limit zero. Thus,

$$\lim_{x \to a} \frac{x^2 + x + 4}{x^3 - 2x^2 + 7} = \frac{a^2 + a + 4}{a^3 - 2a^2 + 7} \qquad \text{provided } a^3 - 2a^2 + 7 \neq 0.$$

(b) The same argument as in (a) shows that $\lim_{x \to 2}(2x + 1) = 2(2) + 1 = 5$. Then the Power Rule (part 6 of Theorem 2) assures us that

$$\lim_{x \to 2} \sqrt{2x + 1} = \sqrt{5}.$$

■

DEFINITION 3

A **polynomial** in the variable x is a function of the form

$$P(x) = a_n x^n + a_{n-1} x^{n-1} + \cdots + a_1 x + a_0,$$

where $a_n, a_{n-1}, \ldots, a_1$, and a_0 are constants and $a_n \neq 0$.
The number n is the **degree** of the polynomial $P(x)$.
A quotient $P(x)/Q(x)$ of two polynomials is called a **rational function**.

For example,

$$3 \qquad \text{is a polynomial of degree 0.}$$
$$2 - x \qquad \text{is a polynomial of degree 1.}$$
$$2x^3 - 17x + 1 \qquad \text{is a polynomial of degree 3.}$$
$$\frac{x}{x^2 + 1} \qquad \text{is a rational function.}$$

The following result is an immediate corollary of Theorem 2.

THEOREM 3

Limits of Polynomials and Rational Functions

1. If $P(x)$ is a polynomial and a is any real number, then

$$\lim_{x \to a} P(x) = P(a).$$

2. If $P(x)$ and $Q(x)$ are polynomials and $Q(a) \neq 0$, then

$$\lim_{x \to a} \frac{P(x)}{Q(x)} = \frac{P(a)}{Q(a)}.$$

The Squeeze Theorem

The following theorem will enable us to calculate some very important limits in subsequent chapters. It is called the *Squeeze Theorem* because it refers to a function g whose values are squeezed between the values of two other functions f and h that have the same limit L at a point a. Being trapped between the values of two functions that approach L, the values of g must also approach L. (See Figure 1.8.)

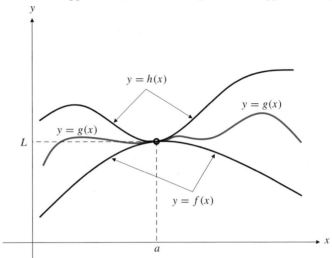

Figure 1.8   The graph of g is squeezed between those of f and h.

THEOREM 4

The Squeeze Theorem

Suppose that $f(x) \leq g(x) \leq h(x)$ hold for all x in some open interval containing a, except possibly at $x = a$ itself. Suppose also that

$$\lim_{x \to a} f(x) = \lim_{x \to a} h(x) = L.$$

Then $\lim_{x \to a} g(x) = L$ also. Similar statements hold for left and right limits.

Example 10   Given that $3 - x^2 \leq u(x) \leq 3 + x^2$ for all $x \neq 0$, find $\lim_{x \to 0} u(x)$.

Solution   Since $\lim_{x \to 0}(3 - x^2) = 3$ and $\lim_{x \to 0}(3 + x^2) = 3$, the Squeeze Theorem implies that $\lim_{x \to 0} u(x) = 3$.

Example 11   Show that if $\lim_{x \to a} |f(x)| = 0$, then $\lim_{x \to a} f(x) = 0$.

Solution   Since $-|f(x)| \leq f(x) \leq |f(x)|$, and $-|f(x)|$ and $|f(x)|$ both have limit 0 as x approaches a, so does $f(x)$ by the Squeeze Theorem.

Exercises 1.2

1. Find: (a) $\lim\limits_{x \to -1} f(x)$, (b) $\lim\limits_{x \to 0} f(x)$, and (c) $\lim\limits_{x \to 1} f(x)$, for the function f whose graph is shown in Figure 1.9.

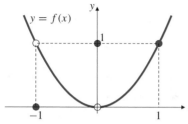

Figure 1.9

2. For the function $y = g(x)$ graphed in Figure 1.10, find each of the following limits or explain why it does not exist.
(a) $\lim\limits_{x \to 1} g(x)$, (b) $\lim\limits_{x \to 2} g(x)$, (c) $\lim\limits_{x \to 3} g(x)$

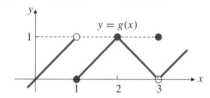

Figure 1.10

In Exercises 3–6, find the indicated one-sided limit of the function g whose graph is given in Figure 1.10.

3. $\lim\limits_{x \to 1-} g(x)$ **4.** $\lim\limits_{x \to 1+} g(x)$

5. $\lim\limits_{x \to 3+} g(x)$ **6.** $\lim\limits_{x \to 3-} g(x)$

In Exercises 7–36, evaluate the limit or explain why it does not exist.

7. $\lim\limits_{x \to 4} (x^2 - 4x + 1)$ **8.** $\lim\limits_{x \to 2} 3(1 - x)(2 - x)$

9. $\lim\limits_{x \to 3} \dfrac{x + 3}{x + 6}$ **10.** $\lim\limits_{t \to -4} \dfrac{t^2}{4 - t}$

11. $\lim\limits_{x \to 1} \dfrac{x^2 - 1}{x + 1}$ **12.** $\lim\limits_{x \to -1} \dfrac{x^2 - 1}{x + 1}$

13. $\lim\limits_{x \to 3} \dfrac{x^2 - 6x + 9}{x^2 - 9}$ **14.** $\lim\limits_{x \to -2} \dfrac{x^2 + 2x}{x^2 - 4}$

15. $\lim\limits_{h \to 2} \dfrac{1}{4 - h^2}$ **16.** $\lim\limits_{h \to 0} \dfrac{3h + 4h^2}{h^2 - h^3}$

17. $\lim\limits_{x \to 9} \dfrac{\sqrt{x} - 3}{x - 9}$ **18.** $\lim\limits_{h \to 0} \dfrac{\sqrt{4 + h} - 2}{h}$

19. $\lim\limits_{x \to \pi} \dfrac{(x - \pi)^2}{\pi x}$ **20.** $\lim\limits_{x \to -2} |x - 2|$

21. $\lim\limits_{x \to 0} \dfrac{|x - 2|}{x - 2}$ **22.** $\lim\limits_{x \to 2} \dfrac{|x - 2|}{x - 2}$

23. $\lim\limits_{t \to 1} \dfrac{t^2 - 1}{t^2 - 2t + 1}$ **24.** $\lim\limits_{x \to 2} \dfrac{\sqrt{4 - 4x + x^2}}{x - 2}$

25. $\lim\limits_{t \to 0} \dfrac{t}{\sqrt{4 + t} - \sqrt{4 - t}}$ **26.** $\lim\limits_{x \to 1} \dfrac{x^2 - 1}{\sqrt{x + 3} - 2}$

27. $\lim\limits_{t \to 0} \dfrac{t^2 + 3t}{(t + 2)^2 - (t - 2)^2}$ **28.** $\lim\limits_{s \to 0} \dfrac{(s + 1)^2 - (s - 1)^2}{s}$

29. $\lim\limits_{y \to 1} \dfrac{y - 4\sqrt{y} + 3}{y^2 - 1}$ **30.** $\lim\limits_{x \to -1} \dfrac{x^3 + 1}{x + 1}$

31. $\lim\limits_{x \to 2} \dfrac{x^4 - 16}{x^3 - 8}$ **32.** $\lim\limits_{x \to 8} \dfrac{x^{2/3} - 4}{x^{1/3} - 2}$

33. $\lim\limits_{x \to 2} \left(\dfrac{1}{x - 2} - \dfrac{4}{x^2 - 4} \right)$ **34.** $\lim\limits_{x \to 2} \left(\dfrac{1}{x - 2} - \dfrac{1}{x^2 - 4} \right)$

35. $\lim\limits_{x \to 0} \dfrac{\sqrt{2 + x^2} - \sqrt{2 - x^2}}{x^2}$ **36.** $\lim\limits_{x \to 0} \dfrac{|3x - 1| - |3x + 1|}{x}$

The limit $\lim\limits_{h \to 0} \dfrac{f(x + h) - f(x)}{h}$ occurs frequently in the study of calculus. (Can you think why?) Evaluate this limit for the functions f in Exercises 37–42.

37. $f(x) = x^2$ **38.** $f(x) = x^3$

39. $f(x) = \dfrac{1}{x}$ **40.** $f(x) = \dfrac{1}{x^2}$

41. $f(x) = \sqrt{x}$ **42.** $f(x) = 1/\sqrt{x}$

Examine the graphs of $\sin x$ and $\cos x$ in Section P.6 to determine the limits in Exercises 43–46.

43. $\lim\limits_{x \to \pi/2} \sin x$ **44.** $\lim\limits_{x \to \pi/4} \cos x$

45. $\lim\limits_{x \to \pi/3} \cos x$ **46.** $\lim\limits_{x \to 2\pi/3} \sin x$

47. Make a table of values of $f(x) = (\sin x)/x$ for a sequence of values of x approaching 0, say ± 1.0, ± 0.1, ± 0.01, ± 0.001, ± 0.0001, and ± 0.00001. Make sure your calculator is set in *radian mode* rather than degree mode. Guess the value of $\lim\limits_{x \to 0} f(x)$.

48. Repeat Exercise 47 for $f(x) = \dfrac{1 - \cos x}{x^2}$.

In Exercises 49–60, find the indicated one-sided limit or explain why it does not exist.

49. $\lim\limits_{x \to 2-} \sqrt{2 - x}$ **50.** $\lim\limits_{x \to 2+} \sqrt{2 - x}$

51. $\lim\limits_{x \to -2-} \sqrt{2 - x}$ **52.** $\lim\limits_{x \to -2+} \sqrt{2 - x}$

53. $\lim\limits_{x \to 0} \sqrt{x^3 - x}$ **54.** $\lim\limits_{x \to 0-} \sqrt{x^3 - x}$

55. $\lim\limits_{x \to 0+} \sqrt{x^3 - x}$ **56.** $\lim\limits_{x \to 0+} \sqrt{x^2 - x^4}$

57. $\lim\limits_{x\to a-} \dfrac{|x-a|}{x^2-a^2}$

58. $\lim\limits_{x\to a+} \dfrac{|x-a|}{x^2-a^2}$

59. $\lim\limits_{x\to 2-} \dfrac{x^2-4}{|x+2|}$

60. $\lim\limits_{x\to 2+} \dfrac{x^2-4}{|x+2|}$

Exercises 61–64 refer to the function

$$f(x) = \begin{cases} x-1 & \text{if } x \leq -1 \\ x^2+1 & \text{if } -1 < x \leq 0 \\ (x+\pi)^2 & \text{if } x > 0. \end{cases}$$

Find the indicated limits.

61. $\lim\limits_{x\to -1-} f(x)$

62. $\lim\limits_{x\to -1+} f(x)$

63. $\lim\limits_{x\to 0+} f(x)$

64. $\lim\limits_{x\to 0-} f(x)$

65. Suppose $\lim_{x\to 4} f(x) = 2$ and $\lim_{x\to 4} g(x) = -3$. Find:

  (a) $\lim\limits_{x\to 4} \big(g(x)+3\big)$ (b) $\lim\limits_{x\to 4} xf(x)$

  (c) $\lim\limits_{x\to 4} \big(g(x)\big)^2$ (d) $\lim\limits_{x\to 4} \dfrac{g(x)}{f(x)-1}$.

66. Suppose $\lim_{x\to a} f(x) = 4$ and $\lim_{x\to a} g(x) = -2$. Find:

  (a) $\lim\limits_{x\to a} \big(f(x)+g(x)\big)$ (b) $\lim\limits_{x\to a} f(x)\cdot g(x)$

  (c) $\lim\limits_{x\to a} 4g(x)$ (d) $\lim\limits_{x\to a} f(x)/g(x)$.

67. If $\lim\limits_{x\to 2} \dfrac{f(x)-5}{x-2} = 3$, find $\lim\limits_{x\to 2} f(x)$.

68. If $\lim\limits_{x\to 0} \dfrac{f(x)}{x^2} = -2$, find $\lim\limits_{x\to 0} f(x)$ and $\lim\limits_{x\to 0} \dfrac{f(x)}{x}$.

Using Graphing Utilities to Find Limits

Graphing calculators or computer software can be used to evaluate limits at least approximately. Simply "zoom" the plot window to show smaller and smaller parts of the graph near the point where the limit is to be found. Find the following limits by graphical techniques. Where you think it justified, give an exact answer. Otherwise, give the answer correct to 4 decimal places. Remember to ensure that your calculator or software is set for radian mode when using trigonometric functions.

69. $\lim\limits_{x\to 0} \dfrac{\sin x}{x}$

70. $\lim\limits_{x\to 0} \dfrac{\sin(2\pi x)}{\sin(3\pi x)}$

71. $\lim\limits_{x\to 1-} \dfrac{\sin\sqrt{1-x}}{\sqrt{1-x^2}}$

72. $\lim\limits_{x\to 0+} \dfrac{x-\sqrt{x}}{\sqrt{\sin x}}$

73. On the same graph plot the three functions $y = x\sin(1/x)$, $y = x$, and $y = -x$ for $-0.2 \leq x \leq 0.2$, $-0.2 \leq y \leq 0.2$. Describe the behaviour of $f(x) = x\sin(1/x)$ near $x = 0$. Does $\lim_{x\to 0} f(x)$ exist, and if so, what is its value? Could you have predicted this before drawing the graph? Why?

Using the Squeeze Theorem

74. If $\sqrt{5-2x^2} \leq f(x) \leq \sqrt{5-x^2}$ for $-1 \leq x \leq 1$, find $\lim\limits_{x\to 0} f(x)$.

75. If $2-x^2 \leq g(x) \leq 2\cos x$ for all x, find $\lim\limits_{x\to 0} g(x)$.

76. (a) Sketch the curves $y = x^2$ and $y = x^4$ on the same graph. Where do they intersect?

  (b) The function $f(x)$ satisfies:

$$\begin{cases} x^2 \leq f(x) \leq x^4 & \text{if } x < -1 \text{ or } x > 1 \\ x^4 \leq f(x) \leq x^2 & \text{if } -1 \leq x \leq 1 \end{cases}$$

  Find   (i) $\lim\limits_{x\to -1} f(x)$,   (ii) $\lim\limits_{x\to 0} f(x)$,   (iii) $\lim\limits_{x\to 1} f(x)$.

77. On what intervals is $x^{1/3} < x^3$? On what intervals is $x^{1/3} > x^3$? If the graph of $y = h(x)$ always lies between the graphs of $y = x^{1/3}$ and $y = x^3$, for what real numbers a can you determine the value of $\lim_{x\to a} h(x)$? Find the limit for each of these values of a.

* **78.** What is the domain of $x \sin \dfrac{1}{x}$? Evaluate $\lim\limits_{x\to 0} x \sin \dfrac{1}{x}$.

* **79.** Suppose $|f(x)| \leq g(x)$ for all x. What can you conclude about $\lim_{x\to a} f(x)$ if $\lim_{x\to a} g(x) = 0$? What if $\lim_{x\to a} g(x) = 3$?

1.3   Limits at Infinity and Infinite Limits

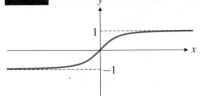

Figure 1.11   The graph of $x/\sqrt{x^2+1}$

In this section we will extend the concept of limit to allow for two situations not covered by the definitions of limit and one-sided limit in the previous section:

  (i) limits at infinity, where x becomes arbitrarily large, positive or negative;

  (ii) infinite limits, which are not really limits at all but provide useful symbolism for describing the behaviour of functions whose values become arbitrarily large, positive or negative.

Limits at Infinity

Consider the function

$$f(x) = \frac{x}{\sqrt{x^2 + 1}}$$

whose graph is shown in Figure 1.11 and for which some values (rounded to 7 decimal places) are given in Table 5. The values of $f(x)$ approach 1 as x takes on larger and larger positive values, and -1 as x takes on negative values that get larger and larger in absolute value. We express this behaviour by writing

$$\lim_{x \to \infty} f(x) = 1 \qquad \text{``$f(x)$ approaches 1 as x approaches infinity.''}$$

$$\lim_{x \to -\infty} f(x) = -1 \qquad \text{``$f(x)$ approaches -1 as x approaches negative infinity.''}$$

The graph of f conveys this limiting behaviour by approaching the horizontal lines $y = 1$ as x moves far to the right, and $y = -1$ as x moves far to the left. These lines are called **horizontal asymptotes** of the graph. In general, if a curve approaches a straight line as it recedes very far away from the origin, that line is called an **asymptote** of the curve.

Table 5.

x	$f(x) = x/\sqrt{x^2 + 1}$
$-1{,}000$	-0.9999995
-100	-0.9999500
-10	-0.9950372
-1	-0.7071068
0	0.0000000
1	0.7071068
10	0.9950372
100	0.9999500
$1{,}000$	0.9999995

DEFINITION 4

Limits at infinity and negative infinity (informal definition)

If the function f is defined on an interval $]a, \infty[$ and if we can ensure that $f(x)$ is as close as we want to the number L by taking x large enough, then we say that $f(x)$ **approaches the limit L as x approaches infinity**, and we write

$$\lim_{x \to \infty} f(x) = L.$$

If f is defined on an interval $]-\infty, b[$ and if we can ensure that $f(x)$ is as close as we want to the number M by taking x negative and large enough in absolute value, then we say that $f(x)$ **approaches the limit M as x approaches negative infinity**, and we write

$$\lim_{x \to -\infty} f(x) = M.$$

Recall that the symbol ∞, called **infinity**, does *not* represent a real number. We cannot use ∞ in arithmetic in the usual way, but we can use the phrase "approaches ∞" to mean "becomes arbitrarily large positive" and the phrase "approaches $-\infty$" to mean "becomes arbitrarily large negative."

Example 1   In Figure 1.12, we can see that $\lim_{x \to \infty} 1/x = \lim_{x \to -\infty} 1/x = 0$. The x-axis is a horizontal asymptote of the graph $y = 1/x$. ∎

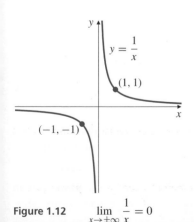

Figure 1.12 $\lim_{x \to \pm\infty} \dfrac{1}{x} = 0$

The theorems of Section 1.2 have suitable counterparts for limits at infinity or negative infinity. In particular, it follows from the example above and from the Product Rule for limits that $\lim_{x \to \pm\infty} 1/x^n = 0$ for any positive integer n. We will use this fact in the following examples. Example 2 shows how to obtain the limits at $\pm\infty$ for the function $x/\sqrt{x^2 + 1}$ by algebraic means, without resorting to making a table of values or drawing a graph, as we did above.

Example 2 Evaluate $\lim\limits_{x\to\infty} f(x)$ and $\lim\limits_{x\to-\infty} f(x)$ for $f(x) = \dfrac{x}{\sqrt{x^2+1}}$.

Solution Rewrite the expression for $f(x)$ as follows:

$$f(x) = \frac{x}{\sqrt{x^2\left(1+\dfrac{1}{x^2}\right)}} = \frac{x}{\sqrt{x^2}\sqrt{1+\dfrac{1}{x^2}}} \qquad \text{Remember } \sqrt{x^2} = |x|.$$

$$= \frac{x}{|x|\sqrt{1+\dfrac{1}{x^2}}}$$

$$= \frac{\operatorname{sgn} x}{\sqrt{1+\dfrac{1}{x^2}}}, \qquad \text{where } \operatorname{sgn} x = \frac{x}{|x|} = \begin{cases} 1 & \text{if } x > 0 \\ -1 & \text{if } x < 0. \end{cases}$$

The factor $\sqrt{1+(1/x^2)}$ approaches 1 as x approaches ∞ or $-\infty$, so $f(x)$ must have the same limits as $x \to \pm\infty$ as does $\operatorname{sgn}(x)$. Therefore,

$$\lim_{x\to\infty} f(x) = 1 \qquad \text{and} \qquad \lim_{x\to-\infty} f(x) = -1.$$

Limits at Infinity for Rational Functions

The only polynomials that have limits at $\pm\infty$ are constant ones, $P(x) = c$. The situation is more interesting for rational functions. Recall that a rational function is a quotient of two polynomials. The following examples show how to render such a function in a form where its limits at infinity and negative infinity (if they exist) are apparent. The way to do this is to *divide the numerator and denominator by the highest power of x appearing in the denominator.* The limits of a rational function at infinity and negative infinity either both fail to exist or both exist and are equal.

Example 3 **(Numerator and denominator of the same degree)** Evaluate $\lim_{x\to\pm\infty} \dfrac{2x^2 - x + 3}{3x^2 + 5}$.

Solution Divide the numerator and the denominator by x^2, the highest power of x appearing in the denominator:

$$\lim_{x\to\pm\infty} \frac{2x^2 - x + 3}{3x^2 + 5} = \lim_{x\to\pm\infty} \frac{2 - (1/x) + (3/x^2)}{3 + (5/x^2)} = \frac{2 - 0 + 0}{3 + 0} = \frac{2}{3}.$$

Example 4 **(Degree of numerator less than degree of denominator)** Evaluate $\lim_{x\to\pm\infty} \dfrac{5x + 2}{2x^3 - 1}$.

Solution Divide the numerator and the denominator by the largest power of x in the denominator, namely, x^3.

$$\lim_{x \to \pm\infty} \frac{5x+2}{2x^3-1} = \lim_{x \to \pm\infty} \frac{(5/x^2)+(2/x^3)}{2-(1/x^3)} = \frac{0+0}{2-0} = 0.$$

∎

Summary of limits at $\pm\infty$ for rational functions

Let $P_m(x) = a_m x^m + \cdots + a_0$ and $Q_n(x) = b_n x^n + \cdots + b_0$ be polynomials of degree m and n, respectively. Then

$$\lim_{x \to \pm\infty} \frac{P_m(x)}{Q_n(x)}$$

(a) equals zero if $m < n$,

(b) equals $\dfrac{a_m}{b_n}$ if $m = n$,

(c) does not exist if $m > n$.

The limiting behaviour of rational functions at infinity and negative infinity is summarized at the left.

The technique used in the previous examples can also be applied to more general kinds of functions. The function in the following example is not rational, and the limit seems to produce a meaningless $\infty - \infty$ until we resolve matters by rationalizing the numerator.

Example 5   Find $\lim_{x \to \infty}(\sqrt{x^2+x} - x)$.

Solution   We are trying to find the limit of the difference of two functions, each of which becomes arbitrarily large as x increases to infinity. We rationalize the expression by multiplying the numerator and the denominator (which is 1) by the conjugate expression $\sqrt{x^2+x} + x$:

$$\lim_{x \to \infty}\left(\sqrt{x^2+x} - x\right) = \lim_{x \to \infty} \frac{\left(\sqrt{x^2+x} - x\right)\left(\sqrt{x^2+x} + x\right)}{\sqrt{x^2+x} + x}$$

$$= \lim_{x \to \infty} \frac{x^2 + x - x^2}{\sqrt{x^2\left(1 + \dfrac{1}{x}\right)} + x}$$

$$= \lim_{x \to \infty} \frac{x}{x\sqrt{1 + \dfrac{1}{x}} + x} = \lim_{x \to \infty} \frac{1}{\sqrt{1 + \dfrac{1}{x}} + 1} = \frac{1}{2}.$$

(Here, $\sqrt{x^2} = x$ because $x > 0$ as $x \to \infty$.)

∎

Remark   The limit $\lim_{x \to -\infty}(\sqrt{x^2+x} - x)$ is not nearly so subtle. Since $-x > 0$ as $x \to -\infty$, we have $\sqrt{x^2+x} - x > \sqrt{x^2+x}$, which grows arbitrarily large as $x \to -\infty$. The limit does not exist.

Infinite Limits

A function whose values grow arbitrarily large can sometimes be said to have an infinite limit. Since infinity is not a number, infinite limits are not really limits at all, but they provide a way of describing the behaviour of functions that grow arbitrarily large positive or negative. A few examples will make the terminology clear.

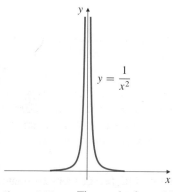

$$y = \frac{1}{x^2}$$

Figure 1.13   The graph of $y = 1/x^2$ (not to scale)

Example 6   (**A two-sided infinite limit**) Describe the behaviour of the function $f(x) = 1/x^2$ near $x = 0$.

Solution   As x approaches 0 from either side, the values of $f(x)$ are positive and grow larger and larger (see Figure 1.13), so the limit of $f(x)$ as x approaches 0 *does not exist*. It is nevertheless convenient to describe the behaviour of f near 0 by saying that $f(x)$ *approaches* ∞ as x approaches zero. We write

$$\lim_{x \to 0} f(x) = \lim_{x \to 0} \frac{1}{x^2} = \infty.$$

Note that in writing this we are *not* saying that $\lim_{x \to 0} 1/x^2$ *exists*. Rather, we are saying that that limit *does not exist because* $1/x^2$ *becomes arbitrarily large near* $x = 0$. Observe how the graph of f approaches the y-axis as x approaches 0. The y-axis is a **vertical asymptote** of the graph. ∎

Example 7   **(One-sided infinite limits)** Describe the behaviour of the function $f(x) = 1/x$ near $x = 0$. (See Figure 1.14.)

Solution   As x approaches 0 from the right, the values of $f(x)$ become larger and larger positive numbers, and we say that f has right-hand limit infinity at $x = 0$:

$$\lim_{x \to 0+} f(x) = \infty.$$

Similarly, the values of $f(x)$ become larger and larger negative numbers as x approaches 0 from the left, so f has left-hand limit $-\infty$ at $x = 0$:

$$\lim_{x \to 0-} f(x) = -\infty.$$

These statements do not say that the one-sided limits *exist*; they do not exist because ∞ and $-\infty$ are not numbers. Since the one-sided limits are not equal even as infinite symbols, all we can say about the two-sided $\lim_{x \to 0} f(x)$ is that it does not exist. ∎

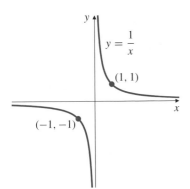

Figure 1.14 $\lim_{x \to 0-} 1/x = -\infty$, $\lim_{x \to 0+} 1/x = \infty$

Example 8   **(Polynomial behaviour at infinity)**
(a) $\lim_{x \to \infty}(3x^3 - x^2 + 2) = \infty$ (b) $\lim_{x \to -\infty}(3x^3 - x^2 + 2) = -\infty$
(c) $\lim_{x \to \infty}(x^4 - 5x^3 - x) = \infty$ (d) $\lim_{x \to -\infty}(x^4 - 5x^3 - x) = \infty$

The highest-degree term of a polynomial dominates the other terms as $|x|$ grows large, so the limits of this term at ∞ and $-\infty$ determine the limits of the whole polynomial. For the polynomial in parts (a) and (b) we have

$$3x^3 - x^2 + 2 = 3x^3\left(1 - \frac{1}{3x} + \frac{2}{3x^3}\right).$$

The factor in the large parentheses approaches 1 as x approaches $\pm\infty$, so the behaviour of the polynomial is just that of its highest-degree term $3x^3$. ∎

We can now say a bit more about the limits at infinity and negative infinity of a rational function whose numerator has higher degree than the denominator. Earlier in this section we said that such a limit *does not exist*. This is true, but we can assign ∞ or $-\infty$ to such limits, as the following example shows.

Example 9   **(Rational functions with numerator of higher degree)** Evaluate
$$\lim_{x \to \infty} \frac{x^3 + 1}{x^2 + 1}.$$

Solution Divide the numerator and the denominator by x^2, the largest power of x in the denominator:

$$\lim_{x \to \infty} \frac{x^3 + 1}{x^2 + 1} = \lim_{x \to \infty} \frac{x + \dfrac{1}{x^2}}{1 + \dfrac{1}{x^2}} = \frac{\lim_{x \to \infty} \left(x + \dfrac{1}{x^2} \right)}{1} = \infty.$$

∎

A polynomial $Q(x)$ of degree $n > 0$ can have at most n *zeros*; that is, there are at most n different real numbers r for which $Q(r) = 0$. If $Q(x)$ is the denominator of a rational function $R(x) = P(x)/Q(x)$, that function will be defined for all x except those finitely many zeros of Q. At each of those zeros $R(x)$ may have limits, infinite limits, or one-sided infinite limits. Here are some examples.

Example 10

(a) $\displaystyle \lim_{x \to 2} \frac{(x-2)^2}{x^2 - 4} = \lim_{x \to 2} \frac{(x-2)^2}{(x-2)(x+2)} = \lim_{x \to 2} \frac{x-2}{x+2} = 0.$

(b) $\displaystyle \lim_{x \to 2} \frac{x-2}{x^2 - 4} = \lim_{x \to 2} \frac{x-2}{(x-2)(x+2)} = \lim_{x \to 2} \frac{1}{x+2} = \frac{1}{4}.$

(c) $\displaystyle \lim_{x \to 2+} \frac{x-3}{x^2 - 4} = \lim_{x \to 2+} \frac{x-3}{(x-2)(x+2)} = -\infty.$ (The values are negative for $x > 2$, x near 2.)

(d) $\displaystyle \lim_{x \to 2-} \frac{x-3}{x^2 - 4} = \lim_{x \to 2-} \frac{x-3}{(x-2)(x+2)} = \infty.$ (The values are positive for $x < 2$, x near 2.)

(e) $\displaystyle \lim_{x \to 2} \frac{x-3}{x^2 - 4} = \lim_{x \to 2} \frac{x-3}{(x-2)(x+2)}$ does not exist.

(f) $\displaystyle \lim_{x \to 2} \frac{2-x}{(x-2)^3} = \lim_{x \to 2} \frac{-(x-2)}{(x-2)^3} = \lim_{x \to 2} \frac{-1}{(x-2)^2} = -\infty.$

In parts (a) and (b) the effect of the zero in the denominator at $x = 2$ is cancelled because the numerator is zero there also. Thus a finite limit exists. This is not true in part (f) because the numerator only vanishes once at $x = 2$, while the denominator vanishes three times there.

∎

Using Maple to Calculate Limits

Maple's `limit` procedure can be easily used to calculate limits, one-sided limits, limits at infinity, and infinite limits. Here is the syntax for calculating

$$\lim_{x \to 2} \frac{x^2 - 4}{x^2 - 5x + 6}, \quad \lim_{x \to 0} \frac{x \sin x}{1 - \cos x}, \quad \lim_{x \to -\infty} \frac{x}{\sqrt{x^2 + 1}}, \quad \lim_{x \to \infty} \frac{x}{\sqrt{x^2 + 1}},$$

$$\lim_{x \to 0} \frac{1}{x}, \quad \lim_{x \to 0-} \frac{1}{x}, \quad \lim_{x \to a-} \frac{x^2 - a^2}{|x - a|}, \quad \text{and} \quad \lim_{x \to a+} \frac{x^2 - a^2}{|x - a|}$$

```
> limit((x^2-4)/(x^2-5*x+6),x=2);
```
$$-4$$
```
> limit(x*sin(x)/(1-cos(x)),x=0);
```
$$2$$

```
>  limit(x/sqrt(x^2+1),x=-infinity);
```
$$-1$$
```
>  limit(x/sqrt(x^2+1),x=infinity);
```
$$1$$
```
>  limit(1/x,x=0);  limit(1/x,x=0,left);
```
undefined
$$-\infty$$
```
>  limit((x^2-a^2)/(abs(x-a)),x=a,left);
```
$$-2\,a$$
```
>  limit((x^2-a^2)/(abs(x-a)),x=a,right);
```
$$2\,a$$

Exercises 1.3

Find the limits in Exercises 1–10.

1. $\displaystyle\lim_{x\to\infty}\frac{x}{2x-3}$

2. $\displaystyle\lim_{x\to\infty}\frac{x}{x^2-4}$

3. $\displaystyle\lim_{x\to\infty}\frac{3x^3-5x^2+7}{8+2x-5x^3}$

4. $\displaystyle\lim_{x\to-\infty}\frac{x^2-2}{x-x^2}$

5. $\displaystyle\lim_{x\to-\infty}\frac{x^2+3}{x^3+2}$

6. $\displaystyle\lim_{x\to\infty}\frac{x^2+\sin x}{x^2+\cos x}$

7. $\displaystyle\lim_{x\to\infty}\frac{3x+2\sqrt{x}}{1-x}$

8. $\displaystyle\lim_{x\to\infty}\frac{2x-1}{\sqrt{3x^2+x+1}}$

9. $\displaystyle\lim_{x\to-\infty}\frac{2x-1}{\sqrt{3x^2+x+1}}$

10. $\displaystyle\lim_{x\to-\infty}\frac{2x-5}{|3x+2|}$

In Exercises 11–34 evaluate the indicated limit. If it does not exist, is the limit ∞, $-\infty$, or neither?

11. $\displaystyle\lim_{x\to3}\frac{1}{3-x}$

12. $\displaystyle\lim_{x\to3}\frac{1}{(3-x)^2}$

13. $\displaystyle\lim_{x\to3-}\frac{1}{3-x}$

14. $\displaystyle\lim_{x\to3+}\frac{1}{3-x}$

15. $\displaystyle\lim_{x\to-5/2}\frac{2x+5}{5x+2}$

16. $\displaystyle\lim_{x\to-2/5}\frac{2x+5}{5x+2}$

17. $\displaystyle\lim_{x\to-(2/5)-}\frac{2x+5}{5x+2}$

18. $\displaystyle\lim_{x\to-(2/5)+}\frac{2x+5}{5x+2}$

19. $\displaystyle\lim_{x\to2+}\frac{x}{(2-x)^3}$

20. $\displaystyle\lim_{x\to1-}\frac{x}{\sqrt{1-x^2}}$

21. $\displaystyle\lim_{x\to1+}\frac{1}{|x-1|}$

22. $\displaystyle\lim_{x\to1-}\frac{1}{|x-1|}$

23. $\displaystyle\lim_{x\to2}\frac{x-3}{x^2-4x+4}$

24. $\displaystyle\lim_{x\to1+}\frac{\sqrt{x^2-x}}{x-x^2}$

25. $\displaystyle\lim_{x\to\infty}\frac{x+x^3+x^5}{1+x^2+x^3}$

26. $\displaystyle\lim_{x\to\infty}\frac{x^3+3}{x^2+2}$

∗ 27. $\displaystyle\lim_{x\to\infty}\frac{x\sqrt{x+1}\left(1-\sqrt{2x+3}\right)}{7-6x+4x^2}$

28. $\displaystyle\lim_{x\to\infty}\left(\frac{x^2}{x+1}-\frac{x^2}{x-1}\right)$

∗ 29. $\displaystyle\lim_{x\to-\infty}\left(\sqrt{x^2+2x}-\sqrt{x^2-2x}\right)$

∗ 30. $\displaystyle\lim_{x\to\infty}\left(\sqrt{x^2+2x}-\sqrt{x^2-2x}\right)$

31. $\displaystyle\lim_{x\to\infty}\frac{1}{\sqrt{x^2-2x}-x}$

32. $\displaystyle\lim_{x\to-\infty}\frac{1}{\sqrt{x^2+2x}-x}$

33. What are the horizontal asymptotes of $y=\dfrac{1}{\sqrt{x^2-2x}-x}$? What are its vertical asymptotes?

34. What are the horizontal and vertical asymptotes of $y=\dfrac{2x-5}{|3x+2|}$?

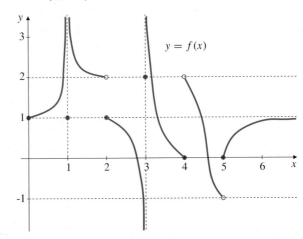

Figure 1.15

The function f whose graph is shown in Figure 1.15 has domain $[0, \infty)$. Find the following limits of f indicated in Exercises 35–45.

35. $\lim\limits_{x \to 0+} f(x)$

36. $\lim\limits_{x \to 1} f(x)$

37. $\lim\limits_{x \to 2+} f(x)$

38. $\lim\limits_{x \to 2-} f(x)$

39. $\lim\limits_{x \to 3-} f(x)$

40. $\lim\limits_{x \to 3+} f(x)$

41. $\lim\limits_{x \to 4+} f(x)$

42. $\lim\limits_{x \to 4-} f(x)$

43. $\lim\limits_{x \to 5-} f(x)$

44. $\lim\limits_{x \to 5+} f(x)$

45. $\lim\limits_{x \to \infty} f(x)$

46. What asymptotes does the graph in Figure 1.15 have?

Exercises 47–52 refer to the greatest integer function $\lfloor x \rfloor$ graphed in Figure 1.16. Find the indicated limit or explain why it does not exist.

47. $\lim\limits_{x \to 3+} \lfloor x \rfloor$

48. $\lim\limits_{x \to 3-} \lfloor x \rfloor$

49. $\lim\limits_{x \to 3} \lfloor x \rfloor$

50. $\lim\limits_{x \to 2.5} \lfloor x \rfloor$

51. $\lim\limits_{x \to 0+} \lfloor 2 - x \rfloor$

52. $\lim\limits_{x \to -3-} \lfloor x \rfloor$

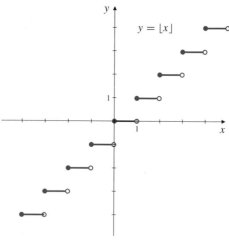

Figure 1.16

53. Parking in a certain parking lot costs $1.50 for each hour or part of an hour. Sketch the graph of the function $C(t)$ representing the cost of parking for t hours. At what values of t does $C(t)$ have a limit? Evaluate $\lim_{t \to t_0-} C(t)$ and $\lim_{t \to t_0+} C(t)$ for an arbitrary number $t_0 > 0$.

54. If $\lim_{x \to 0+} f(x) = L$, find $\lim_{x \to 0-} f(x)$ if (a) f is even, (b) f is odd.

55. If $\lim_{x \to 0+} f(x) = A$ and $\lim_{x \to 0-} f(x) = B$, find

(a) $\lim\limits_{x \to 0+} f(x^3 - x)$

(b) $\lim\limits_{x \to 0-} f(x^3 - x)$

(c) $\lim\limits_{x \to 0-} f(x^2 - x^4)$

(d) $\lim\limits_{x \to 0+} f(x^2 - x^4)$.

1.4 Continuity

When a car is driven along a highway, its distance from its starting point depends on time in a *continuous* way, changing by small amounts over short intervals of time. But not all quantities change in this way. When the car is parked in a parking lot where the rate is quoted as "$2.00 per hour or portion," the parking charges remain at $2.00 for the first hour and then suddenly jump to $4.00 as soon as the first hour has passed. The function relating parking charges to parking time will be called *discontinuous* at each hour. In this section we will define continuity and show how to tell whether a function is continuous. We will also examine some important properties possessed by continuous functions.

Continuity at a Point

Most functions that we encounter have domains that are intervals, or unions of separate intervals. A point P in the domain of such a function is called an **interior point** of the domain if it belongs to some *open* interval contained in the domain. If it is not an interior point, then P is called an **endpoint** of the domain. For example, the domain of the function $f(x) = \sqrt{4 - x^2}$ is the closed interval $[-2, 2]$, which consists of interior points in the interval $]-2, 2[$, a left endpoint -2, and a right endpoint 2. The domain of the function $g(x) = 1/x$ is the union of open intervals

$]-\infty, 0[\cup]0, \infty[$ and consists entirely of interior points. Note that although 0 is an endpoint of each of those intervals, it does not belong to the domain of g.

> **DEFINITION 5**
>
> **Continuity at an interior point**
>
> We say that a function f is **continuous** at an interior point c of its domain if
>
> $$\lim_{x \to c} f(x) = f(c).$$
>
> If either $\lim_{x \to c} f(x)$ fails to exist, or it exists but is not equal to $f(c)$, then we will say that f is **discontinuous** at c.

In graphical terms, f is continuous at an interior point c of its domain if its graph has no break in it at the point $(c, f(c))$; in other words, if you can draw the graph through that point without lifting your pen from the paper. Consider Figure 1.17. In (a), f is continuous at c. In (b), f is discontinuous at c because $\lim_{x \to c} f(x) \neq f(c)$. In (c), f is discontinuous at c because $\lim_{x \to c} f(x)$ does not exist. In both (b) and (c) the graph of f has a break at $x = c$.

Figure 1.17

(a) f is continuous at c.

(b) $\lim_{x \to c} f(x) \neq f(c)$.

(c) $\lim_{x \to c} f(x)$ does not exist.

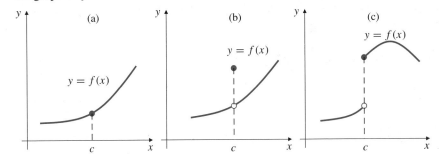

Although a function cannot have a limit at an endpoint of its domain, it can still have a one-sided limit there. We extend the definition of continuity to provide for such situations.

> **DEFINITION 6**
>
> **Right and left continuity**
>
> We say that f is **right continuous** at c if $\lim_{x \to c+} f(x) = f(c)$.
>
> We say that f is **left continuous** at c if $\lim_{x \to c-} f(x) = f(c)$.

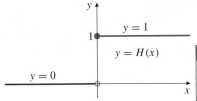

Figure 1.18   The Heaviside function

Example 1   The Heaviside function $H(x)$, whose graph is shown in Figure 1.18, is continuous at every number x except 0. It is right continuous at 0 but is not left continuous or continuous there.

The relationship between continuity and one-sided continuity is summarized in the following theorem.

THEOREM 5

Function f is continuous at c if and only if it is both right continuous and left continuous at c.

Continuity at an endpoint

We say that f is continuous at a left endpoint c of its domain if it is right continuous there.

We say that f is continuous at a right endpoint c of its domain if it is left continuous there.

Example 2   The function $f(x) = \sqrt{4 - x^2}$ has domain $[-2, 2]$. It is continuous at the right endpoint 2 because it is left continuous there, that is, because $\lim_{x \to 2-} f(x) = 0 = f(2)$. It is continuous at the left endpoint -2 because it is right continuous there: $\lim_{x \to -2+} f(x) = 0 = f(-2)$. Of course, f is also continuous at every interior point of its domain. If $-2 < c < 2$, then $\lim_{x \to c} f(x) = \sqrt{4 - c^2} = f(c)$. (See Figure 1.19.)

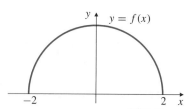

Figure 1.19 $f(x) = \sqrt{4 - x^2}$ is continuous at every point of its domain

Continuity on an Interval

We have defined the concept of continuity at a point. Of greater importance is the concept of continuity on an interval.

Continuity on an interval

We say that function f is **continuous on the interval** I if it is continuous at each point of I. In particular, we will say that f is a **continuous function** if f is continuous at every point of its domain.

Example 3   The function $f(x) = \sqrt{x}$ is a continuous function. Its domain is $[0, \infty[$. It is continuous at the left endpoint 0 because it is right continuous there. Also, f is continuous at every number $c > 0$ since $\lim_{x \to c} \sqrt{x} = \sqrt{c}$.

Example 4   The function $g(x) = 1/x$ is also a continuous function. This may seem wrong to you at first glance because its graph is broken at $x = 0$. (See Figure 1.20.) However, the number 0 is not in the domain of g, so we will prefer to say that g is undefined rather than discontinuous there. (Some authors would say that g is discontinuous at $x - 0$.) If we were to define $g(0)$ to be some number, say 0, then we would say that $g(x)$ is discontinuous at 0. There is no way of defining $g(0)$ so that g becomes continuous at 0.

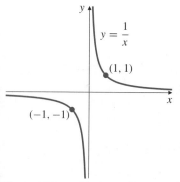

Figure 1.20 $1/x$ is continuous on its domain

Example 5   The greatest integer function $\lfloor x \rfloor$ (see Figure 1.16) is continuous on every interval $[n, n + 1[$, where n is an integer. It is right continuous at each integer n but is not left continuous there, so it is discontinuous at the integers.

$$\lim_{x \to n+} \lfloor x \rfloor = n = \lfloor n \rfloor, \qquad \lim_{x \to n-} \lfloor x \rfloor = n - 1 \neq n = \lfloor n \rfloor.$$

There Are Lots of Continuous Functions

The following functions are continuous wherever they are defined:

(a) all polynomials;

(b) all rational functions;

(c) all rational powers $x^{m/n} = \sqrt[n]{x^m}$;

(d) the sine, cosine, tangent, secant, cosecant, and cotangent functions defined in Section P.6; and

(e) the absolute value function $|x|$.

Theorem 3 of Section 1.2 assures us that every polynomial is continuous everywhere on the real line, and every rational function is continuous everywhere on its domain (which consists of all real numbers except the finitely many where its denominator is zero). If m and n are integers and $n \neq 0$, the rational power function $x^{m/n}$ is defined for all positive numbers x, and also for all negative numbers x if n is odd. The domain includes 0 if and only if $m/n \geq 0$.

The following theorems show that if we combine continuous functions in various ways, the results will be continuous.

THEOREM 6

Combining continuous functions

If the functions f and g are both defined on an interval containing c and both are continuous at c, then the following functions are also continuous at c:

1. the sum $f + g$ and the difference $f - g$;

2. the product fg;

3. the constant multiple kf, where k is any number;

4. the quotient f/g (provided $g(c) \neq 0$); and

5. the nth root $(f(x))^{1/n}$, provided $f(c) > 0$ if n is even.

The proof involves using the various limit rules in Theorem 2 of Section 1.2. For example,

$$\lim_{x \to c} \big(f(x) + g(x) \big) = \lim_{x \to c} f(x) + \lim_{x \to c} g(x) = f(c) + g(c),$$

so $f + g$ is continuous.

THEOREM 7

Composites of continuous functions are continuous

If $f(g(x))$ is defined on an interval containing c, and if f is continuous at L and $\lim_{x \to c} g(x) = L$, then

$$\lim_{x \to c} f(g(x)) = f(L) = f\left(\lim_{x \to c} g(x) \right).$$

In particular, if g is continuous at c (so $L = g(c)$), then the composition $f \circ g$ is continuous at c:

$$\lim_{x \to c} f(g(x)) = f(g(c)).$$

(See Exercise 37 in Section 1.5.)

Example 6   The following functions are continuous everywhere on their respective domains:

(a) $3x^2 - 2x$

(b) $\dfrac{x-2}{x^2-4}$

(c) $|x^2 - 1|$

(d) $\sqrt{x}$

(e) $\sqrt{x^2 - 2x - 5}$

(f) $\dfrac{|x|}{\sqrt{|x+2|}}$

Continuous Extensions and Removable Discontinuities

As we have seen in Section 1.2, a rational function may have a limit even at a point where its denominator is zero. If $f(c)$ is not defined, but $\lim_{x \to c} f(x) = L$ exists, we can define a new function $F(x)$ by

$$F(x) = \begin{cases} f(x) & \text{if } x \text{ is in the domain of } f \\ L & \text{if } x = c. \end{cases}$$

$F(x)$ is continuous at $x = c$. It is called the **continuous extension** of $f(x)$ to $x = c$. For rational functions f, continuous extensions are usually found by cancelling common factors.

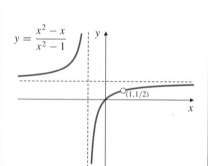

$$y = \frac{x^2 - x}{x^2 - 1}$$

Example 7   Show that $f(x) = \dfrac{x^2 - x}{x^2 - 1}$ has a continuous extension to $x = 1$, and find that extension.

Solution   Although $f(1)$ is not defined, if $x \ne 1$ we have

$$f(x) = \frac{x^2 - x}{x^2 - 1} = \frac{x(x-1)}{(x+1)(x-1)} = \frac{x}{x+1}.$$

The function

$$F(x) = \frac{x}{x+1}$$

is equal to $f(x)$ for $x \ne 1$ but is also continuous at $x = 1$, having there the value $1/2$. The graph of f is shown in Figure 1.21. The continuous extension of $f(x)$ to $x = 1$ is $F(x)$. It has the same graph as $f(x)$ except with no hole at $(1, 1/2)$.

Figure 1.21   This function has a continuous extension to $x = 1$

If a function f is undefined or discontinuous at a point a but can be (re)defined at that *single point* so that it becomes continuous there, then we say that f has a **removable discontinuity** at a. The function f in the above example has a removable discontinuity at $x = 1$. To remove it, define $f(1) = 1/2$.

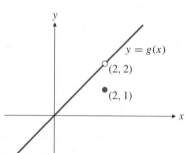

Figure 1.22 g has a removable discontinuity at 2

Example 8   The function $g(x) = \begin{cases} x & \text{if } x \ne 2 \\ 1 & \text{if } x = 2 \end{cases}$ has a removable discontinuity at $x = 2$. To remove it, redefine $g(2) = 2$. (See Figure 1.22.)

Continuous Functions on Closed, Finite Intervals

Continuous functions that are defined on *closed, finite intervals* have special properties that make them particularly useful in mathematics and its applications. We will discuss two of these properties here. Although they may appear obvious, these properties are much more subtle than the results about limits stated earlier in this chapter; their proofs (see Appendix II) require a careful study of the implications of the completeness property of the real numbers.

The first of the properties states that a function $f(x)$ that is continuous on a closed, finite interval $[a, b]$ must have an **absolute maximum value** and an **absolute minimum value**. This means that the values of $f(x)$ at all points of the interval lie between the values of $f(x)$ at two particular points in the interval; the graph of f has a highest point and a lowest point.

THEOREM **8**

The Max-Min Theorem

If $f(x)$ is continuous on the closed, finite interval $[a, b]$, then there exist numbers x_1 and x_2 in $[a, b]$ such that for all x in $[a, b]$,

$$f(x_1) \leq f(x) \leq f(x_2).$$

Thus f has the absolute minimum value $m = f(x_1)$, taken on at the point x_1, and the absolute maximum value $M = f(x_2)$, taken on at the point x_2.

Many important problems in mathematics and its applications come down to having to find maximum and minimum values of functions. Calculus provides some very useful tools for solving such problems. Observe, however, that the theorem above merely asserts that minimum and maximum values *exist;* it doesn't tell us how to find them. In Chapter 4 we will develop techniques for calculating maximum and minimum values of functions. For now, we can solve some simple maximum or minimum value problems involving quadratic functions by completing the square without using any calculus.

Example 9 What is the largest possible area of a rectangular field that can be enclosed by 200 m of fencing?

Solution If the sides of the field are x m and y m (Figure 1.23), then its perimeter is $P = 2x + 2y$ m, and its area is $A = xy$ m^2. We are given that $P = 200$, so $x + y = 100$, and $y = 100 - x$. Neither side can be negative, so x must belong to the closed interval $[0, 100]$. The area of the field can be expressed as a function of x by substituting $100 - x$ for y:

$$A = x(100 - x) = 100x - x^2.$$

We want to find the maximum value of the quadratic function $A(x) = 100x - x^2$ on the interval $[0, 100]$. Theorem 8 assures us that such a maximum exists.

To find the maximum, we complete the square of the function $A(x)$. Note that $x^2 - 100x$ are the first two terms of the square $(x - 50)^2 = x^2 - 100x + 2{,}500$. Thus

$$A(x) = 2{,}500 - (x - 50)^2.$$

Observe that $A(50) = 2{,}500$, and $A(x) < 2{,}500$ if $x \neq 50$, because we are subtracting a positive number $(x - 50)^2$ from $2{,}500$ in this case. Therefore, the maximum value of $A(x)$ is $2{,}500$. The largest field has area $2{,}500$ m^2 and is actually a square with dimensions $x = y = 50$ m.

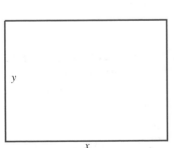

Figure 1.23   Rectangular field: perimeter $= 2x + 2y$, area $= xy$

Theorem 8 implies that a function that is continuous on a closed, finite interval is **bounded**. This means that it cannot take on arbitrarily large positive or negative values; there must exist a number K such that

$$|f(x)| \le K, \qquad \text{that is,} \qquad -K \le f(x) \le K.$$

In fact, for K we can use the larger of the numbers $|f(x_1)|$ and $|f(x_2)|$.

The conclusions of Theorem 8 may fail if the function f is not continuous or if the interval is not closed. See Figures 1.24–1.27 for examples of how such failure can occur.

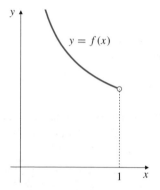

Figure 1.24 $f(x) = 1/x$ is continuous on the open interval $]0, 1[$. It is not bounded and has neither a maximum nor a minimum value

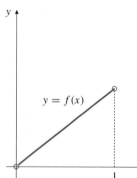

Figure 1.25 $f(x) = x$ is continuous on the open interval $]0, 1[$. It is bounded but has neither a maximum nor a minimum value

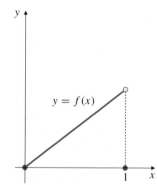

Figure 1.26   This function is defined on the closed interval $[0, 1]$ but is discontinuous at the endpoint $x = 1$. It has a minimum value but no maximum value

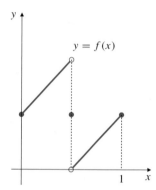

Figure 1.27   This function is discontinuous at an interior point of its domain, the closed interval $[0, 1]$. It is bounded but has neither maximum nor minimum values

Finding Maxima and Minima Graphically

Remark   Graphing utilities can be used to find maximum and minimum values of functions on intervals where they are continuous. In particular, the "zoom box" and "trace" facilities of graphing calculators are helpful. Figure 1.28(a) shows the graph of

$$y = f(x) = \frac{x+1}{x^2+1}$$

on the window $-5 \le x \le 5$, $-2 \le y \le 2$. Observe that f appears to have a maximum value near $x = 0.5$ and a minimum value near $x = -2.5$. Figure 1.28(b) shows the result of expanding the part of the graph in (a) enclosed in the small rectangle (zoom box) to fill the whole screen. Tracing the curve to its highest point gives a more accurate estimate of the maximum value, showing that $f(x)$ has maximum value 1.2071 at $x = 0.4149$, each to 4 significant figures. Further zooming enables us to get any desired degree of accuracy.

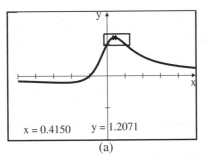
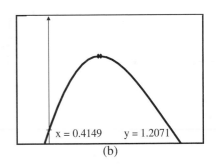

x = 0.4150   y = 1.2071

(a)

x = 0.4149   y = 1.2071

(b)

Figure 1.28   Using a "zoom box" to zoom part of a curve (a) near a maximum value to fill the screen (b) without allowing the curve to become flattened

The second property of a continuous function defined on a closed, finite interval is that the function takes on all real values between any two of its values. This property is called the **intermediate-value property**.

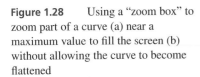

The Intermediate-Value Theorem

If $f(x)$ is continuous on the interval $[a, b]$ and if s is a number between $f(a)$ and $f(b)$, then there exists a number c in $[a, b]$ such that $f(c) = s$.

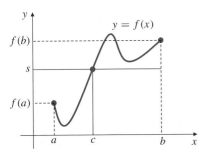

Figure 1.29   The continuous function f takes on the value s at some point c between a and b

In particular, a continuous function defined on a closed interval takes on all values between its minimum value m and its maximum value M, so its range is also a closed interval, $[m, M]$.

Figure 1.29 shows a typical situation. The points $(a, f(a))$ and $(b, f(b))$ are on opposite sides of the horizontal line $y = s$. Being unbroken, the graph $y = f(x)$ must cross this line in order to go from one point to the other. In the figure, it crosses the line only once, at $x = c$. If the line $y = s$ were somewhat higher, there might have been three crossings and three possible values for c.

Theorem 9 is the reason why the graph of a function that is continuous on an interval I cannot have any breaks. It must be **connected**, a single, unbroken curve with no jumps.

Example 10   Determine the intervals on which $f(x) = x^3 - 4x$ is positive and negative.

Solution   Since $f(x) = x(x^2 - 4) = x(x - 2)(x + 2)$, $f(x) = 0$ only at $x = 0$, 2, and -2. Because f is continuous on the whole real line, it must have constant sign on each of the intervals $]-\infty, -2[$, $]-2, 0[$, $]0, 2[$, and $]2, \infty[$. (If there were points a and b in one of those intervals, say in $]0, 2[$, such that $f(a) < 0$ and $f(b) > 0$, then by the Intermediate-Value Theorem there would exist c between a and b, and therefore between 0 and 2, such that $f(c) = 0$. But we know f has no such zero in $]0, 2[$.)

To find whether $f(x)$ is positive or negative throughout each interval, pick a point in the interval and evaluate f at that point.

Since $f(-3) = -15 < 0$, $f(x)$ is negative on $]-\infty, -2[$.

Since $f(-1) = 3 > 0$, $f(x)$ is positive on $]-2, 0[$.

Since $f(1) = -3 < 0$, $f(x)$ is negative on $]0, 2[$.

Since $f(3) = 15 > 0$, $f(x)$ is positive on $]2, \infty[$.

Finding Roots of Equations

Among the many useful tools that calculus will provide are ones that enable us to calculate solutions to equations of the form $f(x) = 0$ to any desired degree of accuracy. Such a solution is called a **root** of the equation, or a **zero** of the function f. Using these tools usually requires previous knowledge that the equation has a solution in some interval. The Intermediate-Value Theorem can provide this information.

Example 11   Show that the equation $x^3 - x - 1 = 0$ has a solution in the interval $[1, 2]$.

Solution   The function $f(x) = x^3 - x - 1$ is a polynomial and is therefore continuous everywhere. Now $f(1) = -1$ and $f(2) = 5$. Since 0 lies between -1 and 5, the Intermediate-Value Theorem assures us that there must be a number c in $[1, 2]$ such that $f(c) = 0$.

One method for finding a zero of a function that is continuous and changes sign on an interval involves bisecting the interval many times, each time determining which half of the previous interval must contain the root, because the function has opposite signs at the two ends of that half. This method is slow. For example, if the original interval has length 1, it will take 11 bisections to cut down to an interval of length less than 0.0005 (because $2^{11} > 2{,}000 = 1/(0.0005)$), and thus to ensure that we have found the root correct to three decimal places. But this method requires no graphics hardware and is easily implemented with a calculator, preferably one into which the formula for the function can be programmed.

Example 12   **(The bisection method)** Solve the equation $x^3 - x - 1 = 0$ of Example 11 correct to three decimal places by successive bisections.

Solution   We start out knowing that there is a root in $[1, 2]$. Table 6 shows the results of the bisections.

Table 6.   The bisection method for $f(x) = x^3 - x - 1 = 0$

Bisection Number	x	$f(x)$	Root in Interval	Midpoint
	1	-1		
	2	5	$[1, 2]$	1.5
1	1.5	0.8750	$[1, 1.5]$	1.25
2	1.25	-0.2969	$[1.25, 1.5]$	1.375
3	1.375	0.2246	$[1.25, 1.375]$	1.3125
4	1.3125	-0.0515	$[1.3125, 1.375]$	1.3438
5	1.3438	0.0826	$[1.3125, 1.3438]$	1.3282
6	1.3282	0.0147	$[1.3125, 1.3282]$	1.3204
7	1.3204	-0.0186	$[1.3204, 1.3282]$	1.3243
8	1.3243	-0.0018	$[1.3243, 1.3282]$	1.3263
9	1.3263	0.0065	$[1.3243, 1.3263]$	1.3253
10	1.3253	0.0025	$[1.3243, 1.3253]$	1.3248
11	1.3248	0.0003	$[1.3243, 1.3248]$	1.3246
12	1.3246	-0.0007	$[1.3246, 1.3248]$	

The root is 1.325, rounded to 3 decimal places. In Section 4.6, calculus will provide us with much faster methods of solving equations such as the one above. ∎

You can use a graphing utility to solve an equation $f(x) = 0$. Just graph the function $f(x)$ over a large enough interval so that you can see roughly where its zeros are. Then select one zero at a time and zoom in on it by successively expanding the part of the viewing window near the zero to fill the whole viewing window. Keep zooming until you can estimate the zero to as many decimal places as you want (or as the calculator or computer will allow).

Many programmable calculators and computer algebra software packages have built-in routines for solving equations. For example, Maple's `fsolve` routine can be used to find the real solution of $x^3 - x - 1 = 0$ in $[1, 2]$. (See Example 11.)

```
>  fsolve(x^3-x-1=0,x=1..2);
```

$$1.324717957$$

Remark The Max-Min Theorem and the Intermediate-Value Theorem are examples of what mathematicians call **existence theorems**. Such theorems assert that something exists without telling you how to find it. Students sometimes complain that mathematicians worry too much about proving that a problem has a solution and not enough about how to find that solution. They argue: "If I can calculate a solution to a problem, then surely I do not need to worry about whether a solution exists." This is, however, false logic. Suppose we pose the problem, "Find the largest positive integer." Of course this problem has no solution; there is no largest positive integer because we can add 1 to any integer and get a larger integer. Suppose, however, that we forget this and try to calculate a solution. We could proceed as follows:

Let N be the largest positive integer.
Since 1 is a positive integer, we must have $N \geq 1$.
Since N^2 is a positive integer, it cannot exceed the largest positive integer.
Therefore, $N^2 \leq N$ and so $N^2 - N \leq 0$.
Thus, $N(N - 1) \leq 0$ and we must have $N - 1 \leq 0$.
Therefore, $N \leq 1$.
We also know that $N \geq 1$; therefore $N = 1$.
Therefore, 1 is the largest positive integer.

The only error we have made here is in the assumption (in the first line) that the problem has a solution. It is partly to avoid logical pitfalls like this that mathematicians prove existence theorems.

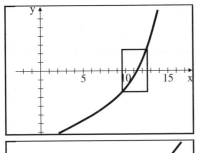

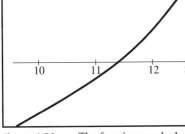

Figure 1.30 The function graphed in the upper window has a root between 11 and 12. The small rectangle (zoom box) is zoomed to fill the screen in the lower window, enabling us to estimate that the root is about 11.4. Successive zooms can provide greater precision

Exercises 1.4

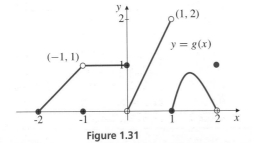

Figure 1.31

Exercises 1–3 refer to the function g defined on $[-2, 2]$, whose graph is shown in Figure 1.31.

1. State whether g is (i) continuous, (ii) left continuous, (iii) right continuous, and (iv) discontinuous at each of the points $-2, -1, 0, 1$, and 2.

2. At what points in its domain does g have a removable discontinuity, and how should g be redefined at each of those points so as to be continuous there?

3. Does g have an absolute maximum value on $[-2, 2]$? an absolute minimum value?

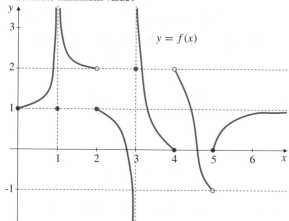

Figure 1.32

4. At what points is the function f, whose graph is shown in Figure 1.32, discontinuous? At which of those points is it left continuous? right continuous?

5. Can the function f graphed in Figure 1.32 be redefined at the single point $x = 1$ so that it becomes continuous there?

6. The function $\text{sgn}(x) = x/|x|$ is neither continuous nor discontinuous at $x = 0$. How is this possible?

In Exercises 7–12, state where in its domain the given function is continuous, where it is left or right continuous, and where it is just discontinuous.

7. $f(x) = \begin{cases} x & \text{if } x < 0 \\ x^2 & \text{if } x \geq 0 \end{cases}$ **8.** $f(x) = \begin{cases} x & \text{if } x < -1 \\ x^2 & \text{if } x \geq -1 \end{cases}$

9. $f(x) = \begin{cases} 1/x^2 & \text{if } x \neq 0 \\ 0 & \text{if } x = 0 \end{cases}$ **10.** $f(x) = \begin{cases} x^2 & \text{if } x \leq 1 \\ 0.987 & \text{if } x > 1 \end{cases}$

11. The least integer function $\lceil x \rceil$ of Example 11 in Section P.5.

12. The parking lot cost function $C(t)$ of Exercise 53 in Section 1.3.

In Exercises 13–16, how should the given function be defined at the given point to be continuous there? Give a formula for the continuous extension to that point.

13. $\dfrac{x^2 - 4}{x - 2}$ at $x = 2$ **14.** $\dfrac{1 + t^3}{1 - t^2}$ at $t = -1$

15. $\dfrac{t^2 - 5t + 6}{t^2 - t - 6}$ at 3 **16.** $\dfrac{x^2 - 2}{x^4 - 4}$ at $\sqrt{2}$

17. Find k so that $f(x) = \begin{cases} x^2 & \text{if } x \leq 2 \\ k - x^2 & \text{if } x > 2 \end{cases}$ is a continuous function.

18. Find m so that $g(x) = \begin{cases} x - m & \text{if } x < 3 \\ 1 - mx & \text{if } x \geq 3 \end{cases}$ is continuous for all x.

19. Does the function x^2 have a maximum value on the open interval $-1 < x < 1$? a minimum value? Explain.

20. The Heaviside function of Example 1 has both absolute maximum and minimum values on the interval $[-1, 1]$, but it is not continuous on that interval. Does this violate the Max-Min Theorem? Why?

Exercises 21–24 involve finding maximum and minimum values of functions. They can all be done by the method of Example 9.

21. The sum of two nonnegative numbers is 8. What is the largest possible value of their product?

22. The sum of two nonnegative numbers is 8. What is (a) the smallest, and (b) the largest possible value for the sum of their squares?

23. A software company estimates that if it assigns x programmers to work on the project, it can develop a new product in T days, where

$$T = 100 - 30x + 3x^2.$$

How many programmers should the company assign in order to complete the development as quickly as possible?

24. It costs a desk manufacturer $\$(245x - 30x^2 + x^3)$ to send a shipment of x desks to its warehouse. How many desks should it include in each shipment to minimize the average shipping cost per desk?

Find the intervals on which the functions $f(x)$ in Exercises 25–28 are positive and negative.

25. $f(x) = \dfrac{x^2 - 1}{x}$ **26.** $f(x) = x^2 + 4x + 3$

27. $f(x) = \dfrac{x^2 - 1}{x^2 - 4}$ **28.** $f(x) = \dfrac{x^2 + x - 2}{x^3}$

29. Show that $f(x) = x^3 + x - 1$ has a zero between $x = 0$ and $x = 1$.

30. Show that the equation $x^3 - 15x + 1 = 0$ has three solutions in the interval $[-4, 4]$.

31. Show that the function $F(x) = (x - a)^2(x - b)^2 + x$ has the value $(a + b)/2$ at some point x.

32. (A fixed-point theorem) Suppose that f is continuous on the closed interval $[0, 1]$ and that $0 \leq f(x) \leq 1$ for every x in $[0, 1]$. Show that there must exist a number c in $[0, 1]$ such that $f(c) = c$. (c is called a fixed point of the function f.) *Hint:* if $f(0) = 0$ or $f(1) = 1$ you are done. If not, apply the Intermediate-Value Theorem to $g(x) = f(x) - x$.

33. If an even function f is right continuous at $x = 0$, show that it is continuous at $x = 0$.

34. If an odd function f is right continuous at $x = 0$, show that it is continuous at $x = 0$ and that it satisfies $f(0) = 0$.

Use a graphing utility to find the maximum and minimum values of the functions in Exercises 35–38 and the points x where they occur. Obtain 3 decimal place accuracy for all answers.

35. $f(x) = \dfrac{x^2 - 2x}{x^4 + 1}$ on $[-5, 5]$

36. $f(x) = \dfrac{\sin x}{6 + x}$ on $[-\pi, \pi]$

37. $f(x) = x^2 + \dfrac{4}{x}$ on $[1, 3]$

38. $f(x) = \sin(\pi x) + x(\cos(\pi x) + 1)$ on $[0, 1]$

Use a graphing utility or a programmable calculator and the Bisection Method to solve the equations in Exercises 39–40 to 3 decimal places. As a first step, try to guess a small interval that you can be sure contains a root.

39. $x^3 + x - 1 = 0$ **40.** $\cos x - x = 0$

Use Maple's `fsolve` routine to solve the equations in 41–42.

41. $\sin x + 1 - x^2 = 0$ (two roots)

42. $x^4 - x - 1 = 0$ (two roots)

43. Investigate the difference between the Maple routines `fsolve(f,x)`, `solve(f,x)`, and `evalf(solve(f,x))`, where `f := x^3-x-1=0`. Note that no interval is specified for x here.

1.5 The Formal Definition of Limit

The material in this section is optional.

The *informal* definition of limit given in Section 1.2 is not precise enough to enable us to prove results about limits such as those given in Theorems 2–4 of Section 1.2. A more precise *formal* definition is based on the idea of controlling the input x of a function f so that the output $f(x)$ will lie in a specific interval.

Example 1   The area of a circular disk of radius r cm is $A = \pi r^2$ cm². A machinist is required to manufacture a circular metal disk having area 400π cm² within an error tolerance of ± 5 cm². How close to 20 cm must the machinist control the radius of the disk to achieve this?

Solution   The machinist wants $|\pi r^2 - 400\pi| < 5$, that is,

$$400\pi - 5 < \pi r^2 < 400\pi + 5,$$

or, equivalently,

$$\sqrt{400 - (5/\pi)} < r < \sqrt{400 + (5/\pi)}$$
$$19.96017 < r < 20.03975.$$

Thus, the machinist needs $|r - 20| < 0.03975$; she must ensure that the radius of the disk differs from 20 cm by less than 0.4 mm so that the area of the disk will lie within the required error tolerance.

When we say that $f(x)$ has limit L as x approaches a, we are really saying that we can ensure that the *error* $|f(x) - L|$ will be less than *any* allowed tolerance, no matter how small, by taking x *close enough* to a (but not equal to a). It is traditional to use ϵ, the Greek letter "epsilon", for the size of the allowable *error* and δ, the Greek letter "delta", for the *difference* $x - a$ that measures how close x must be to a to ensure that the error is within that tolerance. These are the letters that Cauchy and Weierstrass used in their pioneering work on limits and continuity in the nineteenth century.

If ϵ is any positive number, *no matter how small*, we must be able to ensure that $|f(x) - L| < \epsilon$ by restricting x to be *close enough to* (but not equal to) a. How close is close enough? It is sufficient that the distance $|x - a|$ from x to a be less than a positive number δ that depends on ϵ. (See Figure 1.33.) If we can find such a δ for any positive ϵ we are entitled to conclude that $\lim_{x \to a} f(x) = L$.

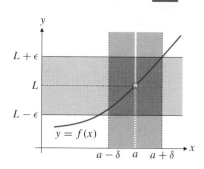

Figure 1.33   If $x \neq a$ and $|x - a| < \delta$, then $|f(x) - L| < \epsilon$

DEFINITION	**9**

A formal definition of limit

We say that $f(x)$ approaches the limit L as x approaches a, and we write

$$\lim_{x \to a} f(x) = L,$$

if the following condition is satisfied:
for every number $\epsilon > 0$ there exists a number $\delta > 0$, depending on ϵ, such that

$$0 < |x - a| < \delta \qquad \text{implies} \qquad |f(x) - L| < \epsilon.$$

The formal definition of limit does not tell you how to find the limit of a function, but it does enable you to verify that a suspected limit is correct. The following examples show how it can be used to verify limit statements for specific functions. The first of these gives a formal verification of the two limits found in Example 3 of Section 1.2.

Example 2   (**Two important limits**) Verify: (a) $\lim_{x \to a} x = a$ and (b) $\lim_{x \to a} k = k$ (k = constant).

Solution

(a) Let $\epsilon > 0$ be given. We must find $\delta > 0$ so that

$$0 < |x - a| < \delta \qquad \text{implies} \qquad |x - a| < \epsilon.$$

Clearly, we can take $\delta = \epsilon$ and the implication above will be true. This proves that $\lim_{x \to a} x = a$.

(b) Let $\epsilon > 0$ be given. We must find $\delta > 0$ so that

$$0 < |x - a| < \delta \qquad \text{implies} \qquad |k - k| < \epsilon.$$

Since $k - k = 0$, we can use any positive number for δ and the implication above will be true. This proves that $\lim_{x \to a} k = k$.

Example 3   Verify that $\lim_{x \to 2} x^2 = 4$.

Solution   Here $a = 2$ and $L = 4$. Let ϵ be a given positive number. We want to find $\delta > 0$ so that if $0 < |x - 2| < \delta$, then $|f(x) - 4| < \epsilon$. Now

$$|f(x) - 4| = |x^2 - 4| = |(x + 2)(x - 2)| = |x + 2||x - 2|.$$

We want the expression above to be less than ϵ. We can make the factor $|x - 2|$ as small as we wish by choosing δ properly, but we need to control the factor $|x + 2|$ so that it does not become too large. If we first assume $\delta \leq 1$ and require that $|x - 2| < \delta$, then we have

$$|x - 2| < 1 \quad \Rightarrow \quad 1 < x < 3 \quad \Rightarrow \quad 3 < x + 2 < 5$$
$$\Rightarrow \quad |x + 2| < 5.$$

Hence,

$$|f(x) - 4| < 5|x - 2| \quad \text{if} \quad |x - 2| < \delta \leq 1.$$

But $5|x - 2| < \epsilon$ if $|x - 2| < \epsilon/5$. Therefore, if we take $\delta = \min\{1, \epsilon/5\}$, the *minimum* (the smaller) of the two numbers 1 and $\epsilon/5$, then

$$|f(x) - 4| < 5|x - 2| < 5 \times \frac{\epsilon}{5} = \epsilon \quad \text{if} \quad |x - 2| < \delta.$$

This proves that $\lim_{x \to 2} f(x) = 4$.

Using the Definition of Limit to Prove Theorems

We do not usually rely on the formal definition of limit to verify specific limits such as those in the two examples above. Rather, we appeal to general theorems about limits, in particular Theorems 2–4 of Section 1.2. The definition is used to prove these theorems. As an example, we prove part 1 of Theorem 2, the *Sum Rule*.

Example 4 (**Proving the rule for the limit of a sum**) If $\lim_{x \to a} f(x) = L$ and $\lim_{x \to a} g(x) = M$, prove that $\lim_{x \to a} \big(f(x) + g(x) \big) = L + M$.

Solution Let $\epsilon > 0$ be given. We want to find a positive number δ such that

$$0 < |x - a| < \delta \quad \Rightarrow \quad \big| \big(f(x) + g(x) \big) - (L + M) \big| < \epsilon.$$

Observe that

$$\big| \big(f(x) + g(x) \big) - (L + M) \big| \qquad \text{Regroup terms.}$$
$$= \big| \big(f(x) - L \big) + \big(g(x) - M \big) \big| \qquad \text{(Use the triangle inequality:}$$
$$|a + b| \leq |a| + |b|).$$
$$\leq |f(x) - L| + |g(x) - M|.$$

Since $\lim_{x \to a} f(x) = L$ and $\epsilon/2$ is a positive number, there exists a number $\delta_1 > 0$ such that

$$0 < |x - a| < \delta_1 \quad \Rightarrow \quad |f(x) - L| < \epsilon/2.$$

Similarly, since $\lim_{x \to a} g(x) = M$, there exists a number $\delta_2 > 0$ such that

$$0 < |x - a| < \delta_2 \quad \Rightarrow \quad |g(x) - M| < \epsilon/2.$$

Let $\delta = \min\{\delta_1, \delta_2\}$, the smaller of δ_1 and δ_2. If $0 < |x - a| < \delta$, then $|x - a| < \delta_1$, so $|f(x) - L| < \epsilon/2$, and $|x - a| < \delta_2$, so $|g(x) - M| < \epsilon/2$. Therefore,

$$\big| \big(f(x) + g(x) \big) - (L + M) \big| < \frac{\epsilon}{2} + \frac{\epsilon}{2} = \epsilon.$$

This shows that $\lim_{x \to a} \big(f(x) + g(x) \big) = L + M$.

Other Kinds of Limits

The formal definition of limit can be modified to give precise definitions of one-sided limits, limits at infinity, and infinite limits. We give some of the definitions here and leave you to supply the others.

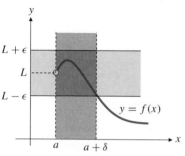

Figure 1.34 If $a < x < a + \delta$, then $|f(x) - L| < \epsilon$

Right limits

We say that $f(x)$ has **right limit** L at a, and we write

$$\lim_{x \to a+} f(x) = L,$$

if the following condition is satisfied:
for every number $\epsilon > 0$ there exists a number $\delta > 0$, depending on ϵ, such that

$$a < x < a + \delta \quad \text{implies} \quad |f(x) - L| < \epsilon.$$

Notice how the condition $0 < |x - a| < \delta$ in the definition of limit becomes $a < x < a + \delta$ in the right limit case (Figure 1.34). The definition for a left limit is formulated in a similar way.

Example 5 Show that $\lim_{x \to 0+} \sqrt{x} = 0$.

Solution Let $\epsilon > 0$ be given. If $x > 0$, then $|\sqrt{x} - 0| = \sqrt{x}$. We can ensure that $\sqrt{x} < \epsilon$ by requiring $x < \epsilon^2$. Thus we can take $\delta = \epsilon^2$ and the condition of the definition will be satisfied:

$$0 < x < \delta = \epsilon^2 \quad \text{implies} \quad |\sqrt{x} - 0| < \epsilon.$$

Therefore $\lim_{x \to 0+} \sqrt{x} = 0$. ∎

To claim that a function f has a limit L at infinity, we must be able to ensure that the error $|f(x) - L|$ is less than any given positive number ϵ by restricting x to be *sufficiently large*, that is, by requiring $x > R$ for some positive number R depending on ϵ.

Limit at infinity

We say that $f(x)$ **approaches the limit** L **as** x **approaches infinity**, and we write

$$\lim_{x \to \infty} f(x) = L,$$

if the following condition is satisfied:
for every number $\epsilon > 0$ there exists a number R, depending on ϵ, such that

$$x > R \quad \text{implies} \quad |f(x) - L| < \epsilon.$$

You are invited to formulate a version of the definition of a limit at negative infinity.

Example 6 Show that $\lim\limits_{x \to \infty} \dfrac{1}{x} = 0$.

Solution Let ϵ be a given positive number. For $x > 0$ we have

$$\left| \frac{1}{x} - 0 \right| = \frac{1}{|x|} = \frac{1}{x} < \epsilon \qquad \text{provided} \qquad x > \frac{1}{\epsilon}.$$

Therefore, the condition of the definition is satisfied with $R = 1/\epsilon$. We have shown that $\lim_{x \to \infty} 1/x = 0$. ∎

To show that $f(x)$ has an infinite limit at a we must ensure that $f(x)$ is larger than any given positive number (say B) by restricting x to a sufficiently small interval centred at a, and requiring that $x \neq a$.

DEFINITION 12

> **Infinite limits**
>
> We say that $f(x)$ approaches infinity as x approaches a and write
>
> $$\lim_{x \to a} f(x) = \infty,$$
>
> if for every positive real number B we can find a positive number δ (depending on B), such that
>
> $$0 < |x - a| < \delta \qquad \text{implies} \qquad f(x) > B.$$

Try to formulate the corresponding definition for the concept $\lim_{x \to a} f(x) = -\infty$. Then try to modify both definitions to cover the case of infinite one-sided limits and infinite limits at infinity.

Example 7 Verify that $\lim\limits_{x \to 0} \dfrac{1}{x^2} = \infty$.

Solution Let B be any positive number. We have

$$\frac{1}{x^2} > B \qquad \text{provided that} \qquad x^2 < \frac{1}{B}.$$

If $\delta = 1/\sqrt{B}$, then

$$0 < |x| < \delta \quad \Rightarrow \quad x^2 < \delta^2 = \frac{1}{B} \quad \Rightarrow \quad \frac{1}{x^2} > B.$$

Therefore $\lim_{x \to 0} 1/x^2 = \infty$. ∎

Exercises 1.5

1. The length L of a metal rod is given in terms of the temperature T (°C) by $L = 39.6 + 0.025T$ cm. Within what range of temperature must the rod be kept if its length must be maintained within ±1 mm of 40 cm?

2. What is the largest tolerable error in the 20 cm edge length of a cubical cardboard box if the volume of the box must be within ±1.2% of 8,000 cm³?

In Exercises 3–6, in what interval must x be confined if $f(x)$ must be within the given distance ϵ of the number L?

3. $f(x) = 2x - 1$, $\quad L = 3$, $\quad \epsilon = 0.02$

4. $f(x) = x^2$, $\quad L = 4$, $\quad \epsilon = 0.1$

5. $f(x) = \sqrt{x}$, $\quad L = 1$, $\quad \epsilon = 0.1$

6. $f(x) = 1/x$, $\quad L = -2$, $\epsilon = 0.01$

In Exercises 7–10, find a number $\delta > 0$ such that if $|x - a| < \delta$, then $|f(x) - L|$ will be less than the given number ϵ.

7. $f(x) = 3x + 1$, $\quad a = 2$, $\quad L = 7$, $\quad \epsilon = 0.03$

8. $f(x) = \sqrt{2x + 3}$, $a = 3$, $\quad L = 3$, $\quad \epsilon = 0.01$

9. $f(x) = x^3$, $\quad a = 2$, $\quad L = 8$, $\quad \epsilon = 0.2$

10. $f(x) = 1/(x + 1)$, $a = 0$, $\quad L = 1$, $\quad \epsilon = 0.05$

In Exercises 11–20, use the formal definition of limit to verify the indicated limit.

11. $\lim\limits_{x \to 1} (3x + 1) = 4$

12. $\lim\limits_{x \to 2} (5 - 2x) = 1$

13. $\lim\limits_{x \to 0} x^2 = 0$

14. $\lim\limits_{x \to 2} \dfrac{x - 2}{1 + x^2} = 0$

15. $\lim\limits_{x \to 1/2} \dfrac{1 - 4x^2}{1 - 2x} = 2$

16. $\lim\limits_{x \to -2} \dfrac{x^2 + 2x}{x + 2} = -2$

17. $\lim\limits_{x \to 1} \dfrac{1}{x + 1} = \dfrac{1}{2}$

18. $\lim\limits_{x \to -1} \dfrac{x + 1}{x^2 - 1} = -\dfrac{1}{2}$

19. $\lim\limits_{x \to 1} \sqrt{x} = 1$

20. $\lim\limits_{x \to 2} x^3 = 8$

Give formal definitions of the limit statements in Exercises 21–26.

21. $\lim\limits_{x \to a-} f(x) = L$

22. $\lim\limits_{x \to -\infty} f(x) = L$

23. $\lim\limits_{x \to a} f(x) = -\infty$

24. $\lim\limits_{x \to \infty} f(x) = \infty$

25. $\lim\limits_{x \to a+} f(x) = -\infty$

26. $\lim\limits_{x \to a-} f(x) = \infty$

Use formal definitions of the various kinds of limits to prove the statements in Exercises 27–30.

27. $\lim\limits_{x \to 1+} \dfrac{1}{x - 1} = \infty$

28. $\lim\limits_{x \to 1-} \dfrac{1}{x - 1} = -\infty$

29. $\lim\limits_{x \to \infty} \dfrac{1}{\sqrt{x^2 + 1}} = 0$

30. $\lim\limits_{x \to \infty} \sqrt{x} = \infty$

Proving Theorems with the Definition of Limit

* 31. Prove that limits are unique; that is, if $\lim_{x \to a} f(x) = L$ and $\lim_{x \to a} f(x) = M$, prove that $L = M$. *Hint:* suppose $L \neq M$ and let $\epsilon = |L - M|/3$.

* 32. If $\lim_{x \to a} g(x) = M$, show that there exists a number $\delta > 0$ such that

$$0 < |x - a| < \delta \quad \Rightarrow \quad |g(x)| < 1 + |M|.$$

(*Hint:* take $\epsilon = 1$ in the definition of limit.) This says that the values of $g(x)$ are **bounded** near a point where g has a limit.

* 33. If $\lim_{x \to a} f(x) = L$ and $\lim_{x \to a} g(x) = M$, prove that $\lim_{x \to a} f(x)g(x) = LM$ (the Product Rule part of Theorem 2). *Hint:* reread Example 4. Let $\epsilon > 0$ and write

$$
\begin{aligned}
|f(x)g(x) - LM| &= |f(x)g(x) - Lg(x) + Lg(x) - LM| \\
&= |(f(x) - L)g(x) + L(g(x) - M)| \\
&\leq |(f(x) - L)g(x)| + |L(g(x) - M)| \\
&= |g(x)||f(x) - L| + |L||g(x) - M|
\end{aligned}
$$

Now try to make each term in the last line less than $\epsilon/2$ by taking x close enough to a. You will need the result of Exercise 32.

* 34. If $\lim_{x \to a} g(x) = M$, where $M \neq 0$, show that there exists a number $\delta > 0$ such that

$$0 < |x - a| < \delta \quad \Rightarrow \quad |g(x)| > |M|/2.$$

* 35. If $\lim_{x \to a} g(x) = M$, where $M \neq 0$, show that

$$\lim\limits_{x \to a} \dfrac{1}{g(x)} = \dfrac{1}{M}.$$

Hint: you will need the result of Exercise 34.

36. Use the facts proved in Exercises 33 and 35 to prove the Quotient Rule (part 5 of Theorem 2): if $\lim_{x \to a} f(x) = L$ and $\lim_{x \to a} g(x) = M$, where $M \neq 0$, then

$$\lim\limits_{x \to a} \dfrac{f(x)}{g(x)} = \dfrac{L}{M}.$$

* 37. Use the definition of limit twice to prove Theorem 7 of Section 1.4; that is, if f is continuous at L and if $\lim_{x \to c} g(x) = L$, then

$$\lim\limits_{x \to c} f(g(x)) = f(L) = f\left(\lim\limits_{x \to c} g(x) \right).$$

* **38.** Prove the Squeeze Theorem (Theorem 4 in Section 1.2).
 Hint: if $f(x) \le g(x) \le h(x)$, then

$$|g(x) - L| = |g(x) - f(x) + f(x) - L|$$
$$\le |g(x) - f(x)| + |f(x) - L|$$
$$\le |h(x) - f(x)| + |f(x) - L|$$
$$= |h(x) - L - (f(x) - L)| + |f(x) - L|$$
$$\le |h(x) - L| + |f(x) - L| + |f(x) - L|$$

Now you can make each term in the last expression less than $\epsilon/3$ and so complete the proof.

Chapter Review

Key Ideas

- **What do the following statements and phrases mean?**

  ◇ the average rate of change of $f(x)$ on $[a, b]$

  ◇ the instantaneous rate of change of $f(x)$ at $x = a$

  ◇ $\lim_{x \to a} f(x) = L$

  ◇ $\lim_{x \to a+} f(x) = L, \quad \lim_{x \to a-} f(x) = L$

  ◇ $\lim_{x \to \infty} f(x) = L, \quad \lim_{x \to -\infty} f(x) = L$

  ◇ $\lim_{x \to a} f(x) = \infty, \quad \lim_{x \to a+} f(x) = -\infty$

  ◇ f is continuous at c.

  ◇ f is left (or right) continuous at c.

  ◇ f has a continuous extension to c.

  ◇ f is a continuous function.

  ◇ f takes on maximum and minimum values on interval I.

  ◇ f is bounded on interval I.

  ◇ f has the intermediate-value property on interval I.

- **State as many "laws of limits" as you can.**

- **What properties must a function have if it is continuous and its domain is a closed, finite interval?**

- **How can you find zeros (roots) of a continuous function?**

Review Exercises

1. Find the average rate of change of x^3 over $[1, 3]$.

2. Find the average rate of change of $1/x$ over $[-2, -1]$.

3. Find the rate of change of x^3 at $x = 2$.

4. Find the rate of change of $1/x$ at $x = -3/2$.

Evaluate the limits in Exercises 5–30 or explain why they do not exist.

5. $\lim_{x \to 1} (x^2 - 4x + 7)$

6. $\lim_{x \to 2} \dfrac{x^2}{1 - x^2}$

7. $\lim_{x \to 1} \dfrac{x^2}{1 - x^2}$

8. $\lim_{x \to 2} \dfrac{x^2 - 4}{x^2 - 5x + 6}$

9. $\lim_{x \to 2} \dfrac{x^2 - 4}{x^2 - 4x + 4}$

10. $\lim_{x \to 2-} \dfrac{x^2 - 4}{x^2 - 4x + 4}$

11. $\lim_{x \to -2+} \dfrac{x^2 - 4}{x^2 + 4x + 4}$

12. $\lim_{x \to 4} \dfrac{2 - \sqrt{x}}{x - 4}$

13. $\lim_{x \to 3} \dfrac{x^2 - 9}{\sqrt{x} - \sqrt{3}}$

14. $\lim_{h \to 0} \dfrac{h}{\sqrt{x + 3h} - \sqrt{x}}$

15. $\lim_{x \to 0+} \sqrt{x - x^2}$

16. $\lim_{x \to 0} \sqrt{x - x^2}$

17. $\lim_{x \to 1} \sqrt{x - x^2}$

18. $\lim_{x \to 1-} \sqrt{x - x^2}$

19. $\lim_{x \to \infty} \dfrac{1 - x^2}{3x^2 - x - 1}$

20. $\lim_{x \to -\infty} \dfrac{2x + 100}{x^2 + 3}$

21. $\lim_{x \to -\infty} \dfrac{x^3 - 1}{x^2 + 4}$

22. $\lim_{x \to \infty} \dfrac{x^4}{x^2 - 4}$

23. $\lim_{x \to 0+} \dfrac{1}{\sqrt{x - x^2}}$

24. $\lim_{x \to 1/2} \dfrac{1}{\sqrt{x - x^2}}$

25. $\lim_{x \to \infty} \sin x$

26. $\lim_{x \to \infty} \dfrac{\cos x}{x}$

27. $\lim_{x \to 0} x \sin \dfrac{1}{x}$

28. $\lim_{x \to 0} \sin \dfrac{1}{x^2}$

29. $\lim_{x \to -\infty} [x + \sqrt{x^2 - 4x + 1}]$

30. $\lim_{x \to \infty} [x + \sqrt{x^2 - 4x + 1}]$

At what, if any, points in its domain is the function f in Exercises 31–38 discontinuous? Is f left or right continuous at these points? In Exercises 35 and 36, H refers to the Heaviside function: $H(x) = 1$ if $x \ge 0$ and $H(x) = 0$ if $x < 0$.

31. $f(x) = x^3 - 4x^2 + 1$

32. $f(x) = \dfrac{x}{x + 1}$

33. $f(x) = \begin{cases} x^2 & \text{if } x > 2 \\ x & \text{if } x \le 2 \end{cases}$

34. $f(x) = \begin{cases} x^2 & \text{if } x > 1 \\ x & \text{if } x \le 1 \end{cases}$

35. $f(x) = H(x - 1)$

36. $f(x) = H(9 - x^2)$

37. $f(x) = |x| + |x + 1|$

38. $f(x) = \begin{cases} |x|/|x + 1| & \text{if } x \ne -1 \\ 1 & \text{if } x = -1 \end{cases}$

Challenging Problems

1. Show that the average rate of change of the function x^3 over the interval $[a, b]$, where $0 < a < b$, is equal to the instantaneous rate of change of x^3 at $x = \sqrt{(a^2 + ab + b^2)/3}$. Is this point to the left or to the right of the midpoint $(a + b)/2$ of the interval $[a, b]$?

2. Evaluate $\lim\limits_{x \to 0} \dfrac{x}{|x - 1| - |x + 1|}$.

3. Evaluate $\lim\limits_{x \to 3} \dfrac{|5 - 2x| - |x - 2|}{|x - 5| - |3x - 7|}$.

4. Evaluate $\lim\limits_{x \to 64} \dfrac{x^{1/3} - 4}{x^{1/2} - 8}$.

5. Evaluate $\lim\limits_{x \to 1} \dfrac{\sqrt{3 + x} - 2}{\sqrt[3]{7 + x} - 2}$.

6. The equation $ax^2 + 2x - 1 = 0$, where a is a constant, has two roots if $a > -1$ and $a \neq 0$:

$$r_+(a) = \frac{-1 + \sqrt{1 + a}}{a} \quad \text{and} \quad r_-(a) = \frac{-1 - \sqrt{1 + a}}{a}.$$

   (a) What happens to the root $r_-(a)$ when $a \to 0$?

   (b) Investigate numerically what happens to the root $r_+(a)$ when $a \to 0$ by trying the values $a = 1, \pm 0.1$, $\pm 0.01, \ldots$. For values such as $a = 10^{-8}$, the limited precision of your calculator may produce some interesting results. What happens, and why?

   (c) Evaluate $\lim\limits_{a \to 0} r_+(a)$ mathematically by using the identity

$$\sqrt{A} - \sqrt{B} = \frac{A - B}{\sqrt{A} + \sqrt{B}}.$$

7. TRUE or FALSE? If TRUE, give reasons; if FALSE, give a counterexample.

   (a) If $\lim\limits_{x \to a} f(x)$ exists but $\lim\limits_{x \to a} g(x)$ does not exist, then $\lim\limits_{x \to a} (f(x) + g(x))$ does not exist.

   (b) If neither $\lim\limits_{x \to a} f(x)$ nor $\lim\limits_{x \to a} g(x)$ exists, then $\lim\limits_{x \to a} (f(x) + g(x))$ does not exist.

   (c) If f is continuous at a, then so is $|f|$.

   (d) If $|f|$ is continuous at a, then so is f.

   (e) If $f(x) < g(x)$ for all x in an interval around a, and if $\lim\limits_{x \to a} f(x)$ and $\lim\limits_{x \to a} g(x)$ both exist, then $\lim\limits_{x \to a} f(x) < \lim\limits_{x \to a} g(x)$.

8. (a) If f is a continuous function defined on a closed interval $[a, b]$, show that $R(f)$ is a closed interval.

   (b) What are the possibilities for $R(f)$ if $D(f)$ is an open interval $]a, b[$?

9. Consider the function $f(x) = \dfrac{x^2 - 1}{|x^2 - 1|}$. Find all points where f is not continuous. Does f have one-sided limits at those points, and if so, what are they?

10. Find the minimum value of $f(x) = 1/(x - x^2)$ on the interval $]0, 1[$. Explain how you know such a minimum value must exist.

* **11.** (a) Suppose f is a continuous function on the interval $[0, 1]$, and $f(0) = f(1)$. Show that $f(a) = f\left(a + \dfrac{1}{2}\right)$ for some $a \in \left[0, \dfrac{1}{2}\right]$.

   Hint: let $g(x) = f\left(x + \dfrac{1}{2}\right) - f(x)$, and use the Intermediate-Value Theorem.

   (b) If n is an integer larger than 2, show that $f(a) = f\left(a + \dfrac{1}{n}\right)$ for some $a \in \left[0, 1 - \dfrac{1}{n}\right]$.

CHAPTER 2

Differentiation

Introduction Two fundamental problems are considered in calculus. The **problem of slopes** is concerned with finding the slope of (the tangent line to) a given curve at a given point on the curve. The **problem of areas** is concerned with finding the area of a plane region bounded by curves and straight lines. The solution of the problem of slopes is the subject of **differential calculus**. As we will see, it has many applications in mathematics and other disciplines. The problem of areas is the subject of **integral calculus**, which we begin in Chapter 5.

2.1 Tangent Lines and Their Slopes

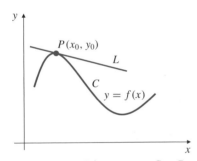

Figure 2.1   *L* is tangent to *C* at *P*

This section deals with the problem of finding a straight line *L* that is tangent to a curve *C* at a point *P*. As is often the case in mathematics, the most important step in the solution of such a fundamental problem is making a suitable definition.

For simplicity, and to avoid certain problems best postponed until later, we will not deal with the most general kinds of curves now, but only with those that are the *graphs of continuous functions*. Let *C* be the graph of $y = f(x)$ and let *P* be the point (x_0, y_0) on *C*, so that $y_0 = f(x_0)$. We assume that *P* is not an endpoint of *C*. Therefore, *C* extends some distance on both sides of *P*. (See Figure 2.1.)

What do we mean when we say that the line *L* is tangent to *C* at *P*? Past experience with tangent lines to circles does not help us to define tangency for more general curves. A tangent line to a circle (Figure 2.2) has the following properties:

(i) It meets the circle at only one point.

(ii) The circle lies on only one side of the line.

(iii) The tangent is perpendicular to the line joining the centre of the circle to the point of contact.

Most curves do not have obvious *centres*, so (iii) is useless for characterizing tangents to them. The curves in Figure 2.3 show that (i) and (ii) cannot be used to define tangency either. In particular, Figure 2.3(d) is not "smooth" at *P* so that curve should not have any tangent line there. A tangent line should have the "same direction" as the curve does at the point of tangency.

A reasonable definition of tangency can be stated in terms of limits. If *Q* is a point on *C* different from *P*, then the line through *P* and *Q* is called a **secant line** to the curve. This line rotates around *P* as *Q* moves along the curve. If *L* is a line through *P* whose slope is the limit of the slopes of these secant lines *PQ* as *Q* approaches *P* along *C* (Figure 2.4), then *L* is tangent to *C* at *P*.

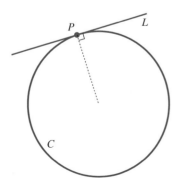

Figure 2.2   *L* is tangent to *C* at *P*

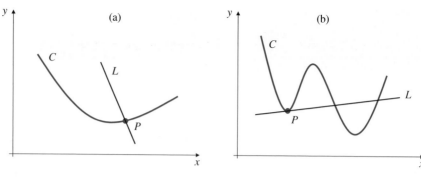

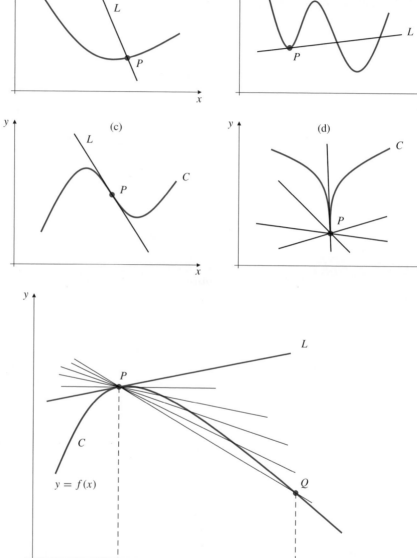

Figure 2.3

(a) L meets C only at P but is not tangent to C

(b) L meets C at several points but is tangent to C at P

(c) L is tangent to C at P but crosses C at P

(d) Many lines meet C only at P but none of them is tangent to C at P

Figure 2.4 Secant lines PQ approach tangent line L as Q approaches P along the curve C

Since C is the graph of the *function* $y = f(x)$, then vertical lines can meet C only once. Since $P = (x_0, f(x_0))$, a different point Q on the graph must have a different x-coordinate, say $x_0 + h$, where $h \neq 0$. Thus $Q = (x_0 + h, f(x_0 + h))$, and the slope of the line PQ is

$$\frac{f(x_0 + h) - f(x_0)}{h}.$$

This expression is called the **Newton quotient** or **difference quotient** for f at x_0. Note that h can be positive or negative, depending on whether Q is to the right or left of P.

DEFINITION **1**

> **Nonvertical tangent lines**
>
> Suppose that the function f is continuous at $x = x_0$ and that
>
> $$\lim_{h \to 0} \frac{f(x_0 + h) - f(x_0)}{h} = m$$
>
> exists. Then the straight line having slope m and passing through the point $P = (x_0, f(x_0))$ is called the **tangent line** (or simply the **tangent**) to the graph of $y = f(x)$ at P. An equation of this tangent is
>
> $$y = m(x - x_0) + y_0.$$

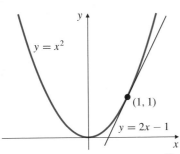

Figure 2.5 The tangent to $y = x^2$ at $(1, 1)$

Example 1 Find an equation of the tangent line to the curve $y = x^2$ at the point $(1, 1)$.

Solution Here $f(x) = x^2$, $x_0 = 1$, and $y_0 = f(1) = 1$. The slope of the required tangent is:

$$m = \lim_{h \to 0} \frac{f(1 + h) - f(1)}{h} = \lim_{h \to 0} \frac{(1 + h)^2 - 1}{h}$$

$$= \lim_{h \to 0} \frac{1 + 2h + h^2 - 1}{h}$$

$$= \lim_{h \to 0} \frac{2h + h^2}{h} = \lim_{h \to 0} 2 + h = 2.$$

Accordingly, the equation of the tangent line at $(1, 1)$ is $y = 2(x - 1) + 1$, or $y = 2x - 1$. See Figure 2.5. ∎

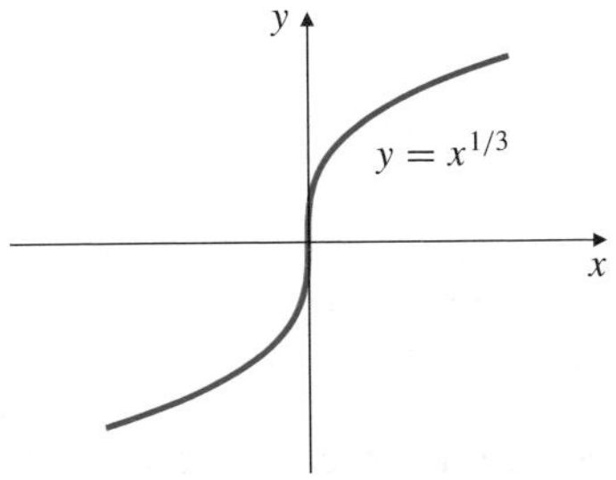

Definition 1 deals only with tangents that have finite slopes and are, therefore, not vertical. It is also possible for the graph of a continuous function to have a *vertical* tangent line.

Figure 2.6 The y-axis is tangent to $y = x^{1/3}$ at the origin

Example 2 Consider the graph of the function $f(x) = \sqrt[3]{x} = x^{1/3}$, which is shown in Figure 2.6. The graph is a smooth curve, and it seems evident that the y-axis is tangent to this curve at the origin. Let us try to calculate the limit of the Newton quotient for f at $x = 0$:

$$\lim_{h \to 0} \frac{f(0 + h) - f(0)}{h} = \lim_{h \to 0} \frac{h^{1/3}}{h} = \lim_{h \to 0} \frac{1}{h^{2/3}} = \infty.$$

Although the limit does not exist, the slope of the secant line joining the origin to another point Q on the curve approaches infinity as Q approaches the origin from either side.

Example 3   On the other hand, the function $f(x) = x^{2/3}$, whose graph is shown in Figure 2.7, does not have a tangent line at the origin because it is not "smooth" there. In this case the Newton quotient is

$$\frac{f(0 + h) - f(0)}{h} = \frac{h^{2/3}}{h} = \frac{1}{h^{1/3}},$$

which has no limit as h approaches zero. (The right limit is ∞; the left limit is $-\infty$.) We say this curve has a **cusp** at the origin. A cusp is an infinitely sharp point; if you were travelling along the curve, you would have to stop and turn $180°$ at the origin.

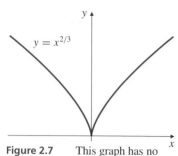

Figure 2.7   This graph has no tangent at the origin

In the light of the two examples above, we extend the definition of tangent line to allow for vertical tangents as follows:

DEFINITION 2

Vertical tangents

If f is continuous at $P = (x_0, y_0)$, where $y_0 = f(x_0)$, and if either

$$\lim_{h \to 0} \frac{f(x_0 + h) - f(x_0)}{h} = \infty \quad \text{or} \quad \lim_{h \to 0} \frac{f(x_0 + h) - f(x_0)}{h} = -\infty,$$

then the vertical line $x = x_0$ is tangent to the graph $y = f(x)$ at P. If the limit of the Newton quotient fails to exist in any other way than by being ∞ or $-\infty$, the graph $y = f(x)$ has no tangent line at P.

Example 4   Does the graph of $y = |x|$ have a tangent line at $x = 0$?

Solution   The Newton quotient here is

$$\frac{|0 + h| - |0|}{h} = \frac{|h|}{h} = \text{sgn } h = \begin{cases} 1, & \text{if } h > 0 \\ -1, & \text{if } h < 0. \end{cases}$$

Since sgn h has different right and left limits at 0 (namely, 1 and -1), the Newton quotient has no limit as $h \to 0$, so $y = |x|$ has no tangent line at $(0, 0)$. (See Figure 2.8.) The graph does not have a cusp at the origin, but it is kinked at that point; *it suddenly changes direction and is not smooth*. Curves have tangents only at points where they are smooth. The graphs of $y = x^{2/3}$ and $y = |x|$ have tangent lines everywhere except at the origin, where they are not smooth.

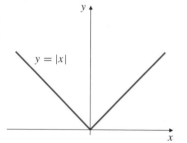

Figure 2.8 $y = |x|$ has no tangent at the origin

DEFINITION 3

The slope of a curve

The **slope** of a curve C at a point P is the slope of the tangent line to C at P if such a tangent line exists. In particular, the slope of the graph of $y = f(x)$ at the point x_0 is

$$\lim_{h \to 0} \frac{f(x_0 + h) - f(x_0)}{h}.$$

Example 5   Find the slope of the curve $y = x/(3x + 2)$ at the point $x = -2$.

Solution   If $x = -2$, then $y = 1/2$, so the required slope is

$$
\begin{aligned}
m &= \lim_{h \to 0} \frac{\dfrac{-2+h}{3(-2+h)+2} - \dfrac{1}{2}}{h} \\
&= \lim_{h \to 0} \frac{-4 + 2h - (-6 + 3h + 2)}{2(-6 + 3h + 2)h} \\
&= \lim_{h \to 0} \frac{-h}{2h(-4 + 3h)} = \lim_{h \to 0} \frac{-1}{2(-4 + 3h)} = \frac{1}{8}.
\end{aligned}
$$

Normals

If a curve C has a tangent line L at point P, then the straight line N through P perpendicular to L is called the **normal** to C at P. If L is horizontal, then N is vertical; if L is vertical, then N is horizontal. If L is neither horizontal nor vertical, then, as shown in Section P.2, the slope of N is the negative reciprocal of the slope of L:

$$
\text{slope of the normal} = \frac{-1}{\text{slope of the tangent}}.
$$

Example 6   Find an equation of the normal to $y = x^2$ at $(1, 1)$.

Solution   By Example 1, the tangent to $y = x^2$ at $(1, 1)$ has slope 2. Hence the normal has slope $-1/2$, and its equation is

$$
y = -\frac{1}{2}(x - 1) + 1 \qquad \text{or} \qquad y = -\frac{x}{2} + \frac{3}{2}.
$$

Example 7   Find equations of the straight lines that are tangent and normal to the curve $y = \sqrt{x}$ at the point $(4, 2)$.

Solution   The slope of the tangent at $(4, 2)$ (Figure 2.9) is

$$
\begin{aligned}
m &= \lim_{h \to 0} \frac{\sqrt{4+h} - 2}{h} = \lim_{h \to 0} \frac{(\sqrt{4+h} - 2)(\sqrt{4+h} + 2)}{h(\sqrt{4+h} + 2)} \\
&= \lim_{h \to 0} \frac{4 + h - 4}{h(\sqrt{4+h} + 2)} \\
&= \lim_{h \to 0} \frac{1}{\sqrt{4+h} + 2} = \frac{1}{4}.
\end{aligned}
$$

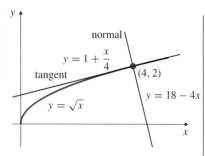

The tangent line has equation

$$y = \frac{1}{4}(x - 4) + 2 \qquad \text{or} \qquad x - 4y + 4 = 0,$$

and the normal has equation

$$y = -4(x - 4) + 2 \qquad \text{or} \qquad y = -4x + 18.$$

Figure 2.9 The tangent and normal to $y = \sqrt{x}$ at $(4, 2)$

Exercises 2.1

In Exercises 1–12, find an equation of the straight line tangent to the given curve at the point indicated.

1. $y = 3x - 1$ at $(1, 2)$

2. $y = x/2$ at $(a, a/2)$

3. $y = 2x^2 - 5$ at $(2, 3)$

4. $y = 6 - x - x^2$ at $x = -2$

5. $y = x^3 + 8$ at $x = -2$

6. $y = \dfrac{1}{x^2 + 1}$ at $(0, 1)$

7. $y = \sqrt{x + 1}$ at $x = 3$

8. $y = \dfrac{1}{\sqrt{x}}$ at $x = 9$

9. $y = \dfrac{2x}{x + 2}$ at $x = 2$

10. $y = \sqrt{5 - x^2}$ at $x = 1$

11. $y = x^2$ at $x = x_0$

12. $y = \dfrac{1}{x}$ at $\left(a, \dfrac{1}{a}\right)$

Do the graphs of the functions f in Exercises 13–17 have tangent lines at the given points? If yes, what is the tangent line?

13. $f(x) = \sqrt{|x|}$ at $x = 0$

14. $f(x) = (x - 1)^{4/3}$ at $x = 1$

15. $f(x) = (x + 2)^{3/5}$ at $x = -2$

16. $f(x) = |x^2 - 1|$ at $x = 1$

17. $f(x) = \begin{cases} \sqrt{x} & \text{if } x \geq 0 \\ -\sqrt{-x} & \text{if } x < 0 \end{cases}$ at $x = 0$

18. Find the slope of the curve $y = x^2 - 1$ at the point $x = x_0$. What is the equation of the tangent line to $y = x^2 - 1$ that has slope -3?

19. (a) Find the slope of $y = x^3$ at the point $x = a$.

(b) Find the equations of the straight lines having slope 3 that are tangent to $y = x^3$.

20. Find all points on the curve $y = x^3 - 3x$ where the tangent line is parallel to the x-axis.

21. Find all points on the curve $y = x^3 - x + 1$ where the tangent line is parallel to the line $y = 2x + 5$.

22. Find all points on the curve $y = 1/x$ where the tangent line is perpendicular to the line $y = 4x - 3$.

23. For what value of the constant k is the line $x + y = k$ normal to the curve $y = x^2$?

24. For what value of the constant k do the curves $y = kx^2$ and $y = k(x - 2)^2$ intersect at right angles? *Hint:* where do the curves intersect? What are their slopes there?

Use a graphics utility to plot the following curves. Where does the curve have a horizontal tangent? Does the curve fail to have a tangent line anywhere?

25. $y = x^3(5 - x)^2$

26. $y = 2x^3 - 3x^2 - 12x + 1$

27. $y = |x^2 - 1| - x$

28. $y = |x + 1| - |x - 1|$

29. $y = (x^2 - 1)^{1/3}$

30. $y = ((x^2 - 1)^2)^{1/3}$

* 31. If line L is tangent to curve C at point P, then the smaller angle between L and the secant line PQ joining P to another point Q on C approaches 0 as Q approaches P along C. Is the converse true: if the angle between PQ and line L (which passes through P) approaches 0, must L be tangent to C?

* 32. Let $P(x)$ be a polynomial. If a is a real number, then $P(x)$ can be expressed in the form

$$P(x) = a_0 + a_1(x - a) + a_2(x - a)^2 + \cdots + a_n(x - a)^n$$

for some $n \geq 0$. If $\ell(x) = m(x - a) + b$, show that the straight line $y = \ell(x)$ is tangent to the graph of $y = P(x)$ at $x = a$ provided $P(x) - \ell(x) = (x - a)^2 Q(x)$, where $Q(x)$ is a polynomial.

2.2 The Derivative

A straight line has the property that its slope is the same at all points. For any other graph, however, the slope may vary from point to point. Thus the slope of the graph of $y = f(x)$ at the point x is itself a function of x. At any point x where the graph has a finite slope we say that f is differentiable, and we call the slope the derivative of f. The derivative is therefore the limit of the Newton quotient.

DEFINITION 4

The **derivative** of a function f is another function f' defined by

$$f'(x) = \lim_{h \to 0} \frac{f(x+h) - f(x)}{h}$$

at all points x for which the limit exists (i.e., is a finite real number). If $f'(x)$ exists, we say that f is **differentiable** at x.

The domain of the derivative f' (read "f prime") is the set of numbers x in the domain of f where the graph of f has a *nonvertical* tangent line, and the value $f'(x_0)$ of f' at such a point x_0 is the slope of the tangent line to $y = f(x)$ there. Thus the equation of the tangent line to $y = f(x)$ at $(x_0, f(x_0))$ is

$$y = f(x_0) + f'(x_0)(x - x_0).$$

The domain $\mathcal{D}(f')$ of f' may be smaller than the domain $\mathcal{D}(f)$ of f because it contains only those points in $\mathcal{D}(f)$ at which f is differentiable. Values of x in $\mathcal{D}(f)$ where f is not differentiable are called **singular points** of f.

Remark   The value of the derivative of f at a particular point x_0 can be expressed as a limit in either of two ways:

$$f'(x_0) = \lim_{h \to 0} \frac{f(x_0 + h) - f(x_0)}{h} = \lim_{x \to x_0} \frac{f(x) - f(x_0)}{x - x_0}.$$

In the second limit $x_0 + h$ is replaced by x, so that $h = x - x_0$ and $h \to 0$ is equivalent to $x \to x_0$.

The process of calculating the derivative f' of a given function f is called **differentiation**. The graph of f' can often be sketched directly from that of f by visualizing slopes, a procedure called **graphical differentiation**. In Figure 2.10 the graphs of f' and g' were obtained by measuring the slopes at the corresponding points in the graphs of f and g lying above them. The height of the graph $y = f'(x)$ at x is the slope of the graph of $y = f(x)$ at x. Note that -1 and 1 are singular points of f. $f(-1)$ and $f(1)$ are defined, but $f'(-1)$ and $f'(1)$ are not defined; the graph of f has no tangent at -1 or at 1.

A function is differentiable on a set S if it is differentiable at every point x in S. Typically, the functions we encounter are defined on intervals or unions of intervals. If f is defined on a closed interval $[a, b]$, Definition 4 does not allow for the existence of a derivative at the endpoints $x = a$ or $x = b$. (Why?)   As

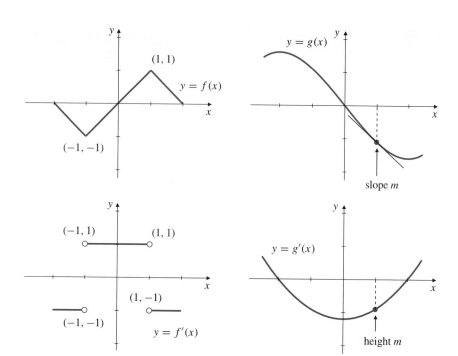

Figure 2.10 Graphical differentiation

we did for continuity in Section 1.4, we extend the definition to allow for a **right derivative** at $x = a$ and a **left derivative** at $x = b$:

$$f'_+(a) = \lim_{h \to 0+} \frac{f(a+h) - f(a)}{h}, \qquad f'_-(b) = \lim_{h \to 0-} \frac{f(b+h) - f(b)}{h}.$$

We now say that f is **differentiable** on $[a, b]$ if $f'(x)$ exists for all x in $]a, b[$ and $f'_+(a)$ and $f'_-(b)$ both exist.

Some Important Derivatives

We now give several examples of the calculation of derivatives algebraically from the definition of derivative. Some of these are the basic building blocks from which more complicated derivatives can be calculated later. They are collected in Table 1 later in this section and should be memorized.

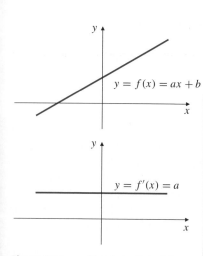

Figure 2.11 The derivative of the linear function $f(x) = ax + b$ is the constant function $f'(x) = a$

Example 1 (**The derivative of a linear function**) Show that if $f(x) = ax + b$, then $f'(x) = a$.

Solution The result is apparent from the graph of f (Figure 2.11), but we will do the calculation using the definition:

$$\begin{aligned}
f'(x) &= \lim_{h \to 0} \frac{f(x+h) - f(x)}{h} \\
&= \lim_{h \to 0} \frac{a(x+h) + b - (ax + b)}{h} \\
&= \lim_{h \to 0} \frac{ah}{h} = a.
\end{aligned}$$

An important special case of Example 1 says that the derivative of a constant function is the zero function:

If $g(x) = c$ (constant), then $g'(x) = 0$.

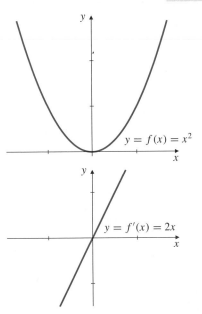

Figure 2.12 The derivative of $f(x) = x^2$ is $f'(x) = 2x$

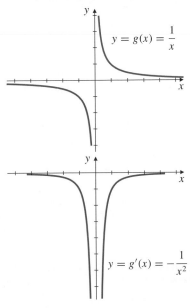

Figure 2.13 The derivative of $g(x) = 1/x$ is $g'(x) = -1/x^2$

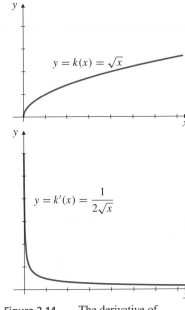

Figure 2.14 The derivative of $k(x) = \sqrt{x}$ is $k'(x) = 1/(2\sqrt{x})$

Example 2 Use the definition of the derivative to calculate the derivatives of the functions:

(a) $f(x) = x^2$, (b) $g(x) = \dfrac{1}{x}$, and (c) $k(x) = \sqrt{x}$.

Solution See Figures 2.12–2.14 for the graphs of these functions and of their derivatives.

(a) $f'(x) = \lim\limits_{h \to 0} \dfrac{f(x+h) - f(x)}{h}$

$\qquad = \lim\limits_{h \to 0} \dfrac{(x+h)^2 - x^2}{h}$

$\qquad = \lim\limits_{h \to 0} \dfrac{2hx + h^2}{h} = \lim\limits_{h \to 0} (2x + h) = 2x.$

(b) $g'(x) = \lim\limits_{h \to 0} \dfrac{g(x+h) - g(x)}{h}$

$\qquad = \lim\limits_{h \to 0} \dfrac{\dfrac{1}{x+h} - \dfrac{1}{x}}{h}$

$\qquad = \lim\limits_{h \to 0} \dfrac{x - (x+h)}{h(x+h)x} = \lim\limits_{h \to 0} -\dfrac{1}{(x+h)x} = -\dfrac{1}{x^2}.$

(c) $k'(x) = \lim\limits_{h \to 0} \dfrac{k(x+h) - k(x)}{h}$

$= \lim\limits_{h \to 0} \dfrac{\sqrt{x+h} - \sqrt{x}}{h}$

$= \lim\limits_{h \to 0} \dfrac{\sqrt{x+h} - \sqrt{x}}{h} \times \dfrac{\sqrt{x+h} + \sqrt{x}}{\sqrt{x+h} + \sqrt{x}}$

$= \lim\limits_{h \to 0} \dfrac{x+h-x}{h(\sqrt{x+h} + \sqrt{x})} = \lim\limits_{h \to 0} \dfrac{1}{\sqrt{x+h} + \sqrt{x}} = \dfrac{1}{2\sqrt{x}}.$

Note that k is not differentiable at $x = 0$. (See Figure 2.14.) Since 0 is in the domain of k, it is a singular point of k.

The three derivative formulas calculated in Example 2 are special cases of the following **General Power Rule**:

If $f(x) = x^r$,   then $f'(x) = r\,x^{r-1}$.

This formula, which we will verify in Section 3.3, is valid for *all values of r and x for which x^{r-1} makes sense as a real number.*

Example 3   (**Differentiating powers**)

If $f(x) = x^{5/3}$, then $f'(x) = \dfrac{5}{3}x^{(5/3)-1} = \dfrac{5}{3}x^{2/3}$ for all real x.

If $g(t) = \dfrac{1}{\sqrt{t}} = t^{-1/2}$, then $g'(t) = -\dfrac{1}{2}t^{-(1/2)-1} = -\dfrac{1}{2}t^{-3/2}$ for $t > 0$.

Eventually we will prove all appropriate cases of the General Power Rule. For the time being, here is a proof of the case $r = n$, a positive integer, based on the *factoring of a difference of nth powers*:

$$a^n - b^n = (a-b)(a^{n-1} + a^{n-2}b + a^{n-3}b^2 + \cdots + ab^{n-2} + b^{n-1}).$$

(Check this formula by multiplying the two factors on the right-hand side.) If $f(x) = x^n$, $a = x + h$ and $b = x$, then $a - b = h$ and,

$f'(x) = \lim\limits_{h \to 0} \dfrac{(x+h)^n - x^n}{h}$

$\overbrace{}^{n \text{ terms}}$

$= \lim\limits_{h \to 0} \dfrac{h\left[(x+h)^{n-1} + (x+h)^{n-2}x + (x+h)^{n-3}x^2 + \cdots + x^{n-1}\right]}{h}$

$= nx^{n-1}.$

An alternative proof based on the product rule and mathematical induction will be given in Section 2.3. The factorization method used above can also be used to demonstrate the General Power Rule for negative integers, $r = -n$, and reciprocals of integers, $r = 1/n$. (See Exercises 50 and 52 at the end of this section.)

Example 4 (**Differentiating the absolute value function**) Verify that:

$$\text{If } f(x) = |x|, \quad \text{then} \quad f'(x) = \frac{x}{|x|} = \text{sgn}\, x.$$

Solution We have

$$f(x) = \begin{cases} x, & \text{if } x \geq 0 \\ -x, & \text{if } x < 0 \end{cases}.$$

Thus, from Example 1 above, $f'(x) = 1$ if $x > 0$ and $f'(x) = -1$ if $x < 0$. Also, Example 4 of Section 2.1 shows that f is not differentiable at $x = 0$, which is a singular point of f. Therefore (see Figure 2.15),

$$f'(x) = \begin{cases} 1, & \text{if } x > 0 \\ -1, & \text{if } x < 0 \end{cases} = \frac{x}{|x|} = \text{sgn}\, x.$$

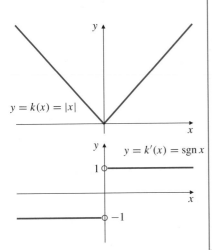

$y = k(x) = |x|$

$y = k'(x) = \text{sgn}\, x$

Figure 2.15 The derivative of $|x|$ is $\text{sgn}\, x = x/|x|$

Table 1 lists a collection of the elementary derivatives calculated above. Beginning in Section 2.3 we will develop general rules for calculating the derivatives of functions obtained by combining simpler functions. Thereafter we will seldom have to revert to the definition of the derivative and to the calculation of limits to evaluate derivatives. It is important, therefore, to remember the derivatives of some elementary functions. Memorize those in Table 1.

Table 1. Some elementary functions and their derivatives

$f(x)$	$f'(x)$
c (constant)	0
x	1
x^2	$2x$
$\dfrac{1}{x}$	$-\dfrac{1}{x^2}$ $(x \neq 0)$
$\sqrt{x}$	$\dfrac{1}{2\sqrt{x}}$ $(x > 0)$
x^r	$r\,x^{r-1}$ $(x^{r-1}$ real)
$|x|$	$\dfrac{x}{|x|} = \text{sgn}\, x$

Leibniz Notation

Because functions can be written in different ways, it is useful to have more than one notation for derivatives. If $y = f(x)$, we can use the dependent variable y to represent the function, and we can denote the derivative of the function with respect to x in any of the following ways:

$$D_x y = y' = \frac{dy}{dx} = \frac{d}{dx} f(x) = f'(x) = Df(x).$$

Often the most convenient way of referring to the derivative of a function given explicitly as an expression in the variable x is to write $\frac{d}{dx}$ in front of that expression. The symbol $\frac{d}{dx}$ is a *differential operator* and should be read "the derivative with respect to x of ..." For example,

$$\frac{d}{dx}x^2 = 2x \quad \text{(the derivative with respect to } x \text{ of } x^2 \text{ is } 2x\text{)}$$

$$\frac{d}{dx}\sqrt{x} = \frac{1}{2\sqrt{x}}$$

$$\frac{d}{dt}t^{100} = 100t^{99}$$

$$\text{if } y = u^3, \text{ then } \frac{dy}{du} = 3u^2.$$

The value of the derivative of a function at a particular number x_0 in its domain can also be expressed in several ways:

Do not confuse the expressions

$$\frac{d}{dx}f(x) \text{ and } \frac{d}{dx}f(x)\bigg|_{x=x_0}.$$

The first expression represents a *function*, $f'(x)$. The second represents a *number*, $f'(x_0)$.

$$D_x y\bigg|_{x=x_0} = y'\bigg|_{x=x_0} = \frac{dy}{dx}\bigg|_{x=x_0} = \frac{d}{dx}f(x)\bigg|_{x=x_0} = f'(x_0) = Df(x_0).$$

The symbol $\bigg|_{x=x_0}$ is called an **evaluation symbol**. It signifies that the expression preceding it should be evaluated at $x = x_0$. Thus,

$$\frac{d}{dx}x^4\bigg|_{x=-1} = 4x^3\bigg|_{x=-1} = 4(-1)^3 = -4.$$

Here is another example in which a derivative is computed from the definition, this time for a somewhat more complicated function.

Example 5 Use the definition of derivative to calculate $\dfrac{d}{dx}\left(\dfrac{x}{x^2+1}\right)\bigg|_{x=2}$.

Solution We could calculate $\dfrac{d}{dx}\left(\dfrac{x}{x^2+1}\right)$ and then substitute $x = 2$, but it is easier to put $x = 2$ in the expression for the Newton quotient before taking the limit:

$$\frac{d}{dx}\left(\frac{x}{x^2+1}\right)\bigg|_{x=2} = \lim_{h\to 0}\frac{\dfrac{2+h}{(2+h)^2+1} - \dfrac{2}{2^2+1}}{h}$$

$$= \lim_{h\to 0}\frac{\dfrac{2+h}{5+4h+h^2} - \dfrac{2}{5}}{h}$$

$$= \lim_{h\to 0}\frac{5(2+h) - 2(5+4h+h^2)}{5(5+4h+h^2)h}$$

$$= \lim_{h\to 0}\frac{-3h - 2h^2}{5(5+4h+h^2)h}$$

$$= \lim_{h\to 0}\frac{-3 - 2h}{5(5+4h+h^2)} = -\frac{3}{25}.$$

The notations dy/dx and $\frac{d}{dx} f(x)$ are called **Leibniz notations** for the derivative, after Gottfried Wilhelm Leibniz (1646–1716), one of the creators of calculus, who used such notations. The main ideas of calculus were developed independently by Leibniz and Isaac Newton (1642–1727); the latter used notations similar to the prime (y') notations we use here.

The Leibniz notation is suggested by the definition of derivative. The Newton quotient $[f(x+h) - f(x)]/h$, whose limit we take to find the derivative dy/dx, can be written in the form $\Delta y/\Delta x$, where $\Delta y = f(x+h) - f(x)$ is the increment in y, and $\Delta x = (x+h) - x = h$ is the corresponding increment in x as we pass from the point $(x, f(x))$ to the point $(x+h, f(x+h))$ on the graph of f. (See Figure 2.16.) Using symbols:

$$\frac{dy}{dx} = \lim_{\Delta x \to 0} \frac{\Delta y}{\Delta x}.$$

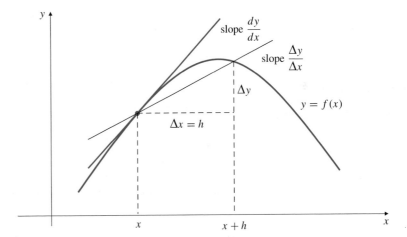

Figure 2.16 $\dfrac{dy}{dx} = \lim\limits_{\Delta x \to 0} \dfrac{\Delta y}{\Delta x}$

Differentials

The Newton quotient $\Delta y/\Delta x$ is actually the quotient of two quantities, Δy and Δx. It is not at all clear, however, that the derivative dy/dx, the limit of $\Delta y/\Delta x$ as Δx approaches zero, can be regarded as a quotient. If y is a continuous function of x, then Δy approaches zero when Δx approaches zero, so dy/dx appears to be the meaningless quantity $0/0$. Nevertheless, it is sometimes useful to be able to refer to quantities dy and dx in such a way that their quotient is the derivative dy/dx. We can justify this by regarding dx as a new *independent* variable (called **the differential of x**) and defining a new *dependent* variable dy (**the differential of y**) as a function of x and dx by

$$dy = \frac{dy}{dx} dx = f'(x)\, dx.$$

For example, if $y = x^2$, we can write $dy = 2x\, dx$ to mean the same thing as $dy/dx = 2x$. This *differential notation* will be used for the interpretation and manipulation of integrals beginning in Chapter 5.

Note that, defined as above, differentials are merely variables that may or may not be small in absolute value. The differentials dy and dx were originally regarded (by Leibniz and his successors) as "infinitesimals" (infinitely small but nonzero) quantities whose quotient dy/dx gave the slope of the tangent line (a secant line meeting the graph of $y = f(x)$ at two points infinitely close together). It can be shown that such "infinitesimal" quantities cannot exist (as real numbers). It is possible to extend the number system to contain infinitesimals and use these to develop calculus, but we will not consider this approach here.

Derivatives Have the Intermediate-Value Property

Is a function f defined on an interval I necessarily the derivative of some other function defined on I? The answer is no; some functions are derivatives and some are not. Although a derivative need not be a continuous function, it must, like a continuous function, have the intermediate-value property: on an interval $[a, b]$, a derivative $f'(x)$ takes on every value between $f'(a)$ and $f'(b)$. An everywhere-defined step function such as the Heaviside $H(x)$ function considered in Example 1 in Section 1.4 does not have this property on, say, the interval $[-1, 1]$, so cannot be the derivative of a function on that interval. This argument does not apply to the signum function, which is the derivative of the absolute value function on any interval (see Example 4), even though it does not have the intermediate-value property on an interval containing the origin. Note, however, that the signum function is not defined at the origin.

If $g(x)$ is continuous on an interval I, then $g(x) = f'(x)$ for some function f that is differentiable on I. We will discuss this further in Chapter 5 and prove it in Appendix III.

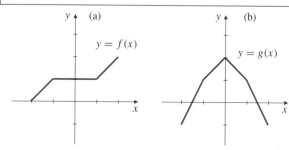

Figure 2.17   This function is not a derivative on $[-1, 1]$; it does not have the intermediate-value property.

Exercises 2.2

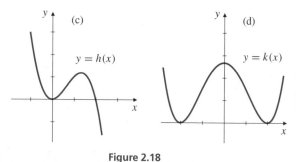

Figure 2.18

Make rough sketches of the graphs of the derivatives of the functions in Exercises 1–4.

1. The function f graphed in Figure 2.18(a).

2. The function g graphed in Figure 2.18(b).

3. The function h graphed in Figure 2.18(c).

4. The function k graphed in Figure 2.18(d).

5. Where is the function f graphed in Figure 2.18(a) differentiable?

6. Where is the function g graphed in Figure 2.18(b) differentiable?

Use a graphics utility with differentiation capabilities to plot the graphs of the following functions and their derivatives. Observe the relationships between the graph of y and that of y' in each case. What features of the graph of y can you infer from the graph of y'?

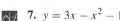

7. $y = 3x - x^2 - 1$ **8.** $y = x^3 - 3x^2 + 2x + 1$

9. $y = |x^3 - x|$ **10.** $y = |x^2 - 1| - |x^2 - 4|$

In Exercises 11–22, calculate the derivative of the given function directly from the definition of derivative.

11. $y = x^2 - 3x$ **12.** $f(x) = 1 + 4x - 5x^2$

13. $f(x) = x^3$ **14.** $s = \dfrac{1}{3 + 4t}$

15. $F(t) = \sqrt{2t + 1}$ **16.** $f(x) = \dfrac{3}{4}\sqrt{2 - x}$

17. $y = x + \dfrac{1}{x}$

18. $z = \dfrac{s}{1+s}$

19. $F(x) = \dfrac{1}{\sqrt{1+x^2}}$

20. $y = \dfrac{1}{x^2}$

21. $y = \dfrac{1}{\sqrt{1+x}}$

22. $f(t) = \dfrac{t^2-3}{t^2+3}$

23. How should the function $f(x) = x \, \text{sgn}\, x$ be defined at $x = 0$ so that it is continuous there? Is it then differentiable there?

24. How should the function $g(x) = x^2 \text{sgn}\, x$ be defined at $x = 0$ so that it is continuous there? Is it then differentiable there?

25. Where does $h(x) = |x^2 + 3x + 2|$ fail to be differentiable?

26. Using a calculator, find the slope of the secant line to $y = x^3 - 2x$ passing through the points corresponding to $x = 1$ and $x = 1 + \Delta x$, for several values of Δx of decreasing size, say $\Delta x = \pm 0.1, \pm 0.01, \pm 0.001, \pm 0.0001$. (Make a table.) Also, calculate $\dfrac{d}{dx}\left(x^3 - 2x\right)\Big|_{x=1}$ using the definition of derivative.

27. Repeat Exercise 26 for the function $f(x) = \dfrac{1}{x}$ and the points $x = 2$ and $x = 2 + \Delta x$.

Using the definition of derivative, find equations for the tangent lines to the curves in Exercises 28–31 at the points indicated.

28. $y = 5 + 4x - x^2$ at the point where $x = 2$

29. $y = \sqrt{x+6}$ at the point $(3, 3)$

30. $y = \dfrac{t}{t^2-2}$ at the point where $t = -2$

31. $y = \dfrac{2}{t^2+t}$ at the point where $t = a$

Calculate the derivatives of the functions in Exercises 32–37 using the General Power Rule. Where is each derivative valid?

32. $f(x) = x^{-17}$

33. $g(t) = t^{22}$

34. $y = x^{1/3}$

35. $y = x^{-1/3}$

36. $t^{-2.25}$

37. $s^{119/4}$

In Exercises 38–48, you may use the formulas for derivatives established in this section.

38. Calculate $\dfrac{d}{ds}\sqrt{s}\,\Big|_{s=9}$.

39. Find $F'(\tfrac{1}{4})$ if $F(x) = \dfrac{1}{x}$.

40. Find $f'(8)$ if $f(x) = x^{-2/3}$.

41. Find $dy/dt\big|_{t=4}$ if $y = t^{1/4}$.

42. Find an equation of the straight line tangent to the curve $y = \sqrt{x}$ at $x = x_0$.

43. Find an equation of the straight line normal to the curve $y = 1/x$ at the point where $x = a$.

44. Show that the curve $y = x^2$ and the straight line $x + 4y = 18$ intersect at right angles at one of their two intersection points. *Hint:* find the product of their slopes at

45. There are two distinct straight lines that pass through the point $(1, -3)$ and are tangent to the curve $y = x^2$. Find their equations. *Hint:* draw a sketch. The points of tangency are not given; let them be denoted (a, a^2).

46. Find equations of two straight lines that have slope -2 and are tangent to the graph of $y = 1/x$.

47. Find the slope of a straight line that passes through the point $(-2, 0)$ and is tangent to the curve $y = \sqrt{x}$.

∗ 48. Show that there are two distinct tangent lines to the curve $y = x^2$ passing through the point (a, b) provided $b < a^2$. How many tangent lines to $y = x^2$ pass through (a, b) if $b = a^2$? If $b > a^2$?

∗ 49. Show that the derivative of an odd differentiable function is even and that the derivative of an even differentiable function is odd.

∗ 50. Prove the case $r = -n$ (n is a positive integer) of the General Power Rule; that is, prove that

$$\dfrac{d}{dx}\, x^{-n} = -n\, x^{-n-1}.$$

Use the factorization of a difference of nth powers given in this section.

∗ 51. Use the factoring of a difference of cubes:

$$a^3 - b^3 = (a-b)(a^2 + ab + b^2),$$

to help you calculate the derivative of $f(x) = x^{1/3}$ directly from the definition of derivative.

∗ 52. Prove the General Power Rule for $\dfrac{d}{dx}x^r$, where $r = 1/n$, n being a positive integer. (*Hint:*

$$\dfrac{d}{dx}x^{1/n} = \lim_{h\to 0} \dfrac{(x+h)^{1/n} - x^{1/n}}{h}$$

$$= \lim_{h\to 0} \dfrac{(x+h)^{1/n} - x^{1/n}}{((x+h)^{1/n})^n - (x^{1/n})^n}.$$

Apply the factorization of the difference of nth powers to the denominator of the latter quotient.)

53. Give a proof of the power rule $\dfrac{d}{dx}x^n = nx^{n-1}$ for positive integers n using the Binomial Theorem:

$$(x+h)^n = x^n + \dfrac{n}{1}x^{n-1}h + \dfrac{n(n-1)}{1\times 2}x^{n-2}h^2$$

$$+ \dfrac{n(n-1)(n-2)}{1\times 2\times 3}x^{n-3}h^3 + \cdots + h^n.$$

∗ 54. Use right and left derivatives, $f'_+(a)$ and $f'_-(a)$, to define the concept of a half-line starting at $(a, f(a))$ being a right or left tangent to the graph of f at $x = a$. Show that the graph has a tangent line at $x = a$ if and only if it has right and left tangents that are opposite halves of the same straight line. What are the left and right tangents to the graphs of $y = x^{1/3}$, $y = x^{2/3}$, and $y = |x|$ at $x = 0$?

2.3 Differentiation Rules

If every derivative had to be calculated directly from the definition of derivative as in the examples of Section 2.2, calculus would indeed be a painful subject. Fortunately there is an easier way. We will develop several general *differentiation rules* that enable us to calculate the derivatives of complicated combinations of functions easily if we already know the derivatives of the elementary functions from which they are constructed. For instance, we will be able to find the derivative of $\dfrac{x^2}{\sqrt{x^2+1}}$ if we know the derivatives of x^2 and $\sqrt{x}$. The rules we develop in this section tell us how to differentiate sums, constant multiples, products, and quotients of functions whose derivatives we already know. In Section 2.4 we will learn how to differentiate composite functions.

Before developing these differentiation rules we need to establish one obvious but very important theorem which states, roughly, that the graph of a function cannot possibly have a break at a point where it is smooth.

THEOREM 1

Differentiability implies continuity

If f is differentiable at x, then f is continuous at x.

PROOF   Since f is differentiable at x, we know that

$$\lim_{h \to 0} \frac{f(x+h) - f(x)}{h} = f'(x)$$

exists. In order to prove that f is continuous at x, we need to show that

$$\lim_{h \to 0} f(x+h) = f(x).$$

Using the limit rules (Theorem 2 of Section 1.2), we have

$$\lim_{h \to 0} f(x+h) = \lim_{h \to 0} \left(f(x) + \frac{f(x+h) - f(x)}{h} \times h \right)$$
$$= f(x) + f'(x) \times 0 = f(x).$$

Sums and Constant Multiples

The derivative of a sum (or difference) of functions is the sum (or difference) of the derivatives of those functions. The derivative of a constant multiple of a function is the same constant multiple of the derivative of the function.

THEOREM 2

Differentiation rules for sums, differences, and constant multiples

If functions f and g are differentiable at x, and if C is a constant, then the functions $f + g$, $f - g$, and Cf are all differentiable at x and

$$(f + g)'(x) = f'(x) + g'(x),$$
$$(f - g)'(x) = f'(x) - g'(x),$$
$$(Cf)'(x) = Cf'(x).$$

PROOF The proofs of all three assertions are straightforward, using the corresponding limit rules from Theorem 2 of Section 1.2. For the sum, we have

$$(f + g)'(x) = \lim_{h \to 0} \frac{(f + g)(x + h) - (f + g)(x)}{h}$$

$$= \lim_{h \to 0} \frac{(f(x + h) + g(x + h)) - (f(x) + g(x))}{h}$$

$$= \lim_{h \to 0} \left(\frac{f(x + h) - f(x)}{h} + \frac{g(x + h) - g(x)}{h} \right)$$

$$= f'(x) + g'(x),$$

because the limit of a sum is the sum of the limits. The proof for the difference $f - g$ is similar. For the constant multiple, we have

$$(Cf)'(x) = \lim_{h \to 0} \frac{Cf(x + h) - Cf(x)}{h}$$

$$= C \lim_{h \to 0} \frac{f(x + h) - f(x)}{h} = Cf'(x).$$

The rule for differentiating sums extends to sums of any finite number of terms:

$$(f_1 + f_2 + \cdots + f_n)' = f_1' + f_2' + \cdots + f_n'. \qquad (*)$$

To see this we can use a technique called **mathematical induction**. (See the note in the margin.) Theorem 2 shows that the case $n = 2$ is true; this is STEP 1. For STEP 2, we must show that if the formula $(*)$ holds for some integer $n = k \geq 2$, then it must also hold for $n = k + 1$. Therefore, assume that

$$(f_1 + f_2 + \cdots + f_k)' = f_1' + f_2' + \cdots + f_k'.$$

Then we have

$$(f_1 + f_2 + \cdots + f_k + f_{k+1})'$$

$$= \left(\underbrace{(f_1 + f_2 + \cdots + f_k)}_{\text{Let this function be } f} + f_{k+1} \right)'$$

$$= (f + f_{k+1})' \qquad \text{(Now use the known case } n = 2.)$$

$$= f' + f_{k+1}'$$

$$= f_1' + f_2' + \cdots + f_k' + f_{k+1}'.$$

Mathematical Induction

Mathematical induction is a technique for proving that a statement about an integer n is true for every integer n greater than or equal to some lowest integer n_0. The proof requires us to carry out two steps:

STEP 1. Prove that the statement is true for $n = n_0$.

STEP 2. Prove that if the statement is true for some integer $n = k$, where $k \geq n_0$, then it is also true for the next larger integer, $n = k + 1$.

Step 2 prevents there from being a smallest integer greater than n_0 for which the statement is false. Being true for n_0, the statement must therefore be true for all larger integers.

With both steps verified, we can claim that $(*)$ holds for any $n \geq 2$ *by induction*. In particular, therefore, the derivative of any polynomial is the sum of the derivatives of its terms.

Example 1 Calculate the derivatives of the functions:

(a) $2x^3 - 5x^2 + 4x + 7$,   (b) $f(x) = 5\sqrt{x} + \dfrac{3}{x} - 18$,   (c) $y = \dfrac{1}{7}t^4 - 3t^{7/3}$.

Solution   Each of these functions is a sum of constant multiples of functions that we already know how to differentiate.

(a) $\dfrac{d}{dx}(2x^3 - 5x^2 + 4x + 7) = 2(3x^2) - 5(2x) + 4(1) + 0 = 6x^2 - 10x + 4.$

(b) $f'(x) = 5\left(\dfrac{1}{2\sqrt{x}}\right) + 3\left(-\dfrac{1}{x^2}\right) - 18(0) = \dfrac{5}{2\sqrt{x}} - \dfrac{3}{x^2}.$

(c) $\dfrac{dy}{dt} = \dfrac{1}{7}(4t^3) - 3\left(\dfrac{7}{3}t^{4/3}\right) = \dfrac{4}{7}t^3 - 7t^{4/3}.$

Example 2   Find an equation of the tangent to the curve $y = \dfrac{3x^3 - 4}{x}$ at the point on the curve where $x = -2$.

Solution   If $x = -2$, then $y = 14$. The slope of the curve at $(-2, 14)$ is

$$\left.\dfrac{dy}{dx}\right|_{x=-2} = \left.\dfrac{d}{dx}\left(3x^2 - \dfrac{4}{x}\right)\right|_{x=-2} = \left.\left(6x + \dfrac{4}{x^2}\right)\right|_{x=-2} = -11.$$

An equation of the tangent line is $y = 14 - 11(x + 2)$, or $y = -11x - 8$.

The Product Rule

The rule for differentiating a product of functions is a little more complicated than that for sums. It is *not* true that the derivative of a product is the product of the derivatives.

THEOREM **3**

The Product Rule

If functions f and g are differentiable at x, then their product fg is also differentiable at x, and

$$(fg)'(x) = f'(x)g(x) + f(x)g'(x).$$

PROOF   We set up the Newton quotient for fg and then add 0 to the numerator in a way that enables us to involve the Newton quotients for f and g separately:

$$(fg)'(x) = \lim_{h \to 0} \frac{f(x+h)g(x+h) - f(x)g(x)}{h}$$

$$= \lim_{h \to 0} \frac{f(x+h)g(x+h) - f(x)g(x+h) + f(x)g(x+h) - f(x)g(x)}{h}$$

$$= \lim_{h \to 0} \left(\frac{f(x+h) - f(x)}{h}g(x+h) + f(x)\frac{g(x+h) - g(x)}{h}\right)$$

$$= f'(x)g(x) + f(x)g'(x).$$

To get the last line we have used the fact that f and g are differentiable and the fact that g is therefore continuous (Theorem 1), as well as limit rules from Theorem 2 of Section 1.2. A graphical proof of the Product Rule is suggested by Figure 2.19.

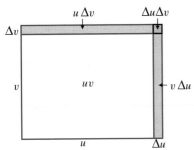

Figure 2.19

A graphical proof of the Product Rule

Here $u = f(x)$ and $v = g(x)$, so that the rectangular area uv represents $f(x)g(x)$. If x changes by an amount Δx, the corresponding increments in u and v are Δu and Δv. The change in the area of the rectangle is

$$\Delta(uv)$$

$$= (u + \Delta u)(v + \Delta v) - uv$$

$$= (\Delta u)v + u(\Delta v) + (\Delta u)(\Delta v),$$

the sum of the three shaded areas. Dividing by Δx and taking the limit as $\Delta x \to 0$, we get

$$\frac{d}{dx}(uv) = \left(\frac{du}{dx}\right)v + u\left(\frac{dv}{dx}\right),$$

since

$$\lim_{\Delta x \to 0} \frac{\Delta u}{\Delta x}\Delta v = \frac{du}{dx} \times 0 = 0.$$

Example 3 Find the derivative of $(x^2 + 1)(x^3 + 4)$ using and without using the Product Rule.

Solution Using the Product Rule with $f(x) = x^2 + 1$ and $g(x) = x^3 + 4$, we calculate

$$\frac{d}{dx}\big((x^2 + 1)(x^3 + 4)\big) = 2x(x^3 + 4) + (x^2 + 1)(3x^2) = 5x^4 + 3x^2 + 8x.$$

On the other hand, we can calculate the derivative by first multiplying the two binomials and then differentiating the resulting polynomial:

$$\frac{d}{dx}\big((x^2 + 1)(x^3 + 4)\big) = \frac{d}{dx}(x^5 + x^3 + 4x^2 + 4) = 5x^4 + 3x^2 + 8x.$$

Example 4 Find $\dfrac{dy}{dx}$ if $y = \left(2\sqrt{x} + \dfrac{3}{x}\right)\left(3\sqrt{x} - \dfrac{2}{x}\right)$.

Solution Applying the Product Rule with f and g being the two functions enclosed in the large parentheses, we obtain

$$\frac{dy}{dx} = \left(\frac{1}{\sqrt{x}} - \frac{3}{x^2}\right)\left(3\sqrt{x} - \frac{2}{x}\right) + \left(2\sqrt{x} + \frac{3}{x}\right)\left(\frac{3}{2\sqrt{x}} + \frac{2}{x^2}\right)$$

$$= 6 - \frac{5}{2x^{3/2}} + \frac{12}{x^3}.$$

Example 5 Let $y = uv$ be the product of the functions u and v. Find $y'(2)$ if $u(2) = 2$, $u'(2) = -5$, $v(2) = 1$, and $v'(2) = 3$.

Solution From the Product Rule we have

$$y' = (uv)' = u'v + uv'.$$

Therefore,

$$y'(2) = u'(2)v(2) + u(2)v'(2) = (-5)(1) + (2)(3) = -5 + 6 = 1.$$

Example 6   Use mathematical induction to verify the formula $\dfrac{d}{dx}x^n = n\,x^{n-1}$ for all positive integers n.

Solution   For $n = 1$ the formula says that $\frac{d}{dx}x^1 = 1 = 1x^0$, so the formula is true in this case. We must show that if the formula is true for $n = k \geq 1$, then it is also true for $n = k + 1$. Therefore, assume that

$$\frac{d}{dx}x^k = kx^{k-1}.$$

Using the Product Rule we calculate

$$\frac{d}{dx}x^{k+1} = \frac{d}{dx}\left(x^k x\right) = (kx^{k-1})(x) + (x^k)(1) = (k+1)x^k = (k+1)x^{(k+1)-1}.$$

Thus the formula is true for $n = k + 1$ also. The formula is true for all integers $n \geq 1$ *by induction*.

The Product Rule can be extended to products of any number of factors, for instance:

$$(fgh)'(x) = f'(x)(gh)(x) + f(x)(gh)'(x)$$
$$= f'(x)g(x)h(x) + f(x)g'(x)h(x) + f(x)g(x)h'(x).$$

In general, the derivative of a product of n functions will have n terms; each term will be the same product but with one of the factors replaced by its derivative:

$$(f_1 f_2 f_3 \cdots f_n)' = f_1' f_2 f_3 \cdots f_n + f_1 f_2' f_3 \cdots f_n + \cdots + f_1 f_2 f_3 \cdots f_n'.$$

This can be proved by mathematical induction. See Exercise 54 at the end of this section.

The Reciprocal Rule

THEOREM 4

The Reciprocal Rule

If f is differentiable at x and $f(x) \neq 0$, then $1/f$ is differentiable at x, and

$$\left(\frac{1}{f}\right)'(x) = \frac{-f'(x)}{(f(x))^2}.$$

PROOF   Using the definition of the derivative, we calculate

$$\frac{d}{dx}\frac{1}{f(x)} = \lim_{h \to 0}\frac{\dfrac{1}{f(x+h)} - \dfrac{1}{f(x)}}{h}$$

$$= \lim_{h \to 0}\frac{f(x) - f(x+h)}{hf(x+h)f(x)}$$

$$= \lim_{h \to 0}\left(\frac{-1}{f(x+h)f(x)}\right)\frac{f(x+h) - f(x)}{h}$$

$$= \frac{-1}{(f(x))^2}f'(x)$$

Again we have to use the continuity of f (Theorem 1) and the limit rules from Section 1.2.

Example 7 Differentiate the functions

(a) $\dfrac{1}{x^2+1}$ and (b) $f(t) = \dfrac{1}{t + \dfrac{1}{t}}$.

Solution Using the Reciprocal Rule:

(a) $\dfrac{d}{dx}\left(\dfrac{1}{x^2+1}\right) = \dfrac{-2x}{(x^2+1)^2}$.

(b) $f'(t) = \dfrac{-1}{\left(t + \dfrac{1}{t}\right)^2}\left(1 - \dfrac{1}{t^2}\right) = \dfrac{-t^2}{(t^2+1)^2}\dfrac{t^2-1}{t^2} = \dfrac{1-t^2}{(t^2+1)^2}$.

We can use the Reciprocal Rule to confirm the General Power Rule for negative integers:

$$\dfrac{d}{dx}x^{-n} = -n\,x^{-n-1},$$

since we have already proved the rule for positive integers. We have

$$\dfrac{d}{dx}x^{-n} = \dfrac{d}{dx}\dfrac{1}{x^n} = \dfrac{-n\,x^{n-1}}{(x^n)^2} = -n\,x^{-n-1}.$$

Example 8 (Differentiating sums of reciprocals)

$$\dfrac{d}{dx}\left(\dfrac{x^2+x+1}{x^3}\right) = \dfrac{d}{dx}\left(\dfrac{1}{x} + \dfrac{1}{x^2} + \dfrac{1}{x^3}\right)$$

$$= \dfrac{d}{dx}(x^{-1} + x^{-2} + x^{-3})$$

$$= -x^{-2} - 2x^{-3} - 3x^{-4} = -\dfrac{1}{x^2} - \dfrac{2}{x^3} - \dfrac{3}{x^4}.$$

The Quotient Rule

The Product Rule and the Reciprocal Rule can be combined to provide a rule for differentiating a quotient of two functions. Observe that

$$\dfrac{d}{dx}\left(\dfrac{f(x)}{g(x)}\right) = \dfrac{d}{dx}\left(f(x)\dfrac{1}{g(x)}\right) = f'(x)\dfrac{1}{g(x)} + f(x)\left(-\dfrac{g'(x)}{(g(x))^2}\right)$$

$$= \dfrac{g(x)f'(x) - f(x)g'(x)}{(g(x))^2}.$$

Thus we have proved the following Quotient Rule.

THEOREM 5

The Quotient Rule

If f and g are differentiable at x, and if $g(x) \neq 0$, then the quotient f/g is differentiable at x and

$$\left(\frac{f}{g}\right)'(x) = \frac{g(x)f'(x) - f(x)g'(x)}{(g(x))^2}.$$

Sometimes students have trouble remembering this rule. (Getting the order of the terms in the numerator wrong will reverse the sign.) Try to remember (and use) the Quotient Rule in the following form:

(quotient)′

$$= \frac{(\text{denominator}) \times (\text{numerator})' - (\text{numerator}) \times (\text{denominator})'}{(\text{denominator})^2}$$

Example 9 Find the derivatives of

(a) $y = \dfrac{1 - x^2}{1 + x^2}$, (b) $\dfrac{\sqrt{t}}{3 - 5t}$, and (c) $f(\theta) = \dfrac{a + b\theta}{m + n\theta}$.

Solution We use the Quotient Rule in each case.

(a) $\dfrac{dy}{dx} = \dfrac{(1 + x^2)(-2x) - (1 - x^2)(2x)}{(1 + x^2)^2} = -\dfrac{4x}{(1 + x^2)^2}.$

(b) $\dfrac{d}{dt}\left(\dfrac{\sqrt{t}}{3 - 5t}\right) = \dfrac{(3 - 5t)\dfrac{1}{2\sqrt{t}} - \sqrt{t}(-5)}{(3 - 5t)^2} = \dfrac{3 + 5t}{2\sqrt{t}(3 - 5t)^2}.$

(c) $f'(\theta) = \dfrac{(m + n\theta)(b) - (a + b\theta)(n)}{(m + n\theta)^2} = \dfrac{mb - na}{(m + n\theta)^2}.$

In all three parts of Example 9 the Quotient Rule yielded fractions with numerators that were complicated but could be simplified algebraically. It is advisable to attempt such simplifications when calculating derivatives; the usefulness of derivatives in applications of calculus often depends on such simplifications.

Example 10 Find equations of any lines that pass through the point $(-1, 0)$ and are tangent to the curve $y = (x - 1)/(x + 1)$.

Solution The point $(-1, 0)$ does not lie on the curve, so it is not the point of tangency. Suppose a line is tangent to the curve at $x = a$, so the point of tangency is $(a, (a - 1)/(a + 1))$. Note that a cannot be -1. The slope of the line must be

$$\left.\frac{dy}{dx}\right|_{x=a} = \left.\frac{(x + 1)(1) - (x - 1)(1)}{(x + 1)^2}\right|_{x=a} = \frac{2}{(a + 1)^2}.$$

If the line also passes through $(-1, 0)$, its slope must also be given by

$$\frac{\dfrac{a-1}{a+1} - 0}{a - (-1)} = \frac{a-1}{(a+1)^2}.$$

Equating these two expressions for the slope, we get an equation to solve for a:

$$\frac{a-1}{(a+1)^2} = \frac{2}{(a+1)^2} \qquad \Longrightarrow \qquad a - 1 = 2.$$

Thus $a = 3$, and the slope of the line is $2/4^2 = 1/8$. There is only one line through $(-1, 0)$ tangent to the given curve, and its equation is

$$y = 0 + \frac{1}{8}(x + 1) \qquad \text{or} \qquad x - 8y + 1 = 0.$$

∎

Remark Derivatives of quotients of functions where the denominator is a monomial, such as in Example 8, are usually easier to do by breaking the quotient into several fractions (as was done in that example) rather than by using the Quotient Rule.

Exercises 2.3

In Exercises 1–32, calculate the derivatives of the given functions. Simplify your answers whenever possible.

1. $y = 3x^2 - 5x - 7$

2. $y = 4x^{1/2} - \dfrac{5}{x}$

3. $f(x) = Ax^2 + Bx + C$

4. $f(x) = \dfrac{6}{x^3} + \dfrac{2}{x^2} - 2$

5. $z = \dfrac{s^5 - s^3}{15}$

6. $y = x^{45} - x^{-45}$

7. $g(t) = t^{1/3} + 2t^{1/4} + 3t^{1/5}$

8. $y = 3\sqrt[3]{t^2} - \dfrac{2}{\sqrt{t^3}}$

9. $u = \dfrac{3}{5}x^{5/3} - \dfrac{5}{3}x^{-3/5}$

10. $F(x) = (3x - 2)(1 - 5x)$

11. $y = \sqrt{x}\left(5 - x - \dfrac{x^2}{3}\right)$

12. $g(t) = \dfrac{1}{2t - 3}$

13. $y = \dfrac{1}{x^2 + 5x}$

14. $y = \dfrac{4}{3 - x}$

15. $f(t) = \dfrac{\pi}{2 - \pi t}$

16. $g(y) = \dfrac{2}{1 - y^2}$

17. $f(x) = \dfrac{1 - 4x^2}{x^3}$

18. $g(u) = \dfrac{u\sqrt{u} - 3}{u^2}$

19. $y = \dfrac{2 + t + t^2}{\sqrt{t}}$

20. $z = \dfrac{x - 1}{x^{2/3}}$

21. $f(x) = \dfrac{3 - 4x}{3 + 4x}$

22. $z = \dfrac{t^2 + 2t}{t^2 - 1}$

23. $s = \dfrac{1 + \sqrt{t}}{1 - \sqrt{t}}$

24. $f(x) = \dfrac{x^3 - 4}{x + 1}$

25. $f(x) = \dfrac{ax + b}{cx + d}$

26. $F(t) = \dfrac{t^2 + 7t - 8}{t^2 - t + 1}$

27. $f(x) = (1 + x)(1 + 2x)(1 + 3x)(1 + 4x)$

28. $f(r) = (r^{-2} + r^{-3} - 4)(r^2 + r^3 + 1)$

29. $y = (x^2 + 4)(\sqrt{x} + 1)(5x^{2/3} - 2)$

30. $y = \dfrac{(x^2 + 1)(x^3 + 2)}{(x^2 + 2)(x^3 + 1)}$

∗ 31. $y = \dfrac{x}{2x + \dfrac{1}{3x + 1}}$

∗ 32. $f(x) = \dfrac{(\sqrt{x} - 1)(2 - x)(1 - x^2)}{\sqrt{x}(3 + 2x)}$

Calculate the derivatives in Exercises 33–36, given that $f(2) = 2$ and $f'(2) = 3$.

33. $\dfrac{d}{dx}\left(\dfrac{x^2}{f(x)}\right)\Big|_{x=2}$

34. $\dfrac{d}{dx}\left(\dfrac{f(x)}{x^2}\right)\Big|_{x=2}$

35. $\dfrac{d}{dx}\left(x^2 f(x)\right)\Big|_{x=2}$

36. $\dfrac{d}{dx}\left(\dfrac{f(x)}{x^2 + f(x)}\right)\Big|_{x=2}$

37. Find

$$\frac{d}{dx}\left(\frac{x^2-4}{x^2+4}\right)\Big|_{x=-2}.$$

38. Find

$$\frac{d}{dt}\left(\frac{t(1+\sqrt{t})}{5-t}\right)\Big|_{t=4}.$$

39. If $f(x) = \dfrac{\sqrt{x}}{x+1}$, find $f'(2)$.

40. Find $\dfrac{d}{dt}\Big((1+t)(1+2t)(1+3t)(1+4t)\Big)\Big|_{t=0}.$

41. Find an equation of the tangent line to $y = \dfrac{2}{3-4\sqrt{x}}$ at the point $(1, -2)$.

42. Find equations of the tangent and normal to $y = \dfrac{x+1}{x-1}$ at $x = 2$.

43. Find the points on the curve $y = x + 1/x$ where the tangent line is horizontal.

44. Find the equations of all horizontal lines that are tangent to the curve $y = x^2(4 - x^2)$.

45. Find the coordinates of all points where the curve $y = \dfrac{1}{x^2+x+1}$ has a horizontal tangent line.

46. Find the coordinates of points on the curve $y = \dfrac{x+1}{x+2}$ where the tangent line is parallel to the line $y = 4x$.

47. Find the equation of the straight line that passes through the point $(0, b)$ and is tangent to the curve $y = 1/x$. Assume $b \neq 0$.

* 48. Show that the curve $y = x^2$ intersects the curve $y = 1/\sqrt{x}$ at right angles.

49. Find two straight lines that are tangent to $y = x^3$ and pass through the point $(2, 8)$.

50. Find two straight lines that are tangent to $y = x^2/(x-1)$ and pass through the point $(2, 0)$.

51. (A Square Root Rule) Show that if f is differentiable at x and $f(x) > 0$, then

$$\frac{d}{dx}\sqrt{f(x)} = \frac{f'(x)}{2\sqrt{f(x)}}.$$

Use this Square Root Rule to find the derivative of $\sqrt{x^2+1}$.

52. Show that $f(x) = |x^3|$ is differentiable at every real number x, and find its derivative.

Mathematical Induction

53. Use mathematical induction to prove that $\dfrac{d}{dx}x^{n/2} = \dfrac{n}{2}x^{(n/2)-1}$ for every positive integer n. Then use the Reciprocal Rule to get the same result for negative integers n.

54. Use mathematical induction to prove the formula for the derivative of a product of n functions given earlier in this section.

2.4 The Chain Rule

Although we can differentiate $\sqrt{x}$ and $x^2 + 1$, we cannot yet differentiate $\sqrt{x^2+1}$. To do this, we need a rule that tells us how to differentiate *composites* of functions whose derivatives we already know. This rule is known as the Chain Rule and is the most often used of all the differentiation rules.

Example 1 The function $\dfrac{1}{x^2-4}$ is the composite $f(g(x))$ of $f(u) = \dfrac{1}{u}$ and $g(x) = x^2 - 4$, which have derivatives

$$f'(u) = \frac{-1}{u^2} \qquad \text{and} \qquad g'(x) = 2x.$$

According to the Reciprocal Rule (which is a special case of the Chain Rule),

$$\frac{d}{dx}f(g(x)) = \frac{d}{dx}\left(\frac{1}{x^2-4}\right) = \frac{-2x}{(x^2-4)^2} = \frac{-1}{(x^2-4)^2}(2x)$$

$$= f'(g(x))g'(x).$$

This example suggests that the derivative of a composite function $f(g(x))$ is the derivative of f evaluated at $g(x)$ multiplied by the derivative of g evaluated at x.

This is the Chain Rule:

$$\frac{d}{dx} f(g(x)) = f'(g(x)) \, g'(x).$$

THEOREM **6**

The Chain Rule

If $f(u)$ is differentiable at $u = g(x)$, and $g(x)$ is differentiable at x, then the composite function $f \circ g(x) = f(g(x))$ is differentiable at x, and

$$(f \circ g)'(x) = f'(g(x))g'(x).$$

In terms of Leibniz notation, if $y = f(u)$ where $u = g(x)$, then $y = f(g(x))$ and:

at u, y is changing $\dfrac{dy}{du}$ times as fast as u is changing;

at x, u is changing $\dfrac{du}{dx}$ times as fast as x is changing.

Therefore, at x, $y = f(u) = f(g(x))$ is changing $\dfrac{dy}{du} \times \dfrac{du}{dx}$ times as fast as x is changing. That is:

$$\frac{dy}{dx} = \frac{dy}{du}\frac{du}{dx}, \qquad \text{where } \frac{dy}{du} \text{ is evaluated at } u = g(x).$$

It appears as though the symbol du cancels from the numerator and denominator, but this is not meaningful because dy/du was not defined as the quotient of two quantities, but rather as a single quantity, the derivative of y with respect to u.

We would like to prove Theorem 6 by writing

$$\frac{\Delta y}{\Delta x} = \frac{\Delta y}{\Delta u}\frac{\Delta u}{\Delta x}$$

and taking the limit as $\Delta x \to 0$. Such a proof is valid for most composite functions but not all. (See Exercise 46 at the end of this section.) A correct proof will be given later in this section, but first we do more examples to give a better idea of how the Chain Rule works.

Example 2 Find the derivative of $y = \sqrt{x^2 + 1}$.

Solution Here $y = f(g(x))$, where $f(u) = \sqrt{u}$ and $g(x) = x^2 + 1$. Since the derivatives of f and g are

$$f'(u) = \frac{1}{2\sqrt{u}} \qquad \text{and} \qquad g'(x) = 2x,$$

the Chain Rule gives

$$\frac{dy}{dx} = \frac{d}{dx} f(g(x)) = f'(g(x)) \cdot g'(x)$$

$$= \frac{1}{2\sqrt{g(x)}} \cdot g'(x) = \frac{1}{2\sqrt{x^2 + 1}} \cdot (2x) = \frac{x}{\sqrt{x^2 + 1}}.$$

Outside and Inside Functions

In the composite $f(g(x))$, the function f is "outside," and the function g is "inside." The Chain Rule says that the derivative of the composite is the derivative f' of the outside function evaluated at the inside function $g(x)$, multiplied by the derivative $g'(x)$ of the inside function:

$$\frac{d}{dx}f(g(x)) = f'(g(x)) \times g'(x).$$

Usually, when applying the Chain Rule, we do not introduce symbols to represent the functions being composed, but rather just proceed to calculate the derivative of the "outside" function and then multiply by the derivative of whatever is "inside." You can say to yourself: "the derivative of f of something is f' of that thing, multiplied by the derivative of that thing."

Example 3   Find derivatives of the following functions:

(a) $(7x - 3)^{10}$,   (b) $f(t) = |t^2 - 1|$,   and   (c) $\left(3x + \dfrac{1}{(2x + 1)^3}\right)^{1/4}$.

Solution

(a) Here, the outside function is the 10th power; it must be differentiated first and the result multiplied by the derivative of the expression $7x - 3$:

$$\frac{d}{dx}(7x - 3)^{10} = 10(7x - 3)^9(7) = 70(7x - 3)^9.$$

(b) Here, we are differentiating the absolute value of something. The derivative is signum of that thing, multiplied by the derivative of that thing:

$$f'(t) = \big(\operatorname{sgn}(t^2-1)\big)(2t) = \frac{2t(t^2 - 1)}{|t^2 - 1|} = \begin{cases} 2t & \text{if } t < -1 \text{ or } t > 1 \\ -2t & \text{if } -1 < t < 1 \\ \text{undefined} & \text{if } t = \pm 1. \end{cases}$$

(c) Here, we will need to use the Chain Rule twice. We begin by differentiating the 1/4 power of something, but the something involves the -3rd power of $2x + 1$, and the derivative of that will also require the Chain Rule:

$$\frac{d}{dx}\left(3x + \frac{1}{(2x + 1)^3}\right)^{1/4} = \frac{1}{4}\left(3x + \frac{1}{(2x + 1)^3}\right)^{-3/4}\frac{d}{dx}\left(3x + \frac{1}{(2x + 1)^3}\right)$$

$$= \frac{1}{4}\left(3x + \frac{1}{(2x + 1)^3}\right)^{-3/4}\left(3 - \frac{3}{(2x + 1)^4}\frac{d}{dx}(2x + 1)\right)$$

$$= \frac{3}{4}\left(1 - \frac{2}{(2x + 1)^4}\right)\left(3x + \frac{1}{(2x + 1)^3}\right)^{-3/4}.$$

When you start to feel comfortable with the Chain Rule, you may want to "save a line or two" by carrying out the whole differentiation in one step:

$$\frac{d}{dx}\left(3x + \frac{1}{(2x + 1)^3}\right)^{1/4} = \frac{1}{4}\left(3x + \frac{1}{(2x + 1)^3}\right)^{-3/4}\left(3 - \frac{3}{(2x + 1)^4}(2)\right)$$

$$= \frac{3}{4}\left(1 - \frac{2}{(2x + 1)^4}\right)\left(3x + \frac{1}{(2x + 1)^3}\right)^{-3/4}.$$

Use of the Chain Rule produces products of factors that do not usually come out in the order you would naturally write them. Often you will want to rewrite the result with the factors in a different order. This is obvious in parts (a) and (c) of the example above. In monomials (expressions that are products of factors), it is common to write the factors in order of increasing complexity from left to right, with numerical factors coming first. One time when you would *not* waste time doing this, or trying to make any other simplification, is when you are going to evaluate the derivative at a particular number. In this case, substitute the number as soon as you have calculated the derivative, before doing any simplification:

$$\frac{d}{dx}(x^2 - 3)^{10}\Big|_{x=2} = 10(x^2 - 3)^9(2x)\Big|_{x=2} = (10)(1^9)(4) = 40.$$

Example 4 Suppose that f is a differentiable function on the real line. In terms of the derivative f' of f, express the derivatives of:

(a) $f(3x)$, (b) $f(x^2)$, (c) $f(\pi f(x))$, and (d) $[f(3 - 2f(x))]^4$.

Solution

(a) $\dfrac{d}{dx} f(3x) = \left(f'(3x)\right)(3) = 3f'(3x).$

(b) $\dfrac{d}{dx} f(x^2) = \left(f'(x^2)\right)(2x) = 2xf'(x^2).$

(c) $\dfrac{d}{dx} f(\pi f(x)) = \left(f'(\pi f(x))\right)(\pi f'(x)) = \pi f'(x) f'(\pi f(x)).$

(d) $\dfrac{d}{dx} [f(3 - 2f(x))]^4 = 4[f(3 - 2f(x))]^3 f'(3 - 2f(x))(-2f'(x))$

$$= -8f'(x)f'(3 - 2f(x))[f(3 - 2f(x))]^3.$$

Finding Derivatives with Maple

Computer algebra systems know the derivatives of elementary functions and can calculate the derivatives of combinations of these functions symbolically, using differentiation rules. Maple's `D` operator can be used to find the derivative function `D(f)` of a function `f` of one variable. Alternatively, you can use `diff` to differentiate an expression with respect to a variable and then use the substitution routine `subs` to evaluate the result at a particular number.

```
>   f := x -> sqrt(1+2*x^2);
```
$$f := x \rightarrow \sqrt{1 + 2x^2}$$
```
>   fprime := D(f);
```
$$fprime := x \rightarrow 2\frac{x}{\sqrt{1 + 2x^2}}$$
```
>   fprime(2);
```
$$\frac{4}{3}$$
```
>   diff(t^2*sin(3*t),t);
```
$$2t \sin(3t) + 3t^2 \cos(3t)$$
```
>   simplify(subs(t=Pi/12, %));
```
$$\frac{1}{12}\pi\sqrt{2} + \frac{1}{96}\pi^2\sqrt{2}$$

Building the Chain Rule into Differentiation Formulas

If u is a differentiable function of x and $y = u^n$, then the Chain Rule gives

$$\frac{d}{dx}u^n = \frac{dy}{dx} = \frac{dy}{du}\frac{du}{dx} = nu^{n-1}\frac{du}{dx}.$$

The formula

$$\frac{d}{dx}u^n = nu^{n-1}\frac{du}{dx}$$

is just the formula $\frac{d}{dx}x^n = nx^{n-1}$ with an application of the Chain Rule built in, so that it applies to functions of x rather than just to x. Some other differentiation rules with built-in Chain Rule applications are:

$$\frac{d}{dx}\left(\frac{1}{u}\right) = \frac{-1}{u^2}\frac{du}{dx} \qquad \text{(the Reciprocal Rule)}$$

$$\frac{d}{dx}\sqrt{u} = \frac{1}{2\sqrt{u}}\frac{du}{dx} \qquad \text{(the Square Root Rule)}$$

$$\frac{d}{dx}u^r = r\,u^{r-1}\frac{du}{dx} \qquad \text{(the General Power Rule)}$$

$$\frac{d}{dx}|u| = \operatorname{sgn}u\,\frac{du}{dx} = \frac{u}{|u|}\frac{du}{dx} \qquad \text{(the Absolute Value Rule)}$$

Proof of the Chain Rule (Theorem 6)

Suppose that f is differentiable at the point $u = g(x)$ and that g is differentiable at x. Let the function $E(k)$ be defined by

$$E(0) = 0,$$

$$E(k) = \frac{f(u+k) - f(u)}{k} - f'(u), \qquad \text{if } k \neq 0.$$

By the definition of derivative, $\lim_{k\to 0} E(k) = f'(u) - f'(u) = 0 = E(0)$, so $E(k)$ is continuous at $k = 0$. Also, whether $k = 0$ or not, we have

$$f(u+k) - f(u) = \big(f'(u) + E(k)\big)k.$$

Now put $u = g(x)$ and $k = g(x+h) - g(x)$, so that $u + k = g(x+h)$, and obtain

$$f(g(x+h)) - f(g(x)) = \big(f'(g(x)) + E(k)\big)(g(x+h) - g(x)).$$

Since g is differentiable at x, $\lim_{h\to 0}[g(x+h) - g(x)]/h = g'(x)$. Also, g is continuous at x by Theorem 1, so $\lim_{h\to 0} k = \lim_{h\to 0}(g(x+h) - g(x)) = 0$. Since E is continuous at 0, $\lim_{h\to 0} E(k) = \lim_{k\to 0} E(k) = E(0) = 0$. Hence

$$\frac{d}{dx}f(g(x)) = \lim_{h\to 0}\frac{f(g(x+h)) - f(g(x))}{h}$$

$$= \lim_{h\to 0}\big(f'(g(x)) + E(k)\big)\frac{g(x+h) - g(x)}{h}$$

$$= \big(f'(g(x)) + 0\big)g'(x) = f'(g(x))g'(x),$$

which was to be proved.

Exercises 2.4

Find the derivatives of the functions in Exercises 1–16.

1. $y = (2x + 3)^6$

2. $y = \left(1 - \dfrac{x}{3}\right)^{99}$

3. $f(x) = (4 - x^2)^{10}$

4. $y = \sqrt{1 - 3x^2}$

5. $F(t) = \left(2 + \dfrac{3}{t}\right)^{-10}$

6. $(1 + x^{2/3})^{3/2}$

7. $\dfrac{3}{5 - 4x}$

8. $(1 - 2t^2)^{-3/2}$

9. $y = |1 - x^2|$

10. $f(t) = |2 + t^3|$

11. $y = 4x + |4x - 1|$

12. $y = (2 + |x|^3)^{1/3}$

13. $y = \dfrac{1}{2 + \sqrt{3x + 4}}$

14. $f(x) = \left(1 + \sqrt{\dfrac{x - 2}{3}}\right)^4$

15. $z = \left(u + \dfrac{1}{u - 1}\right)^{-5/3}$

16. $y = \dfrac{x^5 \sqrt{3 + x^6}}{(4 + x^2)^3}$

17. Sketch the graph of the function in Exercise 10.

18. Sketch the graph of the function in Exercise 11.

Verify that the General Power Rule holds for the functions in Exercises 19–21.

19. $x^{1/4} = \sqrt{\sqrt{x}}$

20. $x^{3/4} = \sqrt{x\sqrt{x}}$

21. $x^{3/2} = \sqrt{(x^3)}$

In Exercises 22–29, express the derivative of the given function in terms of the derivative f' of the differentiable function f.

22. $f(2t + 3)$

23. $f(5x - x^2)$

24. $\left[f\left(\dfrac{2}{x}\right)\right]^3$

25. $\sqrt{3 + 2f(x)}$

26. $f\left(\sqrt{3 + 2t}\right)$

27. $f\left(3 + 2\sqrt{x}\right)$

28. $f\left(2f(3f(x))\right)$

29. $f\left(2 - 3f(4 - 5t)\right)$

30. Find $\dfrac{d}{dx}\left(\dfrac{\sqrt{x^2 - 1}}{x^2 + 1}\right)\Big|_{x=-2}$.

31. Find $\dfrac{d}{dt}\sqrt{3t - 7}\Big|_{t=3}$.

32. If $f(x) = \dfrac{1}{\sqrt{2x + 1}}$, find $f'(4)$.

33. If $y = (x^3 + 9)^{17/2}$, find $y'\big|_{x=-2}$.

34. Find $F'(0)$ if $F(x) = (1 + x)(2 + x)^2(3 + x)^3(4 + x)^4$.

* **35.** Calculate y' if $y = (x + ((3x)^5 - 2)^{-1/2})^{-6}$. Try to do it all in one step.

In Exercises 36–39, find an equation of the tangent line to the given curve at the given point.

36. $y = \sqrt{1 + 2x^2}$ at $x = 2$

37. $y = (1 + x^{2/3})^{3/2}$ at $x = -1$

38. $y = (ax + b)^8$ at $x = b/a$

39. $y = 1/(x^2 - x + 3)^{3/2}$ at $x = -2$

40. Show that the derivative of $f(x) = (x - a)^m (x - b)^n$ vanishes at some point between a and b if m and n are positive integers.

Use Maple or another computer algebra system to evaluate and simplify the derivatives of the functions in Exercises 41–44.

41. $y = \sqrt{x^2 + 1} + \dfrac{1}{(x^2 + 1)^{3/2}}$

42. $y = \dfrac{(x^2 - 1)(x^2 - 4)(x^2 - 9)}{x^6}$

43. $\dfrac{dy}{dt}\Big|_{t=2}$ if $y = (t + 1)(t^2 + 2)(t^3 + 3)(t^4 + 4)(t^5 + 5)$

44. $f'(1)$ if $f(x) = \dfrac{(x^2 + 3)^{1/2}(x^3 + 7)^{1/3}}{(x^4 + 15)^{1/4}}$

45. Does the Chain Rule enable you to calculate the derivatives of $|x|^2$ and $|x^2|$ at $x = 0$? Do these functions have derivatives at $x = 0$? Why?

* **46.** What is wrong with the following "proof" of the Chain Rule? Let $k = g(x + h) - g(x)$. Then $\lim_{h \to 0} k = 0$. Thus

$$\lim_{h \to 0} \frac{f(g(x + h)) - f(g(x))}{h}$$

$$= \lim_{h \to 0} \frac{f(g(x + h)) - f(g(x))}{g(x + h) - g(x)} \cdot \frac{g(x + h) - g(x)}{h}$$

$$= \lim_{h \to 0} \frac{f(g(x) + k) - f(g(x))}{k} \cdot \frac{g(x + h) - g(x)}{h}$$

$$= f'(g(x)) \, g'(x).$$

2.5 Derivatives of Trigonometric Functions

The trigonometric functions, especially sine and cosine, play a very important role in the mathematical modelling of real-world phenomena. In particular, they arise whenever quantities fluctuate in a periodic way. Elastic motions, vibrations, and waves of all kinds naturally involve the trigonometric functions, and many physical and mechanical laws are formulated as differential equations having these functions as solutions.

In this section we will calculate the derivatives of the six trigonometric functions. We only have to work hard for one of them, sine; the others then follow from known identities and the differentiation rules of Section 2.3.

Some Special Limits

First, we have to establish some trigonometric limits that we will need to calculate the derivative of sine. It is assumed throughout that the arguments of the trigonometric functions are measured in radians.

THEOREM **7**

The functions $\sin\theta$ and $\cos\theta$ are continuous at every value of θ. In particular, at $\theta = 0$ we have:

$$\lim_{\theta \to 0} \sin\theta = \sin 0 = 0 \qquad \text{and} \qquad \lim_{\theta \to 0} \cos\theta = \cos 0 = 1.$$

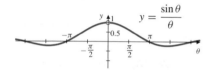

Figure 2.20 It appears that $\lim_{\theta \to 0} (\sin\theta)/\theta = 1$

This result is obvious from the graphs of sine and cosine, so we will not prove it here. A proof can be based on the Squeeze Theorem (Theorem 4 of Section 1.2). The method is suggested in Exercise 62 at the end of this section.

The graph of the function $y = (\sin\theta)/\theta$ is shown in Figure 2.20. Although it is not defined at $\theta = 0$, this function appears to have limit 1 as θ approaches 0.

THEOREM **8**

An important trigonometric limit

$$\lim_{\theta \to 0} \frac{\sin\theta}{\theta} = 1 \qquad \text{(where } \theta \text{ is in radians)}.$$

PROOF Let $0 < \theta < \pi/2$, and represent θ as shown in Figure 2.21. Points $A(1, 0)$ and $P(\cos\theta, \sin\theta)$ lie on the unit circle $x^2 + y^2 = 1$. The area of the circular sector OAP lies between the areas of triangles OAP and OAT:

$$\text{Area } \triangle OAP < \text{Area sector } OAP < \text{Area } \triangle OAT.$$

As shown in Section P.6, the area of a circular sector having central angle θ (radians) and radius 1 is $\theta/2$. The area of a triangle is $(1/2) \times$ base $\times$ height, so

$$\text{Area } \triangle OAP = \frac{1}{2}(1)(\sin\theta) = \frac{\sin\theta}{2},$$

$$\text{Area } \triangle OAT = \frac{1}{2}(1)(\tan\theta) = \frac{\sin\theta}{2\cos\theta}.$$

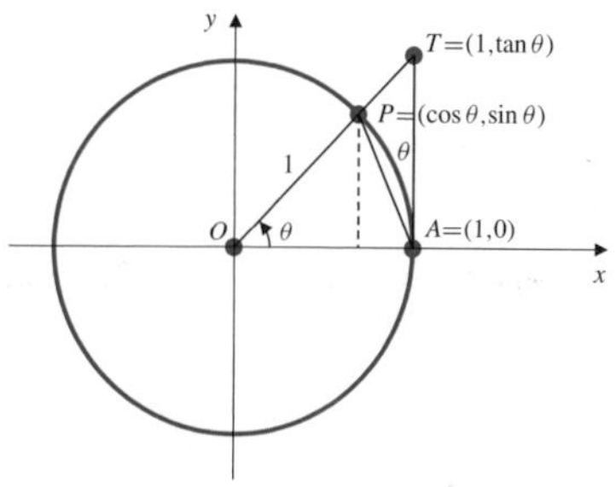

Figure 2.21 Area $\triangle OAP$ < Area sector OAP < Area $\triangle OAT$

Thus

$$\frac{\sin\theta}{2} < \frac{\theta}{2} < \frac{\sin\theta}{2\cos\theta},$$

or, upon multiplication by the positive number $2/\sin\theta$,

$$1 < \frac{\theta}{\sin\theta} < \frac{1}{\cos\theta}.$$

Now take reciprocals, thereby reversing the inequalities:

$$1 > \frac{\sin\theta}{\theta} > \cos\theta.$$

Since $\lim_{\theta\to 0+}\cos\theta = 1$ by Theorem 7, the Squeeze Theorem gives

$$\lim_{\theta\to 0+}\frac{\sin\theta}{\theta} = 1.$$

Finally, note that $\sin\theta$ and θ are *odd functions*. Therefore, $f(\theta) = (\sin\theta)/\theta$ is an *even function*: $f(-\theta) = f(\theta)$, as shown in Figure 2.20. This symmetry implies that the left limit at 0 must have the same value as the right limit:

$$\lim_{\theta\to 0-}\frac{\sin\theta}{\theta} = 1 = \lim_{\theta\to 0+}\frac{\sin\theta}{\theta},$$

so $\lim_{\theta\to 0}(\sin\theta)/\theta = 1$ by Theorem 1 of Section 1.2.

Theorem 8 can be combined with limit rules and known trigonometric identities to yield other trigonometric limits.

Example 1 Show that $\lim\limits_{h\to 0}\dfrac{\cos h - 1}{h} = 0.$

Solution Using the half-angle formula $\cos h = 1 - 2\sin^2(h/2)$, we calculate

$$\lim_{h\to 0}\frac{\cos h - 1}{h} = \lim_{h\to 0} -\frac{2\sin^2(h/2)}{h} \qquad \text{Let } \theta = h/2.$$

$$= -\lim_{\theta\to 0}\frac{\sin\theta}{\theta}\,\sin\theta = -(1)(0) = 0. \quad\blacksquare$$

The Derivatives of Sine and Cosine

To calculate the derivative of $\sin x$ we need the addition formula for sine (see Section P.6):

$$\sin(x + h) = \sin x\,\cos h + \cos x\,\sin h.$$

THEOREM 9 **The derivative of the sine function is the cosine function.**

$$\frac{d}{dx}\sin x = \cos x$$

PROOF We use the definition of derivative, the addition formula for sine, the rules for combining limits, Theorem 8, and the result of Example 1:

$$\frac{d}{dx}\sin x = \lim_{h\to 0}\frac{\sin(x+h)-\sin x}{h}$$

$$= \lim_{h\to 0}\frac{\sin x \cos h + \cos x \sin h - \sin x}{h}$$

$$= \lim_{h\to 0}\frac{\sin x(\cos h - 1) + \cos x \sin h}{h}$$

$$= \lim_{h\to 0}\sin x \cdot \lim_{h\to 0}\frac{\cos h - 1}{h} + \lim_{h\to 0}\cos x \cdot \lim_{h\to 0}\frac{\sin h}{h}$$

$$= (\sin x)\cdot(0) + (\cos x)\cdot(1) = \cos x.$$

THEOREM 10

The derivative of the cosine function is minus the sine function.

$$\frac{d}{dx}\cos x = -\sin x$$

PROOF We could mimic the proof for sine above, using the addition rule for cosine, $\cos(x+h) = \cos x \cos h - \sin x \sin h$. An easier way is to make use of the complementary angle identities, $\sin((\pi/2)-x) = \cos x$ and $\cos((\pi/2)-x) = \sin x$, and the Chain Rule from Section 2.4:

$$\frac{d}{dx}\cos x = \frac{d}{dx}\sin\left(\frac{\pi}{2}-x\right) = (-1)\cos\left(\frac{\pi}{2}-x\right) = -\sin x.$$

Notice the minus sign in the derivative of cosine. The derivative of the sine is the cosine, but the derivative of the cosine is *minus* the sine. This is shown graphically in Figure 2.22.

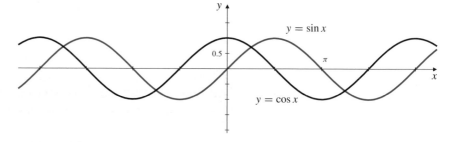

Figure 2.22   The sine and cosine plotted together. The slope of the sine curve at x is $\cos x$; the slope of the cosine curve at x is $-\sin x$.

Example 2   Evaluate the derivatives of the following functions:

(a) $\sin(\pi x) + \cos(3x)$,   (b) $x^2 \sin\sqrt{x}$,   and   (c) $\dfrac{\cos x}{1 - \sin x}$.

Solution

(a) By the Sum Rule and the Chain Rule:

$$\frac{d}{dx}(\sin(\pi x)+\cos(3x)) = \cos(\pi x)(\pi)-\sin(3x)(3) = \pi\cos(\pi x)-3\sin(3x).$$

(b) By the Product and Chain Rules:

$$\frac{d}{dx}(x^2 \sin \sqrt{x}) = 2x \sin \sqrt{x} + x^2 (\cos \sqrt{x}) \frac{1}{2\sqrt{x}} = 2x \sin \sqrt{x} + \frac{1}{2} x^{3/2} \cos \sqrt{x}$$

(c) By the Quotient Rule:

$$\frac{d}{dx}\left(\frac{\cos x}{1 - \sin x}\right) = \frac{(1 - \sin x)(-\sin x) - (\cos x)(0 - \cos x)}{(1 - \sin x)^2}$$

$$= \frac{-\sin x + \sin^2 x + \cos^2 x}{(1 - \sin x)^2}$$

$$= \frac{1 - \sin x}{(1 - \sin x)^2} = \frac{1}{1 - \sin x}$$

We used the identity $\sin^2 x + \cos^2 x = 1$ to simplify the middle line.

Using trigonometric identities can sometimes change the way a derivative is calculated. Carrying out a differentiation in different ways can lead to different-looking answers, but they should be equal if no errors have been made.

Example 3   Use two different methods to find the derivative of the function $f(t) = \sin t \cos t$.

Solution   By the Product Rule:

$$f'(t) = (\cos t)(\cos t) + (\sin t)(-\sin t) = \cos^2 t - \sin^2 t.$$

On the other hand, since $\sin(2t) = 2 \sin t \cos t$, we have

$$f'(t) = \frac{d}{dt}\left(\frac{1}{2} \sin(2t)\right) = \left(\frac{1}{2}\right)(2) \cos(2t) = \cos(2t).$$

The two answers are really the same, since $\cos(2t) = \cos^2 t - \sin^2 t$.

It is very important to remember that the formulas for the derivatives of $\sin x$ and $\cos x$ were obtained under the assumption that x is measured in *radians*. Since $180° = \pi$ radians, $x° = \pi x/180$ radians. By the Chain Rule,

$$\frac{d}{dx}\sin(x°) = \frac{d}{dx}\sin\left(\frac{\pi x}{180}\right) = \frac{\pi}{180}\cos\left(\frac{\pi x}{180}\right) = \frac{\pi}{180}\cos(x°).$$

Similarly, the derivative of $\cos(x°)$ is $-(\pi/180)\sin(x°)$.

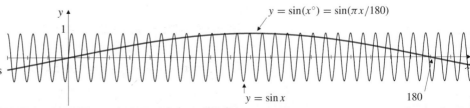

Figure 2.23 $\sin(x°)$ oscillates much more slowly than $\sin x$. Its maximum slope is $\pi/180$

Continuity

The six trigonometric functions are differentiable, and therefore continuous (by Theorem 1), everywhere on their domains. This means that we can calculate the limits of most trigonometric functions as $x \to a$ by evaluating them at $x = a$.

The three "co-" functions (cosine, cotangent, and cosecant) have explicit minus signs in their derivatives.

The Derivatives of the Other Trigonometric Functions

Because $\sin x$ and $\cos x$ are differentiable everywhere, the functions

$$\tan x = \frac{\sin x}{\cos x} \qquad \sec x = \frac{1}{\cos x}$$

$$\cot x = \frac{\cos x}{\sin x} \qquad \csc x = \frac{1}{\sin x}$$

are differentiable at every value of x at which they are defined (i.e., where their denominators are not zero). Their derivatives can be calculated by the Quotient and Reciprocal Rules and are as follows:

$$\frac{d}{dx}\tan x = \sec^2 x \qquad \frac{d}{dx}\sec x = \sec x \tan x$$

$$\frac{d}{dx}\cot x = -\csc^2 x \qquad \frac{d}{dx}\csc x = -\csc x \cot x.$$

Example 4   Verify the derivative formulas for $\tan x$ and $\sec x$.

Solution   We use the Quotient Rule for tangent and the Reciprocal Rule for secant:

$$\frac{d}{dx}\tan x = \frac{d}{dx}\left(\frac{\sin x}{\cos x}\right) = \frac{\cos x \dfrac{d}{dx}(\sin x) - \sin x \dfrac{d}{dx}(\cos x)}{\cos^2 x}$$

$$= \frac{\cos x \cos x - \sin x(-\sin x)}{\cos^2 x} = \frac{\cos^2 x + \sin^2 x}{\cos^2 x}$$

$$= \frac{1}{\cos^2 x} = \sec^2 x.$$

$$\frac{d}{dx}\sec x = \frac{d}{dx}\left(\frac{1}{\cos x}\right) = \frac{-1}{\cos^2 x}\frac{d}{dx}(\cos x)$$

$$= \frac{-1}{\cos^2 x}(-\sin x) = \frac{1}{\cos x}\cdot\frac{\sin x}{\cos x}$$

$$= \sec x \tan x.$$

Example 5

(a) $\dfrac{d}{dx}\left[3x + \cot\left(\dfrac{x}{2}\right)\right] = 3 + \left[-\csc^2\left(\dfrac{x}{2}\right)\right]\dfrac{1}{2} = 3 - \dfrac{1}{2}\csc^2\left(\dfrac{x}{2}\right)$

(b) $\dfrac{d}{dx}\left(\dfrac{3}{\sin(2x)}\right) = \dfrac{d}{dx}(3\csc(2x))$

$$= 3(-\csc(2x)\cot(2x))(2) = -6\csc(2x)\cot(2x).$$

Example 6 Find the tangent and normal lines to the curve $y = \tan(\pi x/4)$ at the point $(1, 1)$.

Solution The slope of the tangent to $y = \tan(\pi x/4)$ at $(1, 1)$ is:

$$\frac{dy}{dx}\Big|_{x=1} = \frac{\pi}{4}\sec^2(\pi x/4)\Big|_{x=1} = \frac{\pi}{4}\sec^2\left(\frac{\pi}{4}\right) = \frac{\pi}{4}\left(\sqrt{2}\right)^2 = \frac{\pi}{2}.$$

The tangent is the line

$$y = 1 + \frac{\pi}{2}(x - 1), \qquad \text{or} \qquad y = \frac{\pi x}{2} - \frac{\pi}{2} + 1.$$

The normal has slope $m = -2/\pi$, so its point-slope equation is

$$y = 1 - \frac{2}{\pi}(x - 1), \qquad \text{or} \qquad y = -\frac{2x}{\pi} + \frac{2}{\pi} + 1.$$

Exercises 2.5

1. Verify the formula for the derivative of $\csc x = 1/(\sin x)$.

2. Verify the formula for the derivative of $\cot x = (\cos x)/(\sin x)$.

Find the derivatives of the functions in Exercises 3–36. Simplify your answers whenever possible. Also be on the lookout for ways you might simplify the given expression before differentiating it.

3. $y = \cos 3x$

4. $y = \sin \dfrac{x}{5}$

5. $y = \tan \pi x$

6. $y = \sec ax$

7. $y = \cot(4 - 3x)$

8. $y = \sin((\pi - x)/3)$

9. $f(x) = \cos(s - rx)$

10. $y = \sin(Ax + B)$

11. $\sin(\pi x^2)$

12. $\cos(\sqrt{x})$

13. $y = \sqrt{1 + \cos x}$

14. $\sin(2 \cos x)$

15. $f(x) = \cos(x + \sin x)$

16. $g(\theta) = \tan(\theta \sin \theta)$

17. $u = \sin^3(\pi x/2)$

18. $y = \sec(1/x)$

19. $F(t) = \sin at \cos at$

20. $G(\theta) = \dfrac{\sin a\theta}{\cos b\theta}$

21. $\sin(2x) - \cos(2x)$

22. $\cos^2 x - \sin^2 x$

23. $\tan x + \cot x$

24. $\sec x - \csc x$

25. $\tan x - x$

26. $\tan(3x)\cot(3x)$

27. $t \cos t - \sin t$

28. $t \sin t + \cos t$

29. $\dfrac{\sin x}{1 + \cos x}$

30. $\dfrac{\cos x}{1 + \sin x}$

31. $x^2 \cos(3x)$

32. $g(t) = \sqrt{(\sin t)/t}$

33. $v = \sec(x^2)\tan(x^2)$

34. $z = \dfrac{\sin \sqrt{x}}{1 + \cos \sqrt{x}}$

35. $\sin(\cos(\tan t))$

36. $f(s) = \cos(s + \cos(s + \cos s))$

37. Given that $\sin 2x = 2 \sin x \cos x$, deduce that $\cos 2x = \cos^2 x - \sin^2 x$.

38. Given that $\cos 2x = \cos^2 x - \sin^2 x$, deduce that $\sin 2x = 2 \sin x \cos x$.

In Exercises 39–42, find equations for the lines that are tangent and normal to the curve $y = f(x)$ at the given point.

39. $y = \sin x$, $(\pi, 0)$

40. $y = \tan(2x)$, $(0, 0)$

41. $y = \sqrt{2}\cos(x/4)$, $(\pi, 1)$

42. $y = \cos^2 x$, $\left(\dfrac{\pi}{3}, \dfrac{1}{4}\right)$

43. Find an equation of the line tangent to the curve $y = \sin(x^\circ)$ at the point where $x = 45$.

44. Find an equation of the straight line normal to $y = \sec(x^\circ)$ at the point where $x = 60$.

45. Find the points on the curve $y = \tan x$, $-\pi/2 < x < \pi/2$, where the tangent is parallel to the line $y = 2x$.

46. Find the points on the curve $y = \tan(2x)$, $-\pi/4 < x < \pi/4$, where the normal is parallel to the line $y = -x/8$.

47. Show that the graphs of $y = \sin x$, $y = \cos x$, $y = \sec x$, and $y = \csc x$ have horizontal tangents.

48. Show that the graphs of $y = \tan x$ and $y = \cot x$ never have horizontal tangents.

Do the graphs of the functions in Exercises 49–52 have any horizontal tangents in the interval $0 \le x \le 2\pi$? If so, where? If not, why not?

49. $y = x + \sin x$

50. $y = 2x + \sin x$

51. $y = x + 2\sin x$

52. $y = x + 2\cos x$

Find the limits in Exercises 53–56.

53. $\lim\limits_{x \to 0} \dfrac{\tan(2x)}{x}$

54. $\lim\limits_{x \to \pi} \sec(1 + \cos x)$

55. $\lim\limits_{x \to 0} (x^2 \csc x \cot x)$

56. $\lim\limits_{x \to 0} \cos\left(\dfrac{\pi - \pi \cos^2 x}{x^2}\right)$

57. Use the method of Example 1 to evaluate $\lim\limits_{h \to 0} \dfrac{1 - \cos h}{h^2}$.

58. Find values of a and b that make

$$f(x) = \begin{cases} ax + b, & x < 0 \\ 2\sin x + 3\cos x, & x \ge 0 \end{cases}$$

differentiable at $x = 0$.

59. How many straight lines that pass through the origin are tangent to $y = \cos x$? Find (to 6 decimal places) the slopes of the two such lines that have the largest positive slopes.

Use Maple or another computer algebra system to evaluate and simplify the derivatives of the functions in Exercises 60–61.

60. $\dfrac{d}{dx} \dfrac{x \cos(x \sin x)}{x + \cos(x \cos x)}\Big|_{x=0}$

61. $\dfrac{d}{dx}\left(\sqrt{2x^2 + 3}\sin(x^2) - \dfrac{(2x^2 + 3)^{3/2}\cos(x^2)}{x}\right)\Big|_{x=\sqrt{\pi}}$

* 62. (The continuity of sine and cosine)

(a) Prove that

$$\lim_{\theta \to 0} \sin\theta = 0 \quad \text{and} \quad \lim_{\theta \to 0} \cos\theta = 1$$

as follows: use the fact that the length of chord AP is less than the length of arc AP in Figure 2.24 to show that

$$\sin^2\theta + (1 - \cos\theta)^2 < \theta^2.$$

Then deduce that $0 \le |\sin\theta| < |\theta|$ and $0 \le |1 - \cos\theta| < |\theta|$. Then use the Squeeze Theorem from Section 1.2.

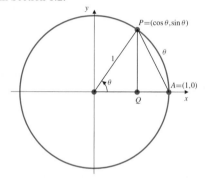

Figure 2.24

(b) Part (a) says that $\sin\theta$ and $\cos\theta$ are continuous at $\theta = 0$. Use the addition formulas to prove that they are therefore continuous at every θ.

2.6 The Mean-Value Theorem

If you set out in a car at 1:00 p.m. and arrive in a town 150 km away from your starting point at 3:00 p.m., then you have travelled at an average speed of $150/2 = 75$ km/h. Although you may not have travelled at constant speed, you must have been going 75 km/h at *at least one instant* during your journey, for if your speed was always less than 75 km/h you would have gone less than 150 km in 2 h, and if your speed was always more than 75 km/h, you would have gone more than 150 km in 2 h. In order to get from a value less than 75 km/h to a value greater than 75 km/h, your speed, which is a continuous function of time, must pass through the value 75 km/h at some intermediate time.

The conclusion that the average speed over a time interval must be equal to the instantaneous speed at some time in that interval is an instance of an important mathematical principle. In geometric terms it says that if A and B are two points on a smooth curve, then there is at least one point C on the curve between A and B where the tangent line is parallel to the chord line AB. See Figure 2.25.

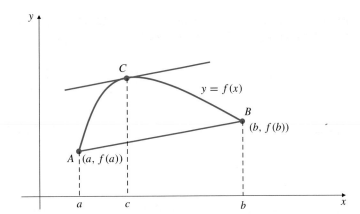

Figure 2.25 There is a point C on the curve where the tangent is parallel to the chord AB.

This principle is stated more precisely in the following theorem.

THEOREM 11

The Mean-Value Theorem

Suppose that the function f is continuous on the closed, finite interval $[a, b]$ and that it is differentiable on the open interval $]a, b[$. Then there exists a point c in the open interval $]a, b[$ such that

$$\frac{f(b) - f(a)}{b - a} = f'(c).$$

This says that the slope of the chord line joining the points $(a, f(a))$ and $(b, f(b))$ is equal to the slope of the tangent line to the curve $y = f(x)$ at the point $(c, f(c))$, so the two lines are parallel.

We will prove the Mean-Value Theorem later in this section. For now we make several observations.

1. The hypotheses of the Mean-Value Theorem are all necessary for the conclusion; if f fails to be continuous at even one point of $[a, b]$ or fails to be differentiable at even one point of $]a, b[$, then there may be no point where the tangent line is parallel to the secant line AB. See Figure 2.26.

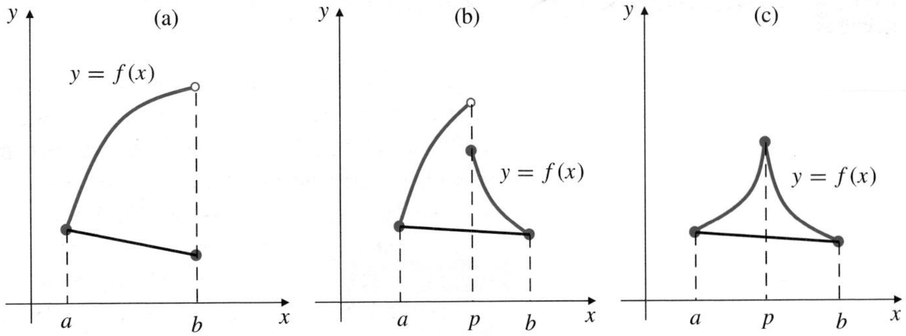

Figure 2.26 Functions that fail to satisfy the hypotheses of the Mean-Value Theorem, and for which the conclusion is false:

(a) f is discontinuous at endpoint b

(b) f is discontinuous at p

(c) f is not differentiable at p

2. The Mean-Value Theorem gives no indication of how many points C there may be on the curve between A and B where the tangent is parallel to AB. If the curve is itself the straight line AB, then every point on the line between A and B has the required property. In general, there may be more than one point (see Figure 2.27); the Mean-Value Theorem asserts only that there must be at least one.

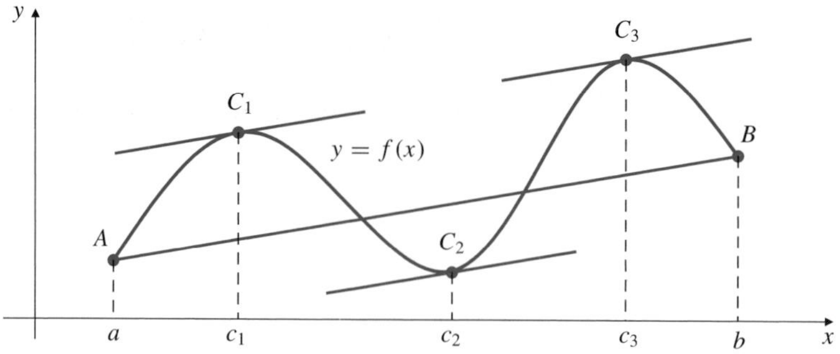

Figure 2.27 For this curve there are three points C where the tangent is parallel to the chord AB

3. The Mean-Value Theorem gives us no information on how to find the point c, which it says must exist. For some simple functions it is possible to calculate c (see the following example), but doing so is usually of no practical value. As we shall see, the importance of the Mean-Value Theorem lies in its use as a theoretical tool. It belongs to a class of theorems called *existence theorems*, as do the Max-Min Theorem and the Intermediate-Value Theorem (Theorems 8 and 9 of Section 1.4).

Example 1 Verify the conclusion of the Mean-Value Theorem for $f(x) = \sqrt{x}$ on the interval $[a, b]$, where $0 \le a < b$.

Solution The theorem says that there must be a number c in the interval $]a, b[$ such that

$$f'(c) = \frac{f(b) - f(a)}{b - a}$$

$$\frac{1}{2\sqrt{c}} = \frac{\sqrt{b} - \sqrt{a}}{b - a} = \frac{\sqrt{b} - \sqrt{a}}{(\sqrt{b} - \sqrt{a})(\sqrt{b} + \sqrt{a})} = \frac{1}{\sqrt{b} + \sqrt{a}}.$$

Thus, $2\sqrt{c} = \sqrt{a} + \sqrt{b}$ and $c = \left(\dfrac{\sqrt{b} + \sqrt{a}}{2}\right)^2$. Since $a < b$ we have

$$a = \left(\frac{\sqrt{a} + \sqrt{a}}{2}\right)^2 < c < \left(\frac{\sqrt{b} + \sqrt{b}}{2}\right)^2 = b,$$

so c lies in the interval $]a, b[$.

The following two examples are more representative of how the Mean-Value Theorem is actually used.

Example 2 Show that $\sin x < x$ for all $x > 0$.

Solution If $x > 2\pi$, then $\sin x \le 1 < 2\pi < x$. If $0 < x \le 2\pi$, then, by the Mean-Value Theorem, there exists c in the open interval $]0, 2\pi[$ such that

$$\frac{\sin x}{x} = \frac{\sin x - \sin 0}{x - 0} = \frac{d}{dx} \sin x \Big|_{x=c} = \cos c < 1.$$

Thus, $\sin x < x$ in this case too.

Example 3 Show that $\sqrt{1+x} < 1 + \dfrac{x}{2}$ for $x > 0$ and for $-1 \le x < 0$.

Solution If $x > 0$, apply the Mean-Value Theorem to $f(x) = \sqrt{1+x}$ on the interval $[0, x]$. There exists c in $]0, x[$ such that

$$\frac{\sqrt{1+x} - 1}{x} = \frac{f(x) - f(0)}{x - 0} = f'(c) = \frac{1}{2\sqrt{1+c}} < \frac{1}{2}.$$

The last inequality holds because $c > 0$. Multiplying by the positive number x and transposing the -1 gives $\sqrt{1+x} < 1 + \dfrac{x}{2}$.

If $-1 \le x < 0$, we apply the Mean-Value Theorem to $f(x) = \sqrt{1+x}$ on the interval $[x, 0]$. There exists c in $]x, 0[$ such that

$$\frac{\sqrt{1+x} - 1}{x} = \frac{1 - \sqrt{1+x}}{-x} = \frac{f(0) - f(x)}{0 - x} = f'(c) = \frac{1}{2\sqrt{1+c}} > \frac{1}{2}$$

(because $0 < 1 + c < 1$). Now we must multiply by the negative number x, which reverses the inequality, $\sqrt{1+x} - 1 < \dfrac{x}{2}$, and the required inequality again follows by transposing the -1. ∎

Increasing and Decreasing Functions

Intervals on which the graph of a function f has positive or negative slope provide useful information about the behaviour of f. The Mean-Value Theorem enables us to determine such intervals by considering the sign of the derivative f'.

DEFINITION 5

> **Increasing and decreasing functions**
>
> Suppose that the function f is defined on an interval I and that x_1 and x_2 are two points of I.
> (a) If $f(x_2) > f(x_1)$ whenever $x_2 > x_1$ we say f is **increasing** on I.
> (b) If $f(x_2) < f(x_1)$ whenever $x_2 > x_1$ we say f is **decreasing** on I.
> (c) If $f(x_2) \ge f(x_1)$ whenever $x_2 > x_1$ we say f is **nondecreasing** on I.
> (d) If $f(x_2) \le f(x_1)$ whenever $x_2 > x_1$ we say f is **nonincreasing** on I.

Figure 2.28 illustrates these terms. Note the distinction between *increasing* and *nondecreasing*. If a function is increasing (or decreasing) on an interval, it must take different values at different points. (Such a function is called **one-to-one**.) A nondecreasing function (or a nonincreasing function) may be constant on all or part of an interval and may therefore not be one-to-one.

THEOREM 12

Let J be an open interval, and let I be an interval consisting of all the points in J and possibly one or both of the endpoints of J. Suppose that f is continuous on I and differentiable on J.

(a) If $f'(x) > 0$ for all x in J, then f is increasing on I.

(b) If $f'(x) < 0$ for all x in J, then f is decreasing on I.

(c) If $f'(x) \ge 0$ for all x in J, then f is nondecreasing on I.

(d) If $f'(x) \le 0$ for all x in J, then f is nonincreasing on I.

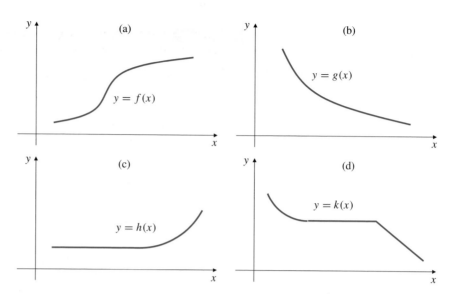

Figure 2.28

(a) Function f is increasing

(b) Function g is decreasing

(c) Function h is nondecreasing

(d) Function k is nonincreasing

PROOF   Let x_1 and x_2 be points in I with $x_2 > x_1$. By the Mean-Value Theorem there exists a point c in $]x_1, x_2[$ (and therefore in J) such that

$$\frac{f(x_2) - f(x_1)}{x_2 - x_1} = f'(c);$$

hence, $f(x_2) - f(x_1) = (x_2 - x_1)\, f'(c)$. Since $x_2 - x_1 > 0$, the difference $f(x_2) - f(x_1)$ has the same sign as $f'(c)$ and may be zero if $f'(c)$ is zero. Thus, all four conclusions follow from the corresponding parts of Definition 5.

Remark   Despite Theorem 12, $f'(x_0) > 0$ at a single point x_0 does *not* imply that f is increasing on *any* interval containing x_0. See Exercise 18 at the end of this section for a counterexample.

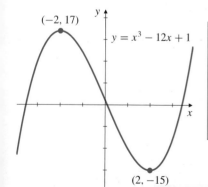

$(-2, 17)$

$y = x^3 - 12x + 1$

$(2, -15)$

Figure 2.29

Example 4   On what intervals is the function $f(x) = x^3 - 12x + 1$ increasing? On what intervals is it decreasing?

Solution   We have $f'(x) = 3x^2 - 12 = 3(x - 2)(x + 2)$. Observe that $f'(x) > 0$ if $x < -2$ or $x > 2$ and $f'(x) < 0$ if $-2 < x < 2$. Therefore, f is increasing on the intervals $]-\infty, -2[$ and $]2, \infty[$ and is decreasing on the interval $]-2, 2[$. See Figure 2.29.   ∎

A function f whose derivative satisfies $f'(x) \geq 0$ on an interval can still be increasing there, rather than just nondecreasing as assured by Theorem 12(c). This will happen if $f'(x) = 0$ only at isolated points, so that f is assured to be increasing on intervals to the left and right of these points.

Example 5   Show that $f(x) = x^3$ is increasing on any interval.

Solution   Let x_1 and x_2 be any two real numbers satsifying $x_1 < x_2$. Since $f'(x) = 3x^2 > 0$ except at $x = 0$, Theorem 12(a) tells us that $f(x_1) < f(x_2)$ if either $x_1 < x_2 \leq 0$ or $0 \leq x_1 < x_2$. If $x_1 < 0 < x_2$ then $f(x_1) < 0 < f(x_2)$. Thus, f is increasing on every interval.   ∎

If a function is constant on an interval, then its derivative is zero on that interval. The Mean-Value Theorem provides a converse of this fact.

THEOREM **13**

If f is continuous on an interval I, and $f'(x) = 0$ at every interior point of I (i.e., at every point of I that is not an endpoint of I), then $f(x) = C$, a constant, on I.

PROOF Pick a point x_0 in I and let $C = f(x_0)$. If x is any other point of I, then the Mean-Value Theorem says that there exists a point c between x_0 and x such that

$$\frac{f(x) - f(x_0)}{x - x_0} = f'(c).$$

The point c must belong to I because an interval contains all points between any two of its points, and c cannot be an endpoint of I since $c \neq x_0$ and $c \neq x$. Since $f'(c) = 0$ for all such points c, we have $f(x) - f(x_0) = 0$ for all x in I, and $f(x) = f(x_0) = C$ as claimed.

We will see how Theorem 13 can be used to establish identities for new functions encountered in later chapters. We will also use it when finding antiderivatives in Section 2.10.

Proof of the Mean-Value Theorem

The Mean-Value Theorem is one of those deep results that is based on the completeness of the real number system via the fact that a continuous function on a closed, finite interval takes on a maximum and minimum value (Theorem 8 of Section 1.4). Before giving the proof we establish two preliminary results.

THEOREM **14**

If f is defined on an open interval $]a, b[$ and achieves a maximum (or minimum) value at the point c in $]a, b[$, and if $f'(c)$ exists, then $f'(c) = 0$. (Values of x where $f'(x) = 0$ are called **critical points** of the function f.)

PROOF Suppose that f has a maximum value at c. Then $f(x) - f(c) \leq 0$ whenever x is in $]a, b[$. If $c < x < b$, then

$$\frac{f(x) - f(c)}{x - c} \leq 0, \qquad \text{so} \quad f'(c) = \lim_{x \to c+} \frac{f(x) - f(c)}{x - c} \leq 0.$$

Similarly, if $a < x < c$, then

$$\frac{f(x) - f(c)}{x - c} \geq 0, \qquad \text{so} \quad f'(c) = \lim_{x \to c-} \frac{f(x) - f(c)}{x - c} \geq 0.$$

Thus $f'(c) = 0$. The proof for a minimum value at c is similar.

THEOREM **15**

Rolle's Theorem

Suppose that the function g is continuous on the closed, finite interval $[a, b]$ and that it is differentiable on the open interval $]a, b[$. If $g(a) = g(b)$, then there exists a point c in the open interval $]a, b[$ such that $g'(c) = 0$.

PROOF If $g(x) = g(a)$ for every x in $[a, b]$, then g is a constant function, so $g'(c) = 0$ for every c in $]a, b[$. Therefore, suppose there exists x_0 in $]a, b[$ such that $g(x_0) \neq g(a)$. Let us assume that $g(x_0) > g(a)$. (If $g(x_0) < g(a)$, the proof is similar.) By the Max-Min Theorem (Theorem 8 of Section 1.4), being continuous on $[a, b]$, g must have a maximum value at some point c in $[a, b]$. Since

$$g(c) \geq g(x_0) > g(a) = g(b),$$

c cannot be either a or b. Therefore, c is in the open interval $]a, b[$, so g is differentiable at c. By Theorem 14, c must be a critical point of g: $g'(c) = 0$.

Remark Rolle's Theorem is a special case of the Mean-Value Theorem in which the chord line has slope 0, so the corresponding parallel tangent line must also have slope 0. We can deduce the Mean-Value Theorem from this special case.

Proof of the Mean-Value Theorem Suppose f satisfies the conditions of the Mean-Value Theorem. Let

$$g(x) = f(x) - \left(f(a) + \frac{f(b) - f(a)}{b - a}(x - a) \right).$$

(For $a \leq x \leq b$, $g(x)$ is the vertical displacement between the curve $y = f(x)$ and the chord line

$$y = f(a) + \frac{f(b) - f(a)}{b - a}(x - a)$$

joining $(a, f(a))$ and $(b, f(b))$. See Figure 2.30.)

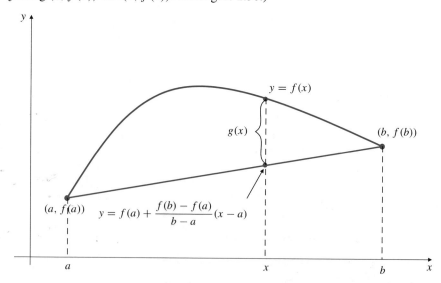

Figure 2.30 $g(x)$ is the vertical distance between the graph of f and the chord line

The function g is also continuous on $[a, b]$ and differentiable on $]a, b[$ because f has these properties. In addition, $g(a) = g(b) = 0$. By Rolle's Theorem, there is some point c in $]a, b[$ such that $g'(c) = 0$. Since

$$g'(x) = f'(x) - \frac{f(b) - f(a)}{b - a},$$

it follows that

$$f'(c) = \frac{f(b) - f(a)}{b - a}.$$

Many of the applications we will make of the Mean-Value Theorem in later chapters will actually use the following generalized version of it.

THEOREM 16

The Generalized Mean-Value Theorem

If functions f and g are both continuous on $[a, b]$ and differentiable on $]a, b[$, and if $g'(x) \neq 0$ for every x in $]a, b[$, then there exists a number c in $]a, b[$ such that

$$\frac{f(b) - f(a)}{g(b) - g(a)} = \frac{f'(c)}{g'(c)}.$$

PROOF   Note that $g(b) \neq g(a)$; otherwise, there would be some number in $]a, b[$ where $g' = 0$. Hence, neither denominator above can be zero. Apply the Mean-Value Theorem to

$$h(x) = \big(f(b) - f(a)\big)\big(g(x) - g(a)\big) - \big(g(b) - g(a)\big)\big(f(x) - f(a)\big).$$

Since $h(a) = h(b) = 0$, there exists c in $]a, b[$ such that $h'(c) = 0$. Thus,

$$\big(f(b) - f(a)\big)g'(c) - \big(g(b) - g(a)\big)f'(c) = 0,$$

and the result follows on division by the g factors.

Exercises 2.6

In Exercises 1–3, illustrate the Mean-Value Theorem by finding any points in the open interval $]a, b[$ where the tangent line to $y = f(x)$ is parallel to the chord line joining $(a, f(a))$ and $(b, f(b))$.

1. $f(x) = x^2$ on $[a, b]$ 　　　 **2.** $f(x) = \dfrac{1}{x}$ on $[1, 2]$

3. $f(x) = x^3 - 3x + 1$ on $[-2, 2]$

* **4.** By applying the Mean-Value Theorem to

$f(x) = \cos x + \dfrac{x^2}{2}$ on the interval $[0, x]$, and using the result of Example 2, show that

$$\cos x > 1 - \frac{x^2}{2}$$

for $x > 0$. This inequality is also true for $x < 0$. Why?

5. Show that $\tan x > x$ for $0 < x < \pi/2$.

6. Let $r > 1$. If $x > 0$ or $-1 \leq x < 0$, show that $(1 + x)^r > 1 + rx$.

7. Let $0 < r < 1$. If $x > 0$ or $-1 \leq x < 0$, show that $(1 + x)^r < 1 + rx$.

Find the intervals of increase and decrease of the functions in Exercises 8–15.

8. $f(x) = x^2 + 2x + 2$ 　　　 **9.** $f(x) = x^3 - 4x + 1$

10. $f(x) = x^3 + 4x + 1$ 　　　 **11.** $f(x) = (x^2 - 4)^2$

12. $f(x) = \dfrac{1}{x^2 + 1}$ 　　　 **13.** $f(x) = x^3(5 - x)^2$

14. $f(x) = x - 2 \sin x$ 　　　 **15.** $f(x) = x + \sin x$

* **16.** If $f(x)$ is differentiable on an interval I and vanishes at $n \geq 2$ distinct points of I, prove that $f'(x)$ must vanish at at least $n - 1$ points in I.

17. What is wrong with the following "proof" of the Generalized Mean-Value Theorem? By the Mean-Value Theorem, $f(b) - f(a) = (b - a)f'(c)$ for some c between a and b and, similarly, $g(b) - g(a) = (b - a)g'(c)$ for some such c. Hence, $(f(b) - f(a))/(g(b) - g(a)) = f'(c)/g'(c)$, as required.

* **18.** Let $f(x) = \begin{cases} x + 2x^2 \sin(1/x) & \text{if } x \neq 0, \\ 0 & \text{if } x = 0. \end{cases}$

(a) Show that $f'(0) = 1$. (*Hint:* use the definition of derivative.)

(b) Show that any interval containing $x = 0$ also contains points where $f'(x) < 0$, so f cannot be increasing on such an interval.

2.7 Using Derivatives

In this section we will look at some examples of ways in which derivatives are used to represent and interpret changes and rates of change in the world around us. It is natural to think of change in terms of dependence on time, such as the velocity of a moving object, but there is no need to be so restrictive. Change with respect to variables other than time can be treated in the same way. For example, a physician may want to know how small changes in dosage can affect the body's response to a drug. An economist may want to study how foreign investment changes with respect to variations in a country's interest rates. These questions can all be formulated in terms of rate of change of a function with respect to a variable.

Approximating Small Changes

If one quantity, say y, is a function of another quantity x, that is

$$y = f(x),$$

we sometimes want to know how a change in the value of x by an amount Δx will affect the value of y. The exact change Δy in y is given by

$$\Delta y = f(x + \Delta x) - f(x),$$

but if the change Δx is small, then we can get a good approximation to Δy by using the fact that $\Delta y / \Delta x$ is approximately the derivative dy/dx. Thus

$$\Delta y = \frac{\Delta y}{\Delta x} \Delta x \approx \frac{dy}{dx} \Delta x = f'(x) \Delta x.$$

Sometimes changes in a quantity are measured with respect to the size of the quantity. The **relative change** in x is the ratio $\Delta x/x$; the **percentage change** in x is the relative change expressed as a percentage:

$$\text{relative change in } x = \frac{\Delta x}{x}$$

$$\text{percentage change in } x = 100 \frac{\Delta x}{x}.$$

Example 1   By approximately what percentage does the area of a circle increase if the radius increases by 2%?

Solution   The area A of a circle is given in terms of the radius r by $A = \pi r^2$. Thus,

$$\Delta A \approx \frac{dA}{dr} \Delta r = 2\pi r \, \Delta r.$$

We divide this approximation by $A = \pi r^2$ to get an approximation that links the relative changes in A and r:

$$\frac{\Delta A}{A} \approx \frac{2\pi r \, \Delta r}{\pi r^2} = 2\frac{\Delta r}{r}.$$

If r increases by 2%, then $\Delta r = \frac{2}{100} r$, so

$$\frac{\Delta A}{A} \approx 2 \times \frac{2}{100} = \frac{4}{100}.$$

Thus, A increases by approximately 4%.

Average and Instantaneous Rates of Change

Recall the concept of average rate of change of a function over an interval, introduced in Section 1.1. The derivative of the function is the limit of this average rate as the length of the interval goes to zero, and so represents the rate of change of the function at a given value of its variable.

DEFINITION 6

> The **average rate of change** of a function $f(x)$ with respect to x over the interval from a to $a + h$ is
>
> $$\frac{f(a + h) - f(a)}{h}.$$
>
> The **(instantaneous) rate of change** of f with respect to x at $x = a$ is the derivative
>
> $$f'(a) = \lim_{h \to 0} \frac{f(a + h) - f(a)}{h},$$
>
> provided the limit exists.

It is conventional to use the word *instantaneous* even when x does not represent time, although the word is frequently omitted. When we say *rate of change* we mean *instantaneous rate of change*.

Example 2   How fast is area A of a circle increasing with respect to its radius when the radius is 5 m?

Solution   The rate of change of the area with respect to the radius is

$$\frac{dA}{dr} = \frac{d}{dr}(\pi r^2) = 2\pi r.$$

When $r = 5$ m, the area is changing at the rate $2\pi \times 5 = 10\pi$ m^2/m. This means that a small change Δr m in the radius when the radius is 5 m would result in a change of about $10\pi \Delta r$ m^2 in the area of the circle. ∎

The above example suggests that the appropriate units for the rate of change of a quantity y with respect to another quantity x are units of y per unit of x.

If $f'(x_0) = 0$, we say that f is **stationary** at x_0 and call x_0 a **critical point** of f. The corresponding point $(x_0, f(x_0))$ on the graph of f is also called a **critical point** of the graph. The graph has a horizontal tangent at a critical point, and f may or may not have a maximum or minimum value there. (See Figure 2.31.) It is still possible for f to be increasing or decreasing on an open interval containing a critical point. (See point a in Figure 2.31.)

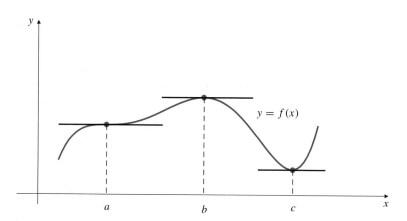

Figure 2.31   Critical points of f

Example 3   Suppose the temperature at a certain location t hours after noon on a certain day is $T°$C (T degrees Celsius), where

$$T = \frac{1}{3}t^3 - 3t^2 + 8t + 10 \qquad (\text{for } 0 \le t \le 5).$$

How fast is the temperature rising or falling at 1:00 p.m.? at 3:00 p.m.? At what instants is the temperature stationary?

Solution   The rate of change of the temperature is given by

$$\frac{dT}{dt} = t^2 - 6t + 8 = (t - 2)(t - 4).$$

If $t = 1$, then $\dfrac{dT}{dt} = 3$, so the temperature is rising at rate $3°$C/h at 1:00 p.m.

If $t = 3$, then $\dfrac{dT}{dt} = -1$, so the temperature is falling at a rate of $1°$C/h at 3:00 p.m.

The temperature is stationary when $\dfrac{dT}{dt} = 0$, that is, at 2:00 p.m. and 4:00 p.m.

Sensitivity to Change

When a small change in x produces a large change in the value of a function $f(x)$, we say that the function is very **sensitive** to changes in x. The derivative $f'(x)$ is a measure of the sensitivity of the dependence of f on x.

Example 4 (**Dosage of a medicine**) A pharmacologist studying a drug that has been developed to lower blood pressure determines experimentally that the average reduction R in blood pressure resulting from a daily dosage of x mg of the drug is given by

$$R = 24.2 \left(1 + \frac{x - 13}{\sqrt{x^2 - 26x + 529}} \right) \quad \text{mm Hg.}$$

(The units are millimetres of mercury (Hg).) Determine the sensitivity of R to dosage x at dosage levels of 5 mg, 15 mg, and 35 mg. At which of these dosage levels would an increase in the dosage have the greatest effect?

Solution The sensitivity of R to x is dR/dx. We have

$$\frac{dR}{dx} = 24.2 \left(\frac{\sqrt{x^2 - 26x + 529}(1) - (x - 13)\dfrac{x - 13}{\sqrt{x^2 - 26x + 529}}}{x^2 - 26x + 529} \right)$$

$$= 24.2 \left(\frac{x^2 - 26x + 529 - (x^2 - 26x + 169)}{(x^2 - 26x + 529)^{3/2}} \right)$$

$$= \frac{8712}{(x^2 - 26x + 529)^{3/2}}.$$

At dosages $x = 5$ mg, 15 mg, and 35 mg we have sensitivities of

$$\left. \frac{dR}{dx} \right|_{x=5} = 0.998 \text{ mm Hg/mg}, \qquad \left. \frac{dR}{dx} \right|_{x=15} = 1.254 \text{ mm Hg/mg},$$

$$\left. \frac{dR}{dx} \right|_{x=35} = 0.355 \text{ mm Hg/mg}.$$

Among these three levels, the greatest sensitivity is at 15 mg. Increasing the dosage from 15 to 16 mg/day could be expected to further reduce average blood pressure by about 1.25 mm Hg.

Derivatives in Economics

Just as physicists use terms such as *velocity* and *acceleration* to refer to derivatives of certain quantities, economists also have their own specialized vocabulary for derivatives. They call them marginals. In economics the term **marginal** denotes the rate of change of a quantity with respect to a variable on which it depends. For example, the **cost of production** $C(x)$ in a manufacturing operation is a function of x, the number of units of product produced. The **marginal cost of production** is the rate of change of C with respect to x, so it is dC/dx. Sometimes the marginal cost of production is loosely defined to be the extra cost of producing one more unit, that is,

$$\Delta C = C(x+1) - C(x).$$

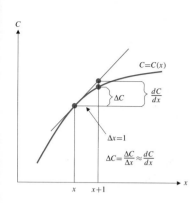

Figure 2.32 The marginal cost dC/dx is approximately the extra cost ΔC of producing $\Delta x = 1$ more unit

To see why this is approximately correct, observe from Figure 2.32 that if the slope of $C = C(x)$ does not change quickly near x, then the difference quotient $\Delta C/\Delta x$ will be close to its limit, the derivative dC/dx, even if $\Delta x = 1$.

Example 5 **(Marginal Tax Rates)** If your marginal income tax rate is 35% and your income increases by \$1,000, you can expect to have to pay an extra \$350 in income taxes. This does not mean that you pay 35% of your entire income in taxes. It just means that at your current income level I, the rate of increase of taxes T with respect to income is $dT/dI = 0.35$. You will pay \$0.35 out of every extra dollar you earn in taxes. Of course, if your income increases greatly, you may land in a higher tax bracket and your marginal rate will increase.

Example 6 **(Marginal Cost of Production)** The cost of producing x tons of coal per day in a mine is $\$C(x)$ where

$$C(x) = 4{,}200 + 5.40x - 0.001x^2 + 0.000002x^3.$$

(a) What is the average cost of producing each ton if the daily production level is 1,000 tons? 2,000 tons?

(b) Find the marginal cost of production if the daily production level is 1,000 tons. 2,000 tons.

(c) If the production level increases slightly from 1,000 tons or from 2,000 tons, what will happen to the average cost per ton?

Solution

(a) The average cost per ton of coal is

$$\frac{C(x)}{x} = \frac{4{,}200}{x} + 5.40 - 0.001x + 0.000002x^2.$$

If $x = 1{,}000$, the average cost per ton is $C(1{,}000)/1{,}000 = \$10.6$ /ton. If $x = 2{,}000$, the average cost per ton is $C(2{,}000)/2{,}000 = \$13.5$ /ton.

(b) The marginal cost of production is

$$C'(x) = 5.40 - 0.002x + 0.000006x^2.$$

If $x = 1,000$, the marginal cost is $C'(1,000) = \$9.4$ /ton. If $x = 2,000$, the marginal cost is $C'(2,000) = \$25.4$ /ton.

(c) If the production level x is increased slightly from $x = 1,000$, then the average cost per ton will drop because the cost is increasing at a rate lower than the average cost. At $x = 2,000$ the opposite is true; an increase in production will increase the average cost per ton.

Economists sometimes prefer to measure relative rates of change that do not depend on the units used to measure the quantities involved. They use the term **elasticity** for such relative rates.

Example 7   **(Elasticity of demand)** The demand y for a certain product (i.e., the amount that can be sold) typically depends on the price p charged for the product: $y = f(p)$. The marginal demand $dy/dp = f'(p)$ (which is typically negative) depends on the units used to measure y and p. The *elasticity of the demand* is the quantity

$$-\frac{p}{y}\frac{dy}{dp} \qquad \text{(the ``$-$'' sign ensures elasticity is positive),}$$

which is independent of units and provides a good measure of the sensitivity of demand to changes in price. To see this, suppose that new units of demand and price are introduced, which are multiples of the old units. In terms of the new units the demand and price are now Y and P, where

$$Y = k_1 y \qquad \text{and} \qquad P = k_2 p.$$

Thus, $Y = k_1 f(P/k_2)$ and $dY/dP = (k_1/k_2)f'(P/k_2) = (k_1/k_2)f'(p)$ by the Chain Rule. It follows that the elasticity has the same value:

$$-\frac{P}{Y}\frac{dY}{dP} = -\frac{k_2 p}{k_1 y}\frac{k_1}{k_2} f'(p) = -\frac{p}{y}\frac{dy}{dp}.$$

Exercises 2.7

In Exercises 1–6, find the approximate percentage changes in the given function $y = f(x)$ that will result from an increase of 2% in the value of x.

1. $y = x^2$

2. $y = 1/x$

3. $y = 1/x^2$

4. $y = x^3$

5. $y = \sqrt{x}$

6. $y = x^{-2/3}$

7. By approximately what percentage will the volume ($V = \frac{4}{3}\pi r^3$) of a ball of radius r increase if the radius increases by 2%?

8. By about what percentage will the edge length of an ice cube decrease if the cube loses 6% of its volume by melting?

9. Find the rate of change of the area of a square with respect to the length of its side when the side is 4 ft.

10. Find the rate of change of the side of a square with respect to the area of the square when the area is 16 m^2.

11. Find the rate of change of the diameter of a circle with respect to its area.

12. Find the rate of change of the area of a circle with respect to its diameter.

13. Find the rate of change of the volume of a sphere (given by $V = \frac{4}{3}\pi r^3$) with respect to its radius r when the radius is 2 m.

14. What is the rate of change of the area A of a square with respect to the length L of the diagonal of the square?

15. What is the rate of change of the circumference C of a circle with respect to the area A of the circle?

16. Find the rate of change of the side s of a cube with respect to the volume V of the cube.

What are the critical points of the functions in Exercises 17–20? On what intervals is each function increasing and decreasing?

17. $f(x) = x^2 - 4$

18. $f(x) = x^3 - 12x + 1$

19. $y = x^3 + 6x^2$

20. $y = 1 - x - x^5$

21. Show that $f(x) = x^3$ is increasing on the whole real line even though $f'(x)$ is not positive at every point.

22. On what intervals is $f(x) = x + 2 \sin x$ increasing?

Use a graphing utility or a computer algebra system to find the critical points of the functions in Exercises 23–26 correct to 6 decimal places.

23. $f(x) = \dfrac{x^2 - x}{x^2 - 4}$

24. $f(x) = \dfrac{2x + 1}{x^2 + x + 1}$

25. $f(x) = x - \sin\left(\dfrac{x}{x^2 + x + 1}\right)$

26. $f(x) = \dfrac{\sqrt{1 - x^2}}{\cos(x + 0.1)}$

27. The volume of water in a tank t min after it starts draining is

$$V(t) = 350(20 - t)^2 \text{ l}.$$

(a) How fast is the water draining out after 5 min? after 15 min?

(b) What is the average rate at which water is draining out during the time interval from 5 to 15 min?

28. **(Poiseuille's Law)** The flow rate F (in litres per minute) of a liquid through a pipe is proportional to the fourth power of the radius of the pipe:

$$F = kr^4.$$

Approximately what percentage increase is needed in the radius of the pipe to increase the flow rate by 10%?

29. **(Gravitational force)** The gravitational force F with which the earth attracts an object in space is given by $F = k/r^2$, where k is a constant and r is the distance from the object to the centre of the earth. If F decreases with respect to r at rate $1 \, pound/mile$ when $r = 4,000$ mi, how fast does F change with respect to r when $r = 8,000$ mi?

30. **(Sensitivity of revenue to price)** The sales revenue $\$R$ from a software product depends on the price $\$p$ charged by the distributor according to the formula

$$R = 4,000p - 10p^2.$$

(a) How sensitive is R to p when $p = \$100$? $p = \$200$? $p = \$300$?

(b) Which of these three is the most reasonable price for the distributor to charge? Why?

31. **(Marginal cost)** The cost of manufacturing x refrigerators is $\$C(x)$, where

$$C(x) = 8,000 + 400x - 0.5x^2.$$

(a) Find the marginal cost if 100 refrigerators are manufactured.

(b) Show that the marginal cost is approximately the difference in cost of manufacturing 101 refrigerators instead of 100.

32. **(Marginal profit)** If a plywood factory produces x sheets of plywood per day, its profit per day will be $\$P(x)$, where

$$P(x) = 8x - 0.005x^2 - 1,000.$$

(a) Find the marginal profit. For what values of x is the marginal profit positive? negative?

(b) How many sheets should be produced each day to generate maximum profits?

33. The cost C (in dollars) of producing n widgets per month in a widget factory is given by

$$C = \frac{80,000}{n} + 4n + \frac{n^2}{100}.$$

Find the marginal cost of production if the number of widgets manufactured each month is (a) 100 and (b) 300.

* 34. In a mining operation the cost C (in dollars) of extracting each tonne of ore is given by

$$C = 10 + \frac{20}{x} + \frac{x}{1,000},$$

where x is the number of tonnes extracted each day. (For small x, C decreases as x increases because of economies of scale, but for large x, C increases with x because of overloaded equipment and labour overtime.) If each tonne of ore can be sold for $\$13$, how many tonnes should be extracted each day to maximize the daily profit of the mine?

* 35. **(Average cost and marginal cost)** If it costs a manufacturer $C(x)$ dollars to produce x items, then his average cost of production is $C(x)/x$ dollars per item. Typically the average cost is a decreasing function of x for small x and an increasing function of x for large x. (Why?) Show that the value of x that minimizes the average cost makes the average cost equal to the marginal cost.

36. **(Constant elasticity)** Show that if demand y is related to price p by the equation $y = Cp^{-r}$, where C and r are positive constants, then the elasticity of demand (see Example 7) is the constant r.

2.8 Higher-Order Derivatives

If the derivative $y' = f'(x)$ of a function $y = f(x)$ is itself differentiable at x, we can calculate *its* derivative, which we call the **second derivative** of f and denote by $y'' = f''(x)$. As is the case for first derivatives, second derivatives can be denoted by various notations depending on the context. Some of the more common ones are

$$y'' = f''(x) = \frac{d^2 y}{dx^2} = \frac{d}{dx}\frac{d}{dx} f(x) = \frac{d^2}{dx^2} f(x) = D_x^2 y = D_x^2 f(x).$$

Similarly, you can consider third-, fourth-, and in general nth-order derivatives. The prime notation is inconvenient for derivatives of high order, so we denote the order by a superscript in parentheses (to distinguish it from an exponent): the nth derivative of $y = f(x)$ is

$$y^{(n)} = f^{(n)}(x) = \frac{d^n y}{dx^n} = \frac{d^n}{dx^n} f(x) = D_x^n y = D_x^n f(x),$$

and it is defined to be the derivative of the $(n-1)$st derivative. For $n = 1$, 2, and 3, primes are still normally used: $f^{(2)}(x) = f''(x)$, $f^{(3)}(x) = f'''(x)$. It is sometimes convenient to denote $f^{(0)}(x) = f(x)$, that is, to regard a function as its own zeroth-order derivative.

Example 1   The **velocity** of a moving object is the (instantaneous) rate of change of the position of the object with respect to time; if the object moves along the x-axis and is at position $x = f(t)$ at time t, then its velocity at that time is

$$v = \frac{dx}{dt} = f'(t).$$

Similarly, the **acceleration** of the object is the rate of change of the velocity. Thus, the acceleration is the *second derivative* of the position:

$$a = \frac{dv}{dt} = \frac{d^2 x}{dt^2} = f''(t).$$

We will investigate the relationships between position, velocity, and acceleration further in Section 2.11.

Example 2   If $y = x^3$, then $y' = 3x^2$, $y'' = 6x$, $y''' = 6$, $y^{(4)} = 0$, and all higher derivatives are zero.

In general, if $f(x) = x^n$ (where n is a positive integer), then

$$f^{(k)}(x) = n(n-1)(n-2)...(n-(k-1)) x^{n-k}$$

$$= \begin{cases} \dfrac{n!}{(n-k)!} x^{n-k} & \text{if } 0 \le k \le n \\ 0 & \text{if } k > n, \end{cases}$$

where $n!$ (called n **factorial**) is defined by:

$$0! = 1$$
$$1! = 0! \times 1 = 1 \times 1 = 1$$
$$2! = 1! \times 2 = 1 \times 2 = 2$$
$$3! = 2! \times 3 = 1 \times 2 \times 3 = 6$$
$$4! = 3! \times 4 = 1 \times 2 \times 3 \times 4 = 24$$
$$\vdots$$
$$n! = (n-1)! \times n = 1 \times 2 \times 3 \times \cdots \times (n-1) \times n.$$

It follows that if P is a polynomial of degree n,

$$P(x) = a_n x^n + a_{n-1} x^{n-1} + \cdots + a_1 x + a_0,$$

where a_n, a_{n-1}, ... , a_1, a_0 are constants, then $P^{(k)}(x) = 0$ for $k > n$. For $k \le n$, $P^{(k)}$ is a polynomial of degree $n - k$; in particular, $P^{(n)}(x) = n! \, a_n$, a constant function.

Example 3   Show that if A, B, and k are constants, then the function

$$y = A\cos(kt) + B\sin(kt)$$

is a solution of the *second-order* **differential equation of simple harmonic motion** (see Section 3.7):

$$\frac{d^2 y}{dt^2} + k^2 y = 0.$$

Solution   To be a solution, the function $y(t)$ must satisfy the differential equation *identically*; that is,

$$\frac{d^2}{dt^2} y(t) + k^2 y(t) = 0$$

must hold for every real number t. We verify this by calculating the first two derivatives of the given function $y(t) = A\cos(kt) + B\sin(kt)$ and observing that the second derivative plus $k^2 y(t)$ is, in fact, zero everywhere:

$$\frac{dy}{dt} = -Ak\sin(kt) + Bk\cos(kt)$$

$$\frac{d^2 y}{dt^2} = -Ak^2 \cos(kt) - Bk^2 \sin(kt) = -k^2 y(t),$$

$$\frac{d^2 y}{dt^2} + k^2 y(t) = 0.$$

Example 4   Find the nth derivative, $y^{(n)}$, of $y = \dfrac{1}{1+x} = (1+x)^{-1}$.

Solution   Begin by calculating the first few derivatives:

$$y' = -(1+x)^{-2}$$
$$y'' = -(-2)(1+x)^{-3} = 2(1+x)^{-3}$$
$$y''' = 2(-3)(1+x)^{-4} = -3!(1+x)^{-4}$$
$$y^{(4)} = -3!(-4)(1+x)^{-5} = 4!(1+x)^{-5}$$

The pattern here is becoming obvious. It seems that

$$y^{(n)} = (-1)^n n! (1+x)^{-n-1}.$$

Note the use of $(-1)^n$ to denote a positive sign if n is even and a negative sign if n is odd.

We have not yet actually proved that the above formula is correct for every n, although it is clearly correct for $n = 1$, 2, 3, and 4. To complete the proof we use mathematical induction (Section 2.3). Suppose that the formula is valid for $n = k$, where k is some positive integer. Consider $y^{(k+1)}$:

$$y^{(k+1)} = \frac{d}{dx} y^{(k)} = \frac{d}{dx} \left((-1)^k k! (1+x)^{-k-1} \right)$$
$$= (-1)^k k! (-k-1)(1+x)^{-k-2} = (-1)^{k+1}(k+1)!(1+x)^{-(k+1)-1}.$$

This is what the formula predicts for the $(k + 1)$st derivative. Therefore, if the formula for $y^{(n)}$ is correct for $n = k$, then it is also correct for $n = k + 1$. Since the formula is known to be true for $n = 1$, it must therefore be true for every integer $n \geq 1$ *by induction*.

◼

Example 5 Find a formula for $f^{(n)}(x)$, given that $f(x) = \sin(ax + b)$.

Solution Begin by calculating several derivatives:

$$f'(x) = a \cos(ax + b)$$
$$f''(x) = -a^2 \sin(ax + b) = -a^2 f(x)$$
$$f'''(x) = -a^3 \cos(ax + b) = -a^2 f'(x)$$
$$f^{(4)}(x) = a^4 \sin(ax + b) = a^4 f(x)$$
$$f^{(5)}(x) = a^5 \cos(ax + b) = a^4 f'(x)$$
$$\vdots$$

The pattern is pretty obvious here. Each new derivative is $-a^2$ times the second previous one. A formula that gives all the derivatives is

$$f^{(n)}(x) = \begin{cases} (-1)^k a^n \sin(ax + b) & \text{if } n = 2k \\ (-1)^k a^n \cos(ax + b) & \text{if } n = 2k + 1 \end{cases} \qquad (k = 0, 1, 2, \ldots),$$

which can also be verified by induction.

◼

Our final example shows that it is not always easy to obtain a formula for the nth derivative of a function.

> **Example 6** Calculate f', f'', and f''' for $f(x) = \sqrt{x^2 + 1}$. Can you see enough of a pattern to predict $f^{(4)}$?
>
> **Solution** Since $f(x) = (x^2 + 1)^{1/2}$ we have
>
> $$f'(x) = \tfrac{1}{2}(x^2 + 1)^{-1/2}(2x) = x(x^2 + 1)^{-1/2},$$
> $$f''(x) = (x^2 + 1)^{-1/2} + x\left(-\tfrac{1}{2}\right)(x^2 + 1)^{-3/2}(2x)$$
> $$= (x^2 + 1)^{-3/2}(x^2 + 1 - x^2) = (x^2 + 1)^{-3/2},$$
> $$f'''(x) = -\tfrac{3}{2}(x^2 + 1)^{-5/2}(2x) = -3x(x^2 + 1)^{-5/2}.$$
>
> Although the expression obtained from each differentiation simplified somewhat, the pattern of these derivatives is not (yet) obvious enough to enable us to predict the formula for $f^{(4)}(x)$ without having to calculate it. In fact,
>
> $$f^{(4)}(x) = 3(4x^2 - 1)(x^2 + 1)^{-7/2},$$
>
> so the pattern (if there is one) doesn't become any clearer at this stage.

Remark Higher-order derivatives can be indicated in Maple by repeating the variable of differentiation or indicating the order by using the $ operator:

```
>   diff(x^5,x,x) + diff(sin(2*x),x$3);
```

$$20\,x^3 - 8\,\cos(2x)$$

The D operator can also be used for higher-order derivatives of a function (as distinct from an expression) by composing it explicitly or using the @@ operator:

```
>   f := x -> x^5; fpp := D(D(f)); (D@@3)(f)(a);
```

$$f := x \to x^5$$
$$fpp := x \to 20\,x^3$$
$$60\,a^2$$

Exercises 2.8

Find y', y'', and y''' for the functions in Exercises 1–12.

1. $y = (3 - 2x)^7$

2. $y = x^2 - \dfrac{1}{x}$

3. $y = \dfrac{6}{(x-1)^2}$

4. $y = \sqrt{ax + b}$

5. $y = x^{1/3} - x^{-1/3}$

6. $y = x^{10} + 2x^8$

7. $y = (x^2 + 3)\sqrt{x}$

8. $y = \dfrac{x - 1}{x + 1}$

9. $y = \tan x$

10. $y = \sec x$

11. $y = \cos(x^2)$

12. $y = \dfrac{\sin x}{x}$

In Exercises 13–23, calculate enough derivatives of the given function to enable you to guess the general formula for $f^{(n)}(x)$. Then verify your guess using mathematical induction.

13. $f(x) = \dfrac{1}{x}$

14. $f(x) = \dfrac{1}{x^2}$

15. $f(x) = \dfrac{1}{2 - x}$

16. $f(x) = \sqrt{x}$

17. $f(x) = \dfrac{1}{a + bx}$

18. $f(x) = x^{2/3}$

19. $f(x) = \cos(ax)$

20. $f(x) = x \cos x$

21. $f(x) = x \sin(ax)$

* 22. $f(x) = \dfrac{1}{|x|}$

* 23. $f(x) = \sqrt{1 - 3x}$

24. If $y = \tan kx$, show that $y'' = 2k^2 y(1 + y^2)$.

25. If $y = \sec kx$, show that $y'' = k^2 y(2y^2 - 1)$.

26. Use mathematical induction to prove that the nth derivative of $y = \sin(ax + b)$ is given by the formula asserted at the end of Example 5.

27. Use mathematical induction to prove that the nth derivative of $y = \tan x$ is of the form $P_{n+1}(\tan x)$, where P_{n+1} is a polynomial of degree $n + 1$.

28. If f and g are twice-differentiable functions, show that $(fg)'' = f''g + 2f'g' + fg''$.

* 29. State and prove the results analogous to that of Exercise 28 but for $(fg)^{(3)}$ and $(fg)^{(4)}$. Can you guess the formula for $(fg)^{(n)}$?

* 30. If $f''(x)$ exists on an interval I and if f vanishes at at least three distinct points of I, prove that f'' must vanish at some point in I.

* 31. Generalize Exercise 30 to a function for which $f^{(n)}$ exists on I and for which f vanishes at at least $n + 1$ distinct points in I.

* 32. Suppose f is twice differentiable on an interval I (i.e., f'' exists on I). Suppose that the points 0 and 2 belong to I and that $f(0) = f(1) = 0$ and $f(2) = 1$. Prove that:

(a) $f'(a) = \dfrac{1}{2}$ for some point a in I.

(b) $f''(b) > \dfrac{1}{2}$ for some point b in I.

(c) $f'(c) = \dfrac{1}{7}$ for some point c in I.

2.9 Implicit Differentiation

We know how to find the slope of a curve which is the graph of a function $y = f(x)$ by calculating the derivative of f. But not all curves are the graphs of such functions. To be the graph of a function $f(x)$, the curve must not intersect any vertical lines at more than one point.

Curves are generally the graphs of *equations* in two variables. Such equations can be written in the form

$$F(x, y) = 0,$$

where $F(x, y)$ denotes an expression involving the two variables x and y. For example, a circle with centre at the origin and radius 5 has equation

$$x^2 + y^2 - 25 = 0,$$

so $F(x, y) = x^2 + y^2 - 25$ for that circle.

Sometimes we can solve an equation $F(x, y) = 0$ for y and so find explicit formulas for one or more functions $y = f(x)$ defined by the equation. Usually, however, we are not able to solve the equation. However, we can still regard it as defining y as one or more functions of x *implicitly*, even it we cannot solve for these functions *explicitly*. Moreover, we still find the derivative dy/dx of these implicit solutions by a technique called **implicit differentiation**. The idea is to differentiate the given equation with respect to x, regarding y as a function of x having derivative dy/dx, or y'.

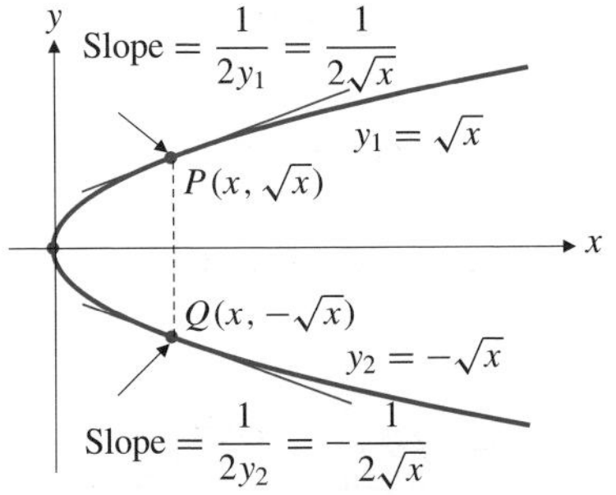

Example 1 Find dy/dx if $y^2 = x$.

Solution The equation $y^2 = x$ defines two differentiable functions of x; in this case we know them explicitly. They are $y_1 = \sqrt{x}$ and $y_2 = -\sqrt{x}$ (Figure 2.33), having derivatives defined for $x > 0$ by

$$\frac{dy_1}{dx} = \frac{1}{2\sqrt{x}} \qquad \text{and} \qquad \frac{dy_2}{dx} = -\frac{1}{2\sqrt{x}}.$$

Slope $= \dfrac{1}{2y_1} = \dfrac{1}{2\sqrt{x}}$

$y_1 = \sqrt{x}$

$P(x, \sqrt{x})$

$Q(x, -\sqrt{x})$

$y_2 = -\sqrt{x}$

Slope $= \dfrac{1}{2y_2} = -\dfrac{1}{2\sqrt{x}}$

Figure 2.33 The equation $y^2 = x$ defines two differentiable functions of x on the interval $x > 0$.

However, we can find the slope of the curve $y^2 = x$ at any point (x, y) satisfying that equation without first solving the equation for y. To find dy/dx we simply differentiate both sides of the equation $y^2 = x$ with respect to x, treating y as a differentiable function of x and using the Chain Rule to differentiate y^2:

$$\frac{d}{dx}(y^2) = \frac{d}{dx}(x) \qquad \left(\text{The Chain Rule gives } \frac{d}{dx}\, y^2 = 2y\frac{dy}{dx}.\right)$$

$$2y\frac{dy}{dx} = 1$$

$$\frac{dy}{dx} = \frac{1}{2y}.$$

Observe that this agrees with the derivatives we calculated above for *both* of the explicit solutions $y_1 = \sqrt{x}$ and $y_2 = -\sqrt{x}$:

$$\frac{dy_1}{dx} = \frac{1}{2y_1} = \frac{1}{2\sqrt{x}} \qquad \text{and} \qquad \frac{dy_2}{dx} = \frac{1}{2y_2} = \frac{1}{2(-\sqrt{x})} = -\frac{1}{2\sqrt{x}}.$$

Example 2   Find the slope of circle $x^2 + y^2 = 25$ at the point $(3, -4)$.

Solution   The circle is not the graph of a single function of x. Again, it combines the graphs of two functions, $y_1 = \sqrt{25 - x^2}$ and $y_2 = -\sqrt{25 - x^2}$ (Figure 2.34). The point $(3, -4)$ lies on the graph of y_2, so we can find the slope by calculating explicitly:

$$\left.\frac{dy_2}{dx}\right|_{x=3} = -\left.\frac{-2x}{2\sqrt{25 - x^2}}\right|_{x=3} = -\frac{-6}{2\sqrt{25 - 9}} = \frac{3}{4}.$$

But we can also solve the problem more easily by differentiating the given equation of the circle implicitly with respect to x:

$$\frac{d}{dx}(x^2) + \frac{d}{dx}(y^2) = \frac{d}{dx}(25)$$

$$2x + 2y\frac{dy}{dx} = 0$$

$$\frac{dy}{dx} = -\frac{x}{y}.$$

The slope at $(3, -4)$ is $\left.-\frac{x}{y}\right|_{(3,-4)} = -\frac{3}{-4} = \frac{3}{4}.$

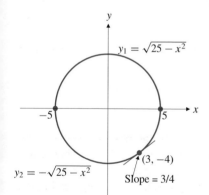

$y_1 = \sqrt{25 - x^2}$

$y_2 = -\sqrt{25 - x^2}$

$(3, -4)$

Slope = 3/4

Figure 2.34   The circle combines the graphs of two functions. The graph of y_2 is the lower semicircle and passes through $(3, -4)$

Example 3   Find $\dfrac{dy}{dx}$ if $y \sin x = x^3 + \cos y$.

Solution   This time we cannot solve the equation for y as an explicit function of x, so we *must* use implicit differentiation.

To find dy/dx by implicit differentiation:

1. Differentiate both sides of the equation with respect to x, regarding y as a function of x and using the Chain Rule to differentiate functions of y.

2. Collect terms with dy/dx on one side of the equation and solve for dy/dx by dividing by its coefficient.

$$\frac{d}{dx}(y \sin x) = \frac{d}{dx}(x^3) + \frac{d}{dx}(\cos y)$$

$\left(\begin{array}{l}\text{Use the Product Rule} \\ \text{on the left side.}\end{array}\right)$

$$(\sin x)\frac{dy}{dx} + y \cos x = 3x^2 - (\sin y)\frac{dy}{dx}$$

$$(\sin x + \sin y)\frac{dy}{dx} = 3x^2 - y \cos x$$

$$\frac{dy}{dx} = \frac{3x^2 - y \cos x}{\sin x + \sin y}$$

In the examples above the derivatives dy/dx calculated by implicit differentiation depend on y, or on both y and x, rather than just on x. This is to be expected because an equation in x and y can define more than one function of x, and the implicitly calculated derivative must apply to each of the solutions. For example, in Example 2, the derivative $dy/dx = -x/y$ also gives the slope $-3/4$ at the point $(3, 4)$ on the circle. When you use implicit differentiation to find the slope of a curve at a point, you will usually have to know both coordinates of the point.

There are subtle dangers involved in calculating derivatives implicitly. When you use the Chain Rule to differentiate an equation involving y with respect to x, you are automatically assuming that the equation defines y as a differentiable function of x. This need not be the case. To see what can happen, consider the problem of finding $y' = dy/dx$ from the equation

$$x^2 + y^2 = K, \qquad\qquad (*)$$

where K is a constant. Just as in Example 2 (where $K = 25$), implicit differentiation gives

$$2x + 2yy' = 0 \qquad \text{or} \qquad y' = -\frac{x}{y}.$$

This formula will give the slope of the curve $(*)$ at any point on the curve where $y \neq 0$. For $K > 0$, $(*)$ represents a circle centred at the origin having radius $\sqrt{K}$. This circle has a finite slope, except at the two points where it crosses the x-axis (where $y = 0$). If $K = 0$, the equation represents only a single point, the origin. The concept of slope of a point is meaningless. For $K < 0$, there are no real points whose coordinates satisfy equation $(*)$, so y' is meaningless here too. The point of this is that being able to calculate y' from a given equation by implicit differentiation does not guarantee that y' actually represents the slope of anything.

If (x_0, y_0) is a point on the graph of the equation $F(x, y) = 0$, there is a theorem that can justify our use of implicit differentiation to find the slope of the graph there. We cannot give a careful statement or proof of this **implicit function theorem** yet (see Section 12.8), but roughly speaking, it says that part of the graph of $F(x, y) = 0$ near (x_0, y_0) is the graph of a function of x that is differentiable at x_0, provided that $F(x, y)$ is a "smooth" function, and that the derivative

$$\left.\frac{d}{dy}F(x_0, y)\right|_{y=y_0} \neq 0.$$

For the circle $x^2 + y^2 - K = 0$ (where $K > 0$) this condition says that $2y_0 \neq 0$, which is the condition that the derivative $y' = -x/y$ should exist at (x_0, y_0).

A useful strategy

When you use implicit differentiation to find the value of a derivative at a particular point, it is best to substitute the coordinates of the point immediately after you carry out the differentiation and before you solve for the derivative dy/dx. It is easier to solve an equation involving numbers than one with algebraic expressions.

Example 4   Find an equation of the tangent to $x^2 + xy + 2y^3 = 4$ at $(-2, 1)$.

Solution   To find the slope of the tangent we differentiate the given equation implicitly with respect to x. Use the Product Rule to differentiate the xy term:

$$2x + y + xy' + 6y^2y' = 0.$$

Substitute the coordinates $x = -2$, $y = 1$, and solve the resulting equation for y':

$$-4 + 1 - 2y' + 6y' = 0 \quad \Rightarrow \quad y' = \frac{3}{4}.$$

The slope of the tangent at $(-2, 1)$ is $3/4$, and its equation is

$$y = \frac{3}{4}(x + 2) + 1 \quad \text{or} \quad 3x - 4y = -10.$$

Example 5   Show that for any constants a and b, the curves $x^2 - y^2 = a$ and $xy = b$ intersect at right angles, that is, at any point where they intersect their tangents are perpendicular.

Solution   The slope at any point on $x^2 - y^2 = a$ is given by $2x - 2yy' = 0$, or $y' = x/y$. The slope at any point on $xy = b$ is given by $y + xy' = 0$, or $y' = -y/x$. If the two curves (they are both hyperbolas if $a \neq 0$ and $b \neq 0$) intersect at (x_0, y_0), then their slopes at that point are x_0/y_0 and $-y_0/x_0$, respectively. Clearly, these slopes are negative reciprocals, so the tangent line to one curve is the normal line to the other at that point. Hence, the curves intersect at right angles. (See Figure 2.35.)

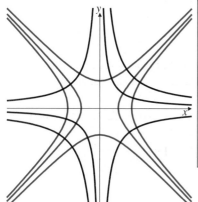

Figure 2.35   Some hyperbolas in the family $x^2 - y^2 = a$ (colour) intersecting some hyperbolas in the family $xy = b$ (black) at right angles

Higher-Order Derivatives

Example 6   Find $y'' = \dfrac{d^2y}{dx^2}$ if $xy + y^2 = 2x$.

Solution   Twice differentiate both sides of the given equation with respect to x:

$$y + xy' + 2yy' = 2$$
$$y' + y' + xy'' + 2(y')^2 + 2yy'' = 0.$$

Now solve these equations for y' and y''.

$$y' = \frac{2 - y}{x + 2y}$$

$$y'' = -\frac{2y' + 2(y')^2}{x + 2y} = -2\,\frac{2 - y}{x + 2y}\,\frac{1 + \dfrac{2 - y}{x + 2y}}{x + 2y}$$

$$= -2\,\frac{(2 - y)(x + y + 2)}{(x + 2y)^3}$$

$$= -2\,\frac{2x - xy + 2y - y^2 + 4 - 2y}{(x + 2y)^3} = -\frac{8}{(x + 2y)^3}.$$

We used the given equation to simplify the numerator in the last line.

Note that Maple uses the symbol ∂ instead of d when expressing the derivative in Leibniz form. This is because the expression it is differentiating can involve more than one variable; $(\partial/\partial x)y$ denotes the derivative of y with respect to the specific variable x rather than any other variables on which y may depend. It is called a **partial derivative**. We will study partial derivatives in Chapter 12. For the time being, just regard ∂ as a d.

Remark   We can use Maple to calculate derivatives implicitly provided we show explicitly which variable depends on which. For example, we can calculate the value of y'' for the curve $xy + y^3 = 3$ at the point $(2, 1)$ as follows. First we differentiate the equation with respect to x, writing $y(x)$ for y to indicate to Maple that it depends on x.

```
>  deq := diff(x*y(x)+(y(x))^3=3, x);
```

$$deq := y(x) + x\left(\frac{\partial}{\partial x}y(x)\right) + 3y(x)^2\left(\frac{\partial}{\partial x}y(x)\right) = 0$$

Now we solve the resulting equation for y':

```
>  yp := solve(deq, diff(y(x),x));
```

$$yp := -\frac{y(x)}{x + 3y(x)^2}$$

We can now differentiate yp with respect to x to get y''.

```
>  ypp := diff(yp,x);
```

$$ypp := -\frac{\frac{\partial}{\partial x}y(x)}{x + 3y(x)^2} + \frac{y(x)\left(1 + 6y(x)\left(\frac{\partial}{\partial x}y(x)\right)\right)}{(x + 3y(x)^2)^2}$$

To get an expression depending only on x and y, we need to substitute the expression obtained for the first derivative into this result. Since the result of this substitution will involve compound fractions, let us simplify the result as well.

```
>  ypp := simplify(subs(diff(y(x),x)=yp, ypp);
```

$$ypp := 2\frac{x\,y(x)}{(x + 3y(x)^2)^3}$$

This is y'' expressed as a function of x and y. Now we want to substitute the coordinates $x = 2$, $y(x) = 1$ to get the value of y'' at $(2, 1)$. However, the order of the substitutions is important. *First* we must replace $y(x)$ with 1 and *then* replace x with 2. (If we replace x first, we would have to then replace $y(2)$ rather than $y(x)$ with 1.) Maple's subs command makes the substitutions in the order they are written.

```
>  subs(y(x)=1, x=2, ypp);
```

$$\frac{4}{125}$$

The General Power Rule

Until now, we have only proven the General Power Rule

$$\frac{d}{dx}x^r = r\,x^{r-1}$$

for integer exponents r and a few special rational exponents such as $r = 1/2$. Using implicit differentiation, we can give the proof for any rational exponent $r = m/n$, where m and n are integers, and $n \neq 0$.

If $y = x^{m/n}$, then $y^n = x^m$. Differentiating implicitly with respect to x, we obtain

$$n\, y^{n-1} \frac{dy}{dx} = m\, x^{m-1}, \qquad \text{so}$$

$$\frac{dy}{dx} = \frac{m}{n} x^{m-1} y^{1-n} = \frac{m}{n} x^{m-1} x^{(m/n)(1-n)} = \frac{m}{n} x^{m-1+(m/n)-m} = \frac{m}{n} x^{(m/n)-1}.$$

Exercises 2.9

In Exercises 1–8, find dy/dx in terms of x and y.

1. $xy - x + 2y = 1$

2. $x^3 + y^3 = 1$

3. $x^2 + xy = y^3$

4. $x^3 y + xy^5 = 2$

5. $x^2 y^3 = 2x - y$

6. $x^2 + 4(y - 1)^2 = 4$

7. $\dfrac{x - y}{x + y} = \dfrac{x^2}{y} + 1$

8. $x\sqrt{x + y} = 8 - xy$

In Exercises 9–16, find an equation of the tangent to the given curve at the given point.

9. $2x^2 + 3y^2 = 5$ at $(1, 1)$

10. $x^2 y^3 - x^3 y^2 = 12$ at $(-1, 2)$

11. $\dfrac{x}{y} + \left(\dfrac{y}{x}\right)^3 = 2$ at $(-1, -1)$

12. $x + 2y + 1 = \dfrac{y^2}{x - 1}$ at $(2, -1)$

13. $2x + y - \sqrt{2}\sin(xy) = \pi/2$ at $\left(\dfrac{\pi}{4}, 1\right)$

14. $\tan(xy^2) = \dfrac{2xy}{\pi}$ at $\left(-\pi, \dfrac{1}{2}\right)$

15. $x\sin(xy - y^2) = x^2 - 1$ at $(1, 1)$

16. $\cos\left(\dfrac{\pi y}{x}\right) = \dfrac{x^2}{y} - \dfrac{17}{2}$ at $(3, 1)$

In Exercises 17–20, find y'' in terms of x and y.

17. $xy = x + y$

18. $x^2 + 4y^2 = 4$

* 19. $x^3 - y^2 + y^3 = x$

* 20. $x^3 - 3xy + y^3 = 1$

21. For $x^2 + y^2 = a^2$ show that $y'' = -\dfrac{a^2}{y^3}$.

22. For $Ax^2 + By^2 = C$ show that $y'' = -\dfrac{AC}{B^2 y^3}$.

Use Maple or another computer algebra program to find the values requested in Exercises 23–26.

23. Find the slope of $x + y^2 + y\sin x = y^3 + \pi$ at $(\pi, 1)$.

24. Find the slope of $\dfrac{x + \sqrt{y}}{y + \sqrt{x}} = \dfrac{3y - 9x}{x + y}$ at the point $(1, 4)$.

25. If $x + y^5 + 1 = y + x^4 + xy^2$, find $d^2 y/dx^2$ at $(1, 1)$.

26. If $x^3 y + xy^3 = 11$, find $d^3 y/dx^3$ at $(1, 2)$.

* 27. Show that the ellipse $x^2 + 2y^2 = 2$ and the hyperbola $2x^2 - 2y^2 = 1$ intersect at right angles.

* 28. Show that the ellipse $x^2/a^2 + y^2/b^2 = 1$ and the hyperbola $x^2/A^2 - y^2/B^2 = 1$ intersect at right angles if $A^2 \le a^2$ and $a^2 - b^2 = A^2 + B^2$. (This says that the ellipse and the hyperbola have the same foci.)

* 29. If $z = \tan\dfrac{x}{2}$, show that

$$\frac{dx}{dz} = \frac{2}{1 + z^2}, \quad \sin x = \frac{2z}{1 + z^2}, \quad \text{and } \cos x = \frac{1 - z^2}{1 + z^2}.$$

* 30. Use implicit differentiation to find y' if $(x - y)/(x + y) = x/y + 1$. Now show that there are, in fact, no points on that curve, so the derivative you calculated is meaningless. This is another example that demonstrates the dangers of calculating something when you don't know whether or not it exists.

2.10   Antiderivatives and Initial-Value Problems

Throughout this chapter we have been concerned with the problem of finding the derivative f' of a given function f. The reverse problem—given the derivative f', find f—is also interesting and important. It is the problem studied in *integral calculus* and is generally harder to solve than the problem of finding a derivative. We will take a preliminary look at this problem in this section and will return to it in more detail in Chapter 5.

Antiderivatives

We begin by defining an antiderivative of a function f to be a function F whose derivative is f. It is appropriate to require that $F'(x) = f(x)$ on an *interval*.

DEFINITION 7

> An **antiderivative** of a function f on an interval I is another function F satisfying
>
> $$F'(x) = f(x) \quad \text{for } x \text{ in } I.$$

Example 1

(a) $F(x) = x$ is an antiderivative of the function $f(x) = 1$ on any interval because $F'(x) = 1 = f(x)$ everywhere.

(b) $G(x) = \frac{1}{2}x^2$ is an antiderivative of the function $g(x) = x$ on any interval because $G'(x) = \frac{1}{2}(2x) = x = g(x)$ everywhere.

(c) $R(x) = -\frac{1}{3}\cos(3x)$ is an antiderivative of $r(x) = \sin(3x)$ on any interval because $R'(x) = -\frac{1}{3}(-3\sin(3x)) = \sin(3x) = r(x)$ everywhere.

(d) $F(x) = -1/x$ is an antiderivative of $f(x) = 1/x^2$ on any interval not containing $x = 0$ because $F'(x) = 1/x^2 = f(x)$ everywhere except at $x = 0$.

Antiderivatives are not unique; indeed, if C is any constant, then $F(x) = x + C$ is an antiderivative of $f(x) = 1$ on any interval. You can always add a constant to an antiderivative F of a function f on an interval and get another antiderivative of f. More importantly, *all* antiderivatives of f on an interval can be obtained by adding constants to any particular one. If F and G are both antiderivatives of f on an interval I, then

$$\frac{d}{dx}\big(G(x) - F(x)\big) = f(x) - f(x) = 0$$

on I, so $G(x) - F(x) = C$ (a constant) on I by Theorem 13 of Section 2.6. Thus $G(x) = F(x) + C$ on I.

Note that neither this conclusion nor Theorem 13 is valid over a set that is not an interval. For example, the derivative of

$$\operatorname{sgn} x = \begin{cases} -1 & \text{if } x < 0 \\ 1 & \text{if } x > 0 \end{cases}$$

is 0 for all $x \neq 0$, but $\operatorname{sgn} x$ is not constant for all $x \neq 0$. $\operatorname{sgn} x$ has *different* constant values on the two intervals $]-\infty, 0[$ and $]0, \infty[$ comprising its domain.

The Indefinite Integral

The *general antiderivative* of a function $f(x)$ on an interval I is $F(x) + C$, where $F(x)$ is any particular antiderivative of $f(x)$ on I and C is a constant. This general antiderivative is called the indefinite integral of $f(x)$ on I and is denoted $\int f(x)\,dx$.

DEFINITION 8

> The **indefinite integral** of $f(x)$ on interval I is
>
> $$\int f(x)\,dx = F(x) + C \qquad \text{on } I,$$
>
> provided $F'(x) = f(x)$ for all x in I.

The symbol $\int$ is called an **integral sign**. It is shaped like an elongated "S" for reasons that will only become apparent when we study the *definite integral* in Chapter 5. Just as you regard dy/dx as a single symbol representing the derivative of y with respect to x, so you should regard $\int f(x)\,dx$ as a single symbol representing the indefinite integral (general antiderivative) of f with respect to x. The constant C is called a **constant of integration**.

Example 2

(a) $\displaystyle\int x\,dx = \frac{1}{2}x^2 + C$ on any interval.

(b) $\displaystyle\int (x^3 - 5x^2 + 7)\,dx = \frac{1}{4}x^4 - \frac{5}{3}x^3 + 7x + C$ on any interval.

(c) $\displaystyle\int \left(\frac{1}{x^2} + \frac{2}{\sqrt{x}}\right)dx = -\frac{1}{x} + 4\sqrt{x} + C$ on any interval to the right of $x = 0$.

All three formulas above can be checked by differentiating the right-hand sides.   ■

Finding antiderivatives is generally more difficult than finding derivatives; many functions do not have antiderivatives that can be expressed as combinations of finitely many elementary functions. However, *every formula for a derivative can be rephrased as a formula for an antiderivative*. For instance,

$$\frac{d}{dx}\sin x = \cos x; \qquad \text{therefore,} \quad \int \cos x\,dx = \sin x + C.$$

We will develop several techniques for finding antiderivatives in later chapters. Until then, we must content ourselves with being able to write a few simple antiderivatives based on the known derivatives of elementary functions:

(a) $\displaystyle\int dx = \int 1\,dx = x + C$ (b) $\displaystyle\int x\,dx = \frac{x^2}{2} + C$

(c) $\displaystyle\int x^2\,dx = \frac{x^3}{3} + C$ (d) $\displaystyle\int \frac{1}{x^2}\,dx = \int \frac{dx}{x^2} = -\frac{1}{x} + C$

(e) $\displaystyle\int \frac{1}{\sqrt{x}}\,dx = 2\sqrt{x} + C$ (f) $\displaystyle\int x^r\,dx = \frac{x^{r+1}}{r+1} + C \ (r \neq -1)$

(g) $\displaystyle\int \sin x\,dx = -\cos x + C$ (h) $\displaystyle\int \cos x\,dx = \sin x + C$

(i) $\displaystyle\int \sec^2 x\,dx = \tan x + C$ (j) $\displaystyle\int \csc^2 x\,dx = -\cot x + C$

(k) $\displaystyle\int \sec x \tan x\,dx = \sec x + C$ (l) $\displaystyle\int \csc x \cot x\,dx = -\csc x + C$

Observe that formulas (a)–(e) are special cases of formula (f). For the moment, r must be rational in (f), but this restriction will be removed later.

The rule for differentiating sums and constant multiples of functions translates into a similar rule for antiderivatives, as reflected in parts (b) and (c) of Example 2 above.

The graphs of the different antiderivatives of the same function on the same interval are vertically displaced versions of the same curve, as shown in Figure 2.36.

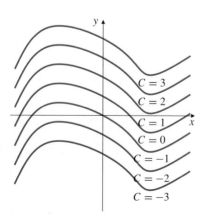

Figure 2.36 Graphs of various antiderivatives of the same function

In general, only one of these curves will pass through any given point, so we can obtain a unique antiderivative of a given function on an interval by requiring the antiderivative to take a prescribed value at a particular point x.

Example 3   Find the function $f(x)$ whose derivative is $f'(x) = 6x^2 - 1$ for all real x and for which $f(2) = 10$.

Solution   Since $f'(x) = 6x^2 - 1$, we have

$$f(x) = \int (6x^2 - 1)\, dx = 2x^3 - x + C$$

for some constant C. Since $f(2) = 10$, we have

$$10 = f(2) = 16 - 2 + C.$$

Thus $C = -4$ and $f(x) = 2x^3 - x - 4$. (By direct calculation we can verify that $f'(x) = 6x^2 - 1$ and $f(2) = 10$.)

Example 4   Find the function $g(t)$ whose derivative is $\dfrac{t+5}{t^{3/2}}$ and whose graph passes through the point $(4, 1)$.

Solution   We have

$$g(t) = \int \frac{t+5}{t^{3/2}}\, dt$$

$$= \int (t^{-1/2} + 5t^{-3/2})\, dt$$

$$= 2t^{1/2} - 10t^{-1/2} + C$$

Since the graph of $y = g(t)$ must pass through $(4, 1)$, we require that

$$1 = g(4) = 4 - 5 + C.$$

Hence, $C = 2$ and

$$g(t) = 2t^{1/2} - 10t^{-1/2} + 2 \qquad \text{for } t > 0.$$

Differential Equations and Initial-Value Problems

A **differential equation** (abbreviated DE) is an equation involving one or more derivatives of an unknown function. Any function whose derivatives satisfy the differential equation *identically on an interval* is called a **solution** of the equation on that interval. For instance, the function $y = x^3 - x$ is a solution of the differential equation

$$\frac{dy}{dx} = 3x^2 - 1$$

on the whole real line. This differential equation has more than one solution; in fact, $y = x^3 - x + C$ is a solution for any value of the constant C.

Example 5   Show that for any constants A and B, the function $y = Ax^3 + B/x$ is a solution of the differential equation $x^2 y'' - xy' - 3y = 0$ on any interval not containing 0.

Solution   If $y = Ax^3 + B/x$, then for $x \neq 0$ we have

$$y' = 3Ax^2 - B/x^2 \quad \text{and} \quad y'' = 6Ax + 2B/x^3.$$

Therefore,

$$x^2 y'' - xy' - 3y = 6Ax^3 + \frac{2B}{x} - 3Ax^3 + \frac{B}{x} - 3Ax^3 - \frac{3B}{x} = 0,$$

provided $x \neq 0$. This is what had to be proved. ∎

The **order** of a differential equation is the order of the highest-order derivative appearing in the equation. The DE in Example 5 is a *second-order* DE since it involves y'' and no higher derivatives of y. Note that the solution verified in Example 5 involves two arbitrary constants, A and B. This solution is called a **general solution** to the equation since it can be shown that every solution is of this form for some choice of the constants A and B. A **particular solution** of the equation is obtained by assigning specific values to these constants. The general solution of an nth-order differential equation typically involves n arbitrary constants.

An **initial-value problem** (abbreviated IVP) is a problem that consists of:

(i) a differential equation (to be solved for an unknown function) and

(ii) prescribed values for the solution and enough of its derivatives at a particular point (the initial point) to determine values for all the arbitrary constants in the general solution of the DE and so yield a particular solution.

Remark   It is common to use the same symbol, say y, to denote both the dependent variable and the function that is the solution to a DE or an IVP; that is, we call the solution function $y = y(x)$ rather than $y = f(x)$.

Remark   The solution of an IVP is valid in the largest interval containing the initial point where the solution function is defined.

Example 6   Use the result of Example 5 to solve the following initial-value problem.

$$\begin{cases} x^2 y'' - xy' - 3y = 0 & (x > 0) \\ y(1) = 2 \\ y'(1) = -6 \end{cases}$$

Solution   As shown in Example 5, the DE $x^2 y'' - xy' - 3y = 0$ has solution $y = Ax^3 + B/x$, which has derivative $y' = 3Ax^2 - B/x^2$. At $x = 1$ we must have $y = 2$ and $y' = -6$. Therefore,

$$A + B = 2$$
$$3A - B = -6.$$

Solving these two linear equations for A and B, we get $A = -1$ and $B = 3$. Hence, $y = -x^3 + 3/x$ for $x > 0$ is the solution of the IVP.   ■

One of the simplest kinds of differential equation is the equation

$$\frac{dy}{dx} = f(x),$$

which is to be solved for y as a function of x. Evidently the solution is

$$y = \int f(x)\, dx.$$

Our ability to find the unknown function $y(x)$ depends on our ability to find an antiderivative of f.

Example 7   Solve the initial-value problem

$$\begin{cases} y' = \dfrac{3 + 2x^2}{x^2} \\ y(-2) = 1. \end{cases}$$

Where is the solution valid?

Solution

$$y = \int \left(\frac{3}{x^2} + 2 \right) dx = -\frac{3}{x} + 2x + C$$

$$1 = y(-2) = \frac{3}{2} - 4 + C$$

Therefore, $C = \frac{7}{2}$ and

$$y = -\frac{3}{x} + 2x + \frac{7}{2}.$$

Although the solution function appears to be defined for all x except 0, it is only a solution of the given initial-value problem for $x < 0$. This is because $]-\infty, 0[$ is the largest interval that contains the initial point -2 but not the point $x = 0$, where the solution y is undefined.   ■

Example 8   Solve the second-order IVP

$$\begin{cases} y'' = \sin x \\ y(\pi) = 2 \\ y'(\pi) = -1. \end{cases}$$

Solution   Since $(y')' = y'' = \sin x$, we have

$$y'(x) = \int \sin x \, dx = -\cos x + C_1.$$

The initial condition for y' gives

$$-1 = y'(\pi) = -\cos \pi + C_1 = 1 + C_1,$$

so that $C_1 = -2$ and $y'(x) = -(\cos x + 2)$. Thus

$$y(x) = -\int (\cos x + 2) \, dx$$
$$= -\sin x - 2x + C_2.$$

The initial condition for y now gives

$$2 = y(\pi) = -\sin \pi - 2\pi + C_2 = -2\pi + C_2,$$

so that $C_2 = 2 + 2\pi$. The solution to the given IVP is

$$y = 2 + 2\pi - \sin x - 2x$$

and is valid for all x.   ∎

Differential equations and initial-value problems are of great importance in applications of calculus, especially for expressing in mathematical form certain laws of nature that involve rates of change of quantities. A large portion of the total mathematical endeavour of the last two hundred years has been devoted to their study. They are usually treated in separate courses on differential equations, but we will discuss them from time to time in this book when appropriate. Throughout this book, exercises about differential equations and initial-value problems are designated with the symbol ◈.

Exercises 2.10

In Exercises 1–14, find the given indefinite integrals.

1. $\displaystyle\int 5 \, dx$

2. $\displaystyle\int x^2 \, dx$

3. $\displaystyle\int \sqrt{x} \, dx$

4. $\displaystyle\int x^{12} \, dx$

5. $\displaystyle\int x^3 \, dx$

6. $\displaystyle\int (x + \cos x) \, dx$

7. $\displaystyle\int \tan x \cos x \, dx$

8. $\displaystyle\int \frac{1 + \cos^3 x}{\cos^2 x} \, dx$

9. $\displaystyle\int (a^2 - x^2) \, dx$

10. $\displaystyle\int (A + Bx + Cx^2) \, dx$

11. $\displaystyle\int (2x^{1/2} + 3x^{1/3}) \, dx$

12. $\displaystyle\int \frac{6(x-1)}{x^{4/3}} \, dx$

13. $\displaystyle\int \left(\frac{x^3}{3} - \frac{x^2}{2} + x - 1 \right) dx$

14. $\displaystyle 105 \int (1 + t^2 + t^4 + t^6) \, dt$

In Exercises 15–22, find the given indefinite integrals. This may require guessing the form of an antiderivative and then checking by differentiation. For instance, you might suspect that

$\int \cos(5x - 2)\, dx = k \sin(5x - 2) + C$ for some k.
Differentiating the answer shows that k must be $1/5$.

15. $\displaystyle\int \cos(2x)\, dx$

16. $\displaystyle\int \sin\left(\frac{x}{2}\right) dx$

* 17. $\displaystyle\int \frac{dx}{(1 + x)^2}$

* 18. $\displaystyle\int \sec(1 - x) \tan(1 - x)\, dx$

* 19. $\displaystyle\int \sqrt{2x + 3}\, dx$

* 20. $\displaystyle\int \frac{4}{\sqrt{x + 1}}\, dx$

21. $\displaystyle\int 2x \sin(x^2)\, dx$

* 22. $\displaystyle\int \frac{2x}{\sqrt{x^2 + 1}}\, dx$

Use trigonometric identities such as $\sec^2 x = 1 + \tan^2 x$,
$\sin(2x) = 2\sin x \cos x$, and
$\cos(2x) = 2\cos^2 x - 1 = 1 - 2\sin^2 x$ to help you evaluate the
indefinite integrals in Exercises 23–26.

* 23. $\displaystyle\int \tan^2 x\, dx$

* 24. $\displaystyle\int \sin x \cos x\, dx$

* 25. $\displaystyle\int \cos^2 x\, dx$

* 26. $\displaystyle\int \sin^2 x\, dx$

Differential equations

In Exercises 27–42, find the solution $y = y(x)$ to the given
initial-value problem. On what interval is the solution valid?
(Note that exercises involving differential equations are prefixed
with the symbol ◈.)

◈ 27. $\begin{cases} y' = x - 2 \\ y(0) = 3 \end{cases}$

◈ 28. $\begin{cases} y' = x^{-2} - x^{-3} \\ y(-1) = 0 \end{cases}$

◈ 29. $\begin{cases} y' = 3\sqrt{x} \\ y(4) = 1 \end{cases}$

◈ 30. $\begin{cases} y' = x^{1/3} \\ y(0) = 5 \end{cases}$

◈ 31. $\begin{cases} y' = Ax^2 + Bx + C \\ y(1) = 1 \end{cases}$

◈ 32. $\begin{cases} y' = x^{-9/7} \\ y(1) = -4 \end{cases}$

◈ 33. $\begin{cases} y' = \cos x \\ y(\pi/6) = 2 \end{cases}$

◈ 34. $\begin{cases} y' = \sin(2x) \\ y(\pi/2) = 1 \end{cases}$

◈ 35. $\begin{cases} y' = \sec^2 x \\ y(0) = 1 \end{cases}$

◈ 36. $\begin{cases} y' = \sec^2 x \\ y(\pi) = 1 \end{cases}$

◈ 37. $\begin{cases} y'' = 2 \\ y'(0) = 5 \\ y(0) = -3 \end{cases}$

◈ 38. $\begin{cases} y'' = x^{-4} \\ y'(1) = 2 \\ y(1) = 1 \end{cases}$

◈ 39. $\begin{cases} y'' = x^3 - 1 \\ y'(0) = 0 \\ y(0) = 8 \end{cases}$

◈ 40. $\begin{cases} y'' = 5x^2 - 3x^{-1/2} \\ y'(1) = 2 \\ y(1) = 0 \end{cases}$

◈ 41. $\begin{cases} y'' = \cos x \\ y(0) = 0 \\ y'(0) = 1 \end{cases}$

◈ 42. $\begin{cases} y'' = x + \sin x \\ y(0) = 2 \\ y'(0) = 0 \end{cases}$

◈ 43. Show that for any constants A and B the function
$y = y(x) = Ax + B/x$ satisfies the *second-order
differential equation* $x^2 y'' + xy' - y = 0$ for $x \neq 0$. Find a
function y satisfying the initial-value problem:

$$\begin{cases} x^2 y'' + xy' - y = 0 \qquad (x > 0) \\ y(1) = 2 \\ y'(1) = 4. \end{cases}$$

◈ 44. Show that for any constants A and B the function
$y = Ax^{r_1} + Bx^{r_2}$ satisfies, for $x > 0$, the differential
equation $ax^2 y'' + bxy' + cy = 0$, provided that r_1 and r_2
are two distinct rational roots of the quadratic equation
$ar(r - 1) + br + c = 0$.

Use the result of Exercise 44 to solve the initial-value problems
in Exercises 45–46 on the interval $x > 0$.

◈ 45. $\begin{cases} 4x^2 y'' + 4xy' - y = 0 \\ y(4) = 2 \\ y'(4) = -2 \end{cases}$

◈ 46. $\begin{cases} x^2 y'' - 6y = 0 \\ y(1) = 1 \\ y'(1) = 1 \end{cases}$

2.11 Velocity and Acceleration

Velocity and Speed

Suppose that an object is moving along a straight line (say the x-axis) so that its
position x is a function of time t, say $x = x(t)$. (We are using x to represent both
the dependent variable and the function.) Suppose we are measuring x in metres
and t in seconds. The **average velocity** of the object over the time interval $[t, t+h]$
is the change in position divided by the change in time, that is, the Newton quotient

$$v_{average} = \frac{\Delta x}{\Delta t} = \frac{x(t+h) - x(t)}{h} \text{ m/s}.$$

The **velocity** $v(t)$ of the object at time t is the limit of this average velocity as $h \to 0$. Thus it is the rate of change (the derivative) of position with respect to time:

$$\text{Velocity:} \quad v(t) = \frac{dx}{dt} = x'(t).$$

Besides telling us how fast the object is moving, the velocity also tells us in which direction it is moving. If $v(t) > 0$, then x is increasing, so the object is moving to the right; if $v(t) < 0$, then x is decreasing, so the object is moving to the left. At a critical point of x, that is, a time t when $v(t) = 0$, the object is instantaneously at rest—at that instant it is not moving in either direction.

We distinguish between the term *velocity* (which involves direction of motion as well as the rate) and **speed**, which only involves the rate, and not the direction. The speed is the absolute value of the velocity:

$$\text{Speed:} \quad s(t) = |v(t)| = \left| \frac{dx}{dt} \right|.$$

A speedometer gives us the speed an automobile is moving; it does not give the velocity. The speedometer does not start to show negative values if the automobile turns around and heads in the opposite direction.

Example 1

(a) Determine the velocity $v(t)$ at time t of an object moving along the x-axis so that at time t its position is given by

$$x = v_0 t + \frac{1}{2} a t^2,$$

where v_0 and a are constants.

(b) Draw the graph of the function $v(t)$, and show that the area under the graph and above the t-axis, over the interval $[t_1, t_2]$, is equal to the distance the object travels in that time interval.

Solution The velocity is given by

$$v(t) = \frac{dx}{dt} = v_0 + at.$$

Its graph is a straight line with slope a and intercept v_0 on the vertical (velocity) axis. The area under the graph (shaded in Figure 2.37) is the sum of the areas of a rectangle and a triangle. Each has base $t_2 - t_1$. The rectangle has height $v(t_1) = v_0 + at_1$, and the triangle has height $a(t_2 - t_1)$. (Why?) Thus the shaded area is equal to

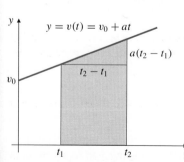

Figure 2.37   The shaded area equals the distance travelled between t_1 and t_2

$$\text{Area} = (t_2 - t_1)(v_0 + at_1) + \frac{1}{2}(t_2 - t_1)[a(t_2 - t_1)]$$

$$= (t_2 - t_1)\left[v_0 + at_1 + \frac{a}{2}(t_2 - t_1)\right]$$

$$= (t_2 - t_1)\left[v_0 + \frac{a}{2}(t_2 + t_1)\right]$$

$$= v_0(t_2 - t_1) + \frac{a}{2}(t_2^2 - t_1^2)$$

$$= x(t_2) - x(t_1),$$

which is the distance travelled by the object between times t_1 and t_2.

∎

Remark   In Example 1 we differentiated the position x to get the velocity v and then used the area under the velocity graph to recover information about the position. It appears that there is a connection between finding areas and finding functions that have given derivatives (i.e., finding antiderivatives). This connection, which we will explore in Chapter 5, is perhaps the most important idea in calculus!

Acceleration

The derivative of the velocity also has a useful interpretation. The rate of change of the velocity with respect to time is the **acceleration** of the moving object. It is measured in units of distance/time^2. The value of the acceleration at time t is

$$\text{Acceleration:} \quad a(t) = v'(t) = \frac{dv}{dt} = \frac{d^2x}{dt^2}.$$

The acceleration is the *second derivative* of the position. If $a(t) > 0$, the velocity is increasing. This does not necessarily mean that the speed is increasing; if the object is moving to the left ($v(t) < 0$) and accelerating to the right ($a(t) > 0$), then it is actually slowing down. The object is speeding up only when the velocity and acceleration have the same sign.

Table 2.   Velocity, acceleration, and speed

If velocity is	and acceleration is	then object is	and speed is
positive	positive	moving right	increasing
positive	negative	moving right	decreasing
negative	positive	moving left	decreasing
negative	negative	moving left	increasing

If $a(t_0) = 0$, then the velocity and the speed are stationary at t_0. If $a(t) = 0$ during an interval of time, then the velocity is unchanging and, therefore, constant over that interval.

Example 2   A point P moves along the x-axis in such a way that its position at time t s is given by

$$x = 2t^3 - 15t^2 + 24t \text{ ft.}$$

(a) Find the velocity and acceleration of P at time t.

(b) In which direction and how fast is P moving at $t = 2$ s? Is it speeding up or slowing down at that time?

(c) When is P instantaneously at rest? When is its speed instantaneously not changing?

(d) When is P moving to the left? to the right?

(e) When is P speeding up? slowing down?

Solution

(a) The velocity and acceleration of P at time t are

$$v = \frac{dx}{dt} = 6t^2 - 30t + 24 = 6(t-1)(t-4) \text{ ft/s} \quad \text{and}$$

$$a = \frac{dv}{dt} = 12t - 30 = 6(2t-5) \text{ ft/s}^2.$$

(b) At $t = 2$ we have $v = -12$ and $a = -6$. Thus P is moving to the left with speed 12 ft/s, and, since the velocity and acceleration are both negative, its speed is increasing.

(c) P is at rest when $v = 0$, that is, when $t = 1$ s or $t = 4$ s. Its speed is unchanging when $a = 0$, that is, at $t = 5/2$ s.

(d) The velocity is continuous for all t so, by the Intermediate-Value Theorem, has a constant sign on the intervals between the points where it is 0. By examining the values of $v(t)$ at $t = 0$, 2, and 5 (or by analyzing the signs of the factors $(t-1)$ and $(t-4)$ in the expression for $v(t)$), we conclude that $v(t) < 0$ (and P is moving to the left) on time interval $(1, 4)$. $v(t) > 0$ (and P is moving to the right) on time intervals $]-\infty, 1[$ and $]4, \infty[$.

(e) The acceleration a is negative for $t < 5/2$ and is positive for $t > 5/2$. Table 3 combines this information with information about v to show where P is speeding up and slowing down.

Table 3.   Data for Example 2

Interval	$v(t)$ is	$a(t)$ is	P is
$]-\infty, 1[$	positive	negative	slowing down
$]1, 5/2[$	negative	negative	speeding up
$]5/2, 4[$	negative	positive	slowing down
$]4, \infty[$	positive	positive	speeding up

The motion of P is shown in Figure 2.38.

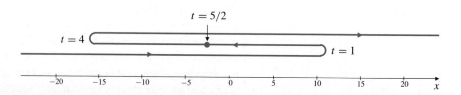

Figure 2.38   The motion of P in Example 2

Example 3 An object is hurled upward from the roof of a building 10 m high. It rises and then falls back; its height above ground t s after it is thrown is

$$y = -4.9\,t^2 + 8t + 10 \text{ m},$$

until it strikes the ground. What is the greatest height above the ground that the object attains? With what speed does the object strike the ground?

Solution Refer to Figure 2.39. The vertical velocity at time t during flight is

$$v(t) = -2(4.9)\,t + 8 = -9.8\,t + 8 \text{ m/s}.$$

The object is rising when $v > 0$, that is, when $0 < t < 8/9.8$, and is falling for $t > 8/9.8$. Thus the object is at its maximum height at time $t = 8/9.8 \approx 0.8163$ s, and this maximum height is

$$y_{\max} = -4.9 \left(\frac{8}{9.8} \right)^2 + 8 \left(\frac{8}{9.8} \right) + 10 \approx 13.27 \text{ m}.$$

The time t at which the object strikes the ground is the positive root of the quadratic equation obtained by setting $y = 0$,

$$-4.9t^2 + 8t + 10 = 0,$$

namely,

$$t = \frac{-8 - \sqrt{64 + 196}}{-9.8} \approx 2.462 \text{ s}.$$

The velocity at this time is $v = -(9.8)(2.462) + 8 \approx -16.12$. Thus the object strikes the ground with a speed of about 16.12 m/s.

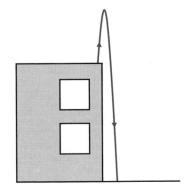

Figure 2.39

Falling Under Gravity

According to Newton's Second Law of Motion, a rock of mass m acted on by an unbalanced force F will experience an acceleration a proportional to and in the same direction as F; with appropriate units of force, $F = ma$. If the rock is sitting on the ground, it is acted on by two forces: the force of gravity acting downward and reaction of the ground acting upward. These forces balance, so there is no resulting acceleration. On the other hand, if the rock is up in the air and is unsupported, the gravitational force on it will be unbalanced and the rock will experience downward acceleration. It will fall.

According to Newton's Universal Law of Gravitation, the force by which the earth attracts the rock is proportional to the mass m of the rock and inversely proportional to the square of its distance r from the centre of the earth: $F = km/r^2$. If the relative change $\Delta r/r$ is small, as will be the case if the rock remains near the surface of the earth, then $F = mg$, where $g = k/r^2$ is approximately constant. It follows that $ma = F = mg$, and the rock experiences *constant* downward acceleration g. Since g does not depend on m, all objects experience the same acceleration when falling near the surface of the earth, provided we ignore air resistance and any other forces that may be acting on them. Newton's laws therefore imply that if the height of such an object at time t is $y(t)$, then

$$\frac{d^2y}{dt^2} = -g.$$

The negative sign is needed because the gravitational acceleration is downward, the opposite direction to that of increasing y. Physical experiments give the following approximate values for g at the surface of the earth:

$$g = 32 \text{ ft/s}^2 \quad \text{or} \quad g = 9.8 \text{ m/s}^2.$$

Example 4 A rock falling freely near the surface of the earth is subject to a constant downward acceleration g, if the effect of air resistance is neglected. If the height and velocity of the rock are y_0 and v_0 at time $t = 0$, find the height $y(t)$ of the rock at any later time t until the rock strikes the ground.

Solution This example asks for a solution $y(t)$ to the second-order initial-value problem:

$$\begin{cases} y''(t) = -g \\ y(0) = y_0 \\ y'(0) = v_0. \end{cases}$$

We have

$$y'(t) = -\int g \, dt = -gt + C_1$$

$$v_0 = y'(0) = 0 + C_1.$$

Thus $C_1 = v_0$.

$$y'(t) = -gt + v_0$$

$$y(t) = \int (-gt + v_0)dt = -\frac{1}{2}gt^2 + v_0 t + C_2$$

$$y_0 = y(0) = 0 + 0 + C_2.$$

Thus $C_2 = y_0$. Finally, therefore,

$$y(t) = -\frac{1}{2}gt^2 + v_0 t + y_0.$$

Example 5 A ball is thrown down with an initial speed of 20 ft/s from the top of a cliff, and it strikes the ground at the bottom of the cliff after 5 s. How high is the cliff?

Solution We will apply the result of Example 4. Here we have $g = 32 \text{ ft/s}^2$, $v_0 = -20$ ft/s, and y_0 is the unknown height of the cliff. The height of the ball t s after it is thrown down is

$$y(t) = -16t^2 - 20t + y_0 \text{ ft.}$$

At $t = 5$ the ball reaches the ground, so $y(5) = 0$:

$$0 = -16(25) - 20(5) + y_0 \qquad \Rightarrow \qquad y_0 = 500.$$

The cliff is 500 ft high.

Example 6 (**Stopping distance**) A car is travelling at 72 km/h. At a certain instant its brakes are applied to produce a constant deceleration of 0.8 m/s^2. How far does the car travel before coming to a stop?

Solution Let $s(t)$ be the distance the car travels in the t seconds after the brakes are applied. Then $s''(t) = -0.8$ (m/s^2), so the velocity at time t is given by

$$s'(t) = \int -0.8 \, dt = -0.8t + C_1 \quad \text{m/s.}$$

Since $s'(0) = 72$ km/h $= 72 \times 1000/3600 = 20$ m/s, we have $C_1 = 20$. Thus,

$$s'(t) = 20 - 0.8t$$

and

$$s(t) = \int (20 - 0.8t) \, dt = 20t - 0.4t^2 + C_2.$$

Since $s(0) = 0$, we have $C_2 = 0$ and $s(t) = 20t - 0.4t^2$. When the car has stopped, its velocity will be 0. Hence, the stopping time is the solution t of the equation

$$0 = s'(t) = 20 - 0.8t,$$

that is, $t = 25$ s. The distance travelled during deceleration is $s(25) = 250$ m.

Exercises 2.11

In Exercises 1–4, a point moves along the x-axis so that its position x at time t is specified by the given function. In each case determine the following:

(a) the time intervals on which the point is moving to the right and (b) to the left;

(c) the time intervals on which the point is accelerating to the right and (d) to the left;

(e) the time intervals when the particle is speeding up and (f) slowing down;

(g) the acceleration at times when the velocity is zero;

(h) the average velocity over the time interval $[0, 4]$.

1. $x = t^2 - 4t + 3$

2. $x = 4 + 5t - t^2$

3. $x = t^3 - 4t + 1$

4. $x = \dfrac{t}{t^2 + 1}$

5. A ball is thrown upward from ground level with an initial speed of 9.8 m/s so that its height in metres after t s is given by $y = 9.8t - 4.9t^2$. What is the acceleration of the ball at any time t? How high does the ball go? How fast is it moving when it strikes the ground?

6. A ball is thrown downward from the top of a 100-metre-high tower with an initial speed of 2 m/s. Its height in metres above the ground t s later is $y = 100 - 2t - 4.9t^2$. How long does it take to reach the ground? What is its average velocity during the fall? At what instant is its velocity equal to its average velocity?

7. **(Takeoff distance)** The distance an aircraft travels along a runway before takeoff is given by $D = t^2$, where D is measured in metres from the starting point, and t is measured in seconds from the time the brake is released. If the aircraft will become airborne when its speed reaches 200 km/h, how long will it take to become airborne, and what distance will it travel in that time?

8. **(Projectiles on Mars)** A projectile fired upward from the surface of the earth falls back to the ground after 10 s. How long would it take to fall back to the surface if it is fired upward on Mars with the same initial velocity? $g_{Mars} = 3.72$ m/s^2.

9. A ball is thrown upward with initial velocity v_0 m/s and reaches a maximum height of h m. How high would it have gone if its initial velocity was $2v_0$? How fast must it be thrown upward to achieve a maximum height of $2h$ m?

10. How fast would the ball in the previous exercise have to be thrown upward on Mars in order to achieve a maximum height of $3h$ m?

11. A rock falls from the top of a cliff and hits the ground at the base of the cliff at a speed of 160 ft/s. How high is the cliff?

12. A rock is thrown down from the top of a cliff with the initial speed of 32 ft/s and hits the ground at the base of the cliff at a speed of 160 ft/s. How high is the cliff?

13. **(Distance travelled while braking)** With full brakes applied, a freight train can decelerate at a constant rate of $1/6$ m/s^2. How far will the train travel while braking to a full stop from an initial speed of 60 km/h?

* 14. Show that if the position x of a moving point is given by a quadratic function of t, $x = At^2 + Bt + C$, then the average velocity over any time interval $[t_1, t_2]$ is equal to the instantaneous velocity at the midpoint of that time interval.

* 15. **(Piecewise motion)** The position of an object moving along the s-axis is given at time t by

$$s = \begin{cases} t^2 & \text{if } 0 \le t \le 2 \\ 4t - 4 & \text{if } 2 < t < 8 \\ -68 + 20t - t^2 & \text{if } 8 \le t \le 10. \end{cases}$$

Determine the velocity and acceleration at any time t. Is the velocity continuous? Is the acceleration continuous? What is the maximum velocity and when is it attained?

(Rocket flight with limited fuel) Figure 2.40 shows the velocity v in ft/s of a small rocket that was fired from the top of a tower at time $t = 0$ (t in seconds), accelerated with constant upward acceleration until its fuel was used up, then fell back to the ground at the foot of the tower. The whole flight lasted 14 s. Exercises 16–19 refer to this rocket.

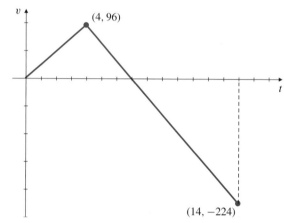

Figure 2.40

16. What was the acceleration of the rocket while its fuel lasted?

17. How long was the rocket rising?

* 18. What is the maximum height above ground that the rocket reached?

* 19. How high was the tower from which the rocket was fired?

20. Redo Example 6 using instead a nonconstant deceleration, $s''(t) = -t$ m/s^2.

Chapter Review

Key Ideas

• **What do the following statements and phrases mean?**

◇ Line L is tangent to curve C at point P.

◇ the Newton quotient of $f(x)$ at $x = a$

◇ the derivative $f'(x)$ of the function $f(x)$

◇ f is differentiable at $x = a$.

◇ the slope of the graph $y = f(x)$ at $x = a$

◇ f is increasing (or decreasing) on interval I.

◇ f is nondecreasing (or nonincreasing) on interval I.

◇ the average rate of change of $f(x)$ on $[a, b]$

◇ the rate of change of $f(x)$ at $x = a$

◇ c is a critical point of $f(x)$.

◇ the second derivative of $f(x)$ at $x = a$

◇ an antiderivative of f on interval I

◇ the indefinite integral of f on interval I

◇ differential equation   ◇ initial-value problem

◇ velocity ◇ speed ◇ acceleration

• **State the following differentiation rules:**

◇ the rule for differentiating a sum of functions

◇ the rule for differentiating a constant multiple of a function

◇ the Product Rule ◇ the Reciprocal Rule

◇ the Quotient Rule ◇ the Chain Rule

• **State the Mean-Value Theorem.**

• **State the Generalized Mean-Value Theorem.**

• **State the derivatives of the following functions:**

◇ x ◇ x^2 ◇ $1/x$ ◇ $\sqrt{x}$

◇ x^n ◇ $|x|$ ◇ $\sin x$ ◇ $\cos x$

◇ $\tan x$ ◇ $\cot x$ ◇ $\sec x$ ◇ $\csc x$

• **What is a proof by mathematical induction?**

Review Exercises

Use the definition of derivative to calculate the derivatives in Exercises 1–4.

1. $\dfrac{dy}{dx}$ if $y = (3x + 1)^2$

2. $\dfrac{d}{dx}\sqrt{1 - x^2}$

3. $f'(2)$ if $f(x) = \dfrac{4}{x^2}$

4. $g'(9)$ if $g(t) = \dfrac{t - 5}{1 + \sqrt{t}}$

5. Find the tangent to $y = \cos(\pi x)$ at $x = 1/6$.

6. Find the normal to $y = \tan(x/4)$ at $x = \pi$.

Calculate the derivatives of the functions in Exercises 7–12.

7. $\dfrac{1}{x - \sin x}$

8. $\dfrac{1 + x + x^2 + x^3}{x^4}$

9. $(4 - x^{2/5})^{-5/2}$

10. $\sqrt{2 + \cos^2 x}$

11. $\tan\theta - \theta\sec^2\theta$

12. $\dfrac{\sqrt{1 + t^2} - 1}{\sqrt{1 + t^2} + 1}$

Evaluate the limits in Exercises 13–16 by interpreting each as a derivative.

13. $\lim\limits_{h \to 0} \dfrac{(x + h)^{20} - x^{20}}{h}$

14. $\lim\limits_{x \to 2} \dfrac{\sqrt{4x + 1} - 3}{x - 2}$

15. $\lim\limits_{x \to \pi/6} \dfrac{\cos(2x) - (1/2)}{x - \pi/6}$

16. $\lim\limits_{x \to -a} \dfrac{(1/x^2) - (1/a^2)}{x + a}$

In Exercises 17–24, express the derivatives of the given functions in terms of the derivatives f' and g' of the differentiable functions f and g.

17. $f(3 - x^2)$

18. $[f(\sqrt{x})]^2$

19. $f(2x)\sqrt{g(x/2)}$

20. $\dfrac{f(x) - g(x)}{f(x) + g(x)}$

21. $f(x + (g(x))^2)$

22. $f\left(\dfrac{g(x^2)}{x}\right)$

23. $f(\sin x)\,g(\cos x)$

24. $\sqrt{\dfrac{\cos f(x)}{\sin g(x)}}$

25. Find the tangent to the curve $x^3y + 2xy^3 = 12$ at the point $(2, 1)$.

26. Find the slope of the curve $3\sqrt{2x}\sin(\pi y) + 8y\cos(\pi x) = 2$ at the point $\left(\frac{1}{3}, \frac{1}{4}\right)$.

Find the indefinite integrals in Exercises 27–30.

27. $\displaystyle\int \dfrac{1 + x^4}{x^2}\, dx$

28. $\displaystyle\int \dfrac{1 + x}{\sqrt{x}}\, dx$

29. $\displaystyle\int \dfrac{2 + 3\sin x}{\cos^2 x}\, dx$

30. $\displaystyle\int (2x + 1)^4\, dx$

31. Find $f(x)$ given that $f'(x) = 12x^2 + 12x^3$ and $f(1) = 0$.

32. Find $g(x)$ if $g'(x) = \sin(x/3) + \cos(x/6)$ and the graph of g passes through the point $(\pi, 2)$.

33. Differentiate $x\sin x + \cos x$ and $x\cos x - \sin x$, and use the results to find the indefinite integrals

$$I_1 = \int x\cos x\, dx \quad \text{and} \quad I_2 = \int x\sin x\, dx.$$

34. Suppose that $f'(x) = f(x)$ for every x, and let $g(x) = x\, f(x)$. Calculate the first several derivatives of g and guess a formula for the nth-order derivative $g^{(n)}(x)$. Verify your guess by induction.

35. Find an equation of the straight line that passes through the origin and is tangent to the curve $y = x^3 + 2$.

36. Find an equation of the straight lines that pass through the point $(0, 1)$ and are tangent to the curve $y = \sqrt{2 + x^2}$.

37. Show that $\dfrac{d}{dx}\left(\sin^n x\sin(nx)\right) = n\sin^{n-1} x\sin((n + 1)x)$. At what points x in $[0, \pi]$ does the graph of $y = \sin^n x\sin(nx)$ have a horizontal tangent? Assume that $n \geq 2$.

38. Find differentiation formulas for $y = \sin^n x\cos(nx)$, $y = \cos^n x\sin(nx)$, and $y = \cos^n x\cos(nx)$ analogous to the one given for $y = \sin^n x\sin(nx)$ in the previous exercise.

39. Let Q be the point $(0, 1)$. Find all points P on the curve $y = x^2$ such that the line PQ is normal to $y = x^2$ at P. What is the shortest distance from Q to the curve $y = x^2$?

40. **(Average and marginal profit)** Figure 2.41 shows the graph of the profit $\$P(x)$ realized by a grain exporter from its sale of x tonnes of wheat. Thus, the average profit per tonne is $\$P(x)/x$. Show that the maximum average profit occurs when the average profit equals the marginal profit. What is the geometric significance of this fact in the figure?

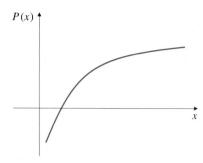

Figure 2.41

41. (Gravitational attraction) The gravitational attraction of the earth on a mass m at distance r from the centre of the earth is a continuous function $F(r)$ given for $r \geq 0$ by

$$F(r) = \begin{cases} \dfrac{mgR^2}{r^2} & \text{if } r \geq R \\ mkr & \text{if } 0 \leq r < R \end{cases}$$

where R is the radius of the earth, and g is the acceleration due to gravity at the surface of the earth.

(a) Find the constant k in terms of g and R.

(b) F decreases as m moves away from the surface of the earth, either upward or downward. Show that F decreases as r increases from R at twice the rate at which F decreases as r decreases from R.

42. (Compressibility of a gas) The isothermal compressibility of a gas is the relative rate of change of the volume V with respect to the pressure P, at a constant temperature T, that is,

$$\frac{1}{V}\frac{dV}{dP}.$$

For a sample of an ideal gas, the temperature, pressure, and volume satisfy the equation $PV = kT$, where k is a constant related to the number of molecules of gas present in the sample. Show that the isothermal compressibility of such a gas is the negative reciprocal of the pressure:

$$\frac{1}{V}\frac{dV}{dP} = -\frac{1}{P}.$$

43. A ball is thrown upward with an initial speed of 10 m/s from the top of a building. A second ball is thrown upward with an initial speed of 20 m/s from the ground. Both balls achieve the same maximum height above the ground. How tall is the building?

44. A ball is dropped from the top of a 60 m high tower at the same instant that a second ball is thrown upward from the ground at the base of the tower. The balls collide at a height of 30 m above the ground. With what initial velocity was the second ball thrown? How fast is each ball moving when they collide?

45. (Braking distance) A car's brakes can decelerate the car at 20 ft/s². How fast can the car travel if it must be able to stop in a distance of 160 ft?

46. (Measuring variations in g) The period P of a pendulum of length L is given by

$$P = 2\pi\sqrt{L/g},$$

where g is the acceleration of gravity.

(a) Assuming that L remains fixed, show that a 1% increase in g results in approximately a 1/2% decrease in the period P. (Variations in the period of a pendulum can be used to detect small variations in g from place to place on the earth's surface.)

(b) For fixed g, what percentage change in L will produce a 1% increase in P?

Challenging Problems

1. René Descartes, the inventor of analytic geometry, calculated the tangent to a parabola (or a circle or other quadratic curve) at a given point (x_0, y_0) on the curve by looking for a straight line through (x_0, y_0) having only one intersection with the given curve. Illustrate his method by writing the equation of a line through (a, a^2), having arbitrary slope m, and then finding the value of m for which the line has only one intersection with the parabola $y = x^2$. Why does the method not work for more general curves?

2. Given that $f'(x) = 1/x$ and $f(2) = 9$, find:

(a) $\displaystyle\lim_{x \to 2} \frac{f(x^2 + 5) - f(9)}{x - 2}$ (b) $\displaystyle\lim_{x \to 2} \frac{\sqrt{f(x)} - 3}{x - 2}$

3. Suppose that $f'(4) = 3$, $g'(4) = 7$, $g(4) = 4$, and $g(x) \neq 4$ for $x \neq 4$. Find:

(a) $\displaystyle\lim_{x \to 4}\left(f(x) - f(4)\right)$ (b) $\displaystyle\lim_{x \to 4} \frac{f(x) - f(4)}{x^2 - 16}$

(c) $\displaystyle\lim_{x \to 4} \frac{f(x) - f(4)}{\sqrt{x} - 2}$ (d) $\displaystyle\lim_{x \to 4} \frac{f(x) - f(4)}{(1/x) - (1/4)}$

(e) $\displaystyle\lim_{x \to 4} \frac{f(x) - f(4)}{g(x) - 4}$ (f) $\displaystyle\lim_{x \to 4} \frac{f(g(x)) - f(4)}{x - 4}$

4. Let $f(x) = \begin{cases} x & \text{if } x = 1,\ 1/2,\ 1/3,\ 1/4,\ \dots \\ x^2 & \text{otherwise.} \end{cases}$

(a) Find all points at which f is continuous. In particular, is it continuous at $x = 0$?

(b) Is the following statement true or false? Justify your answer. For any two real numbers a and b, there is some x between a and b such that $f(x) = (f(a) + f(b))/2$.

(c) Find all points at which f is differentiable. In particular, is it differentiable at $x = 0$?

5. Suppose $f(0) = 0$ and $|f(x)| > \sqrt{|x|}$ for all x. Show that $f'(0)$ does not exist.

6. Suppose that f is a function satisfying the following conditions: $f'(0) = k$, $f(0) \neq 0$, and $f(x + y) = f(x)f(y)$ for all x and y. Show that $f(0) = 1$ and that $f'(x) = k f(x)$ for every x. (We will study functions with these properties in Chapter 3.)

7. Suppose the function g satisfies the conditions: $g'(0) = k$, and $g(x + y) = g(x) + g(y)$ for all x and y. Show that:

  (a) $g(0) = 0$,   (b) $g'(x) = k$ for all x,   and

  (c) $g(x) = kx$ for all x. *Hint:* let $h(x) = g(x) - g'(0)x$.

8. (a) If f is differentiable at x, show that

  (i) $\lim\limits_{h \to 0} \dfrac{f(x) - f(x - h)}{h} = f'(x)$

  (ii) $\lim\limits_{h \to 0} \dfrac{f(x + h) - f(x - h)}{2h} = f'(x)$

  (b) Show that the existence of the limit in (i) guarantees that f is differentiable at x.

  (c) Show that the existence of the limit in (ii) does *not* guarantee that f is differentiable at x. *Hint:* consider the function $f(x) = |x|$ at $x = 0$.

9. Show that there is a line through $(a, 0)$ that is tangent to the curve $y = x^3$ at $x = 3a/2$. If $a \neq 0$, is there any other line through $(a, 0)$ that is tangent to the curve? If (x_0, y_0) is an arbitrary point, what is the maximum number of lines through (x_0, y_0) that can be tangent to $y = x^3$? the minimum number?

10. Make a sketch showing that there are two straight lines, each of which is tangent to both of the parabolas $y = x^2 + 4x + 1$ and $y = -x^2 + 4x - 1$. Find equations of the two lines.

11. Show that if $b > 1/2$, there are three straight lines through $(0, b)$, each of which is normal to the curve $y = x^2$. How many such lines are there if $b = 1/2$? if $b < 1/2$?

12. **(Distance from a point to a curve)** Find the point on the curve $y = x^2$ that is closest to the point $(3, 0)$. *Hint:* the line from $(3, 0)$ to the closest point Q on the parabola is normal to the parabola at Q.

* 13. **(Envelope of a family of lines)** Show that for each value of the parameter m, the line $y = mx - (m^2/4)$ is tangent to the parabola $y = x^2$. (The parabola is called the *envelope* of the family of lines $y = mx - (m^2/4)$.) Find $f(m)$ such that the family of lines $y = mx + f(m)$ has envelope the parabola $y = Ax^2 + Bx + C$.

* 14. **(Common tangents)** Consider the two parabolas with equations $y = x^2$ and $y = Ax^2 + Bx + C$. We assume that $A \neq 0$, and if $A = 1$, then either $B \neq 0$ or $C \neq 0$, so that the two equations do represent different parabolas. Show that:

  (a) the two parabolas are tangent to each other if $B^2 = 4C(A - 1)$;

  (b) the parabolas have two common tangent lines if and only if $A \neq 1$ and $A(B^2 - 4C(A - 1)) > 0$;

  (c) the parabolas have exactly one common tangent line if either $A = 1$ and $B \neq 0$, or $A \neq 1$ and $B^2 = 4C(A - 1)$;

  (d) the parabolas have no common tangent lines if either $A = 1$ and $B = 0$, or $A \neq 1$ and $A(B^2 - 4C(A - 1)) < 0$.

Make sketches illustrating each of the above possibilities.

15. Let C be the graph of $y = x^3$.

  (a) Show that if $a \neq 0$ then the tangent to C at $x = a$ also intersects C at a second point $x = b$.

  (b) Show that the slope of C at $x = b$ is four times its slope at $x = a$.

  (c) Can any line be tangent to C at more than one point?

  (d) Can any line be tangent to the graph of $y = Ax^3 + Bx^2 + Cx + D$ at more than one point?

* 16. Let C be the graph of $y = x^4 - 2x^2$.

  (a) Find all horizontal lines that are tangent to C.

  (b) One of the lines found in (a) is tangent to C at two different points. Show that there are no other lines that have this property.

  (c) Find an equation of a straight line that is tangent to the graph of $y = x^4 - 2x^2 + x$ at two different points. Can there exist more than one such line? Why?

17. **(Double tangents)** A line tangent to the quartic (fourth-degree polynomial) curve C with equation $y = ax^4 + bx^3 + cx^2 + dx + e$ at $x = p$ may intersect C at zero, one, or two other points. If it meets C at only one other point $x = q$, it must be tangent to C at that point also, and it is thus a "double tangent."

  (a) Find the condition that must be satisfied by the coefficients of the quartic to ensure that there does exist such a double tangent, and show that there cannot be more than one such double tangent. Illustrate this by applying your results to $y = x^4 - 2x^2 + x - 1$.

  (b) If the line PQ is tangent to C at two distinct points $x = p$ and $x = q$, show that PQ is parallel to the line tangent to C at $x = (p + q)/2$.

  (c) If the line PQ is tangent to C at two distinct points $x = p$ and $x = q$, show that C has two distinct inflection points R and S and that RS is parallel to PQ.

18. Verify the following formulas for every positive integer n:

  (a) $\dfrac{d^n}{dx^n} \cos(ax) = a^n \cos\left(ax + \dfrac{n\pi}{2}\right)$

  (b) $\dfrac{d^n}{dx^n} \sin(ax) = a^n \sin\left(ax + \dfrac{n\pi}{2}\right)$

  (c) $\dfrac{d^n}{dx^n}\left(\cos^4 x + \sin^4 x\right) = 4^{n-1} \cos\left(4x + \dfrac{n\pi}{2}\right)$

19. **(Rocket with a parachute)** A rocket is fired from the top of a tower at time $t = 0$. It experiences constant upward acceleration until its fuel is used up. Thereafter its acceleration is the constant downward acceleration of gravity until, during its fall, it deploys a parachute that gives it a constant upward acceleration again to slow it down. The rocket hits the ground near the base of the tower. The upward velocity v (in metres per second) is graphed against time in Figure 2.42. From information in the figure answer the following questions:

  (a) How long did the fuel last?

(b) When was the rocket's height maximum?

(c) When was the parachute deployed?

(d) What was the rocket's upward acceleration while its motor was firing?

(e) What was the maximum height achieved by the rocket?

(f) How high was the tower from which the rocket was fired?

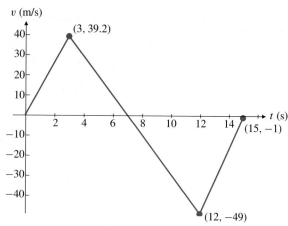

Figure 2.42

CHAPTER 3

Transcendental Functions

Introduction   With the exception of the trigonometric functions, all the functions we have encountered so far have been of three main types: *polynomials, rational functions* (quotients of polynomials), and *algebraic functions* (fractional powers of rational functions). On an interval in its domain, each of these functions can be constructed from real numbers and a single real variable x by using finitely many arithmetic operations (addition, subtraction, multiplication, and division) and by taking finitely many roots (fractional powers). Functions that cannot be so constructed are called **transcendental functions**. The only examples of these that we have seen so far are the trigonometric functions.

Much of the importance of calculus and many of its most useful applications result from its ability to illuminate the behaviour of transcendental functions that arise naturally when we try to model concrete problems in mathematical terms. This chapter is devoted to developing other transcendental functions, including exponential and logarithmic functions and the inverse trigonometric functions.

Some of these functions "undo" what other ones "do," and vice versa. When a pair of functions behaves this way, we call each one the inverse of the other. We begin the chapter by studying inverse functions in general.

3.1 Inverse Functions

Consider the function

$$f(x) = x^3,$$

whose graph is shown in Figure 3.1. Like any function, $f(x)$ has only one value for each x in its domain (the whole real line $\mathbb{R}$). In geometric terms, any *vertical* line meets the graph of f at only one point. For this function f, any *horizontal* line also meets the graph at only one point. This means that different values of x always give different values to $f(x)$. Such a function is said to be *one-to-one*.

DEFINITION 1

A function f is **one-to-one** if $f(x_1) \neq f(x_2)$ whenever x_1 and x_2 belong to the domain of f and $x_1 \neq x_2$ or, equivalently, if

$$f(x_1) = f(x_2) \quad \Longrightarrow \quad x_1 = x_2.$$

A function defined on a single interval is one-to-one there if it is either increasing or decreasing.

Because $f(x) = x^3$ is one-to-one, the equation

$$y = x^3$$

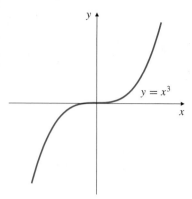

Figure 3.1   The graph of $f(x) = x^3$

Do not confuse the -1 in f^{-1} with an exponent. The inverse f^{-1} is *not* the reciprocal $1/f$. If we want to denote the reciprocal $1/f(x)$ with an exponent we can write it as $\left(f(x)\right)^{-1}$.

Figure 3.2

(a) f is one-to-one and has an inverse. $y = f(x)$ means the same thing as $x = f^{-1}(y)$

(b) g is not one-to-one

has a unique solution x for every given value of y in the range of f. Specifically,

$$x = y^{1/3}.$$

This equation defines x as a function of y. We call this new function the *inverse of f* and denote it f^{-1}. Thus

$$f^{-1}(y) = y^{1/3}.$$

Whenever a function f is one-to-one, for any number y in its range there will always exist a single number x in its domain such that $y = f(x)$. Since x is determined uniquely by y, it is a function of y. We write $x = f^{-1}(y)$ and call f^{-1} the inverse of f. The function f whose graph is shown in Figure 3.2(a) is one-to-one and has an inverse. The function g whose graph is shown in Figure 3.2(b) is not one-to-one and does not have an inverse.

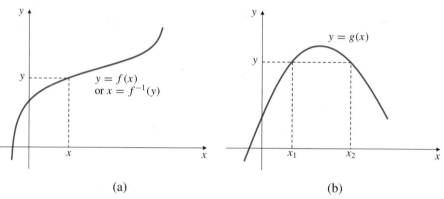

(a)

(b)

We usually like to write functions with the domain variable called x rather than y, so we reverse the roles of x and y and reformulate the above definition as follows.

DEFINITION 2

If f is one-to-one, then it has an **inverse function** f^{-1}. The value of $f^{-1}(x)$ is the unique number y in the domain of f for which $f(y) = x$. Thus,

$$y = f^{-1}(x) \iff x = f(y).$$

As observed above, $y = x^3$ is equivalent to $x = y^{1/3}$, or, reversing the roles of x and y,

$$y = x^{1/3} \iff x = y^3.$$

Example 1   Show that $f(x) = 2x - 1$ is one-to-one, and find its inverse $f^{-1}(x)$.

Solution   Since $f'(x) = 2 > 0$ on $\mathbb{R}$, f is increasing and therefore one-to-one there. Let $y = f^{-1}(x)$. Then

$$x = f(y) = 2y - 1.$$

Solving this equation for y gives $y = \dfrac{x + 1}{2}$. Thus $f^{-1}(x) = \dfrac{x + 1}{2}$.

There are several things you should remember about the relationship between a function f and its inverse f^{-1}. The most important one is that the two equations

$$y = f^{-1}(x) \qquad \text{and} \qquad x = f(y)$$

say the same thing. They are equivalent just as, for example, $y = x + 1$ and $x = y - 1$ are equivalent. Either of the equations can be replaced by the other. This implies that the domain of f^{-1} is the range of f, and vice versa.

The inverse of a one-to-one function is itself one-to-one and so also has an inverse. Not surprisingly, the inverse of f^{-1} is f:

$$y = (f^{-1})^{-1}(x) \quad \Longleftrightarrow \quad x = f^{-1}(y) \quad \Longleftrightarrow \quad y = f(x).$$

We can substitute either of the equations $y = f^{-1}(x)$ or $x = f(y)$ into the other and obtain the **cancellation identities**:

$$f\big(f^{-1}(x)\big) = x, \qquad f^{-1}\big(f(y)\big) = y.$$

The first of these identities holds for all x in the domain of f^{-1} and the second for all y in the domain of f. If S is any set of real numbers and I_S denotes the **identity function** on S, defined by

$$I_S(x) = x \quad \text{for all } x \text{ in } S,$$

then the cancellation identities say that if $\mathcal{D}(f)$ is the domain of f, then

$$f \circ f^{-1} = I_{\mathcal{D}(f^{-1})} \qquad \text{and} \qquad f^{-1} \circ f = I_{\mathcal{D}(f)},$$

where $f \circ g(x)$ denotes the composition $f\big(g(x)\big)$.

If the coordinates of a point $P = (a, b)$ are exchanged to give those of a new point $Q = (b, a)$, then each point is the reflection of the other in the line $x = y$. (To see this, note that the line PQ has slope -1, so it is perpendicular to $y = x$. Also, the midpoint of PQ is $\left(\frac{a+b}{2}, \frac{b+a}{2}\right)$, which lies on $y = x$.) It follows that the graphs of the equations $x = f(y)$ and $y = f(x)$ are reflections of each other in the line $x = y$. Since the equation $x = f(y)$ is equivalent to $y = f^{-1}(x)$, the graphs of the functions f^{-1} and f are reflections of each other in $y = x$. See Figure 3.3.

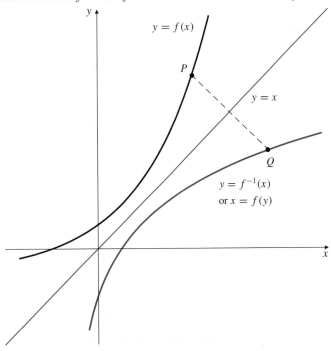

Figure 3.3   The graph of $y = f^{-1}(x)$ is the reflection of the graph of $y = f(x)$ in the line $y = x$

Here is a summary of the properties of inverse functions discussed above:

Properties of inverse functions

1. $y = f^{-1}(x) \iff x = f(y)$.
2. The domain of f^{-1} is the range of f.
3. The range of f^{-1} is the domain of f.
4. $f^{-1}(f(x)) = x$ for all x in the domain of f.
5. $f(f^{-1}(x)) = x$ for all x in the domain of f^{-1}.
6. $(f^{-1})^{-1}(x) = f(x)$ for all x in the domain of f.
7. The graph of f^{-1} is the reflection of the graph of f in the line $x = y$.

Example 2   Show that $g(x) = \sqrt{2x + 1}$ is invertible and find its inverse.

Solution   If $g(x_1) = g(x_2)$, then $\sqrt{2x_1 + 1} = \sqrt{2x_2 + 1}$. Squaring both sides we get $2x_1 + 1 = 2x_2 + 1$, which implies that $x_1 = x_2$. Thus, g is one-to-one and invertible. Let $y = g^{-1}(x)$; then

$$x = g(y) = \sqrt{2y + 1}.$$

It follows that $x \geq 0$ and $x^2 = 2y + 1$. Therefore, $y = \dfrac{x^2 - 1}{2}$ and

$$g^{-1}(x) = \frac{x^2 - 1}{2} \qquad \text{for } x \geq 0.$$

(The restriction $x \geq 0$ applies since the range of g is $[0, \infty[$.) See Figure 3.4(a) for the graphs of g and g^{-1}. ∎

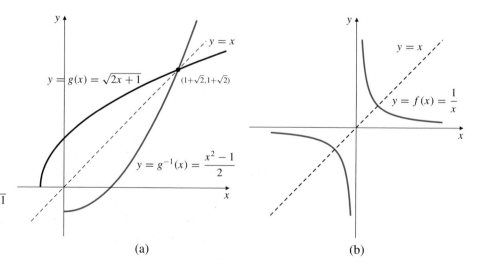

Figure 3.4

(a) The graphs of $g(x) = \sqrt{2x + 1}$ and its inverse

(b) The graph of the self-inverse function $f(x) = 1/x$

(a)

(b)

DEFINITION 3

A function f is **self-inverse** if $f^{-1} = f$, that is, if $f(f(x)) = x$ for every x in the domain of f.

Example 3 The function $f(x) = 1/x$ is self-inverse. If $y = f^{-1}(x)$, then $x = f(y) = 1/y$. Therefore $y = 1/x$, so $f^{-1}(x) = \dfrac{1}{x} = f(x)$. See Figure 3.4(b). The graph of any self-inverse function must be its own reflection in the line $x = y$ and must therefore be symmetric about that line.

◼

Inverting Non-One-to-One Functions

Many important functions such as the trigonometric functions are not one-to-one on their whole domains. It is still possible to define an inverse for such a function, but we have to restrict the domain of the function artificially so that the restricted function is one-to-one.

As an example, consider the function $f(x) = x^2$. Unrestricted, its domain is the whole real line and it is not one-to-one since $f(-a) = f(a)$ for any a. Let us define a new function $F(x)$ equal to $f(x)$ but having a smaller domain, so that it is one-to-one. We can use the interval $[0, \infty[$ as the domain of F:

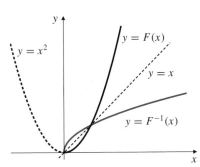

$$F(x) = x^2 \quad \text{for} \quad 0 \le x < \infty.$$

Figure 3.5 The restriction F of x^2 to $[0, \infty[$ and its inverse F^{-1}

The graph of F is shown in Figure 3.5; it is the right half of the parabola $y = x^2$, the graph of f. Evidently F is one-to-one, so it has an inverse F^{-1} which we calculate as follows:

Let $y = F^{-1}(x)$, then $x = F(y) = y^2$ and $y \ge 0$. Thus $y = \sqrt{x}$. Hence $F^{-1}(x) = \sqrt{x}$.

This method of restricting the domain of a non-one-to-one function to make it invertible will be used when we invert the trigonometric functions in Section 3.5.

Derivatives of Inverse Functions

Suppose that the function f is differentiable on an interval $]a, b[$ and that either $f'(x) > 0$ for $a < x < b$ (so that f is increasing on $]a, b[$) or $f'(x) < 0$ for $a < x < b$ (so that f is decreasing on $]a, b[$). In either case f is one-to-one on $]a, b[$ and has an inverse, f^{-1}, defined by

$$y = f^{-1}(x) \quad \Longleftrightarrow \quad x = f(y), \quad (a < y < b).$$

Since we are assuming that the graph $y = f(x)$ has a *nonhorizontal* tangent line at any x in $]a, b[$, its reflection, the graph $y = f^{-1}(x)$, has a *nonvertical* tangent line at any x in the interval between $f(a)$ and $f(b)$. Therefore, f^{-1} is differentiable at any such x. (See Figure 3.6.)

Let $y = f^{-1}(x)$. We want to find dy/dx. Solve the equation $y = f^{-1}(x)$ for $x = f(y)$ and differentiate implicitly with respect to x to obtain

$$1 = f'(y) \frac{dy}{dx}, \qquad \text{so} \qquad \frac{dy}{dx} = \frac{1}{f'(y)} = \frac{1}{f'\left(f^{-1}(x)\right)}.$$

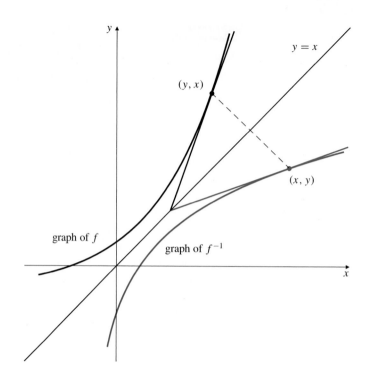

Figure 3.6   Tangents to the graphs of f and f^{-1}

Therefore, the slope of the graph of f^{-1} at (x, y) is the reciprocal of the slope of the graph of f at (y, x) (Figure 3.6) and

$$\frac{d}{dx} f^{-1}(x) = \frac{1}{f'\left(f^{-1}(x)\right)}.$$

In terms of Leibniz notation we have $\dfrac{dy}{dx}\bigg|_{x} = \dfrac{1}{\dfrac{dx}{dy}\bigg|_{y=f^{-1}(x)}}.$

Example 4   Show that $f(x) = x^3 + x$ is one-to-one on the whole real line, and, noting that $f(2) = 10$, find $\left(f^{-1}\right)'(10)$.

Solution   Since $f'(x) = 3x^2 + 1 > 0$ for all real numbers x, f is increasing and therefore one-to-one and invertible. If $y = f^{-1}(x)$, then

$$x = f(y) = y^3 + y \quad \Longrightarrow \quad 1 = (3y^2 + 1)y'$$

$$\Longrightarrow \quad y' = \frac{1}{3y^2 + 1}.$$

Now $x = f(2) = 10$ implies $y = f^{-1}(10) = 2$. Thus,

$$\left(f^{-1}\right)'(10) = \frac{1}{3y^2 + 1}\bigg|_{y=2} = \frac{1}{13}.$$

Exercises 3.1

Show that the functions f in Exercises 1–12 are one-to-one and calculate the inverse functions f^{-1}. Specify the domains and ranges of f and f^{-1}.

1. $f(x) = x - 1$

2. $f(x) = 2x - 1$

3. $f(x) = \sqrt{x - 1}$

4. $f(x) = -\sqrt{x - 1}$

5. $f(x) = x^3$

6. $f(x) = 1 + \sqrt[3]{x}$

7. $f(x) = x^2, \quad x \le 0$

8. $f(x) = (1 - 2x)^3$

9. $f(x) = \dfrac{1}{x + 1}$

10. $f(x) = \dfrac{x}{1 + x}$

11. $f(x) = \dfrac{1 - 2x}{1 + x}$

12. $f(x) = \dfrac{x}{\sqrt{x^2 + 1}}$

In Exercises 13–20, f is a one-to-one function with inverse f^{-1}. Calculate the inverses of the given functions in terms of f^{-1}.

13. $g(x) = f(x) - 2$

14. $h(x) = f(2x)$

15. $k(x) = -3f(x)$

16. $m(x) = f(x - 2)$

17. $p(x) = \dfrac{1}{1 + f(x)}$

18. $q(x) = \dfrac{f(x) - 3}{2}$

19. $r(x) = 1 - 2f(3 - 4x)$

20. $s(x) = \dfrac{1 + f(x)}{1 - f(x)}$

In Exercises 21–23, show that the given function is one-to-one and find its inverse.

21. $f(x) = \begin{cases} x^2 + 1 & \text{if } x \ge 0 \\ x + 1 & \text{if } x < 0 \end{cases}$

22. $g(x) = \begin{cases} x^3 & \text{if } x \ge 0 \\ x^{1/3} & \text{if } x < 0 \end{cases}$

23. $h(x) = x|x| + 1$

24. Find $f^{-1}(2)$ if $f(x) = x^3 + x$.

25. Find $g^{-1}(1)$ if $g(x) = x^3 + x - 9$.

26. Find $h^{-1}(-3)$ if $h(x) = x|x| + 1$.

27. Assume that the function $f(x)$ satisfies $f'(x) = \dfrac{1}{x}$ and that f is one-to-one. If $y = f^{-1}(x)$, show that $dy/dx = y$.

28. Show that $f(x) = \dfrac{4x^3}{x^2 + 1}$ has an inverse and find $\left(f^{-1}\right)'(2)$.

29. Find $\left(f^{-1}\right)'(x)$ if $f(x) = 1 + 2x^3$.

∗ 30. Find $\left(f^{-1}\right)'(-2)$ if $f(x) = x\sqrt{3 + x^2}$.

31. If $f(x) = x^2/(1 + \sqrt{x})$, find $f^{-1}(2)$ correct to 5 decimal places.

32. If $g(x) = 2x + \sin x$, show that g is invertible, and find $g^{-1}(2)$ and $(g^{-1})'(2)$ correct to five decimal places.

33. Show that $f(x) = x \sec x$ is one-to-one on $]-\pi/2, \pi/2[$. What is the domain of $f^{-1}(x)$? Find $(f^{-1})'(0)$.

34. If f and g have respective inverses f^{-1} and g^{-1}, show that the composite function $f \circ g$ has inverse $(f \circ g)^{-1} = g^{-1} \circ f^{-1}$.

∗ 35. For what values of the constants a, b, and c is the function $f(x) = (x - a)/(bx - c)$ self-inverse?

∗ 36. Can an even function be self-inverse? an odd function?

∗ 37. In this section it was claimed that an increasing (or decreasing) function defined on a single interval is necessarily one-to-one. Is the converse of this statement true? Explain.

∗ 38. Repeat the previous exercise with the added assumption that f is continuous on the interval where it is defined.

3.2 Exponential and Logarithmic Functions

This section reviews exponential and logarithmic functions in a form that you are likely to have encountered in your previous mathematical studies. In the following sections we will approach these functions from a different point of view and learn how to find their derivatives.

Exponentials

An **exponential function** is a function of the form $f(x) = a^x$, where the **base** a is a positive constant and the **exponent** x is the variable. Do not confuse such functions with **power** functions like $f(x) = x^a$, where the base is variable and the exponent is constant. The exponential function a^x can be defined for integer and rational exponents x as follows:

DEFINITION **4**

> **Exponential Functions**
>
> If $a > 0$, then
>
> $$a^0 = 1$$
> $$a^n = \underbrace{a \cdot a \cdot a \cdots a}_{n \text{ factors}} \qquad \text{if } n = 1, 2, 3, \ldots$$
> $$a^{-n} = \frac{1}{a^n} \qquad \text{if } n = 1, 2, 3, \ldots$$
> $$a^{m/n} = \sqrt[n]{a^m} \qquad \text{if } n = 1, 2, 3, \ldots \quad \text{and } m = \pm 1, \pm 2, \pm 3, \ldots.$$
>
> In this definition, $\sqrt[n]{a}$ is the number $b > 0$ that satisfies $b^n = a$.

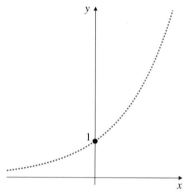

Figure 3.7 $y = 2^r$ for rational r

How should we define a^x if x is not rational? For example, what does 2^π mean? In order to calculate a derivative of a^x we will want the function to be defined for all real numbers x, not just rational ones.

In Figure 3.7 we plot points with coordinates $(x, 2^x)$ for many closely spaced rational values of x. They appear to lie on a smooth curve. The definition of a^x can be extended to irrational x in such a way that a^x becomes a differentiable function of x on the whole real line. We will do so in the next section. For the moment, if x is irrational we can regard a^x as being the limit of values a^r for rational numbers r approaching x:

$$a^x = \lim_{\substack{r \to x \\ r \text{ rational}}} a^r.$$

Example 1   Since the irrational number $\pi = 3.14159265359\ldots$ is the limit of the sequence of rational numbers

$$r_1 = 3, \quad r_2 = 3.1, \quad r_3 = 3.14, \quad r_4 = 3.141, \quad r_5 = 3.1415, \quad \ldots,$$

we can calculate 2^π as the limit of the corresponding sequence

$$2^3 = 8, \quad 2^{3.1} = 8.5741877\ldots, \quad 2^{3.14} = 8.8152409\ldots.$$

This gives $2^\pi = \lim_{n \to \infty} 2^{r_n} = 8.824977827\ldots.$

Exponential functions satisfy several identities called *laws of exponents*:

> **Laws of exponents**
>
> If $a > 0$ and $b > 0$, and x and y are any real numbers, then
>
> (i) $a^0 = 1$ (ii) $a^{x+y} = a^x a^y$
>
> (iii) $a^{-x} = \dfrac{1}{a^x}$ (iv) $a^{x-y} = \dfrac{a^x}{a^y}$
>
> (v) $(a^x)^y = a^{xy}$ (vi) $(ab)^x = a^x b^x$

These identities can be proved for rational exponents using the definitions above. They remain true for irrational exponents, but we can't show that until the next section.

If $a = 1$, then $a^x = 1^x = 1$ for every x. If $a > 1$, then a^x is an increasing function of x; if $0 < a < 1$, then a^x is decreasing. The graphs of some typical exponential functions are shown in Figure 3.8(a). They all pass through the point $(0,1)$ since $a^0 = 1$ for every $a > 0$. Observe that $a^x > 0$ for all $a > 0$ and all real x and that

$$\text{If} \quad a > 1, \qquad \text{then} \qquad \lim_{x \to -\infty} a^x = 0 \quad \text{and} \quad \lim_{x \to \infty} a^x = \infty.$$

$$\text{If} \quad 0 < a < 1, \text{ then} \qquad \lim_{x \to -\infty} a^x = \infty \quad \text{and} \quad \lim_{x \to \infty} a^x = 0.$$

The graph of $y = a^x$ has the x-axis as a horizontal asymptote if $a \neq 1$. It is asymptotic on the left (as $x \to -\infty$) if $a > 1$ and on the right (as $x \to \infty$) if $0 < a < 1$.

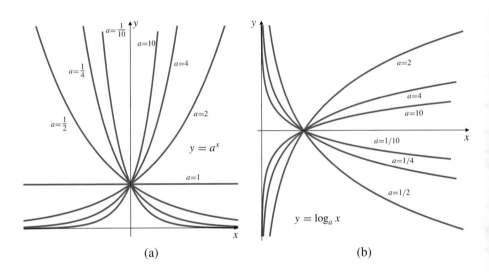

Figure 3.8

(a) Graphs of some exponential functions

(b) Graphs of some logarithmic functions

Logarithms

The function $f(x) = a^x$ is a one-to-one function provided that $a > 0$ and $a \neq 1$. Therefore, f has an inverse which we call a *logarithmic function*.

DEFINITION 5

Logarithms

If $a > 0$ and $a \neq 1$, the function $\log_a x$, called **the logarithm of x to the base a**, is the inverse of the one-to-one function a^x:

$$y = \log_a x \quad \Longleftrightarrow \quad x = a^y, \qquad (a > 0, \quad a \neq 1).$$

Since a^x has domain $]-\infty, \infty[$, $\log_a x$ has range $\infty, \infty[$. Since a^x has range $]0, \infty[$, $\log_a x$ has domain $]0, \infty[$. Since a^x and $\log_a x$ are inverse functions, the following **cancellation identities** hold:

$$\log_a\left(a^x\right) = x \qquad \text{for all real } x$$
$$a^{\log_a x} = x \qquad \text{for all} \quad x > 0.$$

The graphs of some typical logarithmic functions are shown in Figure 3.8(b). They all pass through the point $(1, 0)$. Each graph is the reflection in the line $y = x$ of the corresponding exponential graph in Figure 3.8(a).

Corresponding to the laws of exponents we have several identities involving logarithms:

Laws of logarithms

If $x > 0$, $y > 0$, $a > 0$, $b > 0$, $a \neq 1$, and $b \neq 1$, then

(i) $\log_a 1 = 0$ (ii) $\log_a (xy) = \log_a x + \log_a y$

(iii) $\log_a \left(\dfrac{1}{x}\right) = -\log_a x$ (iv) $\log_a \left(\dfrac{x}{y}\right) = \log_a x - \log_a y$

(v) $\log_a \left(x^y\right) = y \log_a x$ (vi) $\log_a x = \dfrac{\log_b x}{\log_b a}$

These identities can be derived from the laws of exponents.

Example 2   If $a > 0$, $x > 0$, and $y > 0$, verify that $\log_a (xy) = \log_a x + \log_a y$, using laws of exponents.

Solution   Let $u = \log_a x$ and $v = \log_a y$. By the defining property of inverse functions, $x = a^u$ and $y = a^v$. Thus $xy = a^u a^v = a^{u+v}$. Inverting again, we get $\log_a (xy) = u + v = \log_a x + \log_a y$.

Logarithm law (vi) presented above shows that if you know logarithms to a particular base b, you can calculate logarithms to any other base a. Scientific calculators usually have built-in programs for calculating logarithms to base 10 and to base e, a special number that we will discover in Section 3.3. Logarithms to any base can be calculated using either of these functions. For example, computer scientists sometimes need to use logarithms to base 2. Using a scientific calculator, you can readily calculate

$$\log_2 13 = \frac{\log_{10} 13}{\log_{10} 2} = \frac{1.11394335231\ldots}{0.301029995664\ldots} = 3.70043971814\ldots.$$

The laws of logarithms can sometimes be used to simplify complicated expressions.

Example 3   Simplify
(a) $\log_2 10 + \log_2 12 - \log_2 15$,   (b) $\log_{a^2} a^3$,   and   (c) $3^{\log_9 4}$.

Solution

(a) $\log_2 10 + \log_2 12 - \log_2 15 = \log_2 \dfrac{10 \times 12}{15}$ (laws (ii) and (iv))

$= \log_2 8$

$= \log_2 2^3 = 3.$ (cancellation identity)

(b) $\log_{a^2} a^3 = 3 \log_{a^2} a$ (law (v))

$\qquad\qquad = \dfrac{3}{2} \log_{a^2} a^2$ (law (v) again)

$\qquad\qquad = \dfrac{3}{2}.$ (cancellation identity)

(c) $3^{\log_9 4} = 3^{(\log_3 4)/(\log_3 9)}$ (law (vi))

$\qquad\qquad = \left(e^{\log_3 4} \right)^{1/\log_3 9}$

$\qquad\qquad = 4^{1/\log_3 3^2} = 4^{1/2} = 2.$ (cancellation identity)

■

Example 4   Solve the equation $3^{x-1} = 2^x$.

Solution   We can take logarithms of both sides of the equation to any base a and get

$$(x - 1) \log_a 3 = x \log_a 2$$
$$(\log_a 3 - \log_a 2)x = \log_a 3$$
$$x = \frac{\log_a 3}{\log_a 3 - \log_a 2} = \frac{\log_a 3}{\log_a (3/2)}.$$

The numerical value of x can be found using the "log" function on a scientific calculator. (This function is $\log_{10}$.) The value is $x = 2.7095 \ldots$.

■

Corresponding to the asymptotic behaviour of the exponential functions, the logarithmic functions also exhibit asymptotic behaviour. Their graphs are all asymptotic to the y-axis as $x \to 0$ from the right:

If $a > 1$,   then $\displaystyle\lim_{x \to 0+} \log_a x = -\infty$ and $\displaystyle\lim_{x \to \infty} \log_a x = \infty$.

If $0 < a < 1$, then $\displaystyle\lim_{x \to 0+} \log_a x = \infty$ and $\displaystyle\lim_{x \to \infty} \log_a x = -\infty$.

Exercises 3.2

Simplify the expressions in Exercises 1–18.

1. $\dfrac{3^3}{\sqrt{3^5}}$

2. $2^{1/2} 8^{1/2}$

3. $\left(x^{-3} \right)^{-2}$

4. $\left(\dfrac{1}{2} \right)^x 4^{x/2}$

5. $\log_5 125$

6. $\log_4 \left(\dfrac{1}{8} \right)$

7. $\log_{1/3} 3^{2x}$

8. $2^{\log_4 8}$

9. $10^{-\log_{10}(1/x)}$

10. $x^{1/(\log_a x)}$

11. $(\log_a b)(\log_b a)$

12. $\log_x \left(x(\log_y y^2) \right)$

13. $(\log_4 16)(\log_4 2)$

14. $\log_{15} 75 + \log_{15} 3$

15. $\log_6 9 + \log_6 4$

16. $2 \log_3 12 - 4 \log_3 6$

17. $\log_a (x^4 + 3x^2 + 2) + \log_a (x^4 + 5x^2 + 6)$
$\quad -4 \log_a \sqrt{x^2 + 2}$

18. $\log_\pi (1 - \cos x) + \log_\pi (1 + \cos x) - 2 \log_\pi \sin x$

Use the base 10 exponential and logarithm functions 10^x and $\log x$ ($= \log_{10} x$) on a scientific calculator to evaluate the expressions or solve the equations in Exercises 19–24.

19. $3^{\sqrt{2}}$

20. $\log_3 5$

21. $2^{2x} = 5^{x+1}$

22. $x^{\sqrt{2}} = 3$

23. $\log_x 3 = 5$

24. $\log_3 x = 5$

Use the laws of exponents to prove the laws of logarithms in Exercises 25–28.

25. $\log_a \left(\dfrac{1}{x} \right) = -\log_a x$

26. $\log_a \left(\dfrac{x}{y} \right) = \log_a x - \log_a y$

27. $\log_a (x^y) = y \log_a x$

28. $\log_a x = (\log_b x)/(\log_b a)$

29. Solve $\log_4 (x + 4) - 2 \log_4 (x + 1) = \dfrac{1}{2}$ for x.

30. Solve $2 \log_3 x + \log_9 x = 10$ for x.

Evaluate the limits in Exercises 31–34.

31. $\lim\limits_{x \to \infty} \log_x 2$

32. $\lim\limits_{x \to 0+} \log_x (1/2)$

33. $\lim\limits_{x \to 1+} \log_x 2$

34. $\lim\limits_{x \to 1-} \log_x 2$

* 35. Suppose that $f(x) = a^x$ is differentiable at $x = 0$ and that $f'(0) = k$, where $k \neq 0$. Prove that f is differentiable at any real number x and that

$$f'(x) = k \, a^x = k \, f(x).$$

* 36. Continuing Exercise 35, prove that $f^{-1}(x) = \log_a x$ is differentiable at any $x > 0$ and that

$$(f^{-1})'(x) = \dfrac{1}{kx}.$$

3.3 The Natural Logarithm and Exponential

In this section we are going to define a function $\ln x$, called the *natural* logarithm of x, in a way that does not at first seem to have anything to do with the logarithms considered in Section 3.2. We will show, however, that it has the same properties as those logarithms, and in the end we will see that $\ln x = \log_e x$, the logarithm of x to a certain specific base e. We will show that $\ln x$ is a one-to-one function, defined for all positive real numbers. It must therefore have an inverse, e^x, that we will call *the* exponential function. Our final goal is to arrive at a definition of the exponential functions a^x (for any $a > 0$) that is valid for any real number x instead of just rational numbers, and that is known to be continuous and even differentiable without our having to assume those properties as we did in Section 3.2.

The Natural Logarithm

Table 1 lists the derivatives of integer powers of x. Those derivatives are multiples of integer powers of x, but one integer power, x^{-1}, is conspicuously absent from the list of derivatives; we do not yet know a function whose derivative is $x^{-1} = 1/x$. We are going to remedy this situation by defining a function $\ln x$ in such a way that it will have derivative $1/x$.

To get a hint as to how this can be done, review Example 1 of Section 2.11. In that example we showed that the area under the graph of the velocity of a moving object in a time interval was equal to the distance travelled by the object in that time interval. Since the derivative of distance is velocity, measuring the area provided a way of finding a function (the distance) that had a given derivative (the velocity). This relationship between area and derivatives is one of the most important ideas in calculus. It is called the **Fundamental Theorem of Calculus**. We will explore it fully in Chapter 5, but we will make use of the idea now to define $\ln x$, which we want to have derivative $1/x$.

Table 1. Derivatives of integer powers

$f(x)$	$f'(x)$
$\vdots$	$\vdots$
x^3	$3x^2$
x^2	$2x$
x^1	$x^0 = 1$
x^0	0
x^{-1}	$-x^{-2}$
x^{-2}	$-2x^{-3}$
x^{-3}	$-3x^{-4}$
$\vdots$	$\vdots$

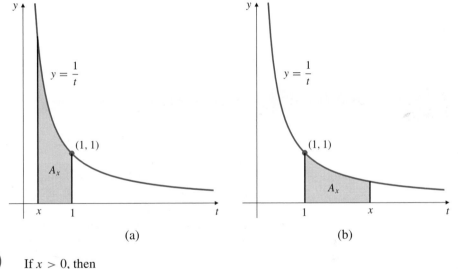

DEFINITION 6

The natural logarithm

For $x > 0$, let A_x be the area of the plane region bounded by the curve $y = 1/t$, the t-axis, and the vertical lines $t = 1$ and $t = x$. The function $\ln x$ is defined by

$$\ln x = \begin{cases} A_x & \text{if } x \geq 1, \\ -A_x & \text{if } 0 < x < 1, \end{cases}$$

as shown in Figure 3.9.

The definition implies that $\ln 1 = 0$, that $\ln x > 0$ if $x > 1$, that $\ln x < 0$ if $0 < x < 1$, and that $\ln$ is a one-to-one function. We now show that if $y = \ln x$, then $y' = 1/x$. The proof of this result is similar to the proof we will give for the Fundamental Theorem of Calculus in Section 5.5.

Figure 3.9

(a) $\ln x = -$area A_x if $0 < x < 1$

(b) $\ln x = $ area A_x if $x \geq 1$

(a)　　　　(b)

THEOREM 1

If $x > 0$, then

$$\frac{d}{dx} \ln x = \frac{1}{x}.$$

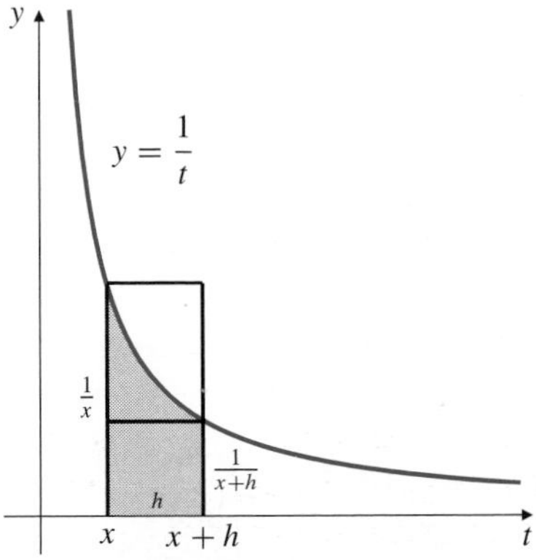

PROOF For $x > 0$ and $h > 0$, $\ln(x + h) - \ln x$ is the area of the plane region bounded by $y = 1/t$, $y = 0$, and the vertical lines $t = x$ and $t = x + h$; it is the shaded area in Figure 3.10. Comparing this area with that of two rectangles, we see that

$$\frac{h}{x + h} < \text{shaded area} = \ln(x + h) - \ln x < \frac{h}{x}.$$

Hence the Newton quotient for $\ln x$ satisfies

$$\frac{1}{x + h} < \frac{\ln(x + h) - \ln x}{h} < \frac{1}{x}.$$

Letting h approach 0 from the right, we obtain (by the Squeeze Theorem applied to one-sided limits)

$$\lim_{h \to 0+} \frac{\ln(x + h) - \ln x}{h} = \frac{1}{x}.$$

Figure 3.10

A similar argument shows that if $0 < x + h < x$, then

$$\frac{1}{x} < \frac{\ln(x + h) - \ln x}{h} < \frac{1}{x + h},$$

so that

$$\lim_{h \to 0-} \frac{\ln(x + h) - \ln x}{h} = \frac{1}{x}.$$

Combining these two one-sided limits we get the desired result:

$$\frac{d}{dx} \ln x = \lim_{h \to 0} \frac{\ln(x + h) - \ln x}{h} = \frac{1}{x}.$$

The two properties $(d/dx) \ln x = 1/x$ and $\ln 1 = 0$ are sufficient to determine the function $\ln x$ completely. We can deduce from these two properties that $\ln x$ satisfies the appropriate laws of logarithms:

THEOREM 2

Properties of the natural logarithm

(i) $\ln(xy) = \ln x + \ln y$ (ii) $\ln\left(\dfrac{1}{x}\right) = -\ln x$

(iii) $\ln\left(\dfrac{x}{y}\right) = \ln x - \ln y$ (iv) $\ln\left(x^r\right) = r \ln x$

Because we do not want to *assume* that exponentials are continuous (as we did in Section 3.2), we should regard (iv) for the moment as only valid for exponents r that are rational numbers.

PROOF   We will only prove part (i) because the other parts are proved by the same method. If $y > 0$ is a constant, then by the Chain Rule,

$$\frac{d}{dx}\big(\ln(xy) - \ln x\big) = \frac{y}{xy} - \frac{1}{x} = 0 \quad \text{for all } x > 0.$$

Theorem 13 of Section 2.6 now tells us that $\ln(xy) - \ln x = C$ (a constant) for $x > 0$. Putting $x = 1$ we get $C = \ln y$ and identity (i) follows.

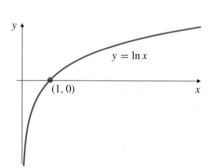

$y = \ln x$

$(1, 0)$

Figure 3.11   The graph of $\ln x$

Part (iv) of Theorem 2 shows that $\ln(2^n) = n \ln 2 \to \infty$ as $n \to \infty$. Therefore, we also have $\ln(1/2)^n = -n \ln 2 \to -\infty$ as $n \to \infty$. Since $(d/dx) \ln x = 1/x > 0$ for $x > 0$, it follows that $\ln x$ is increasing, so we must have (see Figure 3.11)

$$\lim_{x \to \infty} \ln x = \infty, \qquad \lim_{x \to 0+} \ln x = -\infty.$$

Example 1 Show that $\dfrac{d}{dx} \ln |x| = \dfrac{1}{x}$ for any $x \neq 0$. Hence find $\displaystyle\int \dfrac{1}{x}\, dx$.

Solution If $x > 0$, then

$$\frac{d}{dx} \ln |x| = \frac{d}{dx} \ln x = \frac{1}{x}$$

by Theorem 1. If $x < 0$, then, using the Chain Rule,

$$\frac{d}{dx} \ln |x| = \frac{d}{dx} \ln(-x) = \frac{1}{-x}(-1) = \frac{1}{x}.$$

Therefore, $\dfrac{d}{dx} \ln |x| = \dfrac{1}{x}$, and on any interval not containing $x = 0$,

$$\int \frac{1}{x}\, dx = \ln |x| + C.$$

Example 2 Find the derivatives of (a) $\ln |\cos x|$, and (b) $\ln(x + \sqrt{x^2 + 1})$. Simplify your answers as much as possible.

Solution

(a) Using the result of Example 1 and the Chain Rule, we have

$$\frac{d}{dx} \ln |\cos x| = \frac{1}{\cos x}(-\sin x) = -\tan x.$$

(b) $\dfrac{d}{dx} \ln(x + \sqrt{x^2 + 1}) = \dfrac{1}{x + \sqrt{x^2 + 1}} \left(1 + \dfrac{2x}{2\sqrt{x^2 + 1}}\right)$

$$= \frac{1}{x + \sqrt{x^2 + 1}} \frac{\sqrt{x^2 + 1} + x}{\sqrt{x^2 + 1}}$$

$$= \frac{1}{\sqrt{x^2 + 1}}.$$

The Exponential Function

The function $\ln x$ is one-to-one on its domain, the interval $]0, \infty[$, so it has an inverse there. For the moment, let us call this inverse $\exp x$. Thus

$$y = \exp x \quad \Longleftrightarrow \quad x = \ln y \quad (y > 0).$$

Since $\ln 1 = 0$, we have $\exp 0 = 1$. The domain of $\exp$ is $]-\infty, \infty[$, the range of $\ln$. The range of $\exp$ is $]0, \infty[$, the domain of $\ln$. We have cancellation identities

$$\ln(\exp x) = x \quad \text{for all real } x \quad \text{and} \quad \exp(\ln x) = x \quad \text{for } x > 0.$$

We can deduce various properties of $\exp$ from corresponding properties of $\ln$. Not surprisingly, they are properties we would expect an exponential function to have.

THEOREM 3

Properties of the exponential function

(i) $(\exp x)^r = \exp(rx)$

(ii) $\exp(x+y) = (\exp x)(\exp y)$

(iii) $\exp(-x) = \dfrac{1}{\exp(x)}$

(iv) $\exp(x-y) = \dfrac{\exp x}{\exp y}$

For the moment, identity (i) is asserted only for rational numbers r.

PROOF We prove only identity (i); the rest are done similarly. If $u = (\exp x)^r$, then, by Theorem 2(iv), $\ln u = r \ln(\exp x) = rx$. Therefore $u = \exp(rx)$.

Now we make an important definition!

Let $e = \exp(1)$.

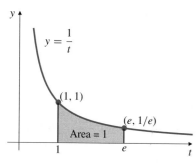

Figure 3.12 The definition of e

The number e satisfies $\ln e = 1$, so the area bounded by the curve $y = 1/t$, the t-axis, and the vertical lines $t = 1$ and $t = e$ must be equal to 1 square unit. See Figure 3.12. The number e is one of the most important numbers in mathematics. Like π, it is irrational and not a zero of any polynomial with rational coefficients. (Such numbers are called **transcendental**.) Its value is between 2 and 3 and begins

$$e = 2.7\ 1828\ 1828\ 45\ 90\ 45 \ldots.$$

Later on we will learn that

$$e = 1 + \frac{1}{1!} + \frac{1}{2!} + \frac{1}{3!} + \frac{1}{4!} + \cdots,$$

a formula from which the value of e can be calculated to any desired precision.

Theorem 3(i) shows that $\exp r = \exp(1r) = (\exp 1)^r = e^r$ holds for any rational number r. Now here is a crucial observation. We only know what e^r means if r is a rational number (if $r = m/n$, then $e^r = \sqrt[n]{e^m}$). But $\exp x$ is defined for all *real x*, rational or not. Since $e^r = \exp r$ when r is rational, we can use $\exp x$ as a *definition* of what e^x means for any real number x, and there will be no contradiction if x happens to be rational.

$$e^x = \exp x \qquad \text{for all real } x.$$

Theorem 3 can now be restated in terms of e^x:

(i) $(e^x)^y = e^{xy}$

(ii) $e^{x+y} = e^x\, e^y$

(iii) $e^{-x} = \dfrac{1}{e^x}$

(iv) $e^{x-y} = \dfrac{e^x}{e^y}$

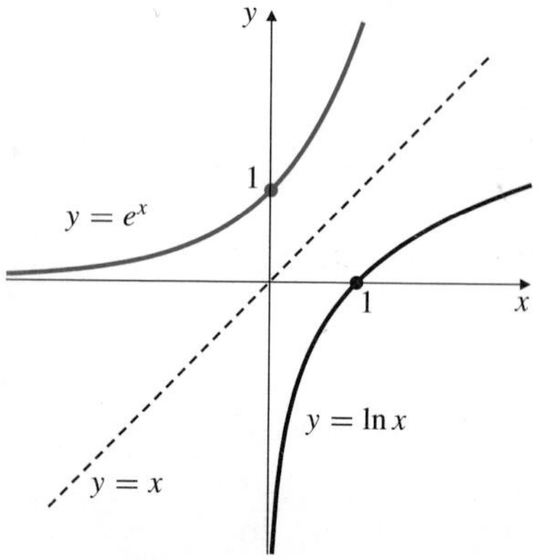

Figure 3.13 The graphs of e^x and $\ln x$

The graph of e^x is the reflection of the graph of its inverse, $\ln x$, in the line $x = y$. Both graphs are shown for comparison in Figure 3.13. Observe that the x-axis is a horizontal asymptote of the graph of $y = e^x$ as $x \to -\infty$. We have

$$\lim_{x \to -\infty} e^x = 0, \qquad \lim_{x \to \infty} e^x = \infty.$$

Since $\exp x = e^x$ actually *is* an exponential function, its inverse must actually *be* a logarithm:

$$\ln x = \log_e x.$$

The derivative of $y = e^x$ is calculated by implicit differentiation:

$$y = e^x \quad \Longrightarrow \quad x = \ln y$$
$$\Longrightarrow \quad 1 = \frac{1}{y} \frac{dy}{dx}$$
$$\Longrightarrow \quad \frac{dy}{dx} = y = e^x.$$

Thus, the exponential function has the remarkable property that it is its own derivative and, therefore, also its own antiderivative:

$$\frac{d}{dx} e^x = e^x, \qquad \int e^x \, dx = e^x + C.$$

Example 3 Find the derivatives of

(a) e^{x^2-3x}, (b) $\sqrt{1 + e^{2x}}$, and (c) $\dfrac{e^x - e^{-x}}{e^x + e^{-x}}$.

Solution

(a) $\dfrac{d}{dx} e^{x^2-3x} = e^{x^2-3x}(2x - 3) = (2x - 3)e^{x^2-3x}$.

(b) $\dfrac{d}{dx} \sqrt{1 + e^{2x}} = \dfrac{1}{2\sqrt{1 + e^{2x}}} \left(e^{2x}(2)\right) = \dfrac{e^{2x}}{\sqrt{1 + e^{2x}}}$.

(c) $\dfrac{d}{dx} \dfrac{e^x - e^{-x}}{e^x + e^{-x}} = \dfrac{(e^x + e^{-x})(e^x - (-e^{-x})) - (e^x - e^{-x})(e^x + (-e^{-x}))}{(e^x + e^{-x})^2}$

$\qquad = \dfrac{(e^x)^2 + 2e^x e^{-x} + (e^{-x})^2 - [(e^x)^2 - 2e^x e^{-x} + (e^{-x})^2]}{(e^x + e^{-x})^2}$

$\qquad = \dfrac{4e^{x-x}}{(e^x + e^{-x})^2} = \dfrac{4}{(e^x + e^{-x})^2}$.

Example 4 Let $f(t) = e^{at}$. Find (a) $f^{(n)}(t)$ and (b) $\int f(t) \, dt$.

Solution

(a) We have $f'(t) = a \, e^{at}$
$$f''(t) = a^2 \, e^{at}$$
$$f'''(t) = a^3 \, e^{at}$$
$$\vdots$$
$$f^{(n)}(t) = a^n \, e^{at}.$$

(b) Also, $\int f(t)\,dt = \int e^{at}\,dt = \dfrac{1}{a}e^{at} + C$, since $\dfrac{d}{dt}\dfrac{1}{a}e^{at} = e^{at}$. ∎

General Exponentials and Logarithms

We can use the fact that e^x is now defined for *all* real x to define the arbitrary exponential a^x (where $a > 0$) for all real x. If r is rational, then $\ln(a^r) = r \ln a$; therefore $a^r = e^{r \ln a}$. However, $e^{x \ln a}$ is defined for all real x, so we can use it as a definition of a^x with no possibility of contradiction arising if x is rational.

DEFINITION **7**

> **The general exponential a^x**
>
> $$a^x = e^{x \ln a}, \qquad (a > 0, \quad x \text{ real}).$$

Example 5 Evaluate 2^π, using the natural logarithm (ln) and exponential (exp or e^x) keys on a scientific calculator, but not using the y^x or $\hat{\ }$ keys.

Solution $2^\pi = e^{\pi \ln 2} = 8.8249778\ldots$. If your calculator has a $\hat{\ }$ key, or an x^y or y^x key, the chances are that it is implemented in terms of the exp and ln functions. ∎

The laws of exponents for a^x as presented in Section 3.2 can now be obtained from those for e^x, as can the derivative:

$$\frac{d}{dx}a^x = \frac{d}{dx}e^{x \ln a} = e^{x \ln a}\ln a = a^x \ln a.$$

We can also verify the General Power Rule for x^a, where a is any real number, provided $x > 0$:

$$\frac{d}{dx}x^a = \frac{d}{dx}e^{a \ln x} = e^{a \ln x}\frac{a}{x} = \frac{a\,x^a}{x} = a\,x^{a-1}.$$

Example 6 Show that the graph of $f(x) = x^\pi - \pi^x$ has a negative slope at $x = \pi$.

Do not confuse x^π, which is a power function of x, and π^x, which is an exponential function of x.

Solution $f'(x) = \pi\,x^{\pi-1} - \pi^x \ln \pi$

$$f'(\pi) = \pi\,\pi^{\pi-1} - \pi^\pi \ln \pi = \pi^\pi(1 - \ln \pi).$$

Since $\pi > 3 > e$, we have $\ln \pi > \ln e = 1$, so $1 - \ln \pi < 0$. Since $\pi^\pi = e^{\pi \ln \pi} > 0$, we have $f'(\pi) < 0$. Thus, the graph $y = f(x)$ has negative slope at $x = \pi$. ∎

Example 7 Find the critical point of $y = x^x$.

Solution We can't differentiate x^x by treating it as a power (like x^a) because the exponent varies. We can't treat it as an exponential (like a^x) because the base varies. We can differentiate it if we first write it in terms of the exponential function, $x^x = e^{x \ln x}$, and then use the Chain Rule and the Product Rule:

$$\frac{dy}{dx} = \frac{d}{dx} e^{x \ln x} = e^{x \ln x} \left(\ln x + x \left(\frac{1}{x} \right) \right) = x^x (1 + \ln x).$$

Now x^x is defined only for $x > 0$, and is itself never 0. (Why?) Therefore the critical point occurs where $1 + \ln x = 0$, that is, $\ln x = -1$, or $x = 1/e$. ∎

Finally, observe that $(d/dx)a^x = a^x \ln a$ is negative for all x if $0 < a < 1$ and is positive for all x if $a > 1$. Thus, a^x is one-to-one and has an inverse function, $\log_a x$, provided $a > 0$ and $a \ne 1$. Its properties follow in the same way as in Section 3.2. If $y = \log_a x$, then $x = a^y$ and, differentiating implicitly with respect to x, we get·

$$1 = a^y \ln a \frac{dy}{dx} = x \ln a \frac{dy}{dx}.$$

Thus, the derivative of $\log_a x$ is given by

$$\frac{d}{dx} \log_a x = \frac{1}{x \ln a}.$$

Since $\log_a x$ can be expressed in terms of logarithms to any other base, say e,

$$\log_a x = \frac{\ln x}{\ln a},$$

we normally use only natural logarithms. Exceptions are found in chemistry, acoustics, and other sciences where "logarithmic scales" are used to measure quantities for which a one unit increase in the measure corresponds to a tenfold increase in the quantity. Logarithms to base 10 are used in defining such scales. In computer science, where powers of 2 play a central role, logarithms to base 2 are often encountered.

Logarithmic Differentiation

Suppose we want to differentiate a function of the form

$$y = (f(x))^{g(x)} \qquad \text{(for } f(x) > 0\text{)}.$$

Since the variable appears in both the base and the exponent, neither the general power rule, $(d/dx)x^a = ax^{a-1}$, nor the exponential rule, $(d/dx)a^x = a^x \ln a$, can be directly applied. One method for finding the derivative of such a function is to express it in the form

$$y = e^{g(x) \ln f(x)}$$

and then differentiate, using the Product Rule to handle the exponent. This is the method used in Example 7.

The derivative in Example 7 can also be obtained by taking natural logarithms of both sides of the equation $y = x^x$ and differentiating implicitly:

$$\ln y = x \ln x$$
$$\frac{1}{y} \frac{dy}{dx} = \ln x + \frac{x}{x} = 1 + \ln x$$
$$\frac{dy}{dx} = y(1 + \ln x) = x^x(1 + \ln x).$$

This latter technique is called **logarithmic differentiation**.

Example 8 Find dy/dt if $y = (\sin t)^{\ln t}$, where $0 < t < \pi$.

Solution We have $\ln y = \ln t \, \ln \sin t$. Thus

$$\frac{1}{y}\frac{dy}{dt} = \frac{1}{t}\ln \sin t + \ln t \frac{\cos t}{\sin t}$$

$$\frac{dy}{dt} = y\left(\frac{\ln \sin t}{t} + \ln t \, \cot t\right)$$

$$= (\sin t)^{\ln t}\left(\frac{\ln \sin t}{t} + \ln t \, \cot t\right).$$

Logarithmic differentiation is also useful for finding the derivatives of functions expressed as products and quotients of many factors. Taking logarithms reduces these products and quotients to sums and differences. This usually makes the calculation easier than it would be using the Product and Quotient Rules, especially if the derivative is to be evaluated at a specific point.

Example 9 Differentiate $y = [(x + 1)(x + 2)(x + 3)]/(x + 4)$.

Solution $\ln|y| = \ln|x + 1| + \ln|x + 2| + \ln|x + 3| - \ln|x + 4|$. Thus,

$$\frac{1}{y}y' = \frac{1}{x + 1} + \frac{1}{x + 2} + \frac{1}{x + 3} - \frac{1}{x + 4}$$

$$y' = \frac{(x + 1)(x + 2)(x + 3)}{x + 4}\left(\frac{1}{x + 1} + \frac{1}{x + 2} + \frac{1}{x + 3} - \frac{1}{x + 4}\right)$$

$$= \frac{(x + 2)(x + 3)}{x + 4} + \frac{(x + 1)(x + 3)}{x + 4} + \frac{(x + 1)(x + 2)}{x + 4}$$

$$- \frac{(x + 1)(x + 2)(x + 3)}{(x + 4)^2}.$$

Example 10 Find $\dfrac{du}{dx}\bigg|_{x=1}$ if $u = \sqrt{(x + 1)(x^2 + 1)(x^3 + 1)}$.

Solution

$$\ln u = \frac{1}{2}\left(\ln(x + 1) + \ln(x^2 + 1) + \ln(x^3 + 1)\right)$$

$$\frac{1}{u}\frac{du}{dx} = \frac{1}{2}\left(\frac{1}{x + 1} + \frac{2x}{x^2 + 1} + \frac{3x^2}{x^3 + 1}\right)$$

At $x = 1$ we have $u = \sqrt{8} = 2\sqrt{2}$. Hence,

$$\frac{du}{dx}\bigg|_{x=1} = \sqrt{2}\left(\frac{1}{2} + 1 + \frac{3}{2}\right) = 3\sqrt{2}.$$

Exercises 3.3

Simplify the expressions given in Exercises 1–10.

1. $e^3/\sqrt{e^5}$

2. $\ln\left(e^{1/2}e^{2/3}\right)$

3. $e^{5\ln x}$

4. $e^{(3\ln 9)/2}$

5. $\ln\dfrac{1}{e^{3x}}$

6. $e^{2\ln\cos x} + \left(\ln e^{\sin x}\right)^2$

7. $3\ln 4 - 4\ln 3$

8. $4\ln\sqrt{x} + 6\ln(x^{1/3})$

9. $2\ln x + 5\ln(x-2)$

10. $\ln(x^2 + 6x + 9)$

Solve the equations in Exercises 11–14 for x.

11. $2^{x+1} = 3^x$

12. $3^x = 9^{1-x}$

13. $\dfrac{1}{2^x} = \dfrac{5}{8^{x+3}}$

14. $2^{x^2-3} = 4^x$

Find the domains of the functions in Exercises 15–16.

15. $\ln\dfrac{x}{2-x}$

16. $\ln(x^2 - x - 2)$

Solve the inequalities in Exercises 17–18.

17. $\ln(2x-5) > \ln(7-2x)$ **18.** $\ln(x^2 - 2) \le \ln x$

In Exercises 19–50, differentiate the given functions. If possible, simplify your answers.

19. $y = e^{5x}$

20. $y = xe^x - x$

21. $y = \dfrac{x}{e^{2x}}$

22. $y = x^2 e^{x/2}$

23. $y = \ln(3x-2)$

24. $y = \ln|3x-2|$

25. $y = \ln(1 + e^x)$

26. $y = 2\ln\sqrt{x^2+2}$

27. $y = \dfrac{e^x + e^{-x}}{2}$

28. $f(x) = e^{(x^2)}$

29. $y = e^{(e^x)}$

30. $x = e^{3t}\ln t$

31. $y = \dfrac{e^x}{1+e^x}$

32. $f(x) = \dfrac{e^x - e^{-x}}{e^x + e^{-x}}$

33. $y = e^x\sin x$

34. $y = e^{-x}\cos x$

35. $y = \ln\ln x$

36. $y = x\ln x - x$

37. $y = x^2\ln x - \dfrac{x^2}{2}$

38. $y = \ln|\sin x|$

39. $y = 5^{2x+1}$

40. $y = 2^{(x^2-3x+8)}$

41. $g(x) = t^x x^t$

42. $h(t) = t^x - x^t$

43. $f(s) = \log_a(bs+c)$

44. $g(x) = \log_x(2x+3)$

45. $y = x^{\sqrt{x}}$

46. $y = (1/x)^{\ln x}$

47. $y = \ln|\sec x + \tan x|$

48. $y = \ln|x + \sqrt{x^2-a^2}|$

49. $y = \ln\left(\sqrt{x^2+a^2} - x\right)$ **50.** $y = (\cos x)^x - x^{\cos x}$

51. Find the nth derivative of $f(x) = xe^{ax}$.

52. Show that the nth derivative of $(ax^2 + bx + c)e^x$ is a function of the same form but with different constants.

53. Find the first four derivatives of e^{x^2}.

54. Find the nth derivative of $\ln(2x+1)$.

55. Differentiate (a) $f(x) = (x^x)^x$ and (b) $g(x) = x^{(x^x)}$. Which function grows more rapidly as x grows large?

∗ 56. Solve the equation $x^{x^{x^{\cdot^{\cdot^{\cdot}}}}} = a$, where $a > 0$. The exponent tower goes on forever.

Use logarithmic differentiation to find the required derivatives in Exercises 57–59.

57. $f(x) = (x-1)(x-2)(x-3)(x-4)$. Find $f'(x)$.

58. $F(x) = \dfrac{\sqrt{1+x}(1-x)^{1/3}}{(1+5x)^{4/5}}$. Find $F'(0)$.

59. $f(x) = \dfrac{(x^2-1)(x^2-2)(x^2-3)}{(x^2+1)(x^2+2)(x^2+3)}$. Find $f'(2)$. Also find $f'(1)$.

60. At what points does the graph $y = x^2 e^{-x^2}$ have a horizontal tangent line?

61. Let $f(x) = xe^{-x}$. Determine where f is increasing and where it is decreasing. Sketch the graph of f.

62. Find the equation of a straight line of slope 4 that is tangent to the graph of $y = \ln x$.

63. Find an equation of the straight line tangent to the curve $y = e^x$ and passing through the origin.

64. Find an equation of the straight line tangent to the curve $y = \ln x$ and passing through the origin.

65. Find an equation of the straight line that is tangent to $y = 2^x$ and that passes through the point $(1, 0)$.

66. For what values of $a > 0$ does the curve $y = a^x$ intersect the straight line $y = x$?

67. Find the slope of the curve $e^{xy}\ln\dfrac{x}{y} = x + \dfrac{1}{y}$ at $(e, 1/e)$.

68. Find an equation of the straight line tangent to the curve $xe^y + y - 2x = \ln 2$ at the point $(1, \ln 2)$.

69. Find the derivative of $f(x) = Ax\cos\ln x + Bx\sin\ln x$. Use the result to help you find the indefinite integrals

$$\int \cos\ln x\, dx \quad\text{and}\quad \int \sin\ln x\, dx.$$

∗ 70. Let $F_{A,B}(x) = Ae^x\cos x + Be^x\sin x$. Show that $(d/dx)F_{A,B}(x) = F_{A+B,B-A}(x)$.

∗ 71. Using the results of Exercise 70, find
(a) $(d^2/dx^2)F_{A,B}(x)$ and (b) $(d^3/dx^3)e^x\cos x$.

∗ 72. Find $\dfrac{d}{dx}(Ae^{ax}\cos bx + Be^{ax}\sin bx)$ and use the answer to help you evaluate
(a) $\int e^{ax}\cos bx\, dx$ and (b) $\int e^{ax}\sin bx\, dx$.

73. Prove identity (ii) of Theorem 2 by examining the derivative of the left side minus the right side as was done in the proof of identity (i).

74. Deduce identity (iii) of Theorem 2 from identities (i) and (ii).

75. Prove identity (iv) of Theorem 2 for rational exponents r by the same method used for Exercise 73.

* **76.** Let $x > 0$, and let $F(x)$ be the area bounded by the curve $y = t^2$, the t-axis, and the vertical lines $t = 0$ and $t = x$. Using the method of the proof of Theorem 1, show that $F'(x) = x^2$. Hence, find an explicit formula for $F(x)$. What is the area of the region bounded by $y = t^2$, $y = 0$, $t = 0$ and $t = 2$?

* **77.** Carry out the following steps to show that $2 < e < 3$. Let

$f(t) = 1/t$ for $t > 0$.

(a) Show that the area under $y = f(t)$, above $y = 0$, and between $t = 1$ and $t = 2$ is less than 1 square unit. Deduce that $e > 2$.

(b) Show that all tangent lines to the graph of f lie below the graph. *Hint:* $f''(t) = 2/t^3 > 0$.

(c) Find the lines T_2 and T_3 that are tangent to $y = f(t)$ at $t = 2$ and $t = 3$, respectively.

(d) Find the area A_2 under T_2, above $y = 0$, and between $t = 1$ and $t = 2$. Also find the area A_3 under T_3, above $y = 0$, and between $t = 2$ and $t = 3$.

(e) Show that $A_2 + A_3 > 1$ square unit. Deduce that $e < 3$.

Growth and Decay

In this section we will study the use of exponential functions to model the growth rates of quantities whose rate of growth is directly related to their size. The growth of such quantities is typically governed by differential equations whose solutions involve exponential functions. Before delving into this topic, we prepare the way by examining the growth behaviour of exponential and logarithmic functions.

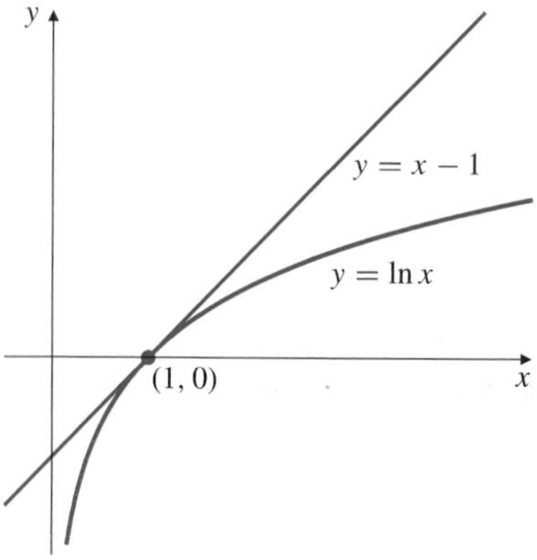

The Growth of Exponentials and Logarithms

Figure 3.14 $\ln x \leq x - 1$ for $x > 0$

In Section 3.3 we showed that both e^x and $\ln x$ grow large (approach infinity) as x grows large. However, e^x increases very rapidly as x increases, and $\ln x$ increases very slowly. In fact, e^x increases, for large x, faster than any positive power of x (no matter how large the power), while $\ln x$ increases more slowly than any positive power of x (no matter how small the power). In order to verify this behaviour we start with an inequality satisfied by $\ln x$. The straight line $y = x - 1$ is tangent to the curve $y = \ln x$ at the point $(1, 0)$. The following theorem asserts that the curve lies below that line. (See Figure 3.14.)

THEOREM **4**

If $x > 0$, then $\ln x \leq x - 1$.

PROOF   Let $g(x) = \ln x - (x - 1)$ for $x > 0$. Then $g(1) = 0$ and

$$g'(x) = \frac{1}{x} - 1 \quad \begin{cases} > 0 & \text{if } 0 < x < 1 \\ < 0 & \text{if } x > 1. \end{cases}$$

As observed in Section 2.6, these inequalities imply that g is increasing on $]0, 1[$ and decreasing on $]1, \infty[$. Thus, $g(x) \leq g(1) = 0$ for all $x > 0$ and $\ln x \leq x - 1$ for all such x.

THEOREM **5**

The growth properties of exp and ln

If $a > 0$, then

(a) $\lim\limits_{x\to\infty} \dfrac{x^a}{e^x} = 0,$ (b) $\lim\limits_{x\to\infty} \dfrac{\ln x}{x^a} = 0,$

(c) $\lim\limits_{x\to-\infty} |x|^a\, e^x = 0,$ (d) $\lim\limits_{x\to 0+} x^a \ln x = 0.$

Each of these limits makes a statement about who "wins" in a contest between an exponential or logarithm and a power. For example, in part (a), the denominator e^x grows large as $x \to \infty$, so it tries to make the fraction x^a/e^x approach 0. On the other hand, if a is a large positive number, the numerator x^a also grows large and tries to make the fraction approach infinity. The assertion of (a) is that in this contest between the exponential and the power, the exponential is stronger and wins; the fraction approaches 0. The content of Theorem 5 can be paraphrased as follows:

In a struggle between a power and an exponential, the exponential wins.
In a struggle between a power and a logarithm, the power wins.

PROOF First we prove part (b). Let $x > 1$, $a > 0$, and let $s = a/2$. Since $\ln(x^s) = s \ln x$, we have, using Theorem 4,

$$0 < s \ln x = \ln(x^s) \le x^s - 1 < x^s.$$

Thus, $0 < \ln x < \dfrac{1}{s} x^s$ and, dividing by $x^a = x^{2s}$,

$$0 < \frac{\ln x}{x^a} < \frac{1}{s}\frac{x^s}{x^{2s}} = \frac{1}{s\,x^s}.$$

Now $1/(s\,x^s) \to 0$ as $x \to \infty$ (since $s > 0$); therefore by the Squeeze Theorem,

$$\lim_{x\to\infty} \frac{\ln x}{x^a} = 0.$$

Next we deduce part (d) from part (b) by substituting $x = 1/t$. As $x \to 0+$, we have $t \to \infty$, so

$$\lim_{x\to 0+} x^a \ln x = \lim_{t\to\infty} \frac{\ln(1/t)}{t^a} = \lim_{t\to\infty} \frac{-\ln t}{t^a} = -0 = 0.$$

Now we deduce (a) from (b). If $x = \ln t$, then $t \to \infty$ as $x \to \infty$, so

$$\lim_{x\to\infty} \frac{x^a}{e^x} = \lim_{t\to\infty} \frac{(\ln t)^a}{t} = \lim_{t\to\infty} \left(\frac{\ln t}{t^{1/a}}\right)^a = 0^a = 0.$$

Finally, (c) follows from (a) via the substitution $x = -t$:

$$\lim_{x\to-\infty} |x|^a\, e^x = \lim_{t\to\infty} |-t|^a\, e^{-t} = \lim_{t\to\infty} \frac{t^a}{e^t} = 0.$$

Exponential Growth and Decay Models

Many natural processes involve quantities that increase or decrease at a rate proportional to their size. For example, the mass of a culture of bacteria growing in a medium supplying adequate nourishment will increase at a rate proportional to that mass. The value of an investment bearing interest that is continuously compounding increases at a rate proportional to that value. The mass of undecayed radioactive material in a sample decreases at a rate proportional to that mass.

All of these phenomena, and others exhibiting similar behaviour, can be modelled mathematically in the same way. If $y = y(t)$ denotes the value of a quantity y at time t, and if y changes at a rate proportional to its size, then

$$\frac{dy}{dt} = ky,$$

where k is the constant of proportionality. The above equation is called the **differential equation of exponential growth or decay** because, for any value of the constant C, the function $y = Ce^{kt}$ satisfies the equation. In fact, if $y(t)$ is any solution of the differential equation $y' = ky$, then

$$\frac{d}{dt}\left(\frac{y(t)}{e^{kt}}\right) = \frac{e^{kt}y'(t) - ke^{kt}y(t)}{e^{2kt}} = \frac{y'(t) - ky(t)}{e^{kt}} = 0 \quad \text{for all } t.$$

Thus $y(t)/e^{kt} = C$, a constant, and $y(t) = Ce^{kt}$. Since $y(0) = Ce^0 = C$,

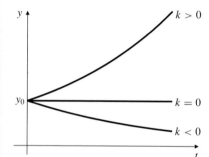

The initial-value problem $\begin{cases} \dfrac{dy}{dt} = ky \\ y(0) = y_0 \end{cases}$ has unique solution $y = y_0 e^{kt}$.

If $y_0 > 0$, then $y(t)$ is an increasing function of t if $k > 0$ and a decreasing function of t if $k < 0$. We say that the quantity y exhibits **exponential growth** if $k > 0$ and **exponential decay** if $k < 0$. (See Figure 3.15.)

Figure 3.15 Solutions of the initial-value problem $dy/dt = ky$, $y(0) = y_0$, for $k > 0$, $k = 0$, and $k < 0$

Example 1 (**Growth of a cell culture**) A certain cell culture grows at a rate proportional to the number of cells present. If the culture contains 500 cells initially and 800 after 24 h, how many cells will be present after a further 12 h?

Solution Let $y(t)$ be the number of cells present t hours after there were 500 cells. Thus $y(0) = 500$ and $y(24) = 800$. Because $dy/dt = ky$, we have

$$y(t) = y(0)e^{kt} = 500e^{kt}.$$

Therefore, $800 = y(24) = 500e^{24k}$, and so $24k = \ln \frac{800}{500} = \ln(1.6)$. It follows that $k = (1/24)\ln(1.6)$ and

$$y(t) = 500e^{(t/24)\ln(1.6)} = 500(1.6)^{t/24}.$$

We want to know y when $t = 36$: $y(36) = 500e^{(36/24)\ln(1.6)} = 500(1.6)^{3/2} \approx 1012$. The cell count grew to about 1,012 in the 12 h after it was 800.

Exponential growth is characterized by a **fixed doubling time**. If T is the time at which y has doubled from its size at $t = 0$, then $2y(0) = y(T) = y(0)e^{kT}$.

Therefore $e^{kT} = 2$. Since $y(t) = y(0)e^{kt}$, we have

$$y(t + T) = y(0)e^{k(t+T)} = e^{kT} y(0)e^{kt} = 2y(t),$$

that is, T units of time are required for y to double from any value. Similarly, exponential decay involves a fixed halving time (usually called the **half-life**). If $y(T) = \frac{1}{2}y(0)$, then $e^{kT} = \frac{1}{2}$ and

$$y(t + T) = y(0)e^{k(t+T)} = \frac{1}{2}y(t).$$

Example 2   (**Radioactive decay**) A radioactive material has a half-life of 1,200 years. What percentage of the original radioactivity of a sample is left after 10 years? How many years are required to reduce the radioactivity by 10%?

Solution   Let $p(t)$ be the percentage of the original radioactivity left after t years. Thus $p(0) = 100$ and $p(1,200) = 50$. Since the radioactivity decreases at a rate proportional to itself, $dp/dt = kp$ and

$$p(t) = 100e^{kt}.$$

Now $50 = p(1,200) = 100e^{1,200k}$, so

$$k = \frac{1}{1,200} \ln \frac{50}{100} = -\frac{\ln 2}{1,200}.$$

The percentage left after 10 years is

$$p(10) = 100e^{10k} = 100e^{-10(\ln 2)/1,200} \approx 99.424.$$

If after t years 90% of the radioactivity is left, then

$$90 = 100e^{kt},$$
$$kt = \ln \frac{90}{100},$$
$$t = \frac{1}{k} \ln(0.9) = -\frac{1,200}{\ln 2} \ln(0.9) \approx 182.4,$$

so it will take a little over 182 years to reduce the radioactivity by 10%.

Sometimes an exponential growth or decay problem will involve a quantity that changes at a rate proportional to the difference between itself and a fixed value:

$$\frac{dy}{dt} = k(y - a).$$

In this case the change of dependent variable $u(t) = y(t) - a$ should be used to convert the differential equation to the standard form. Observe that $u(t)$ changes at the same rate as $y(t)$ (i.e., $du/dt = dy/dt$), so it satisfies

$$\frac{du}{dt} = ku.$$

Example 3   (Newton's law of cooling)   A hot object introduced into a cooler environment will cool at a rate proportional to the excess of its temperature above that of its environment. If a cup of coffee sitting in a room maintained at a temperature of 20°C cools from 80°C to 50°C in five minutes, how much longer will it take to cool to 40°C?

Solution   Let $y(t)$ be the temperature of the coffee t min after it was 80°C. Thus $y(0) = 80$ and $y(5) = 50$. Newton's law says that $dy/dt = k(y - 20)$ in this case, so let $u(t) = y(t) - 20$. Thus, $u(0) = 60$ and $u(5) = 30$. We have

$$\frac{du}{dt} = \frac{dy}{dt} = k(y - 20) = ku.$$

Thus

$$u(t) = 60e^{kt},$$

$$30 = u(5) = 60e^{5k},$$

$$5k = \ln \tfrac{1}{2} = -\ln 2.$$

We want to know t such that $y(t) = 40$, that is, $u(t) = 20$:

$$20 = u(t) = 60e^{-(t/5)\ln 2}$$

$$-\frac{t}{5}\ln 2 = \ln \frac{20}{60} = -\ln 3,$$

$$t = 5\frac{\ln 3}{\ln 2} \approx 7.92.$$

The coffee will take about $7.92 - 5 = 2.92$ min to cool from 50°C to 40°C.

■

Interest on Investments

Suppose that $10,000 is invested at an annual rate of interest of 8%. Thus the value of the investment at the end of 1 year will be $10,000(1.08) = $10,800. If this amount remains invested for a second year at the same rate it will grow to $10,000(1.08)^2 = $11,664$; in general, n years after the original investment was made, it will be worth $10,000(1.08)^n$.

Now suppose that the 8% rate is *compounded semiannually* so that the interest is actually paid at a rate of 4% per 6-month period. After 1 year (2 interest periods) the $10,000 will grow to $10,000(1.04)^2 = $10,816$. This is $16 more than was obtained when the 8% was compounded only once per year. The extra $16 is the interest paid in the second 6-month period on the $400 interest earned in the first 6-month period. Continuing in this way, if the 8% interest is compounded *monthly* (12 periods per year and $\frac{8}{12}$% paid per period) or *daily* (365 periods per year and $\frac{8}{365}$% paid per period), then the original $10,000 would grow in 1 year to $10{,}000\left(1 + \frac{8}{1{,}200}\right)^{12} = $10,830$ or $10{,}000\left(1 + \frac{8}{36{,}500}\right)^{365} = $10,832.78$, respectively.

For any given *nominal* interest rate, the investment grows more if the compounding period is shorter. In general, an original investment of A invested at r%

per annum compounded n times per year grows in one year to

$$\$A \left(1 + \frac{r}{100n}\right)^n.$$

It is natural to ask how well we can do with our investment if we let the number of periods in a year approach infinity, that is, we compound the interest *continuously*. The answer is that in 1 year the $\$A$ will grow to

$$\$A \lim_{n \to \infty} \left(1 + \frac{r}{100n}\right)^n = \$A e^{r/100}.$$

For example, at 8% per annum compounded continuously, our $10,000 will grow in 1 year to $10,000e^{0.08} \approx \$10,832.87$. (Note that this is just a few cents more than we get compounding daily.) To justify this result we need the following theorem.

THEOREM **6**

For every real number x,

$$e^x = \lim_{n \to \infty} \left(1 + \frac{x}{n}\right)^n$$

PROOF If $x = 0$, there is nothing to prove; both sides of the identity are 1. If $x \neq 0$, let $h = x/n$. As n tends to infinity, h approaches 0. Thus,

$$
\begin{aligned}
\lim_{n \to \infty} \ln \left(1 + \frac{x}{n}\right)^n &= \lim_{n \to \infty} n \ln \left(1 + \frac{x}{n}\right) \\
&= \lim_{n \to \infty} x \, \frac{\ln \left(1 + \frac{x}{n}\right)}{\frac{x}{n}} \\
&= x \lim_{h \to 0} \frac{\ln(1 + h)}{h} \qquad \text{(where } h = x/n\text{)} \\
&= x \lim_{h \to 0} \frac{\ln(1 + h) - \ln 1}{h} \qquad \text{(since } \ln 1 = 0\text{)} \\
&= x \left(\frac{d}{dt} \ln t\right)\Big|_{t=1} \qquad \text{(by the definition of derivative)} \\
&= x \, \frac{1}{t}\Big|_{t=1} = x.
\end{aligned}
$$

Since ln is differentiable, it is continuous. Hence, by Theorem 7 of Section 1.4,

$$\ln \left(\lim_{n \to \infty} \left(1 + \frac{x}{n}\right)^n \right) = \lim_{n \to \infty} \ln \left(1 + \frac{x}{n}\right)^n = x.$$

Since the exponential function is the inverse of the natural logarithm, we conclude

$$\lim_{n \to \infty} \left(1 + \frac{x}{n}\right)^n = e^x.$$

In the case $x = 1$ the formula given in Theorem 6 takes the following form:

$$e = \lim_{n \to \infty} \left(1 + \frac{1}{n}\right)^n.$$

Table 2.

n	$\left(1 + \dfrac{1}{n}\right)^n$
1	2
10	$2.59374\cdots$
100	$2.70481\cdots$
1,000	$2.71692\cdots$
10,000	$2.71815\cdots$
100,000	$2.71827\cdots$

We can use this formula to compute approximations to e, as shown in Table 2. In a sense we have cheated in obtaining the numbers in this table; they were produced using the y^x function on a scientific calculator. However, this function is actually computed as $e^{x \ln y}$. In any event, the formula in this table is not a very efficient way to calculate e to any great accuracy. Only 4 decimal places are correct for $n = 100{,}000$. A much better way is to use the series

$$e = 1 + \frac{1}{1!} + \frac{1}{2!} + \frac{1}{3!} + \frac{1}{4!} + \cdots = 1 + 1 + \frac{1}{2} + \frac{1}{6} + \frac{1}{24} + \cdots,$$

which we will establish in Section 4.8.

A final word about interest rates. Financial institutions sometimes quote *effective* rates of interest rather than *nominal* rates. The effective rate tells you what the actual effect of the interest rate will be after one year. Thus, $10,000 invested at an effective rate of 8% will grow to $10,800.00 in one year regardless of the compounding period. A nominal rate of 8% per annum compounded daily is equivalent to an effective rate of about 8.3278%.

Logistic Growth

Few quantities in nature can sustain exponential growth over extended periods of time; the growth is usually limited by external constraints. For example, suppose a small number of rabbits (of both sexes) is introduced to a small island where there were no rabbits previously, and where there are no predators who eat rabbits. By virtue of natural fertility, the number of rabbits might be expected to grow exponentially, but this growth will eventually be limited by the food supply available to the rabbits. Suppose the island can grow enough food to supply a population of L rabbits indefinitely. If there are $y(t)$ rabbits in the population at time t, we would expect $y(t)$ to grow at a rate proportional to $y(t)$ provided $y(t)$ is quite small (much less than L). But as the numbers increase, it will be harder for the rabbits to find enough food, and we would expect the rate of increase to approach 0 as $y(t)$ gets closer and closer to L. One possible model for such behaviour is the differential equation

$$\frac{dy}{dt} = ky\left(1 - \frac{y}{L}\right),$$

which is called the **logistic equation** since it models growth that is limited by the *supply* of necessary resources. Observe that $dy/dt > 0$ if $0 < y < L$ and that this rate is small if y is small (there are few rabbits to reproduce) or if y is close to L (there are almost as many rabbits as the available resources can feed). Observe also that $dy/dt < 0$ if $y > L$; there being more animals than the resources can feed, the rabbits die at a greater rate than they are born. Of course, the steady-state populations $y = 0$ and $y = L$ are solutions of the logistic equation; for both of these $dy/dt = 0$. We will examine techniques for solving differential equations like the logistic equation in Section 7.9. For now, we invite the reader to verify by differentiation that the solution satisfying $y(0) = y_0$, is

$$y = \frac{Ly_0}{y_0 + (L - y_0)e^{-kt}}.$$

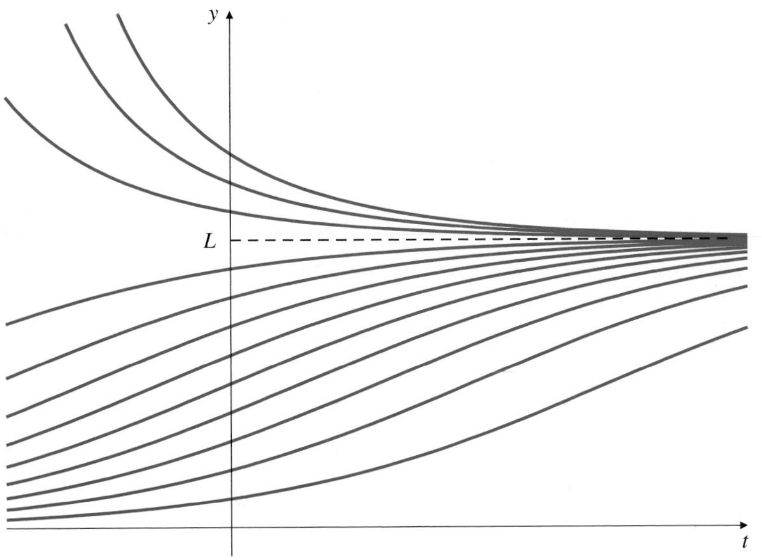

Figure 3.16   Some logistic curves

Observe that, as expected, if $0 < y_0 < L$, then

$$\lim_{t \to \infty} y(t) = L, \qquad \lim_{t \to -\infty} y(t) = 0.$$

The solution given above also holds for $y_0 > L$. However, the solution does not approach 0 as t approaches $-\infty$ in this case. It has a vertical asymptote at a certain negative value of t. (See Exercise 28 below.) The graphs of solutions of the logistic equation for various positive values of y_0 are given in Figure 3.16.

Exercises 3.4

Evaluate the limits in Exercises 1–8.

1. $\lim_{x \to \infty} x^3 e^{-x}$

2. $\lim_{x \to \infty} x^{-3} e^x$

3. $\lim_{x \to \infty} \dfrac{2e^x - 3}{e^x + 5}$

4. $\lim_{x \to \infty} \dfrac{x - 2e^{-x}}{x + 3e^{-x}}$

5. $\lim_{x \to 0+} x \ln x$

6. $\lim_{x \to 0+} \dfrac{\ln x}{x}$

7. $\lim_{x \to 0} x \left(\ln |x| \right)^2$

8. $\lim_{x \to \infty} \dfrac{(\ln x)^3}{\sqrt{x}}$

9. (Bacterial growth) Bacteria grow in a certain culture at a rate proportional to the amount present. If there are 100 bacteria present initially and the amount doubles in 1 h, how many will there be after a further $1\frac{1}{2}$ h?

10. (Dissolving sugar) Sugar dissolves in water at a rate proportional to the amount still undissolved. If there were 50 kg of sugar present initially, and at the end of 5 h only 20 kg is left, how much longer will it take until 90% of the sugar is disssolved?

11. (Radioactive decay) A radioactive substance decays at a rate proportional to the amount present. If 30% of such a substance decays in 15 years, what is the half-life of the substance?

12. (Half-life of radium) If the half-life of radium is 1,690 years, what percentage of the amount present now will be remaining after (a) 100 years, (b) 1,000 years?

13. Find the half-life of a radioactive substance if after 1 year 99.57% of an initial amount still remains.

14. (Bacterial growth) In a certain culture where the rate of growth of bacteria is proportional to the number present, the number triples in 3 days. If at the end of 7 days there are 10 million bacteria present in the culture, how many were present initially?

15. (Weight of a newborn) In the first few weeks after birth, a baby gains weight at a rate proportional to its weight. A baby weighing 4 kg at birth weighs 4.4 kg after 2 weeks. How much did the baby weigh 5 days after birth?

16. (Electric current) When a simple electrical circuit containing inductance and resistance but no capacitance has the electromotive force removed, the rate of decrease of the current is proportional to the current. If the current is $I(t)$ amperes t s after cutoff, and if $I = 40$ when $t = 0$, and $I = 15$ when $t = 0.01$, find a formula for $I(t)$.

17. (Continuously compounding interest) How much money needs to be invested today at a nominal rate of 4% compounded continuously, in order that it should grow to $10,000 in 7 years?

18. (Continuously compounding interest) Money invested at compound interest (with instantaneous compounding) accumulates at a rate proportional to the amount present. If an initial investment of $1,000 grows to $1,500 in exactly 5 years, find (a) the doubling time for the investment and (b) the effective annual rate of interest being paid.

19. (Purchasing power) If the purchasing power of the dollar is decreasing at an effective rate of 9% annually, how long will it take for the purchasing power to be reduced to 25 cents?

∗ **20. (Effective interest rate)** A bank claims to pay interest at an effective rate of 9.5% on an investment account. If the interest is actually being compounded monthly, what is the nominal rate of interest being paid on the account?

Differential equations of the form $y' = a + by$

21. Suppose that $f(x)$ satisfies the differential equation

$$f'(x) = a + bf(x),$$

where a and b are constants.

(a) Solve the differential equation by substituting $u(x) = a + bf(x)$ and solving the simpler differential equation that results for $u(x)$.

(b) Solve the initial-value problem:

$$\begin{cases} \dfrac{dy}{dx} = a + by \\ y(0) = y_0 \end{cases}$$

22. (Drug concentrations in the blood) A drug is introduced into the bloodstream intravenously at a constant rate and breaks down and is eliminated from the body at a rate proportional to its concentration in the blood. The concentration $x(t)$ of the drug in the blood satisfies the differential equation

$$\frac{dx}{dt} = a - bx,$$

where a and b are positive constants.

(a) What is the limiting concentration $\lim_{t \to \infty} x(t)$ of the drug in the blood?

(b) Find the concentration of the drug in the blood at time t, given that the concentration was zero at $t = 0$.

(c) How long after $t = 0$ will it take for the concentration to rise to half its limiting value?

23. (Cooling) Use Newton's law of cooling to determine the reading on a thermometer 5 min after it is taken from an oven at 72°C to the outdoors where the temperature is 20°C, if the reading dropped to 48°C after one min.

24. (Cooling) An object is placed in a freezer maintained at a temperature of −5°C. If the object cools from 45°C to 20°C in 40 min, how many more minutes will it take to cool to 0°C?

25. (Warming) If an object in a room warms up from 5°C to 10°C in 4 min, and if the room is being maintained at 20°C, how much longer will the object take to warm up to 15°C? Assume the object warms at a rate proportional to the difference between its temperature and room temperature.

The logistic equation

∗ **26.** Suppose the quantity $y(t)$ exhibits logistic growth. If the values of $y(t)$ at times $t = 0$, $t = 1$, and $t = 2$ are y_0, y_1, and y_2, respectively, find an equation satisfied by the limiting value L of $y(t)$, and solve it for L. If $y_0 = 3$, $y_1 = 5$, and $y_2 = 6$, find L.

∗ **27.** Show that a solution $y(t)$ of the logistic equation having $0 < y(0) < L$ is increasing most rapidly when its value is $L/2$. (*Hint:* you do not need to use the formula for the solution to see this.)

∗ **28.** If $y_0 > L$, find the interval on which the given solution of the logistic equation is valid. What happens to the solution as t approaches the left endpoint of this interval?

∗ **29.** If $y_0 < 0$, find the interval on which the given solution of the logistic equation is valid. What happens to the solution as t approaches the right endpoint of this interval?

30. (Modelling an epidemic) The number y of persons infected by a highly contagious virus is modelled by a logistic curve

$$y = \frac{L}{1 + Me^{-kt}},$$

where t is measured in months from the time the outbreak was discovered. At that time there were 200 infected persons, and the number grew to 1,000 after 1 month. Eventually, the number levelled out at 10,000. Find the values of the parameters L, M, and k of the model.

31. Continuing the previous exercise, how many people were infected 3 months after the outbreak was discovered, and how fast was the number growing at that time?

3.5 The Inverse Trigonometric Functions

The six trigonometric functions are periodic and, hence, not one-to-one. However, as we did with the function x^2 in Section 3.1, we can restrict their domains in such a way that the restricted functions are one-to-one and invertible.

The Inverse Sine (or Arcsine) Function

Let us define a function $\operatorname{Sin} x$ (note the capital letter) to be $\sin x$, restricted so that its domain is the interval $-\frac{\pi}{2} \le x \le \frac{\pi}{2}$:

DEFINITION 8

> **The restricted function Sin x**
> $$\operatorname{Sin} x = \sin x \qquad \text{if } -\frac{\pi}{2} \le x \le \frac{\pi}{2}.$$

Since its derivative $\cos x$ is positive on the interval $\left]-\frac{\pi}{2}, \frac{\pi}{2}\right[$, the function $\operatorname{Sin} x$ is increasing on its domain, so it is a one-to-one function. It has domain $\left[-\frac{\pi}{2}, \frac{\pi}{2}\right]$ and range $[-1, 1]$. (See Figure 3.17.)

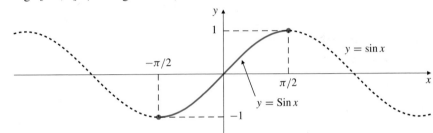

Figure 3.17 The graph of $\operatorname{Sin} x$ forms part of the graph of $\sin x$

Being one-to-one, Sin has an inverse function which is denoted $\sin^{-1}$ (or, in some books and computer programs, by arcsin, Arcsin, or asin) and which is called the **inverse sine** or **arcsine** function.

DEFINITION 9

> **The inverse sine function $\sin^{-1} x$ or $\arcsin x$**
> $$y = \sin^{-1} x \iff x = \operatorname{Sin} y$$
> $$\iff x = \sin y \quad \text{and} \quad -\frac{\pi}{2} \le y \le \frac{\pi}{2}$$

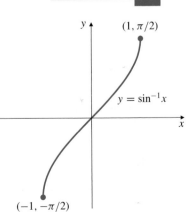

Figure 3.18 The arcsine function

The graph of $\sin^{-1}$ is shown in Figure 3.18; it is the reflection of the graph of Sin in the line $y = x$. The domain of $\sin^{-1}$ is $[-1, 1]$ (the range of Sin), and the range of $\sin^{-1}$ is $\left[-\frac{\pi}{2}, \frac{\pi}{2}\right]$ (the domain of Sin). The cancellation identities for Sin and $\sin^{-1}$ are

$$\sin^{-1}(\operatorname{Sin} x) = \arcsin(\operatorname{Sin} x) = x \qquad \text{for } -\frac{\pi}{2} \le x \le \frac{\pi}{2}$$

$$\operatorname{Sin}(\sin^{-1} x) = \operatorname{Sin}(\arcsin x) = x \qquad \text{for } -1 \le x \le 1$$

Remark As for the general inverse function f^{-1}, be aware that $\sin^{-1} x$ does *not* represent the *reciprocal* $1/\sin x$. (We already have a perfectly good name for the reciprocal of $\sin x$; we call it $\csc x$.) We should think of $\sin^{-1} x$ as "the angle between $-\frac{\pi}{2}$ and $\frac{\pi}{2}$ whose sine is x."

Example 1

(a) $\sin^{-1}\frac{1}{2} = \frac{\pi}{6}$ (because $\sin\frac{\pi}{6} = \frac{1}{2}$ and $-\frac{\pi}{2} < \frac{\pi}{6} < \frac{\pi}{2}$).

(b) $\sin^{-1}\left(-\frac{1}{\sqrt{2}}\right) = -\frac{\pi}{4}$ (because $\sin\left(-\frac{\pi}{4}\right) = -\frac{1}{\sqrt{2}}$ and $-\frac{\pi}{2} < -\frac{\pi}{4} < \frac{\pi}{2}$).

(c) $\sin^{-1}(-1) = -\frac{\pi}{2}$ (because $\sin\left(-\frac{\pi}{2}\right) = -1$).

(d) $\sin^{-1}2$ is not defined. (2 is not in the range of sine.)

Example 2   Find (a) $\sin\left(\sin^{-1}0.7\right)$, (b) $\sin^{-1}\left(\sin 0.3\right)$, (c) $\sin^{-1}\left(\sin\frac{4\pi}{5}\right)$, (d) $\cos\left(\sin^{-1}0.6\right)$.

Solution

(a) $\sin\left(\sin^{-1}0.7\right) = 0.7$ (cancellation identity).

(b) $\sin^{-1}\left(\sin 0.3\right) = 0.3$ (cancellation identity).

(c) The number $\frac{4\pi}{5}$ does not lie in $\left[-\frac{\pi}{2}, \frac{\pi}{2}\right]$ so we can't apply the cancellation identity directly. However, $\sin\frac{4\pi}{5} = \sin\left(\pi - \frac{\pi}{5}\right) = \sin\frac{\pi}{5}$ by the supplementary angle identity. Therefore, $\sin^{-1}\left(\sin\frac{4\pi}{5}\right) = \sin^{-1}\left(\sin\frac{\pi}{5}\right) = \frac{\pi}{5}$ (by cancellation).

(d) Let $\theta = \sin^{-1}0.6$, as shown in the right triangle in Figure 3.19, which has hypotenuse 1 and side opposite θ equal to 0.6. By the Pythagorean Theorem, the side adjacent θ is $\sqrt{1 - (0.6)^2} = 0.8$. Thus $\cos\left(\sin^{-1}0.6\right) = \cos\theta = 0.8$.

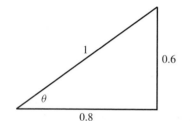

0.6

Figure 3.19

Example 3   Simplify the expression $\tan(\sin^{-1}x)$.

Solution   We want the tangent of an angle whose sine is x. Suppose first that $0 \le x < 1$. As in Example 2, we draw a right triangle (Figure 3.20) with one angle θ, and label the sides so that $\theta = \sin^{-1}x$. The side opposite θ is x, and hypotenuse 1. The remaining side is $\sqrt{1 - x^2}$, and we have

$$\tan(\sin^{-1}x) = \tan\theta = \frac{x}{\sqrt{1 - x^2}}.$$

Because both sides of the above equation are odd functions of x, the same result holds for $-1 < x < 0$.

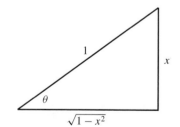

Figure 3.20

Now let us use implicit differentiation to find the derivative of the inverse sine function. If $y = \sin^{-1}x$, then $x = \sin y$ and $-\frac{\pi}{2} \le y \le \frac{\pi}{2}$. Differentiating with respect to x, we obtain

$$1 = (\cos y)\frac{dy}{dx}.$$

Since $-\frac{\pi}{2} \le y \le \frac{\pi}{2}$ we know that $\cos y \ge 0$. Therefore,

$$\cos y = \sqrt{1 - \sin^2 y} = \sqrt{1 - x^2},$$

and $dy/dx = 1/\cos y = 1/\sqrt{1 - x^2}$.

$$\frac{d}{dx}\sin^{-1}x = \frac{d}{dx}\arcsin x = \frac{1}{\sqrt{1-x^2}}.$$

Note that the inverse sine function is differentiable only on the *open* interval $]-1, 1[$; the slope of its graph approaches infinity as $x \to -1+$ or as $x \to 1-$. (See Figure 3.18.)

Example 4   Find the derivative of $\sin^{-1}\left(\dfrac{x}{a}\right)$ and hence evaluate $\displaystyle\int \frac{dx}{\sqrt{a^2-x^2}}$, where $a > 0$.

Solution   By the Chain Rule,

$$\frac{d}{dx}\sin^{-1}\frac{x}{a} = \frac{1}{\sqrt{1-\dfrac{x^2}{a^2}}}\frac{1}{a} = \frac{1}{\sqrt{\dfrac{a^2-x^2}{a^2}}}\frac{1}{a} = \frac{1}{\sqrt{a^2-x^2}} \quad \text{if } a > 0.$$

Hence,

$$\int \frac{1}{\sqrt{a^2-x^2}}\,dx = \sin^{-1}\frac{x}{a} + C \qquad (a > 0).$$

Example 5   Find the solution y of the following initial-value problem:

$$\begin{cases} y' = \dfrac{4}{\sqrt{2-x^2}} & (-\sqrt{2} < x < \sqrt{2}) \\ y(1) = 2\pi. \end{cases}$$

Solution   Using the integral from the previous example, we have

$$y = \int \frac{4}{\sqrt{2-x^2}}\,dx = 4\sin^{-1}\left(\frac{x}{\sqrt{2}}\right) + C$$

for some constant C. Also $2\pi = y(1) = 4\sin^{-1}(1/\sqrt{2}) + C = 4\left(\frac{\pi}{4}\right) + C = \pi + C$. Thus, $C = \pi$ and $y = 4\sin^{-1}(x/\sqrt{2}) + \pi$.

Example 6   (**A sawtooth curve**) Let $f(x) = \sin^{-1}(\sin x)$ for all real x.

(a) Calculate and simplify $f'(x)$.

(b) Where is f differentiable? Where is f continuous?

(c) Use your results from (a) and (b) to sketch the graph of f.

Solution   (a) Using the Chain Rule and the Pythagorean identity we calculate

$$f'(x) = \frac{1}{\sqrt{1-(\sin x)^2}}(\cos x)$$

$$= \frac{\cos x}{\sqrt{\cos^2 x}} = \frac{\cos x}{|\cos x|} = \begin{cases} 1 & \text{if } \cos x > 0 \\ -1 & \text{if } \cos x < 0. \end{cases}$$

(b) f is differentiable at all points where $\cos x \neq 0$, that is, everywhere except at odd multiples of $\pi/2$, namely, $\pm\frac{\pi}{2}, \pm\frac{3\pi}{2}, \pm\frac{5\pi}{2}, \ldots$.
Since sin is continuous everywhere and has values in $[-1, 1]$, and since $\sin^{-1}$ is continuous on $[-1, 1]$, therefore f is continuous on the whole real line.

(c) Since f is continuous, its graph has no breaks. The graph consists of straight line segments of slopes alternating between 1 and -1 on intervals between consecutive odd multiples of $\pi/2$. Since $f'(x) = 1$ on the interval $\left[-\frac{\pi}{2}, \frac{\pi}{2}\right]$ (where $\cos x \geq 0$), the graph must be as shown in Figure 3.21.   ∎

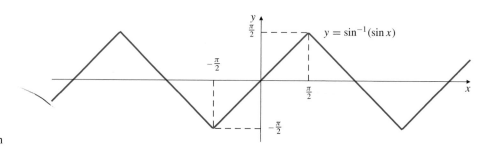

Figure 3.21   A sawtooth graph

The Inverse Tangent (or Arctangent) Function

The inverse tangent function is defined in a manner similar to the inverse sine. We begin by restricting tangent to an interval where it is one-to-one; in this case we use the open interval $\left]-\frac{\pi}{2}, \frac{\pi}{2}\right[$. See Figure 3.22(a).

DEFINITION 10

The restricted function Tan x
$$\text{Tan } x = \tan x \qquad \text{if } -\frac{\pi}{2} < x < \frac{\pi}{2}.$$

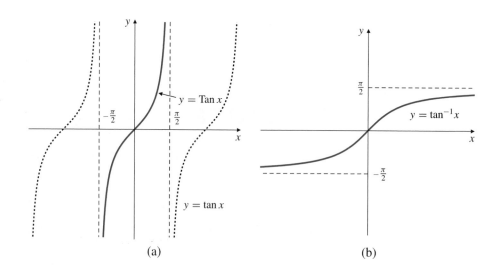

Figure 3.22

(a) The graph of Tan x

(b) The graph of $\tan^{-1} x$

(a)   (b)

The inverse of the function Tan is called the **inverse tangent** function and is denoted $\tan^{-1}$ (or arctan, Arctan, or atan). The domain of $\tan^{-1}$ is the whole real line (the range of Tan). Its range is the open interval $\left]-\frac{\pi}{2}, \frac{\pi}{2}\right[$.

DEFINITION **11**

The inverse tangent function $\tan^{-1}x$ or $\arctan x$

$$y = \tan^{-1} x \iff x = \text{Tan } y$$

$$\iff x = \tan y \quad \text{and} \quad -\frac{\pi}{2} < y < \frac{\pi}{2}$$

The graph of $\tan^{-1}$ is shown in Figure 3.22(b); it is the reflection of the graph of Tan in the line $y = x$.

The cancellation identities for Tan and $\tan^{-1}$ are

$$\tan^{-1}(\text{Tan } x) = \arctan(\text{Tan } x) = x \quad \text{for} \quad -\frac{\pi}{2} < x < \frac{\pi}{2}$$

$$\text{Tan}(\tan^{-1} x) = \text{Tan}(\arctan x) = x \quad \text{for} \quad -\infty < x < \infty$$

Example 7   Evaluate: (a) $\tan(\tan^{-1} 3)$,   (b) $\tan^{-1}\left(\tan \dfrac{3\pi}{4}\right)$, and   (c) $\cos(\tan^{-1} 2)$.

Solution

(a) $\tan(\tan^{-1} 3) = 3$ by cancellation.

(b) $\tan^{-1}\left(\tan \frac{3\pi}{4}\right) = \tan^{-1}(-1) = -\frac{\pi}{4}$.

(c) $\cos(\tan^{-1} 2) = \cos\theta = \frac{1}{\sqrt{5}}$ via the triangle in Figure 3.23. Alternatively, we have $\tan(\tan^{-1} 2) = 2$, so $\sec^2(\tan^{-1} 2) = 1 + 2^2 = 5$. Thus $\cos^2(\tan^{-1} 2) = \frac{1}{5}$. Since cosine is positive on the range of $\tan^{-1}$, we have $\cos(\tan^{-1} 2) = \frac{1}{\sqrt{5}}$. ∎

The derivative of the inverse tangent function is also found by implicit differentiation: if $y = \tan^{-1} x$, then $x = \tan y$ and

$$1 = (\sec^2 y)\frac{dy}{dx} = (1 + \tan^2 y)\frac{dy}{dx} = (1 + x^2)\frac{dy}{dx}.$$

Thus

$$\frac{d}{dx}\tan^{-1} x = \frac{1}{1 + x^2}.$$

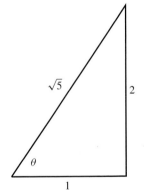

Figure 3.23

Example 8   Find $\dfrac{d}{dx}\tan^{-1}\left(\dfrac{x}{a}\right)$, and hence evaluate $\displaystyle\int \frac{1}{x^2 + a^2}\, dx$.

Solution   We have

$$\frac{d}{dx}\tan^{-1}\left(\frac{x}{a}\right) = \frac{1}{1 + \dfrac{x^2}{a^2}}\frac{1}{a} = \frac{a}{a^2 + x^2};$$

hence

$$\int \frac{dx}{a^2 + x^2} = \frac{1}{a} \, \tan^{-1}\left(\frac{x}{a}\right) + C.$$

Example 9   Prove that $\tan^{-1}\left(\dfrac{x-1}{x+1}\right) = \tan^{-1} x - \dfrac{\pi}{4}$ for $x > -1$.

Solution   Let $f(x) = \tan^{-1}\left(\dfrac{x-1}{x+1}\right) - \tan^{-1} x$. On the interval $]-1, \infty[$ we have, by the Chain Rule and the Quotient Rule,

$$f'(x) = \frac{1}{1 + \left(\dfrac{x-1}{x+1}\right)^2} \frac{(x+1) - (x-1)}{(x+1)^2} - \frac{1}{1+x^2}$$

$$= \frac{(x+1)^2}{(x^2 + 2x + 1) + (x^2 - 2x + 1)} \frac{2}{(x+1)^2} - \frac{1}{1+x^2}$$

$$= \frac{2}{2 + 2x^2} - \frac{1}{1+x^2} = 0.$$

Hence, $f(x) = C$ (constant) on that interval. We can find C by finding $f(0)$:

$$C = f(0) = \tan^{-1}(-1) - \tan^{-1} 0 = -\frac{\pi}{4}.$$

Hence, the given identity holds on $]-1, \infty[$.

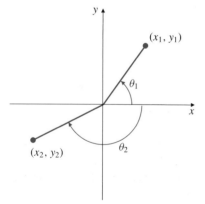

Figure 3.24 $\theta_1 = \tan^{-1}(y_1/x_1)$
$= \text{atan}(y_1/x_1)$
$= \text{atan2}(x_1, y_1)$
$\theta_2 = \text{atan2}(x_2, y_2)$

Remark   Some computer programs, especially spreadsheets, implement two versions of the arctangent function, usually called "atan" and "atan2." The function atan is just the function $\tan^{-1}$ that we have defined; atan(y/x) gives the angle in radians, between the line from the origin to the point (x, y) and the positive x-axis, provided (x, y) lies in quadrants I or IV of the plane. The function atan2 is a function of two variables: atan2(x, y) gives that angle for any point (x, y) not on the y-axis. See Figure 3.24. Some programs, for instance Matlab, reverse the order of the variables x and y in their atan2 function. Maple uses `arctan(x)` and `arctan(y,x)` for the one- and two-variable versions of arctangent.

Other Inverse Trigonometric Functions

The function $\cos x$ is one-to-one on the interval $[0, \pi]$, so we could define the **inverse cosine function**, $\cos^{-1} x$ (or arccos x, or Arccos x, or acos x), so that

$$y = \cos^{-1} x \iff x = \cos y \quad \text{and} \quad 0 \le y \le \pi.$$

However, $\cos y = \sin\left(\frac{\pi}{2} - y\right)$ (the complementary angle identity), and $\frac{\pi}{2} - y$ is in the interval $\left[-\frac{\pi}{2}, \frac{\pi}{2}\right]$ when $0 \le y \le \pi$. Thus, the definition above would lead to

$$y = \cos^{-1} x \iff x = \sin\left(\frac{\pi}{2} - y\right) \iff \sin^{-1} x = \frac{\pi}{2} - y = \frac{\pi}{2} - \cos^{-1} x.$$

It is easier to use this result to define $\cos^{-1} x$ directly:

DEFINITION | **12**

> **The inverse cosine function $\cos^{-1} x$ or $\arccos x$**
>
> $$\cos^{-1} x = \frac{\pi}{2} - \sin^{-1} x \qquad \text{for} \quad -1 \le x \le 1.$$

The cancellation identities for $\cos^{-1} x$ are

$$\cos^{-1}(\cos x) = \arccos(\cos x) = x \qquad \text{for } 0 \le x \le \pi$$
$$\cos(\cos^{-1} x) = \cos(\arccos x) = x \qquad \text{for } -1 \le x \le 1$$

The derivative of $\cos^{-1} x$ is the negative of that of $\sin^{-1} x$ (why?):

$$\frac{d}{dx} \cos^{-1} x = -\frac{1}{\sqrt{1 - x^2}}.$$

The graph of $\cos^{-1}$ is shown in Figure 3.25(a).

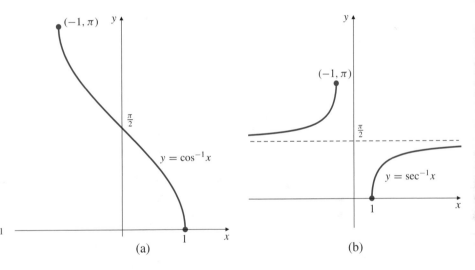

Figure 3.25 The graphs of $\cos^{-1}$ and $\sec^{-1}$

(a)   (b)

Scientific calculators usually implement only the primary trigonometric functions—sine, cosine, and tangent—and the inverses of these three. The secondary functions—secant, cosecant, and cotangent—are calculated using the reciprocal key; to calculate $\sec x$ you calculate $\cos x$ and take the reciprocal of the answer. The inverses of the secondary trigonometric functions are also easily expressed in terms of those of their reciprocal functions. For example, we define:

DEFINITION | **13**

> **The inverse secant function $\sec^{-1} x$ (or $\operatorname{arcsec} x$)**
>
> $$\sec^{-1} x = \cos^{-1}\left(\frac{1}{x}\right) \qquad \text{for} \quad |x| \ge 1.$$

The domain of $\sec^{-1}$ is the union of intervals $]-\infty, -1] \cup [1, \infty[$, and its range is $\left[0, \frac{\pi}{2}\right[\cup \left]\frac{\pi}{2}, \pi\right]$. The graph of $y = \sec^{-1} x$ is shown in Figure 3.25(b). It is the reflection in the line $y = x$ of that part of the graph of $\sec x$ for x between 0 and π.

Observe that

$$\sec(\sec^{-1} x) = \sec\left(\cos^{-1}\left(\frac{1}{x}\right)\right)$$

$$= \frac{1}{\cos\left(\cos^{-1}\left(\frac{1}{x}\right)\right)} = \frac{1}{\frac{1}{x}} = x \qquad \text{for } |x| \geq 1,$$

$$\sec^{-1}(\sec x) = \cos^{-1}\left(\frac{1}{\sec x}\right)$$

$$= \cos^{-1}(\cos x) = x \qquad \text{for } x \text{ in } [0, \pi], \ x \neq \frac{\pi}{2}.$$

We calculate the derivative of $\sec^{-1}$ from that of $\cos^{-1}$:

$$\frac{d}{dx}\sec^{-1} x = \frac{d}{dx}\cos^{-1}\left(\frac{1}{x}\right) = \frac{-1}{\sqrt{1 - \frac{1}{x^2}}}\left(-\frac{1}{x^2}\right)$$

$$= \frac{1}{x^2}\sqrt{\frac{x^2}{x^2 - 1}} = \frac{1}{x^2}\frac{|x|}{\sqrt{x^2 - 1}} = \frac{1}{|x|\sqrt{x^2 - 1}}.$$

Note that we had to use $\sqrt{x^2} = |x|$ in the last line. There are negative values of x in the domain of $\sec^{-1}$. Observe in Figure 3.25(b) that the slope of $y = \sec^{-1}(x)$ is always positive.

Some authors prefer to define $\sec^{-1}$ as the inverse of the restriction of $\sec x$ to the separated intervals $[0, \pi/2[$ and $[\pi, 3\pi/2[$ because this prevents the absolute value from appearing in the formula for the derivative. However, it is much harder to calculate values with that definition. Our definition makes it easy to obtain a value such as $\sec^{-1}(-3)$ from a calculator. Scientific calculators usually have just the inverses of sine, cosine, and tangent built in.

$$\frac{d}{dx}\sec^{-1} x = \frac{1}{|x|\sqrt{x^2 - 1}}.$$

The corresponding integration formula takes different forms on intervals where $x \geq 1$ or $x \leq -1$:

$$\int \frac{1}{x\sqrt{x^2 - 1}}\, dx = \begin{cases} \sec^{-1}x + C & \text{on intervals where } x \geq 1 \\ -\sec^{-1}x + C & \text{on intervals where } x \leq -1 \end{cases}$$

Finally, note that $\csc^{-1}$ and $\cot^{-1}$ are defined similarly to $\sec^{-1}$. They are seldom encountered.

DEFINITION   14

> **The inverse cosecant and inverse cotangent functions**
>
> $$\csc^{-1} x = \sin^{-1}\left(\frac{1}{x}\right), \quad (|x| \geq 1); \qquad \cot^{-1} x = \tan^{-1}\left(\frac{1}{x}\right), \quad (x \neq 0)$$

Exercises 3.5

In Exercises 1–12, evaluate the given expression.

1. $\sin^{-1}\frac{\sqrt{3}}{2}$

2. $\cos^{-1}\left(\frac{-1}{2}\right)$

3. $\tan^{-1}(-1)$

4. $\sec^{-1}\sqrt{2}$

5. $\sin(\sin^{-1} 0.7)$

6. $\cos(\sin^{-1} 0.7)$

7. $\tan^{-1}\left(\tan\frac{2\pi}{3}\right)$

8. $\sin^{-1}(\cos 40°)$

9. $\cos^{-1}(\sin(-0.2))$ **10.** $\sin\left(\cos^{-1}\left(\frac{-1}{3}\right)\right)$

11. $\cos\left(\tan^{-1}\frac{1}{2}\right)$ **12.** $\tan(\tan^{-1}200)$

In Exercises 13–18, simplify the given expression.

13. $\sin(\cos^{-1}x)$ **14.** $\cos(\sin^{-1}x)$

15. $\cos(\tan^{-1}x)$ **16.** $\sin(\tan^{-1}x)$

17. $\tan(\cos^{-1}x)$ **18.** $\tan(\sec^{-1}x)$

In Exercises 19–32, differentiate the given function and simplify the answer whenever possible.

19. $y = \sin^{-1}\left(\dfrac{2x-1}{3}\right)$ **20.** $y = \tan^{-1}(ax+b)$

21. $y = \cos^{-1}\left(\dfrac{x-b}{a}\right)$ **22.** $f(x) = x\sin^{-1}x$

23. $f(t) = t\tan^{-1}t$ **24.** $u = z^2\sec^{-1}(1+z^2)$

25. $F(x) = (1+x^2)\tan^{-1}x$ **26.** $y = \sin^{-1}\dfrac{a}{x}$

27. $G(x) = \dfrac{\sin^{-1}x}{\sin^{-1}2x}$ **28.** $H(t) = \dfrac{\sin^{-1}t}{\sin t}$

29. $f(x) = (\sin^{-1}x^2)^{1/2}$ **30.** $y = \cos^{-1}\dfrac{a}{\sqrt{a^2+x^2}}$

31. $y = \sqrt{a^2-x^2} + a\sin^{-1}\dfrac{x}{a}$ $(a > 0)$

32. $y = a\cos^{-1}\left(1-\dfrac{x}{a}\right) - \sqrt{2ax-x^2}$ $(a > 0)$

33. Find the slope of the curve $\tan^{-1}\left(\dfrac{2x}{y}\right) = \dfrac{\pi x}{y^2}$ at the point $(1, 2)$.

34. Find equations of two straight lines tangent to the graph of $y = \sin^{-1}x$ and having slope 2.

35. Show that, on their respective domains, $\sin^{-1}$ and $\tan^{-1}$ are increasing functions and $\cos^{-1}$ is a decreasing function.

36. The derivative of $\sec^{-1}x$ is positive for every x in the domain of $\sec^{-1}$. Does this imply that $\sec^{-1}$ is increasing on its domain? Why?

37. Sketch the graph of $\csc^{-1}x$ and find its derivative.

38. Sketch the graph of $\cot^{-1}x$ and find its derivative.

39. Show that $\tan^{-1}x + \cot^{-1}x = \frac{\pi}{2}$ for $x > 0$. What is the sum if $x < 0$?

40. Find the derivative of $g(x) = \tan(\tan^{-1}x)$ and sketch the graph of g.

In Exercises 41–44, plot the graphs of the given functions by first calculating and simplifying the derivative of the function. Where is each function continuous? Where is it differentiable?

$*$ **41.** $\cos^{-1}(\cos x)$ $*$ **42.** $\sin^{-1}(\cos x)$

$*$ **43.** $\tan^{-1}(\tan x)$ $*$ **44.** $\tan^{-1}(\cot x)$

45. Show that $\sin^{-1}x = \tan^{-1}\left(\dfrac{x}{\sqrt{1-x^2}}\right)$ if $|x| < 1$.

46. Show that $\sec^{-1}x = \begin{cases} \tan^{-1}\sqrt{x^2-1} & \text{if } x \geq 1 \\ \pi - \tan^{-1}\sqrt{x^2-1} & \text{if } x \leq -1 \end{cases}$

47. Show that $\tan^{-1}x = \sin^{-1}\left(\dfrac{x}{\sqrt{1+x^2}}\right)$ for all x.

48. Show that $\sec^{-1}x = \begin{cases} \sin^{-1}\dfrac{\sqrt{x^2-1}}{x} & \text{if } x \geq 1 \\ \pi - \sin^{-1}\dfrac{\sqrt{x^2-1}}{x} & \text{if } x \leq -1 \end{cases}$

$*$ **49.** Show that the function $f(x)$ of Example 9 is also constant on the interval $]-\infty, -1[$. Find the value of the constant. *Hint:* find $\lim_{x\to-\infty}f(x)$.

$*$ **50.** Find the derivative of $f(x) = x - \tan^{-1}(\tan x)$. What does your answer imply about $f(x)$? Calculate $f(0)$ and $f(\pi)$. Is there a contradiction here?

$*$ **51.** Find the derivative of $f(x) = x - \sin^{-1}(\sin x)$ for $-\pi \leq x \leq \pi$ and sketch the graph of f on that interval.

In Exercises 52–55, solve the initial-value problems.

$\diamond$ **52.** $\begin{cases} y' = \dfrac{1}{1+x^2} \\ y(0) = 1 \end{cases}$ $\diamond$ **53.** $\begin{cases} y' = \dfrac{1}{9+x^2} \\ y(3) = 2 \end{cases}$

$\diamond$ **54.** $\begin{cases} y' = \dfrac{1}{\sqrt{1-x^2}} \\ y(1/2) = 1 \end{cases}$ $\diamond$ **55.** $\begin{cases} y' = \dfrac{4}{\sqrt{25-x^2}} \\ y(0) = 0 \end{cases}$

3.6 Hyperbolic Functions

Any function defined on the real line can be expressed (in a unique way) as the sum of an even function and an odd function. (See Exercise 35 of Section P.5.) The **hyperbolic functions** $\cosh x$ and $\sinh x$ are, respectively, the even and odd functions whose sum is the exponential function e^x.

DEFINITION **15**

> **The hyperbolic cosine and hyperbolic sine functions**
>
> For any real x the **hyperbolic cosine**, $\cosh x$, and the **hyperbolic sine**, $\sinh x$, are defined by
>
> $$\cosh x = \frac{e^x + e^{-x}}{2}, \qquad \sinh x = \frac{e^x - e^{-x}}{2}.$$

(The symbol "sinh" is somewhat hard to pronounce as written. Some people say "shine," and others say "sinch.") Recall that cosine and sine are called *circular functions* because, for any t, the point $(\cos t, \sin t)$ lies on the circle with equation $x^2 + y^2 = 1$. Similarly, cosh and sinh are called *hyperbolic functions* because the point $(\cosh t, \sinh t)$ lies on the rectangular hyperbola with equation $x^2 - y^2 = 1$,

$$\cosh^2 t - \sinh^2 t = 1 \quad \text{for any real } t.$$

To see this, observe that

$$\cosh^2 t - \sinh^2 t = \left(\frac{e^t + e^{-t}}{2}\right)^2 - \left(\frac{e^t - e^{-t}}{2}\right)^2$$
$$= \frac{1}{4}\left(e^{2t} + 2 + e^{-2t} - (e^{2t} - 2 + e^{-2t})\right)$$
$$= \frac{1}{4}(2 + 2) = 1.$$

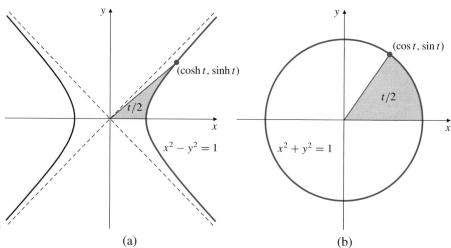

Figure 3.26 Both shaded areas are $t/2$ square units

(a)   (b)

There is no interpretation of t as an arc length or angle as there was in the circular case; however, the *area* of the *hyperbolic sector* bounded by $y = 0$, the hyperbola $x^2 - y^2 = 1$, and the ray from the origin to $(\cosh t, \sinh t)$ is $t/2$ square units (see Exercise 21 of Section 8.4), just as is the area of the circular sector bounded by $y = 0$, the circle $x^2 + y^2 = 1$, and the ray from the origin to $(\cos t, \sin t)$. (See Figure 3.26.)

Observe that, similar to the corresponding values of $\cos x$ and $\sin x$, we have

$$\cosh 0 = 1 \quad \text{and} \quad \sinh 0 = 0,$$

and $\cosh x$, like $\cos x$, is an even function, and $\sinh x$, like $\sin x$, is an odd function:

$$\cosh(-x) = \cosh x, \qquad \sinh(-x) = -\sinh x .$$

Many other properties of the hyperbolic functions resemble those of the corresponding circular functions, sometimes with signs changed.

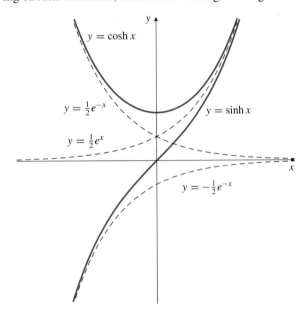

Figure 3.27   The graphs of cosh and sinh and some exponential graphs to which they are asymptotic

Example 1   Show that

$$\frac{d}{dx}\cosh x = \sinh x \quad \text{and} \quad \frac{d}{dx}\sinh x = \cosh x.$$

Solution   We have
$$\frac{d}{dx}\cosh x = \frac{d}{dx}\frac{e^x + e^{-x}}{2} = \frac{e^x + e^{-x}(-1)}{2} = \sinh x$$
$$\frac{d}{dx}\sinh x = \frac{d}{dx}\frac{e^x - e^{-x}}{2} = \frac{e^x - e^{-x}(-1)}{2} = \cosh x.$$

The following addition formulas and double angle formulas can be checked algebraically by using the definition of cosh and sinh and the laws of exponents:

$$\cosh(x + y) = \cosh x \cosh y + \sinh x \sinh y,$$
$$\sinh(x + y) = \sinh x \cosh y + \cosh x \sinh y,$$
$$\cosh(2x) = \cosh^2 x + \sinh^2 x = 1 + 2\sinh^2 x = 2\cosh^2 x - 1,$$
$$\sinh(2x) = 2\sinh x \cosh x.$$

The graphs of cosh and sinh are shown in Figure 3.27. The graph $y = \cosh x$ is called a **catenary**. A chain hanging by its ends will assume the shape of a catenary.

By analogy with the trigonometric functions, four other hyperbolic functions can be defined in terms of cosh and sinh.

DEFINITION | **16**

Other hyperbolic functions

$$\tanh x = \frac{\sinh x}{\cosh x} = \frac{e^x - e^{-x}}{e^x + e^{-x}} \qquad \operatorname{sech} x = \frac{1}{\cosh x} = \frac{2}{e^x + e^{-x}}$$

$$\coth x = \frac{\cosh x}{\sinh x} = \frac{e^x + e^{-x}}{e^x - e^{-x}} \qquad \operatorname{csch} x = \frac{1}{\sinh x} = \frac{2}{e^x - e^{-x}}$$

Multiplying the numerator and denominator of the fraction defining $\tanh x$ by e^{-x} and e^x, respectively, we obtain

$$\lim_{x \to \infty} \tanh x = \lim_{x \to \infty} \frac{1 - e^{-2x}}{1 + e^{-2x}} = 1 \qquad \text{and}$$

$$\lim_{x \to -\infty} \tanh x = \lim_{x \to -\infty} \frac{e^{2x} - 1}{e^{2x} + 1} = -1,$$

so that the graph of $y = \tanh x$ has two horizontal asymptotes. The graph resembles that of the inverse tangent function in shape, as you can see in Figure 3.28.

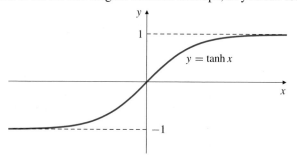

Figure 3.28 The graph of $\tanh x$

The derivatives of the remaining hyperbolic functions

$$\frac{d}{dx} \tanh x = \operatorname{sech}^2 x \qquad \frac{d}{dx} \operatorname{sech} x = -\operatorname{sech} x \, \tanh x$$

$$\frac{d}{dx} \coth x = -\operatorname{csch}^2 x \qquad \frac{d}{dx} \operatorname{csch} x = -\operatorname{csch} x \, \coth x$$

are easily calculated from those of $\cosh x$ and $\sinh x$ using the Reciprocal and Quotient Rules. For example,

$$\frac{d}{dx} \tanh x = \frac{d}{dx} \frac{\sinh x}{\cosh x} = \frac{(\cosh x)(\cosh x) - (\sinh x)(\sinh x)}{\cosh^2 x}$$

$$= \frac{1}{\cosh^2 x} = \operatorname{sech}^2 x.$$

Remark The distinction between trigonometric and hyperbolic functions largely disappears if we allow complex numbers instead of just real numbers as variables. If i is the imaginary unit (so that $i^2 = -1$), then

$$e^{ix} = \cos x + i \sin x \qquad \text{and} \qquad e^{-ix} = \cos x - i \sin x.$$

(See Appendix I.) Therefore,

$$\cosh(ix) = \frac{e^{ix} + e^{-ix}}{2} = \cos x, \qquad \cos(ix) = \cosh(-x) = \cosh x,$$

$$\sinh(ix) = \frac{e^{ix} - e^{-ix}}{2} = i \sin x, \qquad \sin(ix) = \frac{1}{i} \sinh(-x) = i \sinh x.$$

Inverse Hyperbolic Functions

The functions sinh and tanh are increasing and therefore one-to-one and invertible on the whole real line. Their inverses are denoted $\sinh^{-1}$ and $\tanh^{-1}$, respectively:

$$y = \sinh^{-1} x \quad \Longleftrightarrow \quad x = \sinh y,$$
$$y = \tanh^{-1} x \quad \Longleftrightarrow \quad x = \tanh y.$$

Since the hyperbolic functions are defined in terms of exponentials, it is not surprising that their inverses can be expressed in terms of logarithms.

Example 2 Express $\sinh^{-1} x$ and $\tanh^{-1} x$ in terms of logarithms.

Solution Let $y = \sinh^{-1} x$. Then

$$x = \sinh y = \frac{e^y - e^{-y}}{2} = \frac{(e^y)^2 - 1}{2e^y}.$$

(We multiplied the numerator and denominator of the first fraction by e^y to get the second fraction.) Therefore,

$$(e^y)^2 - 2xe^y - 1 = 0.$$

This is a quadratic equation in e^y, and it can be solved by the quadratic formula:

$$e^y = \frac{2x \pm \sqrt{4x^2 + 4}}{2} = x \pm \sqrt{x^2 + 1}.$$

Note that $\sqrt{x^2 + 1} > x$. Since e^y cannot be negative, we need to use the positive square root:

$$e^y = x + \sqrt{x^2 + 1}.$$

Hence $y = \ln\left(x + \sqrt{x^2 + 1}\right)$, and we have

$$\sinh^{-1} x = \ln\left(x + \sqrt{x^2 + 1}\right).$$

Now let $y = \tanh^{-1} x$. Then

$$x = \tanh y = \frac{e^y - e^{-y}}{e^y + e^{-y}} = \frac{e^{2y} - 1}{e^{2y} + 1} \qquad (-1 < x < 1),$$

$$xe^{2y} + x = e^{2y} - 1,$$

$$e^{2y} = \frac{1 + x}{1 - x}, \qquad y = \frac{1}{2} \ln\left(\frac{1 + x}{1 - x}\right).$$

Thus

$$\tanh^{-1} x = \frac{1}{2} \ln\left(\frac{1 + x}{1 - x}\right), \qquad (-1 < x < 1).$$

Since cosh is not one-to-one, its domain must be restricted before an inverse can be defined. Let us define the principal value of cosh to be

$$\text{Cosh}\, x = \cosh x \qquad (x \geq 0).$$

The inverse, $\cosh^{-1}$, is then defined by

$$y = \cosh^{-1} x \iff x = \text{Cosh}\, y$$
$$\iff x = \cosh y \qquad (y \geq 0).$$

As we did for $\sinh^{-1}$, we can obtain the formula

$$\cosh^{-1} x = \ln\left(x + \sqrt{x^2 - 1}\right), \qquad (x \geq 1).$$

Exercises 3.6

1. Verify the formulas for the derivatives of $\text{sech}\, x$, $\text{csch}\, x$, and $\coth x$ given in this section.

2. Verify the addition formulas

$$\cosh(x + y) = \cosh x \, \cosh y + \sinh x \, \sinh y,$$
$$\sinh(x + y) = \sinh x \, \cosh y + \cosh x \, \sinh y.$$

Proceed by expanding the right-hand side of each identity in terms of exponentials. Find similar formulas for $\cosh(x - y)$ and $\sinh(x - y)$.

3. Obtain addition formulas for $\tanh(x + y)$ and $\tanh(x - y)$ from those for sinh and cosh.

4. Sketch the graphs of $y = \coth x$, $y = \text{sech}\, x$, and $y = \text{csch}\, x$, showing any asymptotes.

5. Calculate the derivatives of $\sinh^{-1} x$, $\cosh^{-1} x$, and $\tanh^{-1} x$. Hence express each of the indefinite integrals

$$\int \frac{dx}{\sqrt{x^2 + 1}}, \qquad \int \frac{dx}{\sqrt{x^2 - 1}}, \qquad \int \frac{dx}{1 - x^2}$$

in terms of inverse hyperbolic functions.

6. Calculate the derivatives of the functions $\sinh^{-1}(x/a)$, $\cosh^{-1}(x/a)$, and $\tanh^{-1}(x/a)$ (where $a > 0$), and use your answers to provide formulas for certain indefinite integrals.

7. Simplify the following expressions: (a) $\sinh \ln x$, (b) $\cosh \ln x$, (c) $\tanh \ln x$, (d) $\dfrac{\cosh \ln x + \sinh \ln x}{\cosh \ln x - \sinh \ln x}$.

8. Let $\text{csch}^{-1} x = \sinh^{-1}(1/x)$. Find the domain, range, and derivative of $\text{csch}^{-1} x$, and sketch its graph. Express $\text{csch}^{-1} x$ in terms of logarithms.

9. Do an analogous version of Exercise 8 for $\coth^{-1} x$.

* 10. Define $\text{Sech}\, x$ at a suitably restricted version of $\text{sech}\, x$, and repeat Exercise 8 for the function $\text{Sech}^{-1} x$.

◈ 11. Show that the functions $f_{A,B}(x) = Ae^{kx} + Be^{-kx}$ and $g_{C,D}(x) = C \cosh kx + D \sinh kx$ are both solutions of the differential equation $y'' - k^2 y = 0$. (They are both general solutions.) Express $f_{A,B}$ in terms of $g_{C,D}$, and express $g_{C,D}$ in terms of $f_{A,B}$.

12. Show that $h_{L,M}(x) = L \cosh k(x - a) + M \sinh k(x - a)$ is also a solution of the differential equation in the previous exercise. Express $h_{L,M}$ in terms of the function $f_{A,B}$ above.

13. Solve the initial-value problem $y'' - k^2 y = 0$, $y(a) = y_0$, $y'(a) = v_0$. Express the solution in terms of the function $h_{L,M}$ of Exercise 12.

3.7 Second-Order Linear DEs with Constant Coefficients

A differential equation of the form

$$a\,y'' + b\,y' + cy = 0, \qquad (*)$$

where a, b, and c are constants and $a \neq 0$, is called a **second-order, linear, homogeneous** differential equation with constant coefficients. The *second-order* refers to the presence of a second derivative; the terms *linear* and *homogeneous* refer to the fact that if $y_1(t)$ and $y_2(t)$ are two solutions of the equation, then so is $y(t) = Ay_1(t) + By_2(t)$ for any constants A and B:

If $ay_1''(t) + by_1'(t) + cy_1(t) = 0$ and $ay_2''(t) + by_2'(t) + cy_2(t) = 0$,
and if $y(t) = Ay_1(t) + By_2(t)$, then $ay''(t) + by'(t) + cy(t) = 0$.

(Throughout this section we will assume that the independent variable in our functions is t rather than x, so the prime $(')$ refers to the derivative d/dt. This is because in most applications of such equations the independent variable is time.)

Equations of type $(*)$ arise in many applications of mathematics. In particular, they can model mechanical vibrations such as the motion of a mass suspended from an elastic spring or the current in certain electrical circuits. In most such applications the three constants a, b, and c are positive, though sometimes we may have $b = 0$.

Recipe for Solving $ay'' + by' + cy = 0$

In Section 3.4 we observed that the first-order, constant-coefficient equation $y' = ky$ has solution $y = Ce^{kt}$. Let us try to find a solution of equation $(*)$ having the form $y = e^{rt}$. Substituting this expression into equation $(*)$, we obtain

$$ar^2 e^{rt} + bre^{rt} + ce^{rt} = 0.$$

Since e^{rt} is never zero, $y = e^{rt}$ will be a solution of the differential equation $(*)$ if and only if r satisfies the quadratic **auxiliary equation**

$$ar^2 + br + c = 0, \qquad (**)$$

which has roots given by the quadratic formula:

$$r = \frac{-b \pm \sqrt{b^2 - 4ac}}{2a} = -\frac{b}{2a} \pm \frac{\sqrt{D}}{2a},$$

where $D = b^2 - 4ac$ is called the **discriminant** of the auxiliary equation $(**)$.

There are three cases to consider, depending on whether the discriminant D is positive, zero, or negative.

CASE I Suppose $D = b^2 - 4ac > 0$. Then the auxiliary equation has two different real roots, r_1 and r_2, given by

$$r_1 = \frac{-b - \sqrt{D}}{2a}, \qquad r_2 = \frac{-b + \sqrt{D}}{2a}.$$

(Sometimes these roots can be found easily by factoring the left side of the auxiliary equation.) In this case both $y = y_1(t) = e^{r_1 t}$ and $y = y_2(t) = e^{r_2 t}$ are solutions of the differential equation (∗), and neither is a multiple of the other. As noted above, the function

$$y = A e^{r_1 t} + B e^{r_2 t}$$

is also a solution for any choice of the constants A and B. Since the differential equation is of second order and this solution involves two arbitrary constants, we suspect it is the **general solution**, that is, that every solution of the differential equation can be written in this form. Exercise 18 at the end of this section outlines a way to prove this.

CASE II Suppose $D = b^2 - 4ac = 0$. Then the auxiliary equation has two equal roots, $r_1 = r_2 = -b/(2a) = r$, say. Certainly $y = e^{rt}$ is a solution of (∗). We can find the general solution by letting $y = e^{rt}u(t)$ and calculating the first two derivatives of y:

$$y' = e^{rt}\big(u'(t) + ru(t)\big),$$
$$y'' = e^{rt}\big(u''(t) + 2ru'(t) + r^2u(t)\big).$$

Substituting these expressions into (∗), we obtain

$$e^{rt}\big(au''(t) + (2ar + b)u'(t) + (ar^2 + br + c)u(t)\big) = 0.$$

Since $e^{rt} \neq 0$, $2ar + b = 0$ and r satisfies (∗∗), this equation reduces to $u''(t) = 0$, which has general solution $u(t) = A + Bt$ for arbitrary constants A and B. Thus the general solution of (∗) in this case is

$$y = A e^{rt} + Bt e^{rt}.$$

CASE III Suppose $D = b^2 - 4ac < 0$. Then the auxiliary equation (∗∗) has complex conjugate roots given by

$$r = \frac{-b \pm \sqrt{b^2 - 4ac}}{2a} = k \pm i\omega,$$

where $k = -b/(2a)$, $\omega = \sqrt{4ac - b^2}/(2a)$, and i is the imaginary unit ($i^2 = -1$; see Appendix I). As in Case I, the functions $y_1^*(t) = e^{(k+i\omega)t}$ and $y_2^*(t) = e^{(k-i\omega)t}$ are two independent solutions of (∗), but they are not real-valued. However, since

$$e^{ix} = \cos x + i \sin x \qquad \text{and} \qquad e^{-ix} = \cos x - i \sin x$$

(as noted in the previous section and in Appendix I), we can find two real-valued functions that are solutions of (∗) by suitably combining y_1^* and y_2^*:

$$y_1(t) = \frac{1}{2}y_1^*(t) + \frac{1}{2}y_2^*(t) = e^{kt}\cos(\omega t),$$
$$y_2(t) = \frac{1}{2i}y_1^*(t) - \frac{1}{2i}y_2^*(t) = e^{kt}\sin(\omega t).$$

Therefore, the general solution of (∗) in this case is

$$y = A\, e^{kt} \cos(\omega t) + B\, e^{kt} \sin(\omega t).$$

The following examples illustrate the recipe for solving (∗) in each of the three cases.

Example 1   Find the general solution of $y'' + y' - 2y = 0$.

Solution   The auxiliary equation is $r^2 + r - 2 = 0$, or $(r + 2)(r - 1) = 0$. The auxiliary roots are $r_1 = -2$ and $r_2 = 1$, which are real and unequal. According to Case I, the general solution of the differential equation is

$$y = A\, e^{-2t} + Be^t.$$

Example 2   Find the general solution of $y'' + 6y' + 9y = 0$.

Solution   The auxiliary equation is $r^2 + 6r + 9 = 0$, or $(r + 3)^2 = 0$, which has equal roots $r = -3$. According to Case II, the general solution of the differential equation is

$$y = A\, e^{-3t} + Bt\, e^{-3t}.$$

Example 3   Find the general solution of $y'' + 4y' + 13y = 0$.

Solution   The auxiliary equation is $r^2 + 4r + 13 = 0$, which has solutions

$$r = \frac{-4 \pm \sqrt{16 - 52}}{2} = \frac{-4 \pm \sqrt{-36}}{2} = -2 \pm 3i.$$

Thus $k = -2$ and $\omega = 3$. According to Case III, the general solution of the given differential equation is

$$y = A\, e^{-2t} \cos(3t) + B\, e^{-2t} \sin(3t).$$

Initial-value problems for $ay'' + by' + cy = 0$ specify values for y and y' at an initial point. These values can be used to determine the values of the constants A and B in the general solution, so the initial-value problem has a unique solution.

Example 4   Solve the initial-value problem

$$\begin{cases} y'' + 2y' + 2y = 0 \\ y(0) = 2 \\ y'(0) = -3. \end{cases}$$

Solution   The auxiliary equation is $r^2 + 2r + 2 = 0$, which has roots

$$r = \frac{-2 \pm \sqrt{4-8}}{2} = -1 \pm i.$$

Thus Case III applies, $k = -1$ and $\omega = 1$. Thus, the differential equation has the general solution

$$y = A e^{-t} \cos t + B e^{-t} \sin t.$$

Also,

$$y' = e^{-t}\left(-A \cos t - B \sin t - A \sin t + B \cos t\right)$$
$$= (B - A) e^{-t} \cos t - (A + B) e^{-t} \sin t.$$

Applying the initial conditions $y(0) = 2$ and $y'(0) = -3$, we obtain $A = 2$ and $B - A = -3$. Hence, $B = -1$ and the initial-value problem has the solution

$$y = 2 e^{-t} \cos t - e^{-t} \sin t.$$

Simple Harmonic Motion

Many natural phenomena exhibit periodic behaviour. The swinging of a clock pendulum, the vibrating of a guitar string or drum membrane, the altitude of a rider on a rotating ferris wheel, the motion of an object floating in wavy seas, and the voltage produced by an alternating current generator are but a few examples where quantities depend on time in a periodic way. Being periodic, the circular functions sine and cosine provide a useful model for such behaviour.

It often happens that a quantity displaced from an equilibrium value experiences a restoring force that tends to move it back in the direction of its equilibrium. Besides the obvious examples of elastic motions in physics, one can imagine such a model applying, say, to a biological population in equilibrium with its food supply or the price of a commodity in an elastic economy where increasing price causes decreasing demand and hence decreasing price. In the simplest models, the restoring force is proportional to the amount of displacement from equilibrium. Such a force causes the quantity to oscillate sinusoidally; we say that it executes *simple harmonic motion*.

As a specific example, suppose a mass m is suspended by an elastic spring so that it hangs unmoving in its equilibrium position. If it is displaced vertically by an amount y from this position, a force is exerted by the spring, directed to restore the mass to its equilibrium position. (See Figure 3.29.) This force is proportional to the displacement (Hooke's Law); its magnitude is $-ky$, where k is a positive constant called the **spring constant**. Assuming the spring is weightless, this force imparts to the mass m an acceleration d^2y/dt^2 that satisfies, by Newton's Second Law, $m(d^2y/dt^2) = -ky$ (mass × acceleration = force). Dividing this equation by m, we obtain the equation

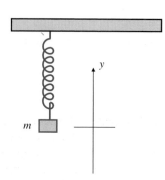

Figure 3.29

$$\frac{d^2y}{dt^2} + \omega^2 y = 0, \qquad \text{where} \quad \omega^2 = \frac{k}{m}.$$

The second-order differential equation

$$\frac{d^2y}{dt^2} + \omega^2 y = 0$$

is called the **equation of simple harmonic motion**. Its auxiliary equation, $r^2 + \omega^2 = 0$, has complex roots $r = \pm i\omega$, so it has general solution

$$y = A\cos\omega t + B\sin\omega t,$$

where A and B are arbitrary constants.

For any values of the constants R and t_0, the function

$$y = R\cos\big(\omega(t - t_0)\big)$$

is also a general solution of the differential equation of simple harmonic motion. If we expand this formula using the addition formula for cosine, we get

$$y = R\cos\omega t_0\cos\omega t + R\sin\omega t_0\sin\omega t$$
$$= A\cos\omega t + B\sin\omega t,$$

where

$$A = R\cos(\omega t_0), \qquad\qquad B = R\sin(\omega t_0),$$
$$R^2 = A^2 + B^2, \qquad\qquad \tan(\omega t_0) = B/A.$$

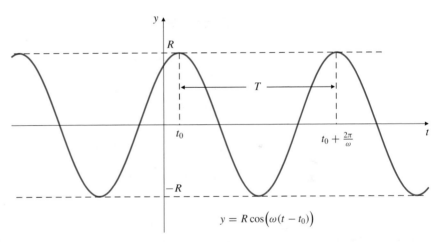

$$y = R\cos\big(\omega(t - t_0)\big)$$

Figure 3.30   Simple harmonic motion

The constants A and B are related to the position y_0 and the velocity v_0 of the mass m at time $t = 0$:

$$y_0 = y(0) = A\cos 0 + B\sin 0 = A,$$
$$v_0 = y'(0) = -A\omega\sin 0 + B\omega\cos 0 = B\omega.$$

The constant $R = \sqrt{A^2 + B^2}$ is called the **amplitude** of the motion. Because $\cos x$ oscillates between -1 and 1, the displacement y varies between $-R$ and R. Note in Figure 3.30 that the graph of the displacement as a function of time is the curve $y = R\cos\omega t$ shifted t_0 units to the right. The number t_0 is called the **time-shift**. (The related quantity ωt_0 is called a **phase-shift**.) The **period** of this curve is $T = 2\pi/\omega$; it is the time interval between consecutive instants when the mass is at the same height moving in the same direction. The reciprocal $1/T$ of the period is called the **frequency** of the motion. It is usually measured in Hertz (Hz), that is, cycles per second. The quantity $\omega = 2\pi/T$ is called the **circular frequency**. It is measured in radians per second since 1 cycle = 1 revolution = 2π radians.

Example 5 Solve the initial-value problem

$$\begin{cases} y'' + 16y = 0 \\ y(0) = -6 \\ y'(0) = 32. \end{cases}$$

Find the amplitude, frequency, and period of the solution.

Solution Here, $\omega^2 = 16$ so $\omega = 4$. The solution is of the form

$$y = A\cos(4t) + B\sin(4t).$$

Since $y(0) = -6$, we have $A = -6$. Also, $y'(t) = -4A\sin(4t) + 4B\cos(4t)$. Since $y'(0) = 32$, we have $4B = 32$, or $B = 8$. Thus, the solution is

$$y = -6\cos(4t) + 8\sin(4t).$$

The amplitude is $\sqrt{(-6)^2 + 8^2} = 10$, the frequency is $\omega/(2\pi) \approx 0.637$ Hz, and the period is $2\pi/\omega \approx 1.57$ seconds.

■

Example 6 **(Spring-mass problem)** Suppose that a 100 gram mass is suspended from a spring and that a force of 3×10^4 dynes (3×10^4 g-cm/s^2) is required to produce a displacement from equilibrium of 1/3 cm. At time $t = 0$ the mass is pulled down 2 cm below equilibrium and flicked upward with a velocity of 60 cm/s. Find its subsequent displacement at any time $t > 0$. Find the frequency, period, amplitude, and time-shift of the motion. Express the position of the mass at time t in terms of the amplitude and the time-shift.

Solution The spring constant k is determined from Hooke's Law, $F = -ky$. Here $F = -3 \times 10^4$ g-cm/s^2 is the force of the spring on the mass displaced 1/3 cm:

$$-3 \times 10^4 = -\frac{1}{3}k,$$

so $k = 9 \times 10^4$ g/s^2. Hence, the circular frequency is $\omega = \sqrt{k/m} = 30$ rad/s, the frequency is $\omega/2\pi = 15/\pi \approx 4.77$ Hz, and the period is $2\pi/\omega \approx 0.209$ s.

Since the displacement at time $t = 0$ is $y_0 = -2$ and the velocity at that time is $v_0 = 60$, the subsequent displacement is $y = A\cos(30t) + B\sin(30t)$, where $A = y_0 = -2$ and $B = v_0/\omega = 60/30 = 2$. Thus

$$y = -2\cos(30t) + 2\sin(30t), \qquad (y \text{ in cm, } t \text{ in seconds}).$$

The amplitude of the motion is $R = \sqrt{(-2)^2 + 2^2} = 2\sqrt{2} \approx 2.83$ cm. The time-shift t_0 must satisfy

$$-2 = A = R\cos(\omega t_0) = 2\sqrt{2}\cos(30t_0),$$
$$2 = B = R\sin(\omega t_0) = 2\sqrt{2}\sin(30t_0),$$

so $\sin(30t_0) = 1/\sqrt{2} = -\cos(30t_0)$. Hence the phase-shift is $30t_0 = 3\pi/4$ radians, and the time-shift is $t_0 = \pi/40 \approx 0.0785$ s. The position of the mass at time $t > 0$ is also given by

$$y = 2\sqrt{2}\cos\left(30(t - \frac{\pi}{40})\right).$$

Damped Harmonic Motion

If a and c are positive and $b = 0$, then equation

$$ay'' + by' + cy = 0$$

is the differential equation of simple harmonic motion and has oscillatory solutions of fixed amplitude as shown above. If $a > 0$, $b > 0$, and $c > 0$, then the roots of the auxiliary equation are either negative real numbers or, if $b^2 < 4ac$, complex numbers $k \pm i\omega$ with negative real parts $k = -b/(2a)$ (Case III). In this latter case the solutions still oscillate, but the amplitude diminishes exponentially as $t \to \infty$ because of the factor $e^{kt} = e^{-(b/2a)t}$. (See Exercise 17 below.) A system whose behaviour is modelled by such an equation is said to exhibit **damped harmonic motion**. If $b^2 = 4ac$ (Case II), the system is said to be **critically damped**, and if $b^2 > 4ac$ (Case I), it is **overdamped**. In these cases the behaviour is no longer oscillatory. (See Figure 3.31. Imagine a mass suspended by a spring in a jar of oil.)

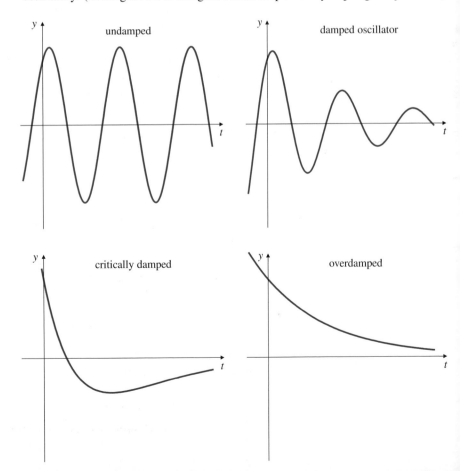

Figure 3.31

Undamped oscillator ($b = 0$)

Damped oscillator ($b > 0$, $b^2 < 4ac$)

Critically damped case ($b > 0$, $b^2 = 4ac$)

Overdamped case ($b > 0$, $b^2 > 4ac$)

Non-homogeneous Equations and Resonance

A second-order, linear, constant coefficient differential equation of the form

$$ay'' + by' + cy = f(t) \qquad (\text{***})$$

is *non-homogeneous* because the term $f(t)$ is not a multiple of y or one of its derivatives like the other terms of the equation. This term can represent an "external" force applied to the system being modelled by the equation. For example, it might represent an outside force being applied to the mass suspended by the spring we considered earlier or an applied voltage in an electric circuit.

If we can find any one solution $y(t) = y_P(t)$ of equation (***) (called a *particular solution*), then the general solution of (***) is $y(t) = y_P(t) + y_H(t)$, where $y_H(t)$ is the general solution of the corresponding homogeneous equation

$$ay'' + by' + cy = 0.$$

Example 7   It is easily seen that $y_P(t) = 1$ is a solution of

$$y'' + 3y' + 2y = 2.$$

(Here $f(t) = 2$.) The corresponding homogeneous equation $y'' + 3y' + 2y = 0$ has auxiliary equation $r^2 + 3r + 2 = 0$, with roots $r = -2$ and $r = -1$. Thus, the non-homogeneous equation has general solution

$$y = 1 + Ae^{-2t} + Be^{-t}$$

with arbitrary constants A and B. ∎

Techniques for solving non-homogeneous equations are beyond the scope of this book, but we can sometimes guess the form of a solution and thus find one. For a sinusoidal forcing term such as $f(t) = \sin(\lambda t)$ we can try

$$y_P(t) = A \cos(\lambda t) + B \sin(\lambda t).$$

This will work as long $\sin(\lambda t)$ is not a solution of the corresponding homogeneous equation. If $\sin(\lambda t)$ is a solution of the corresponding homogeneous equation, try

$$y_P(t) = At \cos(\lambda t) + Bt \sin(\lambda t).$$

Example 8   (**Resonance**) Consider the initial-value problem

$$\begin{cases} y'' + y = \sin(\lambda t) \\ \quad y(0) = 0 \\ \quad y'(0) = 1, \end{cases}$$

where $\lambda \neq 1$. Substituting the trial solution suggested above, we obtain $A = 0$ and $B = 1/(1 - \lambda^2)$. Since the homogeneous equation $y'' + y = 0$ has general solution $y = C \cos t + D \sin t$, the given DE has general solution

$$y = \frac{1}{1 - \lambda^2} \sin(\lambda t) + C \cos t + D \sin t.$$

Applying the two initial conditions leads to the values $C = 0$ and $D = (1 - \lambda - \lambda^2)/(1 - \lambda^2)$, so the IVP has solution

$$y(t) = y_\lambda(t) = \frac{\sin(\lambda t) + (1 - \lambda - \lambda^2)\sin t}{1 - \lambda^2}.$$

For $\lambda = 1$ the nonhomogeneous term in the DE is a solution of the homogeneous equation $y'' + y = 0$, so we must try for a particular solution of another form, namely $y = At \cos t + Bt \sin t$. In this case the solution of the initial-value problem is

$$y_1(t) = \frac{3 \sin t - t \cos t}{2}.$$

(This solution can also be found by calculating $\lim_{\lambda \to 1} y_\lambda(t)$ using l'Hôpital's Rule; see Section 4.9.) Observe that this solution is unbounded; the amplitude of the oscillations becomes larger and larger as t increases. In contrast, the solutions $y_\lambda(t)$ for $\lambda \neq 1$ are bounded for all t, although they can become quite large for some values of t if λ is close to 1. The graphs of the solutions $y_{0.9}(t)$, $y_{0.95}(t)$, and $y_1(t)$ on the interval $-10 \leq t \leq 100$ are shown in Figure 3.32.

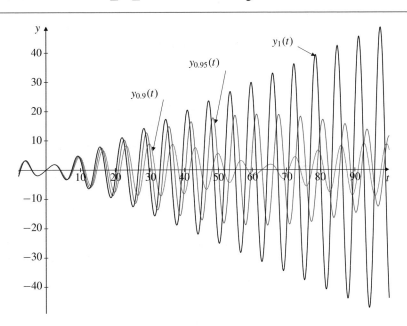

Figure 3.32   Resonance

The phenomenon illustrated in the above example is called **resonance**. Vibrating mechanical systems have natural frequencies at which they will vibrate. If you try to force them to vibrate at a different frequency, the amplitude of the vibrations will themselves vary sinusoidally over time, producing an effect known as **beats**. The amplitudes of the beats can grow quite large, and the period of the beats lengthens as the forcing frequency approaches the natural frequency of the system. If the system has no resistive damping (the one illustrated above has no damping), then forcing vibrations at the natural frequency will cause the system to vibrate at ever increasing amplitudes.

As a concrete example, if you push a child on a swing, the swing will rise highest if your pushes are timed to have the same frequency as the natural frequency of the swing. Resonance is used in the design of tuning circuits of radios; the circuit is tuned (usually by a variable capacitor) so that its natural frequency of oscillation is the frequency of the station being tuned in. The circuit then responds much more strongly to the signal received from that station than to others on different frequencies.

Remark   Maple has a `dsolve` routine for solving (some) differential equations and initial value problems. This routine takes as inputs a DE and, if desired, initial conditions for it. We illustrate for the equation $y'' + 2y' + 5y = 25t + 20$ (assuming that the independent variable is t):

```
>   DE := (D@@2)(y)(t)+2*D(y)(t)+5*y(t)=25*t+20;
```

$$DE := (D^{(2)})(y)(t) + 2D(y)(t) + 5y(t) = 25t + 20$$

```
>   dsolve(DE, y(t));
```

$$y(t) = 2 + 5t + \_C1e^{(-t)} \cos(2t) + \_C2e^{(-t)} \sin(2t)$$

Note Maple's use of $\_C1$ and $\_C2$ for arbitrary constants. For an initial-value problem we supply the DE and its initial conditions to `dsolve` as a single list or set argument enclosed in square brackets or braces:

```
>   dsolve([DE, y(0)=3, D(y)(0)=-2], y(t));
```

$$y(t) = 2 + 5t + e^{(-t)} \cos(2t) - 3e^{(-t)} \sin(2t)$$

You might think that this output indicates that y has been defined as a function of t and you can find a decimal value for, say, $y(1)$ by giving the input `evalf(y(1))`. But this won't work. In fact, the output of the `dsolve` is just an equation with left side the symbol $y(t)$. We can, however, use this output to define y as a function of t as follows:

```
>   y := unapply(op(2,%),t);
```

$$y := t \rightarrow 2 + 5t + e^{(-t)} \cos(2t) - 3e^{(-t)} \sin(2t)$$

The `op(2,%)` in the `unapply` command refers to the second operand of the previous result (i.e., the right side of equation output from the `dsolve`). `unapply(f,t)` converts an expression f to a function of t. To confirm:

```
>   evalf(y(1));
```

$$5.843372646$$

Exercises 3.7

In Exercises 1–12, find the general solutions for the given equations.

1. $y'' + 7y' + 10y = 0$

2. $y'' - 2y' - 3y = 0$

3. $y'' + 2y' = 0$

4. $4y'' - 4y' - 3y = 0$

5. $y'' + 8y' + 16y = 0$

6. $y'' - 2y' + y = 0$

7. $y'' - 6y' + 10y = 0$

8. $9y'' + 6y' + y = 0$

9. $y'' + 2y' + 5y = 0$

10. $y'' - 4y' + 5y = 0$

11. $y'' + 2y' + 3y = 0$

12. $y'' + y' + y = 0$

In Exercises 13–15, solve the given initial-value problems.

13. $\begin{cases} 2y'' + 5y' - 3y = 0 \\ y(0) = 1 \\ y'(0) = 0. \end{cases}$

14. $\begin{cases} y'' + 10y' + 25y = 0 \\ y(1) = 0 \\ y'(1) = 2. \end{cases}$

15. $\begin{cases} y'' + 4y' + 5y = 0 \\ y(0) = 2 \\ y'(0) = 2. \end{cases}$

* 16. Show that if $\epsilon \neq 0$, the function $y_\epsilon(t) = \dfrac{e^{(1+\epsilon)t} - e^t}{\epsilon}$

satisfies the equation $y'' - (2 + \epsilon)y' + (1 + \epsilon)y = 0$. Caclulate $y(t) = \lim_{\epsilon \to 0} y_\epsilon(t)$ and verify that, as expected, it is a solution of $y'' - 2y' + y = 0$.

* 17. If $a > 0$, $b > 0$, and $c > 0$, prove that all solutions of the differential equation $ay'' + by' + cy = 0$ satisfy $\lim_{t \to \infty} y(t) = 0$.

* 18. Prove that the solution given in the discussion of Case I, namely, $y = A e^{r_1 t} + B e^{r_2 t}$, is the general solution for that case as follows: first, let $y = e^{r_1 t} u$ and show that u satisfies the equation

$$u'' - (r_2 - r_1)u' = 0.$$

Then let $v = u'$, so that v must satisfy $v' = (r_2 - r_1)v$. The general solution of this equation is $v = C e^{(r_2 - r_1)t}$, as shown in the discussion of the equation $y' = ky$ at the beginning of Section 4.4. Hence find u and y.

Simple harmonic motion

Exercises 19–22 all refer to the differential equation of simple harmonic motion:

$$\frac{d^2 y}{dt^2} + \omega^2 y = 0, \qquad (\omega \neq 0). \qquad (\dagger)$$

Together they show that $y = A \cos \omega t + B \sin \omega t$ is a *general solution* of this equation, that is, every solution is of this form for some choice of the constants A and B.

* 19. Show that $y = A \cos \omega t + B \sin \omega t$ is a solution of $(\dagger)$.

* 20. If $f(t)$ is any solution of $(\dagger)$, show that $\omega^2 (f(t))^2 + (f'(t))^2$ is constant.

* 21. If $g(t)$ is a solution of $(\dagger)$ satisfying $g(0) = g'(0) = 0$, show that $g(t) = 0$ for all t.

* 22. Suppose that $f(t)$ is any solution of the differential equation $(\dagger)$. Show that $f(t) = A \cos \omega t + B \sin \omega t$, where $A = f(0)$ and $B\omega = f'(0)$.
(*Hint:* let $g(t) = f(t) - A \cos \omega t - B \sin \omega t$.)

* 23. If $b^2 - 4ac < 0$, show that the substitution $y = e^{kt} u(t)$, where $k = -b/(2a)$, transforms $ay'' + by' + cy = 0$ into the equation $u'' + \omega^2 u = 0$, where $\omega^2 = (4ac - b^2)/(4a^2)$. Together with the result of the previous exercise, this confirms the recipe for Case III, in case you didn't feel comfortable with the complex number argument given in the text.

In Exercises 24–25, solve the given initial-value problems. For each problem determine the circular frequency, the frequency, the period, and the amplitude of the solution.

24. $\begin{cases} y'' + 4y = 0 \\ y(0) = 2 \\ y'(0) = -5 \end{cases}$

25. $\begin{cases} y'' + 100y = 0 \\ y(0) = 0 \\ y'(0) = 3 \end{cases}$

* 26. Show that $y = \alpha \cos(\omega(t - c)) + \beta \sin(\omega(t - c))$ is a solution of the differential equation $y'' + \omega^2 y = 0$, and that it satisfies $y(c) = \alpha$ and $y'(c) = \beta\omega$. Express the solution in the form $y = A \cos(\omega t) + B \sin(\omega t)$ for certain values of the constants A and B depending on α, β, c, and ω.

27. Solve $\begin{cases} y'' + y = 0 \\ y(2) = 3 \\ y'(2) = -4 \end{cases}$

28. Solve $\begin{cases} y'' + \omega^2 y = 0 \\ y(a) = A \\ y'(a) = B \end{cases}$

29. What mass should be suspended from the spring in Example 6 to provide a system whose natural frequency of oscillation is 10 Hz? Find the displacement of such a mass from its equilibrium position t s after it is pulled down 1 cm from equilibrium and flicked upward with a speed of 2 cm/s. What is the amplitude of this motion?

30. A mass of 400 g suspended from a certain elastic spring will oscillate with a frequency of 24 Hz. What would be the frequency if the 400 g mass were replaced with a 900 g mass? a 100 g mass?

* 31. Show that if t_0, A, and B are constants and $k = -b/(2a)$ and $\omega = \sqrt{4ac - b^2}/(2a)$, then

$$y = e^{kt} \left[A \cos(\omega(t - t_0)) + B \sin(\omega(t - t_0)) \right]$$

is an alternative to the general solution for $ay'' + by' + cy = 0$ for Case III ($b^2 - 4ac < 0$). This form of the general solution is useful for solving initial-value problems where $y(t_0)$ and $y'(t_0)$ are specified.

* 32. Show that if t_0, A, and B are constants and $k = -b/(2a)$ and $\omega = \sqrt{b^2 - 4ac}/(2a)$, then

$$y = e^{kt} \left[A \cosh(\omega(t - t_0)) + B \sinh(\omega(t - t_0)) \right]$$

is an alternative to the general solution for $ay'' + by' + cy = 0$ for Case I ($b^2 - 4ac > 0$). This form of the general solution is useful for solving initial-value problems where $y(t_0)$ and $y'(t_0)$ are specified.

Use the forms of solution provided by the previous two exercises to solve the initial-value problems in Exercises 33–34.

33. $\begin{cases} y'' + 2y' + 5y = 0 \\ y(3) = 2 \\ y'(3) = 0 \end{cases}$

34. $\begin{cases} y'' + 4y' + 3y = 0 \\ y(3) = 1 \\ y'(3) = 0 \end{cases}$

In Exercises 35–40 try to find a particular solution to the given non-homogeneous equation by guessing its form. Then write the general solution of the equation.

35. $y'' + y' - 2y = 1$

36. $y'' + y' - 2y = t$

37. $y'' + y' - 2y = e^{-t}$

38. $y'' + y' - 2y = 20 \cos(2t)$

39. $y'' + y' - 2y = 10e^t \sin t$

40. $y'' + y' - 2y = e^t$

Euler or equidimensional equations

Exercises 41–46 refer to the **Euler** or **equidimensional** equation

$$at^2 y'' + bt y' + cy = 0, \qquad (*)$$

where a, b, and c are constants and $y(t)$ is defined for $t > 0$. Associated with the DE ($*$) is an **auxiliary** quadratic equation

$$ar(r - 1) + br + c = 0. \qquad (\dagger)$$

* **41.** Show that if r satisfies (†), then $y = t^r$ is a solution of (*). What is the general solution of (*) if (†) has two different real roots r_1 and r_2?

* **42.** If (†) has two equal roots r, show that $y_1 = t^r$ and $y_2 = t^r \ln t$ are both solutions of (*).

* **43.** If (†) has complex conjugate roots $r = \alpha \pm i\beta$, where α and β are real and $\beta \neq 0$, show that $y_1 = t^\alpha \cos(\beta \ln t)$ and $y_2 = t^\alpha \sin(\beta \ln t)$ are both solutions of (*).

44. What is the general solution of $2t^2 y'' - ty' - 2y = 0$?

45. What is the general solution of $t^2 y'' - 3ty' + 4y = 0$?

46. What is the general solution of $t^2 y'' - 3ty' + 13y = 0$?

Chapter Review

Key Ideas

- **State the laws of exponents.**
- **State the laws of logarithms.**
- **What is the significance of the number e?**
- **What do the following statements and phrases mean?**
  - f is one-to-one. $\diamond$ f is invertible.
  - Function f^{-1} is the inverse of function f.
  - $a^b = c$ $\diamond$ $\log_a b = c$
  - the natural logarithm of x
  - logarithmic differentiation
  - the half-life of a varying quantity
  - The quantity y exhibits exponential growth.
  - The quantity y exhibits logistic growth.
  - $y = \sin^{-1} x$ $\diamond$ $y = \tan^{-1} x$
  - The quantity y exhibits simple harmonic motion.
  - The quantity y exhibits damped harmonic motion.
- **Define the functions $\sinh x$, $\cosh x$, and $\tanh x$.**
- **What kinds of functions satisfy second-order differential equations with constant coefficients?**

Review Exercises

1. If $f(x) = 3x + x^3$, show that f has an inverse and find the slope of $y = f^{-1}(x)$ at $x = 0$.

2. Let $f(x) = \sec^2 x \tan x$. Show that f is increasing on the interval $]-\pi/2, \pi/2[$ and, hence, one-to-one and invertible there. What is the domain of f^{-1}? Find $(f^{-1})'(2)$. *Hint:* $f(\pi/4) = 2$.

Exercises 3–5 refer to the function $f(x) = x e^{-x^2}$.

3. Find $\lim_{x \to \infty} f(x)$ and $\lim_{x \to -\infty} f(x)$.

4. On what intervals is f increasing? decreasing?

5. What are the maximum and minimum values of $f(x)$?

6. Find the points on the graph of $y = e^{-x} \sin x$, $(0 \le x \le 2\pi)$ where the graph has a horizontal tangent line.

7. Suppose that a function $f(x)$ satisfies $f'(x) = x f(x)$ for all real x, and $f(2) = 3$. Calculate the derivative of $f(x)/e^{x^2/2}$, and use the result to help you find $f(x)$ explicitly.

8. A lump of modelling clay is being rolled out so that it maintains the shape of a circular cylinder. If the length is increasing at a rate proportional to itself, show that the radius is decreasing at a rate proportional to itself.

9. (a) What nominal interest rate, compounded continuously, will cause an investment to double in 5 years?

   (b) By about how many days will the doubling time in part (a) increase if the nominal interest rate drops by 0.5%?

10. (A poor man's natural logarithm)

   (a) Show that if $a > 0$, then

$$\lim_{h \to 0} \frac{a^h - 1}{h} = \ln a.$$

   Hence show that

$$\lim_{n \to \infty} n(a^{1/n} - 1) = \ln a.$$

   (b) Most calculators, even non-scientific ones, have a square root key. If n is a power of 2, say $n = 2^k$, then $a^{1/n}$ can be calculated by entering a and hitting the square root key k times:

$$a^{1/2^k} = \sqrt{\sqrt{\cdots \sqrt{a}}} \qquad (k \text{ square roots}).$$

   Then you can subtract 1 and multiply by n to get an approximation for $\ln a$. Use $n = 2^{10} = 1024$ and $n = 2^{11} = 2048$ to find approximations for $\ln 2$. Based on the agreement of these two approximations, quote a value of $\ln 2$ to as many decimal places as you feel justified.

11. A nonconstant function f satisfies

$$\frac{d}{dx}(f(x))^2 = (f'(x))^2$$

for all x. If $f(0) = 1$, find $f(x)$.

12. If $f(x) = (\ln x)/x$, show that $f'(x) > 0$ for $0 < x < e$ and $f'(x) < 0$ for $x > e$, so that $f(x)$ has a maximum value at $x = e$. Use this to show that $e^\pi > \pi^e$.

13. Find an equation of a straight line that passes through the origin and is tangent to the curve $y = x^x$.

14. (a) Find $x \neq 2$ such that $\dfrac{\ln x}{x} = \dfrac{\ln 2}{2}$.

(b) Find $b > 1$ such that there is *no* $x \neq b$ with
$\dfrac{\ln x}{x} = \dfrac{\ln b}{b}$.

15. Investment account A bears simple interest at a certain rate. Investment account B bears interest at the same nominal rate but compounded instantaneously. If $1,000 is invested in each account, B produces $10 more in interest after one year than does A. Find the nominal rate both accounts use.

16. Express each of the functions $\cos^{-1} x$, $\cot^{-1} x$, and $\csc^{-1} x$ in terms of $\tan^{-1}$.

17. Express each of the functions $\cos^{-1} x$, $\cot^{-1} x$, and $\csc^{-1} x$ in terms of $\sin^{-1}$.

18. (**A warming problem**) A bottle of milk at $5°C$ is removed from a refrigerator into a room maintained at $20°C$. After 12 min the temperature of the milk is $12°C$. How much longer will it take for the milk to warm up to $18°C$?

19. (**A cooling problem**) A kettle of hot water at $96°C$ is allowed to sit in an air-conditioned room. The water cools to $60°C$ in 10 min and then to $40°C$ in another 10 min. What is the temperature of the room?

20. Show that $e^x > 1 + x$ if $x \neq 0$.

21. Use mathematical induction to show that

$$e^x > 1 + x + \frac{x^2}{2!} + \cdots + \frac{x^n}{n!}$$

if $x > 0$ and n is any positive integer.

Challenging Problems

* 1. (a) Show that the function $f(x) = x^x$ is strictly increasing on $[e^{-1}, \infty[$.

(b) If g is the inverse function to f of part (a), show that

$$\lim_{y \to \infty} \frac{g(y) \ln(\ln y)}{\ln y} = 1$$

Hint: start with the equation $y = x^x$ and take the ln of both sides twice.

Two models for incorporating air resistance into the analysis of the motion of a falling body

2. (**Air resistance proportional to speed**) An object falls under gravity near the surface of the earth, and its motion is impeded by air resistance proportional to its speed. Its velocity v therefore satisfies the equation

$$\frac{dv}{dt} = -g - kv, \qquad (*)$$

where k is a positive constant depending on such factors as the shape and density of the object and the density of the air.

(a) Find the velocity of the object as a function of time t, given that it was v_0 at $t = 0$.

(b) Find the limiting velocity $\lim_{t \to \infty} v(t)$. Observe that this can be done either directly from $(*)$ or from the solution found in (a).

(c) If the object was at height y_0 at time $t = 0$, find its height $y(t)$ at any time during its fall.

* 3. (**Air resistance proportional to the square of speed**) Under certain conditions a better model for the effect of air resistance on a moving object is one where the resistance is proportional to the square of the speed. For an object falling under constant gravitational acceleration g, motion is

$$\frac{dv}{dt} = -g - kv|v|,$$

where $k > 0$. Note that $v|v|$ is used instead of v^2 to ensure that the resistance is always in the opposite direction to the velocity. For an object falling from rest at time $t = 0$, we have $v(0) = 0$ and $v(t) < 0$ for $t > 0$, so the equation of motion becomes

$$\frac{dv}{dt} = -g + kv^2.$$

We are not (yet) in a position to solve this equation. However, we can verify its solution.

(a) Verify that the velocity is given for $t \geq 0$ by

$$v(t) = \sqrt{\frac{g}{k}} \, \frac{1 - e^{2t\sqrt{gk}}}{1 + e^{2t\sqrt{gk}}}.$$

(b) What is the limiting velocity $\lim_{t \to \infty} v(t)$?

(c) Also verify that if the falling object was at height y_0 at time $t = 0$, then its height at subsequent times during its fall is given by

$$y(t) = y_0 + \sqrt{\frac{g}{k}} \, t - \frac{1}{k} \ln\left(\frac{1 + e^{2t\sqrt{gk}}}{2}\right).$$

4. (**A model for the spread of a new technology**) When a new and superior technology is introduced, the percentage p of potential clients that adopt it might be expected to increase logistically with time. However, even newer technologies are continually being introduced, so adoption of a particular one will fall off exponentially over time. The following model exhibits this behaviour:

$$\frac{dp}{dt} = kp\left(1 - \frac{p}{e^{-bt}M}\right).$$

This DE suggests that the growth in p is logistic but that the asymptotic limit is not a constant but rather $e^{-bt} M$, which decreases exponentially with time.

(a) Show that the change of variable $p = e^{-bt} y(t)$ transforms the equation above into a standard logistic equa-

tion, and hence find an explicit formula for $p(t)$ given that $p(0) = p_0$. It will be necessary to assume that $M << 100k/(b+k)$ to ensure that $p(t) < 100$.

(b) If $k = 10$, $b = 1$, $M = 90$, and $p_0 = 1$, how large will $p(t)$ become before it starts to decrease?

CHAPTER 4
Some Applications of Derivatives

Introduction Differential calculus can be used to analyze many kinds of problems and situations that arise in applied disciplines. Calculus has made and will continue to make significant contributions to every field of human endeavour that uses quantitative measurement to further its aims. From economics to physics and from biology to sociology, problems can be found whose solutions can be aided by the use of some calculus.

In this chapter we will examine several kinds of problems to which the techniques we have already learned can be applied. These problems arise both outside and within mathematics. We will deal with the following kinds of problems:

1. Related rates problems, where the rates of change of related quantities are analyzed.
2. Graphing problems, where derivatives are used to illuminate the behaviour of functions.
3. Optimization problems, where a quantity is to be maximized or minimized.
4. Root finding methods, where we try to find numerical solutions of equations.
5. Approximation problems, where complicated functions are approximated by polynomials,
6. Evaluation of limits.

Do not assume that most of the problems we present here are "real-world" problems. Such problems are usually too complex to be treated in a general calculus course. However, the problems we consider, while sometimes artificial, do show how calculus can be applied in concrete situations.

4.1 Related Rates

When two or more quantities that change with time are linked by an equation, that equation can be differentiated with respect to time to produce an equation linking the rates of change of the quantities. Any one of these rates may then be determined when the others, and the values of the quantities themselves, are known. We will consider a couple of examples before formulating a list of procedures for dealing with such problems.

Example 1 An aircraft is flying horizontally at a speed of 600 km/h. How fast is the distance between the aircraft and a radio beacon increasing 1 minute after the aircraft passes 5 km directly above the beacon?

Solution A diagram is useful here; see Figure 4.1. Let C be the point on the aircraft's path directly above the beacon B. Let A be the position of the aircraft t min after it is at C, and let x and s be the distances CA and BA, respectively. From the right triangle BCA we have

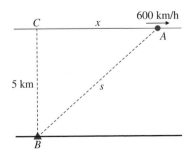

Figure 4.1

$$s^2 = x^2 + 5^2.$$

We differentiate this equation implicitly with respect to t to obtain

$$2s \frac{ds}{dt} = 2x \frac{dx}{dt}.$$

We are given that $dx/dt = 600$ km/h $= 10$ km/min. Therefore, $x = 10$ km at time $t = 1$ min. At that time $s = \sqrt{10^2 + 5^2} = 5\sqrt{5}$ km and is increasing at the rate

$$\frac{ds}{dt} = \frac{x}{s} \frac{dx}{dt} = \frac{10}{5\sqrt{5}}(600) = \frac{1,200}{\sqrt{5}} \approx 536.7 \text{ km/h}.$$

One minute after the aircraft passes over the beacon, its distance from the beacon is increasing at about 537 km/h.

Example 2   How fast is the area of a rectangle changing if one side is 10 cm long and is increasing at a rate of 2 cm/s and the other side is 8 cm long and is decreasing at a rate of 3 cm/s?

Solution   Let the lengths of the sides of the rectangle at time t be x cm and y cm, respectively. Thus the area at time t is $A = xy$ cm^2. (See Figure 4.2.) We want to know the value of dA/dt when $x = 10$ and $y = 8$, given that $dx/dt = 2$ and $dy/dt = -3$. (Note the negative sign to indicate that y is decreasing.) Since all the quantities in the equation $A = xy$ are functions of time, we can differentiate that equation implicitly with respect to time and obtain

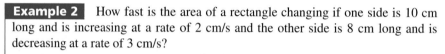

$$\left. \frac{dA}{dt} \right|_{\substack{x=10 \\ y=8}} = \left. \left(\frac{dx}{dt} y + x \frac{dy}{dt} \right) \right|_{\substack{x=10 \\ y=8}} = 2(8) + 10(-3) = -14.$$

At the time in question, the area of the rectangle is decreasing at a rate of 14 cm^2/s.

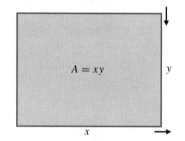

$A = xy$ y

x

Figure 4.2   Rectangle with sides changing

Procedures for Related-Rates Problems

In view of these examples we can formulate a few general procedures for dealing with related-rates problems.

> **How to Solve Related-Rates Problems**
>
> 1. Read the problem very carefully. Try to understand the relationships among the variable quantities. What is given? What is to be found?
> 2. Make a sketch if appropriate.
> 3. Define any symbols you want to use that are not defined in the statement of the problem. Express given and required quantities and rates in terms of these symbols.
> 4. Discover from a careful reading of the problem or consideration of the sketch one or more equations linking the variable quantities. (You will need as many equations as quantities or rates to be found in the problem.)

5. Differentiate the equation(s) implicitly with respect to time, regarding all variable quantities as functions of time. You can manipulate the equation(s) algebraically before the differentiation is performed (for instance, you could solve for the quantities whose rates are to be found), but it is usually easier to differentiate the equations as they are originally obtained and solve for the desired items later.

6. Substitute any given values for the quantities and their rates, then solve the resulting equation(s) for the unknown quantities and rates.

7. Make a concluding statement answering the question asked. Is your answer "reasonable"? If not, check back through your solution to see what went wrong.

Example 3   A lighthouse L is located on a small island 2 km from the nearest point A on a long, straight shoreline. If the lighthouse lamp rotates at 3 revolutions per minute, how fast is the illuminated spot P on the shoreline moving along the shoreline when it is 4 km from A?

Solution   Referring to Figure 4.3, let x be the distance AP and let θ be the angle $\angle PLA$. Then $x = 2 \tan \theta$ and

$$\frac{dx}{dt} = 2 \sec^2 \theta \, \frac{d\theta}{dt}.$$

Now

$$\frac{d\theta}{dt} = 3 \text{ rev/min } \times \ 2\pi \text{ radians/rev} = 6\pi \text{ radians/min.}$$

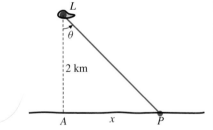

Figure 4.3

When $x = 4$, we have $\tan \theta = 2$ and $\sec^2 \theta = 1 + \tan^2 \theta = 5$. Thus

$$\frac{dx}{dt} = 2 \times 5 \times 6\pi = 60\pi \approx 188.5.$$

The spot of light is moving along the shoreline at a rate of about 188.5 km/min when it is 4 km from A.

(Note that it was essential to convert the rate of change of θ from revolutions per minute to radians per minute. If θ were not measured in radians we could not assert that $(d/d\theta) \tan \theta = \sec^2 \theta$.)

Example 4   A leaky water tank is in the shape of an inverted right circular cone with depth 5 m and top radius 2 m. When the water in the tank is 4 m deep it is leaking out at a rate of $1/12$ m³/min. How fast is the water level in the tank dropping at that time?

Solution   Let r and h denote the surface radius and depth of water in the tank at time t (both measured in metres). Thus, the volume V (in m³) of water in the tank at time t is

$$V = \frac{1}{3}\pi r^2 h.$$

Using similar triangles in Figure 4.4, we can find a relationship between r and h:

$$\frac{r}{h} = \frac{2}{5}, \quad \text{so} \quad r = \frac{2h}{5} \quad \text{and} \quad V = \frac{1}{3}\pi \left(\frac{2h}{5}\right)^2 h = \frac{4\pi}{75}h^3.$$

Differentiating this equation with respect to t we obtain

$$\frac{dV}{dt} = \frac{4\pi}{25}h^2\frac{dh}{dt}.$$

Since $dV/dt = -1/12$ when $h = 4$, we have

$$\frac{-1}{12} = \frac{4\pi}{25}(4^2)\frac{dh}{dt}, \quad \text{so} \quad \frac{dh}{dt} = -\frac{25}{768\pi}.$$

When the water in the tank is 4 m deep, its level is dropping at a rate of $25/(768\pi)$ m/min, or about 1.036 cm/min. ∎

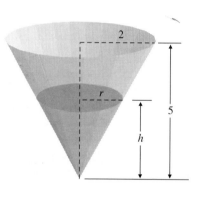

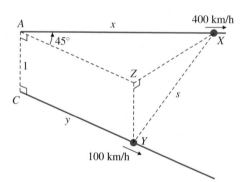

Figure 4.4   The conical tank of Example 4   **Figure 4.5**   Aircraft paths in Example 5

Example 5   At a certain instant an aircraft flying due east at 400 km/h passes directly over a car travelling due southeast at 100 km/h on a straight, level road. If the aircraft is flying at an altitude of 1 km, how fast is the distance between the aircraft and the car increasing 36 s after the aircraft passes directly over the car?

Solution   A good diagram is essential here.   See Figure 4.5.   Let time t be measured in hours from the time the aircraft was at position A directly above the car at position C. Let X and Y be the positions of the aircraft and the car, respectively, at time t. Let x be the distance AX, y be the distance CY, and s the distance XY, all measured in kilometres. Let Z be the point 1 km above Y. Since angle $XAZ = 45°$, the Pythagorean Theorem and Cosine Law yield

$$s^2 = 1 + (ZX)^2 = 1 + x^2 + y^2 - 2xy\cos 45°$$
$$= 1 + x^2 + y^2 - \sqrt{2}xy.$$

Thus,

$$2s \frac{ds}{dt} = 2x \frac{dx}{dt} + 2y \frac{dy}{dt} - \sqrt{2} \frac{dx}{dt} y - \sqrt{2}x \frac{dy}{dt}$$
$$= 400(2x - \sqrt{2}y) + 100(2y - \sqrt{2}x),$$

since $dx/dt = 400$ and $dy/dt = 100$. When $t = 1/100$ (i.e., 36 s after $t = 0$), we have $x = 4$ and $y = 1$. Hence,

$$s^2 = 1 + 16 + 1 - 4\sqrt{2} = 18 - 4\sqrt{2}$$
$$s \approx 3.5133.$$
$$\frac{ds}{dt} = \frac{1}{2s} \left(400(8 - \sqrt{2}) + 100(2 - 4\sqrt{2})\right) \approx 322.86.$$

The aircraft and the car are separating at a rate of about 323 km/h after 36 s.

(Note that it was necessary to convert 36 s to hours in the solution. In general all measurements should be in compatible units.) ∎

Exercises 4.1

1. Find the rate of change of the area of a square whose side is 8 cm long, if the side length is increasing at 2 cm/min.

2. The area of a square is decreasing at 2 ft^2/s. How fast is the side length changing when it is 8 ft?

3. A pebble dropped into a pond causes a circular ripple to expand outward from the point of impact. How fast is the area enclosed by the ripple increasing when the radius is 20 cm and is increasing at a rate of 4 cm/s?

4. The area of a circle is decreasing at a rate of 2 cm^2/min. How fast is the radius of the circle changing when the area is 100 cm^2?

5. The area of a circle is increasing at 1/3 km^2/h. Express the rate of change of the radius of the circle as a function of (a) the radius r and (b) the area A of the circle.

6. At a certain instant the length of a rectangle is 16 m and the width is 12 m. The width is increasing at 3 m/s. How fast is the length changing if the area of the rectangle is not changing?

7. Air is being pumped into a spherical balloon. The volume of the balloon is increasing at a rate of 20 cm^3/s when the radius is 30 cm. How fast is the radius increasing at that time? (The volume of a ball of radius r units is $V = \frac{4}{3}\pi r^3$ cubic units.)

8. When the diameter of a ball of ice is 6 cm, it is decreasing at a rate of 0.5 cm/h due to melting of the ice. How fast is the volume of the ice ball decreasing at that time?

9. How fast is the surface area of a cube changing when the volume of the cube is 64 cm^3 and is increasing at 2 cm^3/s?

10. The volume of a right circular cylinder is 60 cm^3 and is increasing at 2 cm^3/min at a time when the radius is 5 cm and is increasing at 1 cm/min. How fast is the height of the cylinder changing at that time?

11. How fast is the volume of a rectangular box changing when the length is 6 cm, the width is 5 cm, and the depth is 4 cm, if the length and depth are both increasing at a rate of 1 cm/s and the width is decreasing at a rate of 2 cm/s?

12. The area of a rectangle is increasing at a rate of 5 m^2/s while the length is increasing at a rate of 10 m/s. If the length is 20 m and the width is 16 m, how fast is the width changing?

13. A point moves on the curve $y = x^2$. How fast is y changing when $x = -2$ and x is decreasing at a rate 3?

14. A point is moving to the right along the first-quadrant portion of the curve $x^2 y^3 = 72$. When the point has coordinates $(3, 2)$, its horizontal velocity is 2 units/s. What is its vertical velocity?

15. The point P moves so that at time t it is at the intersection of the curves $xy = t$ and $y = tx^2$. How fast is the distance of P from the origin changing at time $t = 2$?

16. **(Radar guns)** A policeman is standing near a highway using a radar gun to catch speeders. (See Figure 4.6.) He aims the gun at a car that has just passed his position and, when the gun is pointing at an angle of 45° to the direction of the highway, notes that the distance between the car and the gun is increasing at a rate of 100 km/h. How fast is the car travelling?

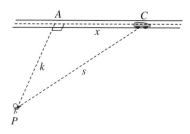

Figure 4.6

17. If the radar gun of Exercise 16 is aimed at a car travelling at 90 km/h along a straight road, what will its reading be at an instant when it is aimed in a direction making an angle of 30° with the road?

18. The top of a ladder 5 m long rests against a vertical wall. If the base of the ladder is being pulled away from the base of the wall at a rate of 1/3 m/s, how fast is the top of the ladder slipping down the wall when it is 3 m above the base of the wall?

19. A man 2 m tall walks toward a lamppost on level ground at a rate of 0.5 m/s. If the lamp is 5 m high on the post, how fast is the length of the man's shadow decreasing when he is 3 m from the post? How fast is the shadow of his head moving at that time?

20. A woman 6 ft tall is walking at 2 ft/s along a straight path on level ground. There is a lamppost 5 ft to the side of the path. A light 15 ft high on the lamppost casts the woman's shadow on the ground. How fast is the length of her shadow changing when the woman is 12 feet from the point on the path closest to the lamppost?

21. **(Cost of production)** It costs a coal mine owner C each day to maintain a production of x tons of coal, where $C = 10,000 + 3x + x^2/8,000$. At what rate is the production increasing when it is 12,000 tons and the daily cost is increasing at $600 per day?

22. **(Distance between ships)** At 1:00 p.m. ship A is 25 km due north of ship B. If ship A is sailing west at a rate of 16 km/h and ship B is sailing south at 20 km/h, find the rate at which the distance between the two ships is changing at 1:30 p.m.

23. What is the first time after 3:00 p.m. that the hands of the clock are together?

24. **(Tracking a balloon)** A balloon released at point A rises vertically with a constant speed of 5 m/s. Point B is level with and 100 m distant from point A. How fast is the angle of elevation of the balloon at B changing when the balloon is 200 m above A?

25. Sawdust is falling onto a pile at a rate of 1/2 m³/min. If the pile maintains the shape of a right circular cone with height equal to half the diameter of its base, how fast is the height of the pile increasing when the pile is 3 m high?

26. **(Conical tank)** A water tank is in the shape of an inverted right circular cone with top radius 10 m and depth 8 m.

Water is flowing in at a rate of 1/10 m³/min. How fast is the depth of water in the tank increasing when the water is 4 m deep?

27. **(Leaky tank)** Repeat Exercise 26 with the added assumption that water is leaking out of the bottom of the tank at a rate of $h^3/1,000$ m³/min when the depth of water in the tank is h m. How full can the tank get in this case?

28. **(Another leaky tank)** Water is pouring into a leaky tank at a rate of 10 m³/h. The tank is a cone with vertex down, 9 m in depth and 6 m in diameter at the top. The surface of water in the tank is rising at a rate of 20 cm/h when the depth is 6 m. How fast is the water leaking out at that time?

29. **(Kite flying)** How fast must you let out line if the kite you are flying is 30 m high, 40 m horizontally away from you, and moving horizontally away from you at a rate of 10 m/min?

30. **(Ferris wheel)** You are riding on a Ferris wheel of diameter 20 m. The wheel is rotating at 1 revolution per minute. How fast are you rising or falling when you are 6 m horizontally away from the vertical line passing through the centre of the wheel?

31. **(Distance between aircraft)** An aircraft is 144 km east of an airport and is travelling west at 200 km/h. At the same time, a second aircraft at the same altitude is 60 km north of the airport and travelling north at 150 km/h. How fast is the distance between the two aircraft changing?

32. **(Production rate)** If a truck factory employs x workers and has daily operating expenses of $$y$, it can produce $P = (1/3)x^{0.6}y^{0.4}$ trucks per year. How fast are the daily expenses decreasing when they are $10,000 and the number of workers is 40, if the number of workers is increasing at 1 per day and production is remaining constant?

33. A lamp is located at point $(3, 0)$ in the xy-plane. An ant is crawling in the first quadrant of the plane and the lamp casts its shadow onto the y-axis. How fast is the ant's shadow moving along the y-axis when the ant is at position $(1, 2)$ and moving so that its x-coordinate is increasing at rate 1/3 units/s and its y-coordinate is decreasing at 1/4 units/s?

34. A straight highway and a straight canal intersect at right angles, the highway crossing over the canal on a bridge 20 m above the water. A boat travelling at 20 km/h passes under the bridge just as a car travelling at 80 km/h passes over it. How fast are the boat and car separating after one minute?

35. **(Filling a trough)** The cross section of a water trough is an equilateral triangle with top edge horizontal. If the trough is 10 m long and 30 cm deep, and if water is flowing in at a rate of 1/4 m³/min, how fast is the water level rising when the water is 20 cm deep at the deepest?

36. **(Draining a pool)** A rectangular swimming pool is 8 m wide and 20 m long. (See Figure 4.7.) Its bottom is a sloping plane, the depth increasing from 1 m at the shallow end to 3 m at the deep end. Water is draining out of the pool at a rate of 1 m³/min. How fast is the surface of the water

falling when the depth of water at the deep end is (a) 2.5 m?
(b) 1 m?

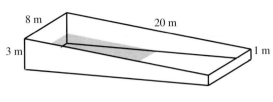

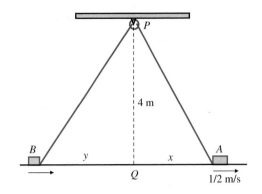

Figure 4.7

Figure 4.9

* **37.** One end of a 10 m long ladder is on the ground and the
ladder is supported partway along its length by resting on
top of a 3 m high fence. (See Figure 4.8.) If the bottom of
the ladder is 4 m from the base of the fence and is being
dragged along the ground away from the fence at a rate of
1/5 m/s, how fast is the free top end of the ladder moving (a)
vertically and (b) horizontally?

* **38.** Two crates, A and B, are on the floor of a warehouse. The
crates are joined by a rope 15 m long, each crate being
hooked at floor level to an end of the rope. The rope is
stretched tight and pulled over a pulley P that is attached to
a rafter 4 m above a point Q on the floor directly between
the two crates. (See Figure 4.9.) If crate A is 3 m from Q
and is being pulled directly away from Q at a rate of
1/2 m/s, how fast is crate B moving toward Q?

39. (**Tracking a rocket**) Shortly after launch, a rocket is 100 km
high and 50 km downrange. If it is travelling at 4 km/s at an
angle of $30°$ above the horizontal, how fast is its angle of
elevation, as measured at the launch site, changing?

40. (**Shadow of a falling ball**) A lamp is 20 m high on a pole.
At time $t = 0$ a ball is dropped from a point level with the
lamp and 10 m away from it. The ball falls under gravity
(acceleration 9.8 m/s^2) until it hits the ground. How fast is
the shadow of the ball moving along the ground (a) 1 s after
it is dropped? (b) just as the ball hits the ground?

41. (**Tracking a rocket**) A rocket blasts off at time $t = 0$ and
climbs vertically with acceleration 10 m/s^2. The progress of
the rocket is monitored by a tracking station located 2 km
horizontally away from the launch pad. How fast is the
tracking station antenna rotating upward 10 s after launch?

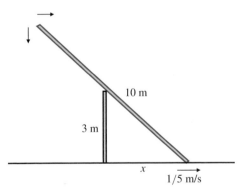

Figure 4.8

4.2 Extreme Values

The first derivative of a function is a source of much useful information about
the behaviour of the function. As we have already seen, the sign of f' tells us
whether f is increasing or decreasing. In this section we use this information
to find maximum and minimum values of functions. In Section 4.5 we will put
the techniques developed here to use in solving problems requiring the finding of
maximum and minimum values.

Maximum and Minimum Values

Recall (from Section 1.4) that a function has a maximum value at x_0 if $f(x) \le f(x_0)$
for all x in the domain of f. The maximum value is $f(x_0)$. To be more precise, we should call such a maximum value an *absolute* or *global* maximum

because it is the largest value that f attains anywhere on its entire domain.

> **Absolute extreme values**
>
> Function f has an **absolute maximum value** $f(x_0)$ at the point x_0 in its domain if $f(x) \le f(x_0)$ holds for every x in the domain of f.
> Similarly, f has an **absolute minimum value** $f(x_1)$ at the point x_1 in its domain if $f(x) \ge f(x_1)$ holds for every x in the domain of f.

A function can have at most one absolute maximum or minimum value, although this value can be assumed at many points. For example, $f(x) = \sin x$ has absolute maximum value 1 occurring at every point of the form $x = (\pi/2) + 2n\pi$ where n is an integer. Of course a function need not have any absolute extreme values. The function $f(x) = 1/x$ becomes arbitrarily large as x approaches 0 from the right, so has no finite absolute maximum. (Remember, ∞ is not a number, so is not a value of f.) Even a bounded function may not have an absolute maximum or minimum value. The function $g(x) = x$ with domain specified to be the *open* interval $]0, 1[$ has neither; the range of g is also the interval $]0, 1[$ and there is no largest or smallest number in this interval. Of course, if the domain of g were extended to be the *closed* interval $[0, 1]$, then g would have both a maximum value, 1, and a minimum value, 0.

Maximum and minimum values of a function are collectively referred to as **extreme values**. The following theorem is a restatement (and slight generalization) of Theorem 8 of Section 1.4. It will prove very useful in some circumstances when we want to find extreme values.

Existence of extreme values

If the domain of the function f is a *closed, finite interval* or a union of finitely many such intervals, and if f is *continuous* on that domain, then f must have an absolute maximum value and an absolute minimum value.

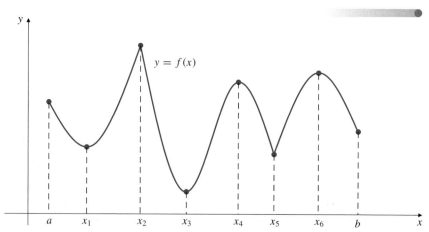

Figure 4.10   Local extreme values

Consider the graph $y = f(x)$ shown in Figure 4.10. Evidently the absolute maximum value of f is $f(x_2)$ and the absolute minimum value is $f(x_3)$. In addition to these extreme values, f has several other "local" maximum and minimum values corresponding to points on the graph that are higher or lower than neighbouring points. Observe that f has *local maximum values* at a, x_2, x_4, and x_6 and local minimum values at x_1, x_3, x_5, and b. The absolute maximum is the highest of the local maxima; the absolute minimum is the lowest of the local minima.

DEFINITION 2

Local extreme values

Function f has a **local maximum value (loc max)** $f(x_0)$ at the point x_0 in its domain provided there exists a number $h > 0$ such that $f(x) \leq f(x_0)$ whenever x is in the domain of f and $|x - x_0| < h$.

Similarly, f has a **local minimum value (loc min)** $f(x_1)$ at the point x_1 in its domain provided there exists a number $h > 0$ such that $f(x) \geq f(x_1)$ whenever x is in the domain of f and $|x - x_1| < h$.

Thus, f has a local maximum (or minimum) value at x if it has an absolute maximum (or minimum) value at x when its domain is restricted to points sufficiently near x. Geometrically, the graph of f is at least as high (or low) at x as it is at nearby points.

Critical Points, Singular Points, and Endpoints

Figure 4.10 suggests that a function $f(x)$ can have local extreme values only at points x of three special types:

(i) **critical points** of f (points x in $\mathcal{D}(f)$ where $f'(x) = 0$),

(ii) **singular points** of f (points x in $\mathcal{D}(f)$ where $f'(x)$ is not defined), and

(iii) **endpoints** of the domain of f (points in $\mathcal{D}(f)$ that do not belong to any open interval contained in $\mathcal{D}(f)$).

In Figure 4.10, x_1, x_3, x_4, and x_6 are critical points, x_2 and x_5 are singular points, and a and b are endpoints.

THEOREM 2

Locating extreme values

If the function f is defined on an interval I and has a local maximum (or local minimum) value at point $x = x_0$ in I, then x_0 must be either a critical point of f, a singular point of f, or an endpoint of I.

PROOF   Suppose that f has a local maximum value at x_0 and that x_0 is neither an endpoint of the domain of f nor a singular point of f. Then for some $h > 0$, $f(x)$ is defined on the open interval $(x_0 - h, x_0 + h)$ and has an absolute maximum (for that interval) at x_0. Also, $f'(x_0)$ exists. By Theorem 14 of Section 2.6, $f'(x_0) = 0$. The proof for the case where f has a local minimum value at x_0 is similar.

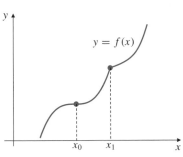

Figure 4.11   A function need not have extreme values at a critical point or a singular point

Although a function cannot have extreme values anywhere other than at endpoints, critical points, and singular points, it need not have extreme values at such points. Figure 4.11 shows the graph of a function with a critical point x_0 and a singular point x_1 at neither of which it has an extreme value. It is more difficult to draw the graph of a function whose domain has an endpoint at which the function fails to have an extreme value. See Exercise 51 at the end of this section for an example of such a function.

Finding Absolute Extreme Values

If a function f is defined on a closed interval or a union of finitely many closed intervals, Theorem 1 assures us that f must have an absolute maximum value and an absolute minimum value. Theorem 2 tells us how to find them. We need only check the values of f at any critical points, singular points, and endpoints.

Example 1   Find the maximum and minimum values of the function

$$g(x) = x^3 - 3x^2 - 9x + 2$$

on the interval $-2 \le x \le 2$.

Solution   Since g is a polynomial, it can have no singular points. For critical points, we calculate

$$g'(x) = 3x^2 - 6x - 9 = 3(x^2 - 2x - 3)$$
$$= 3(x + 1)(x - 3)$$
$$= 0 \quad \text{if} \quad x = -1 \text{ or } x = 3.$$

However, $x = 3$ is not in the domain of g, so we can ignore it. We need to consider only the values of g at the critical point $x = -1$ and at the endpoints $x = -2$ and $x = 2$:

$$g(-2) = 0, \qquad g(-1) = 7, \qquad g(2) = -20.$$

The maximum value of $g(x)$ on $-2 \le x \le 2$ is 7, at the critical point $x = -1$, and the minimum value is -20, at the endpoint $x = 2$. See Figure 4.12.

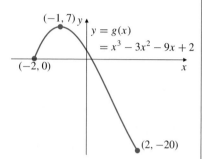

Figure 4.12 g has maximum and minimum values 7 and -20 respectively

Example 2   Find the maximum and minimum values of $h(x) = 3x^{2/3} - 2x$ on the interval $[-1, 1]$.

Solution   The derivative of h is

$$h'(x) = 3\left(\frac{2}{3}\right) x^{-1/3} - 2 = 2(x^{-1/3} - 1).$$

Note that $x^{-1/3}$ is not defined at the point $x = 0$ in $\mathcal{D}(h)$, so $x = 0$ is a singular point of h. Also, h has a critical point where $x^{-1/3} = 1$, that is, at $x = 1$, which also happens to be an endpoint of the domain of h. We must therefore examine the values of h at the points $x = 0$ and $x = 1$, as well as at the other endpoint $x = -1$. We have

$$h(-1) = 5, \qquad h(0) = 0, \qquad h(1) = 1.$$

The function h has maximum value 5 at the endpoint -1 and minimum value 0 at the singular point $x = 0$. See Figure 4.13.

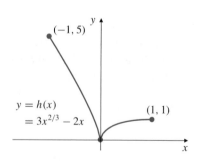

Figure 4.13 h has absolute minimum value 0 at a singular point

The First Derivative Test

Most functions you will encounter in elementary calculus have nonzero derivatives everywhere on their domains except possibly at a finite number of critical points, singular points, and endpoints of their domains. On intervals between these points the derivative exists and is not zero, so the function is either increasing or decreasing there. If f is continuous and increases to the left of x_0 and decreases to the right, then it must have a local maximum value at x_0. The following theorem collects several results of this type together.

| **THEOREM** **3** | **The First Derivative Test** |

PART I. Testing interior critical points and singular points.

Suppose that f is continuous at x_0, and x_0 is not an endpoint of the domain of f.

(a) If there exists an open interval $]a, b[$ containing x_0 such that $f'(x) > 0$ on $]a, x_0[$ and $f'(x) < 0$ on $]x_0, b[$, then f has a local maximum value at x_0.

(b) If there exists an open interval $]a, b[$ containing x_0 such that $f'(x) < 0$ on $]a, x_0[$ and $f'(x) > 0$ on $]x_0, b[$, then f has a local minimum value at x_0.

PART II. Testing endpoints of the domain.

Suppose a is a left endpoint of the domain of f and f is right continuous at a.

(c) If $f'(x) > 0$ on some interval $]a, b[$, then f has a local minimum value at a.

(d) If $f'(x) < 0$ on some interval $]a, b[$, then f has a local maximum value at a.

Suppose b is a right endpoint of the domain of f and f is left continuous at b.

(e) If $f'(x) > 0$ on some interval $]a, b[$, then f has a local maximum value at b.

(f) If $f'(x) < 0$ on some interval $]a, b[$, then f has a local minimum value at b.

Remark If f' is positive (or negative) on *both* sides of a critical or singular point, then f has neither a maximum nor a minimum value at that point.

Example 3 Find the local and absolute extreme values of $f(x) = x^4 - 2x^2 - 3$ on the interval $[-2, 2]$. Sketch the graph of f.

Solution We begin by calculating and factoring the derivative $f'(x)$:

$$f'(x) = 4x^3 - 4x = 4x(x^2 - 1) = 4x(x - 1)(x + 1).$$

The critical points are 0, -1, and 1. The corresponding values are $f(0) = -3$, $f(-1) = f(1) = -4$. There are no singular points. The values of f at the endpoints -2 and 2 are $f(-2) = f(2) = 5$. The factored form of $f'(x)$ is also convenient for determining the sign of $f'(x)$ on intervals between these endpoints and critical points. Where an odd number of the factors of $f'(x)$ are negative, $f'(x)$ will itself be negative; where an even number of factors are negative, $f'(x)$ will be positive. We summarize the positive/negative properties of $f'(x)$ and the implied increasing/decreasing behaviour of $f(x)$ in chart form:

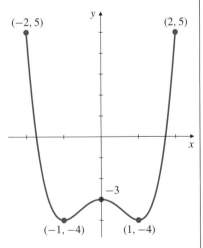

(−2, 5)   (2, 5)

−3

(−1, −4)   (1, −4)

Figure 4.14   The graph
$y = x^4 - 2x^2 - 3$

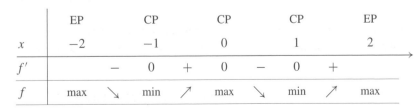

	EP	CP	CP	CP	EP
x	-2	-1	0	1	2
f'	$-$	0 $+$	0 $-$	0 $+$	
f	max ↘	min ↗	max ↘	min ↗	max

Note how the sloping arrows indicate visually the appropriate classification of the endpoints (EP) and critical points (CP) as determined by the First Derivative Test. We will make extensive use of such charts in future sections. The graph of f is shown in Figure 4.14. Since the domain is a closed, finite interval, f must have absolute maximum and minimum values. These are 5 (at ± 2) and -4 (at ± 1).

Example 4 Find and classify the local and absolute extreme values of the function $f(x) = x - x^{2/3}$ with domain $[-1, 2]$. Sketch the graph of f.

Solution $f'(x) = 1 - \frac{2}{3}x^{-1/3} = \left(x^{1/3} - \frac{2}{3}\right)/x^{1/3}$. There is a singular point, $x = 0$, and a critical point, $x = 8/27$. The endpoints are $x = -1$ and $x = 2$. The values of f at these points are $f(-1) = -2$, $f(0) = 0$, $f(8/27) = -4/27$, and $f(2) = 2 - 2^{2/3} \approx 0.4126$ (see Figure 4.15). Another interesting point on the graph is the x-intercept at $x = 1$. Information from f' is summarized in the chart:

	EP		SP		CP		EP
x	-1		0		$8/27$		2
f'		$+$	undef	$-$	0	$+$	
f	min	↗	max	↘	min	↗	max

There are two local minima and two local maxima. The absolute maximum of f is $2 - 2^{2/3}$ at $x = 2$; the absolute minimum is -2 at $x = -1$. ∎

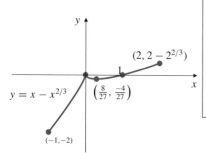

$y = x - x^{2/3}$

$(2, 2 - 2^{2/3})$

$\left(\frac{8}{27}, \frac{-4}{27}\right)$

$(-1, -2)$

Figure 4.15 The graph for Example 4

Functions Not Defined on Closed, Finite Intervals

If the function f is not defined on a closed, finite interval, then Theorem 1 cannot be used to guarantee the existence of maximum and minimum values for f. Of course, f may still have such extreme values. In many applied situations we will want to find extreme values of functions defined on infinite and/or open intervals. The following theorem adapts Theorem 1 to cover some such situations.

THEOREM 4

Existence of extreme values on open intervals

If f is continuous on the open interval $]a, b[$, and if

$$\lim_{x \to a+} f(x) = L \quad \text{and} \quad \lim_{x \to b-} f(x) = M,$$

then the following conclusions hold:

(i) If $f(u) > L$ and $f(u) > M$ for some u in $]a, b[$, then f has an absolute maximum value on $]a, b[$.

(ii) If $f(v) < L$ and $f(v) < M$ for some v in $]a, b[$, then f has an absolute minimum value on $]a, b[$.

In this theorem a may be $-\infty$, in which case $\lim_{x \to a+}$ should be replaced with $\lim_{x \to -\infty}$, and b may be ∞, in which case $\lim_{x \to b-}$ should be replaced with $\lim_{x \to \infty}$. Also, either or both of L and M may be either ∞ or $-\infty$.

PROOF We prove part (i); the proof of (ii) is similar. We are given that there is a number u in $]a, b[$ such that $f(u) > L$ and $f(u) > M$. Here, L and M may be finite numbers or $-\infty$. Since $\lim_{x \to a+} f(x) = L$, there must exist a number x_1 in $]a, u[$ such that

$$f(x) < f(u) \quad \text{whenever} \quad a < x < x_1.$$

Similarly, there must exist a number x_2 in $]u, b[$ such that

$$f(x) < f(u) \quad \text{whenever} \quad x_2 < x < b.$$

(See Figure 4.16.) Thus, $f(x) < f(u)$ at all points of $]a, b[$ that are not in the closed, finite subinterval $[x_1, x_2]$. By Theorem 1, the function f, being continuous on $[x_1, x_2]$, must have an absolute maximum value on that interval, say at the point w. Since u belongs to $[x_1, x_2]$, we must have $f(w) \geq f(u)$, so $f(w)$ is the maximum value of $f(x)$ for all of $]a, b[$.

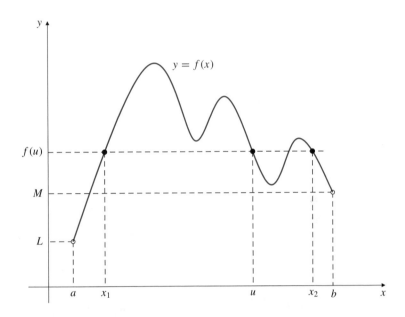

Figure 4.16

Theorem 2 still tells us where to look for extreme values. There are no endpoints to consider in an open interval, but we must still look at the values of the function at any critical points or singular points in the interval.

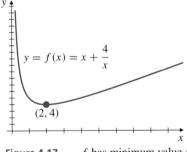

Figure 4.17 f has minimum value 4 at $x = 2$

Example 5   Show that $f(x) = x + (4/x)$ has an absolute minimum value on the interval $]0, \infty[$, and find that minimum value.

Solution   We have

$$\lim_{x \to 0+} f(x) = \infty \qquad \text{and} \qquad \lim_{x \to \infty} f(x) = \infty.$$

Since $f(1) = 5 < \infty$, Theorem 4 guarantees that f must have an absolute minimum value at some point in $]0, \infty[$. To find the minimum value we must check the values of f at any critical points or singular points in the interval. We have

$$f'(x) = 1 - \frac{4}{x^2} = \frac{x^2 - 4}{x^2} = \frac{(x - 2)(x + 2)}{x^2},$$

which equals 0 only at $x = 2$ and $x = -2$. Since f has domain $]0, \infty[$, it has no singular points and only one critical point, namely $x = 2$, where f has the value $f(2) = 4$. This must be the minimum value of f on $]0, \infty[$. (See Figure 4.17.)

Example 6 Let $f(x) = x e^{-x^2}$. Find and classify the critical points of f, evaluate $\lim_{x \to \pm\infty} f(x)$, and use these results to help you sketch the graph of f.

Solution $f'(x) = e^{-x^2}(1 - 2x^2) = 0$ only if $1 - 2x^2 = 0$ since the exponential is always positive. Thus the critical points are $\pm\frac{1}{\sqrt{2}}$. We have $f\left(\pm\frac{1}{\sqrt{2}}\right) = \pm\frac{1}{\sqrt{2e}}$. f' is positive (or negative) when $1 - 2x^2$ is positive (or negative). We summarize the intervals where f is increasing and decreasing in chart form:

x		CP $-1/\sqrt{2}$		CP $1/\sqrt{2}$	
f'	$-$	0	$+$	0	$-$
f	$\searrow$	min	$\nearrow$	max	$\searrow$

Note that $f(0) = 0$ and that f is an odd function ($f(-x) = -f(x)$) so the graph is symmetric about the origin. Also,

$$\lim_{x \to \pm\infty} x e^{-x^2} = \left(\lim_{x \to \pm\infty} \frac{1}{x}\right)\left(\lim_{x \to \pm\infty} \frac{x^2}{e^{x^2}}\right) = 0 \times 0 = 0$$

because $\lim_{x \to \pm\infty} x^2 e^{-x^2} = \lim_{u \to \infty} u e^{-u} = 0$ by Theorem 5 of Section 3.4. Since $f(x)$ is positive at $x = 1/\sqrt{2}$ and is negative at $x = -1/\sqrt{2}$, f must have absolute maximum and minimum values by Theorem 4. These values can only be the values $\pm 1/\sqrt{2e}$ at the two critical points. The graph is shown in Figure 4.18. The x-axis is an asymptote as $x \to \pm\infty$.

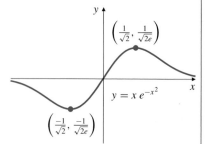

$\left(\frac{1}{\sqrt{2}}, \frac{1}{\sqrt{2e}}\right)$

$y = x e^{-x^2}$

$\left(\frac{-1}{\sqrt{2}}, \frac{-1}{\sqrt{2e}}\right)$

Figure 4.18 The graph for Example 6

Exercises 4.2

In Exercises 1–17, determine whether the given function has any local or absolute extreme values, and find those values if possible.

1. $f(x) = x + 2$ on $[-1, 1]$
2. $f(x) = x + 2$ on $]-\infty, 0]$

3. $f(x) = x + 2$ on $[-1, 1[$
4. $f(x) = x^2 - 1$

5. $f(x) = x^2 - 1$ on $[-2, 3]$
6. $f(x) = x^2 - 1$ on $]2, 3[$

7. $f(x) = x^3 + x - 4$ on $[a, b]$

8. $f(x) = x^3 + x - 4$ on $]a, b[$

9. $f(x) = x^5 + x^3 + 2x$ on $]a, b]$

10. $f(x) = \dfrac{1}{x - 1}$

11. $f(x) = \dfrac{1}{x - 1}$ on $]0, 1[$

12. $f(x) = \dfrac{1}{x - 1}$ on $[2, 3]$
13. $|x - 1|$ on $[-2, 2]$

14. $|x^2 - x - 2|$ on $[-3, 3]$
15. $f(x) = \dfrac{1}{x^2 + 1}$

16. $f(x) = (x + 2)^{2/3}$
17. $f(x) = (x - 2)^{1/3}$

In Exercises 18–42, locate and classify all local extreme values of the given function. Determine whether any of these extreme values are absolute. Sketch the graph of the function.

18. $f(x) = x^2 + 2x$
19. $f(x) = x^3 - 3x - 2$

20. $f(x) = (x^2 - 4)^2$
21. $f(x) = x(x - 1)^2$

22. $f(x) = x^4 + 4x$
23. $f(x) = x^3(x - 1)^2$

24. $f(x) = x^2(x - 1)^2$
25. $f(x) = x(x^2 - 1)^2$

26. $f(x) = \dfrac{x}{x^2 + 1}$
27. $f(x) = \dfrac{x^2}{x^2 + 1}$

28. $f(x) = \dfrac{x}{\sqrt{x^4 + 1}}$
29. $f(x) = x\sqrt{2 - x^2}$

30. $f(x) = x + \sin x$
31. $f(x) = x - 2\sin x$

32. $f(x) = x - 2\tan^{-1} x$
33. $f(x) = 2x - \sin^{-1} x$

34. $f(x) = e^{-x^2/2}$
35. $f(x) = x\, 2^{-x}$

36. $f(x) = x^2 e^{-x^2}$
37. $f(x) = \dfrac{\ln x}{x}$

38. $f(x) = |x + 1|$ 　　　　**39.** $f(x) = |x^2 - 1|$

40. $f(x) = \sin|x|$ 　　　　**41.** $f(x) = |\sin x|$

* **42.** $f(x) = (x - 1)^{2/3} - (x + 1)^{2/3}$

In Exercises 43–48 determine whether the given function has absolute maximum or absolute minimum values. Justify your answers. Find the extreme values if you can.

43. $\dfrac{x}{\sqrt{x^2 + 1}}$ 　　　　**44.** $\dfrac{x}{\sqrt{x^4 + 1}}$

45. $x\sqrt{4 - x^2}$ 　　　　**46.** $\dfrac{x^2}{\sqrt{4 - x^2}}$

* **47.** $\dfrac{1}{x \sin x}$ on $(0, \pi)$ 　　** * 48.** $\dfrac{\sin x}{x}$

49. If a function has an absolute maximum value, must it have any local maximum values? If a function has a local maximum value, must it have an absolute maximum value? Give reasons for your answers.

50. If the function f has an absolute maximum value and $g(x) = |f(x)|$, must g have an absolute maximum value? Justify your answer.

* **51.** **(A function with no max or min at an endpoint)** Let

$$f(x) = \begin{cases} x \sin \dfrac{1}{x} & \text{if } x > 0, \\ 0 & \text{if } x = 0. \end{cases}$$

Show that f is continuous on $[0, \infty[$ and differentiable on $]0, \infty[$ but that it has neither a local maximum nor a local minimum value at the endpoint $x = 0$.

4.3 Concavity and Inflections

Like the first derivative, the second derivative of a function also provides useful information about the behaviour of the function and the shape of its graph; it determines whether the graph is *bending upward* (i.e., has increasing slope) or *bending downward* (i.e., has decreasing slope) as we move along the graph toward the right.

DEFINITION 3

> We say that the function f is **concave up** on an open interval I if it is differentiable there and the derivative f' is an increasing function on I. Similarly, f is **concave down** on I if f' exists and is decreasing on I.

The terms "concave up" and "concave down" are used to describe the graph of the function as well as the function itself.

Note that concavity is defined only for differentiable functions, and even for those, only on intervals on which their derivatives are not constant. According to the above definition, a function is neither concave up nor concave down on an interval where its graph is a straight line segment. We say the function has no concavity on such an interval. We also say a function has opposite concavity on two intervals if it is concave up on one interval and concave down on the other.

The function f whose graph is shown in Figure 4.19 is concave up on the interval $]a, b[$ and concave down on the interval $]b, c[$.

Some geometric observations can be made about concavity:

(i) If f is concave up on an interval, then, on that interval, the graph of f lies above its tangents, and chords joining points on the graph lie above the graph.

(ii) If f is concave down on an interval, then, on that interval, the graph of f lies below its tangents, and chords to the graph lie below the graph.

(iii) If the graph of f has a tangent at a point, and if the concavity of f is opposite on opposite sides of that point, then the graph crosses its tangent at that point.

(This occurs at the point $\big(b, f(b)\big)$ in Figure 4.19. Such a point is called an *inflection point* of the graph of f.)

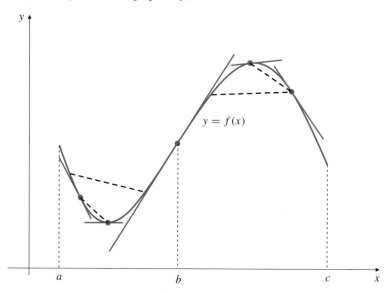

Figure 4.19 f is concave up on $]a, b[$ and concave down on $]b, c[$

Inflection points

We say that the point $\big(x_0, f(x_0)\big)$ *is* an **inflection point** of the curve $y = f(x)$ (or that the function $f(x)$ *has* an **inflection point** at $x = x_0$) if the following two conditions are satisfied:

(a) the graph $y = f(x)$ has a tangent line at x_0, and

(b) the concavity of f is opposite on opposite sides of x_0.

Note that (a) implies that either f is differentiable at x_0 or its graph has a vertical tangent line there, and (b) implies that the graph crosses its tangent line at x_0. An inflection point of a function f is a point on the graph of a function, rather than a point in its domain like a critical point or a singular point. A function may or may not have an inflection point at a critical point or singular point. In general, a point P is an inflection point (or simply *an inflection*) of a curve C (which is not necessarily the graph of a function) if C has a tangent at P and arcs of C extending in opposite directions from P are on opposite sides of that tangent line.

Figures 4.20–4.22 illustrate some situations involving critical and singular points and inflections.

If a function f has a second derivative f'', the sign of that second derivative tells us whether the first derivative f' is increasing or decreasing and hence determines the concavity of f.

THEOREM 5

Concavity and the second derivative

(a) If $f''(x) > 0$ on interval I, then f is concave up on I.

(b) If $f''(x) < 0$ on interval I, then f is concave down on I.

(c) If f has an inflection point at x_0 and $f''(x_0)$ exists, then $f''(x_0) = 0$.

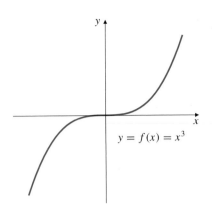

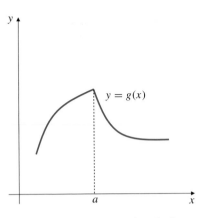

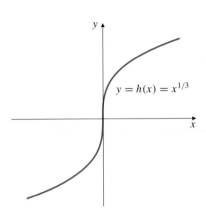

Figure 4.20 $x = 0$ is a critical point of $f(x) = x^3$, and f has an inflection point there

Figure 4.21   The concavity of g is opposite on opposite sides of a, but its graph has no tangent and therefore no inflection point there

Figure 4.22   This graph of h has an inflection point at the origin even though $x = 0$ is a singular point of h

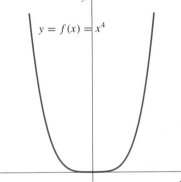

PROOF   Parts (a) and (b) follow from applying Theorem 12 of Section 2.6 to the derivative f' of f. If f has an inflection point at x_0 and $f''(x_0)$ exists, then f must be differentiable in an open interval containing x_0. Since f' is increasing on one side of x_0 and decreasing on the other side, it must have a local maximum or minimum value at x_0. By Theorem 2, $f''(x_0) = 0$.

Figure 4.23 $f''(0) = 0$, but f does not have an inflection point at 0

Theorem 5 tells us that to find (the x-coordinates of) inflection points of a twice differentiable function f we need only look at points where $f''(x) = 0$. However, not every such point has to be an inflection point. For example, $f(x) = x^4$, whose graph is shown in Figure 4.23, does not have an inflection point at $x = 0$ even though $f''(0) = 12x^2|_{x=0} = 0$. In fact, x^4 is concave up on every interval.

Example 1   Determine the intervals of concavity of $f(x) = x^6 - 10x^4$ and the inflection points of its graph.

Solution   We have

$$f'(x) = 6x^5 - 40x^3,$$
$$f''(x) = 30x^4 - 120x^2 = 30x^2(x - 2)(x + 2).$$

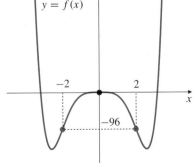

Having factored $f''(x)$ in this manner, we can see that it vanishes only at $x = -2$, $x = 0$, and $x = 2$. On the intervals $]-\infty, -2[$ and $]2, \infty[$, $f''(x) > 0$ so f is concave up. On $]-2, 0[$ and $]0, 2[$, $f''(x) < 0$ so f is concave down. $f''(x)$ changes sign as we pass through -2 and 2. Since $f(\pm 2) = -96$, the graph of f has inflection points at $(\pm 2, -96)$. However, $f''(x)$ does not change sign at $x = 0$, since $x^2 > 0$ for both positive and negative x. Thus there is no inflection point at 0. As was the case for the first derivative, information about the sign of $f''(x)$ and the consequent concavity of f can be conveniently conveyed in a chart:

Figure 4.24   The graph $y = f(x) = x^6 - 10x^4$

x		-2		0		2	
f''	$+$	0	$-$	0	$-$	0	$+$
f	$\smile$	infl	$\frown$		$\frown$	infl	$\smile$

The graph of f is sketched in Figure 4.24.

Example 2 Determine the intervals of increase and decrease, the local extreme values, and the concavity of $f(x) = x^4 - 2x^3 + 1$. Use the information to sketch the graph of f.

Solution

$$f'(x) = 4x^3 - 6x^2 = 2x^2(2x - 3) = 0 \quad \text{at } x = 0 \text{ and } x = 3/2,$$
$$f''(x) = 12x^2 - 12x = 12x(x - 1) = 0 \quad \text{at } x = 0 \text{ and } x = 1.$$

The behaviour of f is summarized in the following chart:

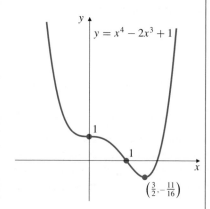

		CP				CP	
x		0		1		$3/2$	
f'	$-$	0	$-$		$-$	0	$+$
f''	$+$	0	$-$	0	$+$		$+$
f	$\searrow$		$\searrow$		$\searrow$	min	$\nearrow$
	$\smile$	infl	$\frown$	infl	$\smile$		$\smile$

Note that f has an inflection at the critical point $x = 0$. We calculate the values of f at the "interesting values of x" in the charts:

$$f(0) = 1, \qquad f(1) = 0, \qquad f\left(\tfrac{3}{2}\right) = -\tfrac{11}{16}.$$

The graph of f is sketched in Figure 4.25.

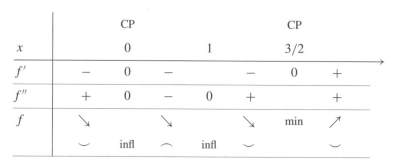

Figure 4.25 The function of Example 2

The Second Derivative Test

A function f will have a local maximum (or minimum) value at a critical point if its graph is concave downward (or upward) in an interval containing that point. In fact, we can often use the value of the second derivative at the critical point to determine whether the function has a local maximum or a local minimum value there.

THEOREM 6

The Second Derivative Test

(a) If $f'(x_0) = 0$ and $f''(x_0) < 0$, then f has a local maximum value at x_0.

(b) If $f'(x_0) = 0$ and $f''(x_0) > 0$, then f has a local minimum value at x_0.

(c) If $f'(x_0) = 0$ and $f''(x_0) = 0$, no conclusion can be drawn; f may have a local maximum at x_0 or a local minimum, or it may have an inflection point instead.

PROOF Suppose that $f'(x_0) = 0$ and $f''(x_0) < 0$. Since

$$\lim_{h \to 0} \frac{f'(x_0 + h)}{h} = \lim_{h \to 0} \frac{f'(x_0 + h) - f'(x_0)}{h} = f''(x_0) < 0,$$

it follows that $f'(x_0 + h) < 0$ for all sufficiently small positive h, and $f'(x_0 + h) > 0$ for all sufficiently small negative h. By the first derivative test (Theorem 3), f must have a local maximum value at x_0. The proof of the local minimum case is similar.

The functions $f(x) = x^4$ (Figure 4.23), $f(x) = -x^4$, and $f(x) = x^3$ (Figure 4.20) all satisfy $f'(0) = 0$ and $f''(0) = 0$. But x^4 has a minimum value at $x = 0$, $-x^4$ has a maximum value at $x = 0$, and x^3 has neither a maximum nor a minimum value at $x = 0$ but has an inflection there. Therefore, we cannot make any conclusion about the nature of a critical point based on knowing that $f''(x) = 0$ there.

Example 3 Find and classify the critical points of $f(x) = x^2 e^{-x}$.

Solution

$$f'(x) = (2x - x^2)e^{-x} = x(2 - x)e^{-x} = 0 \quad \text{at } x = 0 \text{ and } x = 2,$$
$$f''(x) = (2 - 4x + x^2)e^{-x}$$
$$f''(0) = 2 > 0, \qquad f''(2) = -2e^{-2} < 0.$$

Thus, f has a local minimum value at $x = 0$ and a local maximum value at $x = 2$. See Figure 4.26.

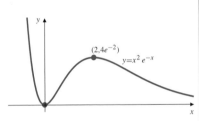

Figure 4.26 The critical points of $f(x) = x^2 e^{-x}$

For many functions the second derivative is more complicated to calculate than the first derivative, so the First Derivative Test is likely to be of more use in classifying critical points than is the Second Derivative Test. Also note that the First Derivative Test can classify local extreme values that occur at endpoints and singular points as well as at critical points.

It is possible to generalize the Second Derivative Test to obtain a higher derivative test to deal with some situations where the second derivative is zero at a critical point. (See Exercise 40 at the end of this section.)

Exercises 4.3

In Exercises 1–22, determine the intervals of constant concavity of the given function and locate any inflection points.

1. $f(x) = \sqrt{x}$

2. $f(x) = 2x - x^2$

3. $f(x) = x^2 + 2x + 3$

4. $f(x) = x - x^3$

5. $f(x) = 10x^3 - 3x^5$

6. $f(x) = 10x^3 + 3x^5$

7. $f(x) = (3 - x^2)^2$

8. $f(x) = (2 + 2x - x^2)^2$

9. $f(x) = (x^2 - 4)^3$

10. $f(x) = \dfrac{x}{x^2 + 3}$

11. $f(x) = \sin x$

12. $f(x) = \cos 3x$

13. $f(x) = x + \sin 2x$

14. $f(x) = x - 2 \sin x$

15. $f(x) = \tan^{-1} x$

16. $f(x) = x \, e^x$

17. $f(x) = e^{-x^2}$

18. $f(x) = \dfrac{\ln(x^2)}{x}$

19. $f(x) = \ln(1 + x^2)$

20. $f(x) = (\ln x)^2$

21. $f(x) = \dfrac{x^3}{3} - 4x^2 + 12x - \dfrac{25}{3}$

22. $f(x) = (x - 1)^{1/3} + (x + 1)^{1/3}$

23. Discuss the concavity of the linear function $f(x) = ax + b$. Does it have any inflections?

Classify the critical points of the functions in Exercises 24–35 using the Second Derivative Test whenever possible.

24. $f(x) = 3x^3 - 36x - 3$

25. $f(x) = x(x-2)^2 + 1$

26. $f(x) = x + \dfrac{4}{x}$

27. $f(x) = x^3 + \dfrac{1}{x}$

28. $f(x) = \dfrac{x}{2^x}$

29. $f(x) = \dfrac{x}{1 + x^2}$

30. $f(x) = xe^x$

31. $f(x) = x \ln x$

32. $f(x) = (x^2 - 4)^2$

33. $f(x) = (x^2 - 4)^3$

34. $f(x) = (x^2 - 3)e^x$

35. $f(x) = x^2 e^{-2x^2}$

36. Let $f(x) = x^2$ if $x \geq 0$ and $f(x) = -x^2$ if $x < 0$. Is 0 a critical point of f? Does f have an inflection point there? Is $f''(0) = 0$? If a function has a nonvertical tangent line at an inflection point, does the second derivative of the function necessarily vanish at that point?

* **37.** Verify that if f is concave up on an interval, then its graph lies above its tangent lines on that interval. *Hint:* suppose f is concave up on an open interval containing x_0. Let $h(x) = f(x) - f(x_0) - f'(x_0)(x - x_0)$. Show that h has a local minimum value at x_0 and hence that $h(x) \geq 0$ on the interval. Show that $h(x) > 0$ if $x \neq x_0$.

* **38.** Verify that the graph $y = f(x)$ crosses its tangent line at an inflection point. *Hint:* consider separately the cases where the tangent line is vertical and nonvertical.

39. Let $f_n(x) = x^n$ and $g_n(x) = -x^n$, $(n = 2, 3, 4, \ldots)$. Determine whether each function has a local maximum, a local minimum, or an inflection point at $x = 0$.

* **40.** (**Higher derivative test**) Use your conclusions from the previous exercise to suggest a generalization of the second derivative test that applies when
$$f'(x_0) = f''(x_0) = \ldots = f^{(k-1)}(x_0) = 0, \; f^{(k)}(x_0) \neq 0,$$

for some $k \geq 2$.

* **41.** This problem shows that no test based solely on the signs of derivatives at x_0 can determine whether every function with a critical point at x_0 has a local maximum or minimum or an inflection point there. Let

$$f(x) = \begin{cases} e^{-1/x^2} & \text{if } x \neq 0, \\ 0 & \text{if } x = 0. \end{cases}$$

Prove the following:

(a) $\lim_{x \to 0} x^{-n} f(x) = 0$ for $n = 0, 1, 2, 3, \ldots$.

(b) $\lim_{x \to 0} P(1/x) f(x) = 0$ for every polynomial P.

(c) For $x \neq 0$, $f^{(k)}(x) = P_k(1/x) f(x) (k = 1, 2, 3, \ldots)$, where P_k is a polynomial.

(d) $f^{(k)}(0)$ exists and equals 0 for $k = 1, 2, 3, \ldots$.

(e) f has a local minimum at $x = 0$; $-f$ has a local maximum at $x = 0$.

(f) If $g(x) = xf(x)$, then $g^{(k)}(0) = 0$ for every positive integer k and g has an inflection point at $x = 0$.

* **42.** A function may have neither a local maximum nor a local minimum nor an inflection at a critical point. Show this by considering the following function:

$$f(x) = \begin{cases} x^2 \sin \dfrac{1}{x} & \text{if } x \neq 0 \\ 0 & \text{if } x = 0 \end{cases}$$

Show that $f'(0) = f(0) = 0$, so the x-axis is tangent to the graph of f at $x = 0$; but $f'(x)$ is not continuous at $x = 0$, so $f''(0)$ does not exist. Show that the concavity of f is not constant on any interval with endpoint 0.

4.4   Sketching the Graph of a Function

When sketching the graph $y = f(x)$ of a function f, we have three sources of useful information:

(i) **the function f itself**, from which we determine the coordinates of some points on the graph, the symmetry of the graph, and any asymptotes;

(ii) **the first derivative, f'**, from which we determine the intervals of increase and decrease and the location of any local extreme values; and

(iii) **the second derivative, f''**, from which we determine the concavity and inflection points, and sometimes extreme values.

Items (ii) and (iii) have been explored in the previous two sections. In this section we consider what we can learn from the function itself about the shape of its graph, and then we illustrate the entire sketching procedure with several examples using all three sources of information.

We could sketch a graph by plotting the coordinates of many points on it and joining them by a suitably smooth curve. This is what computer software and graphics calculators computer software do. When carried out by hand (without a computer or calculator), this simplistic approach is at best tedious and at worst can fail to reveal the most interesting aspects of the graph (singular points, extreme values, and so on). We could also compute the slope at each of the plotted points and, by drawing short line segments through these points with the appropriate slopes, ensure that the sketched graph passes through each plotted point with the correct slope. A more efficient procedure is to obtain the coordinates of only a few points and use qualitative information from the function and its first and second derivatives to determine the *shape* of the graph between these points.

Besides critical and singular points and inflections, a graph may have other "interesting" points. The **intercepts** (points at which the graph intersects the coordinate axes) are usually among these. When sketching any graph it is wise to try to find all such intercepts, that is, all points with coordinates $(x, 0)$ and $(0, y)$ that lie on the graph. Of course, not every graph will have such points, and even when they do exist it may not always be possible to compute them exactly. Whenever a graph is made up of several disconnected pieces (called **components**), the coordinates of *at least one point on each component* must be obtained. It can sometimes be useful to determine the slopes at those points too. Vertical asymptotes (discussed below) usually break the graph of a function into components.

Realizing that a given function possesses some symmetry can aid greatly in obtaining a good sketch of its graph. In Section P.4 we discussed odd and even functions and observed that odd functions have graphs that are symmetric about the origin, while even functions have graphs that are symmetric about the y-axis, as shown in Figure 4.27. These are the symmetries you are most likely to notice, but functions can have other symmetries. For example, the graph of $2 + (x - 1)^2$ will certainly be symmetric about the line $x = 1$, and the graph of $2 + (x - 3)^3$ is symmetric about the point $(3, 2)$.

Asymptotes

Some of the curves we have sketched in previous sections have had **asymptotes**, that is, straight lines to which the curve draws arbitrarily near as it recedes to infinite distance from the origin. Asymptotes are of three types: vertical, horizontal, and oblique.

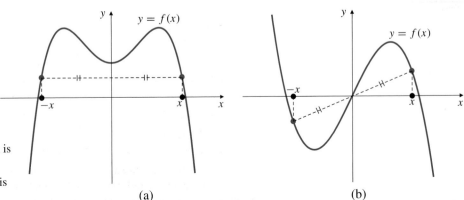

Figure 4.27

(a) The graph of an even function is symmetric about the y-axis

(b) The graph of an odd function is symmetric about the origin

DEFINITION 5

The graph of $y = f(x)$ has a **vertical asymptote** at $x = a$ if

either $\quad \lim_{x \to a-} f(x) = \pm\infty \quad$ or $\quad \lim_{x \to a+} f(x) = \pm\infty, \quad$ or both.

This situation tends to arise when $f(x)$ is a quotient of two expressions and the denominator is zero at $x = a$.

Example 1 Find the vertical asymptotes of $f(x) = \dfrac{1}{x^2 - x}$. How does the graph approach these asymptotes?

Solution The denominator $x^2 - x = x(x - 1)$ approaches 0 as x approaches 0 or 1, so f has vertical asymptotes at $x = 0$ and $x = 1$ (Figure 4.28). Since $x(x - 1)$ is positive on $]-\infty, 0[$ and on $]1, \infty[$ and is negative on $]0, 1[$, we have

$$\lim_{x \to 0-} \frac{1}{x^2 - x} = \infty \qquad\qquad \lim_{x \to 1-} \frac{1}{x^2 - x} = -\infty$$

$$\lim_{x \to 0+} \frac{1}{x^2 - x} = -\infty \qquad\qquad \lim_{x \to 1+} \frac{1}{x^2 - x} = \infty.$$

DEFINITION 6

The graph of $y = f(x)$ has a **horizontal asymptote** $y = L$ if

either $\quad \lim_{x \to \infty} f(x) = L \quad$ or $\quad \lim_{x \to -\infty} f(x) = L, \quad$ or both.

Example 2 Find the horizontal asymptotes of

(a) $f(x) = \dfrac{1}{x^2 - x}$ $\quad$ and $\quad$ (b) $g(x) = \dfrac{x^4 + x^2}{x^4 + 1}$.

Solution

(a) The function f has horizontal asymptote $y = 0$ (Figure 4.28) since

$$\lim_{x \to \pm\infty} \frac{1}{x^2 - x} = \lim_{x \to \pm\infty} \frac{1/x^2}{1 - (1/x)} = \frac{0}{1} = 0.$$

(b) The function $g(x)$ has horizontal asymptote $y = 1$ (Figure 4.29) since

$$\lim_{x \to \pm\infty} \frac{x^4 + x^2}{x^4 + 1} = \lim_{x \to \pm\infty} \frac{1 + (1/x^2)}{1 + (1/x^4)} = \frac{1}{1} = 1.$$

Observe that the graph of g crosses its asymptote twice. (There is a popular misconception among students that curves cannot cross their asymptotes. Exercise 41 below gives an example of a curve that crosses its asymptote infinitely often.)

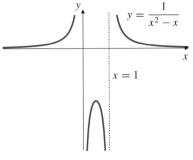

Figure 4.28

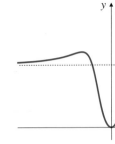

Figure 4.29

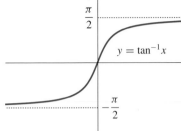

Figure 4.30

The horizontal asymptotes of both functions f and g in Example 2 are **two-sided**, which means that the graphs approach the asymptotes as x approaches both infinity and negative infinity. The function $\tan^{-1} x$ has two **one-sided** asymptotes, $y = \pi/2$ (as $x \to \infty$) and $y = -(\pi/2)$ (as $x \to -\infty$). See Figure 4.30.

It can also happen that the graph of a function $f(x)$ approaches a nonhorizontal straight line as x approaches ∞ or $-\infty$ (or both). Such a line is called an *oblique asymptote* of the graph.

DEFINITION 7

The straight line $y = ax + b$ (where $a \neq 0$), is an **oblique asymptote** of the graph of $y = f(x)$ if

either $\quad \lim_{x \to -\infty} \big(f(x) - (ax + b)\big) = 0 \quad$ or $\quad \lim_{x \to \infty} \big(f(x) - (ax + b)\big) = 0,$

or both.

Example 3 Consider the function

$$f(x) = \frac{x^2 + 1}{x} = x + \frac{1}{x},$$

whose graph is shown in Figure 4.31(a). The straight line $y = x$ is a *two-sided* oblique asymptote of the graph of f because

$$\lim_{x \to \pm\infty} \big(f(x) - x\big) = \lim_{x \to \pm\infty} \frac{1}{x} = 0.$$

Example 4 The graph of $y = \dfrac{x\, e^x}{1 + e^x}$ is shown in Figure 4.31(b). It has a horizontal asymptote $y = 0$ at the left and an oblique asymptote $y = x$ at the right:

$$\lim_{x \to -\infty} \frac{x\, e^x}{1 + e^x} = \frac{0}{1} = 0 \quad \text{and}$$

$$\lim_{x \to \infty} \left(\frac{x\, e^x}{1 + e^x} - x \right) = \lim_{x \to \infty} \frac{x(e^x - 1 - e^x)}{1 + e^x} = \lim_{x \to \infty} \frac{-x}{1}$$

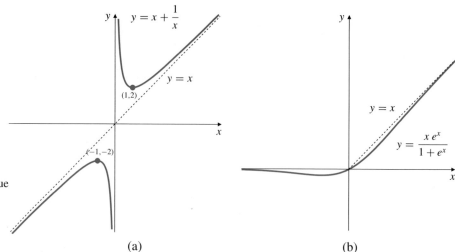

Figure 4.31

(a) $y = f(x)$ has a two-sided oblique asymptote, $y = x$

(b) This graph has a horizontal asymptote at the left and an oblique asymptote at the right

Recall that a **rational function** is a function of the form $f(x) = P(x)/Q(x)$, where P and Q are polynomials. It is possible to be quite specific about the asymptotes of a rational function.

Asymptotes of a rational function

Suppose that $f(x) = \dfrac{P_m(x)}{Q_n(x)}$, where P_m and Q_n are polynomials of degree m and n, respectively. Suppose also that P_m and Q_n have no common linear factors. Then

(a) The graph of f has a vertical asymptote at every position x such that $Q_n(x) = 0$.

(b) The graph of f has a two-sided horizontal asymptote $y = 0$ if $m < n$.

(c) The graph of f has a two-sided horizontal asymptote $y = L$, $(L \neq 0)$ if $m = n$. L is the quotient of the coefficients of the highest degree terms in P_m and Q_n.

(d) The graph of f has a two-sided oblique asymptote if $m = n + 1$. This asymptote can be found by dividing Q_n into P_m to obtain a linear quotient, $ax + b$, and remainder, R, a polynomial of degree at most $n - 1$. That is,

$$f(x) = ax + b + \frac{R(x)}{Q_n(x)}.$$

The oblique asymptote is $y = ax + b$.

(e) The graph of f has no horizontal or oblique asymptotes if $m > n + 1$.

Example 5 Find the oblique asymptote of $y = \dfrac{x^3}{x^2 + x + 1}$.

Solution We can either obtain the quotient by long division:

$$
x^2 + x + 1 \overline{\smash{\big)}\begin{array}{l} x - 1 \\[2pt] x^3 \\ x^3 + x^2 + x \\ \hline {- x^2 - x} \\ - x^2 - x - 1 \\ \hline 1 \end{array}} \qquad \frac{x^3}{x^2 + x + 1} = x - 1 + \frac{1}{x^2 + x + 1}
$$

or we can obtain the same result by "short division":

$$
\frac{x^3}{x^2 + x + 1} = \frac{x^3 + x^2 + x - x^2 - x - 1 + 1}{x^2 + x + 1} = x - 1 + \frac{1}{x^2 + x + 1}.
$$

In any event, we see that the oblique asymptote has equation $y = x - 1$. $\blacksquare$

Examples of Formal Curve Sketching

Here is a checklist of things to consider when you are asked to make a careful sketch of the graph $y = f(x)$. It will, of course, not always be possible to obtain every item of information mentioned in the list.

> **Checklist for curve sketching**
>
> 1. Calculate $f'(x)$ and $f''(x)$, and express the results in factored form.
> 2. Examine $f(x)$ to determine its domain and the following items:
>   (a) any vertical asymptotes. (Look for zeros of denominators.)
>   (b) any horizontal or oblique asymptotes. (Consider $\lim_{x \to \pm\infty} f(x)$.)
>   (c) any obvious symmetry. (Is f even or odd?)
>   (d) any easily calculated intercepts (points with coordinates $(x, 0)$ or $(0, y)$) or endpoints or other "obvious" points. You will add to this list when you know any critical points, singular points, and inflection points. Eventually you should make sure you know the coordinates of at least one point on every component of the graph.
> 3. Examine $f'(x)$ for the following:
>   (a) any critical points.
>   (b) any points where f' is not defined. (These will include singular points, endpoints of the domain of f, and vertical asymptotes.)
>   (c) intervals on which f' is positive or negative. It's a good idea to convey this information in the form of a chart such as those used in the examples. Conclusions about where f is increasing and decreasing and classification of some critical and singular points as local maxima and minima can also be indicated on the chart.
> 4. Examine $f''(x)$ for the following:
>   (a) points where $f''(x) = 0$.
>   (b) points where $f''(x)$ is undefined. (These will include singular points, endpoints, vertical asymptotes, and possibly other points as well, where f' is defined but f'' isn't.)
>   (c) intervals where f'' is positive or negative and where f is therefore concave up or down. Use a chart.
>   (d) any inflection points.

When you have obtained as much of this information as possible, make a careful sketch that reflects *everything* you have learned about the function. Consider where best to place the axes and what scale to use on each so the "interesting features" of the graph show up most clearly. Be alert for seeming inconsistencies in the information—that is a strong suggestion you may have made an error somewhere. For example, if you have determined that $f(x) \to \infty$ as x approaches the vertical asymptote $x = a$ from the right, and also that f is decreasing and concave down on the interval (a, b), then you have very likely made an error. (Try to sketch such a situation to see why.)

Example 6 Sketch the graph of $y = \dfrac{x^2 + 2x + 4}{2x}$.

Solution It is useful to rewrite the function y in the form

$$y = \frac{x}{2} + 1 + \frac{2}{x},$$

since this form not only shows clearly that $y = (x/2) + 1$ is an oblique asymptote, but also makes it easier to calculate the derivatives

$$y' = \frac{1}{2} - \frac{2}{x^2} = \frac{x^2 - 4}{2x^2}, \qquad y'' = \frac{4}{x^3}$$

From y: Domain: all x except 0. Vertical asymptote: $x = 0$,

Oblique asymptote: $y = \dfrac{x}{2} + 1$, $\quad y - \left(\dfrac{x}{2} + 1\right) = \dfrac{2}{x} \to 0$ as $x \to \pm\infty$.

Symmetry: none obvious (y is neither odd nor even).

Intercepts: none. $x^2 + 2x + 4 = (x + 1)^2 + 3 \geq 3$ for all x, and y is not defined at $x = 0$.

From y': Critical points: $x = \pm 2$; points $(-2, -1)$ and $(2, 3)$.

y' not defined at $x = 0$ (vertical asymptote).

From y'': $y'' = 0$ nowhere; y'' undefined at $x = 0$.

x			CP -2		ASY 0		CP 2	
y'	$+$		0	$-$	undef	$-$	0	$+$
y''	$-$			$-$	undef	$+$		$+$
y	↗		max	↘	undef	↘	min	↗
	⌢			⌢		⌣		⌣

The graph is shown in Figure 4.32.

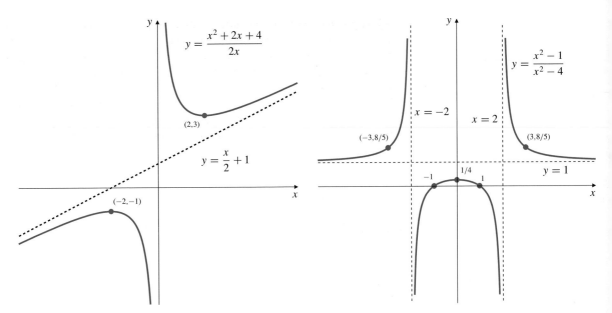

Figure 4.32   **Figure 4.33**

Example 7   Sketch the graph of $f(x) = \dfrac{x^2 - 1}{x^2 - 4}$.

Solution   We have

$$f'(x) = \frac{-6x}{(x^2 - 4)^2}, \qquad f''(x) = \frac{6(3x^2 + 4)}{(x^2 - 4)^3}.$$

From f: Domain: all x except ± 2.   Vertical asymptotes: $x = -2$ and $x = 2$.
Horizontal asymptote: $y = 1$ (as $x \to \pm\infty$).
Symmetry: about the y-axis (y is even).
Intercepts: $(0, 1/4)$, $(-1, 0)$, and $(1, 0)$.
Other points: $(-3, 8/5)$, $(3, 8/5)$.   (The two vertical asymptotes divide the graph into three components; we need points on each.   The outer components require points with $|x| > 2$.)

From f':   Critical point: $x = 0$; f' not defined at $x = 2$ or $x = -2$.

From f'': $f''(x) = 0$ nowhere; f'' not defined at $x = 2$ or $x = -2$.

		ASY		CP		ASY	
x		-2		0		2	
f'	$+$	undef	$+$	0	$-$	undef	$-$
f''	$+$	undef	$-$		$-$	undef	$+$
f	↗	undef	↗	max	↘	undef	↘
	⌣		⌢		⌢		⌣

The graph is shown in Figure 4.33.

Example 8   Sketch the graph of $y = xe^{-x^2/2}$.

Solution   We have $y' = (1 - x^2)e^{-x^2/2}$, $\quad y'' = x(x^2 - 3)e^{-x^2/2}$.
From y: Domain: all x.
              Horizontal asymptote: $y = 0$. Note that if $t = x^2/2$, then
              $|xe^{-x^2/2}| = \sqrt{2t}\,e^{-t} \to 0$ as $t \to \infty$ (hence as $x \to \pm\infty$).
              Symmetry: about the origin (y is odd). Intercepts: $(0, 0)$.
From y':   Critical points: $x = \pm 1$; points $(\pm 1, \pm 1/\sqrt{e}) \approx (\pm 1, \pm 0.61)$.
From y'': $y'' = 0$ at $x = 0$ and $x = \pm\sqrt{3}$;
              points $(0, 0)$, $(\pm\sqrt{3}, \pm\sqrt{3}e^{-3/2}) \approx (\pm 1.73, \pm 0.39)$.

				CP				CP	
x		$-\sqrt{3}$		-1		0		1	$\sqrt{3}$
y'	$-$		$-$	0	$+$		$+$	0 $\quad -$	$-$
y''	$-$	0	$+$		$+$	0	$-$		$-$ $\quad 0$ $\quad +$
y	$\searrow$		$\searrow$	min	$\nearrow$		$\nearrow$	max $\searrow$	$\searrow$
	$\frown$	infl	$\smile$		$\smile$	infl	$\frown$		$\frown$ infl $\smile$

The graph is shown in Figure 4.34.

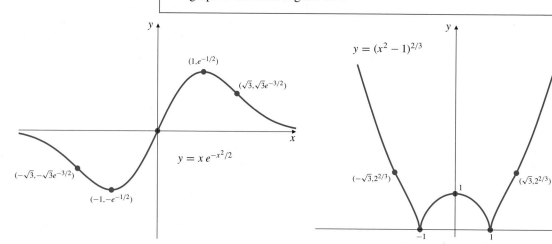

$(1, e^{-1/2})$

$(\sqrt{3}, \sqrt{3}e^{-3/2})$

$y = x\,e^{-x^2/2}$

$(-\sqrt{3}, -\sqrt{3}e^{-3/2})$

$(-1, -e^{-1/2})$

Figure 4.34

$y = (x^2 - 1)^{2/3}$

$(-\sqrt{3}, 2^{2/3})$

$(\sqrt{3}, 2^{2/3})$

Figure 4.35

Example 9   Sketch the graph of $f(x) = (x^2 - 1)^{2/3}$. (See Figure 4.35.)

Solution $f'(x) = \dfrac{4}{3}\dfrac{x}{(x^2 - 1)^{1/3}}$, $f''(x) = \dfrac{4}{9}\dfrac{x^2 - 3}{(x^2 - 1)^{4/3}}$.
From f: Domain: all x.
              Asymptotes: none. ($f(x)$ grows like $x^{4/3}$ as $x \to \pm\infty$.)
              Symmetry: about the y-axis (f is an even function).
              Intercepts: $(\pm 1, 0)$, $(0, 1)$.
From f':   Critical points: $x = 0$; singular points: $x = \pm 1$.
From f'': $f''(x) = 0$ at $x = \pm\sqrt{3}$; points $(\pm\sqrt{3}, 2^{2/3}) \approx (\pm 1.73, 1.59)$;
              $f''(x)$ not defined at $x = \pm 1$.

x		$-\sqrt{3}$		-1 (SP)		0 (CP)		1 (SP)		$\sqrt{3}$	
f'	$-$		$-$	undef	$+$	0	$-$	undef	$+$		$+$
f''	$+$	0	$-$	undef	$-$		$-$	undef	$-$	0	$+$
f	$\searrow$		$\searrow$	min	$\nearrow$	max	$\searrow$	min	$\nearrow$		$\nearrow$
	$\smile$	infl	$\frown$		$\frown$		$\frown$		$\frown$	infl	$\smile$

💻 ***Remark*** **Using a Graphing Utility** The techniques for curve sketching developed above are useful only for graphs of functions that are simple enough to allow you to calculate and analyze their derivatives. In practice you will likely want to use a graphing calculator or a computer to produce the graph quickly and painlessly. To make effective use of such a utility, you have to decide on a viewing window and what horizontal and vertical scales to use. An inappropriate choice of viewing window can cause you to miss significant features of the graph. Here is a Maple command for viewing the graph of the function from Example 6, together with its oblique asymptote; we ask Maple to plot both $(x^2 + 2x + 4)/(2x)$ and $1 + (x/2)$.

```
>  plot({(x^2+2*x+4)/(2*x), 1+(x/2)}, x=-6..6, -7..7);
```

Getting Maple to plot the curve in Example 9 is a bit trickier. Because Maple doesn't want to deal with fractional powers of negative numbers, even when they have positive real values, we must actually plot $|x^2 - 1|^{2/3}$ or else the part of the graph between -1 and 1 will be missing.

```
>  plot((abs(x^2-1))^(2/3), x=-4..4, -1..5);
```

Exercises 4.4

1. Figure 4.36 shows the graphs of a function f, its two derivatives f' and f'', and another function g. Which graph corresponds to each function?

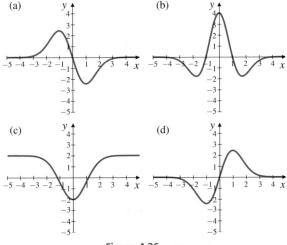

Figure 4.36

2. List, for each function graphed in Figure 4.36, such information that you can determine (approximately) by inspecting the graph (e.g., symmetry, asymptotes, intercepts, intervals of increase and decrease, critical and singular points, local maxima and minima, intervals of constant concavity, inflection points).

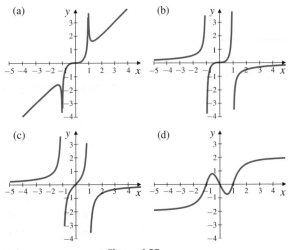

Figure 4.37

3. Figure 4.37 shows the graphs of four functions:

$$f(x) = \frac{x}{1 - x^2} \qquad g(x) = \frac{x^3}{1 - x^4}$$

$$h(x) = \frac{x^3 - x}{\sqrt{x^6 + 1}} \qquad k(x) = \frac{x^3}{\sqrt{|x^4 - 1|}}$$

Which graph corresponds to each function?

4. Repeat Exercise 2 for the graphs in Figure 4.37.

In Exercises 5–6, sketch the graph of a function that has the given properties. Identify any critical points, singular points, local maxima and minima, and inflection points. Assume that f is continuous and its derivatives exist everywhere unless the contrary is implied or explicitly stated.

5. $f(0) = 1$, $f(\pm 1) = 0$, $f(2) = 1$, $\lim_{x \to \infty} f(x) = 2$, $\lim_{x \to -\infty} f(x) = -1$, $f'(x) > 0$ on $]-\infty, 0[$ and on $]1, \infty[$, $f'(x) < 0$ on $]0, 1[$, $f''(x) > 0$ on $]-\infty, 0[$ and on $]0, 2[$, and $f''(x) < 0$ on $]2, \infty[$.

6. $f(-1) = 0$, $f(0) = 2$, $f(1) = 1$, $f(2) = 0$, $f(3) = 1$, $\lim_{x \to \pm\infty}(f(x) + 1 - x) = 0$, $f'(x) > 0$ on $]-\infty, -1[$, $]-1, 0[$ and $]2, \infty[$, $f'(x) < 0$ on $]0, 2[$, $\lim_{x \to -1} f'(x) = \infty$, $f''(x) > 0$ on $]-\infty, -1[$ and on $]1, 3[$, and $f''(x) < 0$ on $]-1, 1[$ and on $]3, \infty[$.

In Exercises 7–39, sketch the graphs of the given functions, making use of any suitable information you can obtain from the function and its first and second derivatives.

7. $y = (x^2 - 1)^3$

8. $y = x(x^2 - 1)^2$

9. $y = \frac{2 - x}{x}$

10. $y = \frac{x - 1}{x + 1}$

11. $y = \frac{x^3}{1 + x}$

12. $y = \frac{1}{4 + x^2}$

13. $y = \frac{1}{2 - x^2}$

14. $y = \frac{x}{x^2 - 1}$

15. $y = \frac{x^2}{x^2 - 1}$

16. $y = \frac{x^3}{x^2 - 1}$

17. $y = \frac{x^3}{x^2 + 1}$

18. $y = \frac{x^2}{x^2 + 1}$

19. $y = \frac{x^2 - 4}{x + 1}$

20. $y = \frac{x^2 - 2}{x^2 - 1}$

21. $y = \frac{x^3 - 4x}{x^2 - 1}$

22. $y = \frac{x^2 - 1}{x^2}$

23. $y = \frac{x^5}{(x^2 - 1)^2}$

24. $y = \frac{(2 - x)^2}{x^3}$

25. $y = \frac{1}{x^3 - 4x}$

26. $y = \frac{x}{x^2 + x - 2}$

27. $y = \frac{x^3 - 3x^2 + 1}{x^3}$

28. $y = x + \sin x$

29. $y = x + 2\sin x$

30. $y = e^{-x^2}$

31. $y = xe^x$

32. $y = e^{-x}\sin x$, $(x \geq 0)$

33. $y = x^2 e^{-x^2}$

34. $y = x^2 e^x$

35. $y = \frac{\ln x}{x}$, $(x > 0)$

36. $y = \frac{\ln x}{x^2}$, $(x > 0)$

37. $y = \frac{1}{\sqrt{4 - x^2}}$

38. $y = \frac{x}{\sqrt{x^2 + 1}}$

39. $y = (x^2 - 1)^{1/3}$

* 40. What is $\lim_{x \to 0+} x \ln x$? $\lim_{x \to 0} x \ln |x|$? If $f(x) = x \ln |x|$ for $x \neq 0$, is it possible to define $f(0)$ in such a way that f is continuous on the whole real line? Sketch the graph of f.

41. What straight line is an asymptote of the curve $y = \frac{\sin x}{1 + x^2}$? At what points does the curve cross this asymptote?

4.5 Extreme-Value Problems

In this section we solve various word problems that, when translated into mathematical terms, require the finding of a maximum or minimum value of a function of one variable. Such problems can range from simple to very complex and difficult; they can be phrased in terminology appropriate to some other discipline or they can be already partially translated into a more mathematical context. We have already encountered a few such problems in earlier chapters.

Let us consider a couple of examples before attempting to formulate any general principles for dealing with such problems.

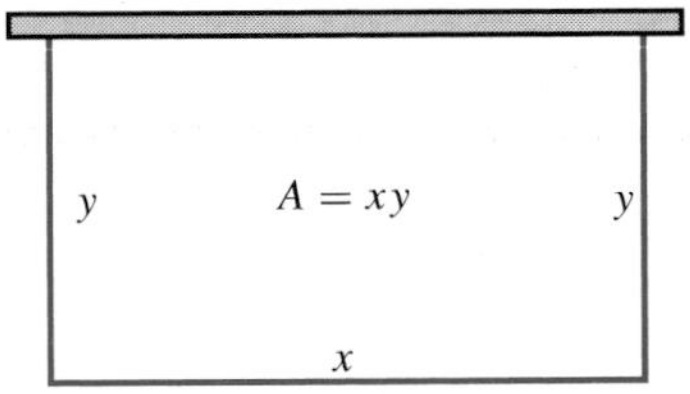

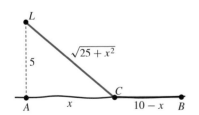

Figure 4.38

Example 1   A rectangular animal enclosure is to be constructed having one side along an existing long wall and the other three sides fenced. If 100 m of fence are available, what is the largest possible area for the enclosure?

Solution   This problem, like many others, is essentially a geometric one. A sketch should be made at the outset, as we have done in Figure 4.38. Let the length and width of the enclosure be x and y m, respectively, and let its area be A m². Thus $A = xy$. Since the total length of the fence is 100 m, we must have $x + 2y = 100$. A appears to be a function of two variables, x and y, but these variables are not independent; they are related by the *constraint* $x + 2y = 100$. This constraint equation can be solved for one variable in terms of the other, and A can therefore be written as a function of only one variable:

$$x = 100 - 2y,$$
$$A = A(y) = (100 - 2y)y = 100y - 2y^2.$$

Evidently we require $y \geq 0$ and $y \leq 50$ (i.e., $x \geq 0$), in order that the area make sense. (It would otherwise be negative.) Thus, we must maximize the function $A(y)$ on the interval $[0, 50]$. Being continuous on this closed, finite interval, A must have a maximum value, by Theorem 1. Clearly, $A(0) = A(50) = 0$ and $A(y) > 0$ for $0 < y < 50$. Hence, the maximum cannot occur at an endpoint. Since A has no singular points, the maximum must occur at a critical point. To find any critical points, we set

$$0 = A'(y) = 100 - 4y.$$

Therefore $y = 25$. Since A must have a maximum value and there is only one possible point where it can be, the maximum must occur at $y = 25$. The greatest possible area for the enclosure is therefore $A(25) = 1{,}250$ m².

Example 2   A lighthouse L is located on a small island 5 km north of a point A on a straight east-west shoreline. A cable is to be laid from L to point B on the shoreline 10 km east of A. The cable will be laid through the water in a straight line from L to a point C on the shoreline between A and B, and from there to B along the shoreline. (See Figure 4.39.) The part of the cable lying in the water costs $5,000/km and the part along the shoreline costs $3,000/km.

(a) Where should C be chosen to minimize the total cost of the cable?

(b) Where should C be chosen if B is only 3 km from A?

Figure 4.39

Solution

(a) Let C be x km from A toward B. Thus $0 \leq x \leq 10$. The length of LC is $\sqrt{25 + x^2}$ km, and the length of CB is $10 - x$ km, as illustrated in Figure 4.39. Hence, the total cost of the cable is $\$T$, where

$$T = T(x) = 5{,}000\sqrt{25 + x^2} + 3{,}000(10 - x), \qquad (0 \leq x \leq 10).$$

T is continuous on the closed, finite interval $[0, 10]$, so it has a minimum value that may occur at one of the endpoints $x = 0$ or $x = 10$ or at a critical point in the interval $]0, 10[$. (T has no singular points.)

To find any critical points, we set

$$0 = \frac{dT}{dx} = \frac{5,000x}{\sqrt{25 + x^2}} - 3,000.$$

Thus
$$5,000x = 3,000\sqrt{25 + x^2}$$
$$25x^2 = 9(25 + x^2)$$
$$16x^2 = 225$$
$$x^2 = \frac{225}{16} = \frac{15^2}{4^2}.$$

The critical points are $x = \pm 15/4$. Only one critical point, $x = 15/4 = 3.75$, lies in the interval $]0, 10[$. Since $T(0) = 55,000$, $T(15/4) = 50,000$, and $T(10) \approx 55,902$, the critical point evidently provides the minimum value for $T(x)$. For minimal cost, C should be 3.75 km from A.

(b) If B is 3 km from A, the corresponding total cost function is

$$T(x) = 5,000\sqrt{25 + x^2} + 3,000(3 - x), \qquad (0 \le x \le 3),$$

which differs from the total cost function $T(x)$ of part (a) only in the added constant (9,000 rather than 30,000). It therefore has the same critical points, $x = \pm 15/4$, neither of which lie in the interval $(0, 3)$. Since $T(0) = 34,000$ and $T(3) \approx 29,155$, in this case we should choose $x = 3$. To minimize the total cost, the cable should go straight from L to B.

Procedure for Solving Extreme-Value Problems

Based on our experience with the examples above we can formulate a checklist of steps involved in solving optimization problems.

> **Solving extreme-value problems**
>
> 1. Read the problem very carefully, perhaps more than once. You must understand clearly what is given and what must be found.
>
> 2. Make a diagram if appropriate. Many problems have a geometric component, and a good diagram can often be an essential part of the solution process.
>
> 3. Define any symbols you wish to use that are not already specified in the statement of the problem.
>
> 4. Express the quantity Q to be maximized or minimized as a function of one or more variables.
>
> 5. If Q depends on n variables, where $n > 1$, find $n - 1$ equations (constraints) linking these variables. (If this cannot be done, the problem cannot be solved by single-variable techniques.)
>
> 6. Use the constraints to eliminate variables and hence express Q as a function of only one variable. Determine the interval(s) in which this variable must lie for the problem to make sense. Alternatively, regard the constraints as implicitly defining $n - 1$ of the variables, and hence Q, as functions of the remaining variable. (It is usually better to avoid this implicit method in an extreme-value problem if you can.)

7. Find the required extreme value of the function Q using the techniques of Section 4.2. Remember to consider any critical points, singular points, and endpoints. Make sure to give a convincing argument that your extreme value is the one being sought; for example, if you are looking for a maximum, the value you have found should not be a minimum.

8. Make a concluding statement answering the question asked. Is your answer for the question "reasonable"? If not, check back through the solution to see what went wrong.

Example 3   Find the length of the shortest ladder that can extend from a vertical wall, over a fence 2 m high located 1 m away from the wall, to a point on the ground outside the fence.

Solution   Let θ be the angle of inclination of the ladder, as shown in Figure 4.40. Using the two right-angled triangles in the figure, we obtain the length L of the ladder as a function of θ:

$$L = L(\theta) = \frac{1}{\cos\theta} + \frac{2}{\sin\theta},$$

where $0 < \theta < \pi/2$. Since

$$\lim_{\theta\to(\pi/2)-} L(\theta) = \infty \quad \text{and} \quad \lim_{\theta\to0+} L(\theta) = \infty,$$

$L(\theta)$ must have a minimum value on $]0, \pi/2[$, occurring at a critical point. (L has no singular points in $]0, \pi/2[$.) To find any critical points, we set

$$0 = L'(\theta) = \frac{\sin\theta}{\cos^2\theta} - \frac{2\cos\theta}{\sin^2\theta} = \frac{\sin^3\theta - 2\cos^3\theta}{\cos^2\theta \sin^2\theta}.$$

Any critical point satisfies $\sin^3\theta = 2\cos^3\theta$, or, equivalently, $\tan^3\theta = 2$. We don't need to solve this equation for $\theta = \tan^{-1}(2^{1/3})$ since it is really the corresponding value of $L(\theta)$ that we want. Observe that

$$\sec^2\theta = 1 + \tan^2\theta = 1 + 2^{2/3}.$$

It follows that

$$\cos\theta = \frac{1}{(1 + 2^{2/3})^{1/2}} \quad \text{and} \quad \sin\theta = \tan\theta\cos\theta = \frac{2^{1/3}}{(1 + 2^{2/3})^{1/2}}.$$

Therefore the minimal value of $L(\theta)$ is

$$\frac{1}{\cos\theta} + \frac{2}{\sin\theta} = (1 + 2^{2/3})^{1/2} + 2\frac{(1 + 2^{2/3})^{1/2}}{2^{1/3}} = \left(1 + 2^{2/3}\right)^{3/2} \approx 4.16.$$

The shortest ladder that can extend from the wall over the fence to the ground outside is about 4.16 m long.

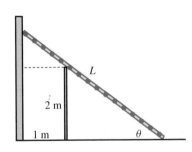

Figure 4.40

L

2 m

1 m

θ

Example 4 Find the most economical shape of a cylindrical tin can.

Solution This problem is stated in a rather vague way. We must consider what is meant by "most economical" and even "shape." Without further information, we can take one of two points of view:

(i) the volume of the tin can is to be regarded as given and we must choose the dimensions to minimize the total surface area, or

(ii) the total surface area is given (we can use just so much metal) and we must choose the dimensions to maximize the volume.

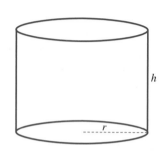

Figure 4.41

We will discuss other possible interpretations later. Since a cylinder is determined by its radius and height (Figure 4.41), its shape is determined by the ratio radius/height. Let r, h, S, and V denote, respectively, the radius, height, total surface area, and volume of the can. The volume of a cylinder is the base area times the height:

$$V = \pi r^2 h.$$

The surface of the can is made up of the cylindrical wall and circular disks for the top and bottom. The disks each have area πr^2, and the cylindrical wall is really just a rolled-up rectangle with base $2\pi r$ (the circumference of the can) and height h. Therefore, the total surface area of the can is

$$S = 2\pi r h + 2\pi r^2.$$

Let us use interpretation (i): V is a given constant, and S is to be minimized. We can use the equation for V to eliminate one of the two variables r and h on which S depends. Say we solve for $h = V/(\pi r^2)$ and substitute into the equation for S to obtain S as a function of r alone:

$$S = S(r) = 2\pi r \frac{V}{\pi r^2} + 2\pi r^2 = \frac{2V}{r} + 2\pi r^2 \qquad (0 < r < \infty).$$

Evidently, $\lim_{r \to 0+} S(r) = \infty$ and $\lim_{r \to \infty} S(r) = \infty$. Being differentiable and therefore continuous on $]0, \infty[$, $S(r)$ must have a minimum value, and it must occur at a critical point. To find any critical points,

$$0 = S'(r) = -\frac{2V}{r^2} + 4\pi r,$$

$$r^3 = \frac{2V}{4\pi} = \frac{1}{2\pi} \pi r^2 h = \frac{1}{2} r^2 h.$$

Thus $h = 2r$ at the critical point of S. Under interpretation (i), the most economical can is shaped so that its height equals the diameter of its base. You are encouraged to show that interpretation (ii) leads to the same conclusion. ∎

Remark There is another way to solve Example 4 that shows directly that interpretations (i) and (ii) must give the same solution. Again, we start from the two equations

$$V = \pi r^2 h \qquad \text{and} \qquad S = 2\pi r h + 2\pi r^2.$$

If we regard h as a function of r and differentiate implicitly, we obtain

$$\frac{dV}{dr} = 2\pi rh + \pi r^2 \frac{dh}{dr},$$

$$\frac{dS}{dr} = 2\pi h + 2\pi r \frac{dh}{dr} + 4\pi r.$$

Under interpretation (i), V is constant and we want a critical point of S; under interpretation (ii), S is constant and we want a critical point of V. In *either* case, $dV/dr = 0$ and $dS/dr = 0$. Hence both interpretations yield

$$2\pi rh + \pi r^2 \frac{dh}{dr} = 0 \qquad \text{and} \qquad 2\pi h + 4\pi r + 2\pi r \frac{dh}{dr} = 0.$$

If we divide the first equation by πr^2 and the second equation by $2\pi r$ and subtract to eliminate dh/dr, we again get $h = 2r$.

Remark   **Modifying Example 4** Given the sparse information provided in the statement of the problem in Example 4, interpretations (i) and (ii) are the best we can do. The problem could be made more meaningful economically (from the point of view, say, of a tin can manufacturer) if more elements were brought into it. For example:

(a) Most cans use thicker material for the cylindrical wall than for the top and bottom disks. If the cylindrical wall material costs $\$A$ per unit area and the material for the top and bottom costs $\$B$ per unit area, we might prefer to minimize the total cost for materials for a can of given volume. What is the optimal shape if $A = 2B$?

(b) Large numbers of cans are to be manufactured. The material is probably being cut out of sheets of metal. The cylindrical walls are made by bending up rectangles, and rectangles can be cut from the sheet with little or no waste. There will, however, always be a proportion of material wasted when the disks are cut out. The exact proportion will depend on how the disks are arranged; two possible arrangements are shown in Figure 4.42. What is the optimal shape of the can if a square packing of disks is used? a hexagonal packing? Any such modification of the original problem will alter the optimal shape to some extent. In "real-world" problems, many factors may have to be taken into account to come up with a "best" strategy.

(c) The problem makes no provision for costs of manufacturing the can other than the cost of sheet metal. There may also be costs for joining the opposite edges of the rectangle to make the cylinder, and for joining the top and bottom disks to the cylinder. These costs may be proportional to the lengths of the joins.

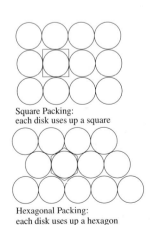

Square Packing:
each disk uses up a square

Hexagonal Packing:
each disk uses up a hexagon

Figure 4.42   Square and hexagonal packing of disks in a plane

In most of the examples above the maximum or minimum value being sought occurred at a critical point. Our final example is one where this is not the case.

Example 5 A man can run twice as fast as he can swim. He is standing at point A on the edge of a circular swimming pool 40 m in diameter, and he wishes to get to the diametrically opposite point B as quickly as possible. He can run around the edge to point C, then swim directly from C to B. Where should C be chosen to minimize the total time taken to get from A to B?

Solution It is convenient to describe the position of C in terms of the angle AOC, where O is the centre of the pool. (See Figure 4.43.) Let θ denote this angle. Clearly $0 \le \theta \le \pi$. (If $\theta = 0$, the man swims the whole way; if $\theta = \pi$, he runs the whole way.) The radius of the pool is 20 m, so arc $AC = 20\theta$. Since angle $BOC = \pi - \theta$, we have angle $BOL = (\pi - \theta)/2$ and chord $BC = 2BL = 40 \sin((\pi - \theta)/2)$.

Suppose the man swims at a rate k m/s and therefore runs at a rate $2k$ m/s. If t is the total time he takes to get from A to B, then

$$t = t(\theta) = \text{time running} + \text{time swimming}$$
$$= \frac{20\theta}{2k} + \frac{40}{k} \sin \frac{\pi - \theta}{2}.$$

(We are assuming that no time is wasted in jumping into the water at C.) The domain of t is $[0, \pi]$ and t has no singular points. Since t is continuous on a closed, finite interval, it must have a minimum value, and that value must occur at a critical point or an endpoint. For critical points,

$$0 = t'(\theta) = \frac{10}{k} - \frac{20}{k} \cos \frac{\pi - \theta}{2}.$$

Thus,

$$\cos \frac{\pi - \theta}{2} = \frac{1}{2}, \qquad \frac{\pi - \theta}{2} = \frac{\pi}{3}, \qquad \theta = \frac{\pi}{3}.$$

This is the only critical value of θ lying in the interval $[0, \pi]$. We have

$$t\left(\frac{\pi}{3}\right) = \frac{10\pi}{3k} + \frac{40}{k} \sin \frac{\pi}{3} = \frac{10}{k}\left(\frac{\pi}{3} + \frac{4\sqrt{3}}{2}\right) \approx \frac{45.11}{k}.$$

We must also look at the endpoints $\theta = 0$ and $\theta = \pi$:

$$t(0) = \frac{40}{k}, \qquad t(\pi) = \frac{10\pi}{k} \approx \frac{31.4}{k}.$$

Evidently $t(\pi)$ is the least of these three times. To get from A to B as quickly as possible, the man should run the entire distance.

Remark This problem shows how important it is to check every candidate point to see whether it gives a maximum or minimum. Here, the critical point $\theta = \pi/3$ yielded the *worst* possible strategy: running one-third of the way around and then swimming the remainder would take the greatest time, not the least.

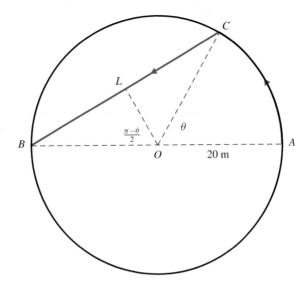

Figure 4.43   Running and swimming to get from A to B

Exercises 4.5

1. Two positive numbers have sum 7. What is the largest possible value for their product?

2. Two positive numbers have product 8. What is the smallest possible value for their sum?

3. Two nonnegative numbers have sum 60. What are the numbers if the product of one of them and the square of the other is maximal?

4. Two numbers have sum 16. What are the numbers if the product of the cube of one and the fifth power of the other is as large as possible?

5. The sum of two nonnegative numbers is 10. What is the smallest value of the sum of the cube of one number and the square of the other?

6. Two nonnegative numbers have sum n. What is the smallest possible value for the sum of their squares?

7. Among all rectangles of given area, show that the square has the least perimeter.

8. Among all rectangles of given perimeter, show that the square has the greatest area.

9. Among all isosceles triangles of given perimeter, show that the equilateral triangle has the greatest area.

10. Find the largest possible area for an isosceles triangle if the the length of each of its two equal sides is 10 m.

11. Find the area of the largest rectangle that can be inscribed in a semicircle of radius R if one side of the rectangle lies along the diameter of the semicircle.

12. Find the largest possible perimeter of a rectangle inscribed in a semicircle of radius R if one side of the rectangle lies

along the diameter of the semicircle. (It is interesting that the rectangle with the largest perimeter has a different shape than the one with the largest area, obtained in Exercise 11.)

13. A rectangle with sides parallel to the coordinate axes is inscribed in the ellipse

$$\frac{x^2}{a^2} + \frac{y^2}{b^2} = 1.$$

Find the largest possible area for this rectangle.

14. Let ABC be a triangle right-angled at C and having area S. Find the maximum area of a rectangle inscribed in the triangle if (a) one corner of the rectangle lies at C, or (b) one side of the rectangle lies along the hypotenuse, AB.

15. (**Designing a billboard**) A billboard is to be made with 100 m^2 of printed area and with margins of 2 m at the top and bottom and 4 m on each side. Find the outside dimensions of the billboard if its total area is to be a minimum.

16. (**Designing a box**) A box is to be made from a rectangular sheet of cardboard 70 cm by 150 cm by cutting equal squares out of the four corners and bending up the resulting four flaps to make the sides of the box. (The box has no top.) What is the largest possible volume of the box?

17. (**Using rebates to maximize profit**) An automobile manufacturer sells 2,000 cars per month, at an average profit of $1,000 per car. Market research indicates that for each $50 of factory rebate the manufacturer offers to buyers it can expect to sell 200 more cars each month. How much of a rebate should it offer to maximize its monthly profit?

18. (Maximizing rental profit) All 80 rooms in a motel will be rented each night if the manager charges $40 or less per room. If he charges $(40 + x)$ per room, then $2x$ rooms will remain vacant. If each rented room costs the manager $10 per day and each unrented room $2 per day in overhead, how much should the manager charge per room to maximize his daily profit?

19. (Minimizing travel time) You are in a dune buggy in the desert 12 km due south of the nearest point A on a straight east-west road. You wish to get to point B on the road 10 km east of A. If your dune buggy can average 15 km/h travelling over the desert and 39 km/h travelling on the road, toward what point on the road should you head in order to minimize your travel time to B?

20. Repeat Exercise 19, but assume that B is only 4 km from A.

21. A one-metre length of stiff wire is cut into two pieces. One piece is bent into a circle, the other piece into a square. Find the length of the part used for the square if the sum of the areas of the circle and the square is (a) maximum and (b) minimum.

22. Find the area of the largest rectangle that can be drawn so that each of its sides passes through a different vertex of a rectangle having sides a and b.

23. What is the length of the shortest line segment having one end on the x-axis, the other end on the y-axis, and passing through the point $(9, \sqrt{3})$?

24. (Getting around a corner) Find the length of the longest beam that can be carried horizontally around the corner from a hallway of width a m to a hallway of width b m. (See Figure 4.44; assume the beam has no width.)

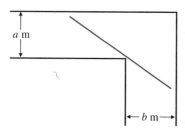

Figure 4.44

25. If the height of both hallways in Exercise 24 is c m, and if the beam need not be carried horizontally, how long can it be and still get around the corner? *Hint:* you can use the result of the previous exercise to do this one easily.

26. The fence in Example 3 is demolished and a new fence is built 2 m away from the wall. How high can the fence be if a 6 m ladder must be able to extend from the wall, over the fence, to the ground outside?

27. Find the shortest distance from the origin to the curve $x^2 y^4 = 1$.

28. Find the shortest distance from the point $(8, 1)$ to the curve $y = 1 + x^{3/2}$.

29. Find the dimensions of the largest right-circular cylinder that can be inscribed in a sphere of radius R.

30. Find the dimensions of the circular cylinder of greatest volume that can be inscribed in a cone of base radius R and height H if the base of the cylinder lies in the base of the cone.

31. A box with square base and no top has a volume of 4 m^3. Find the dimensions of the most economical box.

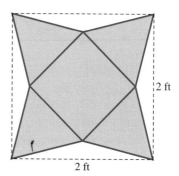

Figure 4.45

32. (Folding a pyramid) A pyramid with a square base and four faces, each in the shape of an isosceles triangle, is made by cutting away four triangles from a 2 ft square piece of cardboard (as shown in Figure 4.45) and bending up the resulting triangles to form the walls of the pyramid. What is the largest volume the pyramid can have? *Hint:* the volume of a pyramid having base area A and height h measured perpendicular to the base is $V = \frac{1}{3}Ah$.

33. (Getting the most light) A window has perimeter 10 m and is in the shape of a rectangle with the top edge replaced by a semicircle. Find the dimensions of the rectangle if the window admits the greatest amount of light.

34. (Fuel tank design) A fuel tank is made of a cylindrical part capped by hemispheres at each end. If the hemispheres are twice as expensive per unit area as the cylindrical wall, and if the volume of the tank is V, find the radius and height of the cylindrical part to minimize the total cost. The surface area of a sphere of radius r is $4\pi r^2$; its volume is $\frac{4}{3}\pi r^3$.

35. (Reflection of light) Light travels in such a way that it requires the minimum possible time to get from one point to another. A ray of light from C reflects off a plane mirror AB at X and then passes through D. (See Figure 4.46.) Show that the rays CX and XD make equal angles with the normal to AB at X. (*Remark:* you may wish to give a proof based on elementary geometry without using any calculus, or you can minimize the travel time on CXD.)

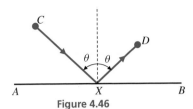

Figure 4.46

* **36. (Snell's Law)** If light travels with speed v_1 in one medium and speed v_2 in a second medium, and if the two media are separated by a plane interface, show that a ray of light passing from point A in one medium to point B in the other is bent at the interface in such a way that

$$\frac{\sin i}{\sin r} = \frac{v_1}{v_2},$$

where i and r are the angles of incidence and refraction, as is shown in Figure 4.47. This is known as Snell's Law. Deduce it from the least-time principle stated in Exercise 35.

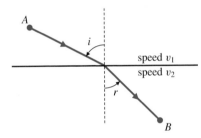

Figure 4.47

37. (Cutting the stiffest beam) The stiffness of a wooden beam of rectangular cross section is proportional to the product of the width and the cube of the depth of the cross section. Find the width and depth of the stiffest beam that can be cut out of a circular log of radius R.

38. Find the equation of the straight line of maximum slope tangent to the curve $y = 1 + 2x - x^3$.

39. A quantity Q grows according to the differential equation

$$\frac{dQ}{dt} = kQ^3(L - Q)^5,$$

where k and L are positive constants. How large is Q when it is growing most rapidly?

* **40.** Find the smallest possible volume of a right-circular cone that can contain a sphere of radius R. (The volume of a cone of base radius r and height h is $\frac{1}{3}\pi r^2 h$.)

* **41. (The best view of a mural)** How far back from a mural should one stand to view it best if the mural is 10 ft high and the bottom of it is 2 ft above eye level? (See Figure 4.48.)

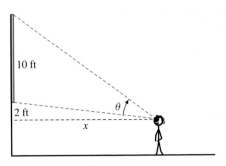

Figure 4.48

* **42. (Improving the enclosure of Example 1)** An enclosure is to be constructed having part of its boundary along an existing straight wall. The other part of the boundary is to be fenced in the shape of an arc of a circle. If 100 m of fencing is available, what is the area of the largest possible enclosure? Into what fraction of a circle is the fence bent?

* **43. (Designing a Dixie cup)** A sector is cut out of a circular disk of radius R, and the remaining part of the disk is bent up so that the two edges join and a cone is formed (Figure 4.49). What is the largest possible volume for the cone?

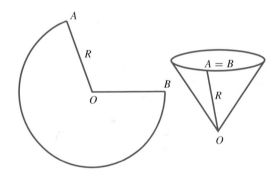

Figure 4.49

* **44. (Minimize the fold)** One corner of a strip of paper a cm wide is folded up so that it lies along the opposite edge (Figure 4.50). Find the least possible length for the fold line.

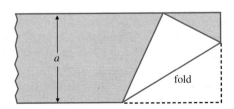

Figure 4.50

4.6 Finding Roots of Equations

Finding solutions (roots) of equations is an important mathematical problem to which calculus can make significant contributions. There are only a few general classes of equations of the form $f(x) = 0$ that we can solve exactly. These include **linear equations**:

$$ax + b = 0 \qquad \Rightarrow \qquad x = -\frac{b}{a},$$

and **quadratic equations**:

$$ax^2 + bx + c = 0 \qquad \Rightarrow \qquad x = \frac{-b \pm \sqrt{b^2 - 4ac}}{2a}.$$

Cubic and quartic (3rd- and 4th-degree polynomial) equations can also be solved, but the formulas are very complicated. We usually solve these and most other equations approximately by using numerical methods, often with the aid of a calculator or computer.

In Section 1.4 we discussed the Bisection Method for approximating a root of an equation $f(x) = 0$. That method uses the Intermediate-Value Theorem and depends only on the continuity of f and our ability to find an interval $[x_1, x_2]$ that must contain the root because $f(x_1)$ and $f(x_2)$ have opposite signs. The method is rather slow; it requires between three and four iterations to gain one significant figure of precision in the root being approximated.

If we know that f is more than just continuous, we can devise better (i.e., faster) methods for finding roots of $f(x) = 0$. We study two such methods in this section:

(a) **Newton's Method**, which requires that f be differentiable and which is usually very efficient, and

(b) **Fixed-Point Iteration**, which is concerned with equations of a different form: $f(x) = x$.

Like the Bisection Method, both of these methods require that we have at the outset a rough idea of where a root can be found, and they generate sequences of approximations that get closer and closer to the root.

Newton's Method

We want to find a **root** of the equation $f(x) = 0$, that is, a number r such that $f(r) = 0$. Such a number is also called a **zero** of the function f. If f is differentiable near the root, then tangent lines can be used to produce a sequence of approximations to the root that approaches the root quite quickly. The idea is as follows. (See Figure 4.51.) Make an initial guess at the root, say $x = x_0$. Draw the tangent line to $y = f(x)$ at $(x_0, f(x_0))$, and find x_1, the x-intercept of this tangent line. Under certain circumstances x_1 will be closer to the root than x_0 was. The process can be repeated over and over to get numbers x_2, x_3, $\ldots$, getting closer and closer to the root r. The number x_{n+1} is the x-intercept of the tangent line to $y = f(x)$ at $(x_n, f(x_n))$.

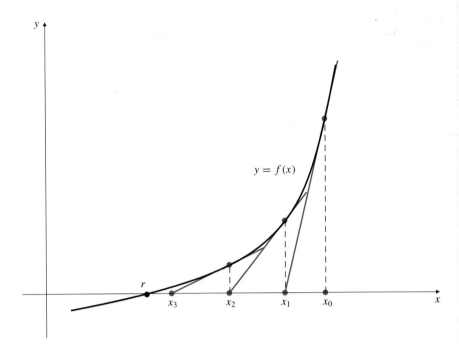

$y = f(x)$

r

x_3 x_2 x_1 x_0

Figure 4.51

The tangent line to $y = f(x)$ at $x = x_0$ has equation

$$y = f(x_0) + f'(x_0)(x - x_0).$$

Since the point $(x_1, 0)$ lies on this line, we have $0 = f(x_0) + f'(x_0)(x_1 - x_0)$. Hence

$$x_1 = x_0 - \frac{f(x_0)}{f'(x_0)}.$$

Similar formulas produce x_2 from x_1, then x_3 from x_2, and so on. The formula producing x_{n+1} from x_n is

$$x_{n+1} = x_n - \frac{f(x_n)}{f'(x_n)}$$

and is known as the **Newton's Method formula**. We usually use a calculator or computer to calculate the successive approximations x_1, x_2, x_3, ..., and observe whether these numbers appear to converge to a limit. If $\lim_{n \to \infty} x_n = r$ exists, and if f/f' is continuous near r, then r must be a root of f because

$$r = \lim_{n \to \infty} x_{n+1} = \lim_{n \to \infty} x_n - \lim_{n \to \infty} \frac{f(x_n)}{f'(x_n)} = r - \frac{f(r)}{f'(r)},$$

from which it follows that $f(r) = 0$. This method is known as **Newton's Method** or **The Newton-Raphson Method**.

Example 1   Use Newton's Method to find the only real root of the equation $x^3 - x - 1 = 0$ correct to 10 decimal places.

Solution   We have $f(x) = x^3 - x - 1$ and $f'(x) = 3x^2 - 1$. Since f is continuous and since $f(1) = -1$ and $f(2) = 5$, the equation has a root in the interval $[1, 2]$. Let us make the initial guess $x_0 = 1.5$. The Newton's Method formula here is

$$x_{n+1} = x_n - \frac{x_n^3 - x_n - 1}{3x_n^2 - 1} = \frac{2x_n^3 + 1}{3x_n^2 - 1},$$

so that, for example, the approximation x_1 is given by

$$x_1 = \frac{2(1.5)^3 + 1}{3(1.5)^2 - 1} \approx 1.347826\ldots.$$

Using a scientific calculator, we calculated the values in Table 1:

Table 1.

n	x_n	$f(x_n)$
0	1.5	$0.875\,000\,000\,000 \cdots$
1	$1.347\,826\,086\,96 \cdots$	$0.100\,682\,173\,091 \cdots$
2	$1.325\,200\,398\,95 \cdots$	$0.002\,058\,361\,917 \cdots$
3	$1.324\,718\,174\,00 \cdots$	$0.000\,000\,924\,378 \cdots$
4	$1.324\,717\,957\,24 \cdots$	$0.000\,000\,000\,000 \cdots$
5	$1.324\,717\,957\,24 \cdots$	

Evidently $r = 1.3247179572$ correctly rounded to 10 decimal places.

Observe the behaviour of the numbers x_n. By the third iteration, x_3, we have apparently achieved a precision of 6 decimal places, and by x_4 over 10 decimal places. It is characteristic of Newton's Method that when you begin to get close to the root the convergence can be very rapid. Compare these results with those obtained for the same equation by the Bisection Method in Example 12 of Section 1.4; there we achieved only 3 decimal place precision after 11 iterations.

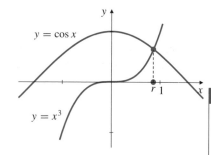

Figure 4.52 Solving $x^3 = \cos x$

Example 2 Solve the equation $x^3 = \cos x$ to 11 decimal places.

Solution We are looking for the x-coordinate r of the intersection of the curves $y = x^3$ and $y = \cos x$. From Figure 4.52 it appears that the curves intersect slightly to the left of $x = 1$. Let us start with the guess $x_0 = 0.8$. If $f(x) = x^3 - \cos x$, then $f'(x) = 3x^2 + \sin x$. The Newton's Method formula for this function is

$$x_{n+1} = x_n - \frac{x_n^3 - \cos x_n}{3x_n^2 + \sin x_n} = \frac{2x_n^3 + x_n \sin x_n + \cos x_n}{3x_n^2 + \sin x_n}.$$

The approximations $x_1, x_2, \ldots$ are given in Table 2:

Table 2.

n	x_n	$f(x_n)$
0	0.8	$-0.184\,706\,709\,347 \cdots$
1	$0.870\,034\,801\,135 \cdots$	$0.013\,782\,078\,762 \cdots$
2	$0.865\,494\,102\,425 \cdots$	$0.000\,006\,038\,051 \cdots$
3	$0.865\,474\,033\,493 \cdots$	$0.000\,000\,001\,176 \cdots$
4	$0.865\,474\,033\,102 \cdots$	$0.000\,000\,000\,000 \cdots$
5	$0.865\,474\,033\,102 \cdots$	

The two curves intersect at $x = 0.86547403310$, rounded to 11 decimal places.

Remark Example 2 shows how useful a sketch can be for determining an initial guess x_0. Even a rough sketch of the graph of $y = f(x)$ can show you how many roots the equation $f(x) = 0$ has, and approximately where they are. Usually, the closer the initial approximation is to the actual root, the smaller the number of iterations needed to achieve the desired precision. Similarly, for an equation of the form $g(x) = h(x)$, making a sketch of the graphs of g and h (on the same set of axes) can suggest starting approximations for any intersection points. In either case, you can then apply Newton's Method to improve the approximations.

Remark   When using Newton's Method to solve an equation that is of the form $g(x) = h(x)$ (such as the one in Example 2), we must rewrite the equation in the form $f(x) = 0$, and apply Newton's Method to f. Usually we just use $f(x) = g(x) - h(x)$, although $f(x) = (g(x)/h(x)) - 1$ is also a possibility.

Remark   If your calculator is programmable, you should learn how to program the Newton's Method formula for a given equation so that generating new iterations requires pressing only a few buttons. If your calculator has graphing capabilities, you can use them to locate a good initial guess.

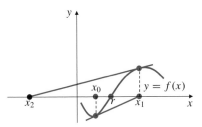

Figure 4.53   Here the Newton's Method iterations do not converge to the root

Newton's Method does not always work as well as it does in the preceding examples. If the first derivative f' is very small near the root, or if the second derivative f'' is very large near the root, a single iteration of the formula can take us from quite close to the root to quite far away. Figure 4.53 illustrates this possibility. (Also see Exercises 15 and 16 at the end of this section.)

The following theorem gives sufficient conditions for the Newton approximations to converge to a root r of the equation $f(x) = 0$ if the initial guess x_0 is sufficiently close to that root.

THEOREM 7

Error bounds for Newton's Method

Suppose that f, f', and f'' are continuous on an interval I containing x_n, x_{n+1}, and a root $x = r$ of $f(x) = 0$. Suppose also that there exist constants K and $L > 0$ such that for all x in I we have

(i) $|f''(x)| \leq K$ and

(ii) $|f'(x)| \geq L$.

Then

(a) $|x_{n+1} - r| \leq \dfrac{K}{2L} |x_{n+1} - x_n|^2$ and

(b) $|x_{n+1} - r| \leq \dfrac{K}{2L} |x_n - r|^2$.

Conditions (i) and (ii) assert that near r the slope of $y = f(x)$ is not too small in size and does not change too rapidly. If $K/(2L) < 1$, the theorem shows that x_n converges quickly to r once n becomes large enough that $|x_n - r| < 1$.

The proof of Theorem 7 depends on the Mean-Value Theorem. We will not give it since the theorem is of little practical use. In practice, we calculate successive approximations using Newton's formula and observe whether they seem to converge to a limit. If they do, and if the values of f at these approximations approach 0, we can be confident that we have located a root.

Fixed-Point Iteration

A number r satisfying the equation $f(r) = r$ is called a **fixed point** of the function f because f leaves that number unchanged. For certain kinds of functions, fixed points can be found by starting with an initial "guess" x_0 and calculating successive approximations $x_1 = f(x_0)$, $x_2 = f(x_1)$, ... In general,

$$x_{n+1} = f(x_n), \qquad \text{for } n = 0, 1, 2, \ldots.$$

Let us begin by investigating a simple example:

Example 3   Find a root of the equation $\cos x = 5x$.

Solution   This equation is of the form $f(x) = x$, where $f(x) = \frac{1}{5}\cos x$. Since $\cos x$ is close to 1 for x near 0, we see that $\frac{1}{5}\cos x$ will be close to $\frac{1}{5}$ when $x = \frac{1}{5}$. This suggests that a reasonable first guess at the fixed point is $x_0 = \frac{1}{5} = 0.2$. The values of subsequent approximations

$$x_1 = \frac{1}{5}\cos(x_0), \quad x_2 = \frac{1}{5}\cos(x_1), \quad x_3 = \frac{1}{5}\cos(x_2), \ldots$$

are presented in Table 3. The root is 0.19616428 to eight decimal places.

Table 3.

n	x_n
0	0.2
1	0.196 013 32
2	0.196 170 16
3	0.196 164 05
4	0.196 164 29
5	0.196 164 28
6	0.196 164 28

Why did the method used in Example 3 work? Will it work for any function f? In order to answer these questions, examine the polygonal line in Figure 4.54. Starting at x_0 it goes vertically to the curve $y = f(x)$, the height there being x_1. Then it goes horizontally to the line $y = x$, meeting that line at a point whose x-coordinate must therefore also be x_1. Then the process repeats; the line goes vertically to the curve $y = f(x)$ and horizontally to $y = x$, arriving at $x = x_2$. The line continues in this way, "spiralling" closer and closer to the intersection of $y = f(x)$ and $y = x$. Each value of x_n is closer to the fixed point r than the previous value.

Now consider the function f whose graph appears in Figure 4.55(a). If we try the same method there, starting with x_0, the polygonal line spirals outward, away from the root, and the resulting values x_n will not "converge" to the root as they did in

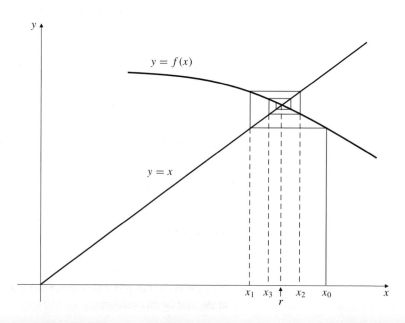

Figure 4.54   Iterations of $x_{n+1} = f(x_n)$ "spiral" toward the fixed point

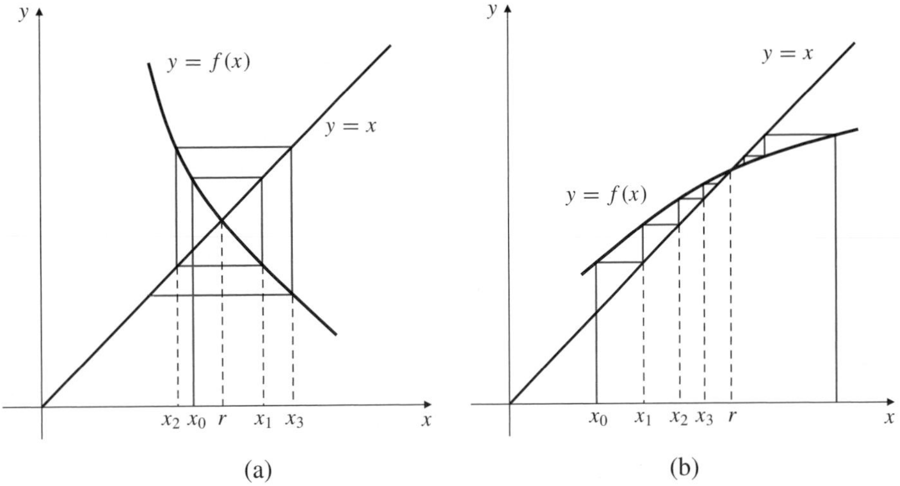

Figure 4.55

(a) A function f for which the iterations $x_{n+1} = f(x_n)$ do not converge

(b) "Staircase" convergence to the fixed point

(a)

(b)

Example 3. To see why the method works for the function in Figure 4.54 but not for the function in Figure 4.55(a), observe the slopes of the two graphs $y = f(x)$, near the fixed point r. Both slopes are negative, but in Figure 4.54 the absolute value of the slope is less than 1 while the absolute value of the slope of f in Figure 4.55(a) is greater than 1. Close consideration of the graphs should convince you that it is this fact that caused the points x_n to get closer to r in Figure 4.54 and farther from r in Figure 4.55(a).

A third example, Figure 4.55(b), shows that the method can be expected to work for functions whose graphs have positive slope near the fixed point r, provided that the slope is less than 1. In this case the polygonal line forms a "staircase" rather than a "spiral" and the successive approximations x_n increase toward the root if $x_0 < r$ and decrease toward it if $x_0 > r$.

The following theorem guarantees that the method of fixed-point iteration will work for a particular class of functions.

THEOREM 8

A fixed-point theorem

Suppose that f is defined on an interval $I = [a, b]$ and satisfies the following two conditions:

(i) $f(x)$ belongs to I whenever x belongs to I and

(ii) there exists a constant K with $0 < K < 1$ such that for every u and v in I,

$$|f(u) - f(v)| \le K|u - v|.$$

Then f has a fixed point r in I, that is, $f(r) = r$, and starting with any number x_0 in I, the iterates

$$x_1 = f(x_0), \quad x_2 = f(x_1), \quad \cdots$$

converge to r.

You are invited to prove this theorem by a method outlined in Exercises 24 and 25 at the end of this section.

"Solve" Routines

Many of the more advanced models of scientific calculators and most computer-based mathematics software have built-in routines for solving general equations numerically or, in a few cases, symbolically. These "Solve" routines assume continuity of the left and right sides of the given equations and often require the user to specify an interval in which to search for the root or an initial guess at the value of the root, or both. Typically the calculator or computer software also has graphing capabilities, and you are expected to use them to get an idea of how many roots the equation has and roughly where they are located before invoking the solving routines. It may also be possible to specify a *tolerance* on the value of the left side − the right side of the equation. For instance, if we want a solution to the equation $f(x) = 0$, it may be more important to us to be sure that an approximate solution $\hat{x}$ satisfies $|f(\hat{x})| < 0.0001$ than it is to be sure that $\hat{x}$ is within any particular distance of the actual root.

The methods used by the solve routines vary from one calculator or software package to another and are frequently very sophisticated, making use of numerical differentiation and other techniques to find roots very quickly, even when the search interval is large.

If you have an advanced scientific calculator and/or computer software with similar capabilities, it is well worth your while to read the manuals that describe how to make effective use of your hardware/software for solving equations. Applications of mathematics to solving "real-world" problems frequently require finding approximate solutions of equations that are intractable by exact methods.

Exercises 4.6

In Exercises 1–10, use Newton's Method to solve the given equations to the precision permitted by your calculator.

1. Find $\sqrt{2}$ by solving $x^2 - 2 = 0$.

2. Find $\sqrt{3}$ by solving $x^2 - 3 = 0$.

3. Find the root of $x^3 + 2x - 1 = 0$ between 0 and 1.

4. Find the root of $x^3 + 2x^2 - 2 = 0$ between 0 and 1.

5. Find the two roots of $x^4 - 8x^2 - x + 16 = 0$ in $[1, 3]$.

6. Find the three roots of $x^3 + 3x^2 - 1 = 0$ in $[-3, 1]$.

7. Solve $\sin x = 1 - x$. Make a sketch to help you make a first guess x_0.

8. Solve $\cos x = x^2$. How many roots are there?

9. How many roots does the equation $\tan x = x$ have? Find the one between $\pi/2$ and $3\pi/2$.

10. Solve $\dfrac{1}{1+x^2} = \sqrt{x}$ by rewriting it in the form $(1+x^2)\sqrt{x} - 1 = 0$.

11. If your calculator has a built-in Solve routine, or if you use computer software with such a routine, use it to solve the equations in the previous 10 exercises.

Find the maximum and minimum values of the functions in Exercises 12–13.

12. $\dfrac{\sin x}{1+x^2}$

13. $\dfrac{\cos x}{1+x^2}$

14. Let $f(x) = x^2$. The equation $f(x) = 0$ clearly has solution $x = 0$. Find the Newton's Method iterations x_1, x_2, and x_3, starting with $x_0 = 1$.

(a) What is x_n?

(b) How many iterations are needed to find the root with error less than 0.0001 in absolute value?

(c) How many iterations are needed to get an approximation x_n for which $|f(x_n)| < 0.0001$?

(d) Why do the Newton's Method iterations converge more slowly here than in the examples done in this section?

15. **(Oscillation)** Apply Newton's Method to

$$f(x) = \begin{cases} \sqrt{x}, & x \geq 0, \\ \sqrt{-x}, & x < 0, \end{cases}$$

starting with the initial guess $x_0 = a > 0$. Calculate x_1 and x_2. What happens? (Make a sketch.) If you ever observed this behaviour when you were using Newton's Method to find a root of an equation, what would you do next?

16. **(Divergent oscillations)** Apply Newton's Method to $f(x) = x^{1/3}$ with $x_0 = 1$. Calculate x_1, x_2, x_3, and x_4. What is happening? Find a formula for x_n.

17. **(Convergent oscillations)** Apply Newton's Method to find $f(x) = x^{2/3}$ with $x_0 = 1$. Calculate $x_1, x_2, x_3,$ and x_4. What is happening? Find a formula for x_n.

Use fixed-point iteration to solve the equations in Exercises 18–22. Obtain 5 decimal place precision.

18. $1 + \frac{1}{4}\sin x = x$

19. $\cos\frac{x}{3} = x$

20. $(x + 9)^{1/3} = x$

21. $\dfrac{1}{2 + x^2} = x$

22. Solve $x^3 + 10x - 10 = 0$ by rewriting it in the form $1 - \frac{1}{10}x^3 = x$.

23. Let $f(x)$ be a differentiable function whose derivative $f'(x)$ is never zero. Let

$$N(x) = x - \frac{f(x)}{f'(x)}.$$

Show that r is a root of $f(x) = 0$ if and only if r is a fixed point of $N(x)$. What are the successive approximations $x_{n+1} = N(x_n)$ starting from x_0 in this case?

Exercises 24–25 constitute a proof of Theorem 8.

* **24.** Condition (ii) of Theorem 8 implies that f is continuous on $I = [a, b]$. Use condition (i) to show that f has a fixed point r on I. *Hint:* apply the Intermediate-Value Theorem to $g(x) = f(x) - x$ on $[a, b]$.

* **25.** Use condition (ii) of Theorem 8 and mathematical induction to show that

$$|x_n - r| \le K^n |x_0 - r|.$$

Since $0 < K < 1$, we know that $K^n \to 0$ as $n \to \infty$. This shows that $\lim_{n\to\infty} x_n = r$.

4.7 Linear Approximations

Many problems in applied mathematics are too difficult to be solved exactly—all we can hope to do is find approximate solutions that are correct to within some acceptably small tolerance. In this section we will examine how knowledge of the values of a function and its first derivative at a point can help us find approximate values for the function at nearby points.

The tangent to the graph $y = f(x)$ at $x = a$ describes the behaviour of that graph near the point $P = (a, f(a))$ better than any other straight line through P, because it goes through P in the same direction as the curve $y = f(x)$. (See Figure 4.56.) We exploit this fact by using the height to the tangent line to calculate approximate values of $f(x)$ for values of x near a. The tangent line has equation $y = f(a) + f'(a)(x - a)$. We call the right side of this equation the linearization of f about $x = a$.

DEFINITION 8

> The **linearization**, or **linear approximation**, of the function f about $x = a$ is the function $L(x)$ defined by
>
> $$L(x) = f(a) + f'(a)(x - a).$$

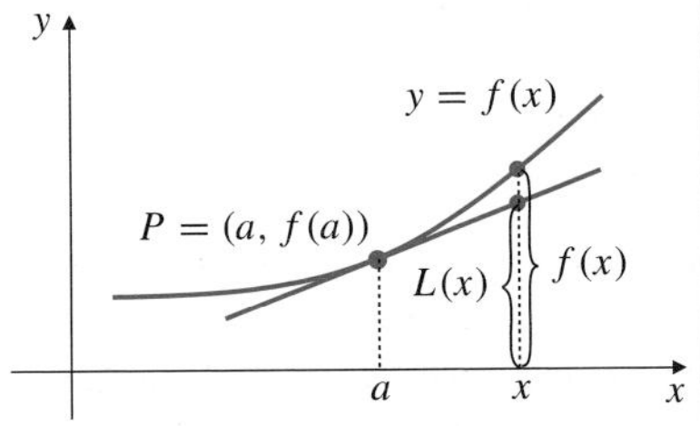

Figure 4.56   The linearization of $f(x)$ about $x = a$

Example 1   Find the linearizations for (a) $f(x) = \sqrt{1 + x}$ about $x = 0$ and (b) $g(x) = 1/x$ about $x = 1/2$.

Solution

(a) Since $f'(x) = 1/(2\sqrt{1 + x})$, we have $f(0) = 1$ and $f'(0) = 1/2$. The linearization of $f(x)$ about $x = 0$ is

$$L(x) = 1 + \frac{1}{2}(x - 0) = 1 + \frac{x}{2}.$$

(b) Since $g'(x) = -1/x^2$, we have $g(1/2) = 2$ and $g'(1/2) = -4$. The linearization of $g(x)$ about $x = 1/2$ is

$$L(x) = 2 - 4\left(x - \frac{1}{2}\right) = 4 - 4x.$$

Approximating Values of Functions

We have already made use of linearization in Section 2.7, where it was disguised as the formula

$$\Delta y \approx \frac{dy}{dx} \Delta x$$

and used to approximate a small change $\Delta y = f(a + \Delta x) - f(a)$ in the values of function f corresponding to the small change in the argument of the function from a to $a + \Delta x$. This is just the linear approximation

$$f(a + \Delta x) \approx f(a) + f'(a)\Delta x.$$

Example 2   A ball of ice melts so that its radius decreases from 5 cm to 4.92 cm. By approximately how much does the volume of the ball decrease?

Solution   The volume V of a ball of radius r is given by $V = \frac{4}{3}\pi r^3$, so

$$\Delta V \approx \frac{4}{3}\pi (3r^2)\, \Delta r = 4\pi r^2\, \Delta r.$$

For $r = 5$ and $\Delta r = -0.08$, we have

$$\Delta V \approx 4\pi (5^2)(-0.08) = -8\pi \approx -25.13.$$

The volume of the ball decreases by about 25 cm^3.

The following example illustrates the use of linearization to find an approximate value of a function near a point where the values of the function and its derivative are known.

Example 3   Use the linearization for $\sqrt{x}$ about $x = 25$ to find an approximate value for $\sqrt{26}$.

Solution   If $f(x) = \sqrt{x}$, then $f'(x) = 1/(2\sqrt{x})$. Since we know that $f(25) = 5$ and $f'(25) = 1/10$, the linearization of $f(x)$ about $x = 25$ is

$$L(x) = 5 + \frac{1}{10}(x - 25).$$

Putting $x = 26$, we get

$$\sqrt{26} = f(26) \approx L(26) = 5 + \frac{1}{10}(26 - 25) = 5.1.$$

If we use the square root function on a calculator we can obtain the "true value" of $\sqrt{26}$ (actually, just another approximation, although presumably a rather better one): $\sqrt{26} = 5.0990195\ldots$, but if we have such a calculator we don't need the approximation in the first place. Approximations are useful when there is no easy way to obtain the true value. However, if we don't know the true value, we would at least like to have some way of determining how good the approximation must be; that is, we want an *estimate for the error*. After all, *any number* is an approximation to $\sqrt{26}$, but the error may be unacceptably large. For instance, the size of the error in the approximation $\sqrt{26} \approx 1,000,000$ is greater than 999,994.

Error Analysis

In any approximation, the **error** is defined by

error = true value − approximate value.

If the linearization of $f(x)$ about $x = a$ is used to approximate $f(x)$ near a, that is,

$$f(x) \approx L(x) = f(a) + f'(a)(x - a),$$

then the error $E(x)$ in this approximation is

$$E(x) = f(x) - L(x) = f(x) - f(a) - f'(a)(x - a).$$

It is the vertical distance at x between the graph of f and the tangent line to that graph at $x = a$, as shown in Figure 4.57. Observe that if x is "near" a, then $E(x)$ is small compared to the horizontal distance between x and a.

The following theorem and its corollaries gives us a way to estimate this error if we know bounds for the *second derivative* of f.

THEOREM 9

An error formula for linearization

If $f''(t)$ exists for all t in an interval containing a and x, then there exists some point X between a and x such that the error $E(x) = f(x) - L(x)$ in the linear approximation $f(x) \approx L(x) = f(a) + f'(a)(x - a)$ satisfies

$$E(x) = \frac{f''(X)}{2}(x - a)^2.$$

PROOF   Let us assume that $x > a$. (The proof for $x < a$ is similar.) Since

$$E(t) = f(t) - f(a) - f'(a)(t - a),$$

we have $E'(t) = f'(t) - f'(a)$. We apply the Generalized Mean-Value Theorem (Theorem 16 of Section 2.6) to the two functions $E(t)$ and $(t - a)^2$ on $[a, x]$. Noting that $E(a) = 0$, we obtain a number c in (a, x) such that

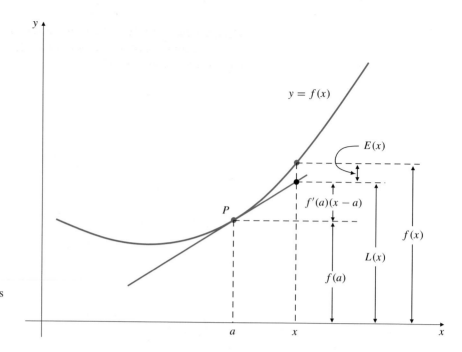

Figure 4.57 $f(x)$ and its linearization $L(x)$ about $x = a$. $E(x)$ is the error in the approximation $f(x) \approx L(x)$

$$\frac{E(x)}{(x-a)^2} = \frac{E(x) - E(a)}{(x-a)^2 - (a-a)^2} = \frac{E'(c)}{2(c-a)} = \frac{f'(c) - f'(a)}{2(c-a)} = \frac{1}{2} f''(X)$$

for some X in (a, c); the latter expression is a consequence of applying the Mean-Value Theorem again, this time to $f'(t)$ on $[a, c]$. Thus,

$$E(x) = \frac{f''(X)}{2}(x-a)^2$$

as claimed.

The following three corollaries are immediate consequences of Theorem 9.

Corollary A. If $f''(t)$ has constant sign (i.e., is always positive or always negative) between a and x, then the error $E(x)$ in the linear approximation $f(x) \approx L(x)$ has that same sign. If $f''(t) > 0$ between a and x, then $f(x) > L(x)$; if $f''(t) < 0$ between a and x, then $f(x) < L(x)$.

Corollary B. If $|f''(t)| < K$ for all t between a and x, then $|E(x)| < (K/2)(x-a)^2$.

Corollary C. If $f''(t)$ satisfies $M < f''(t) < N$ for all t between a and x (where M and N are constants), then

$$L(x) + \frac{M}{2}(x-a)^2 < f(x) < L(x) + \frac{N}{2}(x-a)^2.$$

If M and N have the same sign, a better approximation to $f(x)$ is given by the midpoint of this interval containing $f(x)$:

$$f(x) \approx L(x) + \frac{M+N}{4}(x-a)^2.$$

For this approximation the error is less than half the length of the interval:

$$|\text{Error}| < \frac{N-M}{4}(x-a)^2.$$

Example 4   Determine the sign and estimate the size of the error in the approximation $\sqrt{26} \approx 5.1$ obtained in Example 3. Use these to give an interval that you can be sure contains $\sqrt{26}$.

Solution   For $f(t) = t^{1/2}$, we have

$$f'(t) = \frac{1}{2} t^{-1/2} \qquad \text{and} \qquad f''(t) = -\frac{1}{4} t^{-3/2}.$$

For $25 < t < 26$, we have $f''(t) < 0$, so $\sqrt{26} = f(26) < L(26) = 5.1$. Also, $t^{3/2} > 25^{3/2} = 125$, so $|f''(t)| < (1/4)(1/125) = 1/500$ and

$$|E(26)| < \frac{1}{2} \times \frac{1}{500} \times (26 - 25)^2 = \frac{1}{1,000} = 0.001.$$

Therefore, $f(26) > L(26) - 0.001 = 5.099$, and $\sqrt{26}$ is in the interval $]5.099, 5.1[$. ∎

Remark   We can use Corollary C of Theorem 9 and the fact that $\sqrt{26} < 5.1$ to find a better (i.e., smaller) interval containing $\sqrt{26}$ as follows. If $25 < t < 26$, then $125 = 25^{3/2} < t^{3/2} < 26^{3/2} < 5.1^3$. Thus

$$M = -\frac{1}{4 \times 125} < f''(t) < -\frac{1}{4 \times 5.1^3} = N$$

$$\sqrt{26} \approx L(26) + \frac{M+N}{4} = 5.1 - \frac{1}{4} \left(\frac{1}{4 \times 125} + \frac{1}{4 \times 5.1^3} \right) \approx 5.0990288$$

$$|\text{Error}| < \frac{N - M}{4} = \frac{1}{16} \left(-\frac{1}{5.1^3} + \frac{1}{125} \right) \approx 0.0000288.$$

Thus $\sqrt{26}$ lies in the interval $]5.09900, 5.09906[$.

Example 5   Use a suitable linearization to find an approximate value for $\cos(36°) = \cos(\pi/5)$. Is the true value greater than or less than your approximation? Estimate the size of the error and give an interval that you can be sure contains $\cos(36°)$.

Solution   Let $f(t) = \cos t$, so that $f'(t) = -\sin t$ and $f''(t) = -\cos t$. The value of a nearest to $36°$ for which we know $\cos a$ is $a = 30° = \pi/6$, so we use the linearization at that point:

$$L(x) = \cos \frac{\pi}{6} - \sin \frac{\pi}{6} \left(x - \frac{\pi}{6} \right) = \frac{\sqrt{3}}{2} - \frac{1}{2} \left(x - \frac{\pi}{6} \right).$$

Since $(\pi/5) - (\pi/6) = \pi/30$, our approximation is

$$\cos(36°) = \cos \frac{\pi}{5} \approx L \left(\frac{\pi}{5} \right) = \frac{\sqrt{3}}{2} - \frac{1}{2} \left(\frac{\pi}{30} \right) \approx 0.81367.$$

If $(\pi/6) < t < (\pi/5)$, then $f''(t) < 0$ and $|f''(t)| < \cos(\pi/6) = \sqrt{3}/2$. Therefore, $\cos(36°) < 0.81367$ and

$$|E(36°)| < \frac{\sqrt{3}}{4}\left(\frac{\pi}{30}\right)^2 < 0.00475.$$

Thus, $0.81367 - 0.00475 < \cos(36°) < 0.81367$, so $\cos(36°)$ lies in the interval $]0.80892, 0.81367[$.

■

Remark   The error in the linearization of $f(x)$ about $x = a$ can be interpreted in terms of differentials (see Section 2.2) as follows. If $x - a = \Delta x = dx$, then the change in $f(x)$ as we pass from $x = a$ to $x = a + \Delta x$ is $f(a + \Delta x) - f(a) = \Delta y$, and the corresponding change in the linearization $L(x)$ is $f'(a)(x - a) = f'(a)\,dx$, which is just the value at $x = a$ of the differential $dy = f'(x)\,dx$. Thus

$$E(x) = \Delta y - dy.$$

The error $E(x)$ is small compared with Δx as Δx approaches 0, as seen in Figure 4.57. In fact,

$$\lim_{\Delta x \to 0} \frac{\Delta y - dy}{\Delta x} = \lim_{\Delta x \to 0}\left(\frac{\Delta y}{\Delta x} - \frac{dy}{dx}\right) = \frac{dy}{dx} - \frac{dy}{dx} = 0.$$

If $|f''(t)| \le K$ (constant) near $t = a$, a stronger assertion can be made:

$$\left|\frac{\Delta y - dy}{(\Delta x)^2}\right| = \left|\frac{E(x)}{(\Delta x)^2}\right| \le \frac{K}{2} \qquad \text{so} \qquad |\Delta y - dy| \le \frac{K}{2}(\Delta x)^2.$$

Exercises 4.7

In Exercises 1–10, find the linearization of the given function at the given point.

1. x^2 at $x = 3$ **2.** x^{-3} at $x = 2$

3. $\sqrt{4 - x}$ at $x = 0$ **4.** $\sqrt{3 + x^2}$ at $x = 1$

5. $1/(1 + x)^2$ at $x = 2$ **6.** $1/\sqrt{x}$ at $x = 4$

7. $\sin x$ at $x = \pi$ **8.** $\cos(2x)$ at $x = \pi/3$

9. $\sin^2 x$ at $x = \pi/6$ **10.** $\tan x$ at $x = \pi/4$

11. By approximately how much does the area of a square increase if its side length increases from 10 cm to 10.4 cm?

12. By about how much must the edge length of a cube decrease from 20 cm to reduce the volume of the cube by 12 cm³?

13. A spacecraft orbits the earth at a distance of 4,100 miles from the centre of the earth. By about how much will the circumference of its orbit decrease if the radius decreases by 10 miles?

14. (**Acceleration of gravity**) The acceleration a of gravity at an altitude of h miles above the surface of the earth is given by

$$a = g\left(\frac{R}{R + h}\right)^2,$$

where $g \approx 32$ ft/s² is the acceleration at the surface of the earth, and $R \approx 3960$ miles is the radius of the earth. By about what percentage will a decrease if h increases from 0 to 10 miles?

In Exercises 15–22, use a suitable linearization to approximate the indicated value. Determine the sign of the error and estimate its size. Use this information to specify an interval you can be sure contains the value.

15. $\sqrt{50}$ **16.** $\sqrt{47}$

17. $\sqrt[4]{85}$ **18.** $\dfrac{1}{2.003}$

19. $\cos 46°$ **20.** $\sin \dfrac{\pi}{5}$

21. $\sin(3.14)$ **22.** $\sin 33°$

Use Corollary C of Theorem 9 in the manner suggested in the remark following Example 4 to find better intervals and better approximations to the values in Exercises 23–26.

23. $\sqrt{50}$ as first approximated in Exercise 15.

24. $\sqrt{47}$ as first approximated in Exercise 16.

25. $\cos 36°$ as first approximated in Example 5.

26. $\sin 33°$ as first approximated in Exercise 22.

27. If $f(2) = 4$, $f'(2) = -1$, and $0 \le f''(x) \le 1/x$ for all $x > 0$, find the smallest interval you can that contains $f(3)$.

28. If $f(2) = 4$, $f'(2) = -1$, and $\dfrac{1}{2x} \le f''(x) \le \dfrac{1}{x}$ for $2 \le x \le 3$, find the best approximation you can for $f(3)$.

29. If $g(2) = 1$, $g'(2) = 2$, and $|g''(x)| < 1 + (x-2)^2$ for all $x > 0$, find the best approximation you can for $g(1.8)$. How large can the error be?

30. Show that the linearization of $\sin \theta$ about $\theta = 0$ is $L(\theta) = \theta$. What is the percentage error in the approximation $\sin \theta \approx \theta$ if $|\theta|$ is less than $17°$?

31. A spherical balloon is inflated so that its radius increases from 20.00 cm to 20.20 cm in 1 min. By approximately how much has its volume increased in that minute?

4.8 Taylor Polynomials

The linearization of a function $f(x)$ about $x = a$, namely the linear function

$$P_1(x) = L(x) = f(a) + f'(a)(x - a),$$

describes the behaviour of $f(x)$ near $x = a$ better than any other polynomial of degree 1 because both P_1 and f have the same value and the same derivative at $x = a$:

$$P_1(a) = f(a) \qquad \text{and} \qquad P_1'(a) = f'(a).$$

(We are now using the symbol P_1 instead of L to stress the fact that the linearization is a polynomial of degree 1.)

We can obtain even better approximations to $f(x)$ by using quadratic or higher-degree polynomials and matching more derivatives at $x = a$. For example, if f is twice differentiable near $x = a$, then the quadratic polynomial

$$P_2(x) = f(a) + f'(a)(x - a) + \frac{f''(a)}{2}(x - a)^2$$

satisfies $P_2(a) = f(a)$, $P_2'(a) = f'(a)$, and $P_2''(a) = f''(a)$ and describes the behaviour of $f(x)$ near $x = a$ better than any other polynomial of degree 2.

In general, if $f^{(n)}(x)$ exists in an open interval containing $x = a$, then the polynomial

$$\begin{aligned} P_n(x) = f(a) &+ \frac{f'(a)}{1!}(x - a) + \frac{f''(a)}{2!}(x - a)^2 \\ &+ \frac{f'''(a)}{3!}(x - a)^3 + \cdots + \frac{f^{(n)}(a)}{n!}(x - a)^n \end{aligned}$$

matches f and its first n derivatives at $x = a$,

$$P_n(a) = f(a), \quad P_n'(a) = f'(a), \quad \ldots, \quad P_n^{(n)}(a) = f^{(n)}(a),$$

and so describes $f(x)$ near $x = a$ better than any other polynomial of degree n. P_n is called the **Taylor polynomial of degree n for $f(x)$ about $x = a$**. (If $a = 0$, the Taylor polynomials are sometimes called Maclaurin polynomials.) The Taylor polynomial of degree 0 is just the constant function $P_0(x) = f(a)$.

Example 1 Find the following Taylor polynomials:

(a) $P_2(x)$ for $f(x) = \sqrt{x}$ about $x = 1$.

(b) $P_3(x)$ for $g(x) = \sin x$ about $x = 0$.

(c) $P_n(x)$ for $h(x) = e^x$ about $x = a$.

Solution (a) $f'(x) = (1/2)x^{-1/2}$, $f''(x) = -(1/4)x^{-3/2}$. Thus,

$$P_2(x) = f(1) + f'(1)(x - 1) + \frac{f''(1)}{2!}(x - 1)^2$$

$$= 1 + \frac{1}{2}(x - 1) - \frac{1}{8}(x - 1)^2.$$

(b) $g'(x) = \cos x$, $g''(x) = -\sin x$, $g'''(x) = -\cos x$. Thus,

$$P_3(x) = g(0) + g'(0)x + \frac{g''(0)}{2!}x^2 + \frac{g'''(0)}{3!}x^3$$

$$= x - \frac{1}{6}x^3.$$

(c) Evidently $h^{(n)}(x) = e^x$ for every positive integer n, so

$$P_n(x) = e^a + \frac{e^a}{1!}(x - a) + \frac{e^a}{2!}(x - a)^2 + \cdots + \frac{e^a}{n!}(x - a)^n.$$

Example 2 Use Taylor polynomials for e^x about $x = 0$ to find successive approximations to $e = e^1$. Stop when you think you have 3 decimal places correct.

Solution Since every derivative of e^x is e^x and so is 1 at $x = 0$, the Taylor polynomials for e^x about $x = 0$ are

$$P_n(x) = 1 + \frac{x}{1!} + \frac{x^2}{2!} + \frac{x^3}{3!} + \cdots + \frac{x^n}{n!}.$$

Thus, we have for $x = 1$, adding one more term at each step:

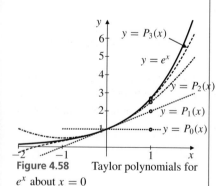

Figure 4.58 Taylor polynomials for e^x about $x = 0$

$$P_0(1) = 1$$

$$P_1(1) = 1 + \frac{1}{1!} = 2$$

$$P_2(1) = P_1(1) + \frac{1}{2!} = P_1(1) + \frac{1}{2} = 2.5$$

$$P_3(1) = P_2(1) + \frac{1}{3!} = P_2(1) + \frac{1}{6} = 2.6666$$

$$P_4(1) = P_3(1) + \frac{1}{4!} = P_3(1) + \frac{1}{24} = 2.7083$$

$$P_5(1) = P_4(1) + \frac{1}{5!} = P_4(1) + \frac{1}{120} = 2.7166$$

$$P_6(1) = P_5(1) + \frac{1}{6!} = P_5(1) + \frac{1}{720} = 2.7180$$

$$P_7(1) = P_6(1) + \frac{1}{7!} = P_6(1) + \frac{1}{5040} = 2.7182$$

It appears that $e \approx 2.718$ to 3 decimal places. We will verify in Example 4 below that $P_7(1)$ does indeed give this much precision. The graphs of e^x and its first four Taylor polynomials are shown in Figure 4.58.

Taylor's Formula

The following theorem provides a formula for the error in a Taylor approximation $f(x) \approx P_n(x)$ similar to that provided for linear approximation by Theorem 9.

THEOREM 10

Taylor's Theorem

If the $(n + 1)$st-order derivative, $f^{(n+1)}(t)$, exists for all t in an interval containing a and x, and if $P_n(x)$ is the Taylor polynomial of degree n for $f(x)$ about $x = a$, that is,

$$P_n(x) = f(a) + f'(a)(x - a) + \frac{f''(a)}{2!}(x - a)^2 + \cdots + \frac{f^{(n)}(a)}{n!}(x - a)^n,$$

then the formula $f(x) = P_n(x) + E_n(x)$ (called **Taylor's formula**) holds where the error term $E_n(x)$ (also called **the Lagrange remainder**) is given by

$$E_n(x) = \frac{f^{(n+1)}(X)}{(n + 1)!}(x - a)^{n+1},$$

where X is some number between a and x.

PROOF $E_n(x) = f(x) - P_n(x)$ is the error in the approximation $f(x) \approx P_n(x)$. Observe that the case $n = 0$ of this theorem, namely,

$$f(x) = P_0(x) + E_0(x) = f(a) + \frac{f'(X)}{1!}(x - a),$$

is just the Mean-Value Theorem

$$\frac{f(x) - f(a)}{x - a} = f'(X) \quad \text{for some } X \text{ between } a \text{ and } x.$$

Also note that the case $n = 1$ is just the error formula for linearization given in Theorem 9.

We will complete the proof for higher n using mathematical induction. Suppose, therefore, that we have proved the case $n = k - 1$, where $k \geq 2$ is an integer. Thus, we are assuming that if f is *any* function whose kth derivative exists on an interval containing a and x, then

$$E_{k-1}(x) = \frac{f^{(k)}(X)}{k!}(x - a)^k,$$

where X is some number between a and x. Let us consider the next higher case: $n = k$. As in the proof of Theorem 9, we assume $x > a$ (the case $x < a$ is similar) and apply the Generalized Mean-Value Theorem to the functions $E_k(t)$ and $(t - a)^{k+1}$ on $[a, x]$. Since $E_k(a) = 0$, we obtain a number c in $]a, x[$ such that

$$\frac{E_k(x)}{(x - a)^{k+1}} = \frac{E_k(x) - E_k(a)}{(x - a)^{k+1} - (a - a)^{k+1}} = \frac{E'_k(c)}{(k + 1)(c - a)^k}.$$

Now

$$E'_k(c) = \frac{d}{dt}\left(f(t) - f(a) - f'(a)(t-a) - \frac{f''(a)}{2!}(t-a)^2 \right.$$

$$\left. - \cdots - \frac{f^{(k)}(a)}{k!}(t-a)^k \right)\Bigg|_{t=c}$$

$$= f'(c) - f'(a) - f''(a)(c-a) - \cdots - \frac{f^{(k)}(a)}{(k-1)!}(c-a)^{k-1}.$$

This last expression is just $E_{k-1}(c)$ for the function $f'(t)$ instead of $f(t)$. By the induction assumption it is equal to

$$\frac{(f')^{(k)}(X)}{k!}(c-a)^k = \frac{f^{(k+1)}(X)}{k!}(c-a)^k,$$

for some X between a and c. Therefore,

$$E_k(x) = \frac{f^{(k+1)}(X)}{(k+1)!}(x-a)^{k+1}.$$

We have shown that the case $n = k$ of Taylor's Theorem is true if the case $n = k - 1$ is true, and the inductive proof is complete.

Example 3   Find the degree 2 Taylor approximation to $\sqrt{26}$ based on values of $f(x) = \sqrt{x}$ and its derivatives at 25. Estimate the size of the error, and specify an interval that you can be sure contains $\sqrt{26}$.

Solution   The first three derivatives of f are:

$$f'(x) = \frac{1}{2}x^{-1/2}, \quad f''(x) = -\frac{1}{4}x^{-3/2}, \quad \text{and} \quad f'''(x) = \tfrac{3}{8}x^{-5/2}.$$

The required approximation is

$$\sqrt{26} = f(26) \approx P_2(26) = f(25) + f'(25)(26-25) + \frac{f''(25)}{2}(26-25)^2$$

$$= 5 + \frac{1}{10} - \frac{1}{2 \times 4 \times 125} = 5.09900.$$

For $25 < t < 26$, we have

$$|f'''(t)| \le \frac{3}{8}\frac{1}{25^{5/2}} = \frac{3}{8 \times 3125} = \frac{3}{25000}.$$

Thus, the error in the approximation satisfies

$$|E_2(26)| \le \frac{3}{25000 \times 6}(26-25)^3 = \frac{1}{50000} = 0.00002.$$

Therefore, $\sqrt{26}$ lies in the interval $]5.09898, 5.09902[$.

Example 4   Use Taylor's Theorem to confirm that the Taylor polynomial $P_7(x)$ for e^x about $x = 0$ is sufficient to give e correct to 3 decimal places as claimed in Example 2.

Solution   The error in the approximation $e^x \approx P_n(x)$ satisfies

$$E_n(x) = \frac{e^X}{(n+1)!} x^{n+1}, \quad \text{for some } X \text{ between } 0 \text{ and } x.$$

If $x = 1$, then $0 < X < 1$, so $e^X < e < 3$ and $0 < E_n(1) < 3/(n+1)!$. To get an approximation for $e = e^1$ correct to 3 decimal places, we need to have $E_n(1) < 0.0005$. Since $3/(8!) = 3/40320 \approx 0.000074$, but $3/(7!) = 3/5040 \approx 0.00059$, we can be sure $n = 7$ will do, but we cannot be sure $n = 6$ will do:

$$e \approx 1 + 1 + \frac{1}{2!} + \frac{1}{3!} + \frac{1}{4!} + \frac{1}{5!} + \frac{1}{6!} + \frac{1}{7!} \approx 2.7183 \approx 2.718$$

to 3 decimal places. ∎

Big-O Notation

DEFINITION 9

We write $f(x) = O\big(u(x)\big)$ as $x \to a$ (read this "$f(x)$ is big-Oh of $u(x)$ as x approaches a") provided that

$$|f(x)| \le K|u(x)|$$

holds for some constant K on some open interval containing $x = a$.
Similarly, $f(x) = g(x) + O\big(u(x)\big)$ as $x \to a$ if $f(x) - g(x) = O\big(u(x)\big)$ as $x \to a$, that is, if

$$|f(x) - g(x)| \le K|u(x)| \quad \text{near } a.$$

For example, $\sin x = O(x)$ as $x \to 0$ because $|\sin x| \le |x|$ near 0.

The following properties of big-O notation follow from the definition:

(i) If $f(x) = O\big(u(x)\big)$ as $x \to a$, then $Cf(x) = O\big(u(x)\big)$ as $x \to a$ for any constant C.

(ii) If $f(x) = O\big(u(x)\big)$ as $x \to a$ and $g(x) = O\big(u(x)\big)$ as $x \to a$, then $f(x) \pm g(x) = O\big(u(x)\big)$ as $x \to a$.

(iii) If $f(x) = O\big((x-a)^k u(x)\big)$ as $x \to a$, then $f(x)/(x-a)^k = O\big(u(x)\big)$ as $x \to a$ for any constant k.

Taylor's Theorem says that if $f^{(n+1)}(t)$ exists on an interval containing a and x, and if $P_n(x)$ is the Taylor polynomial for $f(x)$ about $x = a$, then, as $x \to a$,

$$f(x) = P_n(x) + O\big((x-a)^{n+1}\big).$$

This is a statement about how closely the Taylor polynomial $P_n(x)$ approximates $f(x)$ near $x = a$. The following theorem shows that *only* the Taylor polynomial $P_n(x)$ approximates $f(x)$ this closely.

THEOREM 11

If $f(x) = Q_n(x) + O\big((x - a)^{n+1}\big)$ as $x \to a$, where Q_n is a polynomial of degree at most n, then $Q_n(x) = P_n(x)$, that is, Q_n is the Taylor polynomial for $f(x)$ about $x = a$.

PROOF Let P_n be the Taylor polynomial, then properties (i) and (ii) of big-O imply that $R_n(x) = Q_n(x) - P_n(x) = O\big((x - a)^{n+1}\big)$ as $x \to a$. We want to show that $R_n(x)$ is identically zero so that $Q_n(x) = P_n(x)$ for all x. By replacing x with $a + (x - a)$ and expanding powers, we can write $R_n(x)$ in the form

$$R_n(x) = c_0 + c_1(x - a) + c_2(x - a)^2 + \cdots + c_n(x - a)^n.$$

If $R_n(x)$ is not identically zero, then there is a smallest coefficient c_k $(k \le n)$, such that $c_k \ne 0$, but $c_j = 0$ for $0 \le j \le k - 1$. Thus,

$$R_n(x) = (x - a)^k \big(c_k + c_{k+1}(x - a) + \cdots + c_n(x - a)^{n-k}\big).$$

Therefore, $\lim_{x \to a} R_n(x)/(x - a)^k = c_k \ne 0$. However, by property (iii) above we have $R_n(x)/(x - a)^k = O\big((x - a)^{n+1-k}\big)$. Since $n + 1 - k > 0$, this says $R_n(x)/(x - a)^k \to 0$ as $x \to a$. This contradiction shows that $R_n(x)$ must be identically zero. Therefore $Q_n(x) = P_n(x)$ for all x.

Here is a list of Taylor formulas about $x = 0$ for some elementary functions, with error terms expressed using big-O notation. It is worthwhile remembering these.

$$\frac{1}{1 - x} = 1 + x + x^2 + x^3 + \cdots + x^n + O\big(x^{n+1}\big)$$

$$\ln(1 + x) = x - \frac{x^2}{2} + \frac{x^3}{3} - \cdots + (-1)^{n-1}\frac{x^n}{n} + O\big(x^{n+1}\big)$$

$$e^x = 1 + x + \frac{x^2}{2!} + \frac{x^3}{3!} + \cdots + \frac{x^n}{n!} + O\big(x^{n+1}\big)$$

$$\cos x = 1 - \frac{x^2}{2!} + \frac{x^4}{4!} - \cdots + (-1)^n\frac{x^{2n}}{(2n)!} + O\big(x^{2n+2}\big)$$

$$\sin x = x - \frac{x^3}{3!} + \frac{x^5}{5!} - \cdots + (-1)^n\frac{x^{2n+1}}{(2n + 1)!} + O\big(x^{2n+3}\big)$$

$$\tan^{-1} x = x - \frac{x^3}{3} + \frac{x^5}{5} - \cdots + (-1)^n\frac{x^{2n+1}}{2n + 1} + O\big(x^{2n+3}\big)$$

We can obtain Taylor polynomials for new functions from others already known. As long as the error term is of higher degree than the polynomial obtained, the polynomial must be the Taylor polynomial by Theorem 11. We illustrate this with a few examples.

Example 5   Find the Taylor polynomial of degree $2n$ for $\cosh x$ about $x = 0$.

Solution   Write the Taylor formula for e^x with n replaced by $2n + 1$, and then rewrite that with x replaced by $-x$. We get

$$e^x = 1 + x + \frac{x^2}{2!} + \frac{x^3}{3!} + \cdots + \frac{x^{2n}}{(2n)!} + \frac{x^{2n+1}}{(2n+1)!} + O\left(x^{2n+2}\right),$$

$$e^{-x} = 1 - x + \frac{x^2}{2!} - \frac{x^3}{3!} + \cdots + \frac{x^{2n}}{(2n)!} - \frac{x^{2n+1}}{(2n+1)!} + O\left(x^{2n+2}\right).$$

Now average these two to get

$$\cosh x = \frac{e^x + e^{-x}}{2} = 1 + \frac{x^2}{2!} + \frac{x^4}{4!} + \cdots + \frac{x^{2n}}{(2n)!} + O\left(x^{2n+2}\right).$$

Thus, the Taylor polynomial for $\cosh x$ about $x = 0$ is

$$P_{2n}(x) = 1 + \frac{x^2}{2!} + \frac{x^4}{4!} + \cdots + \frac{x^{2n}}{(2n)!}.$$

Example 6   Obtain the Taylor polynomial of degree 3 for e^{2x} about $x = 1$ from the Taylor polynomial for e^x about $x = 0$.

Solution   Writing $x = 1 + (x - 1)$, we have

$$e^{2x} = e^{2+2(x-1)} = e^2 e^{2(x-1)}$$

$$= e^2 \left[1 + 2(x-1) + \frac{2^2(x-1)^2}{2!} + \frac{2^3(x-1)^3}{3!} + O\left((x-1)^4\right) \right].$$

The Taylor polynomial of degree 3 for e^{2x} about $x = 1$ must be

$$P_3(x) = e^2 + 2e^2(x-1) + 2e^2(x-1)^2 + \frac{4e^2}{3}(x-1)^3.$$

Example 7   Find the Taylor polynomial $P_2(x)$ for $\ln x$ about $x = 2$.

Solution   We replace x with $2 + (x - 2)$.

$$\ln x = \ln(2 + (x - 2)) = \ln\left[2\left(1 + \frac{x-2}{2} \right) \right] = \ln 2 + \ln\left(1 + \frac{x-2}{2} \right)$$

$$= \ln 2 + \frac{x-2}{2} - \frac{1}{2}\left(\frac{x-2}{2} \right)^2 + O\left((x-2)^3\right).$$

Therefore, $P_2(x) = \ln 2 + \frac{1}{2}(x - 2) - \frac{1}{8}(x - 2)^2.$

Exercises 4.8

Calculate the indicated Taylor polynomials for the functions in Exercises 1–6 by using the definition of Taylor polynomial.

1. for e^{-x} about $x = 0$, degree 4.

2. for $\cos x$ about $x = \pi/4$, degree 3.

3. for $\ln x$ about $x = e$, degree 4.

4. for $\sec x$ about $x = 0$, degree 3.

5. for $\sqrt{x}$ about $x = 4$, degree 3.

6. for $1/(2 + x)$ about $x = 1$, degree n.

In Exercises 7–12, use degree 2 Taylor polynomials for the given function near the point specified to approximate the indicated value. Estimate the error and write the smallest interval you can be sure contains the value.

7. $f(x) = x^{1/3}$ near 8; approximate $9^{1/3}$.

8. $f(x) = \sqrt{x}$ near 64; approximate $\sqrt{61}$.

9. $f(x) = \dfrac{1}{x}$ near 1; approximate $\dfrac{1}{1.02}$.

10. $f(x) = \tan^{-1} x$ near 1; approximate $\tan^{-1}(0.97)$.

11. $f(x) = e^x$ near 0; approximate $e^{-0.5}$.

12. $f(x) = \sin x$ near $\pi/4$; approximate $\sin(47°)$.

In Exercises 13–18, write the indicated case of Taylor's formula for the given function. What is the Lagrange remainder in each case?

13. $f(x) = \sin x$, $a = 0$, $n = 7$

14. $f(x) = \cos x$, $a = 0$, $n = 6$

15. $f(x) = \sin x$, $a = \pi/4$, $n = 4$

16. $f(x) = \dfrac{1}{1 - x}$, $a = 0$, $n = 6$

17. $f(x) = \ln x$, $a = 1$, $n = 6$

18. $f(x) = \tan x$, $a = 0$, $n = 3$

Find the requested Taylor polynomials in Exercises 19–24 by using known Taylor polynomials and changing variables as in Examples 6 and 7.

19. $P_3(x)$ for e^{3x} about $x = -1$.

20. $P_8(x)$ for e^{-x^2} about $x = 0$.

21. $P_4(x)$ for $\sin^2 x$ about $x = 0$. *Hint:* $\sin^2 x = \dfrac{1 - \cos 2x}{2}$.

22. $P_5(x)$ for $\sin x$ about $x = \pi$.

23. $P_6(x)$ for $1/(1 + 2x^2)$ about $x = 0$

24. $P_8(x)$ for $\cos(3x - \pi)$ about $x = 0$.

25. Find the Taylor polynomial $P_{2n+1}(x)$ for $\sinh x$ about $x = 0$ by suitably combining polynomials for e^x and e^{-x}.

26. By suitably combining Taylor polynomials for $\ln(1 + x)$ and $\ln(1 - x)$ about $x = 0$, find the Taylor polynomial of degree $2n + 1$ about $x = 0$ for $\tanh^{-1}(x) = \dfrac{1}{2} \ln\left(\dfrac{1 + x}{1 - x}\right)$.

27. Write Taylor's formula for $f(x) = e^{-x}$ with $a = 0$ and use it to calculate $1/e$ to 5 decimal places. (You may use a calculator, but not the e^x function on it.)

* 28. Write the general form of Taylor's formula for $f(x) = \sin x$ about $x = 0$ with Lagrange remainder. How large need n be taken to ensure that the corresponding Taylor polynomial approximation will give the sine of 1 radian correct to 5 decimal places?

29. What is the best degree 2 approximation to the function $f(x) = (x - 1)^2$ near $x = 0$? What is the error in this approximation? Now answer the same questions for $g(x) = x^3 + 2x^2 + 3x + 4$. Can the constant $1/6 = 1/3!$, in the error formula for the degree 2 approximation, be improved (i.e., made smaller)?

4.9 Indeterminate Forms

In Section 2.5 we showed that

$$\lim_{x \to 0} \frac{\sin x}{x} = 1.$$

We could not readily see this by substituting $x = 0$ into the function $(\sin x)/x$ because both $\sin x$ and x are zero at $x = 0$. We call $(\sin x)/x$ an **indeterminate form** of type $[0/0]$ at $x = 0$. The limit of such an indeterminate form can be any number. For instance, each of the quotients kx/x, x/x^3, and x^3/x^2 is an indeterminate form of type $[0/0]$ at $x = 0$, but

$$\lim_{x \to 0} \frac{kx}{x} = k, \qquad \lim_{x \to 0} \frac{x}{x^3} = \infty, \qquad \lim_{x \to 0} \frac{x^3}{x^2} = 0.$$

There are other types of indeterminate forms. Table 4 lists them together with an example of each type.

Table 4.   Types of indeterminate forms

Type	Example
$[0/0]$	$\displaystyle\lim_{x\to0}\frac{\sin x}{x}$
$[\infty/\infty]$	$\displaystyle\lim_{x\to0}\frac{\ln(1/x^2)}{\cot(x^2)}$
$[0\cdot\infty]$	$\displaystyle\lim_{x\to0+} x\ln\frac{1}{x}$
$[\infty-\infty]$	$\displaystyle\lim_{x\to(\pi/2)-}\left(\tan x-\frac{1}{\pi-2x}\right)$
$[0^0]$	$\displaystyle\lim_{x\to0+} x^x$
$[\infty^0]$	$\displaystyle\lim_{x\to(\pi/2)-}(\tan x)^{\cos x}$
$[1^\infty]$	$\displaystyle\lim_{x\to\infty}\left(1+\frac{1}{x}\right)^x$

Indeterminate forms of type $[0/0]$ are the most common.   They can often be evaluated quite easily by using known Taylor formulas.

Example 1   Evaluate $\displaystyle\lim_{x\to0}\frac{2\sin x-\sin(2x)}{2e^x-2-2x-x^2}$.

Solution   Both the numerator and denominator approach 0 as $x\to0$.   Let us replace the trigonometric and exponential functions with their degree 3 Maclaurin polynomials plus error terms written in big-O notation:

$$\lim_{x\to0}\frac{2\sin x-\sin(2x)}{2e^x-2-2x-x^2}$$

$$=\lim_{x\to0}\frac{2\left(x-\dfrac{x^3}{3!}+O(x^5)\right)-\left(2x-\dfrac{2^3x^3}{3!}+O(x^5)\right)}{2\left(1+x+\dfrac{x^2}{2!}+\dfrac{x^3}{3!}+O(x^4)\right)-2-2x-x^2}$$

$$=\lim_{x\to0}\frac{-\dfrac{x^3}{3}+\dfrac{4x^3}{3}+O(x^5)}{\dfrac{x^3}{3}+O(x^4)}$$

$$=\lim_{x\to0}\frac{1+O(x^2)}{\dfrac{1}{3}+O(x)}=\frac{1}{\dfrac{1}{3}}=3.$$

Observe how we used the properties of big-O as listed in the previous section. We needed to use Maclaurin polynomials of degree at least 3 because all lower degree terms cancelled out in the numerator and the denominator.

Example 2 Evaluate $\lim\limits_{x \to 1} \dfrac{\ln x}{x^2 - 1}$.

Solution This is also of type $[0/0]$. We begin by substituting $x = 1 + t$. Note that $x \to 1$ corresponds to $t \to 0$. We can use a known Maclaurin polynomial for $\ln(1 + t)$. For this limit even the degree 1 polynomial $P_1(t) = t$ with error $O(t^2)$ will do.

$$\lim_{x \to 1} \frac{\ln x}{x^2 - 1} = \lim_{t \to 0} \frac{\ln(1 + t)}{(1 + t)^2 - 1} = \lim_{t \to 0} \frac{\ln(1 + t)}{2t + t^2}$$

$$= \lim_{t \to 0} \frac{t + O(t^2)}{2t + t^2} = \lim_{t \to 0} \frac{1 + O(t)}{2 + t} = \frac{1}{2}.$$

l'Hôpital's Rules

You can evaluate many indeterminate forms of type $[0/0]$ with simple algebra, typically by cancelling common factors. Examples can be found in Sections 1.2 and 1.3. Otherwise, you can use the method of Taylor polynomials, if the appropriate polynomials are known or can be calculated easily. We will now develop a third method called **l'Hôpital's Rule**[1] for evaluating limits of indeterminate forms of the types $[0/0]$ and $[\infty/\infty]$. The other types of indeterminate forms can usually be reduced to one of these two by algebraic manipulation and the taking of logarithms.

THEOREM 12

The first l'Hôpital Rule

Suppose the functions f and g are differentiable on the interval $]a, b[$, and $g'(x) \ne 0$ there. Suppose also that

(i) $\lim\limits_{x \to a+} f(x) = \lim\limits_{x \to a+} g(x) = 0$ and

(ii) $\lim\limits_{x \to a+} \dfrac{f'(x)}{g'(x)} = L$ (where L is finite or ∞ or $-\infty$).

Then

$$\lim_{x \to a+} \frac{f(x)}{g(x)} = L.$$

Similar results hold if every occurrence of $\lim_{x \to a+}$ is replaced by $\lim_{x \to b-}$ or even $\lim_{x \to c}$ where $a < c < b$. The cases $a = -\infty$ and $b = \infty$ are also allowed.

PROOF We prove the case involving $\lim_{x \to a+}$ for finite a. Define

$$F(x) = \begin{cases} f(x) & \text{if } a < x < b \\ 0 & \text{if } x = a \end{cases} \qquad \text{and} \qquad G(x) = \begin{cases} g(x) & \text{if } a < x < b \\ 0 & \text{if } x = a \end{cases}$$

Then F and G are continuous on the interval $[a, x]$ and differentiable on the interval (a, x) for every x in (a, b). By the Generalized Mean-Value Theorem (Theorem 16 of Section 2.6) there exists a number c in (a, x) such that

$$\frac{f(x)}{g(x)} = \frac{F(x)}{G(x)} = \frac{F(x) - F(a)}{G(x) - G(a)} = \frac{F'(c)}{G'(c)} = \frac{f'(c)}{g'(c)}.$$

---

[1] The Marquis de l'Hôpital (1661–1704), for whom these rules are named, published the first textbook on calculus. The circumflex (ˆ) did not come into use in the French language until after the French Revolution. The Marquis would have written his name "l'Hospital."

Since $a < c < x$, if $x \to a+$, then necessarily $c \to a+$, so we have

$$\lim_{x \to a+} \frac{f(x)}{g(x)} = \lim_{c \to a+} \frac{f'(c)}{g'(c)} = L.$$

The case involving $\lim_{x \to b-}$ for finite b is proved similarly. The cases where $a = -\infty$ or $b = \infty$ follow from the cases already considered via the change of variable $x = 1/t$:

$$\lim_{x \to \infty} \frac{f(x)}{g(x)} = \lim_{t \to 0+} \frac{f\left(\frac{1}{t}\right)}{g\left(\frac{1}{t}\right)} = \lim_{t \to 0+} \frac{f'\left(\frac{1}{t}\right)\left(\frac{-1}{t^2}\right)}{g'\left(\frac{1}{t}\right)\left(\frac{-1}{t^2}\right)} = \lim_{x \to \infty} \frac{f'(x)}{g'(x)} = L.$$

Example 3   Reevaluate $\lim\limits_{x \to 1} \dfrac{\ln x}{x^2 - 1}$. (See Example 2.)

Solution   We have $\lim\limits_{x \to 1} \dfrac{\ln x}{x^2 - 1}$ $\left[\dfrac{0}{0}\right]$

$$= \lim_{x \to 1} \frac{1/x}{2x} = \lim_{x \to 1} \frac{1}{2x^2} = \frac{1}{2}.$$

Note that in applying l'Hôpital's Rule we calculate the quotient of the derivatives, *not* the derivative of the quotient.

This example illustrates how calculations based on l'Hôpital's Rule are carried out. Having identified the limit as that of a [0/0] indeterminate form, we replace it by the limit of the quotient of derivatives; the existence of this latter limit will justify the equality. It is possible that the limit of the quotient of derivatives may still be indeterminate, in which case a second application of l'Hôpital's Rule can be made. Such applications may be strung out until a limit can finally be extracted, which then justifies all the previous applications of the rule.

Remark   The solution above seems easier than that of Example 2, and we might be tempted to think that l'Hôpital's Rules are easier to use than Taylor polynomials. It was easier here because we only had to apply l'Hôpital's Rule once. If we try to redo Example 1 by l'Hôpital's Rule, we will have to use the rule three times (which corresponds to the fact that degree 3 polynomials were needed in Example 1).

Example 4   Evaluate $\lim\limits_{x \to 0} \dfrac{2 \sin x - \sin(2x)}{2e^x - 2 - 2x - x^2}$.

Solution   We have (using l'Hôpital's Rule three times)

$$\lim_{x \to 0} \frac{2 \sin x - \sin(2x)}{2e^x - 2 - 2x - x^2} \qquad \left[\frac{0}{0}\right]$$

$$= \lim_{x \to 0} \frac{2 \cos x - 2 \cos(2x)}{2e^x - 2 - 2x} \qquad \text{cancel the 2's}$$

$$= \lim_{x \to 0} \frac{\cos x - \cos(2x)}{e^x - 1 - x} \qquad \text{still } \left[\frac{0}{0}\right]$$

$$= \lim_{x \to 0} \frac{-\sin x + 2 \sin(2x)}{e^x - 1} \qquad \text{still } \left[\frac{0}{0}\right]$$

$$= \lim_{x \to 0} \frac{-\cos x + 4 \cos(2x)}{e^x} = \frac{-1 + 4}{1} = 3.$$

Example 5   Evaluate   (a) $\lim\limits_{x \to (\pi/2)-} \dfrac{2x - \pi}{\cos^2 x}$ and   (b) $\lim\limits_{x \to 1+} \dfrac{x}{\ln x}$.

Solution

(a) $\lim\limits_{x \to (\pi/2)-} \dfrac{2x - \pi}{\cos^2 x}$ $\left[\dfrac{0}{0}\right]$

$\qquad = \lim\limits_{x \to (\pi/2)-} \dfrac{2}{-2 \sin x \cos x} = -\infty$

(b) l'Hôpital's Rule cannot be used to evaluate $\lim_{x \to 1+} x/(\ln x)$ because this is not an indeterminate form. The denominator approaches 0 as $x \to 1+$, but the numerator does not approach 0. Since $\ln x > 0$ for $x > 1$, we have, directly,

$$\lim\limits_{x \to 1+} \dfrac{x}{\ln x} = \infty.$$

(Had we tried to apply l'Hôpital's Rule, we would have been led to the erroneous answer $\lim_{x \to 1+}(1/(1/x)) = 1$.)

Example 6   Evaluate $\lim\limits_{x \to 0+} \left(\dfrac{1}{x} - \dfrac{1}{\sin x} \right)$.

Solution   The indeterminate form here is of type $[\infty - \infty]$ to which l'Hôpital's Rule cannot be applied. However, it becomes $[0/0]$ after we combine the fractions into one fraction.

$\lim\limits_{x \to 0+} \left(\dfrac{1}{x} - \dfrac{1}{\sin x} \right)$ $[\infty - \infty]$

$\qquad = \lim\limits_{x \to 0+} \dfrac{\sin x - x}{x \sin x}$ $\left[\dfrac{0}{0}\right]$

$\qquad = \lim\limits_{x \to 0+} \dfrac{\cos x - 1}{\sin x + x \cos x}$ $\left[\dfrac{0}{0}\right]$

$\qquad = \lim\limits_{x \to 0+} \dfrac{- \sin x}{2 \cos x - x \sin x} = \dfrac{-0}{2} = 0.$

A version of l'Hôpital's Rule also holds for indeterminate forms of the type $[\infty/\infty]$.

THEOREM 13

The second l'Hôpital Rule

Suppose that f and g are differentiable on the interval (a, b) and that $g'(x) \neq 0$ there. Suppose also that

(i) $\lim\limits_{x \to a+} g(x) = \pm\infty$ and

(ii) $\lim\limits_{x \to a+} \dfrac{f'(x)}{g'(x)} = L$ (where L is finite, or ∞ or $-\infty$).

Then

$$\lim\limits_{x \to a+} \dfrac{f(x)}{g(x)} = L.$$

Again, similar results hold for $\lim_{x \to b-}$ and for $\lim_{x \to c}$, and the cases $a = -\infty$ and $b = \infty$ are allowed.

The proof of the second l'Hôpital Rule is technically rather more difficult than that of the first Rule and we will not give it here. A sketch of the proof is outlined in Exercise 35 at the end of this section.

Remark   Do *not* try to use l'Hôpital's Rules to evaluate limits that are not indeterminate of type $[0/0]$ or $[\infty/\infty]$; such attempts will almost always lead to false conclusions as observed in Example 5(b) above. (Strictly speaking, the second l'Hôpital Rule can be applied to the form $[a/\infty]$, but there is no point to doing so if a is not infinite, since the limit is obviously 0 in that case.)

Remark   No conclusion about $\lim f(x)/g(x)$ can be made using either l'Hôpital Rule if $\lim f'(x)/g'(x)$ does not exist. Other techniques might still be used. For example, $\lim_{x\to\infty}(\sin x)/x = 0$ by the Squeeze Theorem even though $\lim_{x\to\infty}(\cos x)/1$ does not exist.

Example 7   Evaluate   (a) $\displaystyle\lim_{x\to\infty}\frac{x^2}{e^x}$ and   (b) $\displaystyle\lim_{x\to0+} x^a \ln x$, where $a > 0$.

Solution   Both of these limits are covered by Theorem 5 in Section 3.4. We do them here by l'Hôpital's Rule.

(a) $\displaystyle\lim_{x\to\infty}\frac{x^2}{e^x}$ $\left[\dfrac{\infty}{\infty}\right]$

$\displaystyle = \lim_{x\to\infty}\frac{2x}{e^x}$ still $\left[\dfrac{\infty}{\infty}\right]$

$\displaystyle = \lim_{x\to\infty}\frac{2}{e^x} = 0.$

Similarly, one can show that $\lim_{x\to\infty} x^n/e^x = 0$ for any positive integer n by repeated applications of l'Hôpital's Rule.

(b) $\displaystyle\lim_{x\to0+} x^a \ln x$ $(a > 0)$ $[0\cdot(-\infty)]$

$\displaystyle = \lim_{x\to0+}\frac{\ln x}{x^{-a}}$ $\left[\dfrac{-\infty}{\infty}\right]$

$\displaystyle = \lim_{x\to0+}\frac{1/x}{-ax^{-a-1}} = \lim_{x\to0+}\frac{x^a}{-a} = 0.$

The easiest way to deal with indeterminate forms of types $[0^0]$, $[\infty^0]$, and $[1^\infty]$ is to take logarithms of the expressions involved. The next two examples illustrate the technique.

Example 8   Evaluate $\displaystyle\lim_{x\to0+} x^x$.

Solution   This indeterminate form is of type $[0^0]$. Let $y = x^x$. Then

$$\lim_{x\to0+} \ln y = \lim_{x\to0+} x \ln x = 0,$$

by Example 7(b). Hence $\displaystyle\lim_{x\to0} x^x = \lim_{x\to0+} y = e^0 = 1.$

Example 9 Evaluate $\lim_{x\to\infty}\left(1+\sin\dfrac{3}{x}\right)^x$.

Solution This indeterminate form is of type 1^∞. Let $y=\left(1+\sin\dfrac{3}{x}\right)^x$. Then, taking ln of both sides,

$$\lim_{x\to\infty}\ln y = \lim_{x\to\infty} x\ln\left(1+\sin\frac{3}{x}\right) \qquad [\infty\cdot 0]$$

$$= \lim_{x\to\infty}\frac{\ln\left(1+\sin\dfrac{3}{x}\right)}{\dfrac{1}{x}} \qquad \left[\frac{0}{0}\right]$$

$$= \lim_{x\to\infty}\frac{\dfrac{1}{1+\sin\dfrac{3}{x}}\left(\cos\dfrac{3}{x}\right)\left(-\dfrac{3}{x^2}\right)}{-\dfrac{1}{x^2}} = \lim_{x\to\infty}\frac{3\cos\dfrac{3}{x}}{1+\sin\dfrac{3}{x}} = 3.$$

Hence $\lim_{x\to\infty}\left(1+\sin\dfrac{3}{x}\right)^x = e^3$.

Exercises 4.9

Evaluate the limits in Exercises 1–32.

1. $\lim_{x\to 0}\dfrac{3x}{\tan 4x}$

2. $\lim_{x\to 2}\dfrac{\ln(2x-3)}{x^2-4}$

3. $\lim_{x\to 0}\dfrac{\sin ax}{\sin bx}$

4. $\lim_{x\to 0}\dfrac{1-\cos ax}{1-\cos bx}$

5. $\lim_{x\to 0}\dfrac{\sin^{-1}x}{\tan^{-1}x}$

6. $\lim_{x\to 1}\dfrac{x^{1/3}-1}{x^{2/3}-1}$

7. $\lim_{x\to 0}x\cot x$

8. $\lim_{x\to 0}\dfrac{1-\cos x}{\ln(1+x^2)}$

9. $\lim_{t\to\pi}\dfrac{\sin^2 t}{t-\pi}$

10. $\lim_{x\to 0}\dfrac{10^x-e^x}{x}$

11. $\lim_{x\to\pi/2}\dfrac{\cos 3x}{\pi-2x}$

12. $\lim_{x\to 1}\dfrac{\ln(ex)-1}{\sin\pi x}$

13. $\lim_{x\to\infty}x\sin\dfrac{1}{x}$

14. $\lim_{x\to 0}\dfrac{x-\sin x}{x^3}$

15. $\lim_{x\to 0}\dfrac{x-\sin x}{x-\tan x}$

16. $\lim_{x\to 0}\dfrac{2-x^2-2\cos x}{x^4}$

17. $\lim_{x\to 0+}\dfrac{\sin^2 x}{\tan x-x}$

18. $\lim_{r\to\pi/2}\dfrac{\ln\sin r}{\cos r}$

19. $\lim_{t\to\pi/2}\dfrac{\sin t}{t}$

20. $\lim_{x\to 1-}\dfrac{\arccos x}{x-1}$

21. $\lim_{x\to\infty}x(2\tan^{-1}x-\pi)$

22. $\lim_{t\to(\pi/2)-}(\sec t-\tan t)$

23. $\lim_{t\to 0}\left(\dfrac{1}{t}-\dfrac{1}{te^{at}}\right)$

24. $\lim_{x\to 0+}x^{\sqrt{x}}$

* 25. $\lim_{x\to 0+}(\csc x)^{\sin^2 x}$

* 26. $\lim_{x\to 1+}\left(\dfrac{x}{x-1}-\dfrac{1}{\ln x}\right)$

* 27. $\lim_{t\to 0}\dfrac{3\sin t-\sin 3t}{3\tan t-\tan 3t}$

* 28. $\lim_{x\to 0}\left(\dfrac{\sin x}{x}\right)^{1/x^2}$

* 29. $\lim_{t\to 0}(\cos 2t)^{1/t^2}$

* 30. $\lim_{x\to 0+}\dfrac{\csc x}{\ln x}$

* 31. $\lim_{x\to 1-}\dfrac{\ln\sin\pi x}{\csc\pi x}$

* 32. $\lim_{x\to 0}(1+\tan x)^{1/x}$

33. (A Newton quotient for the second derivative) Evaluate $\lim_{h\to 0}\dfrac{f(x+h)-2f(x)+f(x-h)}{h^2}$ if f is a twice differentiable function.

34. If f has a continuous third derivative, evaluate

$$\lim_{h\to 0}\frac{f(x+3h)-3f(x+h)+3f(x-h)-f(x-3h)}{h^3}$$

* **35. (Proof of the second l'Hôpital Rule)** Fill in the details of the following outline of a proof of the second l'Hôpital Rule (Theorem 13) for the case where a and L are both finite. Let $a < x < t < b$ and show that there exists c in (x, t) such that

$$\frac{f(x) - f(t)}{g(x) - g(t)} = \frac{f'(c)}{g'(c)}.$$

Now juggle the above equation algebraically into the form

$$\frac{f(x)}{g(x)} - L = \frac{f'(c)}{g'(c)} - L + \frac{1}{g(x)}\left(f(t) - g(t)\frac{f'(c)}{g'(c)}\right).$$

It follows that

$$\left|\frac{f(x)}{g(x)} - L\right|$$
$$\leq \left|\frac{f'(c)}{g'(c)} - L\right| + \frac{1}{|g(x)|}\left(|f(t)| + |g(t)|\left|\frac{f'(c)}{g'(c)}\right|\right).$$

Now show that the right side of the above inequality can be made as small as you wish (say less than a positive number ϵ) by choosing first t and then x close enough to a. Remember, you are given that $\lim_{c \to a+}\left(f'(c)/g'(c)\right) = L$ and $\lim_{x \to a+} |g(x)| = \infty$.

Chapter Review

Key Ideas

- **What do the following words, phrases, and statements mean?**
  ◇ critical point of f ◇ singular point of f
  ◇ inflection point of f
  ◇ f has absolute maximum value M.
  ◇ f has a local minimum value at $x = c$.
  ◇ vertical asymptote   ◇ horizontal asymptote
  ◇ oblique asymptote
  ◇ the linearization of $f(x)$ about $x = a$
  ◇ the Taylor polynomial of degree n of $f(x)$ about $x = a$
  ◇ Taylor's formula with Lagrange remainder
  ◇ $f(x) = O\left((x - a)^n\right)$ as $x \to a$.
  ◇ a root of $f(x) = 0$ ◇ a fixed point of $f(x)$
  ◇ an indeterminate form   ◇ l'Hôpital's Rules
- **Describe how to estimate the error in a linear (tangent line) approximation to the value of a function.**
- **Describe how to find a root of an equation $f(x) = 0$ by using Newton's Method. When will this method work well?**

Review Exercises

1. If the radius r of a ball is increasing at a rate of 2 percent per minute, how fast is the volume V of the ball increasing?

2. **(Gravitational attraction)** The gravitational attraction of the earth on a mass m at distance r from the centre of the earth is a continuous function of r for $r \geq 0$, given by

$$F = \begin{cases} \dfrac{mgR^2}{r^2} & \text{if } r \geq R \\ mkr & \text{if } 0 \leq r < R, \end{cases}$$

where R is the radius of the earth, and g is the acceleration due to gravity at the surface of the earth.

(a) Find the constant k in terms of g and R.

(b) F decreases as m moves away from the surface of the earth, either upward or downward. Show that F decreases as r increases from R at twice the rate at which F decreases as r decreases from R.

3. **(Resistors in parallel)** Two variable resistors R_1 and R_2 are connected in parallel so that their combined resistance R is given by

$$\frac{1}{R} = \frac{1}{R_1} + \frac{1}{R_2}.$$

At an instant when $R_1 = 250$ ohms and $R_2 = 1000$ ohms, R_1 is increasing at a rate of 100 ohms/minute. How fast must R_2 be changing at that moment (a) to keep R constant? and (b) to enable R to increase at a rate of 10 ohms/minute?

4. **(Gas law)** The volume V (in m^3), pressure P (in kilopascals, kPa) and temperature T (in kelvin K) for a sample of a certain gas satisfy the equation $pV = 5.0T$.

(a) How rapidly does the pressure increase if the temperature is 400 K and increasing at 4 K/min while the gas is kept confined in a volume of 2.0 m^3?

(b) How rapidly does the pressure decrease if the volume is 2 m^3 and increases at 0.05 m^3/min while the temperature is kept constant at 400 K?

5. **(The size of a print run)** It costs a publisher \$10,000 to set up the presses for a print run of a book and \$8 to cover the material costs for each book printed. In addition, machinery servicing, labour, and warehousing add another $6.25 \times 10^{-7}x^2$ to the cost of each book if x copies are manufactured during the printing. How many copies should the publisher print in order to minimize the average cost per book?

6. **(Maximizing profit)** A bicycle wholesaler must pay the manufacturer \$75 for each bicycle. Market research tells the wholesaler that if she charges her customers \$$x$ per bicycle, she can expect to sell $N(x) = 4.5 \times 10^6/x^2$ of them. What price should she charge to maximize her profit, and how many bicycles should she order from the manufacturer?

7. Find the largest possible volume of a right-circular cone that can be inscribed in a sphere of radius R.

8. **(Minimizing production costs)** The cost $\$C(x)$ of production in a factory varies with the amount x of product manufactured. The cost may rise sharply with x for x small, and more slowly for larger values of x because of economies of scale. However, if x becomes too large, the resources of the factory can be overtaxed, and the cost can begin to rise quickly again. Figure 4.59 shows the graph of a typical such cost function $C(x)$.

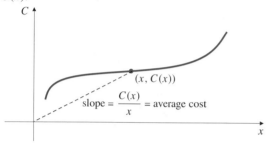

Figure 4.59

If x units are manufactured, the average cost per unit is $\$C(x)/x$, which is the slope of the line from the origin to the point $(x, C(x))$ on the graph.

(a) If it is desired to choose x to minimize this average cost per unit (as would be the case if all units produced could be sold for the same price), show that x should be chosen to make the average cost equal to the marginal cost:

$$\frac{C(x)}{x} = C'(x).$$

(b) Interpret the conclusion of (a) geometrically in the figure.

(c) If the average cost equals the marginal cost for some x, does x necessarily minimize the average cost?

9. **(Box design)** Four squares are cut out of a rectangle of cardboard 50 cm by 80 cm, as shown in Figure 4.60, and the remaining piece is folded into a closed, rectangular box, with two extra flaps tucked in. What is the largest possible volume for such a box?

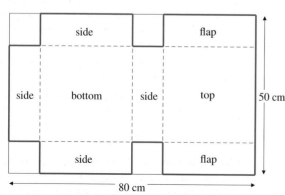

Figure 4.60

10. **(Yield from an orchard)** A certain orchard has 60 trees and produces an average of 800 apples per tree per year. If the density of trees is increased, the yield per tree drops; for each additional tree planted the average yield per tree is reduced by 10 apples per year. How many more trees should be planted to maximize the total annual yield of apples from the orchard?

11. **(Rotation of a tracking antenna)** What is the maximum rate at which the antenna in Exercise 41 of Section 4.1 must be able to turn in order to track the rocket during its entire vertical ascent?

12. An oval table has outer edge in the shape of the curve $x^2 + y^4 = 1/8$, where x and y are measured in metres. What is the width of the narrowest hallway in which the table can be turned horizontally through $180°$?

13. A hollow iron ball whose shell is 2 cm thick weighs half as much as it would if it were solid iron throughout. What is the radius of the ball?

14. **(Range of a cannon fired from a hill)** A cannon ball is fired with a speed of 200 ft/s at an angle of $45°$ above the horizontal from the top of a hill whose height at a horizontal distance x ft from the top is $y = 1,000/(1 + (x/500)^2)$ ft above sea level. How far does the cannon ball travel horizontally before striking the ground?

15. **(Linear approximation for a pendulum)** Because $\sin\theta \approx \theta$ for small values of $|\theta|$, the nonlinear equation of motion of a simple pendulum

$$\frac{d^2\theta}{dt^2} = -\frac{g}{L}\sin\theta,$$

which determines the displacement angle $\theta(t)$ away from vertical at time t for a simple pendulum, is frequently approximated by the simpler linear equation

$$\frac{d^2\theta}{dt^2} = -\frac{g}{L}\theta,$$

when the maximum displacement of the pendulum is not large. What is the percentage error in the right side of the equation if $|\theta|$ does not exceed $20°$?

16. Find the Taylor polynomial of degree 6 for $\sin^2 x$ about $x = 0$ and use it to help you evaluate

$$\lim_{x\to 0} \frac{3\sin^2 x - 3x^2 + x^4}{x^6}.$$

17. Use a degree 2 Taylor polynomial for $\tan^{-1} x$ about $x = 1$ to find an approximate value for $\tan^{-1}(1.1)$. Estimate the size of the error by using Taylor's formula.

18. The line $2y = 10x - 19$ is tangent to $y = f(x)$ at $x = 2$. If an initial approximation $x_0 = 2$ is made for a root of $f(x) = 0$ and Newton's Method is applied once, what will be the new approximation that results?

19. Find all solutions of the equation $\cos x = (x-1)^2$ to 10 decimal places.

20. Find the shortest distance from the point $(2, 0)$ to the curve $y = \ln x$.

21. A car is travelling at night along a level, curved road whose equation is $y = e^x$. At a certain instant its headlights illuminate a signpost located at the point $(1, 1)$. Where is the car at that instant?

Challenging Problems

1. (Growth of a crystal) A single cubical salt crystal is growing in a beaker of salt solution. The crystal's volume V increases at a rate proportional to its surface area and to the amount by which its volume is less than a limiting volume V_0:

$$\frac{dV}{dt} = kx^2(V_0 - V),$$

where x is the edge length of the crystal at time t.

(a) Using $V = x^3$, transform the equation above to one giving the rate of change dx/dt of the edge length x in terms of x.

(b) Show that the growth rate of the edge of the crystal decreases with time but remains positive as long as

$$x < x_0 = V_0^{1/3}.$$

(c) Find the volume of the crystal when its edge length is growing at half the rate it was initially.

2. (A review of calculus!) You are in a tank (the military variety) moving down the y-axis toward the origin. At time $t = 0$ you are 4 km from the origin, and 10 min later you are 2 km from the origin. Your speed is decreasing; it is proportional to your distance from the origin. You know that an enemy tank is waiting somewhere on the positive x-axis, but there is a high wall along the curve $xy = 1$ (all distances in kilometres) preventing you from seeing just where it is. How fast must your gun turret be capable of turning to maximize your chances of surviving the encounter?

3. (The economics of blood testing) Suppose that it is necessary to perform a blood test on a large number N of individuals to detect the presence of a virus. If each test costs $\$C$, then the total cost of the testing program is $\$NC$. If the proportion of people in the population who have the virus is not large, this cost can be greatly reduced by adopting the following strategy. Divide the N samples of blood into N/x groups of x samples each. Pool the blood in each group to make a single sample for that group and test it. If it tests negative, no further testing is necessary for individuals in that group. If the group sample tests positive, test all the individuals in that group.

Suppose that the fraction of individuals in the population infected with the virus is p, so the fraction uninfected is $q = 1 - p$. The probability that a given individual is unaffected is q, so the probability that all x individuals in a group are unaffected is q^x. Therefore, the probability that a pooled sample is infected is $1 - q^x$. Each group requires one test, and the infected groups require an extra x tests. Therefore the expected total number of tests to be performed is

$$T = \frac{N}{x} + \frac{N}{x}(1 - q^x)x = N\left(\frac{1}{x} + 1 - q^x\right).$$

For example, if $p = 0.01$, so that $q = 0.99$ and $x = 20$, then the expected number of tests required is $T = 0.23N$, a reduction of over 75%. But maybe we can do better by making a different choice for x.

(a) For $q = 0.99$, find the number x of samples in a group that minimizes T (i.e., solve $dT/dx = 0$). Show that the minimizing value of x satisfies

$$x = \frac{(0.99)^{-x/2}}{\sqrt{-\ln(0.99)}}.$$

(b) Use the technique of fixed-point iteration (see Section 4.6) to solve the equation in (a) for x. Start with $x = 20$, say.

4. (Measuring variations in g) The period P of a pendulum of length L is given by

$$P = 2\pi\sqrt{L/g},$$

where g is the acceleration of gravity.

(a) Assuming that L remains fixed, show that a 1% increase in g results in approximately a 0.5% decrease in the period P. (Variations in the period of a pendulum can be used to detect small variations in g from place to place on the earth's surface.)

(b) For fixed g, what percentage change in L will produce a 1% increase in P?

5. (Torricelli's Law) The rate at which a tank drains is proportional to the square root of the depth of liquid in the tank above the level of the drain: if $V(t)$ is the volume of liquid in the tank at time t, and $y(t)$ is the height of the surface of the liquid above the drain, then $dV/dt = -k\sqrt{y}$, where k is a constant depending on the size of the drain. For a cylindrical tank with constant cross-sectional area A with drain at the bottom:

(a) Verify that the depth $y(t)$ of liquid in the tank at time t satisfies $dy/dt = -(k/A)\sqrt{y}$.

(b) Verify that if the depth of liquid in the tank at $t = 0$ is y_0, then the depth at subsequent times during the draining process is $y = \left(\sqrt{y_0} - \dfrac{kt}{2A}\right)^2$.

(c) If the tank drains completely in time T, express the depth $y(t)$ at time t in terms of y_0 and T.

(d) In terms of T, how long does it take for half the liquid in the tank to drain out?

6. If a conical tank with top radius R and depth H drains according to Torricelli's Law and empties in time T, show that the depth of liquid in the tank at time t $(0 < t < T)$ is

$$y = y_0 \left(1 - \frac{t}{T}\right)^{2/5},$$

where y_0 is the depth at $t = 0$.

7. Find the largest possible area of a right-angled triangle whose perimeter is P.

8. Find a tangent to the graph of $y = x^3 + ax^2 + bx + c$ that is not parallel to any other tangent.

9. (**Branching angles for electric wires and pipes**)

   (a) The resistance offered by a wire to the flow of electric current through it is proportional to its length and inversely proportional to its cross-sectional area. Thus, the resistance R of a wire of length L and radius r is $R = kL/r^2$, where k is a positive constant.

   A long straight wire of length L and radius r_1 extends from A to B. A second straight wire of smaller radius r_2 is to be connected between a point P on AB and a point C at distance h from B such that CB is perpendicular to AB. (See Figure 4.61.) Find the value of the angle $\theta = \angle BPC$ that minimizes the total resistance of the path APC, that is, the resistance of AP plus the resistance of PC.

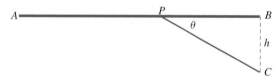

Figure 4.61

   (b) The resistance of a pipe (e.g., a blood vessel) to the flow of liquid through it is, by Poiseuille's Law, proportional to its length and inversely proportional to the *fourth power* of its radius: $R = kL/r^4$. If the situation in part (a) represents pipes instead of wires, find the value of θ that minimizes the total resistance of the path APC. How does your answer relate to the answer for part (a)? Could you have predicted this relationship?

* **10.** (**The range of a spurt**) A cylindrical water tank sitting on a horizontal table has a small hole located on its vertical wall at height h above the bottom of the tank. Water escapes from the tank horizontally through the hole and then curves down under the influence of gravity to strike the table at a distance R from the base of the tank, as shown in Figure 4.62. (We

ignore air resistance.) Torricelli's Law implies that the speed v at which water escapes through the hole is proportional to the square root of the depth of the hole below the surface of the water: if the depth of water in the tank at time t is $y(t) > h$, then $v = k\sqrt{y - h}$, where the constant k depends on the size of the hole.

   (a) Find the range R in terms of v and h.

   (b) For a given depth y of water in the tank, how high should the hole be to maximize R?

   (c) Suppose that the depth of water in the tank at time $t = 0$ is y_0, that the range R of the spurt is R_0 at that time, and that the water level drops to the height h of the hole in T minutes. Find, as a function of t, the range R of the water that escaped through the hole at time t.

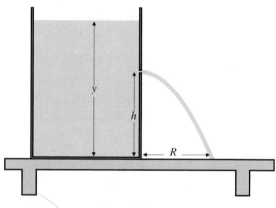

Figure 4.62

* **11.** (**Designing a dustpan**) Equal squares are cut out of two adjacent corners of a square of sheet metal having sides of length 25 cm. The three resulting flaps are bent up, as shown in Figure 4.63, to form the sides of a dustpan. Find the maximum volume of a dustpan made in this way.

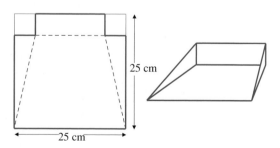

Figure 4.63

CHAPTER 5

Integration

Introduction The second fundamental problem addressed by calculus is the problem of areas, that is, the problem of determining the area of a region of the plane bounded by various curves. Like the problem of tangents considered in Chapter 2, many practical problems in various disciplines require the evaluation of areas for their solution, and the solution of the problem of areas necessarily involves the notion of limits. On the surface the problem of areas appears unrelated to the problem of tangents. However, we will see that the two problems are very closely related; one is the inverse of the other. Finding an area is equivalent to finding an antiderivative or, as we prefer to say, finding an integral. The relationship between areas and antiderivatives is called the Fundamental Theorem of Calculus. When we have proved it we will be able to find areas at will, provided only that we can integrate (i.e., antidifferentiate) the various functions we encounter.

We would like to have at our disposal a set of integration rules similar to the differentiation rules developed in Chapter 2. We can find the derivative of any differentiable function using those differentiation rules. Unfortunately, integration is generally more difficult; indeed, some fairly simple functions are not themselves derivatives of simple functions. For example, e^{x^2} is not the derivative of any finite combination of elementary functions. Nevertheless, we will expend some effort in Section 5.6 and Sections 6.1–6.4 to develop techniques for integrating as many functions as possible. Later in Chapter 6 we will examine how to approximate areas bounded by graphs of functions that we cannot antidifferentiate.

5.1 Sums and Sigma Notation

When we begin calculating areas in the next section we will often encounter sums of values of functions. We need to have a convenient notation for representing sums of arbitrary (possibly large) numbers of terms, and we need to develop techniques for evaluating some such sums.

We use the symbol $\sum$ to represent a sum; it is an enlarged Greek capital letter "S" called *sigma*.

DEFINITION 1

Sigma notation

If m and n are integers with $m \leq n$, and if f is a function defined at the integers $m, m + 1, m + 2, \ldots, n$, the symbol $\sum_{i=m}^{n} f(i)$ represents the sum of the values of f at those integers:

$$\sum_{i=m}^{n} f(i) = f(m) + f(m + 1) + f(m + 2) + \cdots + f(n).$$

The explicit sum appearing on the right side of this equation is the **expansion** of the sum represented in sigma notation on the left side.

Example 1 $$\sum_{i=1}^{5} i^2 = 1^2 + 2^2 + 3^2 + 4^2 + 5^2 = 55.$$

The i that appears in the symbol $\sum_{i=m}^{n} f(i)$ is called an **index of summation**. To evaluate $\sum_{i=m}^{n} f(i)$, replace the index i with the integers m, $m+1$, $\ldots$, n, successively, and sum the results. Observe that the value of the sum does not depend on what we call the index; the index does not appear on the right side of the definition. If we use another letter in place of i in the sum in Example 1, we still get the same value for the sum:

$$\sum_{k=1}^{5} k^2 = 1^2 + 2^2 + 3^2 + 4^2 + 5^2 = 55.$$

The index of summation is a *dummy variable* used to represent an arbitrary point where the function is evaluated to produce a term to be included in the sum. On the other hand, the sum $\sum_{i=m}^{n} f(i)$ does depend on the two numbers m and n, called the **limits of summation**; m is the **lower limit** and n is the **upper limit**.

Example 2 (**Examples of sums using sigma notation**)

$$\sum_{j=1}^{20} j = 1 + 2 + 3 + \cdots + 18 + 19 + 20$$

$$\sum_{i=0}^{n} x^i = x^0 + x^1 + x^2 + \cdots + x^{n-1} + x^n$$

$$\sum_{m=1}^{n} 1 = \underbrace{1 + 1 + 1 + \cdots + 1}_{n \text{ terms}}$$

$$\sum_{k=-2}^{3} \frac{1}{k+7} = \frac{1}{5} + \frac{1}{6} + \frac{1}{7} + \frac{1}{8} + \frac{1}{9} + \frac{1}{10}$$

Sometimes we use a subscripted variable a_i to denote the ith term of a general sum instead of using the functional notation $f(i)$:

$$\sum_{i=m}^{n} a_i = a_m + a_{m+1} + a_{m+2} + \cdots + a_n.$$

In particular, an **infinite series** is such a sum with infinitely many terms:

$$\sum_{n=1}^{\infty} a_n = a_1 + a_2 + a_3 + \cdots$$

When no final term follows the $\cdots$, it is understood that the terms go on forever. We will study infinite series in Chapter 9.

When adding finitely many numbers, the order in which they are added is unimportant; any order will give the same sum. If all the numbers have a common factor, then that factor can be removed from each term and multiplied after the sum is evaluated: $ca + cb = c(a + b)$. These laws of arithmetic translate into the following *linearity* rule for finite sums; if A and B are constants, then

$$\sum_{i=m}^{n}\left(Af(i)+Bg(i)\right)=A\sum_{i=m}^{n}f(i)+B\sum_{i=m}^{n}g(i).$$

Both of the sums $\sum_{j=m}^{m+n} f(j)$ and $\sum_{i=0}^{n} f(i+m)$ have the same expansion, namely, $f(m)+f(m+1)+\cdots+f(m+n)$. Therefore the two sums are equal.

$$\sum_{j=m}^{m+n} f(j)=\sum_{i=0}^{n} f(i+m).$$

This equality can also be derived by substituting $i+m$ for j everywhere j appears on the left side, noting that $i+m=m$ reduces to $i=0$ and $i+m=m+n$ reduces to $i=n$. It is often convenient to make such a **change of index** in a summation.

Example 3   Express $\sum_{j=3}^{17}\sqrt{1+j^2}$ in the form $\sum_{i=1}^{n} f(i)$.

Solution   Let $j=i+2$, Then $j=3$ corresponds to $i=1$ and $j=17$ corresponds to $i=15$. Thus

$$\sum_{j=3}^{17}\sqrt{1+j^2}=\sum_{i=1}^{15}\sqrt{1+(i+2)^2}.$$

Evaluating Sums

When a sum like

$$S=\sum_{i=1}^{n} i=1+2+3+\cdots+n$$

involves a variable or large number of terms, we would like to have a formula giving the value of the sum in **closed form**, that is, not using the ellipsis $(\cdots)$. For the above sum, the formula is

$$S=\sum_{i=1}^{n} i=\frac{n(n+1)}{2}.$$

This can be verified by writing the sum forwards and backwards and adding the two to get

$$
\begin{array}{rccccccccc}
S = & 1 & + & 2 & + & 3 & +\cdots+ & (n-1) & + & n \\
S = & n & + & (n-1) & + & (n-2) & +\cdots+ & 2 & + & 1 \\
\hline
2S = & (n+1) & + & (n+1) & + & (n+1) & +\cdots+ & (n+1) & + & (n+1) = n(n+1)
\end{array}
$$

The formula for S follows when we divide by 2.

It is not usually this easy to evaluate a general sum in closed form. We can only simplify $\sum_{i=m}^{n} f(i)$ for a small class of functions f. The only such formulas we will need in the next sections are collected in Theorem 1.

THEOREM **1** **Summation formulas**

(a) $\displaystyle\sum_{i=1}^{n} 1 = \underbrace{1 + 1 + 1 + \cdots + 1}_{n \text{ terms}} = n$

(b) $\displaystyle\sum_{i=1}^{n} i = 1 + 2 + 3 + \cdots + n = \frac{n(n+1)}{2}$

(c) $\displaystyle\sum_{i=1}^{n} i^2 = 1^2 + 2^2 + 3^2 + \cdots + n^2 = \frac{n(n+1)(2n+1)}{6}$

(d) $\displaystyle\sum_{i=1}^{n} r^{i-1} = 1 + r + r^2 + r^3 + \cdots + r^{n-1} = \frac{r^n - 1}{r - 1}$ if $r \neq 1$.

PROOF Formula (a) is trivial; the sum of n ones is n. One proof of formula (b) was given above. Three others are suggested in Exercises 38–40.

To prove (c) we write n copies of the identity

$$(k+1)^3 - k^3 = 3k^2 + 3k + 1,$$

one for each value of k from 1 to n, and add them up:

$$
\begin{array}{rclcccccc}
2^3 & - & 1^3 & = & 3 \times 1^2 & + & 3 \times 1 & + & 1 \\
3^3 & - & 2^3 & = & 3 \times 2^2 & + & 3 \times 2 & + & 1 \\
4^3 & - & 3^3 & = & 3 \times 3^2 & + & 3 \times 3 & + & 1 \\
\vdots & & \vdots & & \vdots & & \vdots & & \vdots \\
n^3 & - & (n-1)^3 & = & 3(n-1)^2 & + & 3(n-1) & + & 1 \\
(n+1)^3 & - & n^3 & = & 3n^2 & + & 3n & + & 1 \\
\hline
(n+1)^3 & - & 1^3 & = & 3\left(\sum_{i=1}^{n} i^2\right) & + & 3\left(\sum_{i=1}^{n} i\right) & + & n \\
& & & = & 3\left(\sum_{i=1}^{n} i^2\right) & + & \dfrac{3n(n+1)}{2} & + & n.
\end{array}
$$

We used formula (b) in the last line. The final equation can be solved for the desired sum to give formula (c). Note the cancellations that occurred when we added up the left sides of the n equations. The term 2^3 in the first line cancelled the -2^3 in the second line, and so on, leaving us with only two terms, the $(n+1)^3$ from the nth line and the -1^3 from the first line:

$$\sum_{k=1}^{n}\left((k+1)^3 - k^3\right) = (n+1)^3 - 1^3.$$

This is an example of what we call a **telescoping sum**. In general, a sum of the form $\sum_{i=m}^{n}(f(i+1) - f(i))$ telescopes to the closed form $f(n+1) - f(m)$ because all but the first and last terms cancel out.

To prove formula (d), let $s = \sum_{i=1}^{n} r^{i-1}$ and subtract s from rs:

$$(r-1)s = rs - s = (r + r^2 + r^3 + \cdots + r^n) - (1 + r + r^2 + \cdots + r^{n-1})$$
$$= r^n - 1.$$

The result follows on division by $r - 1$.

Example 4 Evaluate $\displaystyle\sum_{k=m+1}^{n} (6k^2 - 4k + 3)$, where $1 \le m < n$.

Solution Using the rules of summation and various summation formulas from Theorem 1 we calculate

$$\sum_{k=1}^{n}(6k^2 - 4k + 3) = 6\sum_{k=1}^{n} k^2 - 4\sum_{k=1}^{n} k + 3\sum_{k=1}^{n} 1$$

$$= 6\,\frac{n(n+1)(2n+1)}{6} - 4\,\frac{n(n+1)}{2} + 3n$$

$$= 2n^3 + n^2 + 2n$$

Thus

$$\sum_{k=m+1}^{n} (6k^2 - 4k + 3) = \sum_{k=1}^{n}(6k^2 - 4k + 3) - \sum_{k=1}^{m}(6k^2 - 4k + 3)$$

$$= 2n^3 + n^2 + 2n - 2m^3 - m^2 - 2m.$$

Remark Maple can find closed form expressions for some sums. For example,

```
>  sum(i^4, i=1..n); factor(%);
```

$$\frac{1}{5}(n+1)^5 - \frac{1}{2}(n+1)^4 + \frac{1}{3}(n+1)^3 - \frac{1}{30}n - \frac{1}{30}$$

$$\frac{1}{30}n(2n+1)(n+1)(3n^2 + 3n - 1)$$

Exercises 5.1

Expand the sums in Exercises 1–8.

1. $\displaystyle\sum_{i=1}^{4} i^3$

2. $\displaystyle\sum_{j=1}^{100} \frac{j}{j+1}$

3. $\displaystyle\sum_{i=1}^{n} 3^i$

4. $\displaystyle\sum_{i=0}^{n-1} \frac{(-1)^i}{i+1}$

5. $\displaystyle\sum_{j=3}^{n} \frac{(-2)^j}{(j-2)^2}$

6. $\displaystyle\sum_{j=1}^{n} \frac{j^2}{n^3}$

7. $\displaystyle\sum_{n=1}^{k} \sin \frac{n\pi}{3k}$

8. $\displaystyle\sum_{k=2}^{n} \frac{e^{-k}}{n+k}$

Write the sums in Exercises 9–16 using sigma notation. (Note that the answers are not unique.)

9. $5 + 6 + 7 + 8 + 9$

10. $2 + 2 + 2 + \cdots + 2$ (200 terms)

11. $2^2 - 3^2 + 4^2 - 5^2 + \cdots - 99^2$

12. $1 + 2x + 3x^2 + 4x^3 + \cdots + 100x^{99}$

13. $1 + x + x^2 + x^3 + \cdots + x^n$

14. $1 - x + x^2 - x^3 + \cdots + x^{2n}$

15. $1 - \dfrac{1}{4} + \dfrac{1}{9} - \dfrac{1}{16} + \cdots + \dfrac{(-1)^{n-1}}{n^2}$

16. $\dfrac{1}{2} + \dfrac{2}{4} + \dfrac{3}{8} + \dfrac{4}{16} + \cdots + \dfrac{n}{2^n}$

Express the sums in Exercises 17–18 in the form $\sum_{i=1}^{n} f(i)$.

17. $\displaystyle\sum_{j=0}^{99} \sin(j)$

18. $\displaystyle\sum_{k=-5}^{m} \frac{1}{k^2 + 1}$

Find closed form values for the sums in Exercises 19–32.

19. $\displaystyle\sum_{i=2}^{6} (i - 1)$

20. $\displaystyle\sum_{j=1}^{1,000} (2j + 3)$

21. $\displaystyle\sum_{i=1}^{n} i^2 + 2i$

22. $\displaystyle\sum_{k=1}^{n-1} (3k^2 - 4)$

23. $\displaystyle\sum_{k=1}^{n} (\pi^k - 3)$

24. $\displaystyle\sum_{i=1}^{n} (2^i - i^2)$

25. $\displaystyle\sum_{m=1}^{n} \ln m$ **26.** $\displaystyle\sum_{i=0}^{n} e^{i/n}$

27. The sum in Exercise 10.   **28.** The sum in Exercise 13.

29. The sum in Exercise 14.

* 30. The sum in Exercise 12. *Hint:* differentiate the sum $\displaystyle\sum_{i=0}^{100} x^i$.

* 31. The sum in Exercise 11. *Hint:* the sum is
$$\sum_{k=1}^{49}\big((2k)^2 - (2k+1)^2\big) = \sum_{k=1}^{49}(-4k - 1).$$

* 32. The sum in Exercise 16. *Hint:* apply the method of proof of Theorem 1(d) to this sum.

33. Verify the formula for the value of a telescoping sum:
$$\sum_{i=m}^{n}\big(f(i+1) - f(i)\big) = f(n+1) - f(m).$$

Why is the word "telescoping" used to describe this sum?

In Exercises 34–36, evaluate the given telescoping sums.

34. $\displaystyle\sum_{n=1}^{10}\big(n^4 - (n-1)^4\big)$ **35.** $\displaystyle\sum_{j=1}^{m}(2^j - 2^{j-1})$

36. $\displaystyle\sum_{i=m}^{2m}\left(\frac{1}{i} - \frac{1}{i+1}\right)$

37. Show that $\dfrac{1}{j(j+1)} = \dfrac{1}{j} - \dfrac{1}{j+1}$, and hence evaluate
$$\sum_{j=1}^{n}\frac{1}{j(j+1)}.$$

38. Figure 5.1 shows a square of side n subdivided into n^2 smaller squares of side 1. How many small squares are shaded? Obtain the closed form expression for $\sum_{i=1}^{n} i$ by considering the sum of the areas of the shaded squares.

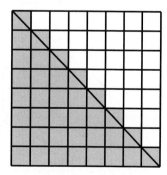

Figure 5.1

39. Write n copies of the identity $(k+1)^2 - k^2 = 2k + 1$, one for each integer k from 1 to n, and add them up to obtain the formula
$$\sum_{i=1}^{n} i = \frac{n(n+1)}{2}.$$

in a manner similar to the proof of Theorem 1(c).

40. Use mathematical induction to prove Theorem 1(b).

41. Use mathematical induction to prove Theorem 1(c).

42. Use mathematical induction to prove Theorem 1(d).

43. Figure 5.2 shows a square of side $\sum_{i=1}^{n} i = n(n+1)/2$ subdivided into a small square of side 1 and $n-1$ L-shaped regions whose short edges are 2, 3, ..., n. Show that the area of the L-shaped region with short side i is i^3, and hence verify that
$$\sum_{i=1}^{n} i^3 = \frac{n^2(n+1)^2}{4}.$$

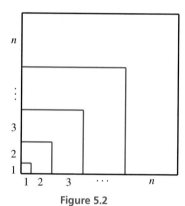

Figure 5.2

* 44. Write n copies of the identity
$$(k+1)^4 - k^4 = 4k^3 + 6k^2 + 4k + 1,$$
one for each integer k from 1 to n, and add them up to obtain the formula
$$\sum_{i=1}^{n} i^3 = \frac{n^2(n+1)^2}{4},$$
in a manner similar to the proof of Theorem 1(c).

45. Use mathematical induction to verify the formula for the sum of cubes given in Exercise 44.

46. Extend the method of Exercise 44 to find a closed form expression for $\sum_{i=1}^{n} i^4$. You will probably want to use Maple or other computer algebra software to do all the algebra.

47. Use Maple or another computer algebra system to find $\sum_{i=1}^{n} i^k$ for $k = 5, 6, 7, 8$. Observe the term involving the highest power of n in each case. Predict the highest power term in $\sum_{i=1}^{n} i^{10}$ and verify your prediction.

5.2   Areas as Limits of Sums

We began the study of derivatives in Chapter 2 by defining what is meant by a tangent line to a curve at a particular point. We would like to begin the study of integrals by defining what is meant by the **area** of a plane region, but a definition of area is much harder to give than a definition of tangency. Let us assume, therefore, that we know intuitively what area means and list some of its properties. (See Figure 5.3.)

(i) The area of a plane region is a nonnegative real number of *square units*.

(ii) The area of a rectangle with width w and height h is $A = wh$.

(iii) The areas of congruent plane regions are equal.

(iv) If region S is contained in region R, then the area of S is less than or equal to that of R.

(v) If region R is a union of (finitely many) non-overlapping regions, then the area of R is the sum of the areas of those regions.

Using these five properties we can calculate the area of any **polygon** (a region bounded by straight line segments). First, we note that properties (iii) and (v) show that the area of a parallelogram is the same as that of a rectangle having the same base width and height. Any triangle can be butted against a congruent copy of itself to form a parallelogram, so a triangle has area half the base width times the height. Finally, any polygon can be subdivided into finitely many nonoverlapping triangles so its area is the sum of the areas of those triangles.

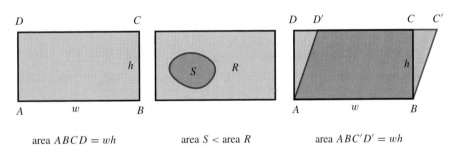

area $ABCD = wh$ area $S <$ area R area $ABC'D' = wh$

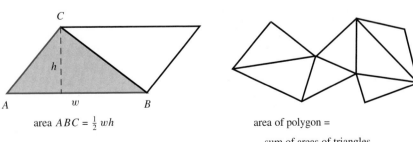

area $ABC = \frac{1}{2}wh$ area of polygon = sum of areas of triangles

Figure 5.3   Properties of area

We can't go beyond polygons without taking limits. If a region has a curved boundary, its area can only be approximated by using rectangles or triangles; calculating the exact area requires the evaluation of a limit. We showed how this could be done for a circle in Section 1.1.

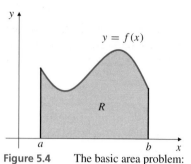

Figure 5.4 The basic area problem: find the area of region R

The Basic Area Problem

In this section we are going to consider how to find the area of a region R lying under the graph $y = f(x)$ of a nonnegative-valued, continuous function f, above the x-axis and between the vertical lines $x = a$ and $x = b$, where $a < b$. (See Figure 5.4.) To accomplish this we proceed as follows. Divide the interval $[a, b]$ into n subintervals by using division points:

$$a = x_0 < x_1 < x_2 < x_3 < \cdots < x_{n-1} < x_n = b.$$

Denote by Δx_i the length of the ith subinterval $[x_{i-1}, x_i]$:

$$\Delta x_i = x_i - x_{i-1}, \qquad (i = 1, 2, 3, \ldots, n).$$

Vertically above each subinterval $[x_{i-1}, x_i]$ build a rectangle whose base has length Δx_i and whose height is $f(x_i)$. The area of this rectangle is $f(x_i)\,\Delta x_i$. Form the sum of these areas:

$$S_n = f(x_1)\,\Delta x_1 + f(x_2)\,\Delta x_2 + f(x_3)\,\Delta x_3 + \cdots + f(x_n)\,\Delta x_n = \sum_{i=1}^{n} f(x_i)\,\Delta x_i.$$

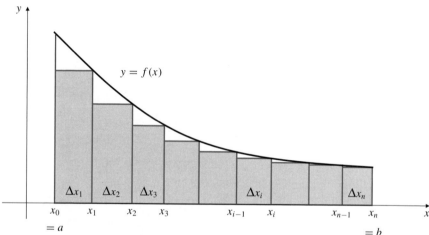

Figure 5.5 Approximating the area under the graph of a decreasing function using rectangles

The rectangles are shown shaded in Figure 5.5 for a decreasing function f. For an increasing function, the tops of the rectangles would lie above the graph of f rather than below it. Evidently S_n is an approximation to the area of the region R, and the approximation gets better as n increases, provided we choose the points $a = x_0 < x_1 < \cdots < x_n = b$ in such a way that the width Δx_i of the widest rectangle approaches zero.

Observe in Figure 5.6, for example, that subdividing a subinterval into two smaller subintervals reduces the error in the approximation by reducing that part of the area under the curve that is not contained in the rectangles. It is reasonable, therefore, to calculate the area of R by finding the limit of S_n as $n \to \infty$ with the restriction that the largest of the subinterval widths Δx_i must approach zero:

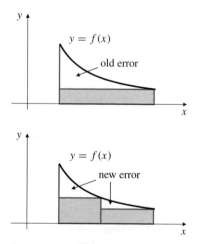

Figure 5.6 Using more rectangles makes the error smaller

$$\text{Area of } R = \lim_{\substack{n \to \infty \\ \max \Delta x_i \to 0}} S_n.$$

Sometimes, but not always, it is useful to choose the points x_i ($0 \le i \le n$) in $[a, b]$ in such a way that the subinterval lengths Δx_i are all equal. In this case we have

$$\Delta x_i = \Delta x = \frac{b-a}{n}, \qquad x_i = a + i\Delta x = a + \frac{i}{n}(b-a).$$

Some Area Calculations

We devote the rest of this section to some examples in which we apply the technique described above for finding areas under graphs of functions by approximating with rectangles. Let us begin with a region for which we already know the area so we can satisfy ourselves that the method does give the correct value.

Example 1   Find the area A of the region lying under the straight line $y = x + 1$, above the x-axis and between the lines $x = 0$ and $x = 2$.

Solution   The region is shaded in Figure 5.7(a). It is a *trapezoid* (a four-sided polygon with one pair of parallel sides) and has area 4 square units. (It can be divided into a rectangle and a triangle, each of area 2 square units.) We will calculate the area as a limit of sums of areas of rectangles constructed as described above. Divide the interval $[0, 2]$ into n subintervals *of equal length* by points

$$x_0 = 0, \ x_1 = \frac{2}{n}, \ x_2 = \frac{4}{n}, \ x_3 = \frac{6}{n}, \ \dots \ x_n = \frac{2n}{n} = 2.$$

The value of $y = x + 1$ at $x = x_i$ is $x_i + 1 = \dfrac{2i}{n} + 1$ and the ith subinterval, $\left[\dfrac{2(i-1)}{n}, \dfrac{2i}{n}\right]$, has length $\Delta x_i = \dfrac{2}{n}$. Observe that $\Delta x_i \to 0$ as $n \to \infty$. The sum of the areas of the approximating rectangles shown in Figure 5.7(a) is

$$
\begin{aligned}
S_n &= \sum_{i=1}^{n} \left(\frac{2i}{n} + 1\right) \frac{2}{n} \\
&= \left(\frac{2}{n}\right)\left[\frac{2}{n} \sum_{i=1}^{n} i + \sum_{i=1}^{n} 1\right] \qquad \text{(Use parts (b) and (a) of Theorem 1.)} \\
&= \left(\frac{2}{n}\right)\left[\frac{2}{n} \frac{n(n+1)}{2} + n\right] \\
&= 2\frac{n+1}{n} + 2.
\end{aligned}
$$

Therefore, the required area A is given by

$$A = \lim_{n \to \infty} S_n = \lim_{n \to \infty}\left(2\frac{n+1}{n} + 2\right) = 2 + 2 = 4 \text{ square units.}$$

Example 2   Find the area of the region bounded by the parabola $y = x^2$ and the straight lines $y = 0$, $x = 0$, and $x = b > 0$.

Solution   The area A of the region is the limit of the sum S_n of areas of the rectangles shown in Figure 5.7(b). Again we have used equal subintervals, each of length b/n. The height of the ith rectangle is $(ib/n)^2$. Thus,

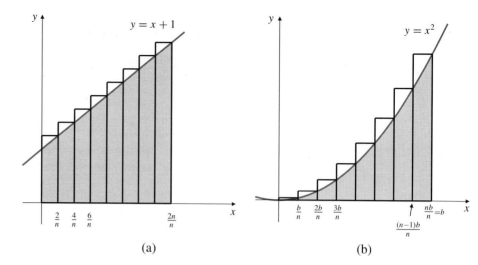

Figure 5.7

(a) The region of Example 1

(b) The region of Example 2

Figure 5.7

(a) The region of Example 1

(b) The region of Example 2

$$S_n = \sum_{i=1}^{n} \left(\frac{ib}{n}\right)^2 \frac{b}{n} = \frac{b^3}{n^3} \sum_{i=1}^{n} i^2 = \frac{b^3}{n^3} \frac{n(n+1)(2n+1)}{6},$$

by formula (c) of Theorem 1. Hence the required area is

$$A = \lim_{n \to \infty} S_n = \lim_{n \to \infty} b^3 \frac{(n+1)(2n+1)}{6n^2} = \frac{b^3}{3} \text{ square units.}$$

Finding an area under the graph of $y = x^k$ over an interval I becomes more and more difficult as k increases if we continue to try to subdivide I into subintervals of equal length. (See Exercise 14 at the end of this section for the case $k = 3$.) It is, however, possible to find the area for arbitrary k if we subdivide the interval I into subintervals whose lengths increase in geometric progression. The next example illustrates this.

Example 3 Let $b > a > 0$ and let k be any real number except -1. Show that the area A of the region bounded by $y = x^k$, $y = 0$, $x = a$, and $x = b$ is

$$A = \frac{b^{k+1} - a^{k+1}}{k+1} \text{ square units.}$$

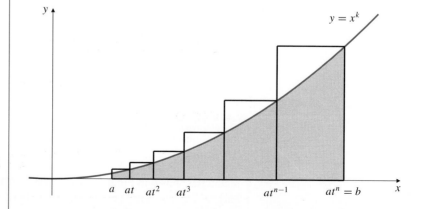

Figure 5.8 For this partition the subinterval lengths increase exponentially

Solution   Let $t = (b/a)^{1/n}$ and let

$$x_0 = a, \; x_1 = at, \; x_2 = at^2, \; x_3 = at^3, \; \ldots \; x_n = at^n = b.$$

These points subdivide the interval $[a, b]$ into n subintervals of which the ith, $[x_{i-1}, x_i]$, has length $\Delta x_i = at^{i-1}(t-1)$. If $f(x) = x^k$, then $f(x_i) = a^k t^{ki}$. The sum of the areas of the rectangles shown in Figure 5.8 is:

$$
\begin{aligned}
S_n &= \sum_{i=1}^{n} f(x_i) \, \Delta x_i \\
&= \sum_{i=1}^{n} a^k t^{ki} \, at^{i-1}(t-1) \\
&= a^{k+1} (t-1) t^k \sum_{i=1}^{n} t^{(k+1)(i-1)} \\
&= a^{k+1} (t-1) t^k \sum_{i=1}^{n} r^{(i-1)} \qquad \text{where } r = t^{k+1} \\
&= a^{k+1} (t-1) t^k \frac{r^n - 1}{r - 1} \qquad \text{(by Theorem 1(d))} \\
&= a^{k+1} (t-1) t^k \frac{t^{(k+1)n} - 1}{t^{k+1} - 1}.
\end{aligned}
$$

Now replace t with its value $(b/a)^{1/n}$ and rearrange factors to obtain

$$
\begin{aligned}
S_n &= a^{k+1} \left(\left(\frac{b}{a}\right)^{1/n} - 1 \right) \left(\frac{b}{a}\right)^{k/n} \frac{\left(\dfrac{b}{a}\right)^{k+1} - 1}{\left(\dfrac{b}{a}\right)^{(k+1)/n} - 1} \\
&= \left(b^{k+1} - a^{k+1} \right) c^{k/n} \frac{c^{1/n} - 1}{c^{(k+1)/n} - 1}, \qquad \text{where } c = \frac{b}{a}.
\end{aligned}
$$

Of the three factors in the final line above, the first does not depend on n, and the second, $c^{k/n}$, approaches $c^0 = 1$ as $n \to \infty$. The third factor is an indeterminate form of type $[0/0]$, which we evaluate using l'Hôpital's Rule. First let $u = 1/n$. Then

$$
\begin{aligned}
\lim_{n \to \infty} \frac{c^{1/n} - 1}{c^{(k+1)/n} - 1} &= \lim_{u \to 0+} \frac{c^u - 1}{c^{(k+1)u} - 1} \qquad \left[\frac{0}{0}\right] \\
&= \lim_{u \to 0+} \frac{c^u \ln c}{(k+1) c^{(k+1)u} \ln c} = \frac{1}{k+1}.
\end{aligned}
$$

Therefore, the required area is

$$A = \lim_{n \to \infty} S_n = \left(b^{k+1} - a^{k+1} \right) \times 1 \times \frac{1}{k+1} = \frac{b^{k+1} - a^{k+1}}{k+1} \quad \text{square units.}$$

It is, as you can see, rather difficult to calculate areas bounded by curves by the methods developed above. Fortunately, there is an easier way, as we will discover in Section 5.5.

Remark   For technical reasons it was necessary to assume $a > 0$ in Example 3. The result is also valid for $a = 0$ provided $k > -1$. In this case we have $\lim_{a\to 0+} a^{k+1} = 0$, so the area under $y = x^k$, above $y = 0$, between $x = 0$ and $x = b > 0$, is $A = b^{k+1}/(k+1)$ square units. For $k = 2$ this agrees with the result of Example 2.

Example 4   Identify the limit $L = \lim_{n\to\infty} \sum_{i=1}^{n} \dfrac{n-i}{n^2}$ as an area, and evaluate it.

Solution   We can rewrite the ith term of the sum so that it depends on i/n:

$$L = \lim_{n\to\infty} \sum_{i=1}^{n} \left(1 - \frac{i}{n}\right)\frac{1}{n}.$$

The terms now appear to be the areas of rectangles of base $1/n$ and heights $1 - x_i$, $(1 \le i \le n)$, where

$$x_1 = \frac{1}{n}, \quad x_2 = \frac{2}{n}, \quad x_3 = \frac{3}{n}, \quad \ldots, \quad x_n = \frac{n}{n}.$$

Thus the limit L is the area under the curve $y = 1 - x$ from $x = 0$ to $x = 1$. (See Figure 5.9.) This region is a triangle having area $1/2$ square unit, so $L = 1/2$. ∎

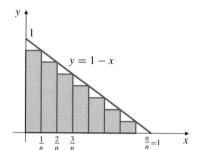

Figure 5.9   Recognizing a sum of areas

Exercises 5.2

Use the techniques of Examples 1 and 2 (with subintervals of equal length) to find the areas of the regions specified in Exercises 1–13.

1. Below $y = 3x$, above $y = 0$, from $x = 0$ to $x = 1$.

2. Below $y = 2x + 1$, above $y = 0$, from $x = 0$ to $x = 3$.

3. Below $y = 2x - 1$, above $y = 0$, from $x = 1$ to $x = 3$.

4. Below $y = 3x + 4$, above $y = 0$, from $x = -1$ to $x = 2$.

5. Below $y = x^2$, above $y = 0$, from $x = 1$ to $x = 3$.

6. Below $y = x^2 + 1$, above $y = 0$, from $x = 0$ to $x = a > 0$.

7. Below $y = x^2 + 2x + 3$, above $y = 0$, from $x = -1$ to $x = 2$.

8. Above $y = x^2 - 1$, below $y = 0$.

9. Above $y = 1 - x$, below $y = 0$, from $x = 2$ to $x = 4$.

10. Above $y = x^2 - 2x$, below $y = 0$.

11. Below $y = 4x - x^2 + 1$, above $y = 1$.

* 12. Below $y = e^x$, above $y = 0$, from $x = 0$ to $x = b > 0$.

* 13. Below $y = 2^x$, above $y = 0$, from $x = -1$ to $x = 1$.

14. Use the formula $\sum_{i=1}^{n} i^3 = n^2(n+1)^2/4$, from Exercises 43–45 of Section 5.1, to find the area of the region lying under $y = x^3$, above the x-axis, and between the vertical lines at $x = 0$ and $x = b > 0$.

15. Use the subdivision of $[a, b]$ given in Example 3 to find the area under $y = 1/x$, above $y = 0$ from $x = a > 0$ to $x = b > a$. Why should your answer not be surprising?

In Exercises 16–19, interpret the given sum S_n as a sum of areas of rectangles approximating the area of a certain region in the plane and hence evaluate $\lim_{n\to\infty} S_n$.

16. $S_n = \sum_{i=1}^{n} \dfrac{2}{n}\left(1 - \dfrac{i}{n}\right)$ **17.** $S_n = \sum_{i=1}^{n} \dfrac{2}{n}\left(1 - \dfrac{2i}{n}\right)$

18. $S_n = \sum_{i=1}^{n} \dfrac{2n + 3i}{n^2}$ *** 19.** $S_n = \sum_{j=1}^{n} \dfrac{1}{n}\sqrt{1 - (j/n)^2}$

5.3 The Definite Integral

In this section we generalize and make more precise the procedure used for finding areas developed in Section 5.2, and we use it to define the *definite integral* of a function f on an interval I. Let us assume, for the time being, that $f(x)$ is defined and continuous on the closed, finite interval $[a, b]$. We no longer assume that the values of f are not negative.

Partitions and Riemann Sums

Let P be a finite set of points arranged in order between a and b on the real line, say

$$P = \{x_0, \, x_1, \, x_2, \, x_3, \, \ldots, \, x_{n-1}, \, x_n\},$$

where $a = x_0 < x_1 < x_2 < x_3 < \cdots < x_{n-1} < x_n = b$. Such a set P is called a **partition** of $[a, b]$; it divides $[a, b]$ into n subintervals of which the ith is $[x_{i-1}, x_i]$. We call these the subintervals of the partition P. The number n depends on the particular partition, so we write $n = n(P)$. The length of the ith subinterval of P is

$$\Delta x_i = x_i - x_{i-1},$$

and we call the greatest of these numbers Δx_i the **norm** of the partition P and denote it $\|P\|$:

$$\|P\| = \max_{1 \le i \le n} \Delta x_i.$$

Since f is continuous on each subinterval $[x_{i-1}, x_i]$ of P, it takes on maximum and minimum values at points of that interval (by Theorem 8 of Section 1.4). Thus there are numbers l_i and u_i in $[x_{i-1}, x_i]$ such that

$$f(l_i) \le f(x) \le f(u_i) \qquad \text{whenever } x_{i-1} \le x \le x_i.$$

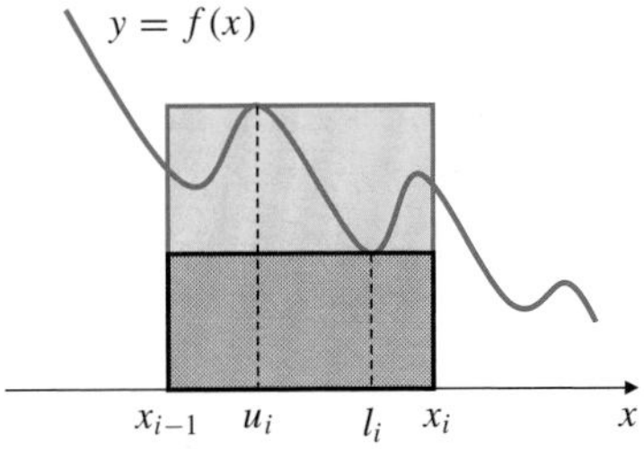

If $f(x) \ge 0$ on $[a, b]$, then $f(l_i) \, \Delta x_i$ and $f(u_i) \, \Delta x_i$ represent the areas of rectangles having the interval $[x_{i-1}, x_i]$ on the x-axis as base, and having tops passing through the lowest and highest points, respectively, on the graph of f on that interval. (See Figure 5.10.) If A_i is that part of the area under $y = f(x)$ and above the x-axis that lies in the vertical strip between $x = x_{i-1}$ and $x = x_i$, then

$$f(l_i) \, \Delta x_i \le A_i \le f(u_i) \, \Delta x_i.$$

If f can have negative values, then one or both of $f(l_i) \, \Delta x_i$ and $f(u_i) \, \Delta x_i$ can be negative and will then represent the negative of the area of a rectangle lying below the x-axis. In any event, we always have $f(l_i) \, \Delta x_i \le f(u_i) \, \Delta x_i$.

$y = f(x)$

$x_{i-1} \quad u_i \qquad l_i \quad x_i \qquad x$

Figure 5.10

DEFINITION 2

> **Upper and lower Riemann sums**
>
> The **lower (Riemann) sum** $L(f, P)$ and the **upper (Riemann) sum** $U(f, P)$ for the function f and the partition P are defined by:
>
> $$L(f, P) = f(l_1) \, \Delta x_1 + f(l_2) \, \Delta x_2 + \cdots + f(l_n) \, \Delta x_n$$
>
> $$= \sum_{i=1}^{n} f(l_i) \, \Delta x_i,$$
>
> $$U(f, P) = f(u_1) \, \Delta x_1 + f(u_2) \, \Delta x_2 + \cdots + f(u_n) \, \Delta x_n$$
>
> $$= \sum_{i=1}^{n} f(u_i) \Delta x_i.$$

Figure 5.11 illustrates these Riemann sums as sums of *signed* areas of rectangles; any such areas that lie below the x-axis are counted as negative.

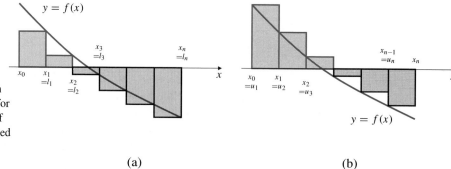

Figure 5.11 (a) A lower Riemann sum and (b) an upper Riemann sum for a decreasing function f. The areas of rectangles shaded in colour are counted as positive; those shaded in grey are counted as negative

(a)

(b)

Example 1 Calculate lower and upper Riemann sums for the function $f(x) = 1/x$ on the interval $[1, 2]$, corresponding to the partition P of $[1, 2]$ into four subintervals of equal length.

Solution The partition P consists of the points $x_0 = 1$, $x_1 = 5/4$, $x_2 = 3/2$, $x_3 = 7/4$, and $x_4 = 2$. Since $1/x$ is decreasing on $[1, 2]$, its minimum and maximum values on the ith subinterval $[x_{i-1}, x_i]$ are $1/x_i$ and $1/x_{i-1}$, respectively. Thus, the lower and upper Riemann sums are

$$L(f, P) = \frac{1}{4}\left(\frac{4}{5} + \frac{2}{3} + \frac{4}{7} + \frac{1}{2}\right) = \frac{533}{840} \approx 0.6345$$

$$U(f, P) = \frac{1}{4}\left(1 + \frac{4}{5} + \frac{2}{3} + \frac{4}{7}\right) = \frac{319}{420} \approx 0.7595.$$

Example 2 Calculate the lower and upper Riemann sums for the function $f(x) = x^2$ on the interval $[0, a]$ (where $a > 0$), corresponding to the partition P_n of $[0, a]$ into n subintervals of equal length.

Solution Each subinterval of P_n has length $\Delta x = a/n$, and the division points are given by $x_i = ia/n$ for $i = 0, 1, 2, \ldots, n$. Since x^2 is increasing on $[0, a]$, its minimum and maximum values over the ith subinterval $[x_{i-1}, x_i]$ occur at $l_i = x_{i-1}$ and $u_i = x_i$, respectively. Thus, the lower Riemann sum of f for P_n is

$$L(f, P_n) = \sum_{i=1}^{n}(x_{i-1})^2 \Delta x = \frac{a^3}{n^3}\sum_{i=1}^{n}(i-1)^2$$

$$= \frac{a^3}{n^3}\sum_{j=0}^{n-1}j^2 = \frac{a^3}{n^3}\frac{(n-1)n(2(n-1)+1)}{6} = \frac{(n-1)(2n-1)a^3}{6n^2},$$

where we have used Theorem 1(c) of Section 5.1 to evaluate the sum of squares.

Similarly, the upper Riemann sum is

$$U(f, P_n) = \sum_{i=1}^{n} (x_i)^2 \Delta x$$

$$= \frac{a^3}{n^3} \sum_{i=1}^{n} i^2 = \frac{a^3}{n^3} \frac{n(n+1)(2n+1)}{6} = \frac{(n+1)(2n+1)a^3}{6n^2}.$$

∎

The Definite Integral

If we calculate $L(f, P)$ and $U(f, P)$ for partitions P having more and more points spaced closer and closer together, we expect that, in the limit, these Riemann sums will converge to a common value that will be the area bounded by $y = f(x)$, $y = 0$, $x = a$, and $x = b$ if $f(x) \geq 0$ on $[a, b]$. This is indeed the case, but we cannot fully prove it yet.

If P_1 and P_2 are two partitions of $[a, b]$ such that every point of P_1 also belongs to P_2, then we say that P_2 is a **refinement** of P_1. It is not difficult to show that in this case

$$L(f, P_1) \leq L(f, P_2) \leq U(f, P_2) \leq U(f, P_1);$$

adding more points to a partition increases the lower sum and decreases the upper sum. (See Exercise 18 at the end of this section.) Given any two partitions, P_1 and P_2, we can form their **common refinement** P, which consists of all of the points of P_1 and P_2. Thus,

$$L(f, P_1) \leq L(f, P) \leq U(f, P) \leq U(f, P_2).$$

Hence, every lower sum is less than or equal to every upper sum. Since the real numbers are complete, there must exist *at least one* real number I such that

$$L(f, P) \leq I \leq U(f, P), \qquad \text{for every partition } P.$$

If there is *only one* such number, we will call it the definite integral of f on $[a, b]$.

DEFINITION 3

> **The definite integral**
>
> Suppose there is exactly one number I such that for every partition P of $[a, b]$ we have
>
> $$L(f, P) \leq I \leq U(f, P).$$
>
> Then we say that the function f is **integrable** on $[a, b]$, and we call I the **definite integral** of f on $[a, b]$. The definite integral is denoted by the symbol
>
> $$I = \int_a^b f(x) \, dx.$$

We stress at once that the definite integral of $f(x)$ over $[a, b]$ is a *number*; it is not a function of x. It depends on the numbers a and b and on the particular function f, but not on the variable x (which is a **dummy variable** like the variable i in the sum $\sum_{i=1}^{n} f(i)$). You can replace x with any other variable without changing the value of the integral:

$$\int_a^b f(x)\, dx = \int_a^b f(t)\, dt.$$

The various parts of the symbol $\int_a^b f(x)\, dx$ have their own names:

(i) $\int$ is called the **integral sign**; it resembles the letter S since it represents the limit of a sum.

(ii) a and b are called the **limits of integration**; a is the **lower limit**, b is the **upper limit**.

(iii) The function f is the **integrand**; x is the **variable of integration**.

(iv) dx is the **differential** of x. It replaces Δx in the Riemann sums. If an integrand depends on more than one variable, the differential tells you which one is the variable of integration.

Example 3 Show that $f(x) = x^2$ is integrable over the interval $[0, a]$, where $a > 0$, and evaluate $\int_0^a x^2\, dx$.

Solution We evaluate the limits as $n \to \infty$ of the lower and upper sums of f over $[0, a]$ obtained in Example 2 above.

$$\lim_{n \to \infty} L(f, P_n) = \lim_{n \to \infty} \frac{(n-1)(2n-1)a^3}{6n^2} = \frac{a^3}{3}$$

$$\lim_{n \to \infty} U(f, P_n) = \lim_{n \to \infty} \frac{(n+1)(2n+1)a^3}{6n^2} = \frac{a^3}{3}$$

If $L(f, P_n) \leq I \leq U(f, P_n)$, we must have $I = a^3/3$. Thus, $f(x) = x^2$ is integrable over $[0, a]$, and

$$\int_0^a f(x)\, dx = \int_0^a x^2\, dx = \frac{a^3}{3}.$$

For all partitions P of $[a, b]$ we have

$$L(f, P) \leq \int_a^b f(x)\, dx \leq U(f, P).$$

If $f(x) \geq 0$ on $[a, b]$, then the area of the region R bounded by the graph of $y = f(x)$, the x-axis, and the lines $x = a$ and $x = b$ is A square units, where $A = \int_a^b f(x)\, dx$. If $f(x) \leq 0$ on $[a, b]$, the area of R is $-\int_a^b f(x)\, dx$ square units. For general f, $\int_a^b f(x)\, dx$ is the area of that part of R lying above the x-axis minus the area of that part lying below the x-axis. (See Figure 5.12.) You can think of $\int_a^b f(x)\, dx$ as a "sum" of "areas" of infinitely many rectangles with heights $f(x)$ and "infinitesimally small widths" dx; it is a limit of the upper and lower Riemann sums.

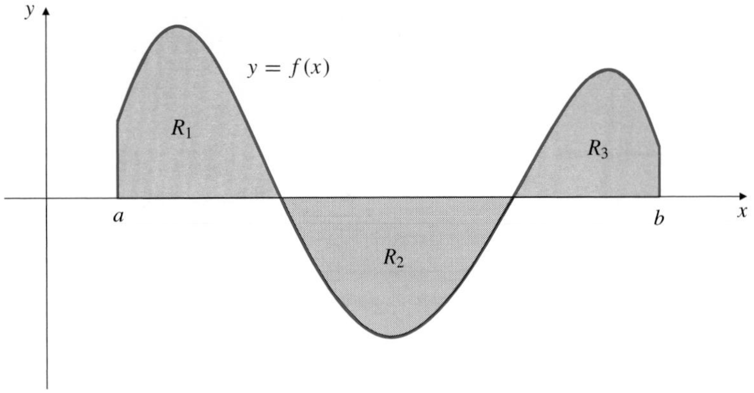

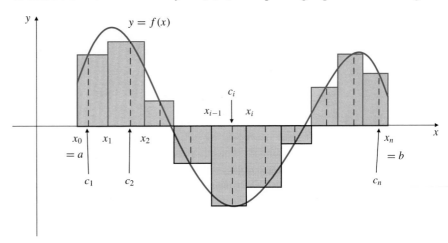

Figure 5.12 $\displaystyle\int_a^b f(x)\,dx$ equals

area R_1 − area R_2 + area R_3

General Riemann Sums

Let $P = \{x_0, x_1, x_2, \ldots, x_n\}$, where $a = x_0 < x_1 < x_2 < \cdots < x_n = b$, be a partition of $[a, b]$ having norm $\|P\| = \max_{1 \le i \le n} \Delta x_i$. In each subinterval $[x_{i-1}, x_i]$ of P pick a point c_i (called a *tag*). Let $c = (c_1, c_2, \ldots, c_n)$ denote the set of these tags. The sum

$$R(f, P, c) = \sum_{i=1}^{n} f(c_i)\,\Delta x_i$$

$$= f(c_1)\,\Delta x_1 + f(c_2)\,\Delta x_2 + f(c_3)\,\Delta x_3 + \cdots + f(c_n)\,\Delta x_n$$

is called the **Riemann sum** of f on $[a, b]$ corresponding to partition P and tags c.

Figure 5.13   The Riemann sum $R(f, P, c)$ is the sum of areas of the rectangles shaded in colour minus the sum of the areas of the rectangles shaded in grey

Note in Figure 5.13 that $R(f, P, c)$ is a sum of *signed* areas of rectangles between the x-axis and the curve $y = f(x)$. For any choice of the tags c, the Riemann sum $R(f, P, c)$ satisfies

$$L(f, P) \le R(f, P, c) \le U(f, P).$$

Therefore, if f is integrable on $[a, b]$, then its integral is the limit of such Riemann sums, where the limit is taken as the number $n(P)$ of subintervals of P increases to infinity in such a way that the lengths of all the subintervals approach zero. That is,

$$\lim_{\substack{n(P)\to\infty \\ \|P\|\to 0}} R(f, P, c) = \int_a^b f(x)\,dx.$$

As we will see in Chapter 7, many applications of integration depend on recognizing that a limit of Riemann sums is a definite integral.

THEOREM **2**

If f is continuous on $[a, b]$, then f is integrable on $[a, b]$.

As noted above, we cannot prove this theorem in its full generality now. The proof makes use of the completeness property of the real numbers and is given in Appendix III. We can, however, make the following observation. In order to prove that f is integrable on $[a, b]$, it is sufficient that, for any given positive number ϵ, we should be able to find a partition P of $[a, b]$ for which $U(f, P) - L(f, P) < \epsilon$. This condition prevents there being more than one number I that is both greater than every lower sum and less than every upper sum. It is not difficult to find such a partition if the function f is nondecreasing (or if it is nonincreasing) on $[a, b]$. (See Exercise 17 at the end of this section.) Therefore, nondecreasing and nonincreasing continuous functions are integrable; so, therefore, is any continuous function that is the sum of a nondecreasing and a nonincreasing function. This class of functions includes any continuous functions we are likely to encounter in concrete applications of calculus but, unfortunately, does not include all continuous functions.

In Section 5.4 we will extend the definition of the definite integral to certain kinds of functions that are not continuous.

Example 4 Express the limit $\displaystyle\lim_{n\to\infty} \sum_{i=1}^n \frac{2}{n}\left(1 + \frac{2i-1}{n}\right)^{1/3}$ as a definite integral.

Solution We want to interpret the sum as a Riemann sum for $f(x) = (1+x)^{1/3}$. The factor $2/n$ suggests that the interval of integration has length 2 and is partitioned into n equal subintervals, each of length $2/n$. Let $c_i = (2i-1)/n$ for $i = 1, 2, 3, \ldots, n$. As $n \to \infty$, $c_1 = 1/n \to 0$ and $c_n = (2n-1)/n \to 2$. Thus, the interval is $[0, 2]$, and the points of the partition are $x_i = 2i/n$. Observe that $x_{i-1} = (2i-2)/n < c_i < 2i/n = x_i$ for each i, so that the sum is indeed a Riemann sum for $f(x)$ over $[0, 2]$. Since f is continuous on that interval, it is integrable there, and

$$\lim_{n\to\infty} \sum_{i=1}^n \frac{2}{n}\left(1 + \frac{2i-1}{n}\right)^{1/3} = \int_0^2 (1+x)^{1/3}\,dx.$$

Exercises 5.3

In Exercises 1–6, let P_n denote the partition of the given interval $[a, b]$ into n subintervals of equal length $\Delta x_i = (b-a)/n$. Evaluate $L(f, P_n)$ and $U(f, P_n)$ for the given functions f and the given values of n.

1. $f(x) = x$ on $[0, 2]$, with $n = 8$

2. $f(x) = x^2$ on $[0, 4]$, with $n = 4$

3. $f(x) = e^x$ on $[-2, 2]$, with $n = 4$

4. $f(x) = \ln x$ on $[1, 2]$, with $n = 5$

5. $f(x) = \sin x$ on $[0, \pi]$, with $n = 6$

6. $f(x) = \cos x$ on $[0, 2\pi]$, with $n = 4$

In Exercises 7–10, calculate $L(f, P_n)$ and $U(f, P_n)$ for the given function f over the given interval $[a, b]$, where P_n is the partition of the interval into n subintervals of equal length $\Delta x = (b - a)/n$. Show that

$$\lim_{n\to\infty} L(f, P_n) = \lim_{n\to\infty} U(f, P_n).$$

Hence f is integrable on $[a, b]$. (Why?) What is $\int_a^b f(x)\,dx$?

7. $f(x) = x$, $[a, b] = [0, 1]$

8. $f(x) = 1 - x$, $[a, b] = [0, 2]$

9. $f(x) = x^3$, $[a, b] = [0, 1]$

10. $f(x) = e^x$, $[a, b] = [0, 3]$

In Exercises 11–16, express the given limit as a definite integral.

11. $\lim_{n\to\infty} \sum_{i=1}^{n} \frac{1}{n}\sqrt{\frac{i}{n}}$

12. $\lim_{n\to\infty} \sum_{i=1}^{n} \frac{1}{n}\sqrt{\frac{i-1}{n}}$

13. $\lim_{n\to\infty} \sum_{i=1}^{n} \frac{\pi}{n}\sin\left(\frac{\pi i}{n}\right)$

14. $\lim_{n\to\infty} \sum_{i=1}^{n} \frac{2}{n}\ln\left(1 + \frac{2i}{n}\right)$

15. $\lim_{n\to\infty} \sum_{i=1}^{n} \frac{1}{n}\tan^{-1}\left(\frac{2i-1}{2n}\right)$

16. $\lim_{n\to\infty} \sum_{i=1}^{n} \frac{n}{n^2 + i^2}$

17. If f is continuous and nondecreasing on $[a, b]$, and P_n is the partition of $[a, b]$ into n subintervals of equal length ($\Delta x_i = (b - a)/n$ for $1 \le i \le n$), show that

$$U(f, P_n) - L(f, P_n) = \frac{(b - a)\big(f(b) - f(a)\big)}{n}.$$

Since we can make the right side as small as we please by choosing n large enough, f must be integrable on $[a, b]$.

18. Let $P = \{a = x_0 < x_1 < x_2 < \cdots < x_n = b\}$ be a partition of $[a, b]$, and let P' be a refinement of P having one more point, x', satisfying, say, $x_{i-1} < x' < x_i$ for some i between 1 and n. Show that

$$L(f, P) \le L(f, P') \le U(f, P') \le U(f, P)$$

for any continuous function f. (Hint: consider the maximum and minimum values of f on the intervals $[x_{i-1}, x_i]$, $[x_{i-1}, x']$, and $[x', x_i]$.) Hence deduce that

$$L(f, P) \le L(f, P'') \le U(f, P'') \le U(f, P) \text{ if } P''$$

is any refinement of P.

5.4 Properties of the Definite Integral

It is convenient to extend the definition of the definite integral $\int_a^b f(x)\,dx$ to allow $a = b$ and $a > b$ as well as $a < b$. The extension still involves partitions P having $x_0 = a$ and $x_n = b$ with intermediate points occurring in order between these end points, so that if $a = b$, then we must have $\Delta x_i = 0$ for every i, and hence the integral is zero. If $a > b$, we have $\Delta x_i < 0$ for each i, so the integral will be negative for positive functions f and vice versa.

Some of the most important properties of the definite integral are summarized in the following theorem.

THEOREM 3

Let f and g be integrable on an interval containing the points a, b, and c. Then

(a) An integral over an interval of zero length is zero.

$$\int_a^a f(x)\,dx = 0.$$

(b) Reversing the limits of integration changes the sign of the integral.

$$\int_b^a f(x)\,dx = -\int_a^b f(x)\,dx.$$

(c) An integral depends linearly on the integrand. If A and B are constants, then

$$\int_a^b \left(Af(x) + Bg(x)\right)dx = A\int_a^b f(x)\,dx + B\int_a^b g(x)\,dx.$$

(d) An integral depends additively on the interval of integration.

$$\int_a^b f(x)\,dx + \int_b^c f(x)\,dx = \int_a^c f(x)\,dx.$$

(e) If $a \le b$ and $f(x) \le g(x)$ for $a \le x \le b$, then

$$\int_a^b f(x)\,dx \le \int_a^b g(x)\,dx.$$

(f) The **triangle inequality** for definite integrals. If $a \le b$, then

$$\left|\int_a^b f(x)\,dx\right| \le \int_a^b |f(x)|\,dx.$$

(g) The integral of an odd function over an interval symmetric about zero is zero. If f is an odd function $(f(-x) = -f(x))$, then

$$\int_{-a}^a f(x)\,dx = 0.$$

(h) The integral of an even function over an interval symmetric about zero is twice the integral over the positive half of the interval. If f is an even function $(f(-x) = f(x))$, then

$$\int_{-a}^a f(x)\,dx = 2\int_0^a f(x)\,dx.$$

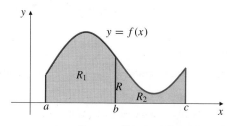

area R_1 + area R_2 = area R

$$\int_a^b f(x)\,dx + \int_b^c f(x)\,dx = \int_a^c f(x)\,dx$$

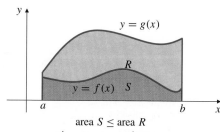

area $S \le$ area R

$$\int_a^b f(x)\,dx \le \int_a^b g(x)\,dx$$

Figure 5.14

(a) Property (d) of Theorem 3

(b) Property (e) of Theorem 3

All these properties can be deduced from the definition of the definite integral. Most of them should appear intuitively reasonable if you regard the integrals as representing (signed) areas. For instance, properties (d) and (e) are, respectively, properties (v) and (iv) of areas mentioned in the first paragraph of Section 5.2. (See Figure 5.14.) Property (f) is a generalization of the triangle inequality for numbers:

$$|x + y| \le |x| + |y|, \quad \text{or more generally,} \quad \left| \sum_{i=1}^{n} x_i \right| \le \sum_{i=1}^{n} |x_i| .$$

It follows from property (e) (assuming that $|f|$ is integrable on $[a, b]$), since $-|f(x)| \le f(x) \le |f(x)|$. The symmetry properties (g) and (h), which are illustrated in Figure 5.15, are particularly useful and should always be kept in mind when evaluating definite integrals because they can save much unnecessary work.

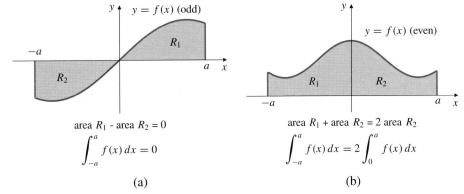

Figure 5.15

(a) Property (g) of Theorem 3

(b) Property (h) of Theorem 3

area R_1 - area $R_2 = 0$

$$\int_{-a}^{a} f(x)\, dx = 0$$

(a)

area R_1 + area $R_2 = 2$ area R_2

$$\int_{-a}^{a} f(x)\, dx = 2 \int_{0}^{a} f(x)\, dx$$

(b)

As yet we have no easy method for evaluating definite integrals. However, some such integrals can be simplified by using various properties in Theorem 3, and others can be interpreted as known areas.

Example 1 Evaluate

(a) $\displaystyle\int_{-2}^{2} (2 + 5x)\, dx$,   (b) $\displaystyle\int_{0}^{3} (2 + x)\, dx$,   and   (c) $\displaystyle\int_{-3}^{3} \sqrt{9 - x^2}\, dx$.

Solution See Figures 5.16–5.18.

(a) By the linearity property (c), $\int_{-2}^{2}(2+5x)\, dx = \int_{-2}^{2} 2\, dx + 5 \int_{-2}^{2} x\, dx$. The first integral on the right represents the area of a rectangle of width 4 and height 2 (Figure 5.16), so it has value 8. The second integral on the right is 0 because its integrand is odd and the interval is symmetric about 0. Thus,

$$\int_{-2}^{2} (2 + 5x)\, dx = 8 + 0 = 8.$$

(b) $\int_{0}^{3}(2 + x)\, dx$ represents the area of the trapezoid in Figure 5.17. Adding the areas of the rectangle and triangle comprising this trapezoid, we get

$$\int_{0}^{3} (2 + x)\, dx = (3 \times 2) + \frac{1}{2}(3 \times 3) = \frac{21}{2}.$$

While areas are measured in squared units of length, definite integrals are numbers and have no units. Even when you use an area to find an integral, do not quote units for the integral.

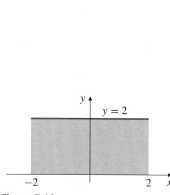

Figure 5.16

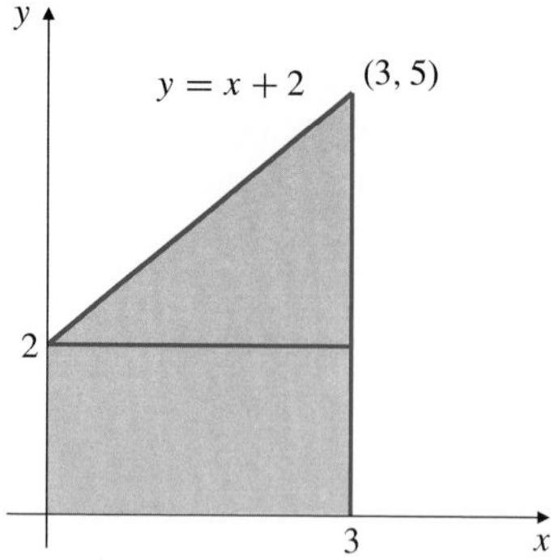

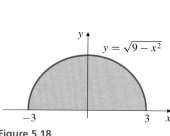

$y = x + 2$ $(3, 5)$

Figure 5.17

$y = \sqrt{9 - x^2}$

Figure 5.18

(c) $\int_{-3}^{3} \sqrt{9 - x^2}\, dx$ represents the area of a semicircle of radius 3 (Figure 5.18), so

$$\int_{-3}^{3} \sqrt{9 - x^2}\, dx = \frac{1}{2}\pi(3^2) = \frac{9\pi}{2}.$$

A Mean-Value Theorem for Integrals

Let f be a function continuous on the interval $[a, b]$. Then f assumes a minimum value m and a maximum value M on the interval, say at points $x = l$ and $x = u$, respectively:

$$m = f(l) \le f(x) \le f(u) = M \qquad \text{for all } x \text{ in } [a, b].$$

For the 2-point partition P of $[a, b]$ having $x_0 = a$ and $x_1 = b$, we have

$$m(b - a) = L(f, P) \le \int_{a}^{b} f(x)\, dx \le U(f, P) = M(b - a).$$

Therefore,

$$f(l) = m \le \frac{1}{b - a} \int_{a}^{b} f(x)\, dx \le M = f(u).$$

By the Intermediate-Value Theorem, $f(x)$ must take on every value between the two values $f(l)$ and $f(u)$ at some point between l and u (Figure 5.19). Hence, there is a number c between l and u such that

$$f(c) = \frac{1}{b - a} \int_{a}^{b} f(x)\, dx,$$

that is, $\int_{a}^{b} f(x)\, dx$ is equal to the area $(b - a)f(c)$ of a rectangle with base width $b - a$ and height $f(c)$ for some c between a and b. This is the Mean-Value Theorem for integrals.

THEOREM **4**

The Mean-Value Theorem for integrals

If f is continuous on $[a, b]$, then there exists a point c in $[a, b]$ such that

$$\int_{a}^{b} f(x)\, dx = (b - a)f(c).$$

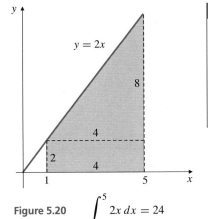

Figure 5.19   Half of the area between $y = f(x)$ and the horizontal line $y = f(c)$ lies above the line and the other half lies below the line

Observe in Figure 5.19 that the area below the curve $y = f(x)$ and above the line $y = f(c)$ is equal to the area above $y = f(x)$ and below $y = f(c)$. In this sense, $f(c)$ is the average value of the function $f(x)$ on the interval $[a, b]$.

DEFINITION 4

> **Average value of a function**
>
> If f is integrable on $[a, b]$, then the **average value** or **mean value** of f on $[a, b]$, denoted by $\bar{f}$, is
>
> $$\bar{f} = \frac{1}{b-a} \int_a^b f(x)\, dx.$$

Example 2   Find the average value of $f(x) = 2x$ on the interval $[1, 5]$.

Solution   The average value (see Figure 5.20) is

$$\bar{f} = \frac{1}{5-1} \int_1^5 2x\, dx = \frac{1}{4}\left(4 \times 2 + \frac{1}{2}(4 \times 8)\right) = 6.$$

Figure 5.20 $\int_1^5 2x\, dx = 24$

Definite Integrals of Piecewise Continuous Functions

The definition of integrability and the definite integral given above can be extended to a wider class than just continuous functions. One simple but very important extension is to the class of *piecewise continuous functions*.

Consider the graph $y = f(x)$ shown in Figure 5.21(a). Although f is not continuous at all points in $[a, b]$ (it is discontinuous at c_1 and c_2), clearly the region lying under the graph and above the x-axis between $x = a$ and $x = b$ does have an area. We would like to represent this area as

$$\int_a^{c_1} f(x)\,dx + \int_{c_1}^{c_2} f(x)\,dx + \int_{c_2}^b f(x)\,dx.$$

This is reasonable because there are continuous functions on $[a, c_1]$, $[c_1, c_2]$ and $[c_2, b]$ equal to $f(x)$ on the corresponding open intervals, $]a, c_1[$, $]c_1, c_2[$ and $]c_2, b[$.

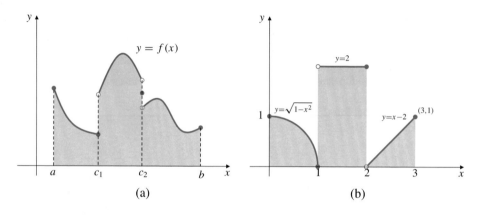

Figure 5.21   Two piecewise continuous functions

(a)   (b)

DEFINITION   **5**

Piecewise continuous functions

Let $c_0 < c_1 < c_2 < \cdots < c_n$ be a finite set of points on the real line. A function f defined on $[c_0, c_n]$ except possibly at some of the points c_i, $(0 \le i \le n)$, is called **piecewise continuous** on that interval if for each i $(1 \le i \le n)$ there exists a function F_i continuous on the *closed* interval $[c_{i-1}, c_i]$ such that

$$f(x) = F_i(x) \qquad \text{on the } \textit{open} \text{ interval} \quad]c_{i-1}, c_i[.$$

In this case, we define the definite integral of f from c_0 to c_n to be

$$\int_{c_0}^{c_n} f(x)\,dx = \sum_{i=1}^{n} \int_{c_{i-1}}^{c_i} F_i(x)\,dx.$$

Example 3   Find $\displaystyle\int_0^3 f(x)\,dx$, where $f(x) = \begin{cases} \sqrt{1-x^2} & \text{if } 0 \le x \le 1 \\ 2 & \text{if } 1 < x \le 2 \\ x - 2 & \text{if } 2 < x \le 3. \end{cases}$

Solution   The value of the integral is the sum of the shaded areas in Figure 5.21(b):

$$\int_0^3 f(x)\,dx = \int_0^1 \sqrt{1-x^2}\,dx + \int_1^2 2\,dx + \int_2^3 (x-2)\,dx$$

$$= \left(\frac{1}{4} \times \pi \times 1^2\right) + (2 \times 1) + \left(\frac{1}{2} \times 1 \times 1\right) = \frac{\pi + 10}{4}.$$

Exercises 5.4

1. Simplify $\displaystyle\int_a^b f(x)\,dx + \int_b^c f(x)\,dx + \int_c^a f(x)\,dx$.

2. Simplify $\displaystyle\int_0^2 3f(x)\,dx + \int_1^3 3f(x)\,dx - \int_0^3 2f(x)\,dx$
$$- \int_1^2 3f(x)\,dx.$$

Evaluate the integrals in Exercises 3–16 by using the properties of the definite integral and interpreting integrals as areas.

3. $\displaystyle\int_{-2}^2 (x+2)\,dx$

4. $\displaystyle\int_0^2 (3x+1)\,dx$

5. $\displaystyle\int_a^b x\,dx$

6. $\displaystyle\int_{-1}^2 (1-2x)\,dx$

7. $\displaystyle\int_{-\sqrt{2}}^{\sqrt{2}} \sqrt{2-t^2}\,dt$

8. $\displaystyle\int_{-\sqrt{2}}^0 \sqrt{2-x^2}\,dx$

9. $\displaystyle\int_{-\pi}^{\pi} \sin(x^3)\,dx$

10. $\displaystyle\int_{-a}^a (a-|s|)\,ds$

11. $\displaystyle\int_{-1}^1 (u^5 - 3u^3 + \pi)\,du$

12. $\displaystyle\int_0^2 \sqrt{2x - x^2}\,dx$

13. $\displaystyle\int_{-4}^4 (e^x - e^{-x})\,dx$

14. $\displaystyle\int_{-3}^3 (2+t)\sqrt{9-t^2}\,dt$

∗ 15. $\displaystyle\int_0^1 \sqrt{4-x^2}\,dx$

∗ 16. $\displaystyle\int_1^2 \sqrt{4-x^2}\,dx$

Given that $\displaystyle\int_0^a x^2\,dx = \frac{a^3}{3}$, evaluate the integrals in Exercises 17–22.

17. $\displaystyle\int_0^2 6x^2\,dx$

18. $\displaystyle\int_2^3 (x^2 - 4)\,dx$

19. $\displaystyle\int_{-2}^2 (4 - t^2)\,dt$

20. $\displaystyle\int_0^2 (v^2 - v)\,dv$

21. $\displaystyle\int_0^1 (x^2 + \sqrt{1-x^2})\,dx$

22. $\displaystyle\int_{-6}^6 x^2(2 + \sin x)\,dx$

The definition of $\ln x$ as an area in Section 3.3 implies that

$$\int_1^x \frac{1}{t}\,dt = \ln x$$

for $x > 0$. Use this to evaluate the integrals in Exercises 23–26.

23. $\displaystyle\int_1^2 \frac{1}{x}\,dx$

24. $\displaystyle\int_2^4 \frac{1}{t}\,dt$

25. $\displaystyle\int_{1/3}^1 \frac{1}{t}\,dt$

26. $\displaystyle\int_{1/4}^3 \frac{1}{s}\,ds$

Find the average values of the functions in Exercises 27–32 over the given intervals.

27. $f(x) = x + 2$ over $[0, 4]$

28. $g(x) = x + 2$ over $[a, b]$

29. $f(t) = 1 + \sin t$ over $[-\pi, \pi]$

30. $k(x) = x^2$ over $[0, 3]$

31. $f(x) = \sqrt{4 - x^2}$ over $[0, 2]$

32. $g(s) = 1/s$ over $[1/2, 2]$

Piecewise continuous functions

33. Evaluate $\displaystyle\int_{-1}^2 \operatorname{sgn} x\,dx$. Recall that $\operatorname{sgn} x$ is 1 if $x > 0$ and -1 if $x < 0$.

34. Find $\displaystyle\int_{-3}^2 f(x)\,dx$, where $f(x) = \begin{cases} 1+x & \text{if } x < 0 \\ 2 & \text{if } x \geq 0. \end{cases}$

35. Find $\displaystyle\int_0^2 g(x)\,dx$, where $g(x) = \begin{cases} x^2 & \text{if } 0 \leq x \leq 1 \\ x & \text{if } 1 < x \leq 2. \end{cases}$

36. Evaluate $\displaystyle\int_0^3 |2 - x|\,dx$.

∗ 37. Evaluate $\displaystyle\int_0^2 \sqrt{4 - x^2}\,\operatorname{sgn}(x - 1)\,dx$.

38. Evaluate $\displaystyle\int_0^{3.5} \lfloor x \rfloor\,dx$, where $\lfloor x \rfloor$ is the greatest integer less than or equal to x. (See Example 10 of section P.5.)

Evaluate the integrals in Exercises 39–40 by inspecting the graphs of the integrands.

39. $\displaystyle\int_{-3}^4 \left(|x+1| - |x-1| + |x+2| \right) dx$

40. $\displaystyle\int_0^3 \frac{x^2 - x}{|x - 1|}\,dx$

41. Find the average value of the function $f(x) = |x+1|\operatorname{sgn} x$ on the interval $[-2, 2]$.

42. If $a < b$ and f is continuous on $[a, b]$, show that
$$\int_a^b \left(f(x) - \bar{f} \right) dx = 0.$$

43. Suppose that $a < b$ and f is continuous on $[a, b]$. Find the constant k that minimizes the integral $\displaystyle\int_a^b \left(f(x) - k \right)^2 dx$.

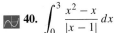

5.5   The Fundamental Theorem of Calculus

In this section we demonstrate the relationship between the definite integral defined in Section 5.3 and the indefinite integral (or general antiderivative) introduced in Section 2.10. A consequence of this relationship is that we will be able to calculate definite integrals of functions whose antiderivatives we can find.

In Section 3.3 we wanted to find a function whose derivative was $1/x$. We solved this problem by defining the desired function $(\ln x)$ in terms of the area under the graph of $y = 1/x$. This idea motivates, and is a special case of, the following theorem.

THEOREM 5

The Fundamental Theorem of Calculus

Suppose that the function f is continuous on an interval I containing the point a.

PART I. Let the function F be defined on I by

$$F(x) = \int_a^x f(t)\,dt.$$

Then F is differentiable on I and $F'(x) = f(x)$ there. Thus, F is an antiderivative of f on I:

$$\frac{d}{dx}\int_a^x f(t)\,dt = f(x).$$

PART II. If $G(x)$ is *any* antiderivative of $f(x)$ on I, so that $G'(x) = f(x)$ on I, then for any b in I we have

$$\int_a^b f(x)\,dx = G(b) - G(a).$$

PROOF   Using the definition of the derivative, we calculate

$$F'(x) = \lim_{h \to 0} \frac{F(x+h) - F(x)}{h}$$

$$= \lim_{h \to 0} \frac{1}{h}\left(\int_a^{x+h} f(t)\,dt - \int_a^x f(t)\,dt\right)$$

$$= \lim_{h \to 0} \frac{1}{h}\int_x^{x+h} f(t)\,dt \qquad \text{by Theorem 3(d)}$$

$$= \lim_{h \to 0} \frac{1}{h}hf(c) \qquad\qquad \text{for some } c = c(h) \text{ (depending on } h\text{)}$$
$$\qquad\qquad\qquad\qquad\qquad \text{between } x \text{ and } x+h \text{ (Theorem 4)}$$

$$= \lim_{c \to x} f(c) \qquad\qquad\quad \text{since } c \to x \text{ as } h \to 0$$

$$= f(x) \qquad\qquad\qquad\quad \text{since } f \text{ is continuous.}$$

Also, if $G'(x) = f(x)$, then $F(x) = G(x) + C$ on I for some constant C (by Theorem 13 of Section 2.6). Hence

$$\int_a^x f(t)\,dt = F(x) = G(x) + C.$$

Let $x = a$ and obtain $0 = G(a) + C$ via Theorem 3(a), so $C = -G(a)$. Now let $x = b$ to get

$$\int_a^b f(t)\,dt = G(b) + C = G(b) - G(a).$$

Of course, we can replace t with x (or any other variable) as the variable of integration on the left-hand side.

Remark   You should remember *both* conclusions of the Fundamental Theorem; they are both useful. Part I concerns the derivative of an integral; it tells you how to differentiate a definite integral with respect to its upper limit. Part II concerns the integral of a derivative; it tells you how to evaluate a definite integral if you can find an antiderivative of the integrand.

DEFINITION 6

To facilitate the evaluation of definite integrals using the Fundamental Theorem, we define the **evaluation symbol**:

$$F(x)\Big|_a^b = F(b) - F(a).$$

Thus

$$\int_a^b f(x)\,dx = \left(\int f(x)\,dx \right)\Big|_a^b,$$

where $\int f(x)\,dx$ denotes the indefinite integral or general antiderivative of f. (See Section 2.10.) When evaluating a definite integral this way, we will omit the constant of integration $(+C)$ from the indefinite integral because it cancels out in the subtraction:

$$(F(x) + C)\Big|_a^b = F(b) + C - (F(a) + C) = F(b) - F(a) = F(x)\Big|_a^b.$$

Any antiderivative of f can be used to calculate the definite integral.

Example 1   Evaluate   (a) $\displaystyle\int_0^a x^2\,dx$ and   (b) $\displaystyle\int_{-1}^2 (x^2 - 3x + 2)\,dx$.

Solution

(a) $\displaystyle\int_0^a x^2\,dx = \frac{1}{3}x^3\Big|_0^a = \frac{1}{3}a^3 - \frac{1}{3}0^3 = \frac{a^3}{3}$ (because $\dfrac{d}{dx}\dfrac{x^3}{3} = x^2$).

(b) $\displaystyle\int_{-1}^2 (x^2 - 3x + 2)\,dx = \left(\frac{1}{3}x^3 - \frac{3}{2}x^2 + 2x \right)\Big|_{-1}^2$

$$= \frac{1}{3}(8) - \frac{3}{2}(4) + 4 - \left(\frac{1}{3}(-1) - \frac{3}{2}(1) + (-2) \right) = \frac{9}{2}.$$

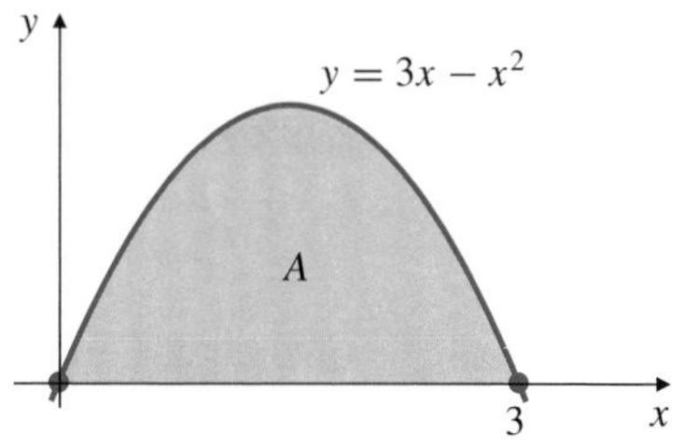

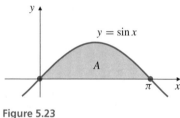

Figure 5.22

Example 2 Find the area A of the plane region lying above the x-axis and under the curve $y = 3x - x^2$.

Solution We need to find the points where the curve $y = 3x - x^2$ meets the x-axis. These are solutions of the equation

$$0 = 3x - x^2 = x(3 - x).$$

The only roots are $x = 0$ and $x = 3$. (See Figure 5.22.) Hence, the area of the region is given by

$$A = \int_0^3 (3x - x^2)\, dx = \left(\frac{3}{2}x^2 - \frac{1}{3}x^3\right)\Big|_0^3$$
$$= \frac{27}{2} - \frac{27}{3} - (0 - 0) = \frac{27}{6} = \frac{9}{2} \text{ square units.}$$

Figure 5.23

Example 3 Find the area under the curve $y = \sin x$, above $y = 0$ from $x = 0$ to $x = \pi$.

Solution The required area, illustrated in Figure 5.23, is

$$A = \int_0^\pi \sin x\, dx = -\cos x\Big|_0^\pi = -(-1 - (1)) = 2 \text{ square units.}$$

Note that while the definite integral is a pure number, an area is a geometric quantity that implicitly involves units. If the units along the x- and y-axes are, for example, metres, the area should be quoted in square metres (m^2). If units of length along the x-axis and y-axis are not specified, areas should be quoted in square units.

Example 4 Find the area of the region R lying above the line $y = 1$ and below the curve $y = 5/(x^2 + 1)$.

Solution The region R is shaded in Figure 5.24. To find the intersections of $y = 1$ and $y = 5/(x^2 + 1)$, we must solve these equations simultaneously:

$$1 = \frac{5}{x^2 + 1}$$

so $x^2 + 1 = 5$, $x^2 = 4$, and $x = \pm 2$.

The area A of the region R is the area under the curve $y = 5/(x^2 + 1)$ and above the x-axis between $x = -2$ and $x = 2$, minus the area of a rectangle of width 4 and height 1.

$$A = \int_{-2}^2 \frac{5}{x^2 + 1}\, dx - 4 = 2\int_0^2 \frac{5}{x^2 + 1}\, dx - 4$$
$$= 10 \tan^{-1} x\Big|_0^2 - 4 = 10 \tan^{-1} 2 - 4 \text{ square units.}$$

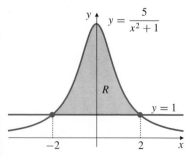

Figure 5.24

Observe the use of symmetry to replace the lower limit of integration by 0. It is easier to substitute 0 into the antiderivative than -2.

Example 5 Find the average value of $f(x) = e^{-x} + \cos x$ on the interval $[-\pi/2, 0]$.

Solution The average value is

$$\bar{f} = \frac{1}{0 - \left(-\dfrac{\pi}{2}\right)} \int_{-(\pi/2)}^{0} (e^{-x} + \cos x)\, dx$$

$$= \frac{2}{\pi} \left(-e^{-x} + \sin x\right)\Big|_{-(\pi/2)}^{0}$$

$$= \frac{2}{\pi} \left(-1 + 0 + e^{\pi/2} - (-1)\right) = \frac{2}{\pi}\, e^{\pi/2}.$$

Beware of integrals of the form $\int_a^b f(x)\, dx$, where f is not continuous at *all* points in the interval $[a, b]$. The Fundamental Theorem does not apply in such cases.

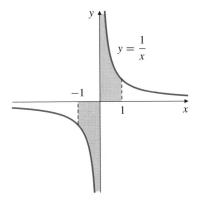

Figure 5.25

Example 6 We know that $\dfrac{d}{dx} \ln|x| = \dfrac{1}{x}$ if $x \neq 0$. It is *incorrect*, however, to state that

$$\int_{-1}^{1} \frac{dx}{x} = \ln|x|\,\Big|_{-1}^{1} = 0 - 0 = 0,$$

even though $1/x$ is an odd function. In fact, $1/x$ undefined and has no limit at $x = 0$, and it is not integrable on $[-1, 0]$ or $[0, 1]$ (Figure 5.25). Observe that

$$\lim_{c \to 0+} \int_c^1 \frac{1}{x}\, dx = \lim_{c \to 0+} -\ln c = \infty,$$

so both shaded regions in Figure 5.25 have infinite area. Integrals of this type are called **improper integrals**. We deal with them in Section 6.5.

We now give some examples illustrating the first conclusion of the Fundamental Theorem.

Example 7 Find the derivatives of the following functions:

(a) $F(x) = \displaystyle\int_x^3 e^{-t^2}\, dt$, (b) $G(x) = x^2 \displaystyle\int_{-4}^{5x} e^{-t^2}\, dt$, (c) $H(x) = \displaystyle\int_{x^2}^{x^3} e^{-t^2}\, dt$.

Solution The solutions involve applying the first conclusion of the Fundamental Theorem together with other differentiation rules.

(a) Observe that $F(x) = -\int_3^x e^{-t^2}\, dt$ (by Theorem 3(b)). Therefore, by the Fundamental Theorem, $F'(x) = -e^{-x^2}$.

(b) By the Product Rule and the Chain Rule,

$$G'(x) = 2x \int_{-4}^{5x} e^{-t^2}\, dt + x^2 \frac{d}{dx} \int_{-4}^{5x} e^{-t^2}\, dt$$

$$= 2x \int_{-4}^{5x} e^{-t^2}\, dt + x^2 e^{-(5x)^2}\,(5)$$

$$= 2x \int_{-4}^{5x} e^{-t^2}\, dt + 5x^2 e^{-25x^2}.$$

(c) Split the integral into a difference of integrals in each of which the variable x appears only in the upper limit.

$$H(x) = \int_0^{x^3} e^{-t^2}\, dt - \int_0^{x^2} e^{-t^2}\, dt$$

$$H'(x) = e^{-(x^3)^2}(3x^2) - e^{-(x^2)^2}(2x)$$

$$= 3x^2\, e^{-x^6} - 2x\, e^{-x^4}.$$

Parts (b) and (c) of Example 7 are examples of the following formulas that build the Chain Rule into the first conclusion of the Fundamental Theorem.

$$\frac{d}{dx}\int_a^{g(x)} f(t)\, dt = f\big(g(x)\big)\, g'(x)$$

$$\frac{d}{dx}\int_{h(x)}^{g(x)} f(t)\, dt = f\big(g(x)\big)\, g'(x) - f\big(h(x)\big)\, h'(x)$$

Example 8   Solve the **integral equation** $f(x) = 2 + 3\int_4^x f(t)\, dt$.

Solution   Differentiate the integral equation to get $f'(x) = 3f(x)$, the DE for exponential growth, having solution $f(x) = Ce^{3x}$. Now put $x = 4$ into the integral equation to get $f(4) = 2$. Hence $2 = Ce^{12}$, so $C = 2e^{-12}$. Therefore, the integral equation has solution $f(x) = 2e^{3x-12}$.

We conclude with an example showing how the Fundamental Theorem can be used to evaluate limits of Riemann sums.

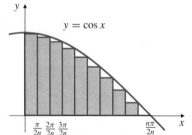

$y = \cos x$

Figure 5.26

Example 9   Evaluate $\displaystyle\lim_{n\to\infty} \frac{1}{n}\sum_{j=1}^{n}\cos\left(\frac{j\pi}{2n}\right)$.

Solution   The sum involves values of $\cos x$ at the right endpoints of the n subintervals of the partition

$$0,\quad \frac{\pi}{2n},\quad \frac{2\pi}{2n},\quad \frac{3\pi}{2n},\quad \ldots,\quad \frac{n\pi}{2n}$$

of the interval $[0, \pi/2]$. Since each of the subintervals of this partition has length $\pi/(2n)$, and since $\cos x$ is continuous on $[0, \pi/2]$, we have, expressing the limit of a Riemann sum as an integral (see Figure 5.26),

$$\lim_{n\to\infty} \frac{\pi}{2n}\sum_{j=1}^{n}\cos\left(\frac{j\pi}{2n}\right) = \int_0^{\pi/2} \cos x\, dx = \sin x\,\Big|_0^{\pi/2} = 1 - 0 = 1.$$

The given sum differs from the Riemann sum above only in that the factor $\pi/2$ is missing. Thus,

$$\lim_{n\to\infty} \frac{1}{n}\sum_{j=1}^{n}\cos\left(\frac{j\pi}{2n}\right) = \frac{2}{\pi}.$$

Exercises 5.5

Evaluate the definite integrals in Exercises 1–20.

1. $\displaystyle\int_0^2 x^3\, dx$

2. $\displaystyle\int_0^4 \sqrt{x}\, dx$

3. $\displaystyle\int_{1/2}^1 \frac{1}{x^2}\, dx$

4. $\displaystyle\int_{-2}^{-1} \left(\frac{1}{x^2} - \frac{1}{x^3}\right) dx$

5. $\displaystyle\int_{-1}^2 (3x^2 - 4x + 2)\, dx$

6. $\displaystyle\int_1^2 \left(\frac{2}{x^3} - \frac{x^3}{2}\right) dx$

7. $\displaystyle\int_{-2}^2 (x^2 + 3)^2\, dx$

8. $\displaystyle\int_4^9 \left(\sqrt{x} - \frac{1}{\sqrt{x}}\right) dx$

9. $\displaystyle\int_{-\pi/4}^{-\pi/6} \cos x\, dx$

10. $\displaystyle\int_0^{\pi/3} \sec^2 \theta\, d\theta$

11. $\displaystyle\int_{\pi/4}^{\pi/3} \sin \theta\, d\theta$

12. $\displaystyle\int_0^{2\pi} (1 + \sin u)\, du$

13. $\displaystyle\int_{-\pi}^{\pi} e^x\, dx$

14. $\displaystyle\int_{-2}^2 \left(e^x - e^{-x}\right) dx$

15. $\displaystyle\int_0^e a^x\, dx \quad (a > 0)$

16. $\displaystyle\int_{-1}^1 2^x\, dx$

17. $\displaystyle\int_{-1}^1 \frac{dx}{1 + x^2}$

18. $\displaystyle\int_0^{1/2} \frac{dx}{\sqrt{1 - x^2}}$

* 19. $\displaystyle\int_{-1}^1 \frac{dx}{\sqrt{4 - x^2}}$

* 20. $\displaystyle\int_{-2}^0 \frac{dx}{4 + x^2}$

Find the area of the region R specified in Exercises 21–32. It is helpful to make a sketch of the region.

21. Bounded by $y = x^4$, $y = 0$, $x = 0$, and $x = 1$

22. Bounded by $y = 1/x$, $y = 0$, $x = e$, and $x = e^2$

23. Above $y = x^2 - 4x$ and below the x-axis

24. Bounded by $y = 5 - 2x - 3x^2$, $y = 0$, $x = -1$, and $x = 1$

25. Bounded by $y = x^2 - 3x + 3$ and $y = 1$

26. Below $y = \sqrt{x}$ and above $y = \dfrac{x}{2}$

27. Above $y = x^2$ and to the right of $x = y^2$

28. Above $y = |x|$ and below $y = 12 - x^2$

29. Bounded by $y = x^{1/3} - x^{1/2}$, $y = 0$, $x = 0$, and $x = 1$

30. Under $y = e^{-x}$ and above $y = 0$ from $x = -a$ to $x = 0$

31. Below $y = 1 - \cos x$ and above $y = 0$ between two consecutive intersections of these graphs

32. Below $y = x^{-1/3}$ and above $y = 0$ from $x = 1$ to $x = 27$

Find the integrals of the piecewise continuous functions in Exercises 33–34.

33. $\displaystyle\int_0^{3\pi/2} |\cos x|\, dx$

34. $\displaystyle\int_1^3 \frac{\operatorname{sgn}(x - 2)}{x^2}\, dx$

In Exercises 35–38, find the average values of the given functions over the intervals specified.

35. $f(x) = 1 + x + x^2 + x^3$ over $[0, 2]$

36. $f(x) = e^{3x}$ over $[-2, 2]$

37. $f(x) = 2^x$ over $[0, 1/\ln 2]$

38. $g(t) = \begin{cases} 0 & \text{if } 0 \le t \le 1 \\ 1 & \text{if } 1 < t \le 3 \end{cases}$ over $[0, 3]$

Find the indicated derivatives in Exercises 39–46.

39. $\displaystyle\frac{d}{dx} \int_2^x \frac{\sin t}{t}\, dt$

40. $\displaystyle\frac{d}{dt} \int_t^3 \frac{\sin x}{x}\, dx$

41. $\displaystyle\frac{d}{dx} \int_{x^2}^0 \frac{\sin t}{t}\, dt$

42. $\displaystyle\frac{d}{dx} x^2 \int_0^{x^2} \frac{\sin u}{u}\, du$

43. $\displaystyle\frac{d}{dt} \int_{-\pi}^t \frac{\cos y}{1 + y^2}\, dy$

44. $\displaystyle\frac{d}{d\theta} \int_{\sin\theta}^{\cos\theta} \frac{1}{1 - x^2}\, dx$

45. $\displaystyle\frac{d}{dx} F(\sqrt{x})$, if $F(t) = \int_0^t \cos(x^2)\, dx$

46. $H'(2)$, if $H(x) = 3x \displaystyle\int_4^{x^2} e^{-\sqrt{t}}\, dt$

47. Solve the integral equation $f(x) = \pi \left(1 + \displaystyle\int_1^x f(t)\, dt\right)$.

48. Solve the integral equation $f(x) = 1 - \displaystyle\int_0^x f(t)\, dt$.

* 49. Criticize the following erroneous calculation:

$$\int_{-1}^1 \frac{dx}{x^2} = -\frac{1}{x}\bigg|_{-1}^1 = -1 + \frac{1}{-1} = -2.$$

Exactly where did the error occur? Why is -2 an unreasonable value for the integral?

* 50. Use a definite integral to define a function $F(x)$ having derivative $\dfrac{\sin x}{1 + x^2}$ for all x and satisfying $F(17) = 0$.

* 51. Does the function $F(x) = \displaystyle\int_0^{2x-x^2} \cos\left(\frac{1}{1 + t^2}\right) dt$ have a maximum or a minimum value? Justify your answer.

Evaluate the limits in Exercises 52–54.

* 52. $\displaystyle\lim_{n\to\infty} \frac{1}{n}\left(\left(1 + \frac{1}{n}\right)^5 + \left(1 + \frac{2}{n}\right)^5 + \cdots + \left(1 + \frac{n}{n}\right)^5\right)$.

* 53. $\displaystyle\lim_{n\to\infty} \frac{\pi}{n}\left(\sin\frac{\pi}{n} + \sin\frac{2\pi}{n} + \sin\frac{3\pi}{n} + \cdots + \sin\frac{n\pi}{n}\right)$.

* 54. $\displaystyle\lim_{n\to\infty} \left(\frac{n}{n^2 + 1} + \frac{n}{n^2 + 4} + \frac{n}{n^2 + 9} + \cdots + \frac{n}{2n^2}\right)$.

5.6 The Method of Substitution

As we have seen, the evaluation of definite integrals is most easily carried out if we can antidifferentiate the integrand. In this section and Sections 6.1–6.4 we develop some *techniques of integration*, that is, methods for finding antiderivatives of functions. Although the techniques we develop can be used for a large class of functions, they will not work for all functions we might want to integrate. If a definite integral involves an integrand whose antiderivative is either impossible or very difficult to find, we may wish, instead, to approximate the definite integral by numerical means. Techniques for doing that will be presented in Sections 6.6–6.8.

Let us begin by assembling a table of some known indefinite integrals. These results have all emerged during our development of differentiation formulas for elementary functions. You should *memorize* them.

Some elementary integrals

1. $\int 1 \, dx = x + C$

2. $\int x \, dx = \frac{1}{2}x^2 + C$

3. $\int x^2 \, dx = \frac{1}{3}x^3 + C$

4. $\int \frac{1}{x^2} \, dx = -\frac{1}{x} + C$

5. $\int \sqrt{x} \, dx = \frac{2}{3}x^{3/2} + C$

6. $\int \frac{1}{\sqrt{x}} \, dx = 2\sqrt{x} + C$

7. $\int x^r \, dx = \frac{1}{r+1}x^{r+1} + C \quad (r \neq -1)$

8. $\int \frac{1}{x} \, dx = \ln|x| + C$

9. $\int \sin ax \, dx = -\frac{1}{a}\cos ax + C$

10. $\int \cos ax \, dx = \frac{1}{a}\sin ax + C$

11. $\int \sec^2 ax \, dx = \frac{1}{a}\tan ax + C$

12. $\int \csc^2 ax \, dx = -\frac{1}{a}\cot ax + C$

13. $\int \sec ax \tan ax \, dx = \frac{1}{a}\sec ax + C$

14. $\int \csc ax \cot ax \, dx = -\frac{1}{a}\csc ax + C$

15. $\int \frac{1}{\sqrt{a^2 - x^2}} \, dx = \sin^{-1}\frac{x}{a} + C \quad (a > 0)$

16. $\int \frac{1}{a^2 + x^2} \, dx = \frac{1}{a}\tan^{-1}\frac{x}{a} + C$

17. $\int e^{ax} \, dx = \frac{1}{a}e^{ax} + C$

18. $\int b^{ax} \, dx = \frac{1}{a \ln b}b^{ax} + C$

19. $\int \cosh ax \, dx = \frac{1}{a}\sinh ax + C$

20. $\int \sinh ax \, dx = \frac{1}{a}\cosh ax + C$

Note that formulas 1–6 are special cases of formula 7, which holds on any interval where x^r makes sense. The linearity formula

$$\int (A f(x) + B g(x)) \, dx = A \int f(x) \, dx + B \int g(x) \, dx$$

makes it possible to integrate sums and constant multiples of functions.

Example 1 (Combining elementary integrals)

(a) $\int (x^4 - 3x^3 + 8x^2 - 6x - 7) \, dx = \frac{x^5}{5} - \frac{3x^4}{4} + \frac{8x^3}{3} - 3x^2 - 7x + C$

(b) $\displaystyle \int \left(5x^{3/5} - \frac{3}{2+x^2} \right) dx = \frac{25}{8} x^{8/5} - \frac{3}{\sqrt{2}} \tan^{-1} \frac{x}{\sqrt{2}} + C$

(c) $\displaystyle \int (4 \cos 5x - 5 \sin 3x) \, dx = \frac{4}{5} \sin 5x + \frac{5}{3} \cos 3x + C$

(d) $\displaystyle \int \left(\frac{1}{\pi x} + a^{\pi x} \right) dx = \frac{1}{\pi} \ln |x| + \frac{1}{\pi \ln a} a^{\pi x} + C, \quad (a > 0).$

Sometimes it is necessary to manipulate an integrand so that the method can be applied.

Example 2
$$\int \frac{(x+1)^3}{x} \, dx = \int \frac{x^3 + 3x^2 + 3x + 1}{x} \, dx$$
$$= \int \left(x^2 + 3x + 3 + \frac{1}{x} \right) dx$$
$$= \frac{1}{3} x^3 + \frac{3}{2} x^2 + 3x + \ln |x| + C.$$

When an integral cannot be evaluated by inspection, as those in Examples 1–2 can, we require one or more special techniques. The most important of these techniques is the **method of substitution**, the integral version of the Chain Rule. If we rewrite the Chain Rule, $\frac{d}{dx} f(g(x)) = f'(g(x)) g'(x)$, in integral form, we obtain

$$\int f'(g(x)) \, g'(x) \, dx = f(g(x)) + C.$$

Observe that the following formalism would produce this latter formula even if we did not already know it was true:

Let $u = g(x)$. Then $du/dx = g'(x)$, or in differential form, $du = g'(x) \, dx$. Thus

$$\int f'(g(x)) \, g'(x) \, dx = \int f'(u) \, du = f(u) + C = f(g(x)) + C.$$

Example 3 (**Examples of substitution**) Find the indefinite integrals:

(a) $\displaystyle \int \frac{x}{x^2 + 1} \, dx,$ (b) $\displaystyle \int \frac{\sin(3 \ln x)}{x} \, dx,$ and (c) $\displaystyle \int e^x \sqrt{1 + e^x} \, dx.$

Solution

(a) $\displaystyle \int \frac{x}{x^2 + 1} \, dx$ Let $u = x^2 + 1$.
Then $du = 2x \, dx$ and
$x \, dx = \frac{1}{2} du..$

$\displaystyle = \frac{1}{2} \int \frac{du}{u} = \frac{1}{2} \ln |u| + C = \frac{1}{2} \ln(x^2 + 1) + C = \ln \sqrt{x^2 + 1} + C.$

(Both versions of the final answer are equally acceptable.)

(b) $\displaystyle \int \frac{\sin(3 \ln x)}{x} \, dx$ Let $u = 3 \ln x$.
Then $du = \dfrac{3}{x} \, dx..$

$\displaystyle = \frac{1}{3} \int \sin u \, du = -\frac{1}{3} \cos u + C = -\frac{1}{3} \cos(3 \ln x) + C.$

(c) $\displaystyle\int e^x \sqrt{1 + e^x}\, dx$ Let $v = 1 + e^x$.

Then $dv = e^x\, dx$..

$$= \int v^{1/2}\, dv = \frac{2}{3} v^{3/2} + C = \frac{2}{3}(1 + e^x)^{3/2} + C.$$

Sometimes the appropriate substitutions are not as obvious as they were in Example 3, and it may be necessary to manipulate the integrand algebraically to put it into a better form for substitution.

Example 4   Evaluate   (a) $\displaystyle\int \frac{1}{x^2 + 4x + 5}\, dx$ and   (b) $\displaystyle\int \frac{dx}{\sqrt{e^{2x} - 1}}$.

Solution

(a) $\displaystyle\int \frac{dx}{x^2 + 4x + 5} = \int \frac{dx}{(x + 2)^2 + 1}$ Let $t = x + 2$.

Then $dt = dx$..

$$= \int \frac{dt}{t^2 + 1}$$

$$= \tan^{-1} t + C = \tan^{-1}(x + 2) + C.$$

(b) $\displaystyle\int \frac{dx}{\sqrt{e^{2x} - 1}} = \int \frac{dx}{e^x \sqrt{1 - e^{-2x}}}$

$$= \int \frac{e^{-x}\, dx}{\sqrt{1 - (e^{-x})^2}}$$ Let $u = e^{-x}$.

Then $du = -e^{-x}\, dx$..

$$= -\int \frac{du}{\sqrt{1 - u^2}}$$

$$= -\sin^{-1} u + C = -\sin^{-1}\left(e^{-x}\right) + C.$$

The method of substitution cannot be *forced* to work. There is no substitution that will do much good with the integral $\int x(2 + x^7)^{1/5}\, dx$, for instance. However, the integral $\int x^6(2 + x^7)^{1/5}\, dx$ will yield to the substitution $u = 2 + x^7$. The substitution $u = g(x)$ is more likely to work if $g'(x)$ is a factor of the integrand.

The following theorem simplifies the use of the method of substitution in definite integrals.

THEOREM   **6**

Substitution in a definite integral

Suppose that g is a differentiable function on $[a, b]$ that satisfies $g(a) = A$ and $g(b) = B$. Also suppose that f is continuous on the range of g. Then

$$\int_a^b f\big(g(x)\big)\, g'(x)\, dx = \int_A^B f(u)\, du.$$

PROOF   Let F be an antiderivative of f; $F'(u) = f(u)$. Then

$$\frac{d}{dx} F\big(g(x)\big) = F'\big(g(x)\big)\, g'(x) = f\big(g(x)\big)\, g'(x).$$

Thus,

$$\int_a^b f\big(g(x)\big)\, g'(x)\, dx = F\big(g(x)\big)\bigg|_a^b = F\big(g(b)\big) - F\big(g(a)\big)$$

$$= F(B) - F(A) = F(u)\bigg|_A^B = \int_A^B f(u)\, du.$$

Example 5   Evaluate the integral $I = \displaystyle\int_0^8 \frac{\cos\sqrt{x+1}}{\sqrt{x+1}}\, dx$.

Solution   Method I. Let $u = \sqrt{x+1}$. Then $du = \dfrac{dx}{2\sqrt{x+1}}$. If $x = 0$, then $u = 1$; if $x = 8$, then $u = 3$. Thus

$$I = 2\int_1^3 \cos u\, du = 2\sin u\bigg|_1^3 = 2\sin 3 - 2\sin 1.$$

Method II. We use the same substitution as in Method I, but we do not transform the limits of integration from x values to u values. Hence we must return to the variable x before substituting in the limits:

$$I = 2\int_{x=0}^{x=8} \cos u\, du = 2\sin u\bigg|_{x=0}^{x=8} = 2\sin\sqrt{x+1}\bigg|_0^8 = 2\sin 3 - 2\sin 1.$$

Note that the limits *must* be written $x = 0$ and $x = 8$ at any stage where the variable is not x. It would have been *wrong* to write

$$I = 2\int_0^8 \cos u\, du$$

because this would imply that u, rather than x, goes from 0 to 8. Method I gives the shorter solution and is therefore preferable. However, in cases where the transformed limits (the u-limits) are very complicated, you might prefer to use Method II.   ∎

Example 6   Find the area of the region bounded by $y = \left(2 + \sin\dfrac{x}{2}\right)^2 \cos\dfrac{x}{2}$, the x-axis, and the lines $x = 0$ and $x = \pi$.

Solution   Because $y \geq 0$ when $0 \leq x \leq \pi$, the required area is

$$A = \int_0^\pi \left(2 + \sin\frac{x}{2}\right)^2 \cos\frac{x}{2}\, dx \qquad\qquad \text{Let } v = 2 + \sin\frac{x}{2}.$$

$$\text{Then } dv = \frac{1}{2}\cos\frac{x}{2}\, dx..$$

$$= 2\int_2^3 v^2\, dv = \frac{2}{3} v^3\bigg|_2^3 = \frac{2}{3}(27 - 8) = \frac{38}{3} \text{ square units.}$$

Remark The condition that f be continuous on the range of the function $u = g(x)$ (for $a \le x \le b$) is essential in Theorem 6. Using the substitution $u = x^2$ in the integral $\int_{-1}^{1} x \csc(x^2)\,dx$ leads to the erroneous conclusion

$$\int_{-1}^{1} x \csc(x^2)\,dx = \frac{1}{2} \int_{1}^{1} \csc u\,du = 0.$$

Although $x \csc(x^2)$ is an odd function, it is not continuous at 0, and it happens that the given integral represents the difference of *infinite* areas. If we assume that f is continuous on an interval containing A and B, then it suffices to know that $u = g(x)$ is one-to-one as well as differentiable. In this case the range of g will lie between A and B, so the condition of Theorem 6 will be satisfied.

Trigonometric Integrals

The method of substitution is often useful for evaluating trigonometric integrals. We begin by listing the integrals of the four trigonometric functions whose integrals we have not yet seen. They arise often in applications and should be memorized.

> **Integrals of tangent, cotangent, secant, and cosecant**
>
> $$\int \tan x\,dx = \ln|\sec x| + C,$$
>
> $$\int \cot x\,dx = \ln|\sin x| + C = -\ln|\csc x| + C,$$
>
> $$\int \sec x\,dx = \ln|\sec x + \tan x| + C,$$
>
> $$\int \csc x\,dx = -\ln|\csc x + \cot x| + C = \ln|\csc x - \cot x| + C.$$

All of these can, of course, be checked by differentiating the right-hand sides. The first two can be evaluated directly by rewriting $\tan x$ or $\cot x$ in terms of $\sin x$ and $\cos x$ and using an appropriate substitution. For example,

$$\int \tan x\,dx = \int \frac{\sin x}{\cos x}\,dx \qquad \text{Let } u = \cos x.$$
$$\text{Then } du = -\sin x\,dx..$$
$$= -\int \frac{du}{u} = -\ln|u| + C$$
$$= -\ln|\cos x| + C = \ln\left|\frac{1}{\cos x}\right| + C = \ln|\sec x| + C.$$

The integral of $\sec x$ can be evaluated by rewriting it in the form

$$\int \sec x\,dx = \int \frac{\sec x(\sec x + \tan x)}{\sec x + \tan x}\,dx$$

and using the substitution $u = \sec x + \tan x$. The integral of $\csc x$ can be evaluated similarly. (Show that the two versions given for that integral are equivalent!)

We now consider integrals of the form

$$\int \sin^m x \cos^n x\,dx.$$

If either m or n is an odd, positive integer, the integral can be done easily by substitution. If, say, $n = 2k + 1$ where k is an integer, then we can use the identity $\sin^2 x + \cos^2 x = 1$ to rewrite the integral in the form

$$\int \sin^m x \, (1 - \sin^2 x)^k \cos x \, dx,$$

which can be integrated using the substitution $u = \sin x$. Similarly, $u = \cos x$ can be used if m is an odd integer.

Example 7 Evaluate: (a) $\displaystyle\int \sin^3 x \cos^8 x \, dx$ and (b) $\displaystyle\int \cos^5 ax \, dx$.

Solution

(a) $\displaystyle\int \sin^3 x \cos^8 x \, dx = \int (1 - \cos^2 x) \cos^8 x \sin x \, dx$ $\qquad$ Let $u = \cos x$,

$$\qquad\qquad\qquad\qquad\qquad\qquad\qquad\qquad\qquad\qquad du = -\sin x \, dx..$$

$$= -\int (1 - u^2) \, u^8 \, du = \int (u^{10} - u^8) \, du$$

$$= \frac{u^{11}}{11} - \frac{u^9}{9} + C = \frac{1}{11} \cos^{11} x - \frac{1}{9} \cos^9 x + C$$

(b) $\displaystyle\int \cos^5 ax \, dx = \int (1 - \sin^2 ax)^2 \cos ax \, dx$ $\qquad$ Let $u = \sin ax$,

$$\qquad\qquad\qquad\qquad\qquad\qquad\qquad\qquad\qquad\qquad du = a \cos ax \, dx..$$

$$= \frac{1}{a} \int (1 - u^2)^2 \, du = \frac{1}{a} \int (1 - 2u^2 + u^4) \, du$$

$$= \frac{1}{a} \left(u - \frac{2}{3} u^3 + \frac{1}{5} u^5 \right) + C$$

$$= \frac{1}{a} \left(\sin ax - \frac{2}{3} \sin^3 ax + \frac{1}{5} \sin^5 ax \right) + C$$

If the powers of $\sin x$ and $\cos x$ are both even, then we can make use of the *double-angle formulas* (see Section P.6):

$$\cos^2 x = \frac{1}{2}(1 + \cos 2x) \qquad \text{and} \qquad \sin^2 x = \frac{1}{2}(1 - \cos 2x).$$

Example 8 **(Integrating even powers of sine and cosine)** Verify the formulas

$$\int \cos^2 x \, dx = \frac{1}{2}(x + \sin x \cos x) + C,$$

$$\int \sin^2 x \, dx = \frac{1}{2}(x - \sin x \cos x) + C.$$

These integrals are encountered frequently and are worth remembering.

Solution Each of the integrals follows from the corresponding double-angle identity. We do the first; the second is similar.

$$\int \cos^2 x \, dx = \frac{1}{2} \int (1 + \cos 2x) \, dx$$

$$= \frac{x}{2} + \frac{1}{4} \sin 2x + C$$

$$= \frac{1}{2}(x + \sin x \cos x) + C \quad \text{(since } \sin 2x = 2 \sin x \cos x\text{)}.$$

Example 9 Evaluate $\int \sin^4 x \, dx$.

Solution We will have to apply the double-angle formula twice.

$$\int \sin^4 x \, dx = \frac{1}{4} \int (1 - \cos 2x)^2 \, dx$$

$$= \frac{1}{4} \int (1 - 2 \cos 2x + \cos^2 2x) \, dx$$

$$= \frac{x}{4} - \frac{1}{4} \sin 2x + \frac{1}{8} \int (1 + \cos 4x) \, dx$$

$$= \frac{x}{4} - \frac{1}{4} \sin 2x + \frac{x}{8} + \frac{1}{32} \sin 4x + C$$

$$= \frac{3}{8} x - \frac{1}{4} \sin 2x + \frac{1}{32} \sin 4x + C$$

(Note that there is no point in inserting the constant of integration C until the last integral has been evaluated.)

Using the identities $\sec^2 x = 1 + \tan^2 x$ and $\csc^2 x = 1 + \cot^2 x$ and one of the substitutions $u = \sec x$, $u = \tan x$, $u = \csc x$, or $u = \cot x$, we can evaluate integrals of the form

$$\int \sec^m x \tan^n x \, dx \qquad \text{or} \qquad \int \csc^m x \cot^n x \, dx,$$

unless m is odd and n is even. (If this is the case, these integrals can be handled by integration by parts; see Section 6.1.)

Example 10 (**Integrals involving secants and tangents**) Evaluate the following integrals:

(a) $\int \tan^2 x \, dx$, (b) $\int \sec^4 t \, dt$, and (c) $\int \sec^3 x \tan^3 x \, dx$.

Solution

(a) $\int \tan^2 x \, dx = \int (\sec^2 x - 1) \, dx = \tan x - x + C$.

(b) $\int \sec^4 t \, dt = \int (1 + \tan^2 t) \sec^2 t \, dt \qquad$ Let $u = \tan t$,

$$du = \sec^2 t \, dt..$$

$$= \int (1 + u^2) \, du = u + \frac{1}{3} u^3 + C = \tan t + \frac{1}{3} \tan^3 t + C.$$

(c) $\displaystyle\int \sec^3 x \, \tan^3 x \, dx$

$\displaystyle = \int \sec^2 x \, (\sec^2 x - 1) \, \sec x \, \tan x \, dx$ 　　　Let $u = \sec x$,

　　　　　　　　　　　　　　　　　　　$du = \sec x \, \tan x \, dx..$

$\displaystyle = \int (u^4 - u^2) \, du = \frac{u^5}{5} - \frac{u^3}{3} + C = \frac{1}{5} \sec^5 x - \frac{1}{3} \sec^3 x + C.$ ∎

Exercises 5.6

Evaluate the integrals in Exercises 1–44. Remember to include a constant of integration with the indefinite integrals. Your answers may appear different from those in the Answers section but may still be correct. For example, evaluating $I = \int \sin x \cos x \, dx$ using the substitution $u = \sin x$ leads to the answer $I = \frac{1}{2} \sin^2 x + C$; using $u = \cos x$ leads to $I = -\frac{1}{2} \cos^2 x + C$; and rewriting $I = \frac{1}{2} \int \sin(2x) \, dx$ leads to $I = -\frac{1}{4} \cos(2x) + C$. These answers are all equivalent up to different choices for the constant of integration:
$\frac{1}{2} \sin^2 x = -\frac{1}{2} \cos^2 + \frac{1}{2} = -\frac{1}{4} \cos(2x) + \frac{1}{4}$.

You can always check your own answer to an indefinite integral by differentiating it to get back to the integrand. This is often easier than comparing your answer with the answer in the back of the book. You may find integrals that you can't do, but you should not make mistakes in those you can do because the answer is so easily checked. (This is a good thing to remember during tests and exams.)

1. $\displaystyle\int e^{5-2x} \, dx$

2. $\displaystyle\int \cos(ax + b) \, dx$

3. $\displaystyle\int \sqrt{3x + 4} \, dx$

4. $\displaystyle\int e^{2x} \sin(e^{2x}) \, dx$

5. $\displaystyle\int \frac{x \, dx}{(4x^2 + 1)^5}$

6. $\displaystyle\int \frac{\sin \sqrt{x}}{\sqrt{x}} \, dx$

7. $\displaystyle\int x \, e^{x^2} \, dx$

8. $\displaystyle\int x^2 \, 2^{x^3+1} \, dx$

9. $\displaystyle\int \frac{\cos x}{4 + \sin^2 x} \, dx$

10. $\displaystyle\int \frac{\sec^2 x}{\sqrt{1 - \tan^2 x}} \, dx$

* 11. $\displaystyle\int \frac{e^x + 1}{e^x - 1} \, dx$

12. $\displaystyle\int \frac{\ln t}{t} \, dt$

13. $\displaystyle\int \frac{ds}{\sqrt{4 - 5s}}$

14. $\displaystyle\int \frac{x + 1}{\sqrt{x^2 + 2x + 3}} \, dx$

15. $\displaystyle\int \frac{t \, dt}{\sqrt{4 - t^4}}$

16. $\displaystyle\int \frac{x^2 \, dx}{2 + x^6}$

* 17. $\displaystyle\int \frac{dx}{e^x + 1}$

* 18. $\displaystyle\int \frac{dx}{e^x + e^{-x}}$

19. $\displaystyle\int \tan x \, \ln \cos x \, dx$

20. $\displaystyle\int \frac{x + 1}{\sqrt{1 - x^2}} \, dx$

21. $\displaystyle\int \frac{dx}{x^2 + 6x + 13}$

22. $\displaystyle\int \frac{dx}{\sqrt{4 + 2x - x^2}}$

23. $\displaystyle\int \sin^3 x \cos^5 x \, dx$

24. $\displaystyle\int \sin^4 t \cos^5 t \, dt$

25. $\displaystyle\int \sin ax \cos^2 ax \, dx$

26. $\displaystyle\int \sin^2 x \cos^2 x \, dx$

27. $\displaystyle\int \sin^6 x \, dx$

28. $\displaystyle\int \cos^4 x \, dx$

29. $\displaystyle\int \sec^5 x \tan x \, dx$

30. $\displaystyle\int \sec^6 x \tan^2 x \, dx$

31. $\displaystyle\int \sqrt{\tan x} \, \sec^4 x \, dx$

32. $\displaystyle\int \sin^{-2/3} x \cos^3 x \, dx$

33. $\displaystyle\int \cos x \sin^4(\sin x) \, dx$

34. $\displaystyle\int \frac{\sin^3 \ln x \cos^3 \ln x}{x} \, dx$

35. $\displaystyle\int \frac{\sin^2 x}{\cos^4 x} \, dx$

36. $\displaystyle\int \frac{\sin^3 x}{\cos^4 x} \, dx$

37. $\displaystyle\int \csc^5 x \cot^5 x \, dx$

38. $\displaystyle\int \frac{\cos^4 x}{\sin^8 x} \, dx$

39. $\displaystyle\int_0^4 x^3 (x^2 + 1)^{-\frac{1}{2}} \, dx$

40. $\displaystyle\int_1^{\sqrt{e}} \frac{\sin(\pi \ln x)}{x} \, dx$

41. $\displaystyle\int_0^{\pi/2} \sin^4 x \, dx$

42. $\displaystyle\int_{\pi/4}^{\pi} \sin^5 x \, dx$

43. $\displaystyle\int_e^{e^2} \frac{dt}{t \ln t}$

44. $\displaystyle\int_{\frac{\pi^2}{16}}^{\frac{\pi^2}{9}} \frac{2^{\sin \sqrt{x}} \cos \sqrt{x}}{\sqrt{x}} \, dx$

* 45. Use the identities $\cos 2\theta = 2\cos^2 \theta - 1 = 1 - 2\sin^2 \theta$ and $\sin \theta = \cos\left(\frac{\pi}{2} - \theta\right)$ to help you evaluate the following:

$$\int_0^{\pi/2} \sqrt{1 + \cos x} \, dx \quad \text{and} \quad \int_0^{\pi/2} \sqrt{1 - \sin x} \, dx$$

46. Find the area of the region bounded by $y = x/(x^2 + 16)$, $y = 0$, $x = 0$, and $x = 2$.

47. Find the area of the region bounded by $y = x/(x^4 + 16)$, $y = 0$, $x = 0$, and $x = 2$.

48. Express the area bounded by the ellipse $(x^2/a^2) + (y^2/b^2) = 1$ as a definite integral. Make a substitution that converts this integral into one representing the area of a circle, and hence evaluate it.

* 49. Use the addition formulas for $\sin(x \pm y)$ and $\cos(x \pm y)$ from Section P.6 to establish the following identities:

$$\cos x \, \cos y = \frac{1}{2}\left(\cos(x - y) + \cos(x + y)\right),$$

$$\sin x \, \sin y = \frac{1}{2}\left(\cos(x - y) - \cos(x + y)\right),$$

$$\sin x \, \cos y = \frac{1}{2}\left(\sin(x + y) + \sin(x - y)\right).$$

* 50. Use the identities established in the previous exercise to calculate the following integrals:

$$\int \cos ax \, \cos bx \, dx, \quad \int \sin ax \, \sin bx \, dx,$$

and $\int \sin ax \, \cos bx \, dx$.

* 51. If m and n are integers, show that:

(i) $\displaystyle\int_{-\pi}^{\pi} \cos mx \, \cos nx \, dx = 0$ if $m \neq n$,

(ii) $\displaystyle\int_{-\pi}^{\pi} \sin mx \, \sin nx \, dx = 0$ if $m \neq n$,

(iii) $\displaystyle\int_{-\pi}^{\pi} \sin mx \, \cos nx \, dx = 0$.

* 52. **(Fourier coefficients)** Suppose that for some positive integer k,

$$f(x) = \frac{a_0}{2} + \sum_{n=1}^{k} (a_n \cos nx + b_n \sin nx)$$

holds for all x in $[-\pi, \pi]$. Use the result of the previous exercise to show that the coefficients a_m ($0 \leq m \leq k$) and b_m ($1 \leq m \leq k$), which are called the Fourier coefficients of f on $[-\pi, \pi]$, are given by

$$a_m = \frac{1}{\pi} \int_{-\pi}^{\pi} f(x) \cos mx \, dx, \quad b_m = \frac{1}{\pi} \int_{-\pi}^{\pi} f(x) \sin mx \, dx.$$

5.7 Areas of Plane Regions

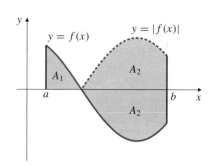

Figure 5.27

In this section we review and extend the use of definite integrals to represent plane areas. Recall that the integral $\int_a^b f(x)\, dx$ measures the area between the graph of f and the x-axis from $x = a$ to $x = b$, but treats as *negative* any part of this area that lies below the x-axis. (We are assuming that $a < b$.) In order to express the total area bounded by $y = f(x)$, $y = 0$, $x = a$, and $x = b$, counting all of the area positively, we should integrate the *absolute value* of f (see Figure 5.27):

$$\int_a^b f(x) \, dx = A_1 - A_2 \quad \text{and} \quad \int_a^b |f(x)| \, dx = A_1 + A_2.$$

There is no "rule" for integrating $\int_a^b |f(x)| \, dx$; one must break the integral into a sum of integrals over intervals where $f(x) > 0$ (so $|f(x)| = f(x)$), and intervals where $f(x) < 0$ (so $|f(x)| = -f(x)$).

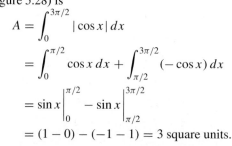

Figure 5.28

Example 1   The area bounded by $y = \cos x$, $y = 0$, $x = 0$, and $x = 3\pi/2$ (see Figure 5.28) is

$$A = \int_0^{3\pi/2} |\cos x| \, dx$$

$$= \int_0^{\pi/2} \cos x \, dx + \int_{\pi/2}^{3\pi/2} (-\cos x) \, dx$$

$$= \sin x \Big|_0^{\pi/2} - \sin x \Big|_{\pi/2}^{3\pi/2}$$

$$= (1 - 0) - (-1 - 1) = 3 \text{ square units.}$$

Areas Between Two Curves

Suppose that a plane region R is bounded by the graphs of two continuous functions, $y = f(x)$ and $y = g(x)$, and the vertical straight lines $x = a$ and $x = b$, as shown in Figure 5.29(a). Assume that $a < b$ and that $f(x) \le g(x)$ on $[a, b]$, so the graph of f lies below that of g. If $f(x) \ge 0$ on $[a, b]$, then the area A of R is the area above the x-axis and under the graph of g minus the area above the x-axis and under the graph of f:

$$A = \int_a^b g(x)\,dx - \int_a^b f(x)\,dx = \int_a^b \big(g(x) - f(x)\big)\,dx.$$

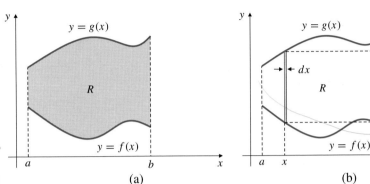

Figure 5.29

(a) The region R lying between two graphs

(b) An area element of the region R

(a)

(b)

It is useful to regard this formula as expressing A as the "sum" (i.e., the integral) of *infinitely many* **area elements**

$$dA = (g(x) - f(x))\,dx,$$

corresponding to values of x between a and b. Each such area element is the area of an infinitely thin vertical rectangle of width dx and height $g(x) - f(x)$ located at position x (see Figure 5.29(b)). Even if f and g can take on negative values on $[a, b]$, this interpretation and the resulting area formula

$$A = \int_a^b \big(g(x) - f(x)\big)\,dx$$

remain valid, provided that $f(x) \le g(x)$ on $[a, b]$ so that all the area elements dA have positive area. Using integrals to represent a quantity as a *sum* of *differential elements* (i.e., a sum of little bits of the quantity) is a very helpful approach. We will do this often in Chapter 7. Of course, what we are really doing is identifying the integral as a *limit* of a suitable Riemann sum.

More generally, if the restriction $f(x) \le g(x)$ is removed, then the vertical rectangle of width dx at position x extending between the graphs of f and g has height $|f(x) - g(x)|$ and hence area

$$dA = |f(x) - g(x)|\,dx.$$

(See Figure 5.30.) Hence the total area lying between the graphs $y = f(x)$ and $y = g(x)$ and between the vertical lines $x = a$ and $x = b > a$ is given by

$$A = \int_a^b \big|f(x) - g(x)\big|\,dx.$$

In order to evaluate this integral, we have to determine the intervals on which $f(x) > g(x)$ or $f(x) < g(x)$, and break the integral into a sum of integrals over each of these intervals.

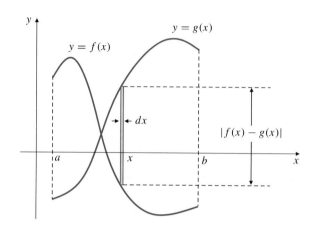

Figure 5.30   An area element for the region between $y = f(x)$ and $y = g(x)$

Example 2   Find the area of the bounded, plane region R lying between the curves $y = x^2 - 2x$ and $y = 4 - x^2$.

Solution   We must find the intersections of the curves, so we solve the equations simultaneously:

$$x^2 - 2x = y = 4 - x^2$$
$$2x^2 - 2x - 4 = 0$$
$$2(x - 2)(x + 1) = 0 \quad \text{so } x = 2 \text{ or } x = -1.$$

The curves are sketched in Figure 5.31, and the bounded (finite) region between them is shaded. (A sketch should always be made in problems of this sort.) Since $4 - x^2 \geq x^2 - 2x$ for $-1 \leq x \leq 2$, the area A of R is given by

$$A = \int_{-1}^{2} \left((4 - x^2) - (x^2 - 2x) \right) dx$$

$$= \int_{-1}^{2} (4 - 2x^2 + 2x) \, dx$$

$$= \left(4x - \frac{2}{3}x^3 + x^2 \right) \Big|_{-1}^{2}$$

$$= 4(2) - \frac{2}{3}(8) + 4 - \left(-4 + \frac{2}{3} + 1 \right) = 9 \text{ square units.}$$

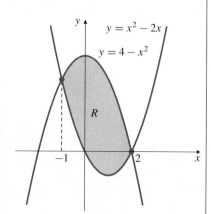

Figure 5.31

Note that in representing the area as an integral we *must subtract the height y to the lower curve from the height y to the upper curve* to get a positive area element dA. Subtracting the wrong way would have produced a negative value for the area.

Example 3   Find the total area A lying between the curves $y = \sin x$ and $y = \cos x$ from $x = 0$ to $x = 2\pi$.

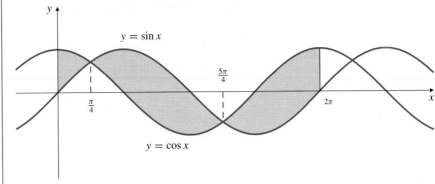

Figure 5.32

Solution   The region is shaded in Figure 5.32. The graphs of sine and cosine cross at $x = \pi/4$ and $x = 5\pi/4$. The required area is

$$
A = \int_0^{\pi/4} (\cos x - \sin x)\, dx + \int_{\pi/4}^{5\pi/4} (\sin x - \cos x)\, dx
$$

$$
+ \int_{5\pi/4}^{2\pi} (\cos x - \sin x)\, dx
$$

$$
= (\sin x + \cos x)\Big|_0^{\pi/4} - (\cos x + \sin x)\Big|_{\pi/4}^{5\pi/4} + (\sin x + \cos x)\Big|_{5\pi/4}^{2\pi}
$$

$$
= (\sqrt{2} - 1) + (\sqrt{2} + \sqrt{2}) + (1 + \sqrt{2}) = 4\sqrt{2} \text{ square units.}
$$

It is sometimes more convenient to use horizontal area elements instead of vertical ones and integrate over an interval of the y-axis instead of the x-axis. This is usually the case if the region whose area we want to find is bounded by curves whose equations are written in terms of functions of y. The region R lying to the right of $x = f(y)$ and to the left of $x = g(y)$, and between the horizontal lines $y = c$ and $y = d > c$ (see Figure 5.33(a)), has area element $dA = \big(g(y) - f(y)\big)\, dy$, and its area is

$$
A = \int_c^d \big(g(y) - f(y)\big)\, dy.
$$

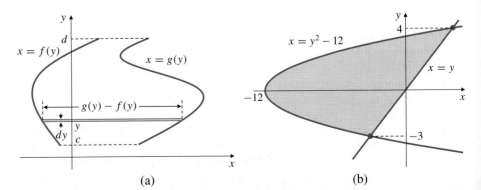

Figure 5.33

(a) A horizontal area element

(b) The finite region bounded by $x = y^2 - 12$ and $x = y$

Example 4   Find the area of the plane region lying to the right of the parabola $x = y^2 - 12$ and to the left of the straight line $y = x$, as illustrated in Figure 5.33(b).

Solution   For the intersections of the curves:

$$y^2 - 12 = x = y$$
$$y^2 - y - 12 = 0$$
$$(y - 4)(y + 3) = 0 \quad \text{so } y = 4 \text{ or } y = -3.$$

Observe that $y^2 - 12 \le y$ for $-3 \le y \le 4$. Thus the area is

$$A = \int_{-3}^{4} \left(y - (y^2 - 12) \right) dy = \left(\frac{y^2}{2} - \frac{y^3}{3} + 12y \right) \Big|_{-3}^{4} = \frac{343}{6} \text{ square units.}$$

Of course, the same result could have been obtained by integrating in the x direction, but the integral would have been more complicated:

$$A = \int_{-12}^{-3} \left(\sqrt{12 + x} - (-\sqrt{12 + x}) \right) dx + \int_{-3}^{4} \left(\sqrt{12 + x} - x \right) dx;$$

different integrals are required over the intervals where the region is bounded below by the parabola and by the straight line. ∎

Exercises 5.7

In Exercises 1–16, sketch and find the area of the plane region bounded by the given curves.

1. $y = x, \quad y = x^2$ **2.** $y = \sqrt{x}, \quad y = x^2$

3. $y = x^2 - 5, \quad y = 3 - x^2$

4. $y = x^2 - 2x, \quad y = 6x - x^2$

5. $2y = 4x - x^2, \quad 2y + 3x = 6$

6. $x - y = 7, \quad x = 2y^2 - y + 3$

7. $y = x^3, \quad y = x$ **8.** $y = x^3, \quad y = x^2$

9. $y = x^3, \quad x = y^2$

10. $x = y^2, \quad x = 2y^2 - y - 2$

11. $y = \dfrac{1}{x}, \quad 2x + 2y = 5$

12. $y = (x^2 - 1)^2, \quad y = 1 - x^2$

13. $y = \dfrac{1}{2}x^2, \quad y = \dfrac{1}{x^2 + 1}$ **14.** $y = \dfrac{4x}{3 + x^2}, \quad y = 1$

15. $y = \dfrac{4}{x^2}, \quad y = 5 - x^2$ **16.** $x = y^2 - \pi^2, \quad x = \sin y$

Find the areas of the regions described in Exercises 17–28. It is helpful to sketch the regions before writing an integral to represent the area.

17. Bounded by $y = \sin x$ and $y = \cos x$, and between two consecutive intersections of these curves

18. Bounded by $y = \sin^2 x$ and $y = 1$, and between two consecutive intersections of these curves

19. Bounded by $y = \sin x$ and $y = \sin^2 x$, between $x = 0$ and $x = \pi/2$

20. Bounded by $y = \sin^2 x$ and $y = \cos^2 x$, and between two consecutive intersections of these curves

21. Under $y = 4x/\pi$ and above $y = \tan x$, between $x = 0$ and the first intersection of the curves to the right of $x = 0$

22. Bounded by $y = x^{1/3}$ and the component of $y = \tan(\pi x/4)$ that passes through the origin

23. Bounded by $y = 2$ and the component of $y = \sec x$ that passes through the point $(0, 1)$

24. Bounded by $y = \sqrt{2} \cos(\pi x/4)$ and $y = |x|$

25. Bounded by $y = \sin(\pi x/2)$ and $y = x$

26. Bounded by $y = e^x$ and $y = x + 2$

27. Find the total area enclosed by the curve $y^2 = x^2 - x^4$.

28. Find the area of the closed loop of the curve $y^2 = x^4(2 + x)$ that lies to the left of the origin.

29. Find the area of the finite plane region that is bounded by the curve $y = e^x$, the line $x = 0$, and the tangent line to $y = e^x$ at $x = 1$.

∗ 30. Find the area of the finite plane region bounded by the curve $y = x^3$ and the tangent line to that curve at the point $(1, 1)$. *Hint:* find the other point at which that tangent line meets the curve.

Chapter Review

Key Ideas

- **What do the following terms and phrases mean?**

  ◇ sigma notation   ◇ a partition of an interval

  ◇ a Riemann sum   ◇ a definite integral

  ◇ an indefinite integral   ◇ an integrable function

  ◇ an area element   ◇ an evaluation symbol

  ◇ the triangle inequality for integrals

  ◇ a piecewise continuous function

  ◇ the average value of function f on $[a, b]$

  ◇ the method of substitution

- **State the Mean-Value Theorem for integrals.**
- **State the Fundamental Theorem of Calculus.**
- **List as many properties of the definite integral as you can.**
- **What is the relationship between the definite integral and the indefinite integral of a function f on an interval $[a, b]$?**
- **What is the derivative of $\int_{f(x)}^{g(x)} h(t)\, dt$ with respect to x?**
- **How can the area between the graphs of two functions be calculated?**

Review Exercises

1. Show that $\dfrac{2j+1}{j^2(j+1)^2} = \dfrac{1}{j^2} - \dfrac{1}{(j+1)^2}$; hence evaluate

$$\sum_{j=1}^{n} \frac{2j+1}{j^2(j+1)^2}.$$

2. **(Stacking balls)** A display of golf balls in a sporting goods store is built in the shape of a pyramid with a rectangular base measuring 40 balls long and 30 balls wide. The next layer up is 39 balls by 29 balls, etc. How many balls are in the pyramid?

3. Let $P_n = \{x_0 = 1, x_1, x_2, \ldots, x_n = 3\}$ be a partition of $[1, 3]$ into n subintervals of equal length, and let $f(x) = x^2 - 2x + 3$. Evaluate $\displaystyle\int_1^3 f(x)\, dx$ by finding $\lim_{n\to\infty} \sum_{i=1}^{n} f(x_i)\, \Delta x_i$.

4. Interpret $R_n = \displaystyle\sum_{i=1}^{n} \frac{1}{n}\sqrt{1 + \frac{i}{n}}$ as a Riemann sum for a certain function f on the interval $[0, 1]$; hence evaluate $\lim_{n\to\infty} R_n$.

Evaluate the integrals in Exercises 5–8 without using the Fundamental Theorem of Calculus. ✗

5. $\displaystyle\int_{-\pi}^{\pi} (2 - \sin x)\, dx$ **6.** $\displaystyle\int_0^{\sqrt{5}} \sqrt{5 - x^2}\, dx$

7. $\displaystyle\int_1^3 \left(1 - \frac{x}{2}\right) dx$ **8.** $\displaystyle\int_0^{\pi} \cos x\, dx$

Find the average values of the functions in Exercises 9–10 over the indicated intervals.

9. $f(x) = 2 - \sin x^3$ on $[-\pi, \pi]$

10. $h(x) = |x - 2|$ on $[0, 3]$

Find the derivatives of the functions in Exercises 11–14.

11. $f(t) = \displaystyle\int_{13}^{t} \sin(x^2)\, dx$ **12.** $f(x) = \displaystyle\int_{-13}^{\sin x} \sqrt{1 + t^2}\, dt$

13. $g(s) = \displaystyle\int_{4s}^{1} e^{\sin u}\, du$ **14.** $g(\theta) = \displaystyle\int_{e^{\sin\theta}}^{e^{\cos\theta}} \ln x\, dx$

15. Solve the integral equation $2f(x) + 1 = 3\displaystyle\int_x^1 f(t)\, dt$.

16. Use the substitution $x = \pi - u$ to show that

$$\int_0^{\pi} x\, f(\sin x)\, dx = \frac{\pi}{2} \int_0^{\pi} f(\sin x)\, dx$$

for any function f continuous on $[0, 1]$.

Find the areas of the finite plane regions bounded by the indicated graphs in Exercises 17–22.

17. $y = 2 + x - x^2$ and $y = 0$

18. $y = (x - 1)^2$, $y = 0$ and $x = 0$

19. $x = y - y^4$ and $x = 0$ **20.** $y = 4x - x^2$ and $y = 3$

21. $y = \sin x$, $y = \cos 2x$, $x = 0$, and $x = \pi/6$

22. $y = 5 - x^2$ and $y = 4/x^2$

Evaluate the integrals in Exercises 23–30.

23. $\displaystyle\int x^2 \cos(2x^3 + 1)\, dx$ **24.** $\displaystyle\int_1^e \frac{\ln x}{x}\, dx$

25. $\displaystyle\int_0^4 \sqrt{9t^2 + t^4}\, dt$ **26.** $\displaystyle\int \sin^3(\pi x)\, dx$

27. $\displaystyle\int_0^{\ln 2} \frac{e^u}{4 + e^{2u}}\, du$ **28.** $\displaystyle\int_1^{\sqrt[4]{e}} \frac{\tan^2 \pi \ln x}{x}\, dx$

29. $\displaystyle\int \frac{\sin\sqrt{2s+1}}{\sqrt{2s+1}}\, ds$ **30.** $\displaystyle\int \cos^2 \frac{t}{5} \sin^2 \frac{t}{5}\, dt$

31. Find the minimum value of $F(x) = \displaystyle\int_0^{x^2 - 2x} \frac{1}{1 + t^2}\, dt$. Does F have a maximum value? Why?

32. Find the maximum value of $\int_a^b (4x - x^2)\, dx$ for intervals $[a, b]$, where $a < b$. How do you know such a maximum value exists?

33. An object moves along the x-axis so that its position at time t is given by the function $x(t)$. In Section 2.11 we defined the average velocity of the object over the time interval $[t_0, t_1]$ to be $v_{av} = \big(x(t_1) - x(t_0)\big)/(t_1 - t_0)$. Show that v_{av} is, in fact, the average value of the velocity function $v(t) = dx/dt$ over the interval $[t_0, t_1]$.

34. If an object falls from rest under constant gravitational acceleration, show that its average height during the time T of its fall is its height at time $T/\sqrt{3}$.

Challenging Problems

1. Evaluate the upper and lower Riemann sums, $U(f, P_n)$ and $L(f, P_n)$, for $f(x) = 1/x$ on the interval $[1, 2]$ for the partition P_n with division points $x_i = 2^{i/n}$ for $0 \le i \le n$. Verify that $\lim_{n\to\infty} U(f, P_n) = \ln 2 = \lim_{n\to\infty} L(f, P_n)$.

∗ 2. (a) Use the addition formulas for $\cos(a+b)$ and $\cos(a-b)$ to show that

$$\cos\left((j+\tfrac{1}{2})t\right) - \cos\left((j-\tfrac{1}{2})t\right) = -2\sin(\tfrac{1}{2}t)\,\sin(jt),$$

and hence deduce that if $t/(2\pi)$ is not an integer, then

$$\sum_{j=1}^{n} \sin(jt) = \frac{\cos\tfrac{t}{2} - \cos\left((n+\tfrac{1}{2})t\right)}{2\sin\tfrac{t}{2}}$$

(b) Use the result to part (a) to evaluate $\int_0^{\pi/2} \sin x \, dx$ as a limit of a Riemann sum.

3. (a) Use the method of the previous exercise to show that if $t/(2\pi)$ is not an integer, then

$$\sum_{j=1}^{n} \cos(jt) = \frac{\sin\left((n+\tfrac{1}{2})t\right) - \sin\tfrac{t}{2}}{2\sin\tfrac{t}{2}}$$

(b) Use the result to part (a) to evaluate $\int_0^{\pi/3} \cos x \, dx$ as a limit of a Riemann sum.

4. Let $f(x) = 1/x^2$ and let $1 = x_0 < x_1 < x_2 < \cdots < x_n = 2$, so that $\{x_0, x_1, x_2, \ldots, x_n\}$ is a partition of $[1, 2]$ into n subintervals. Show that $c_i = \sqrt{x_{i-1}x_i}$ is in the ith subinterval $[x_{i-1}, x_i]$ of the partition, and evaluate the Riemann sum $\sum_{i=1}^{n} f(c_i)\,\Delta x_i$. What does this imply about $\int_1^2 (1/x^2)\,dx$?

∗ 5. (a) Use mathematical induction to verify that for every positive integer k, $\sum_{j=1}^{n} j^k = \dfrac{n^{k+1}}{k+1} + \dfrac{n^k}{2} + P_{k-1}(n)$, where P_{k-1} is a polynomial of degree at most $k-1$. *Hint:* start by iterating the identity

$$(j+1)^{k+1} - j^{k+1} = (k+1)j^k + \frac{(k+1)k}{2}j^{k-1}$$
$$+ \text{lower powers of } j$$

for $j = 1, 2, 3, \ldots, k$ and adding.

(b) Deduce from (a) that $\displaystyle\int_0^a x^k\,dx = \frac{a^{k+1}}{k+1}$.

6. Let C be the cubic curve $y = ax^3 + bx^2 + cx + d$, and let P be any point on C. The tangent to C at P meets C again at point Q. The tangent to C at Q meets C again at R. Show that the area between C and the tangent at Q is 16 times the area between C and the tangent at P.

7. Let C be the cubic curve $y = ax^3 + bx^2 + cx + d$, and let P be any point on C. The tangent to C at P meets C again at point Q. Let R be the inflection point of C. Show that R lies between P and Q on C and that QR divides the area between C and its tangent at P in the ratio 16/11.

8. (Double tangents) Let line PQ be tangent to the graph C of the quartic polynomial $f(x) = ax^4 + bx^3 + cx^2 + dx + e$ at two distinct points: $P = (p, f(p))$ and $Q = (q, f(q))$. Let $U = (u, f(u))$ and $V = (v, f(v))$ be the other two points where the line tangent to C at $T = ((p+q)/2, f((p+q)/2))$ meets C. If A and B are the two inflection points of C, let R and S be the other two points where AB meets C. (See Figure 5.34. Also see Challenging Problem 17 in Chapter 2 for more background.)

(a) Find the ratio of the area bounded by UV and C to the area bounded by PQ and C.

(b) Show that the area bounded by RS and C is divided at A and B into three parts in the ratio $1 : 2 : 1$.

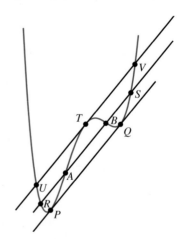

Figure 5.34

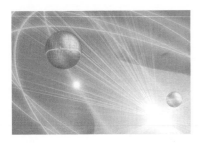

CHAPTER 6
Techniques of Integration

Introduction This chapter is completely concerned with how to evaluate integrals. The first four sections continue our search, begun in Section 5.6, for ways to find antiderivatives and, therefore, definite integrals by the Fundamental Theorem of Calculus. Section 6.5 deals with the problem of finding definite integrals of functions over infinite intervals, or over intervals where the functions are not bounded. The remaining three sections deal with techniques of *numerical integration* that can be used to find approximate values of definite integrals when an antiderivative cannot be found.

It is not necessary to cover the material of this chapter before proceeding to the various applications of integration discussed in Chapter 7, but some of the examples and exercises in that chapter do depend on techniques presented here.

6.1 Integration by Parts

Our next general method for antidifferentiation is called **integration by parts**. Just as the method of substitution can be regarded as inverse to the Chain Rule for differentiation, so the method for integration by parts is inverse to the Product Rule for differentiation.

Suppose that $U(x)$ and $V(x)$ are two differentiable functions. According to the Product Rule,

$$\frac{d}{dx}\big(U(x)V(x)\big) = U(x)\frac{dV}{dx} + V(x)\frac{dU}{dx}.$$

Integrating both sides of this equation and transposing terms, we obtain

$$\int U(x)\frac{dV}{dx}\,dx = U(x)V(x) - \int V(x)\frac{dU}{dx}\,dx$$

or, more simply,

$$\int U\,dV = UV - \int V\,dU.$$

The above formula serves as a *pattern* for carrying out integration by parts, as we will see in the examples below. In each application of the method, we break up the given integrand into a product of two pieces, U and V', where V' is readily integrated and where $\int VU'\,dx$ is usually (but not always) a *simpler* integral than $\int UV'\,dx$. The technique is called integration by parts because it replaces one integral with the sum of an integrated term and another integral that remains to be evaluated. That is, it accomplishes only *part* of the original integration.

Example 1

$$\int x e^x \, dx \qquad\qquad \text{Let} \quad U = x, \qquad dV = e^x \, dx.$$
$$\qquad\qquad\qquad\qquad \text{Then} \ \ dU = dx, \qquad V = e^x.$$

$$= x e^x - \int e^x \, dx \qquad (\text{i.e., } UV - \int V \, dU)$$

$$= x e^x - e^x + C.$$

Note the form in which the integration by parts is carried out. We indicate at the side what choices we are making for U and dV and then calculate dU and V from these. However, we do not actually substitute U and V into the integral; instead, we use the formula $\int U \, dV = UV - \int V \, dU$ as a pattern or mnemonic device to replace the given integral by the equivalent partially integrated form on the second line.

Note also that had we included a constant of integration with V (for example, $V = e^x + K$), that constant would cancel out in the next step:

$$\int x e^x \, dx = x(e^x + K) - \int (e^x + K) \, dx$$
$$= x e^x + Kx - e^x - Kx + C = x e^x - e^x + C.$$

In general, do not include a constant of integration with V or on the right-hand side until the last integral has been evaluated.

Study the various parts of the following example carefully; they show the various ways in which integration by parts is used, and they give some insights into what choices should be made for U and dV in various situations. An improper choice can result in making an integral harder rather than easier. Look for a factor of the integrand that is easily integrated, and include dx with that factor to make up dV. Then U is the remaining factor of the integrand. Sometimes it is necessary to take $dV = dx$ only. When breaking up an integrand using integration by parts, choose U and dV so that, if possible, $V \, dU$ is "simpler" (easier to integrate) than $U \, dV$.

Example 2 Use integration by parts to evaluate:

(a) $\displaystyle\int \ln x \, dx$, (b) $\displaystyle\int x^2 \sin x \, dx$, (c) $\displaystyle\int x \tan^{-1} x \, dx$, (d) $\displaystyle\int \sin^{-1} x \, dx$.

Solution

(a) $\displaystyle\int \ln x \, dx \qquad\qquad \text{Let} \quad U = \ln x, \qquad dV = dx.$
$$\qquad\qquad\qquad\qquad \text{Then} \ \ dU = dx/x, \qquad V = x.$$

$$= x \ln x - \int x \frac{1}{x} \, dx$$

$$= x \ln x - x + C.$$

(b) We have to integrate by parts twice this time:

$$\int x^2 \sin x \, dx \qquad\qquad \text{Let} \quad U = x^2, \qquad dV = \sin x \, dx.$$
$$\text{Then } dU = 2x \, dx, \qquad V = -\cos x.$$

$$= -x^2 \cos x + 2 \int x \cos x \, dx \qquad \text{Let} \quad U = x, \qquad dV = \cos x \, dx.$$
$$\text{Then } dU = dx, \qquad V = \sin x.$$

$$= -x^2 \cos x + 2 \left(x \sin x - \int \sin x \, dx \right)$$

$$= -x^2 \cos x + 2x \sin x + 2 \cos x + C.$$

(c)
$$\int x \tan^{-1} x \, dx \qquad\qquad \text{Let} \quad U = \tan^{-1} x, \qquad dV = x \, dx.$$
$$\text{Then } dU = dx/(1 + x^2), \qquad V = \tfrac{1}{2} x^2.$$

$$= \frac{1}{2} x^2 \tan^{-1} x - \frac{1}{2} \int \frac{x^2}{1 + x^2} \, dx$$

$$= \frac{1}{2} x^2 \tan^{-1} x - \frac{1}{2} \int \left(1 - \frac{1}{1 + x^2} \right) dx$$

$$= \frac{1}{2} x^2 \tan^{-1} x - \frac{1}{2} x + \frac{1}{2} \tan^{-1} x + C.$$

(d)
$$\int \sin^{-1} x \, dx \qquad\qquad \text{Let} \quad U = \sin^{-1} x, \qquad dV = dx.$$
$$\text{Then } dU = dx/\sqrt{1 - x^2}, \qquad V = x.$$

$$= x \sin^{-1} x - \int \frac{x}{\sqrt{1 - x^2}} \, dx \qquad \text{Let } u = 1 - x^2,$$
$$du = -2x \, dx.$$

$$= x \sin^{-1} x + \frac{1}{2} \int u^{-1/2} \, du$$

$$= x \sin^{-1} x + u^{1/2} + C = x \sin^{-1} x + \sqrt{1 - x^2} + C.$$

The following are two useful rules of thumb for choosing U and dV:

(i) If the integrand involves a polynomial multiplied by an exponential, a sine or a cosine, or some other readily integrable function, try $U =$ the polynomial and $dV =$ the rest.

(ii) If the integrand involves a logarithm, an inverse trigonometric function, or some other function that is not readily integrable but whose derivative is readily calculated, try that function for U and let dV equal the rest.

(Of course, these "rules" come with no guarantee. They may fail to be helpful if "the rest" is not of a suitable form. There remain many integrals that cannot be evaluated by any of the standard techniques presented in this chapter.)

The following two examples illustrate a frequently occurring and very useful phenomenon. It may happen after one or two integrations by parts, with the possible application of some known identity, that the original integral reappears on the right-hand side. Unless its coefficient there is 1, we have an equation that can be solved for that integral.

Example 3 Evaluate $I = \int \sec^3 x \, dx$.

Solution Start by integrating by parts:

$$I = \int \sec^3 x \, dx \qquad \text{Let} \quad U = \sec x, \qquad\qquad dV = \sec^2 x \, dx.$$
$$\text{Then } dU = \sec x \tan x \, dx, \quad V = \tan x.$$

$$= \sec x \tan x - \int \sec x \, \tan^2 x \, dx$$

$$= \sec x \tan x - \int \sec x (\sec^2 x - 1) \, dx$$

$$= \sec x \tan x - \int \sec^3 x \, dx + \int \sec x \, dx$$

$$= \sec x \tan x - I + \ln | \sec x + \tan x |.$$

This is an equation that can be solved for the desired integral I: Since $2I = \sec x \tan x + \ln | \sec x + \tan x |$, we have

$$\int \sec^3 x \, dx = I = \frac{1}{2} \sec x \tan x + \frac{1}{2} \ln | \sec x + \tan x | + C.$$

This integral occurs frequently in applications and is worth remembering. ∎

Example 4 Find $I = \int e^{ax} \cos bx \, dx$.

Solution If either $a = 0$ or $b = 0$, the integral is easy to do, so let us assume $a \neq 0$ and $b \neq 0$. We have

$$I = \int e^{ax} \cos bx \, dx \qquad \text{Let} \quad U = e^{ax}, \qquad dV = \cos bx \, dx.$$
$$\text{Then } dU = a \, e^{ax} \, dx, \quad V = (1/b) \sin bx.$$

$$= \frac{1}{b} e^{ax} \sin bx - \frac{a}{b} \int e^{ax} \sin bx \, dx$$

$$\text{Let} \quad U = e^{ax}, \qquad dV = \sin bx \, dx.$$
$$\text{Then } dU = ae^{ax} dx, \quad V = -(\cos bx)/b.$$

$$= \frac{1}{b} e^{ax} \sin bx - \frac{a}{b} \left(-\frac{1}{b} e^{ax} \cos bx + \frac{a}{b} \int e^{ax} \cos bx \, dx \right)$$

$$= \frac{1}{b} e^{ax} \sin bx + \frac{a}{b^2} e^{ax} \cos bx - \frac{a^2}{b^2} I.$$

Thus,

$$\left(1 + \frac{a^2}{b^2} \right) I = \frac{1}{b} e^{ax} \sin bx + \frac{a}{b^2} e^{ax} \cos bx + C_1$$

and

$$\int e^{ax} \cos bx \, dx = I = \frac{b \, e^{ax} \sin bx + a \, e^{ax} \cos bx}{b^2 + a^2} + C.$$

∎

Observe that after the first integration by parts we had an integral that was different from, but no simpler than, the original integral. At this point we might have become discouraged and given up on this method. However, perseverance proved worthwhile; a second integration by parts returned the original integral I in an equation that could be solved for I. Having chosen to let U be the exponential in the first integration by parts (we could have let it be the cosine), we made the same choice for U in the second integration by parts. Had we switched horses in midstream and decided to let U be the trigonometric function the second time, we would have obtained

$$I = \frac{1}{b} e^{ax} \sin bx - \frac{1}{b} e^{ax} \sin bx + I;$$

we would have *undone* what we accomplished in the first step.

If we want to evaluate a definite integral by the method of integration by parts, we must remember to include the appropriate evaluation symbol with the integrated term.

Example 5   (A definite integral)

$$\int_1^e x^3 (\ln x)^2 \, dx \qquad\qquad \text{Let} \quad U = (\ln x)^2, \qquad dV = x^3 \, dx.$$
$$\text{Then } dU = (2 \ln x \, dx)/x, \quad V = x^4/4.$$

$$= \frac{x^4}{4} (\ln x)^2 \Big|_1^e - \frac{1}{2} \int_1^e x^3 \ln x \, dx \qquad \text{Let} \quad U = \ln x, \qquad dV = x^3 \, dx.$$
$$\text{Then } dU = dx/x, \qquad V = x^4/4.$$

$$= \frac{e^4}{4} (1^2) - 0 - \frac{1}{2} \left(\frac{x^4}{4} \ln x \Big|_1^e - \frac{1}{4} \int_1^e x^3 \, dx \right)$$

$$= \frac{e^4}{4} - \frac{e^4}{8} + \frac{1}{8} \frac{x^4}{4} \Big|_1^e = \frac{e^4}{8} + \frac{e^4}{32} - \frac{1}{32} = \frac{5}{32} e^4 - \frac{1}{32}.$$

$\blacksquare$

Don't do

Reduction Formulas

Consider the problem of finding $\int x^4 e^{-x} \, dx$. We can, as in Example 1, proceed by using integration by parts four times. Each time will reduce the power of x by 1. Since this is repetitive and tedious, we prefer the following approach. For $n \geq 0$, let

$$I_n = \int x^n e^{-x} \, dx.$$

We want to find I_4. If we integrate by parts, we obtain a formula for I_n in terms of I_{n-1}:

$$I_n = \int x^n e^{-x} \, dx \qquad\qquad \text{Let} \quad U = x^n, \qquad dV = e^{-x} \, dx.$$
$$\text{Then } dU = nx^{n-1} \, dx, \quad V = -e^{-x}.$$

$$= -x^n e^{-x} + n \int x^{n-1} e^{-x} \, dx = -x^n e^{-x} + n I_{n-1}.$$

The formula

$$I_n = -x^n e^{-x} + n I_{n-1}$$

is called a **reduction formula** because it gives the value of the integral I_n in terms of I_{n-1}, an integral corresponding to a reduced value of the exponent n. Starting with

$$I_0 = \int x^0 e^{-x}\,dx = \int e^{-x}\,dx = -e^{-x} + C,$$

we can apply the reduction formula four times to get

$$I_1 = -xe^{-x} + I_0 = -e^{-x}(x + 1) + C_1$$
$$I_2 = -x^2 e^{-x} + 2I_1 = -e^{-x}(x^2 + 2x + 2) + C_2$$
$$I_3 = -x^3 e^{-x} + 3I_2 = -e^{-x}(x^3 + 3x^2 + 6x + 6) + C_3$$
$$I_4 = -x^4 e^{-x} + 4I_3 = -e^{-x}(x^4 + 4x^3 + 12x^2 + 24x + 24) + C_4.$$

Example 6   Obtain and use a reduction formula to evaluate

$$I_n = \int_0^{\pi/2} \cos^n x\,dx \qquad (n = 0,\ 1,\ 2,\ 3,\ \dots).$$

Solution   Observe first that

$$I_0 = \int_0^{\pi/2} dx = \frac{\pi}{2} \quad \text{and} \quad I_1 = \int_0^{\pi/2} \cos x\,dx = \sin x \Big|_0^{\pi/2} = 1.$$

Now let $n \geq 2$:

$$I_n = \int_0^{\pi/2} \cos^n x\,dx = \int_0^{\pi/2} \cos^{n-1} x \cos x\,dx$$
$$U = \cos^{n-1} x, \qquad dV = \cos x\,dx$$
$$dU = -(n-1)\cos^{n-2} x \sin x\,dx, \qquad V = \sin x$$
$$= \sin x \cos^{n-1} x \Big|_0^{\pi/2} + (n-1) \int_0^{\pi/2} \cos^{n-2} x \sin^2 x\,dx$$
$$= 0 - 0 + (n-1) \int_0^{\pi/2} \cos^{n-2} x\,(1 - \cos^2 x)\,dx$$
$$= (n-1)I_{n-2} - (n-1)I_n.$$

Transposing the term $-(n-1)I_n$, we obtain $nI_n = (n-1)I_{n-2}$, or

$$I_n = \frac{n-1}{n} I_{n-2},$$

which is the required reduction formula. It is valid for $n \geq 2$, which was needed to ensure that $\cos^{n-1}(\pi/2) = 0$. If $n \geq 2$ is an *even integer*, we have

$$I_n = \frac{n-1}{n} I_{n-2} = \frac{n-1}{n} \cdot \frac{n-3}{n-2} I_{n-4} = \cdots$$
$$= \frac{n-1}{n} \cdot \frac{n-3}{n-2} \cdot \frac{n-5}{n-4} \cdots \frac{5}{6} \cdot \frac{3}{4} \cdot \frac{1}{2} \cdot I_0$$
$$= \frac{n-1}{n} \cdot \frac{n-3}{n-2} \cdot \frac{n-5}{n-4} \cdots \frac{5}{6} \cdot \frac{3}{4} \cdot \frac{1}{2} \cdot \frac{\pi}{2}.$$

If $n \geq 3$ is an *odd* integer, we have

$$I_n = \frac{n-1}{n} \cdot \frac{n-3}{n-2} \cdot \frac{n-5}{n-4} \cdots \frac{6}{7} \cdot \frac{4}{5} \cdot \frac{2}{3} \cdot I_1$$

$$= \frac{n-1}{n} \cdot \frac{n-3}{n-2} \cdot \frac{n-5}{n-4} \cdots \frac{6}{7} \cdot \frac{4}{5} \cdot \frac{2}{3}.$$

See Exercise 38 for an interesting consequence of these formulas.

Exercises 6.1

Evaluate the integrals in Exercises 1–28.

1. $\displaystyle\int x \cos x \, dx$

2. $\displaystyle\int (x+3)e^{2x} \, dx$

3. $\displaystyle\int x^2 \cos \pi x \, dx$

4. $\displaystyle\int (x^2 - 2x)e^{kx} \, dx$

5. $\displaystyle\int x^3 \ln x \, dx$

6. $\displaystyle\int x(\ln x)^3 \, dx$

7. $\displaystyle\int \tan^{-1} x \, dx$

8. $\displaystyle\int x^2 \tan^{-1} x \, dx$

9. $\displaystyle\int x \sin^{-1} x \, dx$

10. $\displaystyle\int x^5 e^{-x^2} \, dx$

11. $\displaystyle\int_0^{\pi/4} \sec^5 x \, dx$

12. $\displaystyle\int \tan^2 x \sec x \, dx$

13. $\displaystyle\int e^{2x} \sin 3x \, dx$

14. $\displaystyle\int x e^{\sqrt{x}} \, dx$

* 15. $\displaystyle\int_{1/2}^1 \frac{\sin^{-1} x}{x^2} \, dx$

16. $\displaystyle\int_0^1 \sqrt{x} \sin(\pi \sqrt{x}) \, dx$

17. $\displaystyle\int x \sec^2 x \, dx$

18. $\displaystyle\int x \sin^2 x \, dx$

19. $\displaystyle\int \cos(\ln x) \, dx$

20. $\displaystyle\int_1^e \sin(\ln x) \, dx$

21. $\displaystyle\int \frac{\ln(\ln x)}{x} \, dx$

22. $\displaystyle\int_0^4 \sqrt{x} e^{\sqrt{x}} \, dx$

23. $\displaystyle\int \arccos x \, dx$

24. $\displaystyle\int x \sec^{-1} x \, dx$

25. $\displaystyle\int_1^2 \sec^{-1} x \, dx$

* 26. $\displaystyle\int (\sin^{-1} x)^2 \, dx$

* 27. $\displaystyle\int x(\tan^{-1} x)^2 \, dx$

* 28. $\displaystyle\int x e^x \cos x \, dx$

29. Find the area below $y = e^{-x} \sin x$ and above $y = 0$ from $x = 0$ to $x = \pi$.

30. Find the area of the finite plane region bounded by the curve $y = \ln x$, the line $y = 1$, and the tangent line to $y = \ln x$ at $x = 1$.

Reduction formulas

31. Obtain a reduction formula for $I_n = \int (\ln x)^n \, dx$, and use it to evaluate I_4.

32. Obtain a reduction formula for $I_n = \int_0^{\pi/2} x^n \sin x \, dx$, and use it to evaluate I_6.

33. Obtain a reduction formula for $I_n = \int \sin^n x \, dx$ (where $n \geq 2$), and use it to find I_6 and I_7.

34. Obtain a reduction formula for $I_n = \int \sec^n x \, dx$ (where $n \geq 3$), and use it to find I_6 and I_7.

* 35. By writing

$$I_n = \int \frac{dx}{(x^2 + a^2)^n}$$

$$= \frac{1}{a^2} \int \frac{dx}{(x^2 + a^2)^{n-1}} - \frac{1}{a^2} \int x \frac{x}{(x^2 + a^2)^n} \, dx$$

and integrating the last integral by parts, using $U = x$, obtain a reduction formula for I_n. Use this formula to find I_3.

* 36. If f is twice differentiable on $[a, b]$ and $f(a) = f(b) = 0$, show that

$$\int_a^b (x-a)(b-x)f''(x) \, dx = -2 \int_a^b f(x) \, dx.$$

(*Hint:* use integration by parts on the left-hand side twice.) This formula will be used in Section 6.6 to construct an error estimate for the Trapezoid Rule approximation formula.

* 37. If f and g are two functions having continuous second derivatives on the interval $[a, b]$, and if $f(a) = g(a) = f(b) = g(b) = 0$, show that

$$\int_a^b f(x) g''(x) \, dx = \int_a^b f''(x) g(x) \, dx.$$

What other assumptions about the values of f and g at a and b would give the same result?

* **38. (The Wallis Product)** Let $I_n = \int_0^{\pi/2} \cos^n x \, dx$.

(a) Use the fact that $0 \le \cos x \le 1$ for $0 \le x \le \pi/2$ to show that $I_{2n+2} \le I_{2n+1} \le I_{2n}$, for $n = 0, 1, 2, \ldots$.

(b) Use the reduction formula $I_n = ((n-1)/n)I_{n-2}$ obtained in Example 6, together with the result of (a), to show that

$$\lim_{n \to \infty} \frac{I_{2n+1}}{I_{2n}} = 1.$$

(c) Combine the result of (b) with the explicit formulas

obtained for I_n (for even and odd n) in Example 6 to show that

$$\lim_{n \to \infty} \frac{2}{1} \cdot \frac{2}{3} \cdot \frac{4}{3} \cdot \frac{4}{5} \cdot \frac{6}{5} \cdot \frac{6}{7} \cdots \frac{2n}{2n-1} \cdot \frac{2n}{2n+1} = \frac{\pi}{2}.$$

This interesting product formula for π is due to the seventeenth-century English mathematician John Wallis and is referred to as the Wallis Product.

6.2 Inverse Substitutions

The substitutions considered in Section 5.6 were direct substitutions in the sense that we simplified an integrand by replacing an expression appearing in it with a single variable. In this section we consider the reverse approach; we replace the variable of integration with a function of a new variable. Such substitutions, called *inverse substitutions*, would appear on the surface to make the integral more complicated. That is, substituting $x = g(u)$ in the integral

$$\int_a^b f(x) \, dx$$

leads to the more "complicated" integral

$$\int_{x=a}^{x=b} f\big(g(u)\big) g'(u) \, du.$$

As we will see, however, sometimes such substitutions can actually simplify an integrand, transforming the integral into one that can be evaluated by inspection or to which other techniques can readily be applied.

The Inverse Trigonometric Substitutions

Three very useful inverse substitutions are:

$$x = a \sin \theta, \qquad x = a \tan \theta, \quad \text{and} \quad x = a \sec \theta.$$

These correspond to the direct substitutions:

$$\theta = \sin^{-1} \frac{x}{a}, \qquad \theta = \tan^{-1} \frac{x}{a}, \quad \text{and} \quad \theta = \sec^{-1} \frac{x}{a} = \cos^{-1} \frac{a}{x}.$$

The inverse sine substitution

Integrals involving $\sqrt{a^2 - x^2}$ (where $a > 0$) can frequently be reduced to a simpler form by means of the substitution

$$x = a \sin \theta \quad \text{or, equivalently,} \quad \theta = \sin^{-1} \frac{x}{a}.$$

Observe that $\sqrt{a^2 - x^2}$ makes sense only if $-a \leq x \leq a$, which corresponds to $-\pi/2 \leq \theta \leq \pi/2$. Since $\cos \theta \geq 0$ for such θ, we have

$$\sqrt{a^2 - x^2} = \sqrt{a^2(1 - \sin^2 \theta)} = \sqrt{a^2 \cos^2 \theta} = a \cos \theta.$$

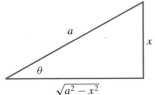

Figure 6.1

(If $\cos \theta$ were not nonnegative, we would have obtained $a|\cos \theta|$ instead.) If needed, the other trigonometric functions of θ can be recovered in terms of x by examining a right-angled triangle labelled to correspond to the substitution. (See Figure 6.1.)

$$\cos \theta = \frac{\sqrt{a^2 - x^2}}{a} \qquad \text{and} \qquad \tan \theta = \frac{x}{\sqrt{a^2 - x^2}}.$$

Example 1 Evaluate $\displaystyle\int \frac{1}{(5 - x^2)^{3/2}} \, dx$.

Solution Refer to Figure 6.2.

$$\int \frac{1}{(5 - x^2)^{3/2}} \, dx \qquad \text{Let } x = \sqrt{5} \sin \theta,$$
$$\qquad\qquad\qquad\qquad\qquad dx = \sqrt{5} \cos \theta \, d\theta.$$

$$= \int \frac{\sqrt{5} \cos \theta \, d\theta}{5^{3/2} \cos^3 \theta}$$

$$= \frac{1}{5} \int \sec^2 \theta \, d\theta = \frac{1}{5} \tan \theta + C = \frac{1}{5} \frac{x}{\sqrt{5 - x^2}} + C$$

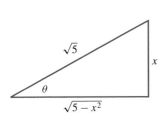

Figure 6.2

Example 2 Find the area of the circular segment shaded in Figure 6.3.

Solution The area is

$$A = 2 \int_b^a \sqrt{a^2 - x^2} \, dx \qquad \text{Let } x = a \sin \theta,$$
$$\qquad\qquad\qquad\qquad\qquad dx = a \cos \theta \, d\theta.$$

$$= 2 \int_{x=b}^{x=a} a^2 \cos^2 \theta \, d\theta$$

$$= a^2 \left(\theta + \sin \theta \cos \theta \right) \Big|_{x=b}^{x=a} \qquad \text{(as in Example 8 of Section 5.6)}$$

$$= a^2 \left(\sin^{-1} \frac{x}{a} + \frac{x\sqrt{a^2 - x^2}}{a^2} \right) \Big|_b^a \qquad \text{(See Figure 6.1.)}$$

$$= \frac{\pi}{2} a^2 - a^2 \sin^{-1} \frac{b}{a} - b\sqrt{a^2 - b^2} \quad \text{square units.}$$

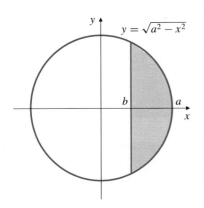

Figure 6.3

The inverse tangent substitution

Integrals involving $\sqrt{a^2 + x^2}$ or $\dfrac{1}{x^2 + a^2}$ (where $a > 0$) are often simplified by the substitution

$$x = a \tan\theta \quad \text{or, equivalently,} \quad \theta = \tan^{-1}\frac{x}{a}.$$

Since x can take any real value, we have $-\pi/2 < \theta < \pi/2$, so $\sec\theta > 0$ and

$$\sqrt{a^2 + x^2} = a\sqrt{1 + \tan^2\theta} = a\sec\theta.$$

Other trigonometric functions of θ can be expressed in terms of x by referring to a right-angled triangle with legs a and x and hypotenuse $\sqrt{a^2 + x^2}$ (see Figure 6.4):

$$\sin\theta = \frac{x}{\sqrt{a^2 + x^2}} \quad \text{and} \quad \cos\theta = \frac{a}{\sqrt{a^2 + x^2}}.$$

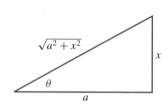

Figure 6.4

Example 3 Evaluate (a) $\displaystyle\int \frac{1}{\sqrt{4 + x^2}}\,dx$ and (b) $\displaystyle\int \frac{1}{(1 + 9x^2)^2}\,dx$.

Solution Figures 6.5 and 6.6 illustrate parts (a) and (b), respectively.

(a) $\displaystyle\int \frac{1}{\sqrt{4 + x^2}}\,dx$ Let $x = 2\tan\theta$,
$$dx = 2\sec^2\theta\,d\theta.$$

$$= \int \frac{2\sec^2\theta}{2\sec\theta}\,d\theta$$

$$= \int \sec\theta\,d\theta$$

$$= \ln|\sec\theta + \tan\theta| + C = \ln\left|\frac{\sqrt{4+x^2}}{2} + \frac{x}{2}\right| + C$$

$$= \ln\left(\sqrt{4 + x^2} + x\right) + C_1, \qquad \text{where } C_1 = C - \ln 2.$$

(Note that $\sqrt{4 + x^2} + x > 0$ for all x, so we do not need an absolute value on it.)

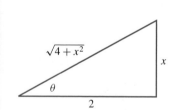

Figure 6.5

(b) $\displaystyle\int \frac{1}{(1 + 9x^2)^2}\,dx$ Let $3x = \tan\theta$,
$$3dx = \sec^2\theta\,d\theta,$$
$$1 + 9x^2 = \sec^2\theta.$$

$$= \frac{1}{3}\int \frac{\sec^2\theta\,d\theta}{\sec^4\theta}$$

$$= \frac{1}{3}\int \cos^2\theta\,d\theta = \frac{1}{6}(\theta + \sin\theta\,\cos\theta) + C$$

$$= \frac{1}{6}\tan^{-1}(3x) + \frac{1}{6}\frac{3x}{\sqrt{1 + 9x^2}}\frac{1}{\sqrt{1 + 9x^2}} + C$$

$$= \frac{1}{6}\tan^{-1}(3x) + \frac{1}{2}\frac{x}{1 + 9x^2} + C$$

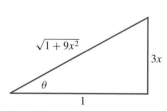

Figure 6.6

The inverse secant substitution

Integrals involving $\sqrt{x^2 - a^2}$ (where $a > 0$) can frequently be simplified by using the substitution

$$x = a \sec\theta \quad \text{or, equivalently,} \quad \theta = \sec^{-1}\frac{x}{a}.$$

We must be more careful with this substitution. Although

$$\sqrt{x^2 - a^2} = a\sqrt{\sec^2\theta - 1} = a\sqrt{\tan^2\theta} = a|\tan\theta|,$$

we cannot always drop the absolute value from the tangent. Observe that $\sqrt{x^2 - a^2}$ makes sense for $x \geq a$ and for $x \leq -a$.

If $x \geq a$, then $0 \leq \theta = \sec^{-1}\frac{x}{a} = \arccos\frac{a}{x} < \frac{\pi}{2}$, and $\tan\theta \geq 0$.

If $x \leq -a$, then $\frac{\pi}{2} < \theta = \sec^{-1}\frac{x}{a} = \arccos\frac{a}{x} \leq \pi$, and $\tan\theta \leq 0$.

In the first case $\sqrt{x^2 - a^2} = a\tan\theta$; in the second case $\sqrt{x^2 - a^2} = -a\tan\theta$.

Example 4 Find $I = \displaystyle\int \frac{dx}{\sqrt{x^2 - a^2}}$, where $a > 0$.

Solution For the moment assume that $x \geq a$. If $x = a\sec\theta$, then $dx = a\sec\theta\,\tan\theta\,d\theta$ and $\sqrt{x^2 - a^2} = a\tan\theta$ (Figure 6.7). Thus,

$$I = \int \sec\theta\,d\theta = \ln|\sec\theta + \tan\theta| + C$$

$$= \ln\left|\frac{x}{a} + \frac{\sqrt{x^2 - a^2}}{a}\right| + C = \ln|x + \sqrt{x^2 - a^2}| + C_1,$$

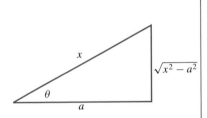

Figure 6.7

where $C_1 = C - \ln a$. If $x \leq -a$, let $u = -x$ so that $u \geq a$ and $du = -dx$. We have

$$I = -\int \frac{du}{\sqrt{u^2 - a^2}} = -\ln|u + \sqrt{u^2 - a^2}| + C_1$$

$$= \ln\left|\frac{1}{-x + \sqrt{x^2 - a^2}} \frac{x + \sqrt{x^2 - a^2}}{x + \sqrt{x^2 - a^2}}\right| + C_1$$

$$= \ln\left|\frac{x + \sqrt{x^2 - a^2}}{-a^2}\right| + C_1 = \ln|x + \sqrt{x^2 - a^2}| + C_2,$$

where $C_2 = C_1 - 2\ln a$. Thus, in either case, we have

$$I = \ln|x + \sqrt{x^2 - a^2}| + C.$$

Completing the Square

Quadratic expressions of the form $Ax^2 + Bx + C$ are often found in integrands. These can be written as sums or differences of squares using the procedure of completing the square. First factor out A so that the remaining expression begins with $x^2 + 2bx$, where $2b = B/A$. These are the first two terms of $(x + b)^2 = x^2 + 2bx + b^2$. Add the third term $b^2 = B^2/4A^2$ and then subtract it again:

$$Ax^2 + Bx + C = A\left(x^2 + \frac{B}{A}x + \frac{C}{A}\right)$$

$$= A\left(x^2 + \frac{B}{A}x + \frac{B^2}{4A^2} + \frac{C}{A} - \frac{B^2}{4A^2}\right)$$

$$= A\left(x + \frac{B}{2A}\right)^2 + \frac{4AC - B^2}{4A}$$

The substitution $u = x + \dfrac{B}{2A}$ should then be made.

Example 5   Evaluate   (a) $\displaystyle\int \frac{1}{\sqrt{2x - x^2}}\, dx$ and (b) $\displaystyle\int \frac{x}{4x^2 + 12x + 13}\, dx$.

Solution

(a) $\displaystyle\int \frac{1}{\sqrt{2x - x^2}}\, dx = \int \frac{dx}{\sqrt{1 - (1 - 2x + x^2)}}$

$$= \int \frac{dx}{\sqrt{1 - (x - 1)^2}} \qquad \text{Let } u = x - 1,$$
$$\qquad\qquad\qquad\qquad\qquad\qquad du = dx.$$

$$= \int \frac{du}{\sqrt{1 - u^2}} = \sin^{-1} u + C = \sin^{-1}(x - 1) + C.$$

(b) $\displaystyle\int \frac{x}{4x^2 + 12x + 13}\, dx = \int \frac{x\, dx}{4\left(x^2 + 3x + \dfrac{9}{4} + 1\right)}$

$$= \frac{1}{4} \int \frac{x\, dx}{\left(x + \dfrac{3}{2}\right)^2 + 1} \qquad \begin{array}{l}\text{Let } u = x + (3/2),\\ du = dx,\\ x = u - (3/2).\end{array}$$

$$= \frac{1}{4} \int \frac{u\, du}{u^2 + 1} - \frac{3}{8} \int \frac{du}{u^2 + 1} \qquad \begin{array}{l}\text{In the first integral}\\ \text{let } v = u^2 + 1,\\ dv = 2u\, du.\end{array}$$

$$= \frac{1}{8} \int \frac{dv}{v} - \frac{3}{8} \tan^{-1} u$$

$$= \frac{1}{8} \ln |v| - \frac{3}{8} \tan^{-1} u + C$$

$$= \frac{1}{8} \ln(4x^2 + 12x + 13) - \frac{3}{8} \tan^{-1}\left(x + \frac{3}{2}\right) + C_1,$$

where $C_1 = C - (\ln 4)/8$.

Other Inverse Substitutions

Integrals involving $\sqrt{ax+b}$ can sometimes be made simpler with the substitution $ax + b = u^2$.

Example 6
$$\int \frac{1}{1 + \sqrt{2x}}\, dx \qquad \text{Let } 2x = u^2,$$
$$2\, dx = 2u\, du.$$
$$= \int \frac{u}{1+u}\, du = \int \frac{1+u-1}{1+u}\, du$$
$$= \int \left(1 - \frac{1}{1+u}\right) du \qquad \text{Let } v = 1+u,$$
$$dv = du.$$
$$= u - \int \frac{dv}{v} = u - \ln|v| + C$$
$$= \sqrt{2x} - \ln\left(1 + \sqrt{2x}\right) + C.$$

Sometimes integrals involving $\sqrt[n]{ax+b}$ will be much simplified by the hybrid substitution $ax + b = u^n$, $a\, dx = n\,u^{n-1}\, du$.

Example 7
$$\int_{-1/3}^{2} \frac{x}{\sqrt[3]{3x+2}}\, dx \qquad \text{Let } 3x + 2 = u^3,$$
$$3\, dx = 3u^2\, du.$$
$$= \int_{1}^{2} \frac{u^3 - 2}{3u}\, u^2\, du$$
$$= \frac{1}{3} \int_{1}^{2} (u^4 - 2u)\, du = \frac{1}{3}\left(\frac{u^5}{5} - u^2\right)\Bigg|_{1}^{2} = \frac{16}{15}.$$

Note that the limits were changed in this definite integral. $u = 1$ when $x = -1/3$, and, coincidentally, $u = 2$ when $x = 2$.

If more than one fractional power is present, it may be possible to eliminate all of them at once.

Example 8 Evaluate $\displaystyle\int \frac{1}{x^{1/2}(1 + x^{1/3})}\, dx.$

Solution We can eliminate both the square root and the cube root by using the inverse substitution $x = u^6$. (The power 6 is chosen because 6 is the least common multiple of 2 and 3.)

$$\int \frac{dx}{x^{1/2}(1 + x^{1/3})} \qquad \text{Let } x = u^6,$$
$$dx = 6u^5\, du.$$
$$= 6\int \frac{u^5\, du}{u^3(1 + u^2)} = 6\int \frac{u^2}{1 + u^2}\, du = 6\int \left(1 - \frac{1}{1+u^2}\right) du$$
$$= 6\left(u - \tan^{-1} u\right) + C = 6\left(x^{1/6} - \tan^{-1} x^{1/6}\right) + C.$$

The tan(θ/2) Substitution

There is a certain special substitution that can transform an integral whose integrand is a rational function of $\sin\theta$ and $\cos\theta$ (i.e., a quotient of polynomials in $\sin\theta$ and $\cos\theta$) into a rational function of x. The substitution is

$$x = \tan\frac{\theta}{2} \quad \text{or, equivalently,} \quad \theta = 2\tan^{-1}x.$$

Observe that

$$\cos^2\frac{\theta}{2} = \frac{1}{\sec^2\dfrac{\theta}{2}} = \frac{1}{1 + \tan^2\dfrac{\theta}{2}} = \frac{1}{1 + x^2},$$

so

$$\cos\theta = 2\cos^2\frac{\theta}{2} - 1 = \frac{2}{1 + x^2} - 1 = \frac{1 - x^2}{1 + x^2}$$

$$\sin\theta = 2\sin\frac{\theta}{2}\cos\frac{\theta}{2} = 2\tan\frac{\theta}{2}\cos^2\frac{\theta}{2} = \frac{2x}{1 + x^2}.$$

Also, $dx = \dfrac{1}{2}\sec^2\dfrac{\theta}{2}\,d\theta$, so

$$d\theta = 2\cos^2\frac{\theta}{2}\,dx = \frac{2\,dx}{1 + x^2}.$$

In summary:

DON'T DO tan(θ/2)

The tan(θ/2) substitution

If $x = \tan(\theta/2)$, then

$$\cos\theta = \frac{1 - x^2}{1 + x^2}, \quad \sin\theta = \frac{2x}{1 + x^2}, \quad \text{and} \quad d\theta = \frac{2\,dx}{1 + x^2}.$$

Note that $\cos\theta$, $\sin\theta$, and $d\theta$ all involve only rational functions of x. We will examine general techniques for integrating rational functions of x in Section 6.3.

Example 9

$$\int \frac{1}{2 + \cos\theta}\,d\theta \qquad \text{Let } x = \tan(\theta/2), \text{ so}$$

$$\cos\theta = \frac{1 - x^2}{1 + x^2},$$

$$d\theta = \frac{2\,dx}{1 + x^2}.$$

$$= \int \frac{\dfrac{2\,dx}{1 + x^2}}{2 + \dfrac{1 - x^2}{1 + x^2}} = 2\int \frac{1}{3 + x^2}\,dx$$

$$= \frac{2}{\sqrt{3}}\tan^{-1}\frac{x}{\sqrt{3}} + C$$

$$= \frac{2}{\sqrt{3}}\tan^{-1}\left(\frac{1}{\sqrt{3}}\tan\frac{\theta}{2}\right) + C.$$

Exercises 6.2

Evaluate the integrals in Exercises 1–36.

1. $\displaystyle \int \frac{dx}{\sqrt{1-4x^2}}$

2. $\displaystyle \int \frac{x^2\,dx}{\sqrt{1-4x^2}}$

3. $\displaystyle \int \frac{x^2\,dx}{\sqrt{9-x^2}}$

4. $\displaystyle \int \frac{dx}{x\sqrt{1-4x^2}}$

5. $\displaystyle \int \frac{dx}{x^2\sqrt{9-x^2}}$

6. $\displaystyle \int \frac{dx}{x\sqrt{9-x^2}}$

7. $\displaystyle \int \frac{x+1}{\sqrt{9-x^2}}\,dx$

8. $\displaystyle \int \frac{dx}{\sqrt{9+x^2}}$

9. $\displaystyle \int \frac{x^3\,dx}{\sqrt{9+x^2}}$

10. $\displaystyle \int \frac{\sqrt{9+x^2}}{x^4}\,dx$

11. $\displaystyle \int \frac{dx}{(a^2-x^2)^{3/2}}$

12. $\displaystyle \int \frac{dx}{(a^2+x^2)^{3/2}}$

13. $\displaystyle \int \frac{x^2\,dx}{(a^2-x^2)^{3/2}}$

14. $\displaystyle \int \frac{dx}{(1+2x^2)^{5/2}}$

15. $\displaystyle \int \frac{dx}{x\sqrt{x^2-4}} \quad (x>2)$

16. $\displaystyle \int \frac{dx}{x^2\sqrt{x^2-a^2}} \quad (x>a>0)$

17. $\displaystyle \int \frac{dx}{x^2+2x+10}$

18. $\displaystyle \int \frac{dx}{x^2+x+1}$

19. $\displaystyle \int \frac{dx}{(4x^2+4x+5)^2}$

20. $\displaystyle \int \frac{x\,dx}{x^2-2x+3}$

21. $\displaystyle \int \frac{x\,dx}{\sqrt{2ax-x^2}}$

22. $\displaystyle \int \frac{dx}{(4x-x^2)^{3/2}}$

23. $\displaystyle \int \frac{x\,dx}{(3-2x-x^2)^{3/2}}$

24. $\displaystyle \int \frac{dx}{(x^2+2x+2)^2}$

25. $\displaystyle \int \frac{dx}{(1+x^2)^3}$

26. $\displaystyle \int \frac{x^2\,dx}{(1+x^2)^2}$

27. $\displaystyle \int \frac{\sqrt{1-x^2}}{x^3}\,dx$

28. $\displaystyle \int \sqrt{9+x^2}\,dx$

29. $\displaystyle \int \frac{dx}{2+\sqrt{x}}$

30. $\displaystyle \int \frac{dx}{1+x^{1/3}}$

31. $\displaystyle \int \frac{1+x^{1/2}}{1+x^{1/3}}\,dx$

32. $\displaystyle \int \frac{x\sqrt{2-x^2}}{\sqrt{x^2+1}}\,dx$

33. $\displaystyle \int_{-\ln 2}^{0} e^x\sqrt{1-e^{2x}}\,dx$

34. $\displaystyle \int_{0}^{\pi/2} \frac{\cos x}{\sqrt{1+\sin^2 x}}\,dx$

35. $\displaystyle \int_{-1}^{\sqrt{3}-1} \frac{dx}{x^2+2x+2}$

36. $\displaystyle \int_{1}^{2} \frac{dx}{x^2\sqrt{9-x^2}}$

In Exercises 37–39, evaluate the integral using the special substitution $x=\tan(\theta/2)$ as in Example 9.

★ 37. $\displaystyle \int \frac{d\theta}{2+\sin\theta}$

★ 38. $\displaystyle \int_{0}^{\pi/2} \frac{d\theta}{1+\cos\theta+\sin\theta}$

★ 39. $\displaystyle \int \frac{d\theta}{3+2\cos\theta}$

40. Find the area of the region bounded by $y=(2x-x^2)^{-1/2}$, $y=0$, $x=1/2$, and $x=1$.

41. Find the area of the region lying below $y=9/(x^4+4x^2+4)$ and above $y=1$.

42. Find the average value of the function $f(x)=(x^2-4x+8)^{-3/2}$ over the interval $[0,4]$.

43. Find the area inside the circle $x^2+y^2=a^2$ and above the line $y=b$, $(-a\le b\le a)$.

44. Find the area inside both of the circles $x^2+y^2=1$ and $(x-2)^2+y^2=4$.

45. Find the area in the first quadrant, above the hyperbola $xy=12$ and inside the circle $x^2+y^2=25$.

46. Find the area to the left of $\dfrac{x^2}{a^2}+\dfrac{y^2}{b^2}=1$ and to the right of the line $x=c$, where $-a\le c\le a$.

★ 47. Find the area of the region bounded by the x-axis, the hyperbola $x^2-y^2=1$, and the straight line from the origin to the point $\left(\sqrt{1+Y^2},\,Y\right)$ on that hyperbola. (Assume $Y>0$.) In particular, show that the area is $t/2$ square units if $Y=\sinh t$.

★ 48. Evaluate the integrals

$$\int \frac{dx}{\sqrt{x^2-a^2}} \quad \text{and} \quad \int \frac{dx}{x^2\sqrt{x^2-a^2}}$$

for $x>a>0$ using the substitution $x=a\cosh u$. (*Hint:* review the properties of the hyperbolic functions in Section 3.6.) This substitution is an alternative to $x=a\sec\theta$ when dealing with $\sqrt{x^2-a^2}$.

6.3 Integrals of Rational Functions

In this section we are concerned with integrals of the form

$$\int \frac{P(x)}{Q(x)} \, dx,$$

where P and Q are polynomials. Recall that a **polynomial** is a function P of the form

$$P(x) = a_n x^n + a_{n-1} x^{n-1} + \cdots + a_2 x^2 + a_1 x + a_0$$

where n is a nonnegative integer, $a_0, a_1, a_2, \ldots, a_n$ are constants, and $a_n \neq 0$. We call n the **degree** of P. A quotient $P(x)/Q(x)$ of two polynomials is called a **rational function**. We need normally concern ourselves only with rational functions $P(x)/Q(x)$ where the degree of P is less than that of Q. If the degree of P equals or exceeds the degree of Q, then we can use long division or some equivalent procedure to express the fraction $P(x)/Q(x)$ as a polynomial plus another fraction $R(x)/Q(x)$, where R, the remainder in the division, has degree less than that of Q.

Example 1 Evaluate $\displaystyle\int \frac{x^3 + 3x^2}{x^2 + 1} \, dx$.

Solution The numerator has degree 3 and the denominator has degree 2 so we need to divide. We use long division:

$$
\begin{array}{r}
x + 3 \\
x^2 + 1 \overline{)\, x^3 + 3x^2 } \\
\underline{x^3 + x} \\
3x^2 - x \\
\underline{3x^2 + 3} \\
- x - 3
\end{array}
$$

$$\frac{x^3 + 3x^2}{x^2 + 1} = x + 3 - \frac{x + 3}{x^2 + 1}.$$

Thus,

$$\int \frac{x^3 + 3x^2}{x^2 + 1} \, dx = \int (x + 3) \, dx - \int \frac{x}{x^2 + 1} \, dx - 3 \int \frac{dx}{x^2 + 1}$$

$$= \frac{1}{2} x^2 + 3x - \frac{1}{2} \ln(x^2 + 1) - 3 \tan^{-1} x + C.$$

Example 2 Evaluate $\displaystyle\int \frac{x}{2x - 1} \, dx$.

Solution The numerator and denominator have the same degree, 1, so division is again required. In this case the division can be carried out by manipulation of the integrand:

$$\frac{x}{2x - 1} = \frac{1}{2} \frac{2x}{2x - 1} = \frac{1}{2} \frac{2x - 1 + 1}{2x - 1} = \frac{1}{2} \left(1 + \frac{1}{2x - 1} \right),$$

a process that we can call *short division*. We have

$$\int \frac{x}{2x - 1} \, dx = \frac{1}{2} \int \left(1 + \frac{1}{2x - 1} \right) dx = \frac{x}{2} + \frac{1}{4} \ln |2x - 1| + C.$$

In the discussion that follows, we always assume that any necessary division has been performed and the quotient polynomial has been integrated. The remaining basic problem with which we will deal in this section is the following:

The basic problem

Evaluate $\displaystyle\int \frac{P(x)}{Q(x)}\, dx$, where the degree of P < the degree of Q.

The complexity of this problem depends on the degree of Q.

Linear and Quadratic Denominators

Suppose that $Q(x)$ has degree 1. Thus $Q(x) = ax + b$, where $a \neq 0$. Then $P(x)$ must have degree 0 and be a constant c. We have $P(x)/Q(x) = c/(ax + b)$. The substitution $u = ax + b$ leads to

$$\int \frac{c}{ax+b}\, dx = \frac{c}{a} \int \frac{du}{u} = \frac{c}{a} \ln|u| + C,$$

so that, for $c = 1$

The case of a linear denominator

$$\int \frac{1}{ax+b}\, dx = \frac{1}{a} \ln|ax+b| + C.$$

Now suppose that $Q(x)$ is quadratic, that is, has degree 2. For purposes of this discussion we can assume that $Q(x)$ is either of the form $x^2 + a^2$ or of the form $x^2 - a^2$, since completing the square and making the appropriate change of variable can always reduce a quadratic denominator to this form, as shown in Section 6.2. Since $P(x)$ can be at most a linear function, $P(x) = Ax + B$, we are led to consider the following four integrals:

$$\int \frac{x\, dx}{x^2 + a^2}, \qquad \int \frac{x\, dx}{x^2 - a^2}, \qquad \int \frac{dx}{x^2 + a^2}, \qquad \text{and} \qquad \int \frac{dx}{x^2 - a^2}.$$

(If $a = 0$, there are only two integrals; each is easily evaluated.) The first two integrals yield to the substitution $u = x^2 \pm a^2$; the third is a known integral. The fourth integral can be done with the substitution $x = a \sin \theta$ if $|x| < |a|$, and with the substitution $x = a \sec \theta$ if $|x| > |a|$, but we will evaluate it by a different method below. The values of all four integrals are given in the following box:

The case of a quadratic denominator

$$\int \frac{x\, dx}{x^2 + a^2} = \frac{1}{2} \ln(x^2 + a^2) + C,$$

$$\int \frac{x\, dx}{x^2 - a^2} = \frac{1}{2} \ln|x^2 - a^2| + C,$$

$$\int \frac{dx}{x^2 + a^2} = \frac{1}{a} \tan^{-1} \frac{x}{a} + C,$$

$$\int \frac{dx}{x^2 - a^2} = \frac{1}{2a} \ln \left| \frac{x - a}{x + a} \right| + C.$$

To obtain the last formula in the box let us try to write the integrand as a sum of two fractions with linear denominators:

$$\frac{1}{x^2 - a^2} = \frac{1}{(x-a)(x+a)} = \frac{A}{x-a} + \frac{B}{x+a} = \frac{Ax + Aa + Bx - Ba}{x^2 - a^2},$$

where we have added the two fractions together again in the last step. If this equation is to hold identically for all x (except $x = \pm a$), then the numerators on the left and right sides must be identical as polynomials in x. The equation $(A + B)x + (Aa - Ba) = 1 = 0x + 1$ can hold for all x only if

$$A + B = 0 \qquad \text{(the coefficient of } x)$$
$$Aa - Ba = 1 \qquad \text{(the constant term)}$$

Solving this pair of linear equations for the unknowns A and B, we get $A = 1/(2a)$ and $B = -1/(2a)$. Therefore,

$$\begin{aligned}
\int \frac{dx}{x^2 - a^2} &= \frac{1}{2a} \int \frac{dx}{x-a} - \frac{1}{2a} \int \frac{dx}{x+a} \\
&= \frac{1}{2a} \ln|x - a| - \frac{1}{2a} \ln|x + a| + C \\
&= \frac{1}{2a} \ln \left| \frac{x-a}{x+a} \right| + C.
\end{aligned}$$

✳ Partial Fractions

The technique used above, involving the writing of a complicated fraction as a sum of simpler fractions, is called the **method of partial fractions**. Suppose that a polynomial $Q(x)$ is of degree n and that its highest degree term is x^n (with coefficient 1). Suppose also that Q factors into a product of n *distinct* linear (degree 1) factors, say

$$Q(x) = (x - a_1)(x - a_2) \cdots (x - a_n),$$

where $a_i \neq a_j$ if $i \neq j$, $1 \leq i, j \leq n$. If $P(x)$ is a polynomial of degree smaller than n, then $P(x)/Q(x)$ has a **partial fraction decomposition** of the form

$$\frac{P(x)}{Q(x)} = \frac{A_1}{x - a_1} + \frac{A_2}{x - a_2} + \cdots + \frac{A_n}{x - a_n}$$

for certain values of the constants $A_1, A_2, \ldots, A_n$. We do not attempt to give any formal proof of this assertion here; such a proof belongs in an algebra course. (See Theorem 1 below for the statement of a more general result.)

Given that $P(x)/Q(x)$ has a partial fraction decomposition as claimed above, there are two methods for determining the constants $A_1, A_2, \ldots, A_n$. The first of these methods, and one that generalizes most easily to the more complicated decompositions considered below, is to add up the fractions in the decomposition, obtaining a new fraction $S(x)/Q(x)$ with numerator $S(x)$, a polynomial of degree one less than that of $Q(x)$. This new fraction will be identical to the original fraction $P(x)/Q(x)$ if S and P are identical polynomials. The constants $A_1, A_2, \ldots, A_n$ are determined by solving the n linear equations resulting from equating the coefficients of like powers of x in the two polynomials S and P.

The second method depends on the following observation: If we multiply the partial fraction decomposition by $x - a_j$, we get

$$(x - a_j)\frac{P(x)}{Q(x)}$$

$$= A_1\frac{x - a_j}{x - a_1} + \cdots + A_{j-1}\frac{x - a_j}{x - a_{j-1}} + A_j + A_{j+1}\frac{x - a_j}{x - a_{j+1}} + \cdots + A_n\frac{x - a_j}{x - a_n}.$$

All terms on the right side are 0 at $x = a_j$ except the jth term, A_j. Hence

$$A_j = \lim_{x \to a_j} (x - a_j)\frac{P(x)}{Q(x)}$$

$$= \frac{P(a_j)}{(a_j - a_1)\cdots(a_j - a_{j-1})(a_j - a_{j+1})\cdots(a_j - a_n)},$$

for $1 \leq j \leq n$. In practice, you can use this method to find each number A_j by cancelling the factor $x - a_j$ from the denominator of $P(x)/Q(x)$ and evaluating the resulting expression at $x = a_j$.

Example 3 Evaluate $\displaystyle\int \frac{(x + 4)}{x^2 - 5x + 6} dx$.

Solution The partial fraction decomposition takes the form

$$\frac{x + 4}{x^2 - 5x + 6} = \frac{x + 4}{(x - 2)(x - 3)} = \frac{A}{x - 2} + \frac{B}{x - 3}.$$

We calculate A and B by both of the methods suggested above.

Method I. Add the partial fractions

$$\frac{x + 4}{x^2 - 5x + 6} = \frac{Ax - 3A + Bx - 2B}{(x - 2)(x - 3)},$$

and equate the coefficient of x and the constant terms in the numerators on both sides to obtain

$$A + B = 1 \quad \text{and} \quad -3A - 2B = 4.$$

Solve these equations to get $A = -6$ and $B = 7$.

Method II. To find A, cancel $x - 2$ from the denominator of the expression $P(x)/Q(x)$ and evaluate the result at $x = 2$. Obtain B similarly.

$$A = \frac{x + 4}{x - 3}\bigg|_{x=2} = -6 \quad \text{and} \quad B = \frac{x + 4}{x - 2}\bigg|_{x=3} = 7.$$

In either case we have

$$\int \frac{(x + 4)}{x^2 - 5x + 6} dx = -6\int \frac{1}{x - 2} dx + 7\int \frac{1}{x - 3} dx$$

$$= -6\ln|x - 2| + 7\ln|x - 3| + C.$$

Example 4   Evaluate $I = \displaystyle\int \frac{x^3 + 2}{x^3 - x}\, dx$.

Solution   Since the numerator does not have degree smaller than the denominator, we must divide:

$$I = \int \frac{x^3 - x + x + 2}{x^3 - x}\, dx = \int \left(1 + \frac{x+2}{x^3 - x}\right) dx = x + \int \frac{x+2}{x^3 - x}\, dx.$$

Now we can use the method of partial fractions.

$$\frac{x+2}{x^3 - x} = \frac{x+2}{x(x-1)(x+1)} = \frac{A}{x} + \frac{B}{x-1} + \frac{C}{x+1}$$

$$= \frac{A(x^2 - 1) + B(x^2 + x) + C(x^2 - x)}{x(x-1)(x+1)}$$

We have

$$\begin{array}{rcll} A + B + C &=& 0 & \text{(coefficient of } x^2) \\ B - C &=& 1 & \text{(coefficient of } x) \\ -A &=& 2 & \text{(constant term).} \end{array}$$

It follows that $A = -2$, $B = 3/2$, and $C = 1/2$. We can also find these values using Method II of the previous example:

$$A = \left.\frac{x+2}{(x-1)(x+1)}\right|_{x=0} = -2, \qquad B = \left.\frac{x+2}{x(x+1)}\right|_{x=1} = \frac{3}{2}, \quad \text{and}$$

$$C = \left.\frac{x+2}{x(x-1)}\right|_{x=-1} = \frac{1}{2}.$$

Finally, we have

$$I = x - 2\int \frac{1}{x}\, dx + \frac{3}{2}\int \frac{1}{x-1}\, dx + \frac{1}{2}\int \frac{1}{x+1}\, dx$$

$$= x - 2\ln|x| + \frac{3}{2}\ln|x-1| + \frac{1}{2}\ln|x+1| + C.$$

Next we consider a rational function whose denominator has a quadratic factor that is equivalent to a sum of squares and cannot, therefore, be further factored into a product of real linear factors.

Example 5   Evaluate $\displaystyle\int \frac{2 + 3x + x^2}{x(x^2 + 1)}\, dx$.

Solution   Note that the numerator has degree 2 and the denominator degree 3, so no division is necessary. If we decompose the integrand as a sum of two simpler fractions, we want one with denominator x and one with denominator $x^2 + 1$. The appropriate form of the decomposition turns out to be

$$\frac{2 + 3x + x^2}{x(x^2 + 1)} = \frac{A}{x} + \frac{Bx + C}{x^2 + 1} = \frac{A(x^2 + 1) + Bx^2 + Cx}{x(x^2 + 1)}.$$

Note that corresponding to the quadratic (degree 2) denominator we use a linear (degree 1) numerator. Equating coefficients in the two numerators, we obtain

$$
\begin{aligned}
A + B &= 1 && \text{(coefficient of } x^2) \\
C &= 3 && \text{(coefficient of } x) \\
A &= 2 && \text{(constant term).}
\end{aligned}
$$

Hence $A = 2$, $B = -1$, and $C = 3$. We have, therefore,

$$
\int \frac{2 + 3x + x^2}{x(x^2 + 1)} \, dx = 2 \int \frac{1}{x} \, dx - \int \frac{x}{x^2 + 1} \, dx + 3 \int \frac{1}{x^2 + 1} \, dx
$$

$$
= 2 \ln|x| - \frac{1}{2} \ln(x^2 + 1) + 3 \tan^{-1} x + C.
$$

We remark that addition of the fractions is the only reasonable real-variable method for determining the constants A, B, and C here. We could determine A by Method II of Example 3, but there is no simple equivalent way of finding B or C without using complex numbers. ∎

Example 6   Evaluate $I = \displaystyle\int \frac{1}{x^3 + 1} \, dx$.

Solution   Here $Q(x) = x^3 + 1 = (x + 1)(x^2 - x + 1)$. The latter factor has no real roots, so it has no real linear subfactors. We have

$$
\frac{1}{x^3 + 1} = \frac{1}{(x + 1)(x^2 - x + 1)} = \frac{A}{x + 1} + \frac{Bx + C}{x^2 - x + 1}
$$

$$
= \frac{A(x^2 - x + 1) + B(x^2 + x) + C(x + 1)}{(x + 1)(x^2 - x + 1)}
$$

$$
\begin{aligned}
A + B &= 0 && \text{(coefficient of } x^2) \\
-A + B + C &= 0 && \text{(coefficient of } x) \\
A + C &= 1 && \text{(constant term).}
\end{aligned}
$$

Hence $A = 1/3$, $B = -1/3$, and $C = 2/3$. We have

$$
I = \frac{1}{3} \int \frac{dx}{x + 1} - \frac{1}{3} \int \frac{x - 2}{x^2 - x + 1} \, dx
$$

$$
= \frac{1}{3} \ln|x + 1| - \frac{1}{3} \int \frac{x - \dfrac{1}{2} - \dfrac{3}{2}}{\left(x - \dfrac{1}{2}\right)^2 + \dfrac{3}{4}} \, dx \qquad \text{Let } u = x - 1/2,
$$
$$
\qquad\qquad\qquad\qquad\qquad\qquad\qquad\qquad\qquad du = dx.
$$

$$
= \frac{1}{3} \ln|x + 1| - \frac{1}{3} \int \frac{u}{u^2 + \dfrac{3}{4}} \, du + \frac{1}{2} \int \frac{1}{u^2 + \dfrac{3}{4}} \, du
$$

$$
= \frac{1}{3} \ln|x + 1| - \frac{1}{6} \ln\left(u^2 + \frac{3}{4}\right) + \frac{1}{2} \frac{2}{\sqrt{3}} \tan^{-1}\left(\frac{2u}{\sqrt{3}}\right) + C
$$

$$
= \frac{1}{3} \ln|x + 1| - \frac{1}{6} \ln(x^2 - x + 1) + \frac{1}{\sqrt{3}} \tan^{-1}\left(\frac{2x - 1}{\sqrt{3}}\right) + C.
$$

∎

We require one final refinement of the method of partial fractions. If any of the linear or quadratic factors of $Q(x)$ is *repeated* (say m times), then the partial fraction decomposition of $P(x)/Q(x)$ requires m distinct fractions corresponding to that factor. The denominators of these fractions have exponents increasing from 1 to m, and the numerators are all constants where the repeated factor is linear or linear where the repeated factor is quadratic. (See Theorem 1 below.)

Example 7 Evaluate $\displaystyle \int \frac{1}{x(x-1)^2}\, dx$.

Solution The appropriate partial fraction decomposition here is

$$
\frac{1}{x(x-1)^2} = \frac{A}{x} + \frac{B}{x-1} + \frac{C}{(x-1)^2}
$$
$$
= \frac{A(x^2 - 2x + 1) + B(x^2 - x) + Cx}{x(x-1)^2}.
$$

Equating coefficients of x^2, x, and 1 in the numerators of both sides, we get

$$
\begin{array}{rcll}
A + B & = 0 & \text{(coefficient of } x^2) \\
-2A - B + C & = 0 & \text{(coefficient of } x) \\
A & = 1 & \text{(constant term).}
\end{array}
$$

Hence $A = 1$, $B = -1$, $C = 1$, and

$$
\int \frac{1}{x(x-1)^2}\, dx = \int \frac{1}{x}\, dx - \int \frac{1}{x-1}\, dx + \int \frac{1}{(x-1)^2}\, dx
$$
$$
= \ln|x| - \ln|x-1| - \frac{1}{x-1} + C
$$
$$
= \ln\left|\frac{x}{x-1}\right| - \frac{1}{x-1} + C.
$$

Example 8 Evaluate $\displaystyle I = \int \frac{x^2 + 2}{4x^5 + 4x^3 + x}\, dx$.

Solution The denominator factors to $x(2x^2 + 1)^2$, so the appropriate partial fraction decomposition is

$$
\frac{x^2 + 2}{x(2x^2 + 1)^2} = \frac{A}{x} + \frac{Bx + C}{2x^2 + 1} + \frac{Dx + E}{(2x^2 + 1)^2}
$$
$$
= \frac{A(4x^4 + 4x^2 + 1) + B(2x^4 + x^2) + C(2x^3 + x) + Dx^2 + Ex}{x(2x^2 + 1)^2}.
$$

Thus

$$
\begin{array}{rcll}
4A + 2B & = 0 & \text{(coefficient of } x^4) \\
2C & = 0 & \text{(coefficient of } x^3) \\
4A + B + D & = 1 & \text{(coefficient of } x^2) \\
C + E & = 0 & \text{(coefficient of } x) \\
A & = 2 & \text{(constant term).}
\end{array}
$$

Solving these equations, we get $A = 2$, $B = -4$, $C = 0$, $D = -3$, and $E = 0$.

$$I = 2 \int \frac{dx}{x} - 4 \int \frac{x\,dx}{2x^2 + 1} - 3 \int \frac{x\,dx}{(2x^2 + 1)^2} \qquad \text{Let } u = 2x^2 + 1,$$
$$du = 4x\,dx.$$

$$= 2 \ln |x| - \int \frac{du}{u} - \frac{3}{4} \int \frac{du}{u^2}$$

$$= 2 \ln |x| - \ln |u| + \frac{3}{4u} + C$$

$$= \ln \left(\frac{x^2}{2x^2 + 1} \right) + \frac{3}{4} \frac{1}{2x^2 + 1} + C.$$

The following theorem summarizes the various aspects of the method of partial fractions.

THEOREM 1

Partial fraction decompositions of rational functions

Let P and Q be real polynomials with real coefficients, and suppose that the degree of P is less than the degree of Q. Then

(a) $Q(x)$ can be factored into the product of a constant K, real linear factors of the form $x - a_i$, and real quadratic factors of the form $x^2 + b_i x + c_i$ having no real roots. The linear and quadratic factors may be repeated:

$$Q(x) = K(x - a_1)^{m_1}(x - a_2)^{m_2} \cdots (x - a_j)^{m_j}(x^2 + b_1 x + c_1)^{n_1}$$
$$\cdots (x^2 + b_k x + c_k)^{n_k}.$$

The degree of Q is $m_1 + m_2 + \cdots + m_j + 2n_1 + 2n_2 + \cdots + 2n_k$.

(b) The rational function $P(x)/Q(x)$ can be expressed as a sum of partial fractions as follows:

(i) corresponding to each factor $(x - a)^m$ of $Q(x)$ the decomposition contains a sum of fractions of the form

$$\frac{A_1}{x - a} + \frac{A_2}{(x - a)^2} + \cdots + \frac{A_m}{(x - a)^m};$$

(ii) corresponding to each factor $(x^2 + bx + c)^n$ of $Q(x)$ the decomposition contains a sum of fractions of the form

$$\frac{B_1 x + C_1}{x^2 + bx + c} + \frac{B_2 x + C_2}{(x^2 + bx + c)^2} + \cdots + \frac{B_n x + C_n}{(x^2 + bx + c)^n}.$$

The constants A_1, A_2, ..., A_m, B_1, B_2, ..., B_n, C_1, C_2, ..., C_n can be determined by adding up the fractions in the decomposition and equating the coefficients of like powers of x in the numerator of the sum with those in $P(x)$.

We will not attempt to prove this theorem here.

Note that part (a) does not tell us how to find the factors of $Q(x)$; it tells us only what form they have. We must know the factors of Q before we can make use of partial fractions to integrate the rational function $P(x)/Q(x)$. Partial fraction decompositions are also used in other mathematical situations, in particular, to solve certain problems involving differential equations.

Exercises 6.3

Evaluate the integrals in Exercises 1–34.

1. $\displaystyle\int \frac{2\,dx}{2x-3}$

2. $\displaystyle\int \frac{dx}{5-4x}$

3. $\displaystyle\int \frac{x\,dx}{\pi x+2}$

4. $\displaystyle\int \frac{x^2}{x-4}\,dx$

5. $\displaystyle\int \frac{1}{x^2-9}\,dx$

6. $\displaystyle\int \frac{dx}{5-x^2}$

7. $\displaystyle\int \frac{dx}{a^2-x^2}$

8. $\displaystyle\int \frac{dx}{b^2-a^2x^2}$

9. $\displaystyle\int \frac{x^2\,dx}{x^2+x-2}$

10. $\displaystyle\int \frac{x\,dx}{3x^2+8x-3}$

11. $\displaystyle\int \frac{x-2}{x^2+x}\,dx$

12. $\displaystyle\int \frac{dx}{x^3+9x}$

13. $\displaystyle\int \frac{dx}{1-6x+9x^2}$

14. $\displaystyle\int \frac{x\,dx}{2+6x+9x^2}$

15. $\displaystyle\int \frac{x^2+1}{6x-9x^2}\,dx$

16. $\displaystyle\int \frac{x^3+1}{12+7x+x^2}\,dx$

17. $\displaystyle\int \frac{dx}{x(x^2-a^2)}$

18. $\displaystyle\int \frac{dx}{x^4-a^4}$

* 19. $\displaystyle\int \frac{x^3\,dx}{x^3-a^3}$

20. $\displaystyle\int \frac{dx}{x^3+2x^2+2x}$

21. $\displaystyle\int \frac{dx}{x^3-4x^2+3x}$

22. $\displaystyle\int \frac{x^2+1}{x^3+8}\,dx$

23. $\displaystyle\int \frac{dx}{(x^2-1)^2}$

24. $\displaystyle\int \frac{x^2\,dx}{(x^2-1)(x^2-4)}$

25. $\displaystyle\int \frac{dx}{x^4-3x^3}$

26. $\displaystyle\int \frac{x\,dx}{(x^2-x+1)^2}$

* 27. $\displaystyle\int \frac{t\,dt}{(t+1)(t^2+1)^2}$

* 28. $\displaystyle\int \frac{dt}{(t-1)(t^2-1)^2}$

* 29. $\displaystyle\int \frac{dx}{x(3+x^2)\sqrt{1-x^2}}$

* 30. $\displaystyle\int \frac{dx}{e^{2x}-4e^x+4}$

* 31. $\displaystyle\int \frac{dx}{x(1+x^2)^{3/2}}$

* 32. $\displaystyle\int \frac{dx}{x(1-x^2)^{3/2}}$

* 33. $\displaystyle\int \frac{dx}{x^2(x^2-1)^{3/2}}$

* 34. $\displaystyle\int \frac{d\theta}{\cos\theta(1+\sin\theta)}$

* 35. Suppose that P and Q are polynomials such that the degree of P is smaller than that of Q. If

$$Q(x) = (x-a_1)(x-a_2)\cdots(x-a_n),$$

where $a_i \neq a_j$ if $i \neq j\,(1 \leq i,\,j \leq n)$, so that $P(x)/Q(x)$ has partial fraction decomposition

$$\frac{P(x)}{Q(x)} = \frac{A_1}{x-a_1} + \frac{A_2}{x-a_2} + \cdots + \frac{A_n}{x-a_n},$$

show that

$$A_j = \frac{P(a_j)}{Q'(a_j)} \qquad (1 \leq j \leq n).$$

This gives yet another method for computing the constants in a partial fraction decomposition if the denominator factors completely into distinct linear factors.

6.4 Integration Using Computer Algebra or Tables

Although anyone who uses calculus should be familiar with the basic techniques of integration, just as anyone who uses arithmetic should be familiar with the techniques of multiplication and division, technology is steadily eroding the necessity for being able to do long, complicated integrals by such methods. In fact, today there are several computer programs that can manipulate mathematical expressions symbolically (rather than just numerically) and that can carry out, with little or no assistance from us, the various algebraic steps and limit calculations that are required to calculate and simplify both derivatives and integrals. Much pain can be avoided and time saved by having the computer evaluate a complicated integral such as

$$\int \frac{1+x+x^2}{(x^4-1)(x^4-16)^2}\,dx$$

rather than doing it by hand using partial fractions. Even without the aid of a computer, we can use tables of standard integrals such as the ones in the back endpapers of this book to help us evaluate complicated integrals. Using computers or tables can nevertheless require that we perform some simplifications beforehand and can make demands on our ability to interpret the answers we get. We give a few examples below.

Using Maple for Integration

Computer algebra systems are capable of evaluating both indefinite and definite integrals symbolically, as well as giving numerical approximations for those definite integrals that have numerical values. The following examples show how to use Maple to evaluate integrals.

We begin by calculating $\int 2^x \sqrt{1 + 4^x}\, dx$ and $\int_0^\pi 2^x \sqrt{1 + 4^x}\, dx$.

We use Maple's "int" command, specifying the function and the variable of integration:

```
>   int(2^x*sqrt(1+4^x),x);
```

$$\frac{1}{2}\frac{e^{(x\ln(2))}\sqrt{1 + (e^{(x\ln(2))})^2}}{\ln(2)} + \frac{1}{2}\frac{\operatorname{arcsinh}(e^{(x\ln(2))})}{\ln(2)}$$

If you don't like the inverse hyperbolic sine, you can convert it to a logarithm:

```
>   convert(%,ln);
```

$$\frac{1}{2}\frac{e^{(x\ln(2))}\sqrt{1 + (e^{(x\ln(2))})^2}}{\ln(2)} + \frac{1}{2}\frac{\ln\left(e^{(x\ln(2))} + \sqrt{1 + (e^{(x\ln(2))})^2}\right)}{\ln(2)}$$

The "%" there refers to the result of the previous calculation. Note how Maple prefers to use $e^{x\ln 2}$ in place of 2^x.

For the definite integral, you specify the interval of values of the variable of integration using two dots between the endpoints as follows:

```
>   int(2^x*sqrt(1+4^x),x=0..Pi);
```

$$\frac{1}{2}\frac{2^\pi\sqrt{1 + (2^\pi)^2} + \ln(2^\pi + \sqrt{1 + 2^{(2\pi)}}) - \sqrt{2} - \ln(1 + \sqrt{2})}{\ln(2)}$$

If you want a decimal approximation to this exact answer, you can ask Maple to evaluate the last result as a floating point number:

```
>   evalf(%);
```

$$56.95542155$$

Remark   Maple defaults to giving 10 significant digits in its floating point numbers unless you request a different precision by declaring a value for the variable "Digits":

```
>   Digits := 20; evalf(Pi);
```

$$3.1415926535897932385$$

Suppose we ask Maple to do an integral that we know we can't do ourselves:

```
>   int(exp(-x^2),x);
```

$$\frac{1}{2}\sqrt{\pi}\,\operatorname{erf}(x)$$

Maple expresses the answer in terms of the **error function** that is defined by

$$\text{erf}(x) = \frac{2}{\sqrt{\pi}} \int_0^x e^{-t^2} \, dt.$$

But observe:

```
>   Int(exp(-x^2),x=-infinity..infinity) = int(exp(-x^2),
x=-infinity..infinity);
```

$$\int_{-\infty}^{\infty} e^{(-x^2)} \, dx = \sqrt{\pi}$$

Note the use of the *inert* Maple command "Int" on the left side to simply print the integral without any evaluation. The active command "int" performs the evaluation.

Computer algebra programs can be used to integrate symbolically many functions, but you may get some surprises when you use them, and you may have to do some of the work to get an answer useful in the context of the problem on which you are working. Such programs, and some of the more sophisticated scientific calculators, are able to evaluate definite integrals numerically to any desired degree of accuracy even if symbolic antiderivatives cannot be found. We will discuss techniques of numerical integration in Sections 6.6–6.8, but note here that Maple's `evalf(Int())` can always be used to get numerical values:

```
>   evalf(Int(sin(cos(x)),x=0..1));
```

$$.7386429980$$

Using Integral Tables

You can get some help evaluating integrals by using an Integral Table, such as the one in the back endpapers of this book. Besides giving the values of the common elementary integrals that you likely remember while you are studying calculus, they also give many more complicated integrals, especially ones representing standard types that often arise in applications. Familiarize yourself with the main headings under which the integrals are classified. Using the tables usually means massaging your integral using simple substitutions until you get it into the form of one of the integrals in the table.

Example 1 Use the table to evaluate $I = \displaystyle\int \frac{t^5}{\sqrt{3 - 2t^4}} \, dt$.

Solution This integral doesn't resemble any in the tables, but there are numerous integrals in the tables involving $\sqrt{a^2 - x^2}$. We can begin to put the integral into this form with the substitution $t^2 = u$, so that $2t \, dt = du$. Thus

$$I = \frac{1}{2} \int \frac{u^2}{\sqrt{3 - 2u^2}} \, du.$$

This is not quite what we want yet; let us get rid of the 2 multiplying the u^2 under the square root. One way to do this is with the change of variable $\sqrt{2}u = x$, so that $du = dx/\sqrt{2}$:

$$I = \frac{1}{4\sqrt{2}} \int \frac{x^2}{\sqrt{3 - x^2}} \, dx.$$

Now the denominator is of the form $\sqrt{a^2 - x^2}$ for $a = \sqrt{3}$. Looking through the part of the table (in the back endpapers) dealing with integrals involving $\sqrt{a^2 - x^2}$ we find the third one, which says that

$$\int \frac{x^2}{\sqrt{a^2 - x^2}}\, dx = -\frac{x}{2}\sqrt{a^2 - x^2} + \frac{a^2}{2}\sin^{-1}\frac{x}{a} + C.$$

Thus

$$I = \frac{1}{4\sqrt{2}}\left(-\frac{x}{2}\sqrt{3 - x^2} + \frac{3}{2}\sin^{-1}\frac{x}{\sqrt{3}}\right) + C_1$$

$$= -\frac{t^2}{8}\sqrt{3 - 2t^4} + \frac{3}{8\sqrt{2}}\sin^{-1}\frac{\sqrt{2}\, t^2}{\sqrt{3}} + C_1.$$

Many of the integrals in the table are reduction formulas. (An integral appears on both sides of the equation.) These can be iterated to simplify integrals as in some of the examples and exercises of Section 6.1.

Example 2 Evaluate $I = \displaystyle\int_0^1 \frac{1}{(x^2 + 1)^3}\, dx$.

Solution The fourth integral in the table of Miscellaneous Algebraic Integrals says that if $n \neq 1$, then

$$\int \frac{dx}{(a^2 \pm x^2)^n} = \frac{1}{2a^2(n - 1)}\left(\frac{x}{(a^2 \pm x^2)^{n-1}} + (2n - 3)\int \frac{dx}{(a^2 \pm x^2)^{n-1}}\right).$$

Using $a = 1$ and the "+" signs, we have

$$\int_0^1 \frac{dx}{(1 + x^2)^n} = \frac{1}{2(n - 1)}\left(\frac{x}{(1 + x^2)^{n-1}}\Bigg|_0^1 + (2n - 3)\int_0^1 \frac{dx}{(1 + x^2)^{n-1}}\right)$$

$$= \frac{1}{2^n(n - 1)} + \frac{2n - 3}{2(n - 1)}\int_0^1 \frac{dx}{(1 + x^2)^{n-1}}.$$

Thus we have

$$I = \frac{1}{16} + \frac{3}{4}\int_0^1 \frac{dx}{(1 + x^2)^2}$$

$$= \frac{1}{16} + \frac{3}{4}\left(\frac{1}{4} + \frac{1}{2}\int_0^1 \frac{dx}{1 + x^2}\right)$$

$$= \frac{1}{16} + \frac{3}{16} + \frac{3}{8}\tan^{-1}x\Bigg|_0^1 = \frac{1}{4} + \frac{3\pi}{32}.$$

Exercises 6.4

1. Use Maple or another computer algebra program to check any of the integrals you have done in the exercises from Sections 5.6 and 6.1–6.3, as well as any of the integrals you have been unable to do.

2. Use Maple or another computer algebra program to evaluate the integral in the opening paragraph of this section.

3. Use Maple or another computer algebra program to reevaluate the integral in Example 1.

4. Use Maple or another computer algebra program to reevaluate the integral in Example 2.

Use the integral tables to help you find the integrals in Exercises 5–14.

5. $\int \dfrac{x^2}{\sqrt{x^2 - 2}}\, dx$

6. $\int \sqrt{(x^2 + 4)^3}\, dx$

7. $\int \dfrac{dt}{t^2\sqrt{3t^2 + 5}}$

8. $\int \dfrac{dt}{t\sqrt{3t - 5}}$

9. $\int x^4 (\ln x)^4\, dx$

10. $\int x^7 e^{x^2}\, dx$

11. $\int x\sqrt{2x - x^2}\, dx$

12. $\int \dfrac{\sqrt{2x - x^2}}{x^2}\, dx$

13. $\int \dfrac{dx}{(\sqrt{4x - x^2})^3}$

14. $\int \dfrac{dx}{(\sqrt{4x - x^2})^4}$

15. Use Maple or another computer algebra program to evaluate the integrals in the previous 10 exercises.

6.5 Improper Integrals

Up to this point we have considered definite integrals of the form

$$I = \int_a^b f(x)\, dx,$$

where the integrand f is *continuous* on the *closed, finite* interval $[a, b]$. Since such a function is necessarily *bounded*, the integral I is necessarily a finite number; for positive f it corresponds to the area of a **bounded region** of the plane, a region contained inside some disk of finite radius with centre at the origin. Such integrals are also called **proper integrals**. We are now going to generalize the definite integral to allow for two possibilities excluded in the situation described above:

(i) We may have $a = -\infty$ or $b = \infty$ or both.

(ii) f may be unbounded as x approaches a or b or both.

Integrals satisfying (i) are called **improper integrals of type I**; integrals satisfying (ii) are called **improper integrals of type II**. Either type of improper integral corresponds (for positive f) to the area of a region in the plane that "extends to infinity" in some direction and therefore is *unbounded*. As we will see, such integrals may or may not have finite values. The ideas involved are best introduced by examples.

Improper Integrals of Type I

Example 1 Find the area of the region A lying under the curve $y = 1/x^2$ and above the x-axis to the right of $x = 1$. (See Figure 6.8(a).)

Solution We would like to calculate the area with an integral

$$A = \int_1^\infty \dfrac{dx}{x^2},$$

Figure 6.8

(a) $A = \int_1^\infty \frac{1}{x^2} \, dx$

(b) $A = \lim_{R \to \infty} \int_1^R \frac{1}{x^2} \, dx$

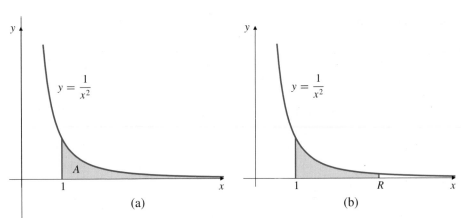

(a)　　　　　　(b)

which is improper of type I since its interval of integration is infinite. It is not immediately obvious whether the area is finite or not; the region has an infinitely long "spike" along the x-axis, but this spike becomes infinitely thin as x approaches ∞. In order to evaluate this improper integral, we interpret it as a limit of proper integrals over intervals $[1, R]$ as $R \to \infty$. (See Figure 6.8(b).)

$$A = \int_1^\infty \frac{dx}{x^2} = \lim_{R \to \infty} \int_1^R \frac{dx}{x^2} = \lim_{R \to \infty} \left(-\frac{1}{x} \right) \Big|_1^R$$
$$= \lim_{R \to \infty} \left(-\frac{1}{R} + 1 \right) = 1$$

Since the limit exists (is finite), we say that the improper integral *converges*. The region has finite area $A = 1$ square unit. ∎

Example 2 Find the area of the region under $y = 1/x$, above $y = 0$, and to the right of $x = 1$. (See Figure 6.9.)

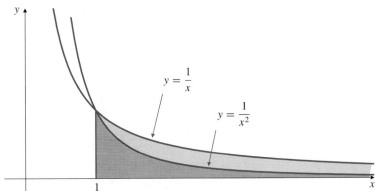

Figure 6.9 The area shaded in colour is infinite

Solution This area is given by the improper integral

$$A = \int_1^\infty \frac{dx}{x} = \lim_{R \to \infty} \int_1^R \frac{dx}{x} = \lim_{R \to \infty} \ln x \Big|_1^R = \lim_{R \to \infty} \ln R = \infty.$$

We say that this improper integral *diverges to infinity*. Observe that the region has a similar shape to the region under $y = 1/x^2$ considered in the above example, but its "spike" is somewhat thicker at each value of $x > 1$. Evidently the extra thickness makes a big difference; this region has *infinite* area. ∎

DEFINITION 1

Improper integrals of type I

If f is continuous on $[a, \infty[$, we define the improper integral of f over $[a, \infty[$ as a limit of proper integrals:

$$\int_a^\infty f(x)\,dx = \lim_{R \to \infty} \int_a^R f(x)\,dx.$$

Similarly, if f is continuous on $]-\infty, b]$, then we define

$$\int_{-\infty}^b f(x)\,dx = \lim_{R \to -\infty} \int_R^b f(x)\,dx.$$

In either case, if the limit exists (is a finite number), we say that the improper integral **converges**; if the limit does not exist, we say that the improper integral **diverges**. If the limit is ∞ (or $-\infty$), we say the improper integral **diverges to infinity** (or **diverges to negative infinity**).

The integral $\int_{-\infty}^\infty f(x)\,dx$ is, for f continuous on the real line, improper of type I at both endpoints. We break it into two separate integrals:

$$\int_{-\infty}^\infty f(x)\,dx = \int_{-\infty}^0 f(x)\,dx + \int_0^\infty f(x)\,dx.$$

The integral on the left converges if and only if *both* integrals on the right converge.

Example 3   Evaluate $\displaystyle\int_{-\infty}^\infty \frac{1}{1+x^2}\,dx.$

Solution   By the (even) symmetry of the integrand (see Figure 6.10), we have

$$\int_{-\infty}^\infty \frac{dx}{1+x^2} = \int_{-\infty}^0 \frac{dx}{1+x^2} + \int_0^\infty \frac{dx}{1+x^2}$$

$$= 2 \lim_{R \to \infty} \int_0^R \frac{dx}{1+x^2}$$

$$= 2 \lim_{R \to \infty} \tan^{-1} R = 2\left(\frac{\pi}{2}\right) = \pi.$$

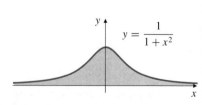
$$y = \frac{1}{1+x^2}$$

Figure 6.10

The use of symmetry here requires some justification. At the time we used it we did not know whether each of the half-line integrals was finite or infinite. However, since both are positive, even if they are infinite, their sum would still be twice one of them. If one had been positive and the other negative, we would not have been justified in cancelling them to get 0 until we knew that they were finite. ($\infty + \infty = \infty$, but $\infty - \infty$ is not defined.) In any event, the given integral converges to π. ∎

Example 4
$$\int_0^\infty \cos x \, dx = \lim_{R\to\infty} \int_0^R \cos x \, dx = \lim_{R\to\infty} \sin R.$$

This limit does not exist (and it is not ∞ or $-\infty$), so all we can say is that the given integral diverges. (See Figure 6.11.) As R increases, the integral alternately adds and subtracts the areas of the hills and valleys but does not approach any unique limit.

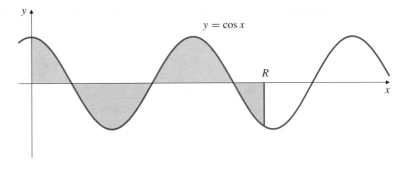

Figure 6.11   Not every divergent improper integral diverges to ∞ or $-\infty$

Improper Integrals of Type II

DEFINITION 2

Improper integrals of type II

If f is continuous on the interval $]a, b]$ and is possibly unbounded near a, we define the improper integral

$$\int_a^b f(x) \, dx = \lim_{c\to a+} \int_c^b f(x) \, dx.$$

Similarly, if f is continuous on $[a, b[$ and is possibly unbounded near b, we define

$$\int_a^b f(x) \, dx = \lim_{c\to b-} \int_a^c f(x) \, dx.$$

These improper integrals may converge, diverge, diverge to infinity, or diverge to negative infinity.

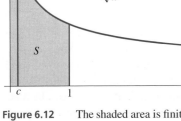

Figure 6.12   The shaded area is finite

Example 5   Find the area of the region S lying under $y = 1/\sqrt{x}$, above the x-axis, between $x = 0$ and $x = 1$.

Solution   The area A is given by

$$A = \int_0^1 \frac{1}{\sqrt{x}} \, dx,$$

which is an improper integral of type II since the integrand is unbounded near $x = 0$. The region S has a "spike" extending to infinity along the y-axis, a vertical asymptote of the integrand, as shown in Figure 6.12. As we did for improper integrals of type I, we express such integrals as limits of proper integrals.

$$A = \lim_{c \to 0+} \int_c^1 x^{-1/2}\, dx = \lim_{c \to 0+} 2x^{1/2}\Big|_c^1 = \lim_{c \to 0+} (2 - 2\sqrt{c}) = 2.$$

This integral converges, and S has a finite area of 2 square units. ∎

While improper integrals of type I are always easily recognized because of the infinite limits of integration, improper integrals of type II can be somewhat harder to spot. You should be alert for singularities of integrands and especially points where they have vertical asymptotes. It may be necessary to break an improper integral into several improper integrals if it is improper at both endpoints or at points inside the interval of integration. For example,

$$\int_{-1}^1 \frac{\ln |x|\, dx}{\sqrt{1 - x}} = \int_{-1}^0 \frac{\ln |x|\, dx}{\sqrt{1 - x}} + \int_0^{1/2} \frac{\ln |x|\, dx}{\sqrt{1 - x}} + \int_{1/2}^1 \frac{\ln |x|\, dx}{\sqrt{1 - x}}.$$

Each integral on the right is improper because of a singularity of its integrand at one endpoint.

Example 6   Evaluate each of the following integrals or show that it diverges:

(a) $\displaystyle\int_0^1 \frac{1}{x}\, dx,$ (b) $\displaystyle\int_0^2 \frac{1}{\sqrt{2x - x^2}}\, dx,$ and   (c) $\displaystyle\int_0^1 \ln x\, dx.$

Solution

(a) $\displaystyle\int_0^1 \frac{1}{x}\, dx = \lim_{c \to 0+} \int_c^1 \frac{1}{x}\, dx = \lim_{c \to 0+} (\ln 1 - \ln c) = \infty.$

This integral diverges to infinity.

(b) $\displaystyle\int_0^2 \frac{1}{\sqrt{2x - x^2}}\, dx = \int_0^2 \frac{1}{\sqrt{1 - (x - 1)^2}}\, dx$ Let $u = x - 1$,

                                                     $du = dx.$

$$= \int_{-1}^1 \frac{1}{\sqrt{1 - u^2}}\, du$$

$$= 2\int_0^1 \frac{1}{\sqrt{1 - u^2}}\, du \qquad \text{(by symmetry)}$$

$$= 2 \lim_{c \to 1-} \int_0^c \frac{1}{\sqrt{1 - u^2}}\, du$$

$$= 2 \lim_{c \to 1-} \sin^{-1} u \Big|_0^c = 2 \lim_{c \to 1-} \sin^{-1} c = \pi.$$

This integral converges to π. Observe how a change of variable can be made even before an improper integral is expressed as a limit of proper integrals.

(c) $\displaystyle\int_0^1 \ln x \, dx = \lim_{c\to 0+} \int_c^1 \ln x \, dx$ (See Example 2(a) of Section 6.1 for the evaluation of the indefinite integral.)

$$= \lim_{c\to 0+} (x \ln x - x)\Big|_c^1$$

$$= \lim_{c\to 0+} (0 - 1 - c \ln c + c)$$

$$= -1 + 0 - \lim_{c\to 0+} \frac{\ln c}{1/c} \qquad \left[\frac{-\infty}{\infty}\right]$$

$$= -1 - \lim_{c\to 0+} \frac{1/c}{-(1/c^2)} \qquad \text{(by l'Hôpital's Rule)}$$

$$= -1 - \lim_{c\to 0+} (-c) = -1 + 0 = -1.$$

The integral converges to -1.

■

The following theorem summarizes the behaviour of improper integrals of types I and II for powers of x.

THEOREM 2

p-integrals

If $0 < a < \infty$, then

(a) $\displaystyle\int_a^\infty x^{-p} \, dx$ $\begin{cases} \text{converges to } \dfrac{a^{1-p}}{p-1} & \text{if } p > 1 \\ \text{diverges to } \infty & \text{if } p \le 1 \end{cases}$

(b) $\displaystyle\int_0^a x^{-p} \, dx$ $\begin{cases} \text{converges to } \dfrac{a^{1-p}}{1-p} & \text{if } p < 1 \\ \text{diverges to } \infty & \text{if } p \ge 1. \end{cases}$

PROOF We prove part (b) only. The proof of part (a) is similar and is left as an exercise. Also, the case $p = 1$ of part (b) is similar to Example 6(a) above, so we need consider only the cases $p < 1$ and $p > 1$. If $p < 1$, then we have

$$\int_0^a x^{-p} \, dx = \lim_{c\to 0+} \int_c^a x^{-p} \, dx$$

$$= \lim_{c\to 0+} \frac{x^{-p+1}}{-p+1}\Big|_c^a$$

$$= \lim_{c\to 0+} \frac{a^{1-p} - c^{1-p}}{1 - p} = \frac{a^{1-p}}{1 - p}$$

because $1 - p > 0$. If $p > 1$, then

$$\int_0^a x^{-p} \, dx = \lim_{c\to 0+} \int_c^a x^{-p} \, dx$$

$$= \lim_{c\to 0+} \frac{x^{-p+1}}{-p+1}\Big|_c^a$$

$$= \lim_{c\to 0+} \frac{c^{-(p-1)} - a^{-(p-1)}}{p - 1} = \infty.$$

The integrals in Theorem 2 are called **p-integrals**. It is very useful to know when they converge and diverge when you have to decide whether certain other improper integrals converge or not and you can't find the appropriate antiderivatives. (See the discussion of estimating convergence below.) Note that $\int_0^\infty x^{-p}\,dx$ does not converge for any value of p.

Remark If f is continuous on the interval $[a, b]$ so that $\int_a^b f(x)\,dx$ is a proper definite integral, then treating the integral as improper will lead to the same value:

$$\lim_{c \to a+} \int_c^b f(x)\,dx = \int_a^b f(x)\,dx = \lim_{c \to b-} \int_a^c f(x)\,dx.$$

This justifies the definition of the definite integral of a piecewise continuous function given in Section 5.4. To integrate a function defined to be different continuous functions on different intervals, we merely add the integrals of the various component functions over their respective intervals. Any of these integrals may be proper or improper; if any are improper, all must converge or the given integral will diverge.

Example 7 Evaluate $\int_0^2 f(x)\,dx$, where $f(x) = \begin{cases} 1/\sqrt{x} & \text{if } 0 < x \le 1 \\ x - 1 & \text{if } 1 < x \le 2. \end{cases}$

Solution The graph of f is shown in Figure 6.13. We have

$$\int_0^2 f(x)\,dx = \int_0^1 \frac{dx}{\sqrt{x}} + \int_1^2 (x - 1)\,dx$$

$$= \lim_{c \to 0+} \int_c^1 \frac{dx}{\sqrt{x}} + \left(\frac{x^2}{2} - x\right)\Big|_1^2 = 2 + \left(2 - 2 - \frac{1}{2} + 1\right) = \frac{5}{2};$$

the first integral on the right is improper but convergent (see Example 5 above) and the second is proper.

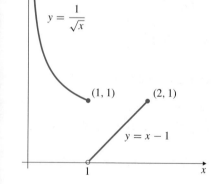

$y = \dfrac{1}{\sqrt{x}}$

$(1, 1)$ $(2, 1)$

$y = x - 1$

Figure 6.13

Estimating Convergence and Divergence

When an improper integral cannot be evaluated by the Fundamental Theorem of Calculus because an antiderivative can't be found, we may still be able to determine whether the integral converges by comparing it with simpler integrals. The following theorem is central to this approach.

THEOREM 3

A comparison theorem for integrals

Let $-\infty \le a < b \le \infty$, and suppose that functions f and g are continuous on the interval $]a, b[$ and satisfy $0 \le f(x) \le g(x)$. If $\int_a^b g(x)\,dx$ converges, then so does $\int_a^b f(x)\,dx$, and

$$\int_a^b f(x)\,dx \le \int_a^b g(x)\,dx.$$

Equivalently, if $\int_a^b f(x)\,dx$ diverges to ∞, then so does $\int_a^b g(x)\,dx$.

PROOF Since both integrands are nonnegative, there are only two possibilities for each integral: it can either converge to a nonnegative number or diverge to ∞. Since $f(x) \le g(x)$ on (a, b), it follows by Theorem 3(e) of Section 5.4 that if $a < r < s < b$, then

$$\int_r^s f(x)\,dx \le \int_r^s g(x)\,dx.$$

This theorem now follows by taking limits as $r \to a+$ and $s \to b-$.

Example 8 Show that $\displaystyle\int_0^\infty e^{-x^2}\,dx$ converges, and find an upper bound for its value.

Solution We can't integrate e^{-x^2}, but we can integrate e^{-x}. We would like to use the inequality $e^{-x^2} \le e^{-x}$, but this is only valid for $x \ge 1$. (See Figure 6.14.) Therefore we break the integral into two parts.

On $[0, 1]$ we have $0 < e^{-x^2} \le 1$, so

$$0 < \int_0^1 e^{-x^2}\,dx \le \int_0^1 dx = 1.$$

On $[1, \infty)$ we have $x^2 \ge x$, so $-x^2 \le -x$ and $0 < e^{-x^2} \le e^{-x}$. Thus,

$$0 < \int_1^\infty e^{-x^2}\,dx \le \int_1^\infty e^{-x}\,dx = \lim_{R\to\infty} \left.\frac{e^{-x}}{-1}\right|_1^R$$

$$= \lim_{R\to\infty} \left(\frac{1}{e} - \frac{1}{e^R}\right) = \frac{1}{e}.$$

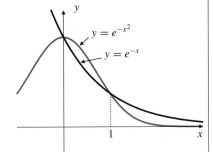

Figure 6.14 Comparing e^{-x^2} and e^{-x}

Hence, $\displaystyle\int_0^\infty e^{-x^2}\,dx$ converges and its value is not greater than $1 + (1/e)$.

We remark that the above integral is, in fact, equal to $\frac{1}{2}\sqrt{\pi}$, although we cannot prove this now. See Section 14.4.

For large or small values of x many integrands behave like powers of x. If so, they can be compared with p-integrals.

Example 9 Determine whether $\displaystyle\int_0^\infty \frac{dx}{\sqrt{x + x^3}}$ converges.

Solution The integral is improper of both types, so we write

$$\int_0^\infty \frac{dx}{\sqrt{x + x^3}} = \int_0^1 \frac{dx}{\sqrt{x + x^3}} + \int_1^\infty \frac{dx}{\sqrt{x + x^3}} = I_1 + I_2.$$

On $(0, 1]$ we have $\sqrt{x + x^3} > \sqrt{x}$, so

$$I_1 < \int_0^1 \frac{dx}{\sqrt{x}} = 2 \qquad \text{(by Theorem 2)}.$$

On $[1, \infty)$ we have $\sqrt{x + x^3} > \sqrt{x^3}$, so

$$I_2 < \int_1^\infty x^{-3/2}\, dx = 2 \qquad \text{(by Theorem 2)}.$$

Hence, the given integral converges and its value is less than 4. ∎

Exercises 6.5

In Exercises 1–22, evaluate the given integral or show that it diverges.

1. $\displaystyle\int_2^\infty \frac{1}{(x-1)^3}\, dx$

2. $\displaystyle\int_3^\infty \frac{1}{(2x-1)^{2/3}}\, dx$

3. $\displaystyle\int_0^\infty e^{-2x}\, dx$

4. $\displaystyle\int_{-\infty}^{-1} \frac{dx}{x^2 + 1}$

5. $\displaystyle\int_{-1}^1 \frac{dx}{(x+1)^{2/3}}$

6. $\displaystyle\int_0^a \frac{dx}{a^2 - x^2}$

7. $\displaystyle\int_0^1 \frac{1}{(1-x)^{1/3}}\, dx$

8. $\displaystyle\int_0^1 \frac{1}{x\sqrt{1-x}}\, dx$

9. $\displaystyle\int_0^{\pi/2} \frac{\cos x\, dx}{(1-\sin x)^{2/3}}$

10. $\displaystyle\int_0^\infty x\,e^{-x}\, dx$

11. $\displaystyle\int_0^1 \frac{dx}{\sqrt{x(1-x)}}$

12. $\displaystyle\int_0^\infty \frac{x}{1+2x^2}\, dx$

13. $\displaystyle\int_0^\infty \frac{x\, dx}{(1+2x^2)^{3/2}}$

14. $\displaystyle\int_0^{\pi/2} \sec x\, dx$

15. $\displaystyle\int_0^{\pi/2} \tan x\, dx$

16. $\displaystyle\int_e^\infty \frac{dx}{x \ln x}$

17. $\displaystyle\int_1^e \frac{dx}{x\sqrt{\ln x}}$

18. $\displaystyle\int_e^\infty \frac{dx}{x(\ln x)^2}$

19. $\displaystyle\int_{-\infty}^\infty \frac{x}{1+x^2}\, dx$

20. $\displaystyle\int_{-\infty}^\infty \frac{x}{1+x^4}\, dx$

21. $\displaystyle\int_{-\infty}^\infty x\,e^{-x^2}\, dx$

22. $\displaystyle\int_{-\infty}^\infty e^{-|x|}\, dx$

23. Find the area below $y = 0$, above $y = \ln x$, and to the right of $x = 0$.

24. Find the area below $y = e^{-x}$, above $y = e^{-2x}$, and to the right of $x = 0$.

25. Find the area of a region that lies above $y = 0$, to the right of $x = 1$, and under the curve $y = \dfrac{4}{2x+1} - \dfrac{2}{x+2}$.

26. Find the area of the plane region that lies under the graph of $y = x^{-2}e^{-1/x}$, above the x-axis, and to the right of the y-axis.

27. Prove Theorem 2(a) by directly evaluating the integrals involved.

28. Evaluate $\int_{-1}^1 (x\operatorname{sgn} x)/(x+2)\, dx$. Recall that $\operatorname{sgn} x = x/|x|$.

29. Evaluate $\int_0^2 x^2 \operatorname{sgn}(x-1)\, dx$.

In Exercises 30–41, state whether the given integral converges or diverges, and justify your claim.

30. $\displaystyle\int_0^\infty \frac{x^2}{x^5 + 1}\, dx$

31. $\displaystyle\int_0^\infty \frac{dx}{1+\sqrt{x}}$

32. $\displaystyle\int_2^\infty \frac{x\sqrt{x}\, dx}{x^2 - 1}$

33. $\displaystyle\int_0^\infty e^{-x^3}\, dx$

34. $\displaystyle\int_0^\infty \frac{dx}{\sqrt{x}+x^2}$

35. $\displaystyle\int_{-1}^1 \frac{e^x}{x+1}\, dx$

36. $\displaystyle\int_0^\pi \frac{\sin x}{x}\, dx$

37. $\displaystyle\int_0^\infty \frac{|\sin x|}{x^2}\, dx$

*38. $\displaystyle\int_0^{\pi^2} \frac{dx}{1-\cos\sqrt{x}}$

*39. $\displaystyle\int_{-\pi/2}^{\pi/2} \csc x\, dx$

*40. $\displaystyle\int_2^\infty \frac{dx}{\sqrt{x}\,\ln x}$

*41. $\displaystyle\int_0^\infty \frac{dx}{xe^x}$

*42. Given that $\int_0^\infty e^{-x^2}\, dx = \dfrac{1}{2}\sqrt{\pi}$, evaluate

(a) $\displaystyle\int_0^\infty x^2 e^{-x^2}\, dx$ and (b) $\displaystyle\int_0^\infty x^4 e^{-x^2}\, dx$.

*43. If f is continuous on $[a, b]$, show that

$$\lim_{c \to a+} \int_c^b f(x)\, dx = \int_a^b f(x)\, dx.$$

Hint: a continuous function on a closed, finite interval is *bounded*: there exists a positive constant K such that $|f(x)| \le K$ for all x in $[a, b]$. Use this fact, together with parts (d) and (f) of Theorem 3 of Section 5.4, to show that

$$\lim_{c \to a+} \left(\int_a^b f(x)\, dx - \int_c^b f(x)\, dx \right) = 0.$$

Similarly, show that

$$\lim_{c \to b-} \int_a^c f(x)\,dx = \int_a^b f(x)\,dx.$$

* **44. (The gamma function)** The gamma function $\Gamma(x)$ is defined by the improper integral

$$\Gamma(x) = \int_0^\infty t^{x-1} e^{-t}\,dt.$$

(Γ is the Greek capital letter gamma.)

(a) Show that the integral converges for $x > 0$.

(b) Use integration by parts to show that $\Gamma(x + 1) = x\Gamma(x)$ for $x > 0$.

(c) Show that $\Gamma(n + 1) = n!$ for $n = 0, 1, 2, \ldots$.

(d) Given that $\int_0^\infty e^{-x^2}\,dx = \frac{1}{2}\sqrt{\pi}$, show that $\Gamma(\frac{1}{2}) = \sqrt{\pi}$ and $\Gamma(\frac{3}{2}) = \frac{1}{2}\sqrt{\pi}$.

In view of (c), $\Gamma(x + 1)$ is often written $x!$ and regarded as a real-valued extension of the factorial function. Some scientific calculators (in particular, HP calculators) with the factorial function $n!$ built in actually calculate the gamma function rather than just the integral factorial. Check whether your calculator does this by asking it for 0.5!. If you get an error message, it's not using the gamma function.

6.6 The Trapezoid and Midpoint Rules

Most of the applications of integration, within and outside of mathematics, involve the definite integral

$$I = \int_a^b f(x)\,dx.$$

Thanks to the Fundamental Theorem of Calculus, we can evaluate such definite integrals by first finding an antiderivative of f. This is why we have spent considerable time on developing techniques of integration. There are, however, two obstacles that can prevent our calculating I in this way:

(i) Finding an antiderivative of f in terms of familiar functions may be impossible, or at least very difficult.

(ii) We may not be given a formula for $f(x)$ as a function of x; for instance, $f(x)$ may be an unknown function whose values at certain points of the interval $[a, b]$ have been determined by experimental measurement.

In the next two sections we investigate the problem of approximating the value of the definite integral I using only the values of $f(x)$ at finitely many points of $[a, b]$. Obtaining such an approximation is called **numerical integration**. Upper and lower sums (or, indeed, any Riemann sum) can be used for this purpose, but these usually require much more calculation to yield a desired precision than the methods we will develop here. We will develop three methods for evaluating definite integrals numerically: the Trapezoid Rule, the Midpoint Rule, and Simpson's Rule. All of these methods can be easily implemented on a small computer or using a scientific calculator. The wide availability of these devices makes numerical integration a steadily more important tool for the user of mathematics. Some of the more advanced calculators have built-in routines for numerical integration.

All the techniques we consider require us to calculate the values of $f(x)$ at a set of equally spaced points in $[a, b]$. The computational "expense" involved in determining an approximate value for the integral I will be roughly proportional to the number of function values required, so that the fewer function evaluations needed to achieve a desired degree of accuracy for the integral, the better we will regard the technique. Time is money, even in the world of computers.

The Trapezoid Rule

We assume that $f(x)$ is continuous on $[a, b]$ and subdivide $[a, b]$ into n subintervals of equal length $h = (b - a)/n$ using the $n + 1$ points

$$x_0 = a, \quad x_1 = a + h, \quad x_2 = a + 2h, \quad \ldots, \quad x_n = a + nh = b.$$

We assume that the value of $f(x)$ at each of these points is known:

$$y_0 = f(x_0), \quad y_1 = f(x_1), \quad y_2 = f(x_2), \quad \ldots, \quad y_n = f(x_n).$$

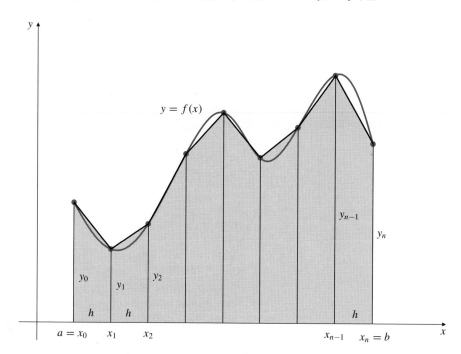

Figure 6.15 The area under $y = f(x)$ is approximated by the sum of the areas of n trapezoids

The Trapezoid Rule approximates $\int_a^b f(x)\,dx$ by using straight line segments between the points (x_{j-1}, y_{j-1}) and (x_j, y_j), $(1 \le j \le n)$, to approximate the graph of f, as shown in Figure 6.15, and summing the areas of the resulting n *trapezoids*. A **trapezoid** is a four-sided polygon with one pair of parallel sides. (For our discussion we assume f is positive so we can talk about "areas," but the resulting formulas apply to any continuous function f.)

The first trapezoid has vertices $(x_0, 0)$, (x_0, y_0), (x_1, y_1), and $(x_1, 0)$. The two parallel sides are vertical and have lengths y_0 and y_1. The perpendicular distance between them is $h = x_1 - x_0$. The area of this trapezoid is h times the average of the parallel sides:

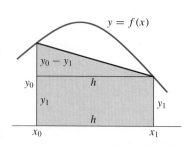

Figure 6.16 The trapezoid has area $y_1 h + \frac{1}{2}(y_0 - y_1)h = \frac{1}{2}h(y_0 + y_1)$

$$h\,\frac{y_0 + y_1}{2} \text{ square units.}$$

This can be seen geometrically by considering the trapezoid as the nonoverlapping union of a rectangle and a triangle; see Figure 6.16. We use this trapezoidal area to approximate the integral of f over the first subinterval $[x_0, x_1]$:

$$\int_{x_0}^{x_1} f(x)\,dx \approx h\,\frac{y_0 + y_1}{2}.$$

We can approximate the integral of f over any subinterval in the same way:

$$\int_{x_{j-1}}^{x_j} f(x)\,dx \approx h\,\frac{y_{j-1}+y_j}{2}, \qquad (1 \le j \le n).$$

It follows that the original integral I can be approximated by the sum of these trapezoidal areas:

$$\int_a^b f(x)\,dx \approx h\left(\frac{y_0+y_1}{2}+\frac{y_1+y_2}{2}+\frac{y_2+y_3}{2}+\cdots+\frac{y_{n-1}+y_n}{2}\right)$$

$$= h\left(\frac{1}{2}\,y_0+y_1+y_2+y_3+\cdots+y_{n-1}+\frac{1}{2}\,y_n\right).$$

DEFINITION 3

> **The Trapezoid Rule**
>
> The n-subinterval **Trapezoid Rule** approximation to $\int_a^b f(x)\,dx$, denoted T_n, is given by $\quad \frac{1}{2}f(x_0)\;+\;f(x_1)\;-\!\sim$
>
> $$T_n = h\left(\frac{1}{2}\,y_0+y_1+y_2+y_3+\cdots+y_{n-1}+\frac{1}{2}\,y_n\right).$$

We now illustrate the Trapezoid Rule by using it to approximate an integral whose value we already know:

$$I = \int_1^2 \frac{1}{x}\,dx = \ln 2 = 0.69314718\ldots.$$

(This value, and those of all the approximations quoted in these sections, were calculated using a scientific calculator.) We will use the same integral to illustrate other methods for approximating definite integrals later.

Example 1 Calculate the Trapezoid Rule approximations T_4, T_8, and T_{16} for

$$I = \int_1^2 \frac{1}{x}\,dx.$$

Solution For $n=4$ we have $h=(2-1)/4=1/4$; for $n=8$ we have $h=1/8$; for $n=16$ we have $h=1/16$. Therefore,

$$T_4 = \frac{1}{4}\left[\frac{1}{2}(1)+\frac{4}{5}+\frac{2}{3}+\frac{4}{7}+\frac{1}{2}\left(\frac{1}{2}\right)\right] = 0.69702381\ldots$$

$$T_8 = \frac{1}{8}\left[\frac{1}{2}(1)+\frac{8}{9}+\frac{4}{5}+\frac{8}{11}+\frac{2}{3}+\frac{8}{13}+\frac{4}{7}+\frac{8}{15}+\frac{1}{2}\left(\frac{1}{2}\right)\right]$$

$$= \frac{1}{8}\left[4\,T_4+\frac{8}{9}+\frac{8}{11}+\frac{8}{13}+\frac{8}{15}\right] = 0.69412185\ldots$$

$$T_{16} = \frac{1}{16}\left[8\,T_8+\frac{16}{17}+\frac{16}{19}+\frac{16}{21}+\frac{16}{23}+\frac{16}{25}+\frac{16}{27}+\frac{16}{29}+\frac{16}{31}\right]$$

$$= 0.69339120\ldots.$$

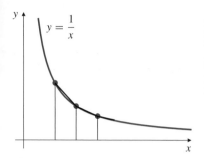

$y = \dfrac{1}{x}$

Figure 6.17 The trapezoid areas are greater than the area under the curve if the curve is concave upward

Note how the function values used to calculate T_4 were reused in the calculation of T_8, and similarly how those in T_8 were reused for T_{16}. When several approximations are needed, it is very useful to double the number of subintervals for each new calculation, so that previously calculated values of f can be reused. ∎

All Trapezoid Rule approximations to $I = \int_1^2 (1/x)\, dx$ are greater than the true value of I. This is because the graph of $y = 1/x$ is concave up on $[1, 2]$, and therefore the tops of the approximating trapezoids lie above the curve. (See Figure 6.17.)

We can calculate the exact errors in the three approximations since we know that $I = \ln 2 = 0.69314718\dots$ (Remember that the error in an approximation is always taken to be the true value minus the approximate value.)

$$I - T_4 = 0.69314718\dots - 0.69702381\dots = -0.00387663\dots$$
$$I - T_8 = 0.69314718\dots - 0.69412185\dots = -0.00097467\dots$$
$$I - T_{16} = 0.69314718\dots - 0.69339120\dots = -0.00024402\dots.$$

Observe that the size of the error decreases to about a quarter of its previous value each time we double n. We will show below that this is to be expected for a "well-behaved" function like $1/x$.

Example 1 is somewhat artificial in the sense that we know the actual value of the integral so we really don't need an approximation. In practical applications of numerical integration we do not know the actual value. It is tempting to calculate several approximations for increasing values of n until the two most recent ones agree to within a prescribed error tolerance. For example, we might be inclined to claim that $\ln 2 \approx 0.69\dots$ from a comparison of T_4 and T_8, and further comparison of T_{16} and T_8 suggests that the third decimal place is probably 3: $I \approx 0.693\dots$. Although this approach cannot be justified in general, it is frequently used in practice.

Figure 6.18 The Midpoint Rule approximation M_n to $\int_a^b f(x)\, dx$ is the Riemann sum based on the heights to the graph of f at the midpoints of the subintervals of the partition

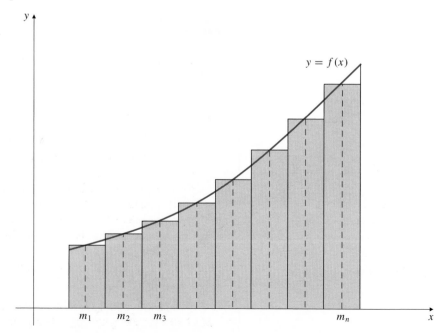

$y = f(x)$

$m_1 \quad m_2 \quad m_3 \qquad\qquad m_n$

The Midpoint Rule

A somewhat simpler approximation to $\int_a^b f(x)\,dx$, based on the partition of $[a, b]$ into n equal subintervals, involves forming a Riemann sum of the areas of rectangles whose heights are taken at the midpoints of the n subintervals. (See Figure 6.18.)

DEFINITION 4

> **The Midpoint Rule**
>
> If $h = (b - a)/n$, let $m_j = a + \left(j - \frac{1}{2}\right) h$ for $1 \le j \le n$. The **Midpoint Rule** approximation to $\int_a^b f(x)\,dx$, denoted M_n, is given by
>
> $$M_n = h\big(f(m_1) + f(m_2) + \cdots + f(m_n)\big) = h \sum_{j=1}^{n} f(m_j).$$

Example 2   Find the Midpoint Rule approximations M_4 and M_8 for the integral $I = \int_1^2 \frac{1}{x}\,dx$ and compare their actual errors with those obtained for the Trapezoid Rule approximations above.

Solution   To find M_4, the interval $[1, 2]$ is divided into four equal subintervals,

$$\left[1, \frac{5}{4}\right], \quad \left[\frac{5}{4}, \frac{3}{2}\right], \quad \left[\frac{3}{2}, \frac{7}{4}\right], \quad \text{and} \quad \left[\frac{7}{4}, 2\right].$$

The midpoints of these intervals are 9/8, 11/8, 13/8, and 15/8, respectively. The midpoints of the subintervals for M_8 are obtained in a similar way. The required Midpoint Rule approximations are

$$M_4 = \frac{1}{4}\left[\frac{8}{9} + \frac{8}{11} + \frac{8}{13} + \frac{8}{15}\right] = 0.69121989\ldots$$

$$M_8 = \frac{1}{8}\left[\frac{16}{17} + \frac{16}{19} + \frac{16}{21} + \frac{16}{23} + \frac{16}{25} + \frac{16}{27} + \frac{16}{29} + \frac{16}{31}\right] = 0.69266055\ldots$$

The errors in these approximations are

$$I - M_4 = 0.69314718\ldots - 0.69121989\ldots = 0.00192729\ldots$$
$$I - M_8 = 0.69314718\ldots - 0.69266055\ldots = 0.00048663\ldots$$

These errors are of opposite sign and about *half the size* of the corresponding Trapezoid Rule errors $I - T_4$ and $I - T_8$. Figure 6.19 suggests the reason for this. The rectangular area $hf(m_j)$ is equal to the area of the trapezoid formed by the tangent line to $y = f(x)$ at $(m_j, f(m_j))$. The shaded region above the curve is the part of the Trapezoid Rule error due to the jth subinterval. The shaded area below the curve is the corresponding Midpoint Rule error.

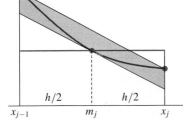

Figure 6.19   The Midpoint Rule error, the area shaded in colour, is opposite in sign and about half the size of the Trapezoid Rule error, the area shaded in grey

One drawback of the Midpoint Rule is that we cannot reuse values of f calculated for M_n when we calculate M_{2n}. However, to calculate T_{2n} we can use the data values already calculated for T_n and M_n. Specifically,

$$T_{2n} = \tfrac{1}{2}(T_n + M_n).$$

A good strategy for using these methods to obtain a value for an integral I to a desired degree of accuracy is to calculate successively:

$$T_n, \quad M_n, \quad T_{2n} = \frac{T_n + M_n}{2}, \quad M_{2n}, \quad T_{4n} = \frac{T_{2n} + M_{2n}}{2}, \quad M_{4n}, \quad \cdots$$

until two consecutive terms agree sufficiently closely. If a single quick approximation is needed, M_n is a better choice than T_n.

Error Estimates

The following theorem provides a bound for the error in the Trapezoid and Midpoint Rule approximations in terms of the second derivative of the integrand.

THEOREM **4**

Error estimates for the Trapezoid and Midpoint Rules

If f has a continuous second derivative on $[a, b]$ and satisfies $|f''(x)| \le K$ there, then

$$\left| \int_a^b f(x)\,dx - T_n \right| \le \frac{K(b-a)}{12}h^2 = \frac{K(b-a)^3}{12n^2},$$

$$\left| \int_a^b f(x)\,dx - M_n \right| \le \frac{K(b-a)}{24}h^2 = \frac{K(b-a)^3}{24n^2},$$

where $h = (b-a)/n$. Note that these error bounds decrease like the square of the subinterval length as n increases.

PROOF We will prove only the Trapezoid Rule error estimate here. (The one for the Midpoint Rule is a little easier to prove; the method is suggested in Exercise 14 below.) The straight line approximating $y = f(x)$ in the first subinterval $[x_0, x_1] = [a, a + h]$ passes through the two points (x_0, y_0) and (x_1, y_1). Its equation is $y = A + B(x - x_0)$, where

$$A = y_0 \quad \text{and} \quad B = \frac{y_1 - y_0}{x_1 - x_0} = \frac{y_1 - y_0}{h}.$$

Let the function $g(x)$ be the vertical distance between the graph of f and this line:

$$g(x) = f(x) - A - B(x - x_0).$$

Since the integral of $A + B(x - x_0)$ over $[x_0, x_1]$ is the area of the first trapezoid, which is $h(y_0 + y_1)/2$ (see Figure 6.20), the integral of $g(x)$ over $[x_0, x_1]$ is the error in the approximation of $\int_{x_0}^{x_1} f(x)\,dx$ by the area of the trapezoid:

$$\int_{x_0}^{x_1} f(x)\,dx - h\,\frac{y_0 + y_1}{2} = \int_{x_0}^{x_1} g(x)\,dx.$$

Now g is twice differentiable, and $g''(x) = f''(x)$. Also $g(x_0) = g(x_1) = 0$. Two integrations by parts (see Exercise 36 of Section 6.1) show that

$$\int_{x_0}^{x_1} (x - x_0)(x_1 - x)\,f''(x)\,dx = \int_{x_0}^{x_1} (x - x_0)(x_1 - x)\,g''(x)\,dx$$

$$= -2\int_{x_0}^{x_1} g(x)\,dx.$$

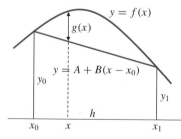

Figure 6.20

By the triangle inequality for definite integrals (Theorem 3(f) of Section 5.4)

$$\left| \int_{x_0}^{x_1} f(x)\,dx - h\,\frac{y_0 + y_1}{2} \right| \leq \frac{1}{2} \int_{x_0}^{x_1} (x - x_0)(x_1 - x)\,|f''(x)|\,dx$$

$$\leq \frac{K}{2} \int_{x_0}^{x_1} \left(-x^2 + (x_0 + x_1)x - x_0 x_1 \right) dx$$

$$= \frac{K}{12}(x_1 - x_0)^3 = \frac{K}{12}h^3.$$

A similar estimate holds on each subinterval $[x_{j-1}, x_j]$ $(1 \leq j \leq n)$. Therefore,

$$\left| \int_a^b f(x)\,dx - T_n \right| = \left| \sum_{j=1}^n \left(\int_{x_{j-1}}^{x_j} f(x)\,dx - h\,\frac{y_{j-1} + y_j}{2} \right) \right|$$

$$\leq \sum_{j=1}^n \left| \int_{x_{j-1}}^{x_j} f(x)\,dx - h\,\frac{y_{j-1} + y_j}{2} \right|$$

$$= \sum_{j=1}^n \frac{K}{12}h^3 = \frac{K}{12}nh^3 = \frac{K(b-a)}{12}h^2,$$

since $nh = b - a$.

We illustrate this error estimate for the approximations of Examples 1 and 2 above.

Example 3 Obtain bounds for the errors for T_4, T_8, T_{16}, M_4, and M_8 for
$$I = \int_1^2 \frac{1}{x}\,dx.$$

Solution If $f(x) = 1/x$, then $f'(x) = -1/x^2$ and $f''(x) = 2/x^3$. On $[1, 2]$ we have $|f''(x)| \leq 2$, so we may take $K = 2$ in the estimate. Thus,

$$|I - T_4| \leq \frac{2(2-1)}{12}\left(\frac{1}{4}\right)^2 = 0.0104\ldots,$$

$$|I - M_4| \leq \frac{2(2-1)}{24}\left(\frac{1}{4}\right)^2 = 0.0052\ldots,$$

$$|I - T_8| \leq \frac{2(2-1)}{12}\left(\frac{1}{8}\right)^2 = 0.0026\ldots,$$

$$|I - M_8| \leq \frac{2(2-1)}{24}\left(\frac{1}{8}\right)^2 = 0.0013\ldots,$$

$$|I - T_{16}| \leq \frac{2(2-1)}{12}\left(\frac{1}{16}\right)^2 = 0.00065\ldots.$$

The actual errors calculated earlier are considerably smaller than these bounds, because $|f''(x)|$ is rather smaller than $K = 2$ over most of the interval $[1, 2]$.

Remark   Error bounds are not usually as easily obtained as they are in Example 3. In particular, if an exact formula for $f(x)$ is not known (as is usually the case if the values of f are obtained from experimental data), then we have no method of calculating $f''(x)$, so we can't determine K. Theorem 4 is of more theoretical than practical importance. It shows us that, for a "well-behaved" function f, the Midpoint Rule error is typically about half as large as the Trapezoid Rule error and that both the Trapezoid Rule and Midpoint Rule errors can be expected to decrease like $1/n^2$ as n increases; in terms of big-O notation,

$$I = T_n + O\left(\frac{1}{n^2}\right) \quad \text{and} \quad I = M_n + O\left(\frac{1}{n^2}\right) \qquad \text{as } n \to \infty.$$

Of course, actual errors are not equal to the error bounds, so they won't always be cut to exactly a quarter of their size when we double n.

Exercises 6.6

In Exercises 1–4, calculate the approximations T_4, M_4, T_8, M_8, and T_{16} for the given integrals. (Use a scientific calculator or computer spreadsheet program.) Also calculate the exact value of each integral, and so determine the exact error in each approximation. Compare these exact errors with the bounds for the size of the error supplied by Theorem 4.

1. $I = \int_0^2 (1 + x^2)\, dx$ **2.** $I = \int_0^1 e^{-x}\, dx$

3. $I = \int_0^{\pi/2} \sin x\, dx$ **4.** $I = \int_0^1 \frac{dx}{1 + x^2}$

5. Figure 6.21 shows the graph of a function f over the interval $[1, 9]$. Using values from the graph, find the Trapezoid Rule estimates T_4 and T_8 for $\int_1^9 f(x)\, dx$.

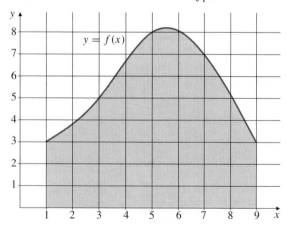

Figure 6.21

6. Obtain the best Midpoint Rule approximation that you can for $\int_1^9 f(x)\, dx$ from the data in Figure 6.21.

7. The map of a region is traced on the grid in Figure 6.22 where 1 unit in both the vertical and horizontal directions

represents 10 km. Use the Trapezoid Rule to obtain two estimates for the area of the region.

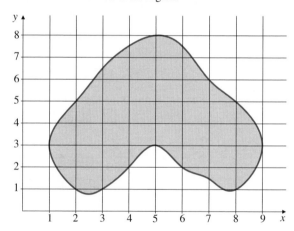

Figure 6.22

8. Find a Midpoint Rule estimate for the area of the region in the previous exercise.

Table 1.

x	$f(x)$	x	$f(x)$
0.0	1.4142	0.1	1.4124
0.2	1.4071	0.3	1.3983
0.4	1.3860	0.5	1.3702
0.6	1.3510	0.7	1.3285
0.8	1.3026	0.9	1.2734
1.0	1.2411	1.1	1.2057
1.2	1.1772	1.3	1.1258
1.4	1.0817	1.5	1.0348
1.6	0.9853		

9. Find T_4, M_4, T_8, M_8, and T_{16} for $\int_0^{1.6} f(x)\, dx$ for the function f whose values are given in Table 1.

10. Find the approximations M_8 and T_{16} for $\int_0^1 e^{-x^2}\,dx$. Quote a value for the integral to as many decimal places as you feel are justified.

11. Repeat Exercise 10 for $\int_0^{\pi/2} \frac{\sin x}{x}\,dx$.
    (Assume the integrand is 1 at $x = 0$.)

12. Compute the actual error in the approximation $\int_0^1 x^2\,dx \approx T_1$ and use it to show that the constant 12 in the estimate of Theorem 4 cannot be improved. That is, show that the absolute value of the actual error is as large as allowed by that estimate.

13. Repeat the previous exercise for M_1.

* 14. Prove the error estimate for the Midpoint Rule in Theorem 4 as follows: If $x_1 - x_0 = h$ and m_1 is the midpoint of $[x_0, x_1]$, use the error estimate for the tangent line

approximation (Theorem 4 of Section 3.5) to show that

$$|f(x) - f(m_1) - f'(m_1)(x - m_1)| \le \frac{K}{2}(x - m_1)^2.$$

Use this inequality to show that

$$\left| \int_{x_0}^{x_1} f(x)\,dx - f(m_1)h \right|$$
$$= \left| \int_{x_0}^{x_1} \big(f(x) - f(m_1) - f'(m_1)(x - m_1)\big)dx \right|$$
$$\le \frac{K}{24}h^3.$$

Complete the proof the same way used for the Trapezoid Rule estimate in Theorem 4.

6.7 Simpson's Rule

The Trapezoid Rule approximation to $\int_a^b f(x)\,dx$ results from approximating the graph of f by straight line segments through adjacent pairs of data points on the graph. Intuitively, we would expect to do better if we approximate the graph by more general curves. Since straight lines are the graphs of linear functions, the simplest obvious generalization is to use the class of quadratic functions, that is, to approximate the graph of f by segments of parabolas. This is the basis of Simpson's Rule.

Suppose that we are given three points in the plane, one on each of three equally spaced vertical lines, spaced, say, h units apart. If we choose the middle of these lines as the y-axis, then the coordinates of the three points will be, say, $(-h, y_L)$, $(0, y_M)$, and (h, y_R), as illustrated in Figure 6.23.

Constants A, B, and C can be chosen so that the parabola $y = A + Bx + Cx^2$ passes through these points; substituting the coordinates of the three points into the equation of the parabola, we get

$$\left.\begin{array}{l} y_L = A - Bh + Ch^2 \\ y_M = A \\ y_R = A + Bh + Ch^2 \end{array}\right\} \quad \Rightarrow \quad A = y_M \quad \text{and} \quad 2Ch^2 = y_L - 2y_M + y_R.$$

Now we have

$$\int_{-h}^{h} (A + Bx + Cx^2)\,dx = \left(Ax + \frac{B}{2}x^2 + \frac{C}{3}x^3 \right)\Big|_{-h}^{h} = 2Ah + \frac{2}{3}Ch^3$$
$$= h\left(2y_M + \frac{1}{3}(y_L - 2y_M + y_R) \right)$$
$$= \frac{h}{3}(y_L + 4y_M + y_R).$$

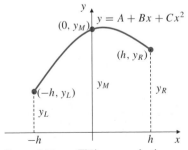

Figure 6.23   Fitting a quadratic graph through three points with equal horizontal spacing

Thus, the area of the plane region bounded by the parabolic arc, the interval of length $2h$ on the x-axis, and the left and right vertical lines is equal to $(h/3)$ times the sum of the heights of the region at the left and right edges and four times the height at the middle. (It is independent of the position of the y-axis.)

Now suppose that we are given the same data for f as we were given for the Trapezoid Rule, that is, we know the values $y_j = f(x_j)$ $(0 \leq j \leq n)$ at $n + 1$ equally spaced points

$$x_0 = a, \quad x_1 = a + h, \quad x_2 = a + 2h, \quad \ldots, \quad x_n = a + nh = b,$$

where $h = (b-a)/n$. We can approximate the graph of f over *pairs* of the subintervals $[x_{j-1}, x_j]$ using parabolic segments, and use the integrals of the corresponding quadratic functions to approximate the integrals of f over these subintervals. Since we need to use the subintervals two at a time, we must assume that n is *even*. Using the integral computed for the parabolic segment above, we have

$$\int_{x_0}^{x_2} f(x)\, dx \approx \frac{h}{3} (y_0 + 4y_1 + y_2)$$

$$\int_{x_2}^{x_4} f(x)\, dx \approx \frac{h}{3} (y_2 + 4y_3 + y_4)$$

$$\vdots$$

$$\int_{x_{n-2}}^{x_n} f(x)\, dx \approx \frac{h}{3} (y_{n-2} + 4y_{n-1} + y_n).$$

Adding these $n/2$ individual approximations we get the Simpson's Rule approximation to the integral $\int_a^b f(x)\, dx$.

DEFINITION   **5**

Simpson's Rule

The **Simpson's Rule** approximation to $\int_a^b f(x)\, dx$ based on a subdivision of $[a, b]$ into an even number n of subintervals of equal length $h = (b - a)/n$ is denoted S_n and is given by:

$$\int_a^b f(x)\, dx \approx S_n$$

$$= \frac{h}{3} \left(y_0 + 4y_1 + 2y_2 + 4y_3 + 2y_4 + \cdots + 2y_{n-2} + 4y_{n-1} + y_n \right)$$

$$= \frac{h}{3} \left(y_{\text{"ends"}} + 4y_{\text{"odds"}} + 2y_{\text{"evens"}} \right).$$

Note that the Simpson's Rule approximation S_n requires no more data than does the Trapezoid Rule approximation T_n; both require the values of $f(x)$ at $n + 1$ equally spaced points. However, Simpson's Rule treats the data differently, weighting successive values either 1/3, 2/3, or 4/3. As we will see, this can produce a much better approximation to the integral of f.

Example 1   Calculate the approximations S_4, S_8, and S_{16} for $I = \displaystyle\int_1^2 \frac{1}{x}\, dx$ and compare them with the actual value $I = \ln 2 = 0.69314718\ldots$, and with the values of T_4, T_8, and T_{16} obtained in Example 1 of Section 6.6.

Solution   We calculate

$$S_4 = \frac{1}{12}\left[1 + 4\left(\frac{4}{5}\right) + 2\left(\frac{2}{3}\right) + 4\left(\frac{4}{7}\right) + \frac{1}{2}\right] = 0.69325397\ldots,$$

$$S_8 = \frac{1}{24}\left[1 + \frac{1}{2} + 4\left(\frac{8}{9} + \frac{8}{11} + \frac{8}{13} + \frac{8}{15}\right)\right.$$
$$\left. + 2\left(\frac{4}{5} + \frac{2}{3} + \frac{4}{7}\right)\right] = 0.69315453\ldots,$$

$$S_{16} = \frac{1}{48}\left[1 + \frac{1}{2}\right.$$
$$+ 4\left(\frac{16}{17} + \frac{16}{19} + \frac{16}{21} + \frac{16}{23} + \frac{16}{25} + \frac{16}{27} + \frac{16}{29} + \frac{16}{31}\right)$$
$$\left. + 2\left(\frac{8}{9} + \frac{4}{5} + \frac{8}{11} + \frac{2}{3} + \frac{8}{13} + \frac{4}{7} + \frac{8}{15}\right)\right] = 0.69314765\ldots.$$

The errors are

$$I - S_4 = 0.69314718\ldots - 0.69325397\ldots = -0.00010679,$$
$$I - S_8 = 0.69314718\ldots - 0.69315453\ldots = -0.00000735,$$
$$I - S_{16} = 0.69314718\ldots - 0.69314765\ldots = -0.00000047.$$

These errors are evidently much smaller than the corresponding errors for the Trapezoid or Midpoint Rule approximations.

Remark   Simpson's Rule S_{2n} makes use of the same $2n + 1$ data values that T_n and M_n together use. It is not difficult to verify that

$$S_{2n} = \frac{T_n + 2M_n}{3}, \qquad S_{2n} = \frac{2T_{2n} + M_n}{3}, \qquad \text{and} \qquad S_{2n} = \frac{4T_{2n} - T_n}{3}.$$

Figure 6.19 and Theorem 4 in Section 6.6 suggest why the first of these formulas ought to yield a particularly good approximation to I.

Obtaining an error estimate for Simpson's Rule is harder than for the Trapezoid Rule. We state the appropriate estimate in the following theorem, but we do not attempt any proof. Proofs can be found in textbooks on numerical analysis.

THEOREM 5

Error estimate for Simpson's Rule

If f has a continuous fourth derivative on the interval $[a, b]$, satisfying $|f^{(4)}(x)| \le K$ there, then

$$\left|\int_a^b f(x)\,dx - S_n\right| \le \frac{K(b-a)}{180}h^4 = \frac{K(b-a)^5}{180n^4},$$

where $h = (b - a)/n$.

Observe that, as n increases, the error decreases as the fourth power of h and, hence, as $1/n^4$. Using the big-O notation we have

$$\int_a^b f(x)\,dx = S_n + O\left(\frac{1}{n^4}\right) \qquad \text{as } n \to \infty.$$

This accounts for the fact that S_n is a much better approximation than is T_n, provided that h is small and $|f^{(4)}(x)|$ is not unduly large compared with $|f''(x)|$. Note also that for any (even) n, S_n gives the exact value of the integral of any *cubic* function $f(x) = A + Bx + Cx^2 + Dx^3$; $f^{(4)}(x) = 0$ identically for such f, so we can take $K = 0$ in the error estimate.

Example 2   Obtain bounds for the absolute values of the errors in the approximations of Example 1.

Solution   If $f(x) = 1/x$, then

$$f'(x) = -\frac{1}{x^2}, \qquad f''(x) = \frac{2}{x^3}, \qquad f^{(3)}(x) = -\frac{6}{x^4}, \qquad f^{(4)}(x) = \frac{24}{x^5}.$$

Clearly, $|f^{(4)}(x)| \le 24$ on $[1, 2]$, so we can take $K = 24$ in the estimate of Theorem 5. We have

$$|I - S_4| \le \frac{24(2-1)}{180}\left(\frac{1}{4}\right)^4 \approx 0.00052083,$$

$$|I - S_8| \le \frac{24(2-1)}{180}\left(\frac{1}{8}\right)^4 \approx 0.00003255,$$

$$|I - S_{16}| \le \frac{24(2-1)}{180}\left(\frac{1}{16}\right)^4 \approx 0.00000203.$$

Again we observe that the actual errors are well within these bounds.

Example 3   A function f satisfies $|f^{(4)}(x)| \le 7$ on the interval $[1, 3]$, and the values $f(1.0) = 0.1860$, $f(1.5) = 0.9411$, $f(2.0) = 1.1550$, $f(2.5) = 1.4511$, and $f(3.0) = 1.2144$. Find the best possible Simpson's Rule approximation to $I = \int_1^3 f(x)\,dx$, based on these data. Give a bound for the size of the error, and specify the smallest interval you can that must contain the value of I.

Solution   We take $n = 4$, so that $h = (3 - 1)/4 = 0.5$, and we obtain

$$I = \int_1^3 f(x)\,dx$$

$$\approx S_4 = \frac{0.5}{3}\left(0.1860 + 4(0.9411 + 1.4511) + 2(1.1550) + 1.2144\right)$$

$$= 2.2132.$$

Since $|f^{(4)}(x)| \le 7$ on $[1, 3]$ we have

$$|I - S_4| \le \frac{7(3-1)}{180}(0.5)^4 < 0.0049.$$

I must therefore satisfy

$$2.2132 - 0.0049 < I < 2.2132 + 0.0049 \quad \text{or} \quad 2.2083 < I < 2.2181.$$

Exercises 6.7

In Exercises 1–4, find Simpson's Rule approximations S_4 and S_8 for the given functions. Compare your results with the actual values of the integrals and with the corresponding Trapezoid Rule approximations obtained in Exercises 1–4 of Section 6.6.

1. $I = \int_0^2 (1 + x^2) \, dx$

2. $I = \int_0^1 e^{-x} \, dx$

3. $I = \int_0^{\pi/2} \sin x \, dx$

4. $I = \int_0^1 \dfrac{dx}{1 + x^2}$

5. Find the Simpson's Rule approximation S_8 for the integral in Exercise 5 of Section 6.6.

6. Find the best Simpson's Rule approximation that you can for the area of the region in Exercise 7 of Section 6.6.

7. Use Theorem 5 to obtain bounds for the errors in the approximations obtained in Exercises 2 and 3 above.

8. Verify that $S_{2n} = \dfrac{T_n + 2M_n}{3} = \dfrac{2T_{2n} + M_n}{3}$, where T_n and M_n refer to the appropriate Trapezoid and Midpoint Rule approximations. Deduce that $S_{2n} = \dfrac{4T_{2n} - T_n}{3}$.

9. Find S_4, S_8, and S_{16} for $\int_0^{1.6} f(x) \, dx$ for the function f whose values are tabulated in Exercise 9 of Section 6.6.

10. Find the Simpson's Rule approximations S_8 and S_{16} for $\int_0^1 e^{-x^2} \, dx$. Quote a value for the integral to the number of decimal places you feel is justified based on comparing the two approximations.

* 11. Compute the actual error in the approximation $\int_0^1 x^4 \, dx \approx S_2$ and use it to show that the constant 180 in the estimate of Theorem 5 cannot be improved.

* 12. Since Simpson's Rule is based on quadratic approximation, it is not surprising that it should give an exact value for an integral of $A + Bx + Cx^2$. It is more surprising that it is exact for a cubic function as well. Verify by direct calculation that $\int_0^1 x^3 \, dx = S_2$.

6.8 Other Aspects of Approximate Integration

The numerical methods described in Sections 6.6 and 6.7 are suitable for finding approximate values for integrals of the form

$$I = \int_a^b f(x) \, dx,$$

where $[a, b]$ is a finite interval and the integrand f is "well-behaved" on $[a, b]$. In particular, I must be a *proper* integral. There are many other methods for dealing with such integrals, some of which we mention later in this section. First, however, we consider what can be done if the function f isn't "well-behaved" on $[a, b]$. We mean by this that either the integral is improper or f doesn't have sufficiently many continuous derivatives on $[a, b]$ to justify whatever numerical methods we want to use.

The ideas of this section are best presented by means of concrete examples.

Example 1 How can you evaluate the integral $I = \int_0^1 \sqrt{x} \, e^x \, dx$ numerically?

Solution Although I is a proper integral, with integrand $f(x) = \sqrt{x} \, e^x$ satisfying $f(x) \to 0$ as $x \to 0+$, nevertheless, the standard numerical methods can be expected to perform poorly for I because the derivatives of f are not bounded near 0. This problem is easily remedied; just make the change of variable $x = t^2$ and rewrite I in the form

$$I = 2 \int_0^1 t^2 \, e^{t^2} \, dt,$$

whose integrand $g(t) = t^2 \, e^{t^2}$ has bounded derivatives near 0. The latter integral can be efficiently approximated by the methods of Sections 6.6 and 6.7.

Approximating Improper Integrals

Example 2   Describe how to evaluate $I = \int_0^1 \dfrac{\cos x}{\sqrt{x}}\, dx$ numerically.

Solution   The integral is improper, but convergent because, on [0, 1],

$$0 < \frac{\cos x}{\sqrt{x}} \leq \frac{1}{\sqrt{x}} \qquad \text{and} \qquad \int_0^1 \frac{dx}{\sqrt{x}} = 2.$$

However, since $\lim_{x \to 0+} \dfrac{\cos x}{\sqrt{x}} = \infty$, we cannot directly apply any of the techniques developed in Sections 6.6 and 6.7. (y_0 is infinite.) The substitution $x = t^2$ removes this difficulty:

$$I = \int_0^1 \frac{\cos t^2}{t} 2t\, dt = 2 \int_0^1 \cos t^2\, dt.$$

The latter integral is not improper and is well-behaved. Numerical techniques can be applied to evaluate it.

■

Example 3   Show how to evaluate $I = \int_0^\infty \dfrac{dx}{\sqrt{2 + x^2 + x^4}}$ by numerical means.

Solution   Here the integral is improper of type I; the interval of integration is infinite. Although there is no singularity at $x = 0$, it is still useful to break the integral into two parts:

$$I = \int_0^1 \frac{dx}{\sqrt{2 + x^2 + x^4}} + \int_1^\infty \frac{dx}{\sqrt{2 + x^2 + x^4}} = I_1 + I_2.$$

I_1 is proper. In I_2 make the change of variable $x = 1/t$:

$$I_2 = \int_0^1 \frac{dt}{t^2 \sqrt{2 + \dfrac{1}{t^2} + \dfrac{1}{t^4}}} = \int_0^1 \frac{dt}{\sqrt{2t^4 + t^2 + 1}}.$$

This is also a proper integral. If desired, I_1 and I_2 can be recombined into a single integral before numerical methods are applied:

$$I = \int_0^1 \left(\frac{1}{\sqrt{2 + x^2 + x^4}} + \frac{1}{\sqrt{2x^4 + x^2 + 1}} \right) dx.$$

■

Example 3 suggests that when an integral is taken over an infinite interval, a change of variable should be made to convert the integral to a finite interval.

Using Taylor's Formula

Taylor's Formula (see Section 4.8) can sometimes be useful for evaluating integrals. Here is an example.

Example 4 Use Taylor's Formula for $f(x) = e^x$, obtained in Section 4.8, to evaluate the integral $\int_0^1 e^{x^2}\,dx$ to within an error of less than 10^{-4}.

Solution In Example 4 of Section 4.8 we showed that

$$f(x) = e^x = 1 + x + \frac{x^2}{2!} + \frac{x^3}{3!} + \cdots + \frac{x^n}{n!} + E_n(x),$$

where

$$E_n(x) = \frac{e^X}{(n+1)!}\,x^{n+1}$$

for some X between 0 and x. If $0 \le x \le 1$, then $0 \le X \le 1$, so $e^X \le e < 3$. Therefore

$$|E_n(x)| \le \frac{3}{(n+1)!}\,x^{n+1}.$$

Now replace x by x^2 in the formula for e^x above and integrate from 0 to 1:

$$\int_0^1 e^{x^2}\,dx = \int_0^1 \left(1 + x^2 + \frac{x^4}{2!} + \cdots + \frac{x^{2n}}{n!}\right) dx + \int_0^1 E_n(x^2)\,dx$$

$$= 1 + \frac{1}{3} + \frac{1}{5 \times 2!} + \cdots + \frac{1}{(2n+1)n!} + \int_0^1 E_n(x^2)\,dx.$$

We want the error to be less than 10^{-4}, so we estimate the remainder term:

$$\left|\int_0^1 E_n(x^2)\,dx\right| \le \frac{3}{(n+1)!}\int_0^1 x^{2(n+1)}\,dx = \frac{3}{(n+1)!(2n+3)} < 10^{-4},$$

provided $(2n+3)(n+1)! > 30{,}000$. Since $13 \times 6! = 9{,}360$ and $15 \times 7! = 75{,}600$, we need $n = 6$. Thus,

$$\int_0^1 e^{x^2}\,dx = 1 + \frac{1}{3} + \frac{1}{5 \times 2!} + \frac{1}{7 \times 3!} + \frac{1}{9 \times 4!} + \frac{1}{11 \times 5!} + \frac{1}{13 \times 6!}$$

$$\approx 1.46264,$$

with error less than 10^{-4}.

Romberg Integration

Using Taylor's Formula, it is possible to verify that for a function f having continuous derivatives up to order $2m+2$ on $[a, b]$ the error $E_n = I - T_n$ in the Trapezoid Rule approximation T_n to $I = \int_a^b f(x)\,dx$ satisfies

$$E_n = I - T_n = \frac{C_1}{n^2} + \frac{C_2}{n^4} + \frac{C_3}{n^6} + \cdots + \frac{C_m}{n^{2m}} + O\left(\frac{1}{n^{2m+2}}\right),$$

where the constants C_j depend on the $2j$th derivative of f. It is possible to use this formula to obtain higher-order approximations to I, starting with Trapezoid Rule approximations. The technique is known as **Romberg integration** or **Richardson extrapolation**.

To begin, suppose we have constructed Trapezoid Rule approximations for values of n that are powers of 2: $n = 1, 2, 4, 8, \ldots$. Accordingly, let us define

$$T_k^0 = T_{2^k}. \qquad \text{Thus} \quad T_0^0 = T_1, \quad T_1^0 = T_2, \quad T_2^0 = T_4, \quad \dots.$$

Using the formula for $T_{2^k} = I - E_{2^k}$ given above, we write

$$T_k^0 = I - \frac{C_1}{4^k} - \frac{C_2}{4^{2k}} - \dots - \frac{C_m}{4^{mk}} + O\left(\frac{1}{4^{(m+1)k}}\right) \qquad (\text{as } k \to \infty).$$

Similarly, replacing k by $k+1$, we get

$$T_{k+1}^0 = I - \frac{C_1}{4^{k+1}} - \frac{C_2}{4^{2(k+1)}} - \dots - \frac{C_m}{4^{m(k+1)}} + O\left(\frac{1}{4^{(m+1)(k+1)}}\right).$$

If we multiply the formula for T_{k+1}^0 by 4 and subtract the formula for T_k^0, the terms involving C_1 will cancel out. The first term on the right will be $4I - I = 3I$, so let us also divide by 3 and define T_{k+1}^1 to be the result. Then as $k \to \infty$, we have

$$T_{k+1}^1 = \frac{4T_{k+1}^0 - T_k^0}{3} = I - \frac{C_2^1}{4^{2k}} - \frac{C_3^1}{4^{3k}} - \dots - \frac{C_m^1}{4^{mk}} + O\left(\frac{1}{4^{(m+1)k}}\right).$$

(The C_i^1 are new constants.)   Unless these constants are much larger than the previous ones, T_{k+1}^1 ought to be a better approximation to I than T_{k+1}^0 since we have eliminated the lowest order (and therefore the largest) of the error terms, $C_1/4^{k+1}$. In fact, Exercise 8 in Section 6.7 shows that $T_{k+1}^1 = S_{2^{k+1}}$, the Simpson's Rule approximation based on 2^{k+1} subintervals.

We can continue the process of eliminating error terms begun above. Replacing $k+1$ by $k+2$ in the expression for T_{k+1}^1 we obtain

$$T_{k+2}^1 = I - \frac{C_2^1}{4^{2(k+1)}} - \frac{C_3^1}{4^{3(k+1)}} - \dots - \frac{C_m^1}{4^{m(k+1)}} + O\left(\frac{1}{4^{(m+1)(k+1)}}\right).$$

To eliminate C_2^1 we can multiply the second formula by 16, subtract the first formula, and divide by 15. Denoting the result T_{k+2}^2, we have, as $k \to \infty$,

$$T_{k+2}^2 = \frac{16T_{k+2}^1 - T_{k+1}^1}{15} = I - \frac{C_3^2}{4^{3k}} - \dots - \dots - \frac{C_m^2}{4^{mk}} + O\left(\frac{1}{4^{(m+1)k}}\right).$$

We can proceed in this way, eliminating one error term after another. In general, for $j < m$ and $k \geq 0$,

$$T_{k+j}^j = \frac{4^j T_{k+j}^{j-1} - T_{k+j-1}^{j-1}}{4^j - 1} = I - \frac{C_{j+1}^j}{4^{(j+1)k}} - \dots - \frac{C_m^j}{4^{mk}} + O\left(\frac{1}{4^{(m+1)k}}\right).$$

The big-O term refers to $k \to \infty$ for fixed j. All this looks very complicated, but it is not difficult to carry out in practice, especially with the aid of a computer spreadsheet. Let $R_j = T_j^j$, called a **Romberg approximation** to I, and calculate the entries in the following scheme in order from left to right and down each column when you come to it:

Scheme for calculating Romberg approximations

$$T_0^0 = T_1 = R_0 \longrightarrow \quad T_1^0 = T_2 \quad \longrightarrow \quad T_2^0 = T_4 \quad \longrightarrow \quad T_3^0 = T_8 \quad \longrightarrow$$

$$\downarrow \qquad\qquad\qquad \downarrow \qquad\qquad\qquad \downarrow$$

$$T_1^1 = S_2 = R_1 \qquad\quad T_2^1 = S_4 \qquad\quad T_3^1 = S_8$$

$$\downarrow \qquad\qquad\qquad \downarrow$$

$$T_2^2 = R_2 \qquad\quad T_3^2$$

$$\downarrow$$

$$T_3^3 = R_3$$

Stop when T_j^{j-1} and R_j differ by less than the acceptable error, and quote R_j as the Romberg approximation to $\int_a^b f(x)\,dx$.

The top line in the scheme is made up of the Trapezoid Rule approximations $T_1, T_2, T_4, T_8, \ldots$. Elements in subsequent rows are calculated by the formulas:

Formulas for calculating Romberg approximations

$$T_1^1 = \frac{4T_1^0 - T_0^0}{3} \qquad T_2^1 = \frac{4T_2^0 - T_1^0}{3} \qquad T_3^1 = \frac{4T_3^0 - T_2^0}{3} \quad \cdots$$

$$T_2^2 = \frac{16T_2^1 - T_1^1}{15} \qquad T_3^2 = \frac{16T_3^1 - T_2^1}{15} \quad \cdots$$

$$T_3^3 = \frac{64T_3^2 - T_2^2}{63} \quad \cdots$$

In general, if $1 \le j \le k$, then $\quad T_k^j = \dfrac{4^j T_k^{j-1} - T_{k-1}^{j-1}}{4^j - 1}$.

Each new entry is calculated from the one above and the one to the left of that one.

Example 5 Calculate the Romberg approximations R_0, R_1, R_2, R_3, and R_4 for the integral $I = \displaystyle\int_1^2 \frac{1}{x}\,dx$.

Solution We will carry all calculations to 8 decimal places. Since we must obtain R_4, we will need to find all the entries in the first five columns of the scheme. First we calculate the first two Trapezoid Rule approximations:

$$R_0 = T_0^0 = T_1 = \frac{1}{2} + \frac{1}{4} = 0.75000000,$$

$$T_1^0 = T_2 = \frac{1}{2}\left[\frac{1}{2}(1) + \frac{2}{3} + \frac{1}{2}\left(\frac{1}{2}\right)\right] = 0.70833333.$$

The remaining required Trapezoid Rule approximations were calculated in Example 1 of Section 6.6, so we will just record them here:

$$T_2^0 = T_4 = 0.69702381,$$

$$T_3^0 = T_8 = 0.69412185,$$

$$T_4^0 = T_{16} = 0.69339120.$$

Now we calculate down the columns from left to right. For the second column:

$$R_1 = S_2 = T_1^1 = \frac{4T_1^0 - T_0^0}{3} = 0.69444444;$$

the third column:

$$S_4 = T_2^1 = \frac{4T_2^0 - T_1^0}{3} = 0.69325397,$$

$$R_2 = T_2^2 = \frac{16T_2^1 - T_1^1}{15} = 0.69317460;$$

the fourth column:

$$S_8 = T_3^1 = \frac{4T_3^0 - T_2^0}{3} = 0.69315453,$$

$$T_3^2 = \frac{16T_3^1 - T_2^1}{15} = 0.69314790,$$

$$R_3 = T_3^3 = \frac{64T_3^2 - T_2^2}{63} = 0.69314748;$$

and the fifth column:

$$S_{16} = T_4^1 = \frac{4T_4^0 - T_3^0}{3} = 0.69314765,$$

$$T_4^2 = \frac{16T_4^1 - T_3^1}{15} = 0.69314719,$$

$$T_4^3 = \frac{64T_4^2 - T_3^2}{63} = 0.69314718,$$

$$R_4 = T_4^4 = \frac{256T_4^3 - T_3^3}{255} = 0.69314718.$$

Since T_4^3 and R_4 agree to the 8 decimal places we are calculating, we conclude that, correct to 8 decimal places,

$$I = \int_1^2 \frac{dx}{x} = \ln 2 \approx 0.69314718\ldots.$$

The various approximations calculated above suggest that for any given value of $n = 2^k$, the Romberg approximation R_n should give the best value obtainable for the integral based on the $n + 1$ data values $y_0, y_1, \ldots, y_n$. This is so only if the derivatives $f^{(n)}(x)$ do not grow too rapidly as n increases.

Other Methods

As developed above, the Trapezoid, Midpoint, Simpson, and Romberg methods all involved using equal subdivisions of the interval $[a, b]$. There are other methods that avoid this restriction. In particular, **Gaussian approximations** involve selecting evaluation points and weights in an optimal way so as to give the most accurate results for "well-behaved" functions. See Exercises 11–13 below. You can consult a text on numerical analysis to learn more about this method.

Finally, we note that even when you apply one of the methods of Sections 6.6 and 6.7, it may be advisable for you to break up the integral into two or more integrals over smaller intervals and then use different subinterval lengths h for each of the different integrals. You will want to evaluate the integrand at more points in an interval where its graph is changing direction erratically than in one where the graph is better behaved.

Exercises 6.8

Rewrite the integrals in Exercises 1–6 in a form to which numerical methods can be readily applied.

1. $\displaystyle\int_0^1 \frac{dx}{x^{1/3}(1+x)}$

2. $\displaystyle\int_0^1 \frac{e^x}{\sqrt{1-x}}\,dx$

3. $\displaystyle\int_{-1}^1 \frac{e^x}{\sqrt{1-x^2}}\,dx$

4. $\displaystyle\int_1^\infty \frac{dx}{x^2+\sqrt{x}+1}$

* 5. $\displaystyle\int_0^{\pi/2} \frac{dx}{\sqrt{\sin x}}$

6. $\displaystyle\int_0^\infty \frac{dx}{x^4+1}$

7. Find T_2, T_4, T_8, and T_{16} for $\int_0^1 \sqrt{x}\,dx$ and find the actual errors in these approximations. Do the errors decrease like $1/n^2$ as n increases? Why?

8. Transform the integral $I = \int_1^\infty e^{-x^2}\,dx$ using the substitution $x = 1/t$ and calculate the Simpson's Rule approximations S_2, S_4, and S_8 for the resulting integral (whose integrand has limit 0 as $t \to 0+$). Quote the value of I to the accuracy you feel is justified. Do the approximations converge as quickly as you might expect? Can you think of a reason why they might not?

9. Evaluate $I = \int_0^1 e^{-x^2}\,dx$, by the Taylor's Formula method of Example 4, to within an error of 10^{-4}.

10. Recall that $\int_0^\infty e^{-x^2}\,dx = \dfrac{1}{2}\sqrt{\pi}$. Combine this fact with the result of the previous exercise to evaluate
$$I = \int_1^\infty e^{-x^2}\,dx \text{ to 3 decimal places.}$$

11. (Gaussian approximation) Find constants A and u, with u between 0 and 1, such that
$$\int_{-1}^1 f(x)\,dx = Af(-u) + Af(u)$$
holds for every cubic polynomial
$f(x) = ax^3 + bx^2 + cx + d$. For a general function $f(x)$ defined on $[-1, 1]$, the approximation
$$\int_{-1}^1 f(x)\,dx \approx Af(-u) + Af(u)$$
is called a *Gaussian* approximation.

12. Use the method of Exercise 11 to approximate the integrals of (a) x^4, (b) $\cos x$, and (c) e^x, over the interval $[-1, 1]$, and find the error in each approximation.

13. (Another Gaussian approximation) Find constants A and B, and u between 0 and 1, such that
$$\int_{-1}^1 f(x)\,dx = Af(-u) + Bf(0) + Af(u)$$

holds for every quintic polynomial
$f(x) = ax^5 + bx^4 + cx^3 + dx^2 + ex + f$.

14. Use the Gaussian approximation
$$\int_{-1}^1 f(x)\,dx \approx Af(-u) + Bf(0) + Af(u),$$
where A, B, and u are as determined in Exercise 13, to find approximations for the integrals of (a) x^6, (b) $\cos x$, and (c) e^x over the interval $[-1, 1]$, and find the error in each approximation.

15. Calculate sufficiently many Romberg approximations $R_1, R_2, R_3, \ldots$ for the integral
$$\int_0^1 e^{-x^2}\,dx$$
to be confident you have evaluated the integral correctly to 6 decimal places.

16. Use the values of $f(x)$ given in the table accompanying Exercise 9 in Section 6.6 to calculate the Romberg approximations R_1, R_2 and R_3 for the integral
$$\int_0^{1.6} f(x)\,dx$$
in that exercise.

* 17. The Romberg approximation R_2 for $\int_a^b f(x)\,dx$ requires five values of f, $y_0 = f(a)$, $y_1 = f(a + h)$, ..., $y_4 = f(x + 4h) = f(b)$, where $h = (b - a)/4$. Write the formula for R_2 explicitly in terms of these five values.

* 18. Explain why the change of variable $x = 1/t$ is not suitable for transforming the integral $\displaystyle\int_\pi^\infty \frac{\sin x}{1+x^2}\,dx$ into a form to which numerical methods can be applied. Try to devise a method whereby this integral could be approximated to any desired degree of accuracy.

* 19. If $f(x) = \dfrac{\sin x}{x}$ for $x \neq 0$ and $f(0) = 1$, show that $f''(x)$ has a finite limit as $x \to 0$. Hence, f'' is bounded on finite intervals $[0, a]$ and Trapezoid Rule approximations T_n to $\int_0^a \dfrac{\sin x}{x}\,dx$ converge suitably quickly as n increases. Higher derivatives are also bounded (Taylor's Formula is useful for showing this) so Simpson's Rule and higher-order approximations can also be used effectively.

Chapter Review

Key Ideas

- **What do the following terms and phrases mean?**

  ◇ integration by parts   ◇ a reduction formula

  ◇ an inverse substitution   ◇ a rational function

  ◇ the method of partial fractions

  ◇ a computer algebra system

  ◇ an improper integral of type I

  ◇ an improper integral of type II

  ◇ a p-integral   ◇ the Trapezoid Rule

  ◇ the Midpoint Rule   ◇ Simpson's Rule

- **Describe the inverse sine and inverse tangent substitutions.**
- **What is the significance of the comparison theorem for improper integrals?**
- **When is numerical integration necessary?**

Summary of Techniques of Integration

Students sometimes have difficulty deciding which method to use to evaluate a given integral. Often no one method will suffice to produce the whole solution, but one method may lead to a different, possibly simpler, integral that can then be dealt with on its own merits. Here are a few guidelines:

1. First, and always, be alert for simplifying substitutions. Even when these don't accomplish the whole integration, they can lead to integrals to which some other method can be applied.

2. If the integral involves a quadratic expression $Ax^2 + Bx + C$ with $A \neq 0$ and $B \neq 0$, complete the square. A simple substitution then reduces the quadratic expression to a sum or difference of squares.

3. Integrals of products of trigonometric functions can sometimes be evaluated or rendered simpler by the use of appropriate trigonometric identities such as:

$$\sin^2 x + \cos^2 x = 1$$
$$\sec^2 x = 1 + \tan^2 x$$
$$\csc^2 x = 1 + \cot^2 x$$
$$\sin x \cos x = \tfrac{1}{2} \sin 2x$$
$$\sin^2 x = \tfrac{1}{2}(1 - \cos 2x)$$
$$\cos^2 x = \tfrac{1}{2}(1 + \cos 2x).$$

4. Integrals involving $(a^2 - x^2)^{1/2}$ can be transformed using $x = a \sin \theta$. Integrals involving $(a^2 + x^2)^{1/2}$ or $1/(a^2 + x^2)$ may yield to $x = a \tan \theta$. Integrals involving $(x^2 - a^2)^{1/2}$ can be transformed using $x = a \sec \theta$ or $x = a \cosh \theta$.

5. Use integration by parts for integrals of functions such as products of polynomials and transcendental functions, and for inverse trigonometric functions and logarithms. Be alert for ways of using integration by parts to obtain formulas representing complicated integrals in terms of simpler ones.

6. Use partial fractions to integrate rational functions whose denominators can be factored into real linear and quadratic factors. Remember to divide the polynomials first, if necessary, to reduce the fraction to one whose numerator has degree smaller than that of its denominator.

7. There is a table of integrals at the back of this book. If you can't do an integral directly, try to use the methods above to convert it to the form of one of the integrals in the table.

8. If you can't find any way to evaluate a definite integral for which you need a numerical value, consider using a computer or calculator and one of the numerical methods presented in Sections 6.6–6.8.

Review Exercises on Techniques of Integration

Here is an opportunity to get more practice evaluating integrals. Unlike the exercises in Sections 5.6 and 6.1–6.3, which used only the technique of the particular section, these exercises are grouped randomly so you will have to decide which techniques to use.

1. $\displaystyle \int \frac{x\,dx}{2x^2 + 5x + 2}$

2. $\displaystyle \int \frac{x\,dx}{(x - 1)^3}$

3. $\displaystyle \int \sin^3 x \cos^3 x \, dx$

4. $\displaystyle \int \frac{(1 + \sqrt{x})^{1/3}}{\sqrt{x}} \, dx$

5. $\displaystyle \int \frac{3\,dx}{4x^2 - 1}$

6. $\displaystyle \int (x^2 + x - 2) \sin 3x \, dx$

7. $\displaystyle \int \frac{\sqrt{1 - x^2}}{x^4} \, dx$

8. $\displaystyle \int x^3 \cos(x^2) \, dx$

9. $\displaystyle \int \frac{x^2 \, dx}{(5x^3 - 2)^{2/3}}$

10. $\displaystyle \int \frac{dx}{x^2 + 2x - 15}$

11. $\displaystyle \int \frac{dx}{(4 + x^2)^2}$

12. $\displaystyle \int (\sin x + \cos x)^2 \, dx$

13. $\displaystyle \int 2^x \sqrt{1 + 4^x} \, dx$

14. $\displaystyle \int \frac{\cos x}{1 + \sin^2 x} \, dx$

15. $\displaystyle \int \frac{\sin^3 x}{\cos^7 x} \, dx$

16. $\displaystyle \int \frac{x^2 \, dx}{(3 + 5x^2)^{3/2}}$

17. $\displaystyle \int e^{-x} \sin(2x) \, dx$

18. $\displaystyle \int \frac{2x^2 + 4x - 3}{x^2 + 5x} \, dx$

19. $\displaystyle \int \cos(3 \ln x) \, dx$

20. $\displaystyle \int \frac{dx}{4x^3 + x}$

21. $\displaystyle \int \frac{x \ln(1 + x^2)}{1 + x^2} \, dx$

22. $\displaystyle \int \sin^2 x \cos^4 x \, dx$

23. $\displaystyle \int \frac{x^2}{\sqrt{2 - x^2}} \, dx$

24. $\displaystyle \int \tan^4 x \sec x \, dx$

25. $\displaystyle\int \frac{x^2\,dx}{(4x+1)^{10}}$

26. $\displaystyle\int x\sin^{-1}\frac{x}{2}\,dx$

27. $\displaystyle\int \sin^5(4x)\,dx$

28. $\displaystyle\int \frac{dx}{x^5-2x^3+x}$

29. $\displaystyle\int \frac{dx}{2+e^x}$

30. $\displaystyle\int x^3 3^x\,dx$

31. $\displaystyle\int \frac{\sin^2 x\cos x}{2-\sin x}\,dx$

32. $\displaystyle\int \frac{x^2+1}{x^2+2x+2}\,dx$

33. $\displaystyle\int \frac{dx}{x^2\sqrt{1-x^2}}$

34. $\displaystyle\int x^3(\ln x)^2\,dx$

35. $\displaystyle\int \frac{x^3}{\sqrt{1-4x^2}}\,dx$

36. $\displaystyle\int \frac{e^{1/x}\,dx}{x^2}$

37. $\displaystyle\int \frac{x+1}{\sqrt{x^2+1}}\,dx$

38. $\displaystyle\int e^{(x^{1/3})}\,dx$

39. $\displaystyle\int \frac{x^3-3}{x^3-9x}\,dx$

40. $\displaystyle\int \frac{10^{\sqrt{x+2}}}{\sqrt{x+2}}\,dx$

41. $\displaystyle\int \sin^5 x\cos^9 x\,dx$

42. $\displaystyle\int \frac{x^2\,dx}{\sqrt{x^2-1}}$

43. $\displaystyle\int \frac{x\,dx}{x^2+2x-1}$

44. $\displaystyle\int \frac{2x-3}{\sqrt{4-3x+x^2}}\,dx$

45. $\displaystyle\int x^2\sin^{-1}(2x)\,dx$

46. $\displaystyle\int \frac{\sqrt{3x^2-1}}{x}\,dx$

47. $\displaystyle\int \cos^4 x\sin^4 x\,dx$

48. $\displaystyle\int \sqrt{x-x^2}\,dx$

49. $\displaystyle\int \frac{dx}{(4+x)\sqrt{x}}$

50. $\displaystyle\int x\tan^{-1}\frac{x}{3}\,dx$

51. $\displaystyle\int \frac{x^4-1}{x^3+2x^2}\,dx$

52. $\displaystyle\int \frac{dx}{x(x^2+4)^2}$

53. $\displaystyle\int \frac{\sin(2\ln x)}{x}\,dx$

54. $\displaystyle\int \frac{\sin(\ln x)}{x^2}\,dx$

55. $\displaystyle\int \frac{e^{2\tan^{-1}x}}{1+x^2}\,dx$

56. $\displaystyle\int \frac{x^3+x-2}{x^2-7}\,dx$

57. $\displaystyle\int \frac{\ln(3+x^2)}{3+x^2}x\,dx$

58. $\displaystyle\int \cos^7 x\,dx$

59. $\displaystyle\int \frac{\sin^{-1}(x/2)}{(4-x^2)^{1/2}}\,dx$

60. $\displaystyle\int \tan^4(\pi x)\,dx$

61. $\displaystyle\int \frac{(x+1)\,dx}{\sqrt{x^2+6x+10}}$

62. $\displaystyle\int e^x(1-e^{2x})^{5/2}\,dx$

63. $\displaystyle\int \frac{x^3\,dx}{(x^2+2)^{7/2}}$

64. $\displaystyle\int \frac{x^2}{2x^2-3}\,dx$

65. $\displaystyle\int \frac{x^{1/2}}{1+x^{1/3}}\,dx$

66. $\displaystyle\int \frac{dx}{x(x^2+x+1)^{1/2}}$

67. $\displaystyle\int \frac{1+x}{1+\sqrt{x}}\,dx$

68. $\displaystyle\int \frac{x\,dx}{4x^4+4x^2+5}$

69. $\displaystyle\int \frac{x\,dx}{(x^2-4)^2}$

70. $\displaystyle\int \frac{dx}{x^3+x^2+x}$

71. $\displaystyle\int x^2\tan^{-1}x\,dx$

72. $\displaystyle\int e^x\sec(e^x)\,dx$

73. $\displaystyle\int \frac{dx}{4\sin x-3\cos x}$

74. $\displaystyle\int \frac{dx}{x^{1/3}-1}$

75. $\displaystyle\int \frac{dx}{\tan x+\sin x}$

76. $\displaystyle\int \frac{x\,dx}{\sqrt{3-4x-4x^2}}$

77. $\displaystyle\int \frac{\sqrt{x}}{1+x}\,dx$

78. $\displaystyle\int \sqrt{1+e^x}\,dx$

79. $\displaystyle\int \frac{x^4\,dx}{x^3-8}$

80. $\displaystyle\int xe^x\cos x\,dx$

Other Review Exercises

1. Evaluate $I=\int x\,e^x\cos x\,dx$ and $J=\int x\,e^x\sin x\,dx$ by differentiating $e^x\big((ax+b)\cos x+(cx+d)\sin x\big)$ and examining coefficients.

2. For which real numbers r is the following reduction formula (obtained using integration by parts) valid?

$$\int_0^\infty x^r e^{-x}\,dx = r\int_0^\infty x^{r-1}e^{-x}\,dx$$

Evaluate the integrals in Exercises 3–6 or show that they diverge.

3. $\displaystyle\int_0^{\pi/2}\csc x\,dx$

4. $\displaystyle\int_1^\infty \frac{1}{x+x^3}\,dx$

5. $\displaystyle\int_0^1 \sqrt{x}\ln x\,dx$

6. $\displaystyle\int_{-1}^1 \frac{dx}{x\sqrt{1-x^2}}$

7. Show that the integral $I=\int_0^\infty (1/(\sqrt{x}\,e^x))\,dx$ converges and that its value satisfies $I<(2e+1)/e$.

8. By measuring the areas enclosed by contours on a topographic map, a geologist determines the cross-sectional areas A (m^2) through a 60 m high hill at various heights h (m) given in Table 2.

Table 2.

h	0	10	20	30	40	50	60
A	10,200	9,200	8,000	7,100	4,500	2,400	100

If she uses the Trapezoid Rule to estimate the volume of the hill (which is $V=\int_0^{60}A(h)\,dh$), what will be her estimate, to the nearest 1,000 m^3?

9. What will be the geologist's estimate of the volume of the hill in Exercise 8 if she uses Simpson's Rule instead of the Trapezoid Rule?

10. Find the Trapezoid Rule and Midpoint Rule approximations T_4 and M_4 for the integral $I = \int_0^1 \sqrt{2 + \sin(\pi x)}\, dx$. Quote the results to 5 decimal places. Quote a value of I to as many decimal places as you feel are justified by these approximations.

11. Use the results of Exercise 10 to calculate the Trapezoid Rule approximation T_8 and the Simpson's Rule approximation S_8 for the integral I in that exercise. Quote a value of I to as many decimal places as you feel are justified by these approximations.

12. Devise a way to evaluate $I = \int_{1/2}^{\infty} x^2/(x^5 + x^3 + 1)\, dx$ numerically, and use it to find I correct to 3 decimal places.

13. You want to approximate the integral $I = \int_0^4 f(x)\, dx$ of an unknown function $f(x)$, and you measure the following values of f:

Table 3.

x	0	1	2	3	4
$f(x)$	0.730	1.001	1.332	1.729	2.198

(a) What are the approximations T_4 and S_4 to I that you calculate with these data.

(b) You then decide to make more measurements in order to calculate T_8 and S_8. You obtain $T_8 = 5.5095$. What do you obtain for S_8?

(c) You have theoretical reasons to believe that $f(x)$ is, in fact, a polynomial of degree 3. Do your calculations support this theory? Why or why not?

Challenging Problems

1. (a) Some people think that $\pi = 22/7$. Prove that this is not so by showing that

$$\int_0^1 \frac{x^4(1 - x)^4}{x^2 + 1}\, dx = \frac{22}{7} - \pi.$$

(b) If $I = \int_0^1 x^4(1 - x)^4\, dx$, show that

$$\frac{22}{7} - I < \pi < \frac{22}{7} - \frac{I}{2}.$$

(c) Evaluate I and hence determine an explicit small interval containing π.

2. (a) Find a reduction formula for $\int (1 - x^2)^n\, dx$.

(b) Show that if n is a positive integer, then

$$\int_0^1 (1 - x^2)^n\, dx = \frac{2^{2n}(n!)^2}{(2n + 1)!}.$$

(c) Use your reduction formula to evaluate $\int (1 - x^2)^{-3/2}\, dx$.

3. (a) Show that $x^4 + x^2 + 1$ factors into a product of two real quadratics, and evaluate $\int (x^2 + 1)/(x^4 + x^2 + 1)\, dx$. Hint: $x^4 + x^2 + 1 = (x^2 + 1)^2 - x^2$.

(b) Use the same method to find $\int (x^2 + 1)/(x^4 + 1)\, dx$.

4. Let $I_{m,n} = \int_0^1 x^m (\ln x)^n\, dx$.

(a) Show that $I_{m,n} = (-1)^n \int_0^{\infty} x^n e^{-(m+1)x}\, dx$.

(b) Show that $I_{m,n} = \dfrac{(-1)^n n!}{(m + 1)^{n+1}}$.

5. Let $I_n = \int_0^1 x^n e^{-x}\, dx$.

(a) Show that $0 < I_n < \dfrac{1}{n + 1}$ and hence that $\lim_{n \to \infty} I_n = 0$.

(b) Show that $I_n = nI_{n-1} - \dfrac{1}{e}$ for $n \geq 1$, and $I_0 = 1 - \dfrac{1}{e}$.

(c) Verify by induction that $I_n = n!\left(1 - \dfrac{1}{e}\sum_{j=0}^{n}\dfrac{1}{j!}\right)$.

(d) Deduce from (a) and (c) that $\lim_{n \to \infty} \sum_{j=0}^{n}\dfrac{1}{j!} = e$.

6. If K is very large, which of the approximations T_{100} (Trapezoidal Rule), M_{100} (Midpoint Rule), and S_{100} (Simpson's Rule) will be closest to the true value for $\int_0^1 e^{-Kx}\, dx$? Which will be farthest? Justify your answers. (*Caution*: this is trickier than it sounds!)

7. Simpson's Rule gives the exact definite integral for a cubic f. Suppose you want a numerical integration rule that gives the exact answer for a polynomial of degree 5. You might approximate the integral over the subinterval $[m - h, m + h]$ by something of the form $2h\Big(af(m - h) + bf(m - \dfrac{h}{2}) + f(m) + bf(m + \dfrac{h}{2}) + af(m + h)\Big)$, for some constants a, b, and c.

(a) Determine a, b, and c for which this will work. (*Hint*: take $m = 0$ to make things simple.)

(b) Use this method to approximate $\int_0^1 e^{-x}\, dx$ using first one and then two of these intervals (thus evaluating the integrand at nine points).

8. The convergence of improper integrals can be a more delicate matter when the integrand changes sign. Here is one method that can be used to prove convergence in some cases where the comparison theorem fails.

(a) Suppose that $f(x)$ is differentiable on $[1, \infty)$, $f'(x)$ is continuous there, $f'(x) < 0$, and $\lim_{x \to \infty} f(x) = 0$. Show that $\int_1^{\infty} f'(x)\cos(x)\, dx$ converges. *Hint*: what is $\int_1^{\infty} |f'(x)|\, dx$?

(b) Under the same hypotheses, show that $\int_1^{\infty} f(x)\sin x\, dx$ converges. *Hint*: integrate by parts and use (a).

(c) Show that $\int_1^{\infty} \dfrac{\sin x}{x}\, dx$ converges but $\int_1^{\infty} \dfrac{|\sin x|}{x}\, dx$ diverges. *Hint*: $|\sin x| \geq \sin^2 x = \dfrac{1 - \cos(2x)}{2}$. Note that (b) would work just as well with $\sin x$ replaced by $\cos(2x)$.

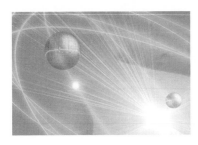

CHAPTER 7

Applications of Integration

Introduction Numerous quantities in mathematics, physics, economics, biology, and indeed any quantitative science can be conveniently represented by integrals. In addition to measuring plane areas, the problem that motivated the definition of the definite integral, we can use these integrals to express volumes of solids, lengths of curves, areas of surfaces, forces, work, energy, pressure, probabilities, dollar values of a stream of payments, and a variety of other quantities that are in one sense or another equivalent to areas under graphs.

In addition, as we have seen previously, many of the basic principles that govern the behaviour of our world are expressed in terms of differential equations and initial-value problems. Indefinite integration is a key tool in the solution of such problems.

In this chapter we examine some of these applications. For the most part they are independent of one another, and for that reason some of the later sections in this chapter can be regarded as optional material. The material of Sections 7.1–7.3, however, should be regarded as core because these ideas will arise again in the study of multivariable calculus.

7.1 Volumes of Solids of Revolution

In this section we show how volumes of certain three-dimensional regions (or *solids*) can be expressed as definite integrals and thereby determined. We will not attempt to give a definition of *volume* but will rely on our intuition and experience with solid objects to provide enough insight for us to specify the volumes of certain simple solids. For example, if the base of a rectangular box is a rectangle of length l and width w (and therefore area $A = lw$), and if the box has height h, then its volume is $V = Ah = lwh$. If l, w, and h are measured in *units* (e.g., centimetres), then the volume is expressed in *cubic units* (cubic centimetres).

A rectangular box is a special case of a solid called a **prism** or **cylinder**. (See Figure 7.1.) Such a solid has a flat base, occupying a plane region having area A. Every cross-section of the solid in a plane parallel to the base is congruent to the base and so has the same area. If the solid has height h (so that its top is in a plane parallel to the base and h units above it), then the volume of the solid is $V = Ah$. Such solids are usually called *prisms* if the base is bounded by straight lines and *cylinders* if the base is bounded by curves. In particular, if the base is a circular disk of radius r and the top of the solid is directly above the base, then the solid is a **right-circular cylinder**. If it has height h, then its volume is $V = \pi r^2 h$ cubic units. Cylinders and prisms are said to be **right** if their side walls are perpendicular to their bases; otherwise they are **oblique**. Obliqueness has no effect on the volume $V = Ah$, for which h is always measured in a direction perpendicular to the base.

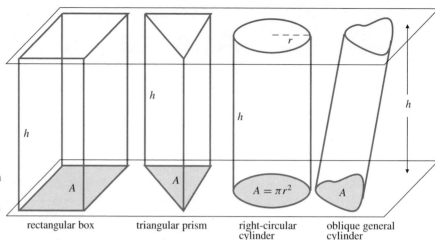

Figure 7.1 The volume of any prism or cylinder is the base area times the height (measured perpendicularly to the base): $V = Ah$

rectangular box   triangular prism   right-circular cylinder   oblique general cylinder

Volumes by Slicing

Knowing the volume of a cylinder enables us to determine the volumes of some more general solids. We can divide solids into thin "slices" by parallel planes. (Think of a loaf of sliced bread.) Each slice is approximately a cylinder of very small "height"; the height is the thickness of the slice. See Figure 7.2, where the height is measured horizontally in the direction of the x-axis. If we know the cross-sectional area of each slice, we can determine its volume and sum these volumes to find the volume of the solid.

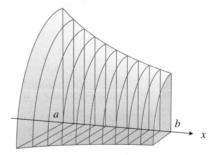

Figure 7.2 Slicing a solid perpendicularly to an axis

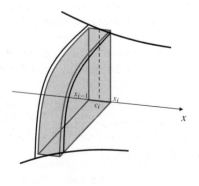

Figure 7.3 The volume of a slice

To be specific, suppose that the solid S lies between planes perpendicular to the x-axis at positions $x = a$ and $x = b$ and that the cross-sectional area of S in the plane perpendicular to the x-axis at x is a known function $A(x)$, for $a \le x \le b$. We assume that $A(x)$ is continuous on $[a, b]$. If $a = x_0 < x_1 < x_2 < \cdots < x_{n-1} < x_n = b$, then $P = \{x_0, x_1, x_2, \ldots, x_{n-1}, x_n\}$ is a partition of $[a, b]$ into n subintervals, and the planes perpendicular to the x-axis at the $x_1, x_2, \ldots, x_{n-1}$ divide the solid into n slices of which the ith has thickness $\Delta x_i = x_i - x_{i-1}$. The volume ΔV_i of that slice lies between the maximum and minimum values of $A(x) \Delta x_i$ for values of x in $[x_{i-1}, x_i]$ (Figure 7.3), so

$$\Delta V_i = A(c_i) \Delta x_i$$

for some c_i in $[x_{i-1}, x_i]$, by the Intermediate-Value Theorem. The volume of the solid is therefore given by the Riemann sum

$$V = \sum_{i=1}^{n} \Delta V_i = \sum_{i=1}^{n} A(c_i)\,\Delta x_i.$$

Letting n approach infinity in such a way that max Δx_i approaches 0, we obtain the definite integral of $A(x)$ over $[a, b]$ as the limit of this Riemann sum. Therefore:

> The volume V of a solid between $x = a$ and $x = b$ having cross-sectional area $A(x)$ at position x is
>
> $$V = \int_a^b A(x)\,dx.$$

There is another way to obtain this formula and others of a similar nature. Consider a slice of the solid between the planes perpendicular to the x-axis at positions x and $x + \Delta x$. Since $A(x)$ is continuous, it doesn't change much in a short interval, so if Δx is small, then the slice has volume ΔV approximately equal to the volume of a cylinder of base area $A(x)$ and height Δx:

$$\Delta V \approx A(x)\,\Delta x.$$

The error in this approximation is small compared to the size of ΔV. This suggests, correctly, that the **volume element**, that is, the volume of an infinitely thin slice of thickness dx is $dV = A(x)\,dx$, and that the volume of the solid is the "sum" (i.e., the integral) of these volume elements between the two ends of the solid, $x = a$ and $x = b$ (see Figure 7.4):

$$V = \int_{x=a}^{x=b} dV, \qquad \text{where} \qquad dV = A(x)\,dx.$$

We will use this *differential element* approach to model other applications that result in integrals rather than setting up explicit Riemann sums each time. Even though this argument does not constitute a proof of the formula, you are strongly encouraged to think of the formula this way; the volume is the integral of the volume elements.

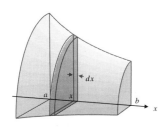

Figure 7.4   The volume element

Solids of Revolution

Many common solids have circular cross-sections in planes perpendicular to some axis. Such solids are called **solids of revolution** because they can be generated by rotating a plane region about an axis in that plane so that it sweeps out the solid. For example, a solid ball is generated by rotating a half-disk about the diameter of that half-disk (Figure 7.5(a)). Similarly, a solid right-circular cone is generated by rotating a right-angled triangle about one of its legs (Figure 7.5(b)).

If the region R bounded by $y = f(x)$, $y = 0$, $x = a$, and $x = b$ is rotated about the x-axis, then the cross-section of the solid generated in the plane perpendicular to the x-axis at x is a circular disk of radius $|f(x)|$. The area of this cross-section is $A(x) = \pi\big(f(x)\big)^2$, so the volume of the solid of revolution is

$$V = \pi \int_a^b (f(x))^2\,dx.$$

Example 1 (**The volume of a ball**) Find the volume of a solid ball of radius a.

Solution The ball can be generated by rotating the half-disk, $0 \le y \le \sqrt{a^2 - x^2}$, $-a \le x \le a$ about the x-axis. See the cutaway view in Figure 7.5(a). Therefore its volume is

$$V = \pi \int_{-a}^{a} (\sqrt{a^2 - x^2})^2 \, dx = 2\pi \int_0^a (a^2 - x^2) \, dx$$

$$= 2\pi \left(a^2 x - \frac{x^3}{3} \right) \Big|_0^a = 2\pi \left(a^3 - \frac{1}{3}a^3 \right) = \frac{4}{3}\pi a^3 \text{ cubic units.}$$

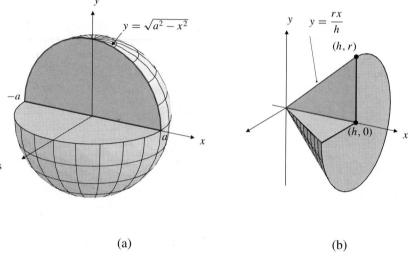

Figure 7.5

(a) The ball is generated by rotating the half-disk $0 \le y \le \sqrt{a^2 - x^2}$ (shown in colour) about the x-axis

(b) The cone of base radius r and height h is generated by rotating the triangle $0 \le x \le h$, $0 \le y \le rx/h$ (in colour) about the x-axis

(a)

(b)

Example 2 (**The volume of a right-circular cone**) Find the volume of the right-circular cone of base radius r and height h that is generated by rotating the triangle with vertices $(0, 0)$, $(h, 0)$, and (h, r) about the x-axis.

Solution The line from $(0, 0)$ to (h, r) has equation $y = rx/h$. Thus the volume of the cone (see the cutaway view in Figure 7.5(b)) is

$$V = \pi \int_0^h \left(\frac{rx}{h} \right)^2 \, dx = \pi \left(\frac{r}{h} \right)^2 \frac{x^3}{3} \Big|_0^h = \frac{1}{3}\pi r^2 h \text{ cubic units.}$$

Improper integrals can represent volumes of unbounded solids. If the improper integral converges, the unbounded solid has a finite volume.

Example 3 Find the volume of the infinitely long horn that is generated by rotating the region bounded by $y = 1/x$ and $y = 0$ and lying to the right of $x = 1$ about the x-axis. The horn is illustrated in Figure 7.6.

Solution   The volume of the horn is

$$V = \pi \int_1^\infty \left(\frac{1}{x}\right)^2 dx = \pi \lim_{R \to \infty} \int_1^R \frac{1}{x^2}\, dx$$

$$= -\pi \lim_{R \to \infty} \frac{1}{x}\Big|_1^R = -\pi \lim_{R \to \infty} \left(\frac{1}{R} - 1\right) = \pi \text{ cubic units.}$$

It is interesting to note that this finite volume arises from rotating a region that itself has infinite area: $\int_1^\infty dx/x = \infty$. We have a paradox: it takes an infinite amount of paint to paint the region but only a finite amount to fill the horn obtained by rotating the region. (How can you resolve this paradox?)

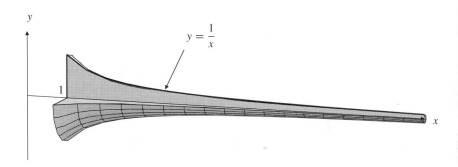

Figure 7.6   Cutaway view of an infinitely long horn

The following example shows how to deal with a problem where the axis of rotation is not the x-axis. Just rotate a suitable area element about the axis to form a volume element.

Example 4   A ring-shaped solid is generated by rotating the finite plane region R bounded by the curve $y = x^2$ and the line $y = 1$ about the line $y = 2$. Find its volume.

Solution   First we solve the pair of equations $y = x^2$ and $y = 1$ to obtain the intersections at $x = -1$ and $x = 1$. The solid lies between these two values of x. The area element of R at position x is a vertical strip of width dx extending upward from $y = x^2$ to $y = 1$. When R is rotated about the line $y = 2$, this area element sweeps out a thin, washer-shaped volume element of thickness dx and radius $2-x^2$, having a hole of radius 1 through the middle. (See Figure 7.7.) The cross-sectional area of this element is the area of a circle of radius $2 - x^2$ minus the area of the hole, a circle of radius 1. Thus,

$$dV = \left(\pi(2 - x^2)^2 - \pi(1)^2\right) dx = \pi(3 - 4x^2 + x^4)\, dx.$$

Since the solid extends from $x = -1$ to $x = 1$, its volume is

$$V = \pi \int_{-1}^1 (3 - 4x^2 + x^4)\, dx = 2\pi \int_0^1 (3 - 4x^2 + x^4)\, dx$$

$$= 2\pi \left(3x - \frac{4x^3}{3} + \frac{x^5}{5}\right)\Big|_0^1 = 2\pi\left(3 - \frac{4}{3} + \frac{1}{5}\right) = \frac{56\pi}{15} \text{ cubic units.}$$

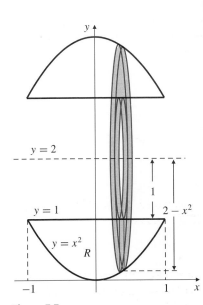

Figure 7.7

Sometimes we want to rotate a region bounded by curves with equations of the form $x = g(y)$ about the y-axis. In this case, the roles of x and y are reversed, and we use horizontal slices instead of vertical ones.

Example 5 Find the volume of the solid generated by rotating the region to the right of the y-axis and to the left of the curve $x = 2y - y^2$ about the y-axis.

Solution For intersections of $x = 2y - y^2$ and $x = 0$, we have

$$2y - y^2 = 0 \quad \Longrightarrow \quad y = 0 \quad \text{or} \quad y = 2.$$

The solid lies between the horizontal planes at $y = 0$ and $y = 2$. A horizontal area element at height y and having thickness dy rotates about the y-axis to generate a thin disk-shaped volume element of radius $2y - y^2$ and thickness dy. (See Figure 7.8.) Its volume is

$$dV = \pi(2y - y^2)^2 \, dy = \pi(4y^2 - 4y^3 + y^4) \, dy.$$

Thus, the volume of the solid is

$$
\begin{aligned}
V &= \pi \int_0^2 (4y^2 - 4y^3 + y^4) \, dy \\
&= \pi \left(\frac{4y^3}{3} - y^4 + \frac{y^5}{5} \right) \Big|_0^2 \\
&= \pi \left(\frac{32}{3} - 16 + \frac{32}{5} \right) = \frac{16\pi}{15} \text{ cubic units.}
\end{aligned}
$$

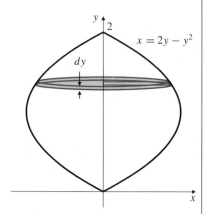

Figure 7.8

Cylindrical Shells

Suppose that the region R bounded by $y = f(x) \geq 0$, $y = 0$, $x = a \geq 0$, and $x = b > a$ is rotated about the y-axis to generate a solid of revolution. In order to find the volume of the solid using (plane) slices, we would need to know the cross-sectional area $A(y)$ in each plane of height y, and this would entail solving the equation $y = f(x)$ for one or more solutions of the form $x = g(y)$. In practice this can be inconvenient or impossible.

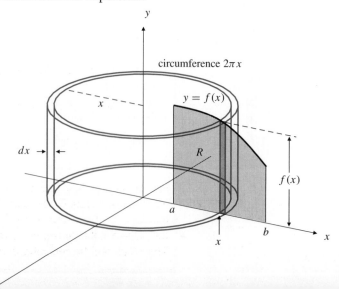

Figure 7.9 When rotated around the y-axis, the area element of width dx under $y = f(x)$ at x generates a cylindrical shell of height $f(x)$, circumference $2\pi x$, and hence volume $dV = 2\pi x \, f(x) \, dx$

The standard area element of R at position x is a vertical strip of width dx, height $f(x)$, and area $dA = f(x)\,dx$. When R is rotated about the y-axis, this strip sweeps out a volume element in the shape of a circular **cylindrical shell** having radius x, height $f(x)$, and thickness dx. (See Figure 7.9.) Regard this shell as a rolled-up rectangular slab with dimensions $2\pi x$, $f(x)$, and dx; evidently it has volume

$$dV = 2\pi x\, f(x)\, dx.$$

The volume of the solid of revolution is the sum *(integral)* of the volumes of such shells with radii ranging from a to b:

The volume of the solid obtained by rotating the plane region $0 \le y \le f(x)$, $0 \le a < x < b$ about the y-axis is

$$V = 2\pi \int_a^b x\, f(x)\, dx.$$

Example 6 (**The volume of a torus**) A disk of radius a has centre at the point $(b, 0)$, where $b > a > 0$. The disk is rotated about the y-axis to generate a **torus** (a doughnut-shaped solid), illustrated in Figure 7.10. Find its volume.

Solution The circle with centre at $(b, 0)$ and having radius a has equation $(x - b)^2 + y^2 = a^2$, so its upper semicircle is the graph of the function

$$f(x) = \sqrt{a^2 - (x - b)^2}.$$

We will double the volume of the upper half of the torus, which is generated by rotating the half-disk $0 \le y \le \sqrt{a^2 - (x - b)^2}$, $b - a \le x \le b + a$ about the y-axis. The volume of the complete torus is

$$V = 2 \times 2\pi \int_{b-a}^{b+a} x\, \sqrt{a^2 - (x - b)^2}\, dx \qquad \text{Let } u = x - b,$$
$$\hspace{9cm} du = dx..$$
$$= 4\pi \int_{-a}^{a} (u + b)\sqrt{a^2 - u^2}\, du$$
$$= 4\pi \int_{-a}^{a} u\, \sqrt{a^2 - u^2}\, du + 4\pi b \int_{-a}^{a} \sqrt{a^2 - u^2}\, du$$
$$= 0 + 4\pi b \frac{\pi a^2}{2} = 2\pi^2 a^2 b \text{ cubic units.}$$

(The first of the final two integrals is 0 because the integrand is odd and the interval is symmetric about 0; the second is the area of a semicircle of radius a.) Note that the volume of the torus is $(\pi a^2)(2\pi b)$, that is, the area of the disk being rotated times the distance travelled by the centre of that disk as it rotates about the y-axis. This result will be generalized by Pappus's Theorem in Section 7.5.

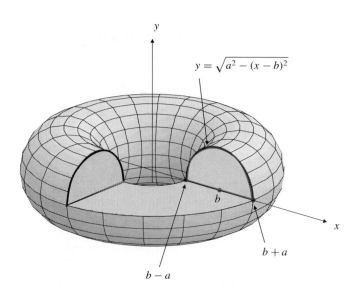

Figure 7.10   Cutaway view of a torus

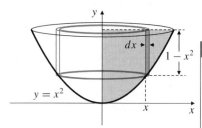

Figure 7.11   A parabolic bowl

Example 7   Find the volume of a bowl obtained by revolving the parabolic arc $y = x^2, 0 \le x \le 1$ about the y-axis.

Solution   The interior of the bowl corresponds to revolving the region given by $x^2 \le y \le 1, 0 \le x \le 1$ about the y-axis. The area element at position x has height $1 - x^2$ and generates a cylindrical shell of volume $dV = 2\pi x(1 - x^2)\,dx$. (See Figure 7.11.) Thus the volume of the bowl is

$$V = 2\pi \int_0^1 x(1 - x^2)\,dx$$

$$= 2\pi \left(\frac{x^2}{2} - \frac{x^4}{4}\right)\Big|_0^1 = \frac{\pi}{2} \text{ cubic units.}$$

We have described two methods for determining the volume of a solid of revolution, slicing and cylindrical shells. The choice of method for a particular solid is usually dictated by the form of the equations defining the region being rotated and by the axis of rotation. The volume element dV can always be determined by rotating a suitable area element dA about the axis of rotation. If the region is bounded by vertical lines and one or more graphs of the form $y = f(x)$, the appropriate area element is a vertical strip of width dx. If the rotation is about the x-axis or any other horizontal line, this strip generates a disk- or washer-shaped slice of thickness dx. If the rotation is about the y-axis or any other vertical line, the strip generates a cylindrical shell of thickness dx. On the other hand, if the region being rotated is bounded by horizontal lines and one or more graphs of the form $x = g(y)$, it is easier to use a horizontal strip of width dy as the area element, and this generates a slice if the rotation is about a vertical line and a cylindrical shell if the rotation is about a horizontal line. For very simple regions either method can be made to work easily.

Table 1. Volumes of solids of revolution

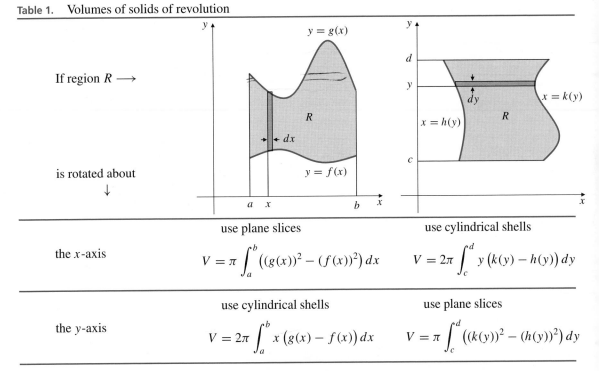

	use plane slices	use cylindrical shells
the x-axis	$V = \pi \displaystyle\int_a^b \left((g(x))^2 - (f(x))^2\right) dx$	$V = 2\pi \displaystyle\int_c^d y\left(k(y) - h(y)\right) dy$

	use cylindrical shells	use plane slices
the y-axis	$V = 2\pi \displaystyle\int_a^b x\left(g(x) - f(x)\right) dx$	$V = \pi \displaystyle\int_c^d \left((k(y))^2 - (h(y))^2\right) dy$

Our final example involves rotation about a vertical line other than the y-axis.

Example 8 The triangular region bounded by $y = x$, $y = 0$, and $x = a > 0$ is rotated about the line $x = b > a$. (See Figure 7.12.) Find the volume of the solid so generated.

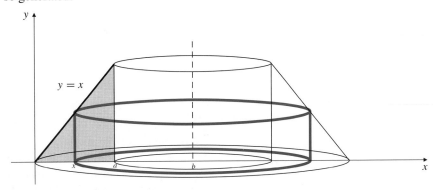

Figure 7.12

Solution Here the vertical area element at x generates a cylindrical shell of radius $b - x$, height x, and thickness dx. Its volume is $dV = 2\pi(b - x)x\,dx$, and the volume of the solid is

$$V = 2\pi \int_0^a (b-x)\,x\,dx = 2\pi \left(\frac{bx^2}{2} - \frac{x^3}{3}\right)\Big|_0^a = \pi \left(a^2 b - \frac{2a^3}{3}\right) \text{ cubic units.}$$

Exercises 7.1

Find the volume of each solid S in Exercises 1–4 in two ways, using the method of slicing and the method of cylindrical shells.

1. S is generated by rotating about the x-axis the region bounded by $y = x^2$, $y = 0$, and $x = 1$.

2. S is generated by rotating the region of Exercise 1 about the y-axis.

3. S is generated by rotating about the x-axis the region bounded by $y = x^2$ and $y = \sqrt{x}$ between $x = 0$ and $x = 1$.

4. S is generated by rotating the region of Exercise 3 about the y-axis.

Find the volumes of the solids obtained if the plane regions R described in Exercises 5–10 are rotated about (a) the x-axis and (b) the y-axis.

5. R is bounded by $y = x(2 - x)$ and $y = 0$ between $x = 0$ and $x = 2$.

6. R is the finite region bounded by $y = x$ and $y = x^2$.

7. R is the finite region bounded by $y = x$ and $x = 4y - y^2$.

8. R is bounded by $y = 1 + \sin x$ and $y = 1$ from $x = 0$ to $x = \pi$.

9. R is bounded by $y = 1/(1 + x^2)$, $y = 2$, $x = 0$, and $x = 1$.

10. R is the finite region bounded by $y = 1/x$ and $3x + 3y = 10$.

11. The triangular region with vertices $(0, -1)$, $(1, 0)$, and $(0, 1)$ is rotated about the line $x = 2$. Find the volume of the solid so generated.

12. Find the volume of the solid generated by rotating the region $0 \le y \le 1 - x^2$ about the line $y = 1$.

13. What percentage of the volume of a ball of radius 2 is removed if a hole of radius 1 is drilled through the centre of the ball?

14. A cylindrical hole is bored through the centre of a ball of radius R. If the length of the hole is L, show that the volume of the remaining part of the ball depends only on L and not on R.

15. A cylindrical hole of radius a is bored through a solid right-circular cone of height h and base radius $b > a$. If the axis of the hole lies along that of the cone, find the volume of the remaining part of the cone.

16. Find the volume of the solid obtained by rotating a circular disk about one of its tangent lines.

17. A plane slices a ball of radius a into two pieces. If the plane passes b units away from the centre of the ball (where $b < a$), find the volume of the smaller piece.

18. Water partially fills a hemispherical bowl of radius 30 cm so that the maximum depth of the water is 20 cm. What volume of water is in the bowl?

19. Find the volume of the ellipsoid of revolution obtained by rotating the ellipse $(x^2/a^2) + (y^2/b^2) = 1$ about the x-axis.

20. Recalculate the volume of the torus of Example 6 by slicing perpendicular to the y-axis rather than using cylindrical shells.

21. The region R bounded by $y = e^{-x}$ and $y = 0$ and lying to the right of $x = 0$ is rotated (a) about the x-axis and (b) about the y-axis. Find the volume of the solid of revolution generated in each case.

22. The region R bounded by $y = x^{-k}$ and $y = 0$ and lying to the right of $x = 1$ is rotated about the x-axis. Find all real values of k for which the solid so generated has finite volume.

23. Repeat Exercise 22 with rotation about the y-axis.

24. The region shaded in Figure 7.13 is rotated about the x-axis. Use Simpson's Rule to find the volume of the resulting solid.

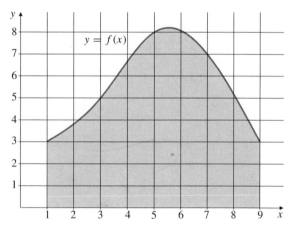

Figure 7.13

25. The region shaded in Figure 7.13 is rotated about the y-axis. Use Simpson's Rule to find the volume of the resulting solid.

26. The region shaded in Figure 7.13 is rotated about the line $x = -1$. Use Simpson's Rule to find the volume of the resulting solid.

* 27. Find the volume of the solid generated by rotating the finite region in the first quadrant bounded by the coordinate axes and the curve $x^{2/3} + y^{2/3} = 4$ about either of the coordinate axes. (Both volumes are the same. Why?)

* 28. Given that the surface area of a sphere of radius r is kr^2 for some constant k, express the volume of a ball of radius R as an integral of volume elements that are the volumes of spherical shells of varying radii and thickness dr. Hence find k.

The following problems are *very difficult*. You will need some ingenuity and a lot of hard work to solve them by the techniques available to you now.

* **29.** A wine glass in the shape of a right-circular cone of height h and semivertical angle α (see Figure 7.14) is filled with wine. Slowly a ball is lowered into the glass, displacing wine and causing it to overflow. Find the radius R of the ball that causes the greatest volume of wine to overflow out of the glass.

* **30.** The finite plane region bounded by the curve $xy = 1$ and the straight line $2x + 2y = 5$ is rotated about that line to generate a solid of revolution. Find the volume of that solid.

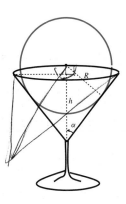

Figure 7.14

7.2   Other Volumes by Slicing

The method of slicing introduced in Section 7.1 can be used to determine volumes of solids that are not solids of revolution. All we need to know is the area of cross-section of the solid in every plane perpendicular to some fixed axis. If that axis is the x-axis, if the solid lies between the planes at $x = a$ and $x = b > a$, and if the cross-sectional area in the plane at x is the continuous (or even piecewise continuous) function $A(x)$, then the volume of the solid is

$$V = \int_a^b A(x)\,dx.$$

In this section we consider some examples that are not solids of revolution.

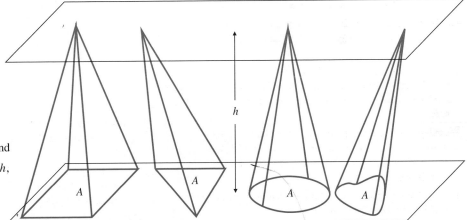

Figure 7.15   Some pyramids and cones. Each has volume $V = \dfrac{1}{3}Ah$, where A is the area of the base, and h is the height measured perpendicular to the base

Pyramids and **cones** are solids consisting of all points on line segments that join a fixed point, the **vertex**, to all the points in a region lying in a plane not containing the vertex. The region is called the **base** of the pyramid or cone. Some pyramids and cones are shown in Figure 7.15. If the base is bounded by straight lines, the solid is called a **pyramid**; if the base has a curved boundary the solid is called a cone. All pyramids and cones have volume

$$V = \frac{1}{3}Ah,$$

where A is the area of the base region and h is the height from the vertex to the plane of the base, measured in the direction perpendicular to that plane. We will give a very simple proof of this fact in Section 16.4. For the time being we verify it for the case of a rectangular base.

Example 1   Verify the formula for the volume of a pyramid with rectangular base of area A and height h.

Solution   Cross-sections of the pyramid in planes parallel to the base are similar rectangles. If the origin is at the vertex of the pyramid and the x-axis is perpendicular to the base, then the cross-section at position x is a rectangle whose dimensions are x/h times the corresponding dimensions of the base. For example, in Figure 7.16(a), the length LM is x/h times the length PQ, as can be seen from the similar triangles OLM and OPQ. Thus, the area of the rectangular cross-section at x is

$$A(x) = \left(\frac{x}{h}\right)^2 A.$$

The volume of the pyramid is therefore

$$V = \int_0^h \left(\frac{x}{h}\right)^2 A\, dx = \frac{A}{h^2}\frac{x^3}{3}\bigg|_0^h = \frac{1}{3}Ah \text{ cubic units.}$$

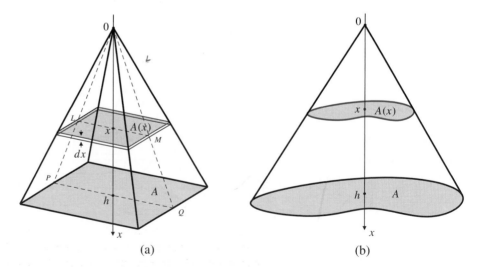

Figure 7.16

(a) A rectangular pyramid

(b) A general cone

(a)   (b)

A similar argument, resulting in the same formula for the volume, holds for a cone, that is, a pyramid with a more general (curved) shape to its base, such as that in Figure 7.16(b). Although it is not as obvious as in the case of the pyramid, the cross-section at x still has area $(x/h)^2$ times that of the base.

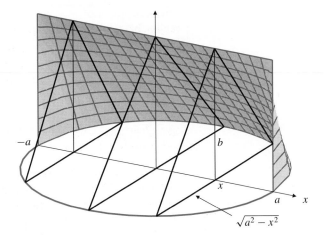

Figure 7.17   The tent of Example 2
with the front covering removed to
show the shape more clearly

Example 2   A tent has a circular base of radius a metres and is supported by a
horizontal ridge bar held at height b metres above a diameter of the base by vertical
supports at each end of the diameter. The material of the tent is stretched tight so
that each cross-section perpendicular to the ridge bar is an isosceles triangle. (See
Figure 7.17.) Find the volume of the tent.

Solution   Let the x-axis be the diameter of the base under the ridge bar. The
cross-section at position x has base length $2\sqrt{a^2 - x^2}$, so its area is

$$A(x) = \frac{1}{2}\left(2\sqrt{a^2 - x^2}\right)b = b\sqrt{a^2 - x^2}.$$

Thus, the volume of the solid is

$$V = \int_{-a}^{a} b\sqrt{a^2 - x^2}\,dx = b\int_{-a}^{a} \sqrt{a^2 - x^2}\,dx = b\frac{\pi a^2}{2} = \frac{\pi}{2}a^2 b \text{ m}^3.$$

Note that we evaluated the last integral by inspection. It is the area of a half-disk of
radius a. ∎

Example 3   Two circular cylinders, each having radius a, intersect so that their
axes meet at right angles. Find the volume of the region lying inside both cylinders.

Solution   We represent the cylinders in a three-dimensional Cartesian coordinate
system where the plane containing the x- and y-axes is horizontal and the z-
axis is vertical. One-eighth of the solid is represented in Figure 7.18, that part
corresponding to all three coordinates being positive. The two cylinders have axes
along the x- and y-axes, respectively. The cylinder with axis along the x-axis
intersects the plane of the y- and z-axes in a circle of radius a.

Similarly, the other cylinder meets the plane of the x- and z-axes in a circle of radius a. It follows that if the region lying inside both cylinders (and having $x \geq 0$, $y \geq 0$, and $z \geq 0$) is sliced horizontally, then the slice at height z above the xy-plane is a square of side $\sqrt{a^2 - z^2}$ and has area $A(z) = a^2 - z^2$. The volume V of the whole region, being eight times that of the part shown, is

$$V = 8 \int_0^a (a^2 - z^2)\, dz = 8 \left(a^2 z - \frac{z^3}{3} \right) \Big|_0^a = \frac{16}{3} a^3 \text{ cubic units.}$$

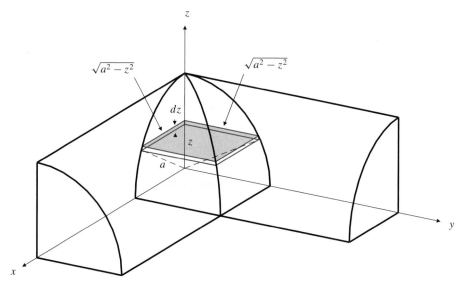

Figure 7.18 One-eighth of the solid lying inside two perpendicular cylindrical pipes. The horizontal slice shown is square

Exercises 7.2

1. A solid is 2 m high. The cross-section of the solid at height x above its base has area $3x$ square metres. Find the volume of the solid.

2. The cross-section at height z of a solid of height h is a rectangle with dimensions z and $h - z$. Find the volume of the solid.

3. Find the volume of a solid of height 1 whose cross-section at height z is an ellipse with semi-axes z and $\sqrt{1 - z^2}$.

4. A solid extends from $x = 1$ to $x = 3$. The cross-section of the solid in the plane perpendicular to the x-axis at x is a square of side x. Find the volume of the solid.

5. A solid is 6 ft high. Its horizontal cross-section at height z ft above its base is a rectangle with length $2 + z$ ft and width $8 - z$ ft. Find the volume of the solid.

6. A solid extends along the x-axis from $x = 1$ to $x = 4$. Its cross-section at position x is an equilateral triangle with edge length $\sqrt{x}$. Find the volume of the solid.

7. Find the volume of a solid that is h cm high if its horizontal cross-section at any height y above its base is a circular

sector having radius a cm and angle $2\pi \left(1 - (y/h) \right)$ radians.

8. The opposite ends of a solid are at $x = 0$ and $x = 2$. The area of cross-section of the solid in a plane perpendicular to the x-axis at x is kx^3 square units. The volume of the solid is 4 cubic units. Find k.

9. Find the cross-sectional area of a solid in any horizontal plane at height z above its base if the volume of that part of the solid lying below any such plane is z^3 cubic units.

10. All the cross-sections of a solid in horizontal planes are squares. The volume of the part of the solid lying below any plane of height z is $4z$ cubic units, where $0 < z < h$, the height of the solid. Find the edge length of the square cross-section at height z for $0 < z < h$.

11. A solid has a circular base of radius r. All sections of the solid perpendicular to a particular diameter of the base are squares. Find the volume of the solid.

12. Repeat Exercise 11 but with sections that are equilateral triangles instead of squares.

13. The base of a solid is an isosceles right-angled triangle with

equal legs measuring 12 cm. Each cross-section perpendicular to one of these legs is half of a circular disk. Find the volume of the solid.

14. **(Cavalieri's Principle)** Two solids have equal cross-sectional areas at equal heights above their bases. If both solids have the same height, show that they both have the same volume.

15. The top of a circular cylinder of radius r is a plane inclined at an angle to the horizontal. (See Figure 7.19.) If the lowest and highest points on the top are at heights a and b, respectively, above the base, find the volume of the cylinder. (Note that there is an easy geometric way to get the answer, but you should also try to do it by slicing. You can use either rectangular or trapezoidal slices.)

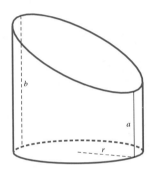

Figure 7.19

* 16. **(Volume of an ellipsoid)** Find the volume enclosed by the ellipsoid

$$\frac{x^2}{a^2} + \frac{y^2}{b^2} + \frac{z^2}{c^2} = 1.$$

Hint: this is not a solid of revolution. As in Example 3, the z-axis is perpendicular to the plane of the x- and y-axes. Each horizontal plane $z = k$ $(-c \le k \le c)$ intersects the ellipsoid in an ellipse $(x/a)^2 + (y/b)^2 = 1 - (k/c)^2$. Thus $dV = dz \times$ the area of this ellipse. The area of the ellipse $(x/a)^2 + (y/b)^2 = 1$ is πab.

Figure 7.20

* 17. **(Notching a log)** A $45°$ notch is cut to the centre of a cylindrical log having radius 20 cm, as shown in Figure 7.20. One plane face of the notch is perpendicular to the axis of the log. What volume of wood was removed from the log by cutting the notch?

18. **(A smaller notch)** Repeat Exercise 17, but assume that the notch penetrates only one quarter way (10 cm) into the log.

19. What volume of wood is removed from a 3 in thick board if a circular hole of radius 2 in is drilled through it with the axis of the hole tilted at an angle of $45°$ to board?

* 20. **(More intersecting cylinders)** The axes of two circular cylinders intersect at right angles. If the radii of the cylinders are a and b $(a > b > 0)$, show that the region lying inside both cylinders has volume

$$V = 8 \int_0^b \sqrt{b^2 - z^2} \sqrt{a^2 - z^2}\, dz.$$

Hint: review Example 3. Try to make a similar diagram, showing only one-eighth of the region. The integral is not easily evaluated.

21. A circular hole of radius 2 cm is drilled through the middle of a circular log of radius 4 cm, with the axis of the hole perpendicular to the axis of the log. Find the volume of wood removed from the log. *Hint:* this is very similar to Exercise 20. You will need to use numerical methods or a calculator with a numerical integration function to get the answer.

7.3 Arc Length and Surface Area

In this section we consider how integrals can be used to find the lengths of curves and the areas of the surfaces of solids of revolution.

Arc Length

If A and B are two points in the plane, let $|AB|$ denote the distance between A and

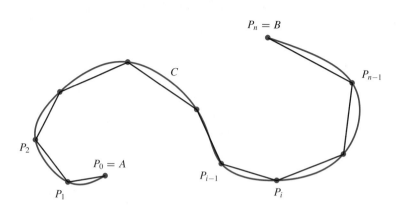

Figure 7.21 A polygonal approximation to a curve C

Given a curve C joining the two points A and B, we would like to define what is meant by the *length* of the curve C from A to B. Suppose we choose points $A = P_0$, P_1, P_2, ..., P_{n-1}, and $P_n = B$ in order along the curve, as shown in Figure 7.21. The polygonal line $P_0 P_1 P_2 \ldots P_{n-1} P_n$ constructed by joining adjacent pairs of these points with straight line segments forms a *polygonal approximation* to C, having length

$$L_n = |P_0 P_1| + |P_1 P_2| + \cdots + |P_{n-1} P_n| = \sum_{i=1}^{n} |P_{i-1} P_i|.$$

Intuition tells us that the shortest curve joining two points is a straight line segment, so the length L_n of any such polygonal approximation to C cannot exceed the length of C. If we increase n by adding more vertices to the polygonal line between existing vertices, L_n cannot get smaller and may increase. If there exists a finite number K such that $L_n \leq K$ for every polygonal approximation to C, then there will be a smallest such number K (by the completeness of the real numbers), and we call this smallest K the arc length of C.

DEFINITION 1

The **arc length** of the curve C from A to B is the smallest real number s such that the length L_n of every polygonal approximation to C satisfies $L_n \leq s$.

A curve with a finite arc length is said to be **rectifiable**. Its arc length s is the limit of the lengths L_n of polygonal approximations as $n \to \infty$ in such a way that the maximum segment length $|P_{i-1} P_i| \to 0$.

It is possible to construct continuous curves that are bounded (they do not go off to infinity anywhere) but are not rectifiable; they have infinite length. To avoid such pathological examples, we will assume that our curves are **smooth**; they will be defined by functions having continuous derivatives.

The Arc Length of the Graph of a Function

Let f be a function defined on a closed, finite interval $[a, b]$ and having a continuous derivative f' there. If C is the graph of f, that is, the graph of the equation $y = f(x)$, then any partition of $[a, b]$ provides a polygonal approximation to C. For the partition

$$\{a = x_0 < x_1 < x_2 < \cdots < x_n = b\},$$

let P_i be the point $(x_i, f(x_i))$, $(0 \le i \le n)$. The length of the polygonal line $P_0 P_1 P_2 \ldots P_{n-1} P_n$ is

$$L_n = \sum_{i=1}^{n} |P_{i-1} P_i| = \sum_{i=1}^{n} \sqrt{(x_i - x_{i-1})^2 + \left(f(x_i) - f(x_{i-1})\right)^2}$$

$$= \sum_{i=1}^{n} \sqrt{1 + \left(\frac{f(x_i) - f(x_{i-1})}{x_i - x_{i-1}}\right)^2} \, \Delta x_i,$$

where $\Delta x_i = x_i - x_{i-1}$. By the Mean-Value Theorem there exists a number c_i in the interval $[x_{i-1}, x_i]$ such that

$$\frac{f(x_i) - f(x_{i-1})}{x_i - x_{i-1}} = f'(c_i),$$

so we have $L_n = \sum_{i=1}^{n} \sqrt{1 + \left(f'(c_i)\right)^2} \, \Delta x_i.$

Thus L_n is a Riemann sum for $\int_a^b \sqrt{1 + (f'(x))^2} \, dx$. Being the limit of such Riemann sums as $n \to \infty$ in such a way that $\max(\Delta x_i) \to 0$, that integral is the length of the curve C.

> The arc length s of the curve $y = f(x)$ from $x = a$ to $x = b$ is given by
>
> $$s = \int_a^b \sqrt{1 + \left(f'(x)\right)^2} \, dx = \int_a^b \sqrt{1 + \left(\frac{dy}{dx}\right)^2} \, dx.$$

You can regard the integral formula above as giving the arc length s of C as a "sum" of **arc length elements**

> $$s = \int_{x=a}^{x=b} ds, \qquad \text{where} \qquad ds = \sqrt{1 + \left(f'(x)\right)^2} \, dx.$$

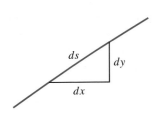

Figure 7.22   A differential triangle

Figure 7.22 provides a convenient way to remember this; it also suggests how we can arrive at similar formulas for arc length elements of other kinds of curves. The *differential triangle* in the figure suggests that

$$(ds)^2 = (dx)^2 + (dy)^2.$$

Dividing this equation by $(dx)^2$, taking the square root, and then multiplying by dx, we get

$$\left(\frac{ds}{dx}\right)^2 = 1 + \left(\frac{dy}{dx}\right)^2$$

$$\frac{ds}{dx} = \sqrt{1 + \left(\frac{dy}{dx}\right)^2}$$

$$ds = \sqrt{1 + \left(\frac{dy}{dx}\right)^2} \, dx = \sqrt{1 + \left(f'(x)\right)^2} \, dx.$$

A similar argument shows that for a curve specified by an equation of the form $x = g(y)$, $(c \le y \le d)$, the arc length element is

$$ds = \sqrt{1 + \left(\frac{dx}{dy}\right)^2} \, dy = \sqrt{1 + \left(g'(y)\right)^2} \, dy.$$

Example 1 Find the length of the curve $y = x^{2/3}$ from $x = 1$ to $x = 8$.

Solution Since $dy/dx = \frac{2}{3}x^{-1/3}$ is continuous between $x = 1$ and $x = 8$ and $x^{1/3} > 0$ there, the length of the curve is given by

$$s = \int_1^8 \sqrt{1 + \frac{4}{9}x^{-2/3}}\,dx = \int_1^8 \sqrt{\frac{9x^{2/3} + 4}{9x^{2/3}}}\,dx$$

$$= \int_1^8 \frac{\sqrt{9x^{2/3} + 4}}{3x^{1/3}}\,dx \qquad \text{Let } u = 9x^{2/3} + 4,$$
$$du = 6x^{-1/3}\,dx.$$

$$= \frac{1}{18}\int_{13}^{40} u^{1/2}\,du = \frac{1}{27}u^{3/2}\Big|_{13}^{40} = \frac{40\sqrt{40} - 13\sqrt{13}}{27} \text{ units.}$$

Example 2 Find the length of the curve $y = x^4 + \dfrac{1}{32x^2}$ from $x = 1$ to $x = 2$.

Solution Here $\dfrac{dy}{dx} = 4x^3 - \dfrac{1}{16x^3}$ and

$$1 + \left(\frac{dy}{dx}\right)^2 = 1 + \left(4x^3 - \frac{1}{16x^3}\right)^2$$

$$= 1 + (4x^3)^2 - \frac{1}{2} + \left(\frac{1}{16x^3}\right)^2$$

$$= (4x^3)^2 + \frac{1}{2} + \left(\frac{1}{16x^3}\right)^2 = \left(4x^3 + \frac{1}{16x^3}\right)^2.$$

The expression in the last set of parentheses is positive for $1 \le x \le 2$, so the length of the curve is

$$s = \int_1^2 \left(4x^3 + \frac{1}{16x^3}\right)dx = \left(x^4 - \frac{1}{32x^2}\right)\Big|_1^2$$

$$= 16 - \frac{1}{128} - \left(1 - \frac{1}{32}\right) = 15 + \frac{3}{128} \text{ units.}$$

The examples above are deceptively simple; the curves were chosen so that the arc length integrals could be easily evaluated. For instance, the number 32 in the curve in Example 2 was chosen so the expression $1 + (dy/dx)^2$ would turn out to be a perfect square and its square root would cause no problems. Because of the square root in the formula, arc length problems for most curves lead to integrals that are difficult or impossible to evaluate without using numerical techniques.

Example 3 (**Manufacturing corrugated panels**) Flat rectangular sheets of metal 2 m wide are to be formed into corrugated roofing panels 2 m wide by bending them into the sinusoidal shape shown in Figure 7.23. The period of the cross-sectional sine curve is 20 cm. Its amplitude is 5 cm, so the panel is 10 cm thick. How long should the flat sheets be cut if the resulting panels must be 5 m long?

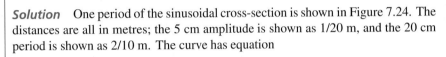

Figure 7.23 A corrugated roofing panel

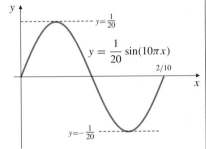

Figure 7.24

Solution One period of the sinusoidal cross-section is shown in Figure 7.24. The distances are all in metres; the 5 cm amplitude is shown as 1/20 m, and the 20 cm period is shown as 2/10 m. The curve has equation

$$y = \frac{1}{20}\sin(10\pi x).$$

Note that 25 periods are required to produce a 5 m long panel. The length of the flat sheet required is 25 times the length of one period of the sine curve:

$$s = 25\int_0^{2/10}\sqrt{1 + \left(\frac{\pi}{2}\cos(10\pi x)\right)^2}\,dx \qquad \text{Let } t = 10\pi x,$$
$$dt = 10\pi\,dx.$$
$$= \frac{5}{2\pi}\int_0^{2\pi}\sqrt{1 + \frac{\pi^2}{4}\cos^2 t}\,dt = \frac{10}{\pi}\int_0^{\pi/2}\sqrt{1 + \frac{\pi^2}{4}\cos^2 t}\,dt.$$

The integral can be evaluated numerically using the techniques of the previous chapter or by using the definite integral function on an advanced scientific calculator. The value is $s \approx 7.32$. The flat metal sheet should be about 7.32 m long to yield a 5 m long finished panel.

If integrals needed for standard problems such as arc lengths of simple curves cannot be evaluated exactly, they are sometimes used to define new functions whose values are tabulated or built in to computer programs. An example of this is the complete elliptic integral function that arises in the next example.

Example 4 (**The circumference of an ellipse**) Find the circumference of the ellipse

$$\frac{x^2}{a^2} + \frac{y^2}{b^2} = 1,$$

where $a \geq b > 0$. See Figure 7.25.

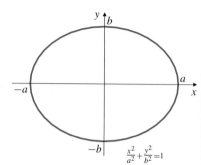

Figure 7.25

Solution   The upper half of the ellipse has equation $y = b\sqrt{1 - \dfrac{x^2}{a^2}} = \dfrac{b}{a}\sqrt{a^2 - x^2}$. Hence,

$$\frac{dy}{dx} = -\frac{b}{a}\frac{x}{\sqrt{a^2 - x^2}},$$

so

$$1 + \left(\frac{dy}{dx}\right)^2 = 1 + \frac{b^2}{a^2}\frac{x^2}{a^2 - x^2}$$

$$= \frac{a^4 - (a^2 - b^2)x^2}{a^2(a^2 - x^2)}.$$

The circumference of the ellipse is four times the arc length of the part lying in the first quadrant, so

$$s = 4\int_0^a \frac{\sqrt{a^4 - (a^2 - b^2)x^2}}{a\sqrt{a^2 - x^2}}\,dx \qquad \begin{aligned} &\text{Let } x = a\sin t, \\ &dx = a\cos t\,dt. \end{aligned}$$

$$= 4\int_0^{\pi/2} \frac{\sqrt{a^4 - (a^2 - b^2)a^2\sin^2 t}}{a(a\cos t)}\,a\cos t\,dt$$

$$= 4\int_0^{\pi/2} \sqrt{a^2 - (a^2 - b^2)\sin^2 t}\,dt$$

$$= 4a\int_0^{\pi/2} \sqrt{1 - \frac{a^2 - b^2}{a^2}\sin^2 t}\,dt$$

$$= 4a\int_0^{\pi/2} \sqrt{1 - \varepsilon^2 \sin^2 t}\,dt \text{ units,}$$

where $\varepsilon = (\sqrt{a^2 - b^2})/a$ is the *eccentricity* of the ellipse. (See Section 8.1 for a discussion of ellipses.) Note that $0 \le \varepsilon < 1$. The function $E(\varepsilon)$, defined by

$$E(\varepsilon) = \int_0^{\pi/2} \sqrt{1 - \varepsilon^2 \sin^2 t}\,dt,$$

is called the **complete elliptic integral of the second kind**. The integral cannot be evaluated by elementary techniques for general ε, although numerical methods can be applied to find approximate values for any given value of ε. Tables of values of $E(\varepsilon)$ for various values of ε can be found in collections of mathematical tables. As shown above, the circumference of the ellipse is given by $4aE(\varepsilon)$. Note that for $a = b$ we have $\varepsilon = 0$, and the formula returns the circumference of a circle; $s = 4a(\pi/2) = 2\pi a$ units. ∎

Areas of Surfaces of Revolution

When a plane curve is rotated (in three dimensions) about a line in the plane of the curve, it sweeps out a **surface of revolution**. For instance, a sphere of radius a is generated by rotating a semicircle of radius a about the diameter of that semicircle. The area of a surface of revolution can be found by integrating an area element

dS constructed by rotating the arc length element ds of the curve about the given line. If the radius of rotation of an arc length element ds is r, then it generates, on rotation, a circular band of width ds and length (circumference) $2\pi r$. The area of this band is, therefore,

$$dS = 2\pi r\, ds,$$

as shown in Figure 7.26. The areas of surfaces of revolution around various lines can be obtained by integrating dS with appropriate choices of r. Here are some important special cases.

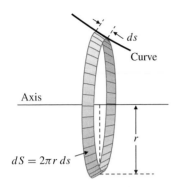

$dS = 2\pi r\, ds$

Figure 7.26 The circular band generated by rotating arc length element ds about the axis

---

Area of a surface of revolution

If $f'(x)$ is continuous on $[a, b]$ and the curve $y = f(x)$ is rotated about the x-axis, the area of the surface of revolution so generated is

$$S = 2\pi \int_{x=a}^{x=b} |y|\, ds = 2\pi \int_a^b |f(x)|\sqrt{1 + (f'(x))^2}\, dx.$$

If the rotation is about the y-axis, the surface area is

$$S = 2\pi \int_{x=a}^{x=b} |x|\, ds = 2\pi \int_a^b |x|\sqrt{1 + (f'(x))^2}\, dx.$$

If $g'(y)$ is continuous on $[c, d]$ and the curve $x = g(y)$ is rotated about the x-axis, the area of the surface of revolution so generated is

$$S = 2\pi \int_{y=c}^{y=d} |y|\, ds = 2\pi \int_c^d |y|\sqrt{1 + (g'(y))^2}\, dy.$$

If the rotation is about the y-axis, the surface area is

$$S = 2\pi \int_{y=c}^{y=d} |x|\, ds = 2\pi \int_c^d |g(y)|\sqrt{1 + (g'(y))^2}\, dy.$$

Remark   Students sometimes wonder whether such complicated formulas are actually necessary. Why not just use $dS = 2\pi |y|\, dx$ for the area element when $y = f(x)$ is rotated about the x-axis instead of the more complicated area element $dS = 2\pi |y|\, ds$? After all, we are regarding dx and ds as both being infinitely small, and we certainly used dx for the width of the disk-shaped volume element when we rotated the region under $y = f(x)$ about the x-axis to generate a solid of revolution. The reason is somewhat subtle. For small thickness Δx, the volume of a slice of the solid of revolution is only approximately $\pi y^2 \Delta x$, but the error is *small compared to the volume of this slice*. On the other hand, if we use $2\pi |y| \Delta x$ as an approximation to the area of a thin band of the surface of revolution corresponding to an x interval of width Δx, the error is *not small compared to the area of that band*. If, for instance, the curve $y = f(x)$ has slope 1 at x, then the width of the band is really $\Delta s = \sqrt{2}\, \Delta x$, so that the area of the band is $\Delta S = 2\pi \sqrt{2}|y| \Delta x$, not just $2\pi |y| \Delta x$. Always use the appropriate arc length element along the curve when you rotate a curve to find the area of a surface of revolution.

Example 5 **(Surface area of a sphere)** Find the area of the surface of a sphere of radius a.

Solution Such a sphere can be generated by rotating the semicircle with equation $y = \sqrt{a^2 - x^2}$, $(-a \leq x \leq a)$, about the x-axis. (See Figure 7.27.) Since

$$\frac{dy}{dx} = -\frac{x}{\sqrt{a^2 - x^2}} = -\frac{x}{y},$$

the area of the sphere is given by

$$S = 2\pi \int_{-a}^{a} y \sqrt{1 + \left(\frac{x}{y}\right)^2} \, dx$$

$$= 4\pi \int_{0}^{a} \sqrt{y^2 + x^2} \, dx$$

$$= 4\pi \int_{0}^{a} \sqrt{a^2} \, dx = 4\pi a x \Big|_{0}^{a} = 4\pi a^2 \text{ square units.}$$

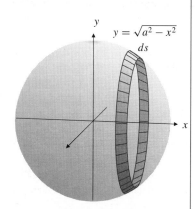

Figure 7.27 An area element on a sphere

Example 6 **(Surface area of a parabolic dish)** Find the surface area of a parabolic reflector whose shape is obtained by rotating the parabolic arc $y = x^2$, $(0 \leq x \leq 1)$, about the y-axis, as illustrated in Figure 7.28.

Solution The arc length element for the parabola $y = x^2$ is $ds = \sqrt{1 + 4x^2} \, dx$, so the required surface area is

$$S = 2\pi \int_{0}^{1} x \sqrt{1 + 4x^2} \, dx \qquad \text{Let } u = 1 + 4x^2,$$
$$\qquad\qquad\qquad\qquad\qquad\qquad du = 8x \, dx.$$

$$= \frac{\pi}{4} \int_{1}^{5} u^{1/2} \, du$$

$$= \frac{\pi}{6} u^{3/2} \Big|_{1}^{5} = \frac{\pi}{6} (5\sqrt{5} - 1) \text{ square units.}$$

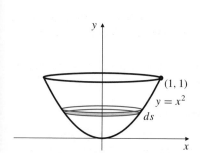

Figure 7.28 The area element is a horizontal band here

Exercises 7.3

In Exercises 1–14, find the lengths of the given curves.

1. $y = 2x - 1$ from $x = 1$ to $x = 3$

2. $y = ax + b$ from $x = A$ to $x = B$

3. $y = \frac{2}{3} x^{3/2}$ from $x = 0$ to $x = 8$

4. $y^2 = (x - 1)^3$ from $(1, 0)$ to $(2, 1)$

5. $y^3 = x^2$ from $(-1, 1)$ to $(1, 1)$

6. $2(x + 1)^3 = 3(y - 1)^2$ from $(-1, 1)$ to $(0, 1 + \sqrt{2/3})$

7. $y = \frac{x^3}{12} + \frac{1}{x}$ from $x = 1$ to $x = 4$

8. $y = \frac{x^3}{3} + \frac{1}{4x}$ from $x = 1$ to $x = 2$

9. $4y = 2 \ln x - x^2$ from $x = 1$ to $x = e$

10. $y = x^2 - \frac{\ln x}{8}$ from $x = 1$ to $x = 2$

11. $y = \frac{e^x + e^{-x}}{2}$ $(= \cosh x)$ from $x = 0$ to $x = a$

12. $y = \ln \cos x$ from $x = \pi/6$ to $x = \pi/4$

* **13.** $y = x^2$ from $x = 0$ to $x = 2$

* **14.** $y = \ln \dfrac{e^x - 1}{e^x + 1}$ from $x = 2$ to $x = 4$

15. Find the circumference of the closed curve
$x^{2/3} + y^{2/3} = a^{2/3}$. *Hint:* the curve is symmetric about both
coordinate axes (why?), so one-quarter of it lies in the first
quadrant.

Use numerical methods (or a calculator with integration
function) to find the lengths of the curves in Exercises 16–19 to 4
decimal places.

16. $y = x^4$ from $x = 0$ to $x = 1$

17. $y = x^{1/3}$ from $x = 1$ to $x = 2$

18. the circumference of the ellipse $3x^2 + y^2 = 3$

19. the shorter arc of the ellipse $x^2 + 2y^2 = 2$ between $(0, 1)$
and $(1, 1/\sqrt{2})$

In Exercises 20–27, find the areas of the surfaces obtained by
rotating the given curve about the indicated lines.

20. $y = x^2$, $(0 \le x \le 2)$, about the y-axis

21. $y = x^3$, $(0 \le x \le 1)$, about the x-axis

22. $y = x^{3/2}$, $(0 \le x \le 1)$, about the x-axis

23. $y = x^{3/2}$, $(0 \le x \le 1)$, about the y-axis

24. $y = e^x$, $(0 \le x \le 1)$, about the x-axis

25. $y = \sin x$, $(0 \le x \le \pi)$, about the x-axis

26. $y = \dfrac{x^3}{12} + \dfrac{1}{x}$, $(1 \le x \le 4)$, about the x-axis

27. $y = \dfrac{x^3}{12} + \dfrac{1}{x}$, $(1 \le x \le 4)$, about the y-axis

28. (**Surface area of a cone**) Find the area of the curved surface
of a right-circular cone of base radius r and height h by
rotating the straight line segment from $(0, 0)$ to (r, h) about
the y-axis.

29. (**How much icing on a doughnut?**) Find the surface area of

the torus (doughnut) obtained by rotating the circle
$(x - b)^2 + y^2 = a^2$ about the y-axis.

30. (**Area of a prolate spheroid**) Find the area of the surface
obtained by rotating the ellipse $x^2 + 4y^2 = 4$ about the
x-axis.

31. (**Area of an oblate spheroid**) Find the area of the surface
obtained by rotating the ellipse $x^2 + 4y^2 = 4$ about the
y-axis.

* **32.** The ellipse of Example 4 is rotated about the line $y = c > b$
to generate a doughnut with elliptical cross-sections.
Express the surface area of this doughnut in terms of the
complete elliptic integral function $E(\varepsilon)$ introduced in that
example.

* **33.** Express the integral formula obtained for the length of the
metal sheet in Example 3 in terms of the complete elliptic
integral function $E(\epsilon)$ introduced in Example 4.

34. (**An interesting property of spheres**) If two parallel planes
intersect a sphere, show that the surface area of that part of
the sphere lying between the two planes depends only on the
radius of the sphere and the distance between the planes, and
not on the position of the planes.

35. For what real values of k does the surface generated by
rotating the curve $y = x^k$, $(0 < x \le 1)$, about the y-axis
have a finite surface area?

* **36.** The curve $y = \ln x$, $(0 < x \le 1)$, is rotated about the y-axis.
Find the area of the area of the horn-shaped surface so
generated.

37. A hollow container in the shape of an infinitely long horn is
generated by rotating the curve $y = 1/x$, $(1 \le x < \infty)$,
about the x-axis.

(a) Find the volume of the container.

(b) Show that the container has infinite surface area.

(c) How do you explain the "paradox" that the container
can be filled with a finite volume of paint but requires
infinitely much paint to cover its surface?

7.4 Mass, Moments, and Centre of Mass

Many quantities of interest in physics, mechanics, ecology, finance, and other
disciplines are described in terms of densities over regions of space, the plane,
or even the real line. To determine the total value of such a quantity we must
add up (integrate) the contributions from the various places where the quantity is
distributed.

Mass and Density

If a solid object is made of a homogeneous material, we would expect different parts
of the solid that have the same volume to have the same mass as well. We express

this homogeneity by saying that the object has constant density, that density being the mass divided by the volume for the whole object or for any part of it. Thus, a rectangular brick with dimensions 20 cm, 10 cm, and 8 cm would have volume $V = 20 \times 10 \times 8 = 1{,}600$ cm^3, and if it was made of material having constant density $\delta = 3$ g/cm^3, it would have mass $m = \delta V = 3 \times 1{,}600 = 4{,}800$ g. (We will use the lowercase Greek letter delta (δ) to represent density.)

If the density of the material constituting a solid object is not constant but varies from point to point in the object, no such simple relationship exists between mass and volume. If the density $\delta = \delta(P)$ is a *continuous* function of position P, we can subdivide the solid into many small volume elements and, by regarding δ as approximately constant over each such element, determine the masses of all the elements and add them up to get the mass of the solid. The mass Δm of a volume element ΔV containing the point P would satisfy

$$\Delta m \approx \delta(P)\,\Delta V,$$

so the mass m of the solid can be approximated:

$$m = \sum \Delta m \approx \sum \delta(P)\,\Delta V.$$

Such approximations become exact as we pass to the limit of differential mass and volume elements, $dm = \delta(P)\,dV$, so we expect to be able to calculate masses as integrals, that is, as the limits of such sums:

$$m = \int dm = \int \delta(P)\,dV.$$

Example 1 The density of a solid vertical cylinder of height H cm and base area A cm^2 is $\delta = \delta_0(1 + h)$ g/cm^3, where h is the height in centimetres above the base and δ_0 is a constant. Find the mass of the cylinder.

Solution See Figure 7.29(a). A slice of the solid at height h above the base and having thickness dh is a circular disk of volume $dV = A\,dh$. Since the density is constant over this disk, the mass of the volume element is

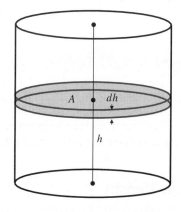

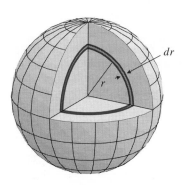

Figure 7.29

(a) A solid cylinder whose density varies with height

(b) Cutaway view of a planet whose density depends on distance from the centre

(a)

(b)

$$dm = \delta \, dV = \delta_0 (1 + h) \, A \, dh.$$

Therefore, the mass of the whole cylinder is

$$m = \int_0^H \delta_0 A (1 + h) \, dh = \delta_0 A \left(H + \frac{H^2}{2} \right) \text{ g.}$$

■

Example 2 **(Using spherical shells)** The density of a spherical planet of radius R m varies with distance r from the centre according to the formula

$$\delta = \frac{\delta_0}{1 + r^2} \text{ kg/m}^3.$$

Find the mass of the planet.

Solution Recall that the surface area of a sphere of radius r is $4\pi r^2$. The planet can be regarded as being composed of concentric spherical shells having radii between 0 and R. The volume of a shell of radius r and thickness dr (see Figure 7.29(b)) is equal to its surface area times its thickness, and its mass is its volume times its density:

$$dV = 4\pi r^2 \, dr; \qquad dm = \delta \, dV = 4\pi \delta_0 \frac{r^2}{1 + r^2} \, dr.$$

We add the masses of these shells to find the mass of the whole planet:

$$m = 4\pi \delta_0 \int_0^R \frac{r^2}{1 + r^2} \, dr = 4\pi \delta_0 \int_0^R \left(1 - \frac{1}{1 + r^2} \right) dr$$

$$= 4\pi \delta_0 (r - \tan^{-1} r) \Big|_0^R = 4\pi \delta_0 (R - \tan^{-1} R) \text{ kg.}$$

■

Similar techniques can be applied to find masses of one- and two-dimensional objects, such as wires and thin plates, that have variable densities of the forms mass/unit length (**line density**) and mass/unit area (**areal density**).

Example 3 A wire of variable composition is stretched along the x-axis from $x = 0$ to $x = L$ cm. Find the mass of the wire if the line density at position x is $\delta(x) = kx$ g/cm, where k is a positive constant.

Solution The mass of a length element dx of the wire located at position x is given by $dm = \delta(x) \, dx = kx \, dx$. Thus, the mass of the wire is

$$m = \int_0^L kx \, dx = \left(\frac{kx^2}{2} \right) \Big|_0^L = \frac{kL^2}{2} \text{ g.}$$

Example 4   Find the mass of a disk of radius a cm whose centre is at the origin in the xy-plane if the areal density at position (x, y) is $\delta = k(2a + x)$ g/cm^2. Here k is a constant.

Solution   The density depends only on the horizontal coordinate x, so it is constant along vertical lines on the disk. This suggests that thin vertical strips should be used as area elements. A vertical strip of thickness dx at x has area $dA = 2\sqrt{a^2 - x^2}\, dx$ (see Figure 7.30); its mass is therefore

$$dm = \delta\, dA = 2k(2a + x)\sqrt{a^2 - x^2}\, dx.$$

Hence, the mass of the disk is

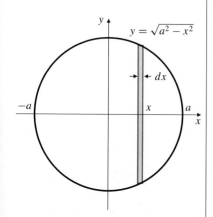

$$m = \int_{x=-a}^{x=a} dm = 2k \int_{-a}^{a} (2a + x)\sqrt{a^2 - x^2}\, dx$$

$$= 4ak \int_{-a}^{a} \sqrt{a^2 - x^2}\, dx + 2k \int_{-a}^{a} x\sqrt{a^2 - x^2}\, dx$$

$$= 4ak \frac{\pi a^2}{2} + 0 = 2\pi ka^3 \text{ g}.$$

Figure 7.30

We used the area of a semicircle to evaluate the first integral. The second integral is zero because the integrand is odd and the interval is symmetric about $x = 0$. ∎

Distributions of mass along one-dimensional structures (lines or curves) necessarily lead to integrals of functions of one variable, but distributions of mass on a surface or in space can lead to integrals involving functions of more than one variable. Such integrals are studied in multivariable calculus. (See, for example, Section 14.7.) In the examples above, the given densities were functions of only one variable, so these problems, although higher dimensional in nature, led to integrals of functions of only one variable and could be solved by the methods at hand.

Moments and Centres of Mass

A mass m located at position x on the x-axis is said to have **moment** xm about the point $x = 0$ or, more generally, moment $(x - x_0)m$ about the point $x = x_0$. If the x-axis is a horizontal arm hinged at x_0, the moment about x_0 measures the tendency of the weight of the mass m to cause the arm to rotate. If several masses m_1, m_2, m_3, ..., m_n are located at the points x_1, x_2, x_3, ..., x_n, respectively, then the total moment of the system of masses about the point $x = x_0$ is the sum of the individual moments (see Figure 7.31):

$$M_{x=x_0} = (x_1 - x_0)m_1 + (x_2 - x_0)m_2 + \cdots + (x_n - x_0)m_n = \sum_{j=1}^{n} (x_j - x_0)m_j.$$

Figure 7.31   A system of discrete masses on a line

The **centre of mass** of the system of masses is the point $\bar{x}$ about which the total moment of the system is zero. Thus

$$0 = \sum_{j=1}^{n}(x_j - \bar{x})m_j = \sum_{j=1}^{n}x_j m_j - \bar{x}\sum_{j=1}^{n}m_j.$$

The centre of mass of the system is therefore given by

$$\bar{x} = \frac{\displaystyle\sum_{j=1}^{n}x_j m_j}{\displaystyle\sum_{j=1}^{n}m_j} = \frac{M_{x=0}}{m},$$

where m is the total mass of the system and $M_{x=0}$ is the total moment about $x = 0$. If you think of the x-axis as being a weightless wire supporting the masses, then $\bar{x}$ is the point at which the wire could be supported and remain in perfect balance (equilibrium), not tipping either way. Even if the axis represents a nonweightless support, say a seesaw, supported at $x = \bar{x}$, it will remain balanced after the masses are added, provided it was balanced beforehand. For many purposes a system of masses behaves as though its total mass were concentrated at its centre of mass.

Now suppose that a one-dimensional distribution of mass with continuously variable line density $\delta(x)$ lies along the interval $[a, b]$ of the x-axis. An element of length dx at position x contains mass $dm = \delta(x)\,dx$, so its moment is $dM_{x=0} = x\,dm = x\delta(x)\,dx$ about $x = 0$. The total moment about $x = 0$ is the *sum* (integral) of these moment elements:

$$M_{x=0} = \int_a^b x\delta(x)\,dx.$$

Since the total mass is

$$m = \int_a^b \delta(x)\,dx,$$

we obtain the following formula for the centre of mass.

The centre of mass of a distribution of mass with line density $\delta(x)$ on the interval $[a, b]$ is given by

$$\bar{x} = \frac{M_{x=0}}{m} = \frac{\displaystyle\int_a^b x\delta(x)\,dx}{\displaystyle\int_a^b \delta(x)\,dx}.$$

Example 5   At what point can the wire of Example 3 be suspended so that it will balance?

Solution   In Example 3 we evaluated the mass of the wire to be $kL^2/2$ g. Its moment about $x = 0$ is

$$M_{x=0} = \int_0^L x\delta(x)\,dx$$

$$= \int_0^L kx^2\,dx = \left(\frac{kx^3}{3}\right)\Big|_0^L = \frac{kL^3}{3}\ \text{g·cm.}$$

(Note that the appropriate units for the moment are units of mass times units of distance: in this case gram-centimetres.) The centre of mass of the wire is

$$\bar{x} = \frac{kL^3/3}{kL^2/2} = \frac{2L}{3}.$$

The wire will be balanced if suspended at position $x = 2L/3$ cm.

Two- and Three-Dimensional Examples

The system of mass considered in Example 5 is one-dimensional and lies along a straight line. If mass is distributed in a plane or in space, similar considerations prevail. For a system of masses m_1 at (x_1, y_1), m_2 at (x_2, y_2), ..., m_n at (x_n, y_n), the **moment about** $x = 0$ is

$$M_{x=0} = x_1 m_1 + x_2 m_2 + \cdots + x_n m_n = \sum_{j=1}^n x_j m_j,$$

and the **moment about** $y = 0$ is

$$M_{y=0} = y_1 m_1 + y_2 m_2 + \cdots + y_n m_n = \sum_{j=1}^n y_j m_j.$$

The **centre of mass** is the point $(\bar{x}, \bar{y})$ where

$$\bar{x} = \frac{M_{x=0}}{m} = \frac{\displaystyle\sum_{j=1}^n x_j m_j}{\displaystyle\sum_{j=1}^n m_j} \qquad \text{and} \qquad \bar{y} = \frac{M_{y=0}}{m} = \frac{\displaystyle\sum_{j=1}^n y_j m_j}{\displaystyle\sum_{j=1}^n m_j}.$$

For continuous distributions of mass, the sums become appropriate integrals.

Example 6 Find the centre of mass of a rectangular plate that occupies the region $0 \le x \le a$, $0 \le y \le b$, if the areal density of the material in the plate at position (x, y) is ky.

Solution Since the density is independent of x and the rectangle is symmetric about the line $x = a/2$, the x-coordinate of the centre of mass must be $\bar{x} = a/2$. A thin horizontal strip of width dy at height y (see Figure 7.32) has mass $dm = aky\,dy$. The moment of this strip about $y = 0$ is $dM_{y=0} = y\,dm = kay^2\,dy$. Hence, the mass and moment about $y = 0$ of the whole plate are

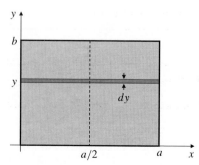

Figure 7.32

$$m = ka \int_0^b y \, dy = \frac{kab^2}{2},$$

$$M_{y=0} = ka \int_0^b y^2 \, dy = \frac{kab^3}{3}.$$

Therefore, $\bar{y} = M_{y=0}/m = 2b/3$, and the centre of mass of the plate is $(a/2, 2b/3)$. The plate would be balanced if supported at this point.

For distributions of mass in three-dimensional space one defines, analogously, the moments $M_{x=0}$, $M_{y=0}$, and $M_{z=0}$ of the system of mass about the planes $x = 0$, $y = 0$, and $z = 0$, respectively. The centre of mass is $(\bar{x}, \bar{y}, \bar{z})$ where

$$\bar{x} = \frac{M_{x=0}}{m}, \qquad \bar{y} = \frac{M_{y=0}}{m}, \qquad \text{and} \qquad \bar{z} = \frac{M_{z=0}}{m},$$

m being the total mass: $m = m_1 + m_2 + \cdots + m_n$. Again, the sums are replaced with integrals for continuous distributions of mass.

Example 7 Find the centre of mass of a solid hemisphere of radius R ft if its density at height z ft above the base plane of the hemisphere is $\delta_0 z$ lb/ft^3.

Solution The solid is symmetric about the vertical axis (let us call it the z-axis), and the density is constant in planes perpendicular to this axis. Therefore the centre of mass must lie somewhere on this axis. A slice of the solid at height z above the base, and having thickness dz, is a disk of radius $\sqrt{R^2 - z^2}$. (See Figure 7.33.) Its volume is $dV = \pi(R^2 - z^2)\,dz$, and its mass is $dm = \delta_0 z\,dV = \delta_0 \pi (R^2 z - z^3)\,dz$. Its moment about the base plane $z = 0$ is $dM_{z=0} = z\,dm = \delta_0 \pi (R^2 z^2 - z^4)\,dz$. The mass of the solid is

$$m = \delta_0 \pi \int_0^R (R^2 z - z^3)\,dz = \delta_0 \pi \left(\frac{R^2 z^2}{2} - \frac{z^4}{4} \right)\Bigg|_0^R = \frac{\pi}{4}\delta_0 R^4 \text{ lb}.$$

The moment of the hemisphere about the plane $z = 0$ is

$$M_{z=0} = \delta_0 \pi \int_0^R (R^2 z^2 - z^4)\,dz = \delta_0 \pi \left(\frac{R^2 z^3}{3} - \frac{z^5}{5} \right)\Bigg|_0^R = \frac{2\pi}{15}\delta_0 R^5 \text{ lb·ft}.$$

The centre of mass therefore lies along the axis of symmetry of the hemisphere at height $\bar{z} = M_{z=0}/m = 8R/15$ ft above the base of the hemisphere.

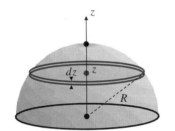

Figure 7.33 Mass element of a solid hemisphere with density depending on height

Example 8 Find the centre of mass of a plate that occupies the region $a \le x \le b$, $0 \le y \le f(x)$, if the density at any point (x, y) is $\delta(x)$.

Solution The appropriate area element is shown in Figure 7.34. It has area $f(x)\,dx$, mass

$$dm = \delta(x)f(x)\,dx,$$

and moment about $x = 0$

$$dM_{x=0} = x\delta(x)f(x)\,dx.$$

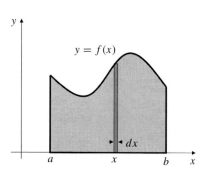

Figure 7.34 Mass element of a plate

Since the density depends only on x, the mass element dm has constant density, so the y-coordinate of *its* centre of mass is at its midpoint: $\bar{y}_{dm} = \frac{1}{2} f(x)$. Therefore, the moment of the mass element dm about $y = 0$ is

$$dM_{y=0} = \bar{y}_{dm}\, dm = \frac{1}{2}\delta(x)\big(f(x)\big)^2 dx.$$

The coordinates of the centre of mass of the plate are $\bar{x} = \dfrac{M_{x=0}}{m}$ and $\bar{y} = \dfrac{M_{y=0}}{m}$, where

$$m = \int_a^b \delta(x) f(x)\, dx,$$

$$M_{x=0} = \int_a^b x\delta(x) f(x)\, dx,$$

$$M_{y=0} = \frac{1}{2}\int_a^b \delta(x)\big(f(x)\big)^2 dx.$$

Remark Similar formulas can be obtained if the density depends on y instead of x, provided that the region admits a suitable horizontal area element (e.g., the region might be specified by $c \le y \le d$, $0 \le x \le g(y)$). Finding centres of mass for plates that occupy regions specified by functions of x, but where the density depends on y, generally requires the use of "double integrals." Such problems are therefore studied in multivariable calculus. (See Section 14.7.)

Exercises 7.4

Find the mass and centre of mass for the systems in Exercises 1–16. Be alert for symmetries.

1. A straight wire of length L cm, where the density at distance s cm from one end is $\delta(s) = \sin \pi s/L$ g/cm

2. A straight wire along the x-axis from $x = 0$ to $x = L$ if the density is constant δ_0, but the cross-sectional radius of the wire varies so that its value at x is $a + bx$

3. A quarter-circular plate having radius a, constant areal density δ_0, and occupying the region $x^2 + y^2 \le a^2$, $x \ge 0$, $y \ge 0$

4. A quarter-circular plate of radius a occupying the region $x^2 + y^2 \le a^2$, $x \ge 0$, $y \ge 0$, having areal density $\delta(x) = \delta_0 x$

5. A plate occupying the region $0 \le y \le 4 - x^2$ if the areal density at (x, y) is ky

6. A right-triangular plate with legs 2 m and 3 m if the areal density at any point P is $5h$ kg/m^2, h being the distance of P from the shorter leg

7. A square plate of edge a cm if the areal density at P is kx g/cm^2, where x is the distance from P to one edge of the square

8. The plate in Exercise 7, but with areal density kr g/cm^2, where r is the distance (in centimetres) from P to one of the diagonals of the square

9. A plate of density $\delta(x)$ occupying the region $a \le x \le b$, $f(x) \le y \le g(x)$

10. A rectangular brick with dimensions 20 cm, 10 cm, and 5 cm, if the density at P is kx g/cm^3, where x is the distance from P to one of the 10×5 faces

11. A solid ball of radius R m if the density at P is z kg/m^3, where z is the distance from P to a plane at distance $2R$ m from the centre of the ball

12. A right-circular cone of base radius a cm and height b cm if the density at point P is kz g/cm^3, where z is the distance of P from the base of the cone

* 13. The solid occupying the quarter of a ball of radius a centred at the origin having as base the region $x^2 + y^2 \le a^2$, $x \ge 0$ in the xy-plane, if the density at height z above the base is $\delta_0 z$

* 14. The cone of Exercise 12, but with density at P equal to kx g/cm^3, where x is the distance of P from the axis of symmetry of the cone. *Hint:* use a cylindrical shell centred on the axis of symmetry as volume element. This element has constant density, so its centre of mass is known, and its moment can be determined from its mass.

* 15. A semicircular plate occupying the region $x^2 + y^2 \le a^2$, $y \ge 0$, if the density at distance s from the origin is ks g/cm^2

* **16.** The wire in Exercise 1 if it is bent in a semicircle

17. It is estimated that the density of matter in the neighbourhood of a gas giant star is given by $\delta(r) = Ce^{-kr^2}$, where C and k are positive constants, and r is the distance from the centre of the star. The radius of the star is indeterminate but can be taken to be infinite since

$\delta(r)$ decreases very rapidly for large r. Find the approximate mass of the star in terms of C and k.

18. Find the average distance $\bar{r}$ of matter in the star of Exercise 17 from the centre of the star. $\bar{r}$ is given by $\int_0^\infty r \, dm / \int_0^\infty dm$, where dm is the mass element at distance r from the centre of the star.

7.5 Centroids

If matter is distributed uniformly in a system so that the density δ is constant, then that density cancels out of the numerator and denominator in sum or integral expressions for coordinates of the centre of mass. In such cases the centre of mass depends only on the *shape* of the object, that is, on geometric properties of the region occupied by the object, and we call it the **centroid** of the region.

Centroids are calculated using the same formulas as those used for centres of mass, except that the density (being constant) is taken to be unity, so the mass is just the length, area, or volume of the region, and the moments are referred to as **moments of the region**, rather than of any mass occupying the region. If we set $\delta(x) = 1$ in the formulas obtained in Example 8 of Section 7.4, we obtain the following result:

> **The centroid of a standard plane region**
>
> The centroid of the plane region $a \le x \le b$, $0 \le y \le f(x)$, is $(\bar{x}, \bar{y})$, where $\bar{x} = \dfrac{M_{x=0}}{A}$, $\bar{y} = \dfrac{M_{y=0}}{A}$, and
>
> $$A = \int_a^b f(x) \, dx, \quad M_{x=0} = \int_a^b x f(x) \, dx, \quad M_{y=0} = \frac{1}{2} \int_a^b \left(f(x) \right)^2 dx.$$

Thus, for example, $\bar{x}$ is the *average value* of the function x over the region.

The centroids of some regions are obvious by symmetry. The centroid of a circular disk or an elliptical disk is at the centre of the disk. The centroid of a rectangle is at the centre also; the centre is the point of intersection of the diagonals. The centroid of any region lies on any axes of symmetry of the region.

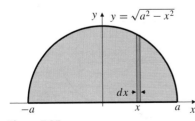

Figure 7.35

Example 1   What is the average value of y over the half-disk $-a \le x \le a$, $0 \le y \le \sqrt{a^2 - x^2}$? Find the centroid of the half-disk.

Solution   By symmetry, the centroid lies on the y-axis, so its x-coordinate is $\bar{x} = 0$. (See Figure 7.35.) Since the area of the half-disk is $A = \frac{1}{2} \pi a^2$, the average value of y over the half-disk is

$$\bar{y} = \frac{M_{y=0}}{A} = \frac{2}{\pi a^2} \frac{1}{2} \int_{-a}^a (a^2 - x^2) \, dx = \frac{2}{\pi a^2} \frac{2a^3}{3} = \frac{4a}{3\pi}.$$

The centroid of the half-disk is $\left(0, \dfrac{4a}{3\pi} \right)$.

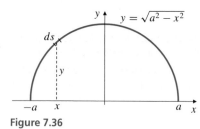

Figure 7.36

Example 2 Find the centroid of the semicircle $y = \sqrt{a^2 - x^2}$.

Solution Here, the "region" is a one-dimensional curve, having length rather than area. Again $\bar{x} = 0$ by symmetry. A short arc of length ds at height y on the semicircle has moment $dM_{y=0} = y\,ds$ about $y = 0$. (See Figure 7.36.) Since

$$ds = \sqrt{1 + \left(\frac{dy}{dx}\right)^2}\,dx = \sqrt{1 + \frac{x^2}{a^2 - x^2}}\,dx = \frac{a\,dx}{\sqrt{a^2 - x^2}},$$

and since $y = \sqrt{a^2 - x^2}$ on the semicircle, we have

$$M_{y=0} = \int_{-a}^{a} \sqrt{a^2 - x^2}\,\frac{a\,dx}{\sqrt{a^2 - x^2}} = a\int_{-a}^{a} dx = 2a^2.$$

Since the length of the semicircle is πa, we have $\bar{y} = \dfrac{M_{y=0}}{\pi a} = \dfrac{2a}{\pi}$, and the centroid of the semicircle is $\left(0, \dfrac{2a}{\pi}\right)$. Note that the centroid of a semicircle of radius a is not the same as that of half-disk of radius a. Note also that the centroid of the semicircle does not lie on the semicircle itself. ∎

THEOREM 1

The centroid of a triangle

The centroid of a triangle is the point at which all three medians of the triangle intersect.

PROOF Recall that a median of a triangle is a straight line joining one vertex of the triangle to the midpoint of the opposite side. Given any median of a triangle, we will show that the centroid lies on that median. Thus, the centroid must lie on all three medians.

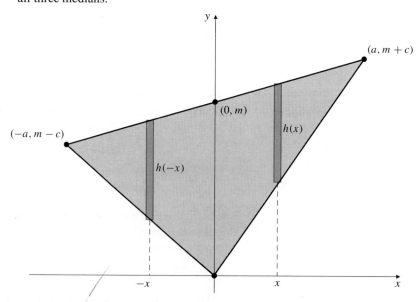

Figure 7.37

Adopt a coordinate system where the median in question lies along the y-axis and such that a vertex of the triangle is at the origin. (See Figure 7.37.) Let the midpoint of the opposite side be $(0, m)$. Then the other two vertices of the triangle must have coordinates of the form $(-a, m - c)$ and $(a, m + c)$ so that $(0, m)$ will be the midpoint between them. The two vertical area elements shown in the figure are at the same distance on opposite sides of the y-axis, so they have the same heights $h(-x) = h(x)$ (by similar triangles) and the same area. The sum of the moments about $x = 0$ of these area elements is

$$dM_{x=0} = -xh(-x)\,dx + xh(x)\,dx = 0,$$

so the moment of the whole triangle about $x = 0$ is

$$M_{x=0} = \int_{x=-a}^{x=a} dM_{x=0} = 0.$$

Therefore, the centroid of the triangle lies on the y-axis.

Remark By solving simultaneously the equations of any two medians of a triangle, we can verify the following formula:

> **Coordinates of the centroid of a triangle**
>
> The coordinates of the centroid of a triangle are the averages of the corresponding coordinates of the three vertices of the triangle. The triangle with vertices (x_1, y_1), (x_2, y_2), and (x_3, y_3) has centroid
>
> $$(\bar{x}, \bar{y}) = \left(\frac{x_1 + x_2 + x_3}{3}, \frac{y_1 + y_2 + y_3}{3} \right).$$

If a region is a union of nonoverlapping subregions, then any moment of the region is the sum of the corresponding moments of the subregions. This fact enables us to calculate the centroid of the region if we know the centroids and areas of all the subregions.

Example 3 Find the centroid of the trapezoid with vertices $(0, 0)$, $(1, 0)$, $(1, 2)$, and $(0, 1)$.

Solution The trapezoid is the union of a square and a (nonoverlapping) triangle, as shown in Figure 7.38. By symmetry, the square has centroid $(\bar{x}_S, \bar{y}_S) = \left(\frac{1}{2}, \frac{1}{2} \right)$, and its area is $A_S = 1$. The triangle has area $A_T = \frac{1}{2}$, and its centroid is $(\bar{x}_T, \bar{y}_T)$, where

$$\bar{x}_T = \frac{0 + 1 + 1}{3} = \frac{2}{3} \quad \text{and} \quad \bar{y}_T = \frac{1 + 1 + 2}{3} = \frac{4}{3}.$$

Continuing to use subscripts S and T to denote the square and triangle, respectively, we calculate

$$M_{x=0} = M_{S;x=0} + M_{T;x=0} = A_S\bar{x}_S + A_T\bar{x}_T = 1 \times \frac{1}{2} + \frac{1}{2} \times \frac{2}{3} = \frac{5}{6},$$

$$M_{y=0} = M_{S;y=0} + M_{T;y=0} = A_S\bar{y}_S + A_T\bar{y}_T = 1 \times \frac{1}{2} + \frac{1}{2} \times \frac{4}{3} = \frac{7}{6}.$$

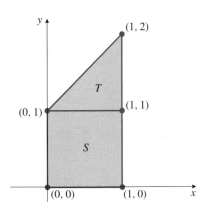

Figure 7.38

Since the area of the trapezoid is $A = A_S + A_T = \frac{3}{2}$, its centroid is

$$(\bar{x}, \bar{y}) = \left(\frac{5}{6} \Big/ \frac{3}{2}, \frac{7}{6} \Big/ \frac{3}{2}\right) = \left(\frac{5}{9}, \frac{7}{9}\right).$$

■

Example 4 Find the centroid of the solid region obtained by rotating about the y-axis the first quadrant region lying between the x-axis and the parabola $y = 4-x^2$.

Solution By symmetry, the centroid of the parabolic solid will lie on its axis of symmetry, the y-axis. A thin, disk-shaped slice of the solid at height y and having thickness dy (see Figure 7.39) has volume

$$dV = \pi x^2\, dy = \pi(4 - y)\, dy$$

and moment about the base plane

$$dM_{y=0} = y\, dV = \pi(4y - y^2)\, dy.$$

Hence, the volume of the solid is

$$V = \pi \int_0^4 (4 - y)\, dy = \pi \left(4y - \frac{y^2}{2}\right)\Bigg|_0^4 = \pi(16 - 8) = 8\pi,$$

and its moment about $y = 0$ is

$$M_{y=0} = \pi \int_0^4 (4y - y^2)\, dy = \pi \left(2y^2 - \frac{y^3}{3}\right)\Bigg|_0^4 = \pi\left(32 - \frac{64}{3}\right) = \frac{32}{3}\pi.$$

Hence, the centroid is located at $\bar{y} = \dfrac{32\pi}{3} \times \dfrac{1}{8\pi} = \dfrac{4}{3}$.

■

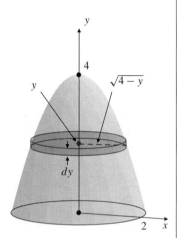

Figure 7.39

Pappus's Theorem

The following theorem relates volumes or surface areas of revolution to the centroid of the region or curve being rotated.

THEOREM 2

Pappus's Theorem

(a) If a plane region R lies on one side of a line L in that plane and is rotated about L to generate a solid of revolution, then the volume V of that solid is the product of the area of R and the distance travelled by the centroid of R under the rotation; that is,

$$V = 2\pi \bar{r} A,$$

where A is the area of R, and $\bar{r}$ is the perpendicular distance from the centroid of R to L.

(b) If a plane curve C lies on one side of a line L in that plane and is rotated about that line to generate a surface of revolution, then the area S of that surface is the length of C times the distance travelled by the centroid of C:

$$S = 2\pi \bar{r} s,$$

where s is the length of the curve C and $\bar{r}$ is the perpendicular distance from the centroid of C to the line L.

PROOF We prove part (a). The proof of (b) is similar and is left as an exercise. Let us take L to be the y-axis and suppose that R lies between $x = a$ and $x = b$ where $0 \le a < b$. Thus $\bar{r} = \bar{x}$, the x-coordinate of the centroid of R. Let dA denote the area of a thin strip of R at position x and having width dx. (See Figure 7.40.) This strip generates, on rotation about L, a cylindrical shell of volume $dV = 2\pi x \, dA$, so the volume of the solid of revolution is

$$V = 2\pi \int_{x=a}^{x=b} x \, dA = 2\pi M_{x=0} = 2\pi \bar{x} A = 2\pi \bar{r} A.$$

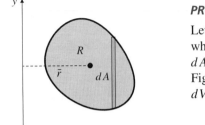

Figure 7.40

As the following examples illustrate, Pappus's Theorem can be used in two ways: either the centroid can be determined when the appropriate volume or surface area is known or the volume or surface area can be determined if the centroid of the rotating region or curve is known.

Example 5 Use Pappus's Theorem to find the centroid of the semicircle $y = \sqrt{a^2 - x^2}$.

Solution The centroid of the semicircle lies on its axis of symmetry, the y-axis, so it is located at a point with coordinates $(0, \bar{y})$. Since the semicircle has length πa units and generates, on rotation about the x-axis, a sphere having area $4\pi a^2$ square units, we obtain, using part (b) of Pappus's Theorem,

$$4\pi a^2 = 2\pi (\pi a) \bar{y}.$$

Thus $\bar{y} = 2a/\pi$, as shown previously in Example 2.

Example 6 Use Pappus's Theorem to find the volume and surface area of the torus (doughnut) obtained by rotating the disk $(x - b)^2 + y^2 \le a^2$ about the y-axis. Here $0 < a < b$. (See Figure 7.10 in Section 7.1.)

Solution The centroid of the disk is at $(b, 0)$, which is at distance $\bar{r} = b$ units from the axis of rotation. Since the disk has area πa^2 square units, the volume of the torus is

$$V = 2\pi b(\pi a^2) = 2\pi^2 a^2 b \text{ cubic units.}$$

To find the surface area S of the torus (in case you want to have icing on the doughnut), rotate the circular boundary of the disk, which has length $2\pi a$, about the y-axis and obtain

$$S = 2\pi b(2\pi a) = 4\pi^2 ab \text{ square units.}$$

Exercises 7.5

Find the centroids of the geometric structures in Exercises 1–21. Be alert for symmetries and opportunities to use Pappus's Theorem.

1. The quarter-disk $x^2 + y^2 \leq r^2$, $x \geq 0$, $y \geq 0$

2. The region $0 \leq y \leq 9 - x^2$

3. The region $0 \leq x \leq 1$, $0 \leq y \leq \dfrac{1}{\sqrt{1 + x^2}}$

4. The circular disk sector $x^2 + y^2 \leq r^2$, $0 \leq y \leq x$

5. The circular disk segment $0 \leq y \leq \sqrt{4 - x^2} - 1$

6. The semi-elliptic disk $0 \leq y \leq b\sqrt{1 - (x/a)^2}$

7. The quadrilateral with vertices (in clockwise order) $(0, 0)$, $(3, 1)$, $(4, 0)$, and $(2, -2)$

8. The region bounded by the semicircle $y = \sqrt{1 - (x - 1)^2}$, the y-axis, and the line $y = x - 2$.

9. A hemispherical surface of radius r

10. A solid half ball of radius r

11. A solid cone of base radius r and height h

12. A conical surface of base radius r and height h

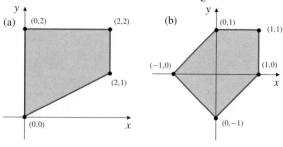
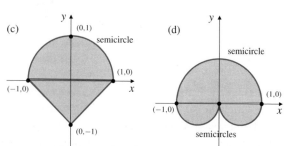

Figure 7.41

13. The region in Figure 7.41(a)

14. The region in Figure 7.41(b)

15. The region in Figure 7.41(c)

16. The region in Figure 7.41(d)

17. The plane region $0 \leq y \leq \sin x$, $0 \leq x \leq \pi$

18. The plane region $0 \leq y \leq \cos x$, $0 \leq x \leq \pi/2$

19. The quarter-circle arc $x^2 + y^2 = r^2$, $x \geq 0$, $y \geq 0$

20. The solid obtained by rotating the region in Figure 7.41(a)

about the y-axis

21. The solid obtained by rotating the plane region $0 \leq y \leq 2x - x^2$ about the line $y = -2$.

22. The line segment from $(1, 0)$ to $(0, 1)$ is rotated about the line $x = 2$ to generate part of a conical surface. Find the area of that surface.

23. The triangle with vertices $(0, 0)$, $(1, 0)$, and $(0, 1)$ is rotated about the line $x = 2$ to generate a certain solid. Find the volume of that solid.

24. An equilateral triangle of edge s cm is rotated about one of its edges to generate a solid. Find the volume and surface area of that solid.

25. Find to 5 decimal places the coordinates of the centroid of the region $0 \leq x \leq \pi/2$, $0 \leq y \leq \sqrt{x}\cos x$.

26. Find to 5 decimal places the coordinates of the centroid of the region $0 < x \leq \pi/2$, $\ln(\sin x) \leq y \leq 0$.

27. Find the centroid of the infinitely long spike-shaped region lying between the x-axis and the curve $y = (x + 1)^{-3}$ and to the right of the y-axis.

28. Show that the curve $y = e^{-x^2}$ ($-\infty < x < \infty$) generates a surface of finite area when rotated about the x-axis. What does this imply about the location of the centroid of this infinitely long curve?

29. Obtain formulas for the coordinates of the centroid of the plane region $c \leq y \leq d$, $0 < f(y) \leq x \leq g(y)$.

30. Prove part (b) of Pappus's Theorem (Theorem 2).

31. **(Stability of a floating object)** Determining the orientation that a floating object will assume is a problem of critical importance to ship designers. Boats must be designed to float stably in an upright position; if the boat tilts somewhat from upright, the forces on it must be such as to right it again. The two forces on a floating object that need to be taken into account are its weight $\mathbf{W}$ and the balancing buoyant force $\mathbf{B} = -\mathbf{W}$. The weight $\mathbf{W}$ must be treated for mechanical purposes as being applied at the centre of mass (CM) of the object. The buoyant force, however, acts at the *centre of buoyancy* (CB), which is the centre of mass of the water displaced by the object, and is therefore the centroid of the "hole in the water" made by the object.

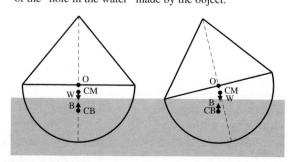

Figure 7.42

For example, consider a channel marker buoy consisting of a hemispherical hull surmounted by a conical tower supporting a navigation light. The buoy has a vertical axis of symmetry. If it is upright, both the CM and the CB lie on this line, as shown in Figure 7.42(left).

Is this upright flotation of the buoy stable? It is if the CM lies below the centre O of the hemispherical hull, as shown in the figure. To see why, imagine the buoy tilted slightly from the vertical as shown in the right figure. Observe that the CM still lies on the axis of symmetry of the buoy, but the CB lies on the vertical line through O. The forces **W** and **B** no longer act along the same line, but their torques are such as to rotate the buoy back to a vertical upright position. If CM had been above O in the left figure, the torques would have been such as to tip the buoy over once it was displaced even slightly from the vertical.

A wooden beam has a square cross-section and specific gravity 0.5, so that it will float with half of its volume submerged. (See Figure 7.43.) Assuming it will float horizontally in the water, what is the stable orientation of the square cross section with respect to the surface of the water? In particular, will the beam float with a flat face upward, or an edge upward? Prove your assertions. You may find

Maple or another symbolic algebra program useful.

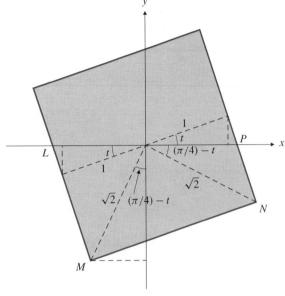

Figure 7.43

7.6 Other Physical Applications

In this section we present some examples of the use of integration to calculate quantities arising in physics and mechanics.

Hydrostatic Pressure

The **pressure** p at depth h beneath the surface of a liquid is the *force per unit area* exerted on a horizontal plane surface at that depth due to the weight of the liquid above it. Hence p is given by

$$p = \delta g h,$$

where δ is the density of the liquid, and g is the acceleration produced by gravity where the fluid is located. (See Figure 7.44.) For water at the surface of the earth we have, approximately, $\delta = 1{,}000$ kg/m^3 and $g = 9.8$ m/s^2, so the pressure at depth h m is

$$p = 9{,}800h \text{ N/m}^2.$$

The unit of force used here is the newton (N); 1 N = 1 kg·m/s^2, the force that imparts an acceleration of 1 m/s^2 to a mass of 1 kg.

The molecules in a liquid interact in such a way that the pressure at any depth acts equally in all directions; the pressure against a vertical surface is the same as that against a horizontal surface at the same depth. This is **Pascal's principle**.

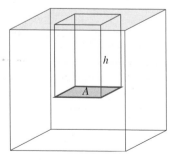

Figure 7.44   The volume of liquid above the area A is $V = Ah$. The weight of this liquid is $\delta V g = \delta g h A$, so the pressure (force per unit area) at depth h is $p = \delta g h$

The total force exerted by a liquid on a horizontal surface (say, the bottom of a tank holding the liquid) is found by multiplying the area of that surface by the pressure at the depth of the surface below the top of the liquid. For nonhorizontal surfaces, however, the pressure is not constant over the whole surface, and the total force cannot be determined so easily. In this case we divide the surface into area elements dA, each at some particular depth h, and we then sum (i.e., integrate) the corresponding force elements $dF = \delta gh \, dA$ to find the total force.

Example 1 One vertical wall of a water trough is a semicircular plate of radius R m with curved edge downward. If the trough is full, so that the water comes up to the top of the plate, find the total force of the water on the plate.

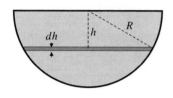

Figure 7.45

Solution A horizontal strip of the surface of the plate at depth h m and having width dh m (see Figure 7.45) has length $2\sqrt{R^2 - h^2}$ m; hence, its area is $dA = 2\sqrt{R^2 - h^2} \, dh$ m². The force of the water on this strip is

$$dF = \delta gh \, dA = 2\delta gh\sqrt{R^2 - h^2} \, dh.$$

Thus, the total force on the plate is

$$F = \int_{h=0}^{h=R} dF = 2\delta g \int_0^R h\sqrt{R^2 - h^2} \, dh \qquad \text{Let } u = R^2 - h^2,$$
$$du = -2h \, dh.$$

$$= \delta g \int_0^{R^2} u^{1/2} \, du = \delta g \, \frac{2}{3}u^{3/2}\Big|_0^{R^2}$$

$$\approx \frac{2}{3} \times 9{,}800R^3 \approx 6{,}533R^3 \text{ N.} \qquad \blacksquare$$

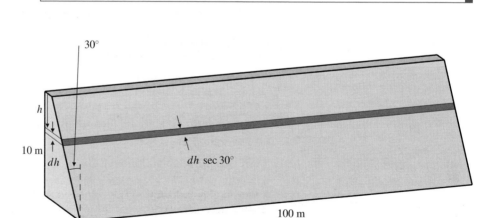

Figure 7.46

Example 2 (Force on a dam) Find the total force on a section of a dam 100 m long and having a vertical height of 10 m, if the surface holding back the water is inclined at an angle of 30° to the vertical and the water comes up to the top of the dam.

Solution   The water in a horizontal layer of thickness dh m at depth h m makes contact with the dam along a slanted strip of width $dh \sec 30° = (2/\sqrt{3})\,dh$ m. (See Figure 7.46.) The area of this strip is $dA = (200/\sqrt{3})\,dh$ m^2, and the force of water against the strip is

$$dF = \delta g h\, dA = \frac{200}{\sqrt{3}} \times 1{,}000 \times 9.8 h\, dh \approx 1{,}131{,}600 h\, dh \text{ N}.$$

The total force on the dam section is therefore

$$F \approx 1{,}131{,}600 \int_0^{10} h\, dh = 1{,}131{,}600 \times \frac{10^2}{2} \approx 5.658 \times 10^7 \text{ N}.$$

Work

When a force acts on an object to move that object, it is said to have done **work** on the object. The amount of work done by a constant force is measured by the product of the force and the distance through which it moves the object. This assumes that the force is in the direction of the motion.

$$\text{Work} = \text{Force} \times \text{Distance}$$

Work is always related to a particular force. If other forces acting on an object cause it to move in a direction opposite to the force F, then work is said to have been done *against* the force F.

Suppose that a force in the direction of the x-axis moves an object from $x = a$ to $x = b$ on that axis and that the force varies continuously with the position x of the object; that is, $F = F(x)$ is a continuous function. The element of work done by the force in moving the object through a very short distance from x to $x + dx$ is $dW = F(x)\,dx$, so the total work done by the force is

$$W = \int_{x=a}^{x=b} dW = \int_a^b F(x)\,dx.$$

Example 3   (Stretching or compressing a spring) By **Hooke's Law**, the force $F(x)$ required to extend (or compress) an elastic spring to x units longer (or shorter) than its natural length is proportional to x:

$$F(x) = kx,$$

where k is the **spring constant** for the particular spring. If a force of 2,000 N is required to extend a certain spring to 4 cm longer than its natural length, how much work must be done to extend it that far?

Solution   Since $F(x) = kx = 2{,}000$ N when $x = 4$ cm, we must have $k = 2{,}000/4 = 500$ N/cm. The work done in extending the spring 4 cm is

$$W = \int_0^4 kx\, dx = k \qquad \text{N} \times \frac{4^2 \text{ cm}^2}{2} = 4{,}000 \text{ N·cm} = 40 \text{ N·m}.$$

40 newton-metres (joul   the spring 4 cm.

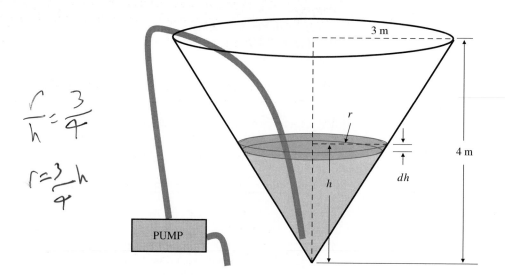

$$\frac{r}{h} = \frac{3}{4}$$

$$r = \frac{3}{4} h$$

Figure 7.47

Example 4 **(Work done to pump out a tank)** Water fills a tank in the shape of a right-circular cone with top radius 3 m and depth 4 m. How much work must be done (against gravity) to pump all the water out of the tank over the top edge of the tank?

Solution A thin, disk-shaped slice of water at height h above the vertex of the tank has radius r (see Figure 7.47), where $r = \frac{3}{4} h$ by similar triangles. The volume of this slice is

$$dV = \pi r^2 \, dh = \frac{9}{16} \pi h^2 \, dh,$$

and its *weight* (the force of gravity on the mass of water in the slice) is

$$dF = \delta g \, dV = \frac{9}{16} \delta g \, \pi h^2 \, dh.$$

The water in this disk must be raised (against gravity) a distance $(4 - h)$ m by the pump. The work required to do this is

$$dW = \frac{9}{16} \delta g \, \pi (4 - h) h^2 \, dh.$$

The total work that must be done to empty the tank is the sum (integral) of all these elements of work for disks at depths between 0 and 4 m:

$$W = \int_0^4 \frac{9}{16} \delta g \, \pi (4h^2 - h^3) \, dh$$

$$= \frac{9}{16} \delta g \, \pi \left(\frac{4h^3}{3} - \frac{h^4}{4} \right) \Big|_0^4$$

$$000 \times 9.8 \times \frac{64}{3} \approx 3.69 \times 10^5 \text{ N·m.}$$

Example 5 (**Work to raise material into orbit**) The gravitational force of the earth on a mass m located at height h above the surface of the earth is given by

$$F(h) = \frac{Km}{(R+h)^2},$$

where R is the radius of the earth and K is a constant that is independent of m and h. Determine, in terms of K and R, the work that must be done against gravity to raise an object from the surface of the earth to:

(a) a height H above the surface of the earth, and

(b) an infinite height above the surface of the earth.

Solution The work done to raise the mass m from height h to height $h + dh$ is

$$dW = \frac{Km}{(R+h)^2}\, dh.$$

(a) The total work to raise it from height $h = 0$ to height $h = H$ is

$$W = \int_0^H \frac{Km}{(R+h)^2}\, dh = \frac{-Km}{R+h}\Big|_0^H = Km\left(\frac{1}{R} - \frac{1}{R+H}\right).$$

If R and H are measured in metres and F is measured in newtons, then W is measured in newton-metres (N·m), or joules.

(b) The total work necessary to raise the mass m to an infinite height is

$$W = \int_0^\infty \frac{Km}{(R+h)^2}\, dh = \lim_{H\to\infty} Km\left(\frac{1}{R} - \frac{1}{R+H}\right) = \frac{Km}{R}.$$

Potential and Kinetic Energy

The units of work (force × distance) are the same as those of energy. Work done against a force may be regarded as storing up energy for future use or for conversion to other forms. Such stored energy is called **potential energy** (P.E.). For instance, in extending or compressing an elastic spring, we are doing work against the tension in the spring and hence storing energy in the spring. When work is done against a (variable) force $F(x)$ to move an object from $x = a$ to $x = b$, the energy stored is

$$\text{P.E.} = -\int_a^b F(x)\, dx.$$

Since the work is being done against F, the signs of $F(x)$ and $b - a$ are opposite, so the integral is negative; the explicit negative sign is included so that the calculated potential energy will be positive.

One of the forms of energy into which potential energy can be converted is **kinetic energy** (K.E.), the energy of motion. If an object of mass m is moving with velocity v, it has kinetic energy

$$\text{K.E.} = \frac{1}{2}m v^2.$$

For example, if an object is raised and then dropped, it accelerates downward under gravity as more and more of the potential energy stored in it when it was raised is converted to kinetic energy.

Consider the change in potential energy stored in a mass m as it moves along the x-axis from a to b under the influence of a force $F(x)$ depending only on x:

$$\text{P.E.}(b) - \text{P.E.}(a) = -\int_a^b F(x)\,dx.$$

(The change in P.E. is negative if m is moving in the direction of F.) According to Newton's second law of motion, the force $F(x)$ causes the mass m to accelerate, with acceleration dv/dt given by

$$F(x) = m\frac{dv}{dt} \qquad (\text{force} = \text{mass} \times \text{acceleration}).$$

By the Chain Rule we can rewrite dv/dt in the form

$$\frac{dv}{dt} = \frac{dv}{dx}\frac{dx}{dt} = v\frac{dv}{dx},$$

so $F(x) = mv\dfrac{dv}{dx}$. Hence,

$$\begin{aligned}
\text{P.E.}(b) - \text{P.E.}(a) &= -\int_a^b mv\frac{dv}{dx}\,dx \\
&= -m\int_{x=a}^{x=b} v\,dv \\
&= -\frac{1}{2}mv^2\bigg|_{x=a}^{x=b} \\
&= \text{K.E.}(a) - \text{K.E.}(b).
\end{aligned}$$

It follows that

$$\text{P.E.}(b) + \text{K.E.}(b) = \text{P.E.}(a) + \text{K.E.}(a).$$

This shows that the total energy (potential + kinetic) remains constant as the mass m moves under the influence of a force F, *depending only on position*. Such a force is said to be **conservative**, and the above result is called the **law of conservation of energy**.

Example 6 **(Escape velocity)** Use the result of Example 5 together with the following known values,

(a) the radius R of the earth is about $6{,}400$ km, or 6.4×10^6 m,

(b) the acceleration of gravity g at the surface of the earth is about 9.8 m/s^2,

to determine the constant K in the gravitational force formula of Example 5, and hence to determine the escape velocity for a projectile fired vertically from the surface of the earth. The **escape velocity** is the (minimum) speed that such a projectile must have at firing to ensure that it will continue to move farther and farther away from the earth and not fall back.

Solution   According to the formula of Example 5, the force of gravity on a mass m kg at the surface of the earth ($h = 0$) is

$$F = \frac{Km}{(R+0)^2} = \frac{Km}{R^2}.$$

According to Newton's second law of motion, this force is related to the acceleration of gravity (g) there by the equation $F = mg$. Thus,

$$\frac{Km}{R^2} = mg \qquad \text{and} \quad K = gR^2.$$

According to the law of conservation of energy, the projectile must have sufficient kinetic energy at firing to do the work necessary to raise the mass m to infinite height. By the result of Example 5, this required energy is Km/R. If the initial velocity of the projectile is v, we want

$$\frac{1}{2}mv^2 \geq \frac{Km}{R}.$$

Thus v must satisfy

$$v \geq \sqrt{\frac{2K}{R}} = \sqrt{2gR} \approx \sqrt{2 \times 9.8 \times 6.4 \times 10^6} \approx 1.12 \times 10^4 \text{ m/s}.$$

Thus, the escape velocity is approximately 11.2 km/s and is independent of the mass m. In this calculation we have neglected any air resistance near the surface of the earth. Such resistance depends on velocity rather than on position, so it is not a conservative force. The effect of such resistance would be to use up (convert to heat) some of the initial kinetic energy and so raise the escape velocity. ∎

Exercises 7.6

1. A tank has a square base 2 m on each side and vertical sides 6 m high. If the tank is filled with water, find the total force exerted by the water (a) on the bottom of the tank and (b) on one of the four vertical walls of the tank.

2. A swimming pool 20 m long and 8 m wide has a sloping plane bottom so that the depth of the pool is 1 m at one end and 3 m at the other end. Find the total force exerted on the bottom if the pool is full of water.

3. A dam 200 m long and 24 m high presents a sloping face of 26 m slant height to the water in a reservoir behind the dam (Figure 7.48). If the surface of the water is level with the top of the dam, what is the total force of the water on the dam?

4. A pyramid with a square base, 4 m on each side and four equilateral triangular faces, sits on the level bottom of a lake at a place where the lake is 10 m deep. Find the total force of the water on each of the triangular faces.

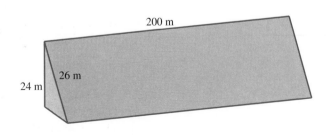

Figure 7.48

5. A lock on a canal has a gate in the shape of a vertical rectangle 5 m wide and 20 m high. If the water on one side of the gate comes up to the top of the gate, and the water on the other side comes only 6 m up the gate, find the total force that must be exerted to hold the gate in place.

6. If 100 N·cm of work must be done to compress an elastic spring to 3 cm shorter than its natural length, how much work must be done to compress it 1 cm further?

7. Find the total work that must be done to pump all the water in the tank of Exercise 1 out over the top of the tank.

8. Find the total work that must be done to pump all the water in the swimming pool of Exercise 2 out over the top edge of the pool.

9. Find the work that must be done to pump all the water in a full hemispherical bowl of radius a m to a height h m above the top of the bowl.

* **10.** A bucket is raised vertically from ground level at a constant speed of 2 m/min by a winch. If the bucket weighs 1 kg and contains 15 kg of water when it starts up but loses water by leakage at a rate of 1 kg/min thereafter, how much work must be done by the winch to raise the bucket to a height of 10 m?

7.7 Applications in Business, Finance, and Ecology

If the rate of change $f'(x)$ of a function $f(x)$ is known, the change in value of the function over an interval from $x = a$ to $x = b$ is just the integral of f' over $[a, b]$:

$$f(b) - f(a) = \int_a^b f'(x)\, dx.$$

For example, if the speed of a moving car at time t is $v(t)$ km/h, then the distance travelled by the car during the time interval $[0, T]$ (hours) is $\int_0^T v(t)\, dt$ km.

Similar situations arise naturally in business and economics, where the rates of change are often called marginals.

Example 1   (**Finding total revenue from marginal revenue**)   A supplier of calculators realizes a marginal revenue of $15 - 5e^{-x/50}$ per calculator when she has sold x calculators. What will be her total revenue from the sale of 100 calculators?

Solution   The marginal revenue is the rate of change of revenue with respect to the number of calculators sold. Thus, the revenue from the sale of dx calculators after x have already been sold is

$$dR = (15 - 5e^{-x/50})\, dx$$

dollars. The total revenue from the sale of the first 100 calculators is R, where

$$\begin{aligned} R = \int_{x=0}^{x=100} dR &= \int_0^{100} (15 - 5e^{-x/50})\, dx \\ &= \left. \left(15x + 250e^{-x/50}\right)\right|_0^{100} \\ &= 1{,}500 + 250e^{-2} - 250 \approx 1{,}283.83, \end{aligned}$$

that is, about \$1,284.

The Present Value of a Stream of Payments

Suppose that you have a business that generates income continuously at a variable rate $P(t)$ dollars per year at time t and that you expect this income to continue for the next T years. How much is the business worth today?

The answer surely depends on interest rates. One dollar to be received t years from now is worth less than one dollar received today, which could be invested at interest to yield more than one dollar t years from now. The higher the interest rate, the lower the value today of a payment that is not due until sometime in the future.

To analyze this situation, suppose that the nominal interest rate is $r\%$ per annum, but is compounded continuously. Let $\delta = r/100$. As shown in Section 3.4, an investment of $1 today will grow to

$$\lim_{n \to \infty} \left(1 + \frac{\delta}{n}\right)^{nt} = e^{\delta t}$$

dollars after t years. Therefore, a payment of $1 after t years must be worth only $\$e^{-\delta t}$ today. This is called the *present value* of the future payment. When viewed this way, the interest rate δ is frequently called a *discount rate*; it represents the amount by which future payments are discounted.

Returning to the business income problem, in the short time interval from t to $t + dt$, the business produces income $\$P(t)\,dt$, of which the present value is $\$e^{-\delta t} P(t)\,dt$. Therefore, the present value $\$V$ of the income stream over the time interval $[0, T]$ is the "sum" of these contributions:

$$V = \int_0^T e^{-\delta t} P(t)\,dt.$$

Example 2   What is the present value of a constant, continual stream of payments at a rate of $10,000 per year, to continue forever, starting now? Assume an interest rate of 6% per annum, compounded continuously.

Solution   The required present value is

$$V = \int_0^\infty e^{-0.06t} 10,000\,dt = 10,000 \lim_{R \to \infty} \left. \frac{e^{-0.06t}}{-0.06} \right|_0^R \approx \$166,667.$$

The Economics of Exploiting Renewable Resources

As noted in Section 3.4, the rate of increase of a biological population sometimes conforms to a logistic model[1]

$$\frac{dx}{dt} = kx\left(1 - \frac{x}{L}\right).$$

Here, $x = x(t)$ is the size (or biomass) of the population at time t, k is the natural rate at which the population would grow if its food supply were unlimited, and L is the natural limiting size of the population—the carrying capacity of its environment. Such models are thought to apply, for example, to the Antarctic blue whale and to several species of fish and trees. If the resource is harvested (say, the fish are caught) at a rate $h(t)$ units per year at time t, then the population grows at a slower rate:

$$\frac{dx}{dt} = kx\left(1 - \frac{x}{L}\right) - h(t). \tag{$*$}$$

---

[1]   This example was suggested by Professor C. W. Clark, of the University of British Columbia.

In particular, if we harvest the population at its current rate of growth,

$$h(t) = kx \left(1 - \frac{x}{L}\right),$$

then $dx/dt = 0$, and the population will maintain a constant size. Assume that each unit of harvest produces an income of $\$p$ for the fishing industry. The total annual income from harvesting the resource at its current rate of growth will be

$$T = ph(t) = pkx \left(1 - \frac{x}{L}\right).$$

Considered as a function of x, this total annual income is quadratic and has a maximum value when $x = L/2$, the value that ensures $dT/dx = 0$. The industry can maintain a stable maximum annual income by ensuring that the population level remains at half the maximal size of the population with no harvesting.

The analysis above, however, does not take into account the discounted value of future harvests. If the discount rate is δ, compounded continuously, then the present value of the income $\$ph(t)\,dt$ due between t and $t + dt$ years from now is $e^{-\delta t} ph(t)\,dt$. The total present value of all income from the fishery in future years is

$$T = \int_0^\infty e^{-\delta t} ph(t)\,dt.$$

What fishing strategy will maximize T? If we substitute for $h(t)$ from equation (∗) governing the growth rate of the population, we get

$$T = \int_0^\infty pe^{-\delta t}\left[kx\left(1 - \frac{x}{L}\right) - \frac{dx}{dt}\right]\,dt$$

$$= \int_0^\infty kpe^{-\delta t}x\left(1 - \frac{x}{L}\right)\,dt - \int_0^\infty pe^{-\delta t}\frac{dx}{dt}\,dt.$$

Integrate by parts in the last integral above, taking $U = pe^{-\delta t}$ and $dV = \dfrac{dx}{dt}\,dt$:

$$T = \int_0^\infty kpe^{-\delta t}x\left(1 - \frac{x}{L}\right)\,dt - \left[pe^{-\delta t}x\Big|_0^\infty + \int_0^\infty p\delta e^{-\delta t}x\,dt\right]$$

$$= px(0) + \int_0^\infty pe^{-\delta t}\left[kx\left(1 - \frac{x}{L}\right) - \delta x\right]\,dt.$$

To make this expression as large as possible, we should choose the population size x to maximize the quadratic expression

$$Q(x) = kx\left(1 - \frac{x}{L}\right) - \delta x$$

at as early a time t as possible, and keep the population size constant at that level thereafter. The maximum occurs where $Q'(x) = k - (2kx/L) - \delta = 0$, that is, where

$$x = \frac{L}{2} - \frac{\delta L}{2k} = (k - \delta)\frac{L}{2k}.$$

The maximum present value of the fishery is realized if the population level x is held at this value. Note that this population level is smaller than the optimal level $L/2$ we obtained by ignoring the discount rate. The higher the discount rate δ, the smaller will be the income-maximizing population level. More unfortunately, if $\delta \geq k$, the model predicts greatest income from fishing the species to *extinction* immediately! (See Figure 7.49.)

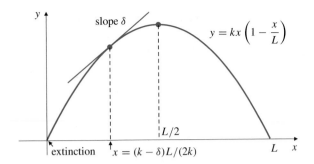

Figure 7.49   The greater the discount rate δ, the smaller the population size x that will maximize the present value of future income from harvesting. If $\delta \geq k$, the model predicts fishing the species to extinction

Of course, this model fails to take into consideration other factors that may affect the fishing strategy, such as the increased cost of harvesting when the population level is small and the effect of competition among various parts of the fishing industry. Nevertheless, it does explain the regrettable fact that, under some circumstances, an industry based on a renewable resource can find it in its best interest to destroy the resource. This is especially likely to happen when the natural growth rate k of the resource is low, as it is for the case of whales and most trees. There is good reason not to allow economics alone to dictate the management of the resource.

Exercises 7.7

1. **(Cost of production)** The marginal cost of production in a coal mine is $\$6 - 2 \times 10^{-3}x + 6 \times 10^{-6}x^2$ per ton after the first x tons are produced each day. In addition, there is a fixed cost of $\$4,000$ per day to open the mine. Find the total cost of production on a day when 1,000 tons are produced.

2. **(Total sales)** The sales of a new computer chip are modelled by $s(t) = te^{-t/10}$, where $s(t)$ is the number of thousands of chips sold per week, t weeks after the chip was introduced to the market. How many chips were sold in the first year?

3. **(Internet connection rates)** An internet service provider charges clients at a continuously decreasing marginal rate of $\$4/(1 + \sqrt{t})$ per hour when the client has already used t hours during a month. How much will be billed to a client who uses x hours in a month? (x need not be an integer.)

4. **(Total revenue from declining sales)** The price per kilogram of maple syrup in a store rises at a constant rate from $\$10$ at the beginning of the year to $\$15$ at the end of the year. As the price rises, the quantity sold decreases; the sales rate is $400/(1 + 0.1t)$ kg/year at time t years, $(0 \leq t \leq 1)$. What total revenue does the store obtain from sales of the syrup during the year?

(Stream of payment problems) Find the present value of a continuous stream of payments of $\$1,000$ per year for the periods and discount rates given in Exercises 5–10. In each case the discount rate is compounded continuously.

5. 10 years at a discount rate of 2%

6. 10 years at a discount rate of 5%

7. 10 years beginning 2 years from now at a discount rate of 8%

8. 25 years beginning 10 years from now at a discount rate of 5%

9. For all future time at a discount rate of 2%

10. Beginning in 10 years and continuing forever after at a discount rate of 5%

11. Find the present value of a continuous stream of payments over a 10-year period beginning at a rate of $\$1,000$ per year now and increasing steadily at $\$100$ per year. The discount rate is 5%.

12. Find the present value of a continuous stream of payments over a 10-year period beginning at a rate of $\$1,000$ per year now and increasing steadily at 10% per year. The discount rate is 5%.

13. Money flows continuously into an account at a rate of $\$5,000$ per year. If the account earns interest at a rate of 5% compounded continuously, how much will be in the account after 10 years?

14. Money flows continuously into an account beginning at a rate of $\$5,000$ per year and increasing at 10% per year. Interest causes the account to grow at a real rate of 6% (so that $\$1$ grows to $\$1.06^t$ in t years). How long will it take for the balance in the account to reach $\$1,000,000$?

15. If the discount rate δ varies with time, say $\delta = \delta(t)$, show that the present value of a payment of $\$P$ due t years from now is $\$Pe^{-\lambda(t)}$, where

$$\lambda(t) = \int_0^t \delta(\tau)\,d\tau.$$

What is the value of a stream of payments due at a rate $P(t)$ at time t, from $t = 0$ to $t = T$?

16. **(Discount rates and population models)** Suppose that the growth rate of a population is a function of the population size: $dx/dt = F(x)$. (For the logistic model, $F(x) = kx(1 - (x/L))$.) If the population is harvested at rate $h(t)$ at time t, then $x(t)$ satisfies

$$\frac{dx}{dt} = F(x) - h(t).$$

Show that the value of x that maximizes the present value of all future harvests satisfies $F'(x) = \delta$, where δ is the (continuously compounded) discount rate. *Hint:* mimic the argument used above for the logistic case.

17. **(Managing a fishery)** The carrying capacity of a certain lake is $L = 80,000$ of a certain species of fish. The natural growth rate of this species is 12% per year ($k = 0.12$). Each fish is worth $6. The discount rate is 5%. What population of fish should be maintained in the lake to maximize the present value of all future revenue from harvesting the fish? What is the annual revenue resulting from maintaining this population level?

18. **(Blue whales)** It is speculated that the natural growth rate of the Antarctic blue whale population is about 2% per year ($k = 0.02$) and that the carrying capacity of its habitat is about $L = 150,000$. One blue whale is worth, on average, $10,000. Assuming that the blue whale population satisfies a logistic model, and using the data above, find the following:

(a) the maximum sustainable annual harvest of blue whales.

(b) the annual revenue resulting from the maximum annual sustainable harvest.

(c) the annual interest generated if the whale population (assumed to be at the level $L/2$ supporting the maximum sustainable harvest) is exterminated and the proceeds invested at 2%. (d) at 5%.

(e) the total present value of all future revenue if the population is maintained at the level $L/2$ and the discount rate is 5%.

* 19. The model developed above does not allow for the costs of harvesting. Try to devise a way to alter the model to take this into account. Typically, the cost of catching a fish goes up as the number of fish goes down.

7.8 Probability

Probability theory is a very important field of application of the definite integral. This subject cannot, of course, be developed thoroughly here—an adequate presentation requires one or more whole courses—but we can give a brief introduction that suggests some of the ways integrals are used in probability theory.

The **probability** of an event occurring is a real number between 0 and 1 that measures the proportion of times the event can be expected to occur in a large number of trials. If the occurrence of an event is certain, its probability is 1; if the event cannot possibly occur, its probability is 0. For example, the probability that a tossed coin will land heads is 1/2 because we would expect it to land heads about half the time if it were tossed a great many times. In such a tossing of a coin there are only two possible outcomes, heads or tails, each equally likely, that is, each having probability 1/2. (We are assuming the coin won't ever land standing on its edge.) For any toss, let $X = 0$ if the outcome is heads, and let $X = 1$ if the outcome is tails. X is called a **discrete random variable**. The probability that $X = 0$ is 1/2 and the probability that $X = 1$ is 1/2, so we write

$$\Pr(X = 0) = \frac{1}{2} \quad \text{and} \quad \Pr(X = 1) = \frac{1}{2}.$$

Note that $\Pr(X = 0) + \Pr(X = 1) = 1$, since it is certain that the coin will land either heads or tails.

Example 1   A single die is rolled so that it will show one of the numbers 1 to 6 on top when it stops. If X denotes the number showing on any roll, then X is a discrete random variable. Assuming no one value of X is any more likely than any other, the probability that the number showing is x must be 1/6 for each possible value of x; that is,

$$\Pr(X = x) = \frac{1}{6} \quad \text{for each } x \text{ in } \{1, 2, 3, 4, 5, 6\}.$$

The discrete random variable X is therefore said to be distributed **uniformly**. Again we note that

$$\sum_{n=1}^{6} \Pr(X = n) = 1,$$

reflecting the fact that the rolled die must certainly give one of the six possible outcomes. The probability that a roll will produce a value from 1 to 4 is

$$\Pr(1 \le X \le 4) = \sum_{n=1}^{4} \Pr(X = n) = \frac{1}{6} + \frac{1}{6} + \frac{1}{6} + \frac{1}{6} = \frac{2}{3}.$$

Now we consider an example with a continuous range of possible outcomes.

Figure 7.50

(a) X is the number of degrees in the acute angle the needle makes with the line

(b) The probability density function f of the random variable X

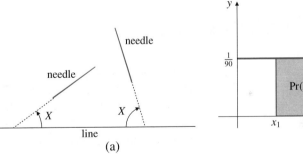
(a)

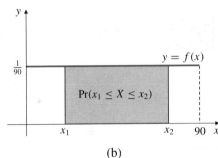
(b)

Example 2   Suppose that a needle is dropped at random on a flat table with a straight line drawn on it. For each drop, let X be the number of degrees in the (acute) angle that the needle makes with the line. (See Figure 7.50(a).) Evidently X can take any real value in the interval $[0, 90]$; therefore X is called a **continuous random variable**. The probability that X takes on any particular real value is 0. (There are infinitely many real numbers in $[0, 90]$ and none is more likely than any other.) However, the probability that X lies in some interval, say $[10, 20]$, is the same as the probability that it lies in any other interval of the same length. Since the interval has length 10 and the interval of all possible values of X has length 90, this probability is

$$\Pr(10 \le X \le 20) = \frac{10}{90} = \frac{1}{9}.$$

More generally, if $0 \le x_1 \le x_2 \le 90$, then

$$\Pr(x_1 \leq X \leq x_2) = \frac{1}{90}(x_2 - x_1).$$

This situation can be conveniently represented as follows: Let $f(x)$ be defined on the interval [0, 90], taking at each point the constant value 1/90:

$$f(x) = \frac{1}{90}, \qquad 0 \leq x \leq 90.$$

The area under the graph of f is 1, and $\Pr(x_1 \leq X \leq x_2)$ is equal to the area under that part of the graph lying over the interval $[x_1, x_2]$. (See Figure 7.50(b).) The function $f(x)$ is called the **probability density function** for the random variable X. Since $f(x)$ is constant on its domain, X is said to be **uniformly distributed**.

DEFINITION 2

Probability density functions

A function defined on an interval $[a, b]$ is a probability density function for a continuous random variable X distributed on $[a, b]$ if, whenever x_1 and x_2 satisfy $a \leq x_1 \leq x_2 \leq b$, we have

$$\Pr(x_1 \leq X \leq x_2) = \int_{x_1}^{x_2} f(x)\, dx.$$

In order to be such a probability density function, f must satisfy two conditions:

(a) $f(x) \geq 0$ on $[a, b]$ (probability cannot be negative)   and

(b) $\int_a^b f(x)\, dx = 1$ ($\Pr(a \leq X \leq b) = 1$).

These ideas extend to random variables distributed on semi-infinite or infinite intervals, but the integrals appearing will be improper in those cases.

In the example of the dropping needle, the probability density function has a horizontal straight line graph, and we termed such a probability distribution uniform. The uniform probability density function on the interval $[a, b]$ is

$$f(x) = \begin{cases} \dfrac{1}{b - a} & \text{if } a \leq x \leq b \\ 0 & \text{otherwise.} \end{cases}$$

Many other functions are commonly encountered as density functions for continuous random variables.

Example 3   (**The exponential distribution**)   The length of time T that any particular atom in a radioactive sample survives before decaying is a random variable taking values in $[0, \infty[$. It has been observed that the proportion of atoms that survive to time t becomes small exponentially as t increases; thus

$$\Pr(T \geq t) = Ce^{-kt}.$$

Let f be the probability density function for the random variable T. Then

$$\int_t^\infty f(x)\,dx = \Pr(T \geq t) = Ce^{-kt}.$$

Differentiating this equation with respect to t (using the Fundamental Theorem of Calculus), we obtain $-f(t) = -Cke^{-kt}$, so $f(t) = Cke^{-kt}$. C is determined by the requirement that $\int_0^\infty f(t)\,dt = 1$. We have

$$1 = Ck \int_0^\infty e^{-kt}\,dt = \lim_{R\to\infty} Ck \int_0^R e^{-kt}\,dt = -C \lim_{R\to\infty} (e^{-kR} - 1) = C.$$

Thus $C = 1$ and $f(t) = ke^{-kt}$. Note that $\Pr(T \geq (\ln 2)/k) = e^{-k(\ln 2)/k} = 1/2$, reflecting the fact that the half-life of such a radioactive sample is $(\ln 2)/k$.

Example 4   For what value of C is $f(x) = C(1 - x^2)$ a probability density function on $[-1, 1]$? If X is a random variable with this density what is the probability that $X \leq 1/2$?

Solution   Observe that $f(x) \geq 0$ on $[-1, 1]$ if $C \geq 0$. Since

$$\int_{-1}^1 f(x)\,dx = C \int_{-1}^1 (1 - x^2)\,dx = 2C\left(x - \frac{x^3}{3}\right)\Big|_0^1 = \frac{4C}{3},$$

$f(x)$ will be a probability density function if $C = 3/4$. In this case

$$\Pr\left(X \leq \frac{1}{2}\right) = \frac{3}{4} \int_{-1}^{1/2} (1 - x^2)\,dx = \frac{3}{4}\left(x - \frac{x^3}{3}\right)\Big|_{-1}^{1/2}$$

$$= \frac{3}{4}\left(\frac{1}{2} - \frac{1}{24} - (-1) + \frac{-1}{3}\right) = \frac{27}{32}.$$

Expectation, Mean, Variance, and Standard Deviation

Consider a simple gambling game in which the player pays the house C dollars for the privilege of rolling a single die and in which he wins X dollars, where X is the number showing on top of the rolled die. In each game the possible winnings are 1, 2, 3, 4, 5, or 6 dollars, each with probability 1/6. In n games the player can expect to win about $n/6 + 2n/6 + 3n/6 + 4n/6 + 5n/6 + 6n/6 = 21n/6 = 7n/2$ dollars, so that his expected *average winnings per game* are 7/2 dollars, \$3.50. If $C > 3.5$, the player can expect, on average, to lose money. The amount 3.5 is called the **expectation**, or **mean**, of the discrete random variable X. The mean is usually denoted by μ, the Greek letter *mu* (pronounced "mew").

In general, if a random variable can take on values x_1 with probability p_1, x_2 with probability p_2, ..., and x_n with probability p_n (where $p_1 + p_2 + \cdots + p_n = 1$), the mean μ or expectation $E(X)$ of that random variable X is given by

$$\mu = E(X) = \sum_{i=1}^n x_i p_i.$$

We formulate an analogous definition for the mean or expectation of a continuous random variable as follows:

Mean or expectation

If X is a continuous random variable on $[a, b]$ with probability density function $f(x)$, the **mean** (denoted μ), or **expectation** of X (denoted $E(X)$), is

$$\mu = E(X) = \int_a^b x f(x)\, dx.$$

Note that in this usage $E(X)$ does not define a function of X but a constant (parameter) associated with the probability distribution of X. Note also that if $f(x)$ were a mass density such as that studied in Section 7.4, then μ would be the moment of the mass about 0 and, since the total mass would be $\int_a^b f(x)\, dx = 1$, μ would in fact be the centre of mass.

More generally, it can be shown that the **expectation** of any function $g(X)$ of the random variable X is

$$E(g(X)) = \int_a^b g(x) f(x)\, dx.$$

Variance

The **variance** of a random variable X with density $f(x)$ on $[a, b]$ is the expectation of the square of the distance of X from its mean μ. The variance is denoted σ^2 or $\text{Var}(X)$.

$$\sigma^2 = \text{Var}(X) = E((X - \mu)^2) = \int_a^b (x - \mu)^2 f(x)\, dx.$$

The symbol σ is the lowercase Greek letter *sigma*. (The symbol Σ used for summation is an uppercase sigma.) Since $\int_a^b f(x)\, dx = 1$, the expression above for the variance can be rewritten as follows:

$$\sigma^2 = \text{Var}(X) = \int_a^b x^2 f(x)\, dx - 2\mu \int_a^b x f(x)\, dx + \mu^2 \int_a^b f(x)\, dx$$

$$= \int_a^b x^2 f(x)\, dx - 2\mu^2 + \mu^2 = E(X^2) - \mu^2,$$

that is,

$$\sigma^2 = \text{Var}(X) = E(X^2) - \mu^2 = E(X^2) - (E(X))^2.$$

DEFINITION 5

Standard deviation

The **standard deviation** of the random variable X is σ, the square root of the variance. Thus, it is the square root of the mean square deviation of X from its mean:

$$\sigma = \left(\int_a^b (x - \mu)^2 f(x)\, dx \right)^{1/2} = \sqrt{E(X^2) - \mu^2}.$$

The standard deviation gives a measure of how spread out the probability distribution of X is. The smaller the standard deviation, the more concentrated is the area under the density curve around the mean, and so the smaller is the probability that a value of X will be far away from the mean. (See Figure 7.51.)

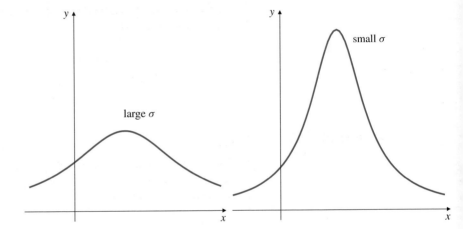

Figure 7.51 Densities with large and small standard deviations

Example 5 Find the mean μ and the standard deviation σ of a random variable X distributed uniformly on the interval $[a, b]$. Find $\Pr(\mu - \sigma \leq X \leq \mu + \sigma)$.

Solution The probability density function is $f(x) = 1/(b - a)$ on $[a, b]$, so the mean is given by

$$\mu = E(X) = \int_a^b \frac{x}{b - a}\, dx = \frac{1}{b - a} \frac{x^2}{2}\bigg|_a^b = \frac{1}{2} \frac{b^2 - a^2}{b - a} = \frac{b + a}{2}.$$

Hence, the mean is, as might have been anticipated, the midpoint of $[a, b]$. The expectation of X^2 is given by

$$E(X^2) = \int_a^b \frac{x^2}{b - a}\, dx = \frac{1}{b - a} \frac{x^3}{3}\bigg|_a^b = \frac{1}{3} \frac{b^3 - a^3}{b - a} = \frac{b^2 + ab + a^2}{3}.$$

Hence, the variance is

$$\sigma^2 = E(X^2) - \mu^2 = \frac{b^2 + ab + a^2}{3} - \frac{b^2 + 2ab + a^2}{4} = \frac{(b-a)^2}{12},$$

and the standard deviation is

$$\sigma = \frac{b-a}{2\sqrt{3}} \approx 0.29(b-a).$$

Finally,

$$\Pr(\mu - \sigma \le X \le \mu + \sigma) = \int_{\mu-\sigma}^{\mu+\sigma} \frac{dx}{b-a} = \frac{1}{b-a} \frac{2(b-a)}{2\sqrt{3}} = \frac{1}{\sqrt{3}} \approx 0.577.$$

∎

Example 6 Find the mean μ and the standard deviation σ of a random variable X distributed exponentially with density function $f(x) = ke^{-kx}$ on the interval $[0, \infty[$. Find $\Pr(\mu - \sigma \le X \le \mu + \sigma)$.

Solution We use integration by parts to find the mean:

$$\mu = E(X) = k \int_0^\infty xe^{-kx}\, dx$$

$$= \lim_{R \to \infty} k \int_0^R xe^{-kx}\, dx \qquad \text{Let} \quad U = x, \qquad dV = e^{-kx}\, dx.$$
$$\qquad\qquad\qquad\qquad\qquad\qquad \text{Then} \quad dU = dx, \qquad V = -e^{-kx}/k.$$

$$= \lim_{R \to \infty} \left(-xe^{-kx}\Big|_0^R + \int_0^R e^{-kx}\, dx \right)$$

$$= \lim_{R \to \infty} \left(-Re^{-kR} - \frac{1}{k}\left(e^{-kR} - 1\right) \right) = \frac{1}{k}, \qquad \text{since } k > 0.$$

Thus, the mean of the exponential distribution is $1/k$. This fact can be quite useful in determining the value of k for an exponentially distributed random variable. A similar integration by parts enables us to evaluate

$$E(X^2) = k \int_0^\infty x^2 e^{-kx}\, dx = 2 \int_0^\infty xe^{-kx}\, dx = \frac{2}{k^2},$$

so the variance of the exponential distribution is

$$\sigma^2 = E(X^2) - \mu^2 = \frac{1}{k^2},$$

and the standard deviation is equal to the mean

$$\sigma = \mu = \frac{1}{k}.$$

Now we have

$$\Pr(\mu - \sigma \le X \le \mu + \sigma) = \Pr(0 \le X \le 2/k)$$
$$= k \int_0^{2/k} e^{-kx} \, dx$$
$$= -e^{-kx} \Big|_0^{2/k}$$
$$= 1 - e^{-2} \approx 0.86,$$

which is independent of the value of k. Exponential densities for small and large values of k are graphed in Figure 7.52.

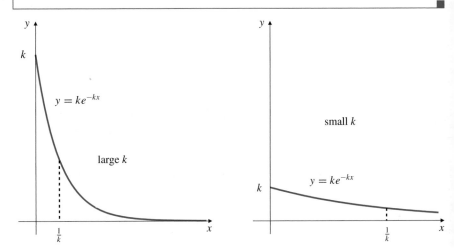

Figure 7.52   Exponential density functions

The Normal Distribution

The most important probability distributions are the so-called **normal** or **Gaussian** distributions. Such distributions govern the behaviour of many interesting random variables, in particular, those associated with random errors in measurements. There is a family of normal distributions, all related to the particular normal distribution called the **standard normal distribution**, which has the following probability density function:

DEFINITION 6

The standard normal probability density

$$f(z) = \frac{1}{\sqrt{2\pi}} e^{-z^2/2}, \qquad -\infty < z < \infty.$$

It is common to use z to denote the random variable in the standard normal distribution; the other normal distributions are obtained from this one by a change of variable. The graph of the standard normal density has a pleasant bell shape, as shown in Figure 7.53.

As we have noted previously, the function e^{-z^2} has no elementary antiderivative, so the improper integral

$$I = \int_{-\infty}^{\infty} e^{-z^2/2} \, dz$$

cannot be evaluated using the Fundamental Theorem of Calculus, although it is a convergent improper integral. The integral can be evaluated using techniques of

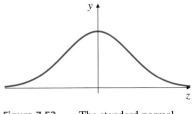

Figure 7.53   The standard normal density function $f(z) = \dfrac{1}{\sqrt{2\pi}} e^{-z^2/2}$

multivariable calculus involving double integrals of functions of two variables. (We do so in Section 14.4.) The value is $I = \sqrt{2\pi}$, which ensures that the above-defined standard normal density $f(z)$ is indeed a probability density function:

$$\int_{-\infty}^{\infty} f(z)\,dz = \frac{1}{\sqrt{2\pi}} \int_{-\infty}^{\infty} e^{-z^2/2}\,dz = 1.$$

Since $ze^{-z^2/2}$ is an odd function of z and its integral on $]-\infty, \infty[$ converges, the mean of the standard normal distribution is 0:

$$\mu = E(Z) = \frac{1}{\sqrt{2\pi}} \int_{-\infty}^{\infty} ze^{-z^2/2}\,dz = 0.$$

We calculate the variance of the standard normal distribution using integration by parts as follows:

$$
\begin{aligned}
\sigma^2 &= E(Z^2) \\
&= \frac{1}{\sqrt{2\pi}} \int_{-\infty}^{\infty} z^2 e^{-z^2/2}\,dz \\
&= \frac{1}{\sqrt{2\pi}} \lim_{R\to\infty} \int_{-R}^{R} z^2 e^{-z^2/2}\,dz \qquad \text{Let} \quad U = z, \quad dV = ze^{-z^2/2}\,dz. \\
&\qquad\qquad\qquad\qquad\qquad\qquad\qquad\quad \text{Then } dU = dz, \quad V = -e^{-z^2/2}. \\
&= \frac{1}{\sqrt{2\pi}} \lim_{R\to\infty} \left(-ze^{-z^2/2}\Big|_{-R}^{R} + \int_{-R}^{R} e^{-z^2/2}\,dz \right) \\
&= \frac{1}{\sqrt{2\pi}} \lim_{R\to\infty} (-2Re^{-R^2/2}) + \frac{1}{\sqrt{2\pi}} \int_{-\infty}^{\infty} e^{-z^2/2}\,dz \\
&= 0 + 1 = 1.
\end{aligned}
$$

Hence, the standard deviation of the standard normal distribution is 1.

Other normal distributions are obtained from the standard normal distribution by a change of variable.

DEFINITION 7

The general normal distribution

A random variable X on $]-\infty, \infty[$ is said to be *normally distributed with mean μ and standard deviation σ* (where μ is any real number and $\sigma > 0$) if its probability density function $f_{\mu,\sigma}$ is given in terms of the standard normal density f by

$$f_{\mu,\sigma}(x) = \frac{1}{\sigma} f\left(\frac{x - \mu}{\sigma}\right) = \frac{1}{\sigma\sqrt{2\pi}} e^{-(x-\mu)^2/(2\sigma^2)}.$$

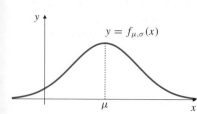

$y = f_{\mu,\sigma}(x)$

Figure 7.54   A general normal density with mean μ

(See Figure 7.54.) Using the change of variable $z = (x - \mu)/\sigma$, $dz = dx/\sigma$, we can verify that

$$\int_{-\infty}^{\infty} f_{\mu,\sigma}(x)\,dx = \int_{-\infty}^{\infty} f(z)\,dz = 1,$$

so $f_{\mu,\sigma}(x)$ is indeed a probability density function. Using the same change of variable, we can show that

$$E(X) = \mu \qquad \text{and} \qquad E((X - \mu)^2) = \sigma^2.$$

Hence, $f_{\mu,\sigma}(x)$ does indeed have a mean μ and a standard deviation σ.

Because $e^{-z^2/2}$ cannot be easily antidifferentiated, we cannot determine normal probabilities (i.e., areas) by using the Fundamental Theorem of Calculus. Numerical integrations can be performed, or one can consult a book of statistical tables for computed areas under the standard normal curve. Specifically, these tables usually provide values for what is called the **cumulative distribution function** of a random variable with standard normal distribution. This is the function

$$F(z) = \frac{1}{\sqrt{2\pi}} \int_{-\infty}^{z} e^{-x^2/2}\, dx = \Pr(Z \le z),$$

which represents the area under the standard normal density function from $-\infty$ up to z, as shown in Figure 7.55.

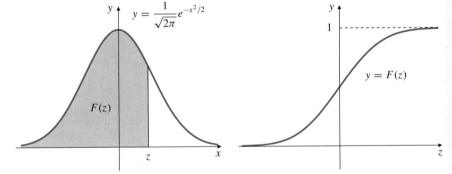

Figure 7.55   The cumulative distribution function $F(z)$ for the standard normal distribution is the area under the standard normal density function from $-\infty$ to z

For use in the following examples and exercises, we include here an abbreviated version of such a table.

Table 2.   Values of the standard normal distribution function $F(z)$ (rounded to 3 decimal places)

z	0.0	0.1	0.2	0.3	0.4	0.5	0.6	0.7	0.8	0.9
−3.0	0.001	0.001	0.001	0.000	0.000	0.000	0.000	0.000	0.000	0.000
−2.0	0.023	0.018	0.014	0.011	0.008	0.006	0.005	0.003	0.003	0.002
−1.0	0.159	0.136	0.115	0.097	0.081	0.067	0.055	0.045	0.036	0.029
−0.0	0.500	0.460	0.421	0.382	0.345	0.309	0.274	0.242	0.212	0.184
0.0	0.500	0.540	0.579	0.618	0.655	0.691	0.726	0.758	0.788	0.816
1.0	0.841	0.864	0.885	0.903	0.919	0.933	0.945	0.955	0.964	0.971
2.0	0.977	0.982	0.986	0.989	0.992	0.994	0.995	0.997	0.997	0.998
3.0	0.999	0.999	0.999	1.000	1.000	1.000	1.000	1.000	1.000	1.000

Example 7   If Z is a standard normal random variable, find
(a) $\Pr(-1.2 \le Z \le 2.0)$ and   (b) $\Pr(Z \ge 1.5)$.

Solution   Using values from the table we obtain

$$\Pr(-1.2 \le Z \le 2.0) = \Pr(Z \le 2.0) - \Pr(Z < -1.2)$$
$$= F(2.0) - F(-1.2) \approx 0.977 - 0.115$$
$$= 0.862$$

$$\Pr(Z \ge 1.5) = 1 - \Pr(Z < 1.5)$$
$$= 1 - F(1.5) \approx 1 - 0.933 = 0.067.$$

Example 8   A random variable X is distributed normally with mean 2 and standard deviation 0.4. Find

(a) $\Pr(1.8 \le X \le 2.4)$ and   (b) $\Pr(X > 2.4)$.

Solution   Since X is distributed normally with mean 2 and standard deviation 0.4, $Z = (X - 2)/0.4$ is distributed according to the standard normal distribution (with mean 0 and standard deviation 1). Accordingly,

$$\Pr(1.8 \le X \le 2.4) = \Pr(-0.5 \le Z \le 1)$$
$$= F(1) - F(-0.5) \approx 0.841 - 0.309 = 0.532,$$
$$\Pr(X > 2.4) = \Pr(Z > 1) = 1 - \Pr(Z \le 1)$$
$$= 1 - F(1) \approx 1 - 0.841 = 0.159.$$

Exercises 7.8

For each function $f(x)$ in Exercises 1–7, find the following:

(a) the value of C for which f is a probability density on the given interval,

(b) the mean μ, variance σ^2, and standard deviation σ of the probability density f, and

(c) $\Pr(\mu - \sigma \le X \le \mu + \sigma)$, that is, the probability that the random variable X is no further than one standard deviation away from its mean.

1. $f(x) = Cx$ on $[0, 3]$ 2. $f(x) = Cx$ on $[1, 2]$

3. $f(x) = Cx^2$ on $[0, 1]$ 4. $f(x) = C \sin x$ on $[0, \pi]$

5. $f(x) = C(x - x^2)$ on $[0, 1]$

6. $f(x) = C\,xe^{-kx}$ on $[0, \infty)$, $(k > 0)$

7. $f(x) = C\,e^{-x^2}$ on $[0, \infty)$. *Hint:* use properties of the standard normal density to show that $\int_0^\infty e^{-x^2}\,dx = \sqrt{\pi}/2$.

8. Is it possible for a random variable to be uniformly distributed on the whole real line? Explain why.

9. Carry out the calculations to show that the normal density $f_{\mu,\sigma}(x)$ defined in the text is a probability density function and has mean μ and standard deviation σ.

* 10. Show that $f(x) = \dfrac{2}{\pi(1 + x^2)}$ is a probability density on

$[0, \infty)$. Find the expectation of X for this density. If a machine generates values of a random variable X distributed with density $f(x)$, how much would you be willing to pay, per game, to play a game in which you operate the machine to produce a value of X and win X dollars? Explain.

11. Calculate $\Pr(|X - \mu| \ge 2\sigma)$ for

(a) the uniform distribution on $[a, b]$,

(b) the exponential distribution with density $f(x) = ke^{-kx}$ on $[0, \infty)$, and

(c) the normal distribution with density $f_{\mu,\sigma}(x)$.

12. The length of time T (in hours) between malfunctions of a computer system is an exponentially distributed random variable. If the average length of time between successive malfunctions is 20 hours, find the probability that the system, having just had a malfunction corrected, will operate without malfunction for at least 12 hours.

13. The number X of metres of cable produced any day by a cable-making company is a normally distributed random variable with mean 5,000 and standard deviation 200. On what fraction of the days the company operates will the number of metres of cable produced exceed 5,500?

7.9   First-Order Differential Equations

This final section on applications of integration concentrates on application of the indefinite integral rather than of the definite integral. We can use the techniques of integration developed in Chapters 5 and 6 to solve certain kinds of first-order differential equations that arise in a variety of modelling situations. We have already seen some examples of applications of differential equations to modelling growth and decay phenomena in Section 3.4.

Separable Equations

Consider the logistic equation introduced in Section 3.4 to model the growth of an animal population with a limited food supply:

$$\frac{dy}{dt} = ky\left(1 - \frac{y}{L}\right),$$

where $y(t)$ is the size of the population at time t, k is a positive constant related to the fertility of the population, and L is the steady-state population size that can be sustained by the available food supply. This equation is an example of a class of first-order differential equations called **separable equations** because when they are written in terms of differentials, they can be separated with only the dependent variable on one side of the equation and only the independent variable on the other. The logistic equation can be written in the form

$$\frac{L\,dy}{y(L-y)} = k\,dt,$$

and solved by integrating both sides. Expanding the left side in partial fractions and integrating, we get

$$\int\left(\frac{1}{y} + \frac{1}{L-y}\right) dy = kt + C.$$

Assuming that $0 < y < L$, we therefore obtain

$$\ln y - \ln(L - y) = kt + C,$$

$$\ln\left(\frac{y}{L-y}\right) = kt + C.$$

We can solve this equation for y by taking exponentials of both sides:

$$\frac{y}{L-y} = e^{kt+C} = C_1 e^{kt}$$

$$y = (L-y)C_1 e^{kt}$$

$$y = \frac{C_1 L e^{kt}}{1 + C_1 e^{kt}},$$

where $C_1 = e^C$.

Generally, separable equations are of the form

$$\frac{dy}{dx} = f(x)g(y).$$

We solve them by rewriting them in the form

$$\frac{dy}{g(y)} = f(x)\,dx$$

and integrating both sides.

Example 1   Solve the equation $\dfrac{dy}{dx} = \dfrac{x}{y}$.

Solution   We rewrite the equation in the form $y\,dy = x\,dx$ and integrate both sides to get

$$\frac{1}{2}y^2 = \frac{1}{2}x^2 + C,$$

or $y^2 - x^2 = C_1$, where $C_1 = 2C$ is an arbitrary constant. The solution curves are rectangular hyperbolas. Their asymptotes $y = x$ and $y = -x$ are also solutions corresponding to $C = 0$.

Example 2   Solve the initial-value problem

$$\begin{cases} \dfrac{dy}{dx} = x^2 y^3 \\ y(1) = 3. \end{cases}$$

Solution   Separating the differential equation gives $\dfrac{dy}{y^3} = x^2\,dx$. Thus,

$$\int \frac{dy}{y^3} = \int x^2\,dx, \qquad \text{so} \qquad \frac{-1}{2y^2} = \frac{x^3}{3} + C.$$

Since $y = 3$ when $x = 1$, we have $-\frac{1}{18} = \frac{1}{3} + C$ and $C = -\frac{7}{18}$. Substituting this value into the above solution and solving for y, we obtain

$$y(x) = \frac{3}{\sqrt{7 - 6x^3}}.$$

This solution is valid for $x < \left(\frac{7}{6}\right)^{1/3}$.

Example 3   Solve the **integral equation** $y(x) = 3 + 2\displaystyle\int_1^x t\,y(t)\,dt$.

Solution   Differentiating the integral equation with respect to x gives

$$\frac{dy}{dx} = 2x\,y(x) \qquad \text{or} \qquad \frac{dy}{y} = 2x\,dx.$$

Thus $\ln|y(x)| = x^2 + C$, and solving for y, $y(x) = C_1 e^{x^2}$. Putting $x = 1$ in the integral equation provides an initial value: $y(1) = 3 + 0 = 3$, so $C_1 = 3/e$ and

$$y(x) = 3e^{x^2 - 1}.$$

Example 4 (**A solution concentration problem**) Initially a tank contains 1,000 L of brine with 50 kg of dissolved salt. If brine containing 10 g of salt per litre is flowing into the tank at a constant rate of 10 L/min, if the contents of the tank are kept thoroughly mixed at all times, and if the solution also flows out at 10 L/min, how much salt remains in the tank at the end of 40 min?

Solution Let $x(t)$ be the number of kilograms of salt in solution in the tank after t min. Thus $x(0) = 50$. Salt is coming into the tank at a rate of 10 g/L $\times$ 10 L/min = 100 g/min = 1/10 kg/min. At all times the tank contains 1,000 L of liquid, so the concentration of salt in the tank at time t is $x/1,000$ kg/L. Since the contents flow out at 10 L/min, salt is being removed at a rate of $10x/1,000 = x/100$ kg/min. Therefore,

$$\frac{dx}{dt} = \text{rate in} - \text{rate out} = \frac{1}{10} - \frac{x}{100} = \frac{10-x}{100}$$

or

$$\frac{dx}{10-x} = \frac{dt}{100}.$$

Integrating both sides of this equation, we obtain

$$-\ln|10-x| = \frac{t}{100} + C.$$

Observe that $x(t) \neq 10$ for any finite time t (since $\ln 0$ is not defined). Since $x(0) = 50 > 10$, it follows that $x(t) > 10$ for all $t > 0$. ($x(t)$ is necessarily continuous so it cannot take any value less than 10 without somewhere taking the value 10 by the Intermediate-Value Theorem.) Hence, we can drop the absolute value from the solution above and obtain

$$\ln(x-10) = -\frac{t}{100} - C.$$

Since $x(0) = 50$, we have $-C = \ln 40$ and

$$x = x(t) = 10 + 40e^{-t/100}.$$

After 40 min there will be $10 + 40e^{-0.4} \approx 36.8$ kg of salt in the tank.

∎

Example 5 (**A rate of reaction problem**) In a chemical reaction that goes to completion in solution, one molecule of each of two reactants, A and B, combine to form each molecule of the product C. According to the law of mass action, the reaction proceeds at a rate proportional to the product of the concentrations of A and B in the solution. Thus, if there were initially present a molecules/cm^3 of A and b molecules/cm^3 of B, then the number $x(t)$ of molecules/cm^3 of C present at time t thereafter is determined by the differential equation

$$\frac{dx}{dt} = k(a-x)(b-x).$$

We solve this equation by the technique of partial fraction decomposition under the assumption that $b \neq a$:

$$\int \frac{dx}{(a-x)(b-x)} = k \int dt = kt + C.$$

Since

$$\frac{1}{(a-x)(b-x)} = \frac{1}{b-a} \left(\frac{1}{a-x} - \frac{1}{b-x} \right),$$

and since necessarily $x \leq a$ and $x \leq b$, we have

$$\frac{1}{b-a} \left(-\ln(a-x) + \ln(b-x) \right) = kt + C,$$

or

$$\ln \left(\frac{b-x}{a-x} \right) = (b-a)\, kt + C_1, \quad \text{where } C_1 = (b-a)C.$$

By assumption, $x(0) = 0$, so $C_1 = \ln(b/a)$ and

$$\ln \frac{a(b-x)}{b(a-x)} = (b-a)\, kt.$$

This equation can be solved for x to yield $x = x(t) = \dfrac{ab(e^{(b-a)kt} - 1)}{be^{(b-a)kt} - a}.$

∎

Example 6 Find a family of curves, each of which intersects every parabola with equation of the form $y = Cx^2$ at right angles.

Solution The family of parabolas $y = Cx^2$ satisfies the differential equation

$$\frac{d}{dx} \left(\frac{y}{x^2} \right) = \frac{d}{dx} C = 0,$$

that is,

$$x^2 \frac{dy}{dx} - 2xy = 0 \quad \text{or} \quad \frac{dy}{dx} = \frac{2y}{x}.$$

Any curve that meets the parabolas $y = Cx^2$ at right angles must, at any point (x, y) on it, have slope equal to the negative reciprocal of the slope of the particular parabola passing through that point. Thus such a curve must satisfy

$$\frac{dy}{dx} = -\frac{x}{2y}.$$

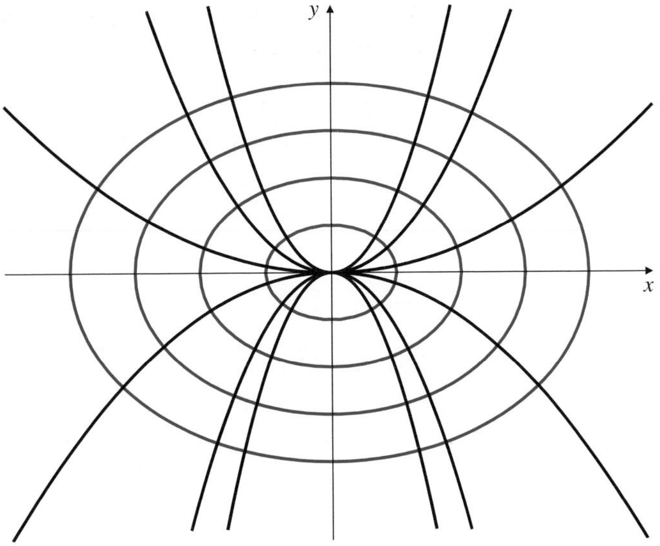

Figure 7.56 The parabolas $y = C_1 x^2$ and the ellipses $x^2 + 2y^2 = C_2$ intersect at right angles

Separation of the variables leads to $2y\,dy = -x\,dx$, and integration of both sides then yields $y^2 = -\frac{1}{2}x^2 + C_1$ or $x^2 + 2y^2 = C$, where $C = 2C_1$. This equation represents a family of ellipses centred at the origin. Each ellipse meets each parabola at right angles, as shown in Figure 7.56. When the curves of one family intersect the curves of a second family at right angles, each family is called the family of **orthogonal trajectories** of the other family. ∎

First-Order Linear Equations

A first-order **linear** differential equation is one of the type

$$\frac{dy}{dx} + p(x)y = q(x),$$

where $p(x)$ and $q(x)$ are given functions, which we assume to be continuous. We can solve such equations (i.e., find y as a function of x) by the following procedure.

Let $\mu(x)$ be any antiderivative of $p(x)$:

$$\mu(x) = \int p(x)\,dx \quad \text{and} \quad \frac{d\mu}{dx} = p(x).$$

If $y = y(x)$ satisfies the given equation, then we calculate, using the Product Rule,

$$\frac{d}{dx}\left(e^{\mu(x)}y(x)\right) = e^{\mu(x)}\frac{dy}{dx} + e^{\mu(x)}\frac{d\mu}{dx}y(x)$$
$$= e^{\mu(x)}\left(\frac{dy}{dx} + p(x)y\right) = e^{\mu(x)}q(x).$$

Therefore $e^{\mu(x)}y(x) = \int e^{\mu(x)}q(x)\,dx$, or

$$y(x) = e^{-\mu(x)} \int e^{\mu(x)} q(x)\, dx.$$

We reuse this method, rather than the final formula, in the examples below. $e^{\mu(x)}$ is called an **integrating factor** of the given differential equation because, if we multiply the equation by $e^{\mu(x)}$, the left side becomes the derivative of $e^{\mu(x)} y(x)$.

Example 7 Solve $\dfrac{dy}{dx} + \dfrac{y}{x} = 1$ for $x > 0$.

Solution Here, $p(x) = 1/x$, so $\mu(x) = \int p(x)\, dx = \ln x$ (for $x > 0$) and $e^{\mu(x)} = x$. We calculate

$$\frac{d}{dx}(xy) = x\frac{dy}{dx} + y = x\left(\frac{dy}{dx} + \frac{y}{x}\right) = x$$

and

$$xy = \int x\, dx = \frac{1}{2}x^2 + C.$$

Finally,

$$y = \frac{1}{x}\left(\frac{1}{2}x^2 + C\right) = \frac{x}{2} + \frac{C}{x}.$$

This is a solution of the given equation for any value of the constant C.

Example 8 Solve $\dfrac{dy}{dx} + xy = x^3$.

Solution Here, $p(x) = x$, so $\mu(x) = x^2/2$ and $e^{\mu(x)} = e^{x^2/2}$. We calculate

$$\frac{d}{dx}\left(e^{x^2/2}y\right) = e^{x^2/2}\frac{dy}{dx} + e^{x^2/2}xy = e^{x^2/2}\left(\frac{dy}{dx} + xy\right) = x^3 e^{x^2/2}.$$

Thus,

$$e^{x^2/2}\, y = \int x^3\, e^{x^2/2}\, dx \qquad \text{Let} \quad U = x^2, \qquad dV = x\, e^{x^2/2}\, dx.$$
$$\text{Then } dU = 2x\, dx, \qquad V = e^{x^2/2}.$$
$$= x^2 e^{x^2/2} - 2\int x\, e^{x^2/2}\, dx$$
$$= x^2 e^{x^2/2} - 2 e^{x^2/2} + C,$$

and, finally, $y = x^2 - 2 + Ce^{-x^2/2}$.

Our final example reviews a typical *stream of payments* problem of the sort considered in Section 7.7. This time we treat the problem as an initial-value problem for a differential equation.

Example 9 A savings account is opened with a deposit of A dollars. At any time t years thereafter, money is being continually deposited into the account at a rate of $(C + Dt)$ dollars per year. If interest is also being paid into the account at a nominal rate of $100R\%$ per year, compounded continuously, find the balance $B(t)$ dollars in the account after t years. Illustrate the solution for the data $A = 5,000$, $C = 1,000$, $D = 200$, $R = 0.13$, and $t = 5$.

Solution As noted in Section 3.4, continuous compounding of interest at a nominal rate of $100R\%$ causes \$1.00 to grow to $\$e^{Rt}$ in t years. Without subsequent deposits, the balance in the account would grow according to the differential equation of exponential growth:

$$\frac{dB}{dt} = RB.$$

Allowing for additional growth due to the continual deposits, we observe that B must satisfy the differential equation

$$\frac{dB}{dt} = RB + (C + Dt)$$

or, equivalently, $dB/dt - RB = C + Dt$. This is a linear equation for B having $p(t) = -R$. Hence, we may take $\mu(t) = -Rt$ and $e^{\mu(t)} = e^{-Rt}$. We now calculate

$$\frac{d}{dt}\left(e^{-Rt} B(t)\right) = e^{-Rt}\frac{dB}{dt} - Re^{-Rt} B(t) = (C + Dt)\, e^{-Rt}$$

and

$$e^{-Rt} B(t) = \int (C + Dt)e^{-Rt}\, dt \qquad \text{Let} \quad U = C + Dt, \quad dV = e^{-Rt}\, dt.$$
$$\text{Then } dU = D\, dt, \qquad V = -e^{-Rt}/R.$$
$$= -\frac{C + Dt}{R} e^{-Rt} + \frac{D}{R}\int e^{-Rt}\, dt$$
$$= -\frac{C + Dt}{R} e^{-Rt} - \frac{D}{R^2} e^{-Rt} + K, \qquad (K = \text{constant}).$$

Hence

$$B(t) = -\frac{C + Dt}{R} - \frac{D}{R^2} + Ke^{Rt}.$$

Since $A = B(0) = -\dfrac{C}{R} - \dfrac{D}{R^2} + K$, we have $K = A + \dfrac{C}{R} + \dfrac{D}{R^2}$ and

$$B(t) = \left(A + \frac{C}{R} + \frac{D}{R^2}\right)e^{Rt} - \frac{C + Dt}{R} - \frac{D}{R^2}.$$

For the illustration $A = 5,000$, $C = 1,000$, $D = 200$, $R = 0.13$, and $t = 5$, we obtain, using a calculator, $B(5) = 19,762.82$. The account will contain \$19,762.82, after 5 years, under these circumstances. ∎

Exercises 7.9

Solve the differential equations in Exercises 1–16.

1. $\dfrac{dy}{dx} = \dfrac{y}{2x}$

2. $\dfrac{dy}{dx} = \dfrac{3y-1}{x}$

3. $\dfrac{dy}{dx} = \dfrac{x^2}{y^2}$

4. $\dfrac{dy}{dx} = x^2 y^2$

5. $\dfrac{dY}{dt} = tY$

6. $\dfrac{dx}{dt} = e^x \sin t$

7. $\dfrac{dy}{dx} = 1 - y^2$

8. $\dfrac{dy}{dx} = 1 + y^2$

9. $\dfrac{dy}{dt} = 2 + e^y$

10. $\dfrac{dy}{dx} = y^2(1-y)$

11. $\dfrac{dy}{dx} - \dfrac{2y}{x} = x^2$

12. $\dfrac{dy}{dx} + \dfrac{2y}{x} = \dfrac{1}{x^2}$

13. $\dfrac{dy}{dx} + 2y = 3$

14. $\dfrac{dy}{dx} + y = e^x$

15. $\dfrac{dy}{dx} + y = x$

16. $\dfrac{dy}{dx} + 2e^x y = e^x$

Solve the integral equations in Exercises 17–20.

17. $y(x) = 2 + \displaystyle\int_0^x \dfrac{t}{y(t)}\,dt$

18. $y(x) = 1 + \displaystyle\int_0^x \dfrac{\big(y(t)\big)^2}{1+t^2}\,dt$

19. $y(x) = 1 + \displaystyle\int_1^x \dfrac{y(t)\,dt}{t(t+1)}$

20. $y(x) = 3 + \displaystyle\int_0^x e^{-y(t)}\,dt$

21. Why is the solution given for the chemical reaction rate problem in Example 5 not valid for $a = b$? Find the solution for the case $a = b$.

22. An object of mass m falling near the surface of the earth is retarded by air resistance proportional to its velocity so that, according to Newton's Second Law of Motion,

$$m\dfrac{dv}{dt} = mg - kv,$$

where $v = v(t)$ is the velocity of the object at time t, and g is the acceleration of gravity near the surface of the earth. Assuming that the object falls from rest at time $t = 0$, that is, $v(0) = 0$, find the velocity $v(t)$ for any $t > 0$ (up until the object strikes the ground). Show $v(t)$ approaches a limit as $t \to \infty$. Do you need the explicit formula for $v(t)$ to determine this limiting velocity?

23. Repeat Exercise 22 except assuming that the air resistance is proportional to the square of the velocity so that the equation of motion is

$$m\dfrac{dv}{dt} = mg - kv^2.$$

24. Find the amount in a savings account after one year if the initial balance in the account was $1,000, if the interest is paid continuously into the account at a nominal rate of 10% per annum, compounded continuously, and if the account is being continuously depleted (by taxes, say) at a rate of $y^2/1,000,000$ dollars per year, where $y = y(t)$ is the balance in the account after t years. How large can the account grow? How long will it take the account to grow to half this balance?

25. Find the family of curves each of which intersects all of the hyperbolas $xy = C$ at right angles.

26. Repeat the solution concentration problem in Example 4, changing the rate of inflow of brine into the tank to 12 L/min but leaving all the other data as they were in that example. Note that the volume of liquid in the tank is no longer constant as time increases.

Chapter Review

Key Ideas

- **What do the following phrases mean?**
  - ◇ a solid of revolution
  - ◇ a volume element
  - ◇ the arc length of a curve
  - ◇ the moment of a point mass m about $x = 0$
  - ◇ the centre of mass of a distribution of mass
  - ◇ the centroid of a plane region
  - ◇ a first-order separable differential equation
  - ◇ a first-order linear differential equation

- Let D be the plane region $0 \le y \le f(x)$, $a \le x \le b$. Use integrals to represent the following:
  - ◇ the volume generated by revolving D about the x-axis
  - ◇ the volume generated by revolving D about the y-axis
  - ◇ the moment of D about the y-axis
  - ◇ the moment of D about the x-axis
  - ◇ the centroid of D

- Let C be the curve $y = f(x)$, $a \le x \le b$. Use integrals to represent the following:
  - ◇ the length of C
  - ◇ the area of the surface generated by revolving C about the

x-axis

◇ the area of the surface generated by revolving *C* about the *y*-axis

Review Exercises

1. Figure 7.57 shows cross-sections along the axes of two circular spools. The left spool will hold 1,000 metres of thread if wound full with no bulging. How many metres of thread of the same size will the right spool hold?

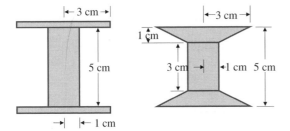

Figure 7.57

2. Water sitting in a bowl evaporates at a rate proportional to its surface area. Show that the depth of water in the bowl decreases at a constant rate, regardless of the shape of the bowl.

3. A barrel is 4 ft high and its volume is 16 cubic feet. Its top and bottom are circular disks of radius 1 ft, and its side wall is obtained by rotating the parabola $x = a - by^2$, $-2 \le y \le 2$, about the *y*-axis. Find, approximately, the values of the positive constants *a* and *b*.

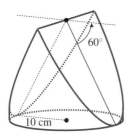

Figure 7.58

4. The solid in Figure 7.58 is cut from a vertical cylinder of radius 10 cm by two planes making angles of 60° with the horizontal. Find its volume.

5. Find to 4 decimal places the value of the positive constant *a* for which the curve $y = (1/a)\cosh ax$ has arc length 2 units between $x = 0$ and $x = 1$.

6. Find the area of the surface obtained by rotating the curve $y = \sqrt{x}$, $(0 \le x \le 6)$, about the *x*-axis.

7. Find the centroid of the plane region $x \ge 0$, $y \ge 0$, $x^2 + 4y^2 \le 4$.

8. A thin plate in the shape of a circular disk has radius 3 ft and constant areal density. A circular hole of radius 1 ft is cut out

of the disk, centred 1 ft from the centre of the disk. Find the centre of mass of the remaining part of the disk.

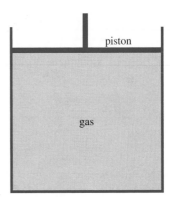

Figure 7.59

9. According to Boyle's Law, the product of the pressure and volume of a gas remains constant if the gas expands or is compressed isothermally. The cylinder in Figure 7.59 is filled with a gas that exerts a force of 1,000 N on the piston when the piston is 20 cm above the base of the cylinder. How much work is done by the piston if it compresses the gas isothermally by descending to a height of 5 cm above the base?

10. Suppose two functions *f* and *g* have the following property: for any $a > 0$, the solid produced by revolving the region of the *xy*-plane bounded by $y = f(x)$, $y = g(x)$, $x = 0$, and $x = a$ about the *x*-axis has the same volume as the solid produced by revolving the same region about the *y*-axis. What can you say about *f* and *g*?

11. Find the equation of a curve that passes through the point $(2, 4)$ and has slope $3y/(x - 1)$ at any point (x, y) on it.

12. Find a family of curves that intersect every ellipse of the form $3x^2 + 4y^2 = C$ at right angles.

13. The income and expenses of a seasonal business result in deposits and withdrawals from its bank account that correspond to a flow rate into the account of $\$P(t)$/year at time *t* years, where $P(t) = 10,000\sin(2\pi t)$. If the account earns interest at an instantaneous rate of 4% per year, and has \$8,000 in it at time $t = 0$, how much is in the account two years later?

Challenging Problems

1. The curve $y = e^{-kx}\sin x$, $(x \ge 0)$, is revolved about the *x*-axis to generate a string of "beads" whose volumes decrease to the right if $k > 0$.

   (a) Show that the ratio of the volume of the $(n + 1)$st bead to that of the *n*th bead depends on *k*, but not on *n*.

   (b) For what value of *k* is the ratio in part (a) equal to 1/2?

   (c) Find the total volume of all the beads as a function of $k > 0$.

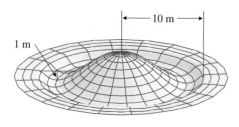

Figure 7.60

2. (Conservation of earth) A landscaper wants to create on level ground a ring-shaped pool having an outside radius of 10 m and a maximum depth of 1 m surrounding a hill that will be built up using all the earth excavated from the pool. (See Figure 7.60.) She decides to use a fourth-degree polynomial to determine the cross-sectional shape of the hill and pool bottom: at distance r metres from the centre of the development the height above or below normal ground level will be

$$h(r) = a(r^2 - 100)(r^2 - k^2) \text{ metres,}$$

for some $a > 0$, where k is the inner radius of the pool. Find k and a so that the requirements given above are all satisfied. How much earth must be moved from the pool to build the hill?

3. (Rocket design) The nose of a rocket is a solid of revolution of base radius r and height h that must join smoothly to the cylindrical body of the rocket. (See Figure 7.61.) Taking the origin at the tip of the nose and the x-axis along the central axis of the rocket, various nose shapes can be obtained by revolving the cubic curve

$$y = f(x) = ax + bx^2 + cx^3$$

about the x-axis. The cubic curve must have slope 0 at $x = h$, and its slope must be positive for $0 < x < h$. Find the particular cubic curve that maximizes the volume of the nose. Also show that this choice of the cubic makes the slope dy/dx at the origin as large as possible and, hence, corresponds to the bluntest nose.

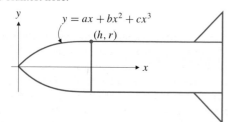

Figure 7.61

4. (Quadratic splines) Let $A = (x_1, y_1)$, $B = (x_2, y_2)$, and $C = (x_3, y_3)$ be three points with $x_1 < x_2 < x_3$. A function $f(x)$ whose graph passes through the three points is a *quadratic spline* if $f(x)$ is a quadratic function on $[x_1, x_2]$ and a possibly different quadratic function on $[x_2, x_3]$, and the two quadratics have the same slope at x_2. For this problem, take $A = (0, 1)$, $B = (1, 2)$, and $C = (3, 0)$.

(a) Find a one parameter family $f(x, m)$ of quadratic splines through A, B, and C, having slope m at B.

(b) Find the value of m for which the length of the graph $y = f(x, m)$ between $x = 0$ and $x = 3$ is minimum. What is this minimum length? Compare it with the length of the polygonal line ABC.

5. A concrete wall in the shape of a circular ring must be built to have maximum height 2 m, inner radius 15 m, and width 1 m at ground level, so that its outer radius is 16 m. (See Figure 7.62.) Built on level ground, the wall will have a curved top with height at distance $15 + x$ metres from the centre of the ring given by the cubic function

$$f(x) = x(1 - x)(ax + b) \text{ m,}$$

which must not vanish anywhere in the open interval $(0, 1)$. Find the values of a and b that minimize the total volume of concrete needed to build the wall.

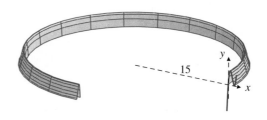

Figure 7.62

6. (The volume of an n-dimensional ball) Euclidean n-dimensional space consists of *points* $(x_1, x_2, \ldots, x_n)$ with n real coordinates. By analogy with the 3-dimensional case, we call the set of such points that satisfy the inequality $x_1^2 + x_2^2 + \cdots + x_n^2 \le r^2$ the n-dimensional *ball* centred at the origin. For example, the 1-dimensional ball is the interval $-r \le x_1 \le r$, which has *volume* (i.e., *length*), $V_1(r) = 2r$. The 2-dimensional ball is the disk $x_1^2 + x_2^2 \le r^2$, which has *volume* (i.e., *area*),

$$V_2(r) = \pi r^2 = \int_{-r}^{r} 2\sqrt{r^2 - x^2}\, dx$$

$$= \int_{-r}^{r} V_1\left(\sqrt{r^2 - x^2}\right) dx.$$

The 3-dimensional ball $x_1^2 + x_2^2 + x_3^2 \le r^2$ has volume

$$V_3(r) = \frac{4}{3}\pi r^3 = \int_{-r}^{r} \pi \left(\sqrt{r^2 - x^2}\right)^2 dx$$

$$= \int_{-r}^{r} V_2\left(\sqrt{r^2 - x^2}\right) dx.$$

By analogy with these formulas, the volume $V_n(r)$ of the n-dimensional ball of radius r is the integral of the volume of the $(n-1)$-dimensional ball of radius $\sqrt{r^2 - x^2}$ from $x = -r$ to $x = r$:

$$V_n(r) = \int_{-r}^{r} V_{n-1}\left(\sqrt{r^2 - x^2}\right) dx.$$

Using a computer algebra program, calculate $V_4(r)$, $V_5(r)$, ..., $V_{10}(r)$, and guess formulas for $V_{2n}(r)$ (the even-dimensional balls) and $V_{2n+1}(r)$ (the odd-dimensional balls). If your computer algebra software is sufficiently powerful, you may be able to verify your guesses by induction. Otherwise, use them to predict $V_{11}(r)$ and $V_{12}(r)$, then check your predictions by starting from $V_{10}(r)$.

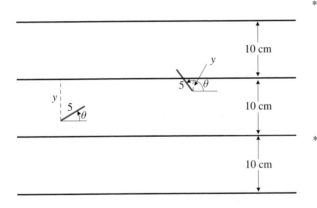

Figure 7.63

* **7. (Buffon's needle problem)** A horizontal flat surface is ruled with parallel lines 10 cm apart, as shown in Figure 7.63. A needle 5 cm long is dropped at random onto the surface. Find the probability that the needle intersects one of the lines. *Hint:* Let the "lower" end of the needle (the end further down the page in the figure) be considered the reference point. (If both ends are the same height, use the left end.) Let y be the distance from the reference point to the nearest line above it, and let θ be the angle between the needle and the line extending to the right of the reference point in the figure. What are the possible values of y and θ? In a plane with Cartesian coordinates θ and y sketch the region consisting of all points (θ, y) corresponding to possible positions of the needle. Also sketch the region corresponding to those positions for which the needle crosses one of the parallel lines. The required probability is the area of the second region divided by the area of the first.

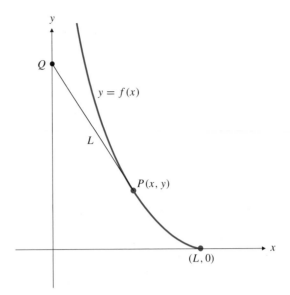

Figure 7.64

* **8. (The path of a trailer)** Find the equation $y = f(x)$ of a curve in the first quadrant of the xy-plane, starting from the point $(L, 0)$, and having the property that if the tangent line to the curve at P meets the y-axis at Q, then the length of PQ is the constant L. (See Figure 7.64. This curve is called a **tractrix** after the Latin participle *tractus* meaning *dragged*. It is the path of the rear end P of a trailer of length L, originally lying along the x-axis, as the trailer is pulled (dragged) by a tractor Q moving along the y-axis away from the origin.)

* **9. (Approximating the surface area of an ellipsoid)** A physical geographer studying the flow of streams around oval stones needed to calculate the surface areas of many such stones which he modelled as ellipsoids:

$$\frac{x^2}{a^2} + \frac{y^2}{b^2} + \frac{z^2}{c^2} = 1.$$

He wanted a simple formula for the surface area so that he could implement it in a spreadsheet containing the measurements a, b, and c of the stones. Unfortunately, there is no exact formula for the area of a general ellipsoid in terms of elementary functions. However, there are such formulas for ellipsoids of revolution, where two of the three semi-axes are equal. These ellipsoids are called spheroids; an *oblate spheroid* (like the earth) has its two longer semi-axes equal; a *prolate spheroid* (like an American football) has its two shorter semi-axes equal. A reasonable approximation to the area of a general ellipsoid can be obtained by linear interpolation between these two.

To be specific, assume the semi-axes are arranged in decreasing order $a \geq b \geq c$, and let the surface area be $S(a, b, c)$.

(a) Calculate $S(a, a, c)$, the area of an oblate spheroid.

(b) Calculate $S(a, c, c)$, the area of a prolate spheroid.

(c) Construct an approximation for $S(a, b, c)$ that divides the interval from $S(a, a, c)$ to $S(a, c, c)$ in the same ratio

that b divides the interval from a to c.

(d) Approximate the area of the ellipsoid

$$\frac{x^2}{9} + \frac{y^2}{4} + z^2 = 1$$

using the above method.

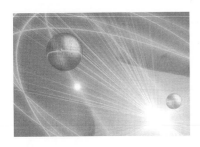

CHAPTER 8

Conics, Parametric Curves, and Polar Curves

Introduction Until now, most curves we have encountered have been graphs of functions, and they provided useful visual information about the behaviour of the functions. In this chapter we begin to look at plane curves as interesting objects in their own right. First, we examine conic sections, curves with quadratic equations obtained by intersecting a plane with a right-circular cone. Then we consider curves that can be described by two parametric equations that give the coordinates of points on the curve as functions of a parameter. If this parameter is time, the equations describe the path of a moving point in the plane. Finally, we consider curves described by equations in a new coordinate system called polar coordinates, in which a point is located by giving its distance and direction from the origin. In Chapter 11 we will expand our study of curves to three dimensions.

8.1 Conics

Circles, ellipses, parabolas, and hyperbolas are called **conic sections** (or, more simply, just **conics**) because they are curves in which planes intersect right-circular cones.

To be specific, suppose that a line A is fixed in space, and V is a point fixed on A. The **right-circular cone** having **axis** A, **vertex** V, and **semi-vertical angle** α is the surface consisting of all points on straight lines through V that make angle α with the line A. (See Figure 8.1.) The cone has two halves (called **nappes**) lying on opposite sides of the vertex V. Any plane P that does not pass through V will intersect the cone (one or both nappes) in a curve $\mathcal{C}$. (See Figure 8.2.) If a line normal (i.e., perpendicular) to P makes angle θ with the axis A of the cone, where $0 \le \theta \le \pi/2$, then

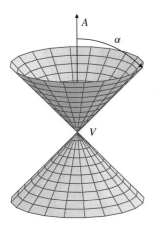

$\mathcal{C}$ is a circle if $\qquad \theta = 0$

$\mathcal{C}$ is an ellipse if $\qquad 0 < \theta < \dfrac{\pi}{2} - \alpha$

$\mathcal{C}$ is a parabola if $\qquad \theta = \dfrac{\pi}{2} - \alpha$

$\mathcal{C}$ is a hyperbola if $\qquad \dfrac{\pi}{2} - \alpha < \theta \le \dfrac{\pi}{2}.$

Figure 8.1 A cone with vertex V, axis A, and semi-vertical angle α

In Sections 10.4 and 10.5 it is shown that planes are represented by first-degree equations and cones by second-degree equations. Therefore, all conics can be represented analytically (in terms of Cartesian coordinates x and y in the plane of the conic) by a second-degree equation of the general form

$$Ax^2 + Bxy + Cy^2 + Dx + Ey + F = 0,$$

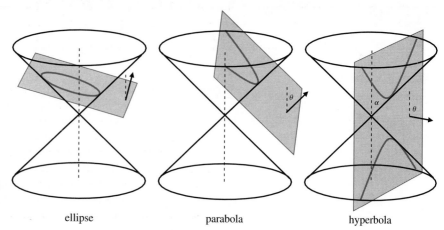

Figure 8.2 Planes intersecting cones in an ellipse, a parabola, and a hyperbola

ellipse   parabola   hyperbola

where A, B, ..., F are constants. However, such an equation can also represent the empty set, a single point, or one or two straight lines if the left-hand side factors into linear factors:

$$(A_1x + B_1y + C_1)(A_2x + B_2y + C_2) = 0.$$

After straight lines the conic sections are the simplest of plane curves. They have many properties that make them useful in applications of mathematics; that is why we include a discussion of them here. Much of this material is optional from the point of view of a calculus course, but familiarity with the properties of conics can be very important in some applications. Most of the properties of conics were discovered by the Greek geometer Apollonius of Perga, around 200 BC. It is remarkable that he was able to obtain these properties using only the techniques of classical Euclidean geometry; today, most of these properties are expressed more conveniently using analytic geometry and specific coordinate systems.

Parabolas

DEFINITION 1

Parabolas

A **parabola** consists of points in the plane that are equidistant from a given point (the **focus** of the parabola) and a given straight line (the **directrix** of the parabola). The line through the focus perpendicular to the directrix is called the **principal axis** (or simply **the axis**) of the parabola. The **vertex** of the parabola is the point where the parabola crosses its principal axis. It is on the axis halfway between the focus and the directrix.

Example 1 Find an equation of the parabola whose focus is the point $F = (a, 0)$ and whose directrix is the line L with equation $x = -a$.

Solution The parabola has axis along the x-axis and vertex at the origin. (See Figure 8.3.) If $P = (x, y)$ is any point on the parabola, then the distance from P to

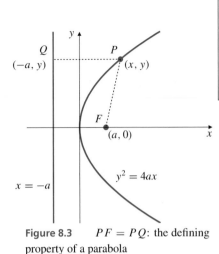

Figure 8.3 $PF = PQ$: the defining property of a parabola

F is equal to the distance from P to the nearest point Q on L. Thus

$$\sqrt{(x-a)^2 + y^2} = x + a$$
$$\text{or } x^2 - 2ax + a^2 + y^2 = x^2 + 2ax + a^2,$$

or, upon simplification, $y^2 = 4ax$.

Similarly, we can obtain standard equations for parabolas with vertices at the origin and foci at $(-a, 0)$, $(0, a)$ and $(0, -a)$:

Table 1. Standard equations of parabolas

Focus	Directrix	Equation
$(a, 0)$	$x = -a$	$y^2 = 4ax$
$(-a, 0)$	$x = a$	$y^2 = -4ax$
$(0, a)$	$y = -a$	$x^2 = 4ay$
$(0, -a)$	$y = a$	$x^2 = -4ay$

The Focal Property of a Parabola

All of the conic sections have interesting and useful focal properties relating to the way in which surfaces of revolution they generate reflect light if the surfaces are mirrors. For instance, a circle will clearly reflect back along the same path any ray of light incident along a line passing through its centre. The focal properties of parabolas, ellipses, and hyperbolas can be derived from the reflecting property of a straight line (i.e., a plane mirror) by elementary geometrical arguments.

Light travels in straight lines in a medium of constant optical density (one where the speed of light is constant). This is a consequence of the physical Principle of Least Action, which asserts that in travelling between two points, light takes the path requiring the minimum travel time. Given a straight line L in a plane and two points A and B in the plane on the same side of L, the point P on L for which the sum of the distances $AP + PB$ is minimum is such that AP and PB make equal angles with L, or equivalently, with the normal to L at P. (See Figure 8.4.) If B' is the point such that L is the right bisector of the line segment BB', then P is the intersection of L and AB'. Since one side of a triangle cannot exceed the sum of the other two sides,

$$AP + PB = AP + PB' = AB' \le AQ + QB' = AQ + QB.$$

> **Reflection by a straight line**
>
> The point P on L at which a ray from A reflects so as to pass through B is the point that minimizes the sum of the distances $AP + PB$.

Now consider a parabola with focus F and directrix D. Let P be on the parabola and let T be the line tangent to the parabola at P. (See Figure 8.5.) Let Q be any point on T. Then FQ meets the parabola at a point X between F and Q. Let M and N be points on D such that MX and NP are perpendicular to D, and let A be a point on the line through N and P that lies on the same side of the parabola as F. We have

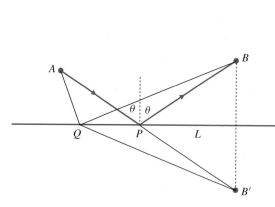

Figure 8.4 Reflection by a straight line

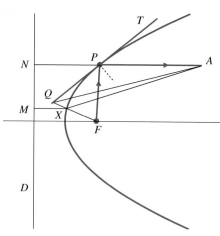

Figure 8.5 Reflection by a parabola

$$FP + PA = NP + PA = NA \leq MX + XA = FX + XA$$
$$\leq FX + XQ + QA = FQ + QA.$$

Thus, among all points Q on the line T, $Q = P$ is the one that minimizes the sum of distances $FQ + QA$. By the observation made for straight lines above, FP and PA make equal angles with T and so also with the normal to the parabola at P. (The parabola and the tangent line have the same normal at P.)

> **Reflection by a parabola**
>
> Any ray from the focus will be reflected parallel to the axis of the parabola. Equivalently, any incident ray parallel to the axis of the parabola will be reflected through the focus.

Ellipses

DEFINITION 2

> **Ellipses**
>
> An **ellipse** consists of all points in the plane, the sum of whose distances from two fixed points (the **foci**) is constant.

Example 2 Find the ellipse with foci at the points $(-c, 0)$ and $(c, 0)$ if the sum of the distances from any point P on the ellipse to these two foci is $2a$.

Solution The ellipse passes through the four points $(a, 0)$, $(-a, 0)$, $(0, b)$, and $(0, -b)$, where $b^2 = a^2 - c^2$. (See Figure 8.6.) Also, if $P = (x, y)$ is on the ellipse, then

$$\sqrt{(x - c)^2 + y^2} + \sqrt{(x + c)^2 + y^2} = 2a.$$

Transposing one term from the left side to the right side and squaring, we get

$$(x - c)^2 + y^2 = 4a^2 - 4a\sqrt{(x + c)^2 + y^2} + (x + c)^2 + y^2.$$

Now we expand the squares, cancel terms, transpose, and square again:

$$a\sqrt{(x+c)^2 + y^2} = a^2 + cx$$
$$a^2(x^2 + 2cx + c^2 + y^2) = a^4 + 2a^2cx + c^2x^2$$
$$(a^2 - c^2)x^2 + a^2y^2 = a^2(a^2 - c^2).$$

Finally, replace $a^2 - c^2$ with b^2 and divide by a^2b^2 to get the standard equation of the ellipse:

$$\frac{x^2}{a^2} + \frac{y^2}{b^2} = 1.$$

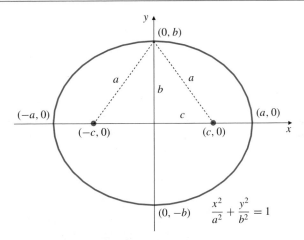

Figure 8.6   An ellipse

The following quantities describe this ellipse:

a	is the **semi-major axis**
b	is the **semi-minor axis**
$c = \sqrt{a^2 - b^2}$	is the **semi-focal separation**.

The point halfway between the foci is called the **centre** of the ellipse. In the example above it is the origin. Note that $a > b$ in this example. If $a < b$, then the ellipse has its foci at $(0, c)$ and $(0, -c)$, where $c = \sqrt{b^2 - a^2}$. The line containing the foci (the **major axis**) and the line through the centre perpendicular to that line (the **minor axis**) are called the **principal axes** of the ellipse.

The **eccentricity** of an ellipse is the ratio of the semi-focal separation to the semi-major axis. We denote the eccentricity ε. For the ellipse $\frac{x^2}{a^2} + \frac{y^2}{b^2} = 1$ with $a > b$,

$$\varepsilon = \frac{c}{a} = \frac{\sqrt{a^2 - b^2}}{a}.$$

Note that $\varepsilon < 1$ for any ellipse; the greater the value of ε, the more elongated (less circular) is the ellipse. If $\varepsilon = 0$ so that $a = b$ and $c = 0$, the two foci coincide and the ellipse is a circle.

The Focal Property of an Ellipse

Let P be any point on an ellipse having foci F_1 and F_2. The normal to the ellipse at P bisects the angle between the lines F_1P and F_2P.

> **Reflection by an ellipse**
>
> Any ray coming from one focus of an ellipse will be reflected through the other focus.

To see this, observe that if Q is any point on the line T tangent to the ellipse at P, then F_1Q meets the ellipse at a point X between F_1 and Q (see Figure 8.7), so

$$F_1P + PF_2 = F_1X + XF_2 \le F_1X + XQ + QF_2 = F_1Q + QF_2.$$

Among all points on T, P is the one that minimizes the sum of the distances to F_1 and F_2. This implies that the normal to the ellipse at P bisects the angle F_1PF_2.

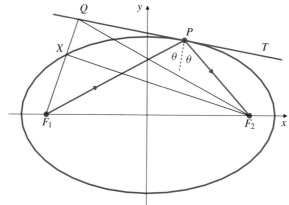

Figure 8.7   A ray from one focus of an ellipse is reflected to the other focus

The Directrices of an Ellipse

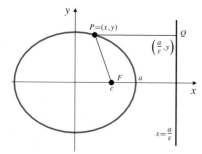

Figure 8.8   A focus and corresponding directrix of an ellipse

If $a > b > 0$, each of the lines $x = a/\varepsilon$ and $x = -a/\varepsilon$ is called a **directrix** of the ellipse $\dfrac{x^2}{a^2} + \dfrac{y^2}{b^2} = 1$. If P is on the ellipse, then the ratio of the distance from P to a focus to its distance from the corresponding directrix is equal to the eccentricity ε. If $P = (x, y)$, F is the focus $(c, 0)$, Q is on the corresponding directrix $x = a/\varepsilon$, and PQ is perpendicular to the directrix, then (see Figure 8.8)

$$PF^2 = (x - c)^2 + y^2$$

$$= x^2 - 2cx + c^2 + b^2\left(1 - \frac{x^2}{a^2}\right)$$

$$= x^2\left(\frac{a^2 - b^2}{a^2}\right) - 2cx + a^2 - b^2 + b^2$$

$$= \varepsilon^2 x^2 - 2\varepsilon a x + a^2 \qquad \text{(because } c = \varepsilon a\text{)}.$$

$$= (a - \varepsilon x)^2.$$

Thus $PF = a - \varepsilon x$. Also, $QP = (a/\varepsilon) - x = (a - \varepsilon x)/\varepsilon$. Therefore $PF/QP = \varepsilon$, as asserted.

A parabola may be considered as the limiting case of an ellipse whose eccentricity has increased to 1. The distance between the foci is infinite, so the centre, one focus, and its corresponding directrix have moved off to infinity leaving only one focus and its directrix in the finite plane.

Hyperbolas

Hyperbolas

A **hyperbola** consists of all points in the plane, the difference of whose distances from two fixed points (the **foci**) is constant.

Example 3 If the foci of a hyperbola are $F_1 = (c, 0)$ and $F_2 = (-c, 0)$, and the difference of the distances from a point $P = (x, y)$ on the hyperbola to these foci is $2a$ (where $a < c$), then

$$PF_2 - PF_1 = \sqrt{(x+c)^2 + y^2} - \sqrt{(x-c)^2 + y^2} = \begin{cases} 2a & \text{(right branch)} \\ -2a & \text{(left branch)}. \end{cases}$$

(See Figure 8.9.) Simplifying this equation by squaring and transposing as was done for the ellipse in Example 2, we obtain the standard equation for the hyperbola:

$$\frac{x^2}{a^2} - \frac{y^2}{b^2} = 1,$$

where $b^2 = c^2 - a^2$.

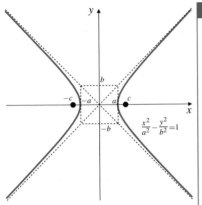

Figure 8.9

The points $(a, 0)$ and $(-a, 0)$ (called the **vertices**) lie on the hyperbola, one on each branch. (The two branches correspond to the intersections of the plane of the hyperbola with the two nappes of a cone.) Some parameters used to describe the hyperbola are

a	the **semi-transverse axis**
b	the **semi-conjugate axis**
$c = \sqrt{a^2 + b^2}$	the **semi-focal separation**.

The midpoint of the line segment $F_1 F_2$ (in this case the origin) is called the **centre** of the hyperbola. The line through the centre, the vertices, and the foci is the **transverse axis**. The line through the centre perpendicular to the transverse axis is the **conjugate axis**. The conjugate axis does not intersect the hyperbola. If a rectangle with sides $2a$ and $2b$ is drawn centred at the centre of the hyperbola and with two sides tangent to the hyperbola at the vertices, the two diagonal lines of the rectangle are **asymptotes** of the hyperbola. They have equations $(x/a) \pm (y/b) = 0$; that is, they are solutions of the degenerate equation

$$\frac{x^2}{a^2} - \frac{y^2}{b^2} = 0.$$

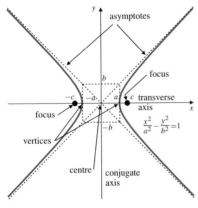

Figure 8.10 Terms associated with a hyperbola

The hyperbola approaches arbitrarily close to these lines as it recedes from the origin. (See Figure 8.10.) A **rectangular** hyperbola is one whose asymptotes are perpendicular lines. (This is so if $b = a$.)

The eccentricity of the hyperbola is

$$\varepsilon = \frac{c}{a} = \frac{\sqrt{a^2 + b^2}}{a}.$$

Note that $\varepsilon > 1$. The lines $x = \pm(a/\varepsilon)$ are called the **directrices** of the hyperbola $(x^2/a^2) - (y^2/b^2) = 1$. (See Figure 8.11.) In a manner similar to that used for the ellipse, you can show that if P is on the hyperbola, then

$$\frac{\text{distance from } P \text{ to a focus}}{\text{distance from } P \text{ to the corresponding directrix}} = \varepsilon.$$

The eccentricity of a rectangular hyperbola is $\sqrt{2}$.

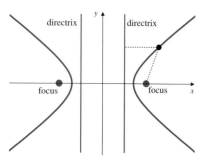

Figure 8.11   The directrices of a hyperbola

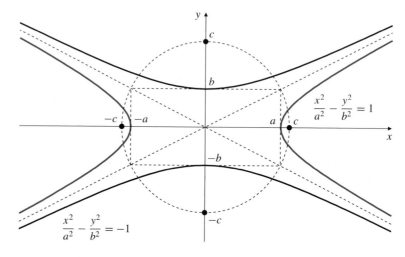

Figure 8.12   Two conjugate hyperbolas and their common asymptotes

A hyperbola with the same asymptotes as $x^2/a^2 - y^2/b^2 = 1$, but with transverse axis along the y-axis, vertices at $(0, b)$ and $(0, -b)$ and foci at $(0, c)$ and $(0, -c)$ is represented by the equation

$$\frac{x^2}{a^2} - \frac{y^2}{b^2} = -1.$$

The two hyperbolas are said to be **conjugate** to one another. (See Figure 8.12.) The *conjugate axis* of a hyperbola is the *transverse axis* of the conjugate hyperbola. Together, the transverse and conjugate axes of a hyperbola are called its **principal axes**.

The Focal Property of a Hyperbola

Let P be any point on a hyperbola with foci F_1 and F_2. Then the tangent line to the hyperbola at P bisects the angle between the lines $F_1 P$ and $F_2 P$.

> **Reflection by a hyperbola**
>
> A ray from one focus of a hyperbola is reflected by the hyperbola so that it appears to have come from the other focus.

To see this, let P be on the right branch, let T be the line tangent to the hyperbola at P, and let C be a circle of large radius centred at F_2. (See Figure 8.13.) Let $F_2 P$ intersect this circle at D. Let Q be any point on T. Then $Q F_1$ meets the hyperbola at X between Q and F_1, and $F_2 X$ meets C at E. Since X is on the radial line $F_2 E$,

!BEWARE

Check the equalities and inequalities in this chain one at a time to make sure you understand why it is true.

it is closer to E than it is to other points on C. That is, $XE \leq XD$. Thus

$$
\begin{aligned}
F_1P + PD &= F_1P + F_2D - F_2P \\
&= F_2D - (F_2P - F_1P) \\
&= F_2E - (F_2X - F_1X) \\
&= F_1X + F_2E - F_2X \\
&= F_1X + XE \\
&\leq F_1X + XD \\
&\leq F_1X + XQ + QD = F_1Q + QD.
\end{aligned}
$$

P is the point on T which minimizes the sum of distances to F_1 and D; therefore the normal to the hyperbola at P bisects the angle F_1PD. Therefore, T bisects the angle F_1PF_2.

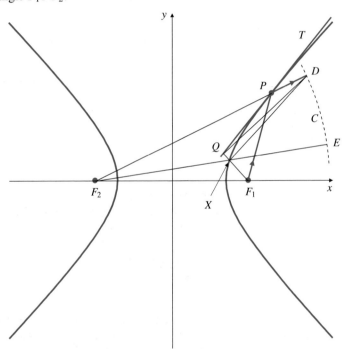

Figure 8.13 A ray from one focus is reflected along a line from the other focus

Classifying General Conics

A second-degree equation in two variables,

$$Ax^2 + Bxy + Cy^2 + Dx + Ey + F = 0, \qquad (A^2 + B^2 + C^2 > 0),$$

generally represents a conic curve, but in certain degenerate cases it may represent two straight lines ($x^2 - y^2 = 0$ represents the lines $x = y$ and $x = -y$), one straight line ($x^2 = 0$ represents the line $x = 0$), a single point ($x^2 + y^2 = 0$ represents the origin), or no points at all ($x^2 + y^2 = -1$ is not satisfied by any point in the plane).

The nature of the set of points represented by a given second-degree equation can be determined by rewriting the equation in a form that can be recognized as one of the standard types. If $B = 0$, this rewriting can be accomplished by completing the squares in the x and y terms.

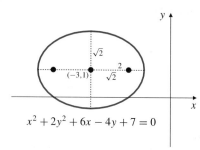

$x^2 + 2y^2 + 6x - 4y + 7 = 0$

Figure 8.14

Example 4 Describe the curve with equation $x^2 + 2y^2 + 6x - 4y + 7 = 0$.

Solution We complete the squares in the x and y terms, and rewrite the equation in the form

$$x^2 + 6x + 9 + 2(y^2 - 2y + 1) = 9 + 2 - 7 = 4$$

$$\frac{(x+3)^2}{4} + \frac{(y-1)^2}{2} = 1.$$

Therefore, it represents an ellipse with centre at $(-3, 1)$, semi-major axis $a = 2$, and semi-minor axis $b = \sqrt{2}$. Since $c = \sqrt{a^2 - b^2} = \sqrt{2}$, the foci are $(-3 \pm \sqrt{2}, 1)$. See Figure 8.14. ∎

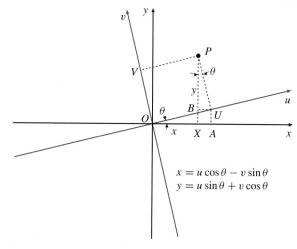

$$x = u \cos \theta - v \sin \theta$$
$$y = u \sin \theta + v \cos \theta$$

Figure 8.15 Rotation of axes

If $B \neq 0$, the equation has an xy term, and it cannot represent a circle. To see what it does represent, we can rotate the coordinate axes to produce an equation with no xy term. Let new coordinate axes (a u-axis and a v-axis) have the same origin but be rotated an angle θ from the x- and y-axes, respectively. (See Figure 8.15.) If point P has coordinates (x, y) with respect to the old axes and coordinates (u, v) with respect to the new axes, then an analysis of triangles in the figure shows that

$$x = OA - XA = OU \cos \theta - OV \sin \theta = u \cos \theta - v \sin \theta,$$
$$y = XB + BP = OU \sin \theta + OV \cos \theta = u \sin \theta + v \cos \theta.$$

Substituting these expressions into the equation

$$Ax^2 + Bxy + Cy^2 + Dx + Ey + F = 0, \qquad (A^2 + B^2 + C^2 > 0)$$

leads to a new equation,

$$A'u^2 + B'uv + C'v^2 + D'u + E'v + F = 0,$$

where

⚠ BEWARE

A lengthy calculation is needed here. The details have been omitted.

$$A' = \frac{1}{2}\Big(A(1 + \cos 2\theta) + B \sin 2\theta + C(1 - \cos 2\theta)\Big)$$
$$B' = (C - A) \sin 2\theta + B \cos 2\theta$$
$$C' = \frac{1}{2}\Big(A(1 - \cos 2\theta) - B \sin 2\theta + C(1 + \cos 2\theta)\Big)$$
$$D' = D \cos \theta + E \sin \theta$$
$$E' = -D \sin \theta + E \cos \theta.$$

Note that F remains unchanged. If we choose θ so that

$$\tan 2\theta = \frac{B}{A - C}, \qquad \text{or} \qquad \theta = \frac{\pi}{4} \text{ if } A = C, \ B \neq 0,$$

then $B' = 0$, and the new equation can then be analyzed as described previously.

Example 5   Identify the curve with equation $xy = 1$.

Solution   The reader is likely well aware that the given equation represents a rectangular hyperbola with the coordinate axes as asymptotes. Since the given equation involves $A = C = D = E = 0$ and $B = 1$, it is appropriate to rotate the axes through angle $\pi/4$ so that

$$x = \frac{1}{\sqrt{2}}(u - v), \qquad y = \frac{1}{\sqrt{2}}(u + v).$$

The transformed equation is $u^2 - v^2 = 2$, which is, as suspected, a rectangular hyperbola with vertices at $u = \pm\sqrt{2}, v = 0$, foci at $u = \pm 2, v = 0$, and asymptotes $u = \pm v$. Hence, $xy = 1$ represents a rectangular hyperbola with coordinate axes as asymptotes, vertices at $(1, 1)$ and $(-1, -1)$, and foci at $(\sqrt{2}, \sqrt{2})$ and $(-\sqrt{2}, -\sqrt{2})$.

Example 6   Show that the curve $2x^2 + xy + y^2 = 2$ is an ellipse, and find the lengths of its semi-major and semi-minor axes.

Solution   Here, $A = 2$, $B = C = 1$, $D = E = 0$, and $F = -2$. We rotate the axes through angle θ where $\tan 2\theta = B/(A - C) = 1$. Thus $B' = 0$, $2\theta = \pi/4$, and $\sin 2\theta = \cos 2\theta = 1/\sqrt{2}$. We have

$$A' = \frac{1}{2}\left[2\left(1 + \frac{1}{\sqrt{2}}\right) + \frac{1}{\sqrt{2}} + \left(1 - \frac{1}{\sqrt{2}}\right)\right] = \frac{3 + \sqrt{2}}{2}$$

$$C' = \frac{1}{2}\left[2\left(1 - \frac{1}{\sqrt{2}}\right) - \frac{1}{\sqrt{2}} + \left(1 + \frac{1}{\sqrt{2}}\right)\right] = \frac{3 - \sqrt{2}}{2}.$$

The transformed equation is $(3 + \sqrt{2})u^2 + (3 - \sqrt{2})v^2 = 4$, which represents an ellipse with semi-major axis $2/\sqrt{3 - \sqrt{2}}$ and semi-minor axis $2/\sqrt{3 + \sqrt{2}}$. (We will discover another way to do a question like this in Section 13.3.)

Exercises 8.1

Find equations of the conics specified in Exercises 1–6.

1. ellipse with foci at $(0, \pm 2)$ and semi-major axis 3.

2. ellipse with foci at $(0, 1)$ and $(4, 1)$ and eccentricity $1/2$.

3. parabola with focus at $(2, 3)$ and vertex at $(2, 4)$.

4. parabola passing through the origin and having focus at $(0, -1)$ and axis along $y = -1$.

5. hyperbola with foci at $(0, \pm 2)$ and semi-transverse axis 1.

6. hyperbola with foci at $(\pm 5, 1)$ and asymptotes

$$x = \pm(y - 1).$$

In Exercises 7–15, identify and sketch the set of points in the plane satisfying the given equation. Specify the asymptotes of any hyperbolas.

7. $x^2 + y^2 + 2x = -1$

8. $x^2 + 4y^2 - 4y = 0$

9. $4x^2 + y^2 - 4y = 0$

10. $4x^2 - y^2 - 4y = 0$

11. $x^2 + 2x - y = 3$

12. $x + 2y + 2y^2 = 1$

13. $x^2 - 2y^2 + 3x + 4y = 2$

14. $9x^2 + 4y^2 - 18x + 8y = -13$

15. $9x^2 + 4y^2 - 18x + 8y = 23$

16. Identify and sketch the curve that is the graph of the equation $(x - y)^2 - (x + y)^2 = 1$.

∗ 17. Light rays in the xy-plane coming from the point $(3, 4)$ reflect in a parabola so that they form a beam parallel to the x-axis. The parabola passes through the origin. Find its equation. (There are two possible answers.)

18. Light rays in the xy-plane coming from the origin are reflected by an ellipse so that they converge at the point $(3, 0)$. Find all possible equations for the ellipse.

In Exercises 19–22, identify the conic and find its centre, principal axes, foci, and eccentricity. Specify the asymptotes of any hyperbolas.

19. $xy + x - y = 2$

∗ 20. $x^2 + 2xy + y^2 = 4x - 4y + 4$

∗ 21. $8x^2 + 12xy + 17y^2 = 20$

∗ 22. $x^2 - 4xy + 4y^2 + 2x + y = 0$

23. The *focus-directrix definition of a conic* defines a conic as a set of points P in the plane that satisfy the condition

$$\frac{\text{distance from } P \text{ to } F}{\text{distance from } P \text{ to } D} = \varepsilon,$$

where F is a fixed point, D a fixed straight line, and ε a fixed positive number. The conic is an ellipse, a parabola, or a hyperbola according to whether $\varepsilon < 1$, $\varepsilon = 1$, or $\varepsilon > 1$. Find the equation of the conic if F is the origin and D is the line $x = -p$.

Another parameter associated with conics is the **semi-latus rectum**, usually denoted ℓ. For a circle it is equal to the radius.

For other conics it is half the length of the chord through a focus and perpendicular to the axis (for a parabola), the major axis (for an ellipse), or the transverse axis (for a hyperbola). That chord is called the **latus rectum** of the conic.

24. Show that the semi-latus rectum of the parabola is twice the distance from the vertex to the focus.

25. Show that the semi-latus rectum for an ellipse with semi-major axis a and semi-minor axis b is $\ell = b^2/a$.

26. Show that the formula in the above exercise also gives the semi-latus rectum of a hyperbola with semi-transverse axis a and semi-conjugate axis b.

∗ 27. Suppose a plane intersects a right-circular cone in an ellipse and that two spheres (one on each side of the plane) are inscribed between the cone and the plane so that each is tangent to the cone around a circle and is also tangent to the plane at a point. Show that the points where these two spheres touch the plane are the foci of the ellipse. *Hint:* all tangent lines drawn to a sphere from a given point outside the sphere are equal in length. The distance between the two circles in which the spheres intersect the cone, measured along generators of the cone (i.e., straight lines lying on the cone), is the same for all generators.

∗ 28. State and prove a result analogous to that in the above exercise but pertaining to a hyperbola.

∗ 29. Suppose a plane intersects a right-circular cone in a parabola with vertex at V. Suppose that a sphere is inscribed between the cone and the plane as in the previous exercises and is tangent to the plane of the parabola at point F. Show that the chord to the parabola through F which is perpendicular to FV has length equal to that of the latus rectum of the parabola. Therefore, F is the focus of the parabola.

8.2 Parametric Curves

Suppose that an object moves around in the xy-plane so that the coordinates of its position at any time t are continuous functions of the variable t:

$$x = f(t), \qquad y = g(t).$$

The path followed by the object is a curve C in the plane that is specified by the two equations above. We call these equations *parametric equations* of C. A curve specified by a particular pair of parametric equations is called a *parametric curve*.

DEFINITION 4

Parametric curves

A **parametric curve** C in the plane consists of an ordered pair (f, g) of continuous functions each defined on the same interval I. The equations

$$x = f(t), \qquad y = g(t), \qquad \text{for } t \text{ in } I,$$

are called **the parametric equations** of the curve C. The independent variable t is called the **parameter**.

Note that the parametric curve C was *not* defined as a set of points in the plane, but rather as the ordered pair of functions whose range is that set of points. Different pairs of functions can give the same set of points in the plane, but we may still want to regard them as different parametric curves. Nevertheless, we will often refer to the set of points (the path traced out by (x, y) as t traverses I) as the curve C. The axis (real line) of the parameter t is distinct from the coordinate axes of the plane of the curve. (See Figure 8.16.) We will usually denote the parameter by t; in many applications the parameter represents time, but this need not always be the case. Because f and g are assumed to be continuous, the curve $x = f(t)$, $y = g(t)$ has no breaks in it. A parametric curve has a *direction* (indicated, say, by arrowheads), namely, the direction corresponding to increasing values of the parameter t, as shown in Figure 8.16.

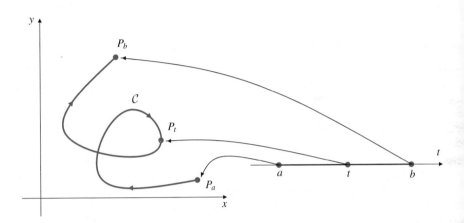

Figure 8.16   A parametric curve

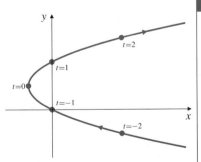

Figure 8.17   The parabola defined parametrically by $x = t^2 - 1$, $y = t + 1$, $(-\infty < t < \infty)$

Example 1   Sketch and identify the parametric curve

$$x = t^2 - 1, \qquad y = t + 1 \qquad (-\infty < t < \infty).$$

Solution   We could construct a table of values of x and y for various values of t, thus getting the coordinates of a number of points on a curve. However, for this example it is easier to *eliminate the parameter* from the pair of parametric equations, thus producing a single equation in x and y whose graph is the desired curve:

$$t = y - 1, \qquad x = t^2 - 1 = (y - 1)^2 - 1 = y^2 - 2y.$$

All points on the curve lie on the parabola $x = y^2 - 2y$. Since $y \to \pm\infty$ as $t \to \pm\infty$, the parametric curve is the whole parabola. (See Figure 8.17.) ∎

Although the curve in Example 1 is more easily identified when the parameter is eliminated, there is a loss of information in going to the nonparametric form. Specifically, we lose the sense of the curve as the path of a moving point and hence also the direction of the curve. If the t in the parametric form denotes the time at which an object is at the point (x, y), the nonparametric equation $x = y^2 - 2y$ no longer tells us where the object is at any particular time t.

Example 2 (**Parametric equations of a straight line**) The straight line passing through the two points $P_0 = (x_0, y_0)$ and $P_1 = (x_1, y_1)$ (see Figure 8.18) has parametric equations

$$\begin{cases} x = x_0 + t(x_1 - x_0) \\ y = y_0 + t(y_1 - y_0) \end{cases} \quad (-\infty < t < \infty).$$

To see that these equations represent a straight line, note that

$$\frac{y - y_0}{x - x_0} = \frac{y_1 - y_0}{x_1 - x_0} = \text{constant} \quad (\text{assuming } x_1 \neq x_0).$$

The point $P = (x, y)$ is at position P_0 when $t = 0$ and at P_1 when $t = 1$. If $t = 1/2$, then P is the midpoint between P_0 and P_1. Note that the line segment from P_0 to P_1 corresponds to values of t between 0 and 1.

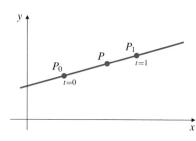

Figure 8.18

Example 3 (**An arc of a circle**) Sketch and identify the curve $x = 3\cos t$, $y = 3\sin t$, $(0 \leq t \leq 3\pi/2)$.

Solution Since $x^2 + y^2 = 9\cos^2 t + 9\sin^2 t = 9$, all points on the curve lie on the circle $x^2 + y^2 = 9$. As t increases from 0 through $\pi/2$ and π to $3\pi/2$, the point (x, y) moves from $(3, 0)$ through $(0, 3)$ and $(-3, 0)$ to $(0, -3)$. The parametric curve is three-quarters of the circle. See Figure 8.19. The parameter t has geometric significance in this example. If P_t is the point on the curve corresponding to parameter value t, then t is the angle at the centre of the circle corresponding to the arc from the initial point to P_t.

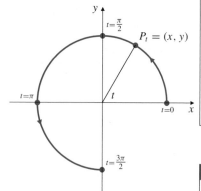

Figure 8.19

Example 4 (**Parametric equations of an ellipse**) Sketch and identify the curve $x = a\cos t$, $y = b\sin t$, $(0 \leq t \leq 2\pi)$, where $a > b > 0$.

Solution Observe that

$$\frac{x^2}{a^2} + \frac{y^2}{b^2} = \cos^2 t + \sin^2 t = 1.$$

Therefore, the curve is all or part of an ellipse with major axis from $(-a, 0)$ to $(a, 0)$ and minor axis from $(0, -b)$ to $(0, b)$. As t increases from 0 to 2π, the point (x, y) moves counterclockwise around the ellipse starting from $(a, 0)$ and returning to the same point. Thus the curve is the whole ellipse.

Figure 8.20(a) shows how the parameter t can be interpreted as an angle and how the points on the ellipse can be obtained using circles of radii a and b. Since the curve starts and ends at the same point, it is called a **closed curve**.

Example 5 Sketch the parametric curve

$$x = t^3 - 3t, \quad y = t^2 \quad (-2 \leq t \leq 2).$$

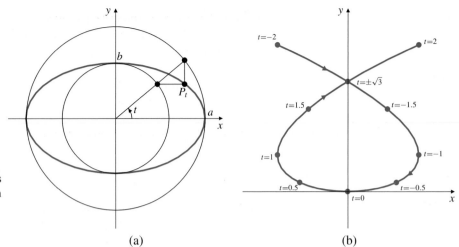

Figure 8.20

(a) An ellipse parametrized in terms
of an angle and constructed with
the help of two circles

(b) A self-intersecting parametric
curve

Solution   We could eliminate the parameter and obtain

$$x^2 = t^2(t^2 - 3)^2 = y(y - 3)^2,$$

but this doesn't help much since we do not recognize this curve from its Cartesian
equation. Instead, let us calculate the coordinates of some points:

Table 2.   Coordinates of some points on the curve of Example 5

t	-2	$-\dfrac{3}{2}$	-1	$-\dfrac{1}{2}$	0	$\dfrac{1}{2}$	1	$\dfrac{3}{2}$	2
x	-2	$\dfrac{9}{8}$	2	$\dfrac{11}{8}$	0	$-\dfrac{11}{8}$	-2	$-\dfrac{9}{8}$	2
y	4	$\dfrac{9}{4}$	1	$\dfrac{1}{4}$	0	$\dfrac{1}{4}$	1	$\dfrac{9}{4}$	4

Note that the curve is symmetric about the y-axis because x is an odd function of
t and y is an even function of t. (At t and $-t$, x has opposite values but y has the
same value.)

The curve intersects itself on the y-axis. (See Figure 8.20(b).) To find this
self-intersection, set $x = 0$:

$$0 = x = t^3 - 3t = t(t - \sqrt{3})(t + \sqrt{3}).$$

For $t = 0$ the curve is at $(0, 0)$, but for $t = \pm\sqrt{3}$ the curve is at $(0, 3)$. The
self-intersection occurs because the curve passes through the same point for two
different values of the parameter. ∎

Remark   Here is how to get Maple to plot the parametric curve in the example
above. Note the square brackets enclosing the two functions $t^3 - 3t$ and t^2, and the
parameter interval, followed by the ranges of x and y for the plot.

```
>  plot([t^3-3*t, t^2, t=-2..2], x=-3..3, y=-1..5);
```

General Plane Curves and Parametrizations

According to Definition 4, a parametric curve always involves a particular set of parametric equations; it is not just a set of points in the plane. When we are interested in considering a curve solely as a set of points (a *geometric object*), we need not be concerned with any particular pair of parametric equations representing that curve. In this case we call the curve simply a *plane curve*.

DEFINITION 5

Plane curves

A **plane curve** is a set of points (x, y) in the plane such that $x = f(t)$ and $y = g(t)$ for some t in an interval I, where f and g are continuous functions defined on I. Any such interval I and function pair (f, g) that generate the points of C is called a **parametrization** of C.

Since a plane curve does not involve any specific parametrization, it has no specific direction.

Example 6   The circle $x^2 + y^2 = 1$ is a plane curve. Each of the following is a possible parametrization of C:

(i) $x = \cos t$, $y = \sin t$, $(0 \le t \le 2\pi)$,

(ii) $x = \sin s^2$, $y = \cos s^2$, $(0 \le s \le \sqrt{2\pi})$,

(iii) $x = \cos(\pi u + 1)$, $y = \sin(\pi u + 1)$, $(-1 \le u \le 1)$,

(iv) $x = 1 - t^2$, $y = t\sqrt{2 - t^2}$, $(-\sqrt{2} \le t \le \sqrt{2})$.

To verify that any of these represents the circle, substitute the appropriate functions for x and y in the expression $x^2 + y^2$, and show that the result simplifies to the value 1. This shows that the parametric curve lies on the circle. Then examine the ranges of x and y as the parameter varies over its domain. For example, for (iv) we have

$$x^2 + y^2 = (1 - t^2)^2 + (t\sqrt{2 - t^2})^2 = 1 - 2t^2 + t^4 + 2t^2 - t^4 = 1,$$

and (x, y) moves from $(-1, 0)$ through $(0, -1)$ to $(1, 0)$ as t increases from $-\sqrt{2}$ through -1 to 0, and then continues on through $(0, 1)$ back to $(-1, 0)$ as t continues to increase from 0 through 1 to $\sqrt{2}$.

There are, of course, infinitely many other possible parametrizations of this curve. ∎

Example 7   If f is a continuous function on an interval I, then the graph of f is a plane curve. One obvious parametrization of this curve is

$$x = t,   y = f(t),   (t \text{ in } I).$$

Some Interesting Plane Curves

We complete this section by parametrizing two curves that arise in the physical world.

The brachistochrone and tautochrone problems

Suppose a wire is bent into a curve from point A to a lower point B and a bead can slide without friction along the wire. If the bead is released at A, it will fall toward B. What curve should be used to minimize the time it takes to fall from A to B? This problem, known as the *brachistochrone* (Greek for "shortest time") problem, has as its solution part of an upside down arch of a cycloid. Moreover, it takes the same amount of time for the bead to slide from any point on the curve to the lowest point B, making the cycloid the solution of the *tautochrone* ("equal time") problem as well. We will examine these matters further in the Challenging Exercises at the end of Chapter 11.

Example 8 (A cycloid) If a circle rolls without slipping along a straight line, find the path followed by a point fixed on the circle. This path is called a **cycloid**.

Solution Suppose that the line on which the circle rolls is the x-axis, that the circle has radius a and lies above the line, and that the point whose motion we follow is originally at the origin O. See Figure 8.21. After the circle has rolled through an angle t, it is tangent to the line at T, and the point whose path we are trying to find has moved to position P, as shown in the figure. Since no slipping occurs,

$$\text{segment } OT = \text{ arc } PT = at.$$

Let PQ be perpendicular to TC, as shown in the figure. If P has coordinates (x, y), then

$$x = OT - PQ = at - a\sin(\pi - t) = at - a\sin t,$$
$$y = TC + CQ = a + a\cos(\pi - t) = a - a\cos t.$$

The parametric equations of the cycloid are therefore

$$x = a(t - \sin t), \qquad y = a(1 - \cos t).$$

Observe that the cycloid has a cusp at the points where it returns to the x-axis, that is, at points corresponding to $t = 2n\pi$, where n is an integer. Even though the functions x and y are everywhere differentiable functions of t, the curve is not smooth everywhere. We shall consider such matters in the next section. ∎

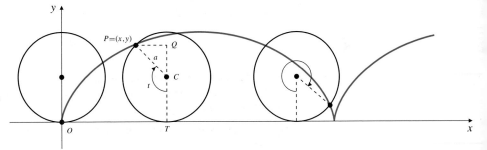

Figure 8.21 Each arch of the cycloid is traced out by P as the wheel rolls through one complete revolution

Example 9 (An involute of a circle) A string is wound around a fixed circle. One end is unwound in such a way that the part of the string not lying on the circle is extended in a straight line. The curve followed by this free end of the string is called an **involute** of the circle. (The involute of any curve is the path traced out by the end of the curve as the curve is straightened out beginning at that end.)

Suppose the circle has equation $x^2 + y^2 = a^2$, and suppose the end of the string being unwound starts at the point $A = (a, 0)$. At some subsequent time during the unwinding let P be the position of the end of the string, and let T be the point where the string leaves the circle. The line PT must be tangent to the circle at T.

We parametrize the path of P in terms of the angle AOT, which we denote by t. Let points R on OA and S on TR be as shown in Figure 8.22. TR is perpendicular to OA and to PS. Note that

$$OR = OT\cos t = a\cos t, \qquad RT = OT\sin t = a\sin t.$$

Since angle OTP is $90°$, we have angle $STP = t$. Since $PT = \text{arc } AT = at$ (because the string does not stretch or slip on the circle), we have

$$SP = TP\sin t = at\sin t, \qquad ST = TP\cos t = at\cos t.$$

If P has coordinates (x, y), then $x = OR + SP$, and $y = RT - ST$:

$$x = a\cos t + at\sin t, \qquad y = a\sin t - at\cos t, \qquad (t \geq 0).$$

These are parametric equations of the involute.

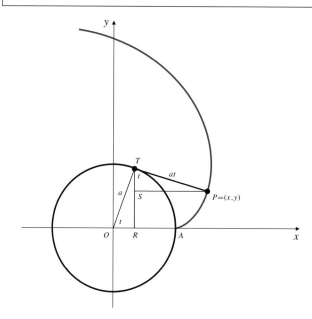

Figure 8.22   An involute of a circle

Exercises 8.2

In Exercises 1–10, sketch the given parametric curve, showing its direction with an arrow. Eliminate the parameter to give a Cartesian equation in x and y whose graph contains the parametric curve.

1. $x = 1 + 2t$, $y = t^2$, $(-\infty < t < \infty)$

2. $x = 2 - t$, $y = t + 1$, $(0 \leq t < \infty)$

3. $x = \dfrac{1}{t}$, $y = t - 1$, $(0 < t < 4)$

4. $x = \dfrac{1}{1+t^2}$, $y = \dfrac{t}{1+t^2}$, $(-\infty < t < \infty)$

5. $x = 3\sin 2t$, $y = 3\cos 2t$, $\left(0 \leq t \leq \dfrac{\pi}{3}\right)$

6. $x = a\sec t$, $y = b\tan t$, $\left(-\dfrac{\pi}{2} < t < \dfrac{\pi}{2}\right)$

7. $x = 3\sin \pi t$, $y = 4\cos \pi t$, $(-1 \leq t \leq 1)$

8. $x = \cos\sin s$, $y = \sin\sin s$, $(-\infty < s < \infty)$

9. $x = \cos^3 t$, $y = \sin^3 t$, $(0 \leq t \leq 2\pi)$

10. $x = 1 - \sqrt{4 - t^2}$, $y = 2 + t$, $(-2 \leq t \leq 2)$

11. Describe the parametric curve $x = \cosh t$, $y = \sinh t$, and find its Cartesian equation.

12. Describe the parametric curve $x = 2 - 3\cosh t$, $y = -1 + 2\sinh t$.

13. Describe the curve $x = t\cos t$, $y = t\sin t$, $(0 \le t \le 4\pi)$.

14. Show that each of the following sets of parametric equations represents a different arc of the parabola with equation $2(x + y) = 1 + (x - y)^2$.

    (i) $x = \cos^4 t$, $y = \sin^4 t$

    (ii) $x = \sec^4 t$, $y = \tan^4 t$

    (iii) $x = \tan^4 t$, $y = \sec^4 t$

15. Find a parametrization of the parabola $y = x^2$ using as parameter the slope of the tangent line at the general point.

16. Find a parametrization of the circle $x^2 + y^2 = R^2$ using as parameter the slope m of the line joining the general point to the point $(R, 0)$. Does the parametrization fail to give any point on the circle?

17. A circle of radius a is centred at the origin O. T is a point on the circle such that OT makes angle t with the positive x-axis. The tangent to the circle at T meets the x-axis at X. The point $P = (x, y)$ is at the intersection of the vertical line through X and the horizontal line through T. Find, in terms of the parameter t, parametric equations for the curve C traced out by P as T moves around the circle. Also, eliminate t and find an equation for C in x and y. Sketch C.

18. Repeat Exercise 17 with the following modification. OT meets a second circle of radius b centred at O at the point Y. $P = (x, y)$ is at the intersection of the vertical line through X and the horizontal line through Y.

* 19. (**The folium of Descartes**) Eliminate the parameter from the parametric equations

$$x = \frac{3t}{1 + t^3}, \qquad y = \frac{3t^2}{1 + t^3} \qquad (t \ne -1),$$

and hence find an ordinary equation in x and y for this curve. The parameter t can be interpreted as the slope of the line joining the general point (x, y) to the origin. Sketch the curve and show that the line $x + y = -1$ is an asymptote.

* 20. (**A prolate cycloid**) A railroad wheel has a flange extending below the level of the track on which the wheel rolls. If the radius of the wheel is a and that of the flange is $b > a$, find parametric equations of the path of a point P at the circumference of the flange as the wheel rolls along the track. (Note that for a portion of each revolution of the wheel, P is moving backward.) Try to sketch the graph of this prolate cycloid.

* 21. (**Hypocycloids**) If a circle of radius b rolls, without slipping, around the inside of a fixed circle of radius $a > b$, a point on the circumference of the rolling circle traces a curve called a hypocycloid. If the fixed circle is centred at the origin and the point tracing the curve starts at $(a, 0)$, show that the

hypocycloid has parametric equations

$$x = (a - b)\cos t + b\cos\left(\frac{a - b}{b}t\right),$$

$$y = (a - b)\sin t - b\sin\left(\frac{a - b}{b}t\right),$$

where t is the angle between the positive x-axis and the line from the origin to the point at which the rolling circle touches the fixed circle.

If $a = 2$ and $b = 1$, show that the hypocycloid becomes a straight line segment.

If $a = 4$ and $b = 1$, show that the parametric equations of the hypocycloid simplify to $x = 4\cos^3 t$, $y = 4\sin^3 t$. This curve is called a hypocycloid of four cusps or an **astroid**. (See Figure 8.23.) It has Cartesian equation $x^{2/3} + y^{2/3} = 4^{2/3}$.

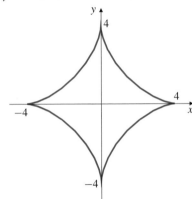

Figure 8.23   The astroid $x^{2/3} + y^{2/3} = 4^{2/3}$

Hypocycloids resemble the curves produced by a popular children's toy called Spirograph, but Spirograph curves result from following a point inside the disc of the rolling circle rather than on its circumference, and they therefore do not have sharp cusps.

* 22. (**The witch of Agnesi**)

(a) Show that the curve traced out by the point P constructed from a circle as shown in Figure 8.24 has parametric equations $x = \tan t$, $y = \cos^2 t$ in terms of the angle t shown. (*Hint:* you will need to make extensive use of similar triangles.)

(b) Use a trigonometric identity to eliminate t from the parametric equations, and hence find an ordinary Cartesian equation for the curve.

This curve is named for the Italian mathematician Maria Agnesi (1718-1799), one of the foremost women scholars of her century and author of an important calculus text. The term *witch* is due to a mistranslation of the Italian word *versiera* ("turning curve"), which she used to describe the curve. The word is similar to *avversiera* ("wife of the devil" or "witch").

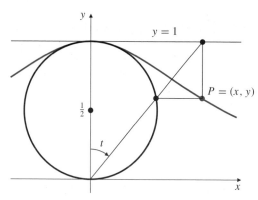

Figure 8.24 The witch of Agnesi

In Exercises 23–26, obtain a graph of the curve $x = \sin(mt)$, $y = \sin(nt)$ for the given values of m and n. Such curves are called **Lissajous figures**. They arise in the analysis of electrical signals using an oscilloscope. A signal of fixed but unknown frequency is applied to the vertical input, and a control signal is applied to the horizontal input. The horizontal frequency is varied until a stable Lissajous figure is observed. The (known) frequency of the control signal and the shape of the figure then determine the unknown frequency.

23. $m = 1$, $n = 2$ **24.** $m = 1$, $n = 3$

25. $m = 2$, $n = 3$ **26.** $m = 2$, $n = 5$

27. (Epicycloids) Use a graphing calculator or computer graphing program to investigate the behaviour of curves with equations of the form

$$x = \left(1 + \frac{1}{n}\right)\cos t - \frac{1}{n}\cos(nt)$$

$$y = \left(1 + \frac{1}{n}\right)\sin t - \frac{1}{n}\sin(nt)$$

for various integer and fractional values of $n \geq 3$. Can you formulate any principles governing the behaviour of such curves?

28. (More hypocycloids) Use a graphing calculator or computer graphing program to investigate the behaviour of curves with equations of the form

$$x = \left(1 + \frac{1}{n}\right)\cos t + \frac{1}{n}\cos((n-1)t)$$

$$y = \left(1 + \frac{1}{n}\right)\sin t - \frac{1}{n}\sin((n-1)t)$$

for various integer and fractional values of $n \geq 3$. Can you formulate any principles governing the behaviour of these curves?

8.3 Smooth Parametric Curves and Their Slopes

We say that a plane curve is *smooth* if it has a tangent line at each point P and this tangent turns in a continuous way as P moves along the curve. (That is, the angle between the tangent line at P and some fixed line, the x-axis say, is a continuous function of the position of P.)

If the curve $\mathcal{C}$ is the graph of function f, then $\mathcal{C}$ is certainly smooth on any interval where the derivative $f'(x)$ exists and is a continuous function of x. It may also be smooth on intervals containing isolated singular points; for example, the curve $y = x^{1/3}$ is smooth everywhere even though dy/dx does not exist at $x = 0$.

For parametric curves $x = f(t)$, $y = g(t)$, the situation is more complicated. Even if f and g have continuous derivatives everywhere, such curves may fail to be smooth at certain points, specifically points where $f'(t) = g'(t) = 0$.

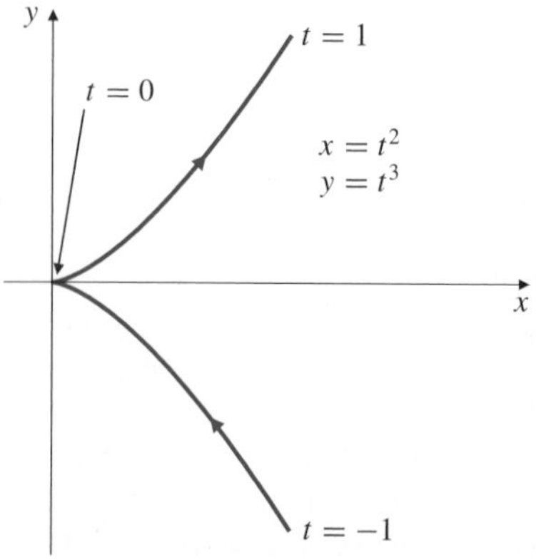

Example 1 Consider the parametric curve $x = f(t) = t^2$, $y = g(t) = t^3$. Even though $f'(t) = 2t$ and $g'(t) = 3t^2$ are continuous for all t, the curve is not smooth at $t = 0$. (See Figure 8.25.) Observe that both f' and g' vanish at $t = 0$: $f'(0) = g'(0) = 0$. If we regard the parametric equations as specifying the position at time t of a moving point P, then the horizontal velocity is $f'(t)$ and the vertical velocity is $g'(t)$. Both velocities are 0 at $t = 0$, so P has come to a stop at that instant. When it starts moving again, it need not move in the direction it was going before it stopped. The cycloid of Example 8 of Section 8.2 is another example where a parametric curve is not smooth at points where dx/dt and dy/dt both vanish.

Figure 8.25 This curve is not smooth at the origin but has a cusp there

(figure labels: $t = 0$, $t = 1$, $x = t^2$, $y = t^3$, $t = -1$)

The Slope of a Parametric Curve

The following theorem confirms that a parametric curve is smooth at points where the derivatives of its coordinate functions are continuous and not both zero.

THEOREM 1

Let C be the parametric curve $x = f(t)$, $y = g(t)$, where $f'(t)$ and $g'(t)$ are continuous on an interval I. If $f'(t) \neq 0$ on I, then C is smooth and has at each t a tangent line with slope

$$\frac{dy}{dx} = \frac{g'(t)}{f'(t)}.$$

If $g'(t) \neq 0$ on I, then C is smooth and has at each t a normal line with slope

$$-\frac{dx}{dy} = -\frac{f'(t)}{g'(t)}.$$

Thus, C is smooth except possibly at points where $f'(t)$ and $g'(t)$ are both 0.

PROOF   If $f'(t) \neq 0$ on I, then f is either increasing or decreasing on I and so is one-to-one and invertible. The part of C corresponding to values of t in I has ordinary equation $y = g\big(f^{-1}(x)\big)$ and hence slope

$$\frac{dy}{dx} = g'\big(f^{-1}(x)\big) \frac{d}{dx} f^{-1}(x) = \frac{g'\big(f^{-1}(x)\big)}{f'\big(f^{-1}(x)\big)} = \frac{g'(t)}{f'(t)}.$$

We have used here the formula

$$\frac{d}{dx} f^{-1}(x) = \frac{1}{f'\big(f^{-1}(x)\big)}$$

for the derivative of an inverse function obtained in Section 3.1. This slope is a continuous function of t, so the tangent to C turns continuously for t in I. The proof for $g'(t) \neq 0$ is similar. In this case the slope of the normal is a continuous function of t, so the normal turns continuously. Therefore so does the tangent.

If f' and g' are continuous, and both vanish at some point t_0, then the curve $x = f(t)$, $y = g(t)$ *may or may not* be smooth around t_0. Example 1 was an example of a curve that was not smooth at such a point.

Example 2   The curve with parametrization $x = t^3$, $y = t^6$ is just the parabola $y = x^2$, so it is smooth everywhere, although $dx/dt = 3t^2$ and $dy/dt = 6t^5$ both vanish at $t = 0$.

Tangents and normals to parametric curves

If f' and g' are continuous and not both 0 at t_0, then the parametric equations

$$\begin{cases} x = f(t_0) + f'(t_0)(t - t_0) \\ y = g(t_0) + g'(t_0)(t - t_0) \end{cases} \qquad (-\infty < t < \infty)$$

represent the tangent line to the parametric curve $x = f(t)$, $y = g(t)$ at the point $\big(f(t_0), g(t_0)\big)$. The normal line there has parametric equations

$$\begin{cases} x = f(t_0) + g'(t_0)(t - t_0) \\ y = g(t_0) - f'(t_0)(t - t_0) \end{cases} \qquad (-\infty < t < \infty).$$

Both lines pass through $\big(f(t_0), g(t_0)\big)$ when $t = t_0$.

Example 3   Find equations of the tangent and normal lines to the parametric curve $x = t^2 - t$, $y = t^2 + t$ at the point where $t = 2$.

Solution   At $t = 2$ we have $x = 2$, $y = 6$ and

$$\frac{dx}{dt} = 2t - 1 = 3, \qquad \frac{dy}{dt} = 2t + 1 = 5.$$

Hence, the tangent and the normal lines have parametric equations

Tangent: $\begin{cases} x = 2 + 3(t - 2) = 3t - 4 \\ y = 6 + 5(t - 2) = 5t - 4 \end{cases}$

Normal: $\begin{cases} x = 2 + 5(t - 2) = 5t - 8 \\ y = 6 - 3(t - 2) = -3t + 12 \end{cases}$

The concavity of a parametric curve can be determined using the second derivatives of the parametric equations. The procedure is just to calculate d^2y/dx^2 using the chain rule:

$$\frac{d^2y}{dx^2} = \frac{d}{dx}\frac{dy}{dx} = \frac{d}{dx}\frac{g'(t)}{f'(t)} = \frac{d}{dt}\left(\frac{g'(t)}{f'(t)}\right)\frac{dt}{dx}$$
$$= \frac{f'(t)g''(t) - g'(t)f''(t)}{(f'(t))^2}\frac{1}{f'(t)}.$$

Concavity of a parametric curve

On an interval where $f'(t) \neq 0$, the parametric curve $x = f(t)$, $y = g(t)$ has concavity determined by

$$\frac{d^2y}{dx^2} = \frac{f'(t)g''(t) - g'(t)f''(t)}{(f'(t))^3}.$$

Sketching Parametric Curves

As in the case of graphs of functions, derivatives provide useful information about the shape of a parametric curve. At points where $dy/dt = 0$ but $dx/dt \neq 0$, the tangent is horizontal; at points where $dx/dt = 0$ but $dy/dt \neq 0$, the tangent is vertical. For points where $dx/dt = dy/dt = 0$, anything can happen; it is wise to calculate left- and right-hand limits of the slope dy/dx as the parameter t approaches one of these points. Concavity can be determined using the formula obtained above. We illustrate these ideas by reconsidering a parametric curve encountered in the previous section.

Example 4   Use slope and concavity information to sketch the graph of the parametric curve

$$x = f(t) = t^3 - 3t, \qquad y = g(t) = t^2, \qquad (-2 \le t \le 2)$$

previously encountered in Example 5 of Section 8.2.

Solution   We have

$$f'(t) = 3(t^2 - 1) = 3(t-1)(t+1), \qquad g'(t) = 2t.$$

The curve has a horizontal tangent at $t = 0$, that is, at $(0,0)$, and vertical tangents at $t = \pm 1$, that is, at $(2, 1)$ and $(-2, 1)$. Directional information for the curve between these points is summarized in the following chart.

t	-2		-1		0		1		2
$f'(t)$		$+$	0	$-$	$-$	$-$	0	$+$	
$g'(t)$		$-$	$-$	$-$	0	$+$	$+$	$+$	
x		$\rightarrow$	$\cdot$	$\leftarrow$	$\leftarrow$	$\leftarrow$	$\cdot$	$\rightarrow$	
y		$\downarrow$	$\downarrow$	$\downarrow$	$\cdot$	$\uparrow$	$\uparrow$	$\uparrow$	
curve		$\searrow$	$\downarrow$	$\swarrow$	$\leftarrow$	$\nwarrow$	$\uparrow$	$\nearrow$	

For concavity we calculate the second derivative d^2y/dx^2 by the formula obtained above. Since $f''(t) = 6t$ and $g''(t) = 2$, we have

$$\frac{d^2y}{dx^2} = \frac{f'(t)g''(t) - g'(t)f''(t)}{(f'(t))^3}$$

$$= \frac{3(t^2-1)(2) - 2t(6t)}{[3(t^2-1)]^3} = -\frac{2}{9}\frac{t^2+1}{(t^2-1)^3},$$

which is never zero but which fails to be defined at $t = \pm 1$. Evidently the curve is concave upward for $-1 < t < 1$ and concave downward elsewhere. The curve is sketched in Figure 8.26.

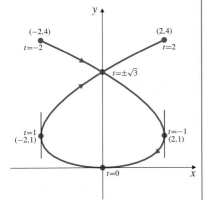

Figure 8.26

Exercises 8.3

In Exercises 1–8, find the coordinates of the points at which the given parametric curve has (a) a horizontal tangent and (b) a vertical tangent.

1. $x = t^2 + 1$, $y = 2t - 4$ **2.** $x = t^2 - 2t$, $y = t^2 + 2t$

3. $x = t^2 - 2t$, $y = t^3 - 12t$

4. $x = t^3 - 3t$, $y = 2t^3 + 3t^2$

5. $x = te^{-t^2/2}$, $y = e^{-t^2}$

6. $x = \sin t$, $y = \sin t - t \cos t$

7. $x = \sin 2t$, $y = \sin t$ **8.** $x = \dfrac{3t}{1+t^3}$, $y = \dfrac{3t^2}{1+t^3}$

Find the slopes of the curves in Exercises 9–12 at the points indicated.

9. $x = t^3 + t$, $y = 1 - t^3$, at $t = 1$

10. $x = t^4 - t^2$, $y = t^3 + 2t$, at $t = -1$

11. $x = \cos 2t$, $y = \sin t$, at $t = \pi/6$

12. $x = e^{2t}$, $y = te^{2t}$, at $t = -2$

Find parametric equations of the tangents to the curves in Exercises 13–14 at the indicated points.

13. $x = t^3 - 2t$, $y = t + t^3$, at $t = 1$

14. $x = t - \cos t$, $y = 1 - \sin t$, at $t = \pi/4$

15. Show that the curve $x = t^3 - t$, $y = t^2$ has two different tangent lines at the point $(0, 1)$ and find their slopes.

16. Find the slopes of two lines that are tangent to $x = \sin t$, $y = \sin 2t$ at the origin.

Where, if anywhere, do the curves in Exercises 17–20 fail to be smooth?

17. $x = t^3$, $y = t^2$

18. $x = (t - 1)^4$, $y = (t - 1)^3$

19. $x = t \sin t$, $y = t^3$ **20.** $x = t^3$, $y = t - \sin t$

In Exercises 21–25, sketch the graphs of the given parametric curves, making use of information from the first two derivatives. Unless otherwise stated, the parameter interval for each curve is the whole real line.

21. $x = t^2 - 2t$, $y = t^2 - 4t$ **22.** $x = t^3$, $y = 3t^2 - 1$

23. $x = t^3 - 3t$, $y = \dfrac{2}{1 + t^2}$

24. $x = t^3 - 3t - 2$, $y = t^2 - t - 2$

25. $x = \cos t + t \sin t$, $y = \sin t - t \cos t$, $(t \geq 0)$. (See Example 9 of Section 8.2.)

8.4 Arc Lengths and Areas for Parametric Curves

In this section we look at the problems of finding lengths of curves defined parametrically, areas of surfaces of revolution obtained by rotating parametric curves, and areas of plane regions bounded by parametric curves.

Arc Lengths and Surface Areas

Let $\mathcal{C}$ be a smooth parametric curve with equations

$$x = f(t), \qquad y = g(t), \qquad (a \leq t \leq b).$$

(We assume that $f'(t)$ and $g'(t)$ are continuous on the interval $[a, b]$ and are never both zero.) From the differential triangle with legs dx and dy and hypotenuse ds (see Figure 8.27), we obtain $(ds)^2 = (dx)^2 + (dy)^2$, so we have

The arc length element for a parametric curve

$$ds = \frac{ds}{dt}\,dt = \sqrt{\left(\frac{ds}{dt}\right)^2}\,dt = \sqrt{\left(\frac{dx}{dt}\right)^2 + \left(\frac{dy}{dt}\right)^2}\,dt$$

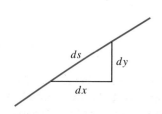

Figure 8.27   A differential triangle

The length of the curve $\mathcal{C}$ is given by

$$s = \int_{t=a}^{t=b} ds = \int_a^b \sqrt{\left(\frac{dx}{dt}\right)^2 + \left(\frac{dy}{dt}\right)^2}\,dt.$$

Example 1   Find the length of the parametric curve

$$x = e^t \cos t, \qquad y = e^t \sin t, \qquad (0 \le t \le 2).$$

Solution   We have

$$\frac{dx}{dt} = e^t(\cos t - \sin t), \qquad \frac{dy}{dt} = e^t(\sin t + \cos t).$$

Squaring these formulas, adding and simplifying, we get

$$\left(\frac{ds}{dt}\right)^2 = e^{2t}(\cos t - \sin t)^2 + e^{2t}(\sin t + \cos t)^2$$

$$= e^{2t}\left(\cos^2 t - 2\cos t \sin t + \sin^2 t + \sin^2 t + 2\sin t \cos t + \cos^2 t\right)$$

$$= 2e^{2t}.$$

The length of the curve is therefore

$$s = \int_0^2 \sqrt{2e^{2t}}\, dt = \sqrt{2} \int_0^2 e^t\, dt = \sqrt{2}\,(e^2 - 1) \text{ units.}$$

Parametric curves can be rotated around various axes to generate surfaces of revolution. The areas of these surfaces can be found by the same procedure used for graphs of functions, with the appropriate version of ds. If the curve

$$x = f(t), \qquad y = g(t), \qquad (a \le t \le b)$$

is rotated about the x-axis, the area S of the surface so generated is given by

$$S = 2\pi \int_{t=a}^{t=b} |y|\, ds = 2\pi \int_a^b |g(t)|\sqrt{(f'(t))^2 + (g'(t))^2}\, dt.$$

If the rotation is about the y-axis, then the area is

$$S = 2\pi \int_{t=a}^{t=b} |x|\, ds = 2\pi \int_a^b |f(t)|\sqrt{(f'(t))^2 + (g'(t))^2}\, dt.$$

Example 2   Find the area of the surface of revolution obtained by rotating the astroid curve

$$x = a \cos^3 t, \qquad y = a \sin^3 t$$

(where $a > 0$) about the x-axis.

Solution   The curve is symmetric about both coordinate axes. (See Figure 8.28.) The entire surface will be generated by rotating the upper half of the curve, and, in fact, we need only rotate the first quadrant part and multiply by 2. The first quadrant part of the curve corresponds to $0 \le t \le \pi/2$. We have

$$\frac{dx}{dt} = -3a \cos^2 t \sin t, \qquad \frac{dy}{dt} = 3a \sin^2 t \cos t.$$

Accordingly, the arc length element is

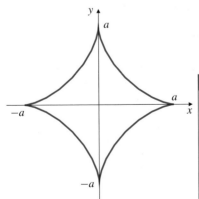

Figure 8.28

$$ds = \sqrt{9a^2 \cos^4 t \sin^2 t + 9a^2 \sin^4 t \cos^2 t}\, dt$$
$$= 3a \cos t \sin t \sqrt{\cos^2 t + \sin^2 t}\, dt$$
$$= 3a \cos t \sin t\, dt.$$

Therefore, the required surface area is

$$S = 2 \times 2\pi \int_0^{\pi/2} a \sin^3 t\, 3a \cos t \sin t\, dt$$
$$= 12\pi a^2 \int_0^{\pi/2} \sin^4 t \cos t\, dt \qquad \text{Let } u = \sin t,$$
$$\hspace{8cm} du = \cos t\, dt.$$
$$= 12\pi a^2 \int_0^1 u^4\, du = \frac{12\pi a^2}{5} \text{ square units.}$$

Areas Bounded by Parametric Curves

Consider the parametric curve $\mathcal{C}$ with equations $x = f(t)$, $y = g(t)$, $(a \le t \le b)$, where f is differentiable and g is continuous on $[a, b]$. For the moment, let us also assume that $f'(t) \ge 0$ and $g(t) \ge 0$ on $[a, b]$, so $\mathcal{C}$ has no points below the x-axis and is traversed from left to right as t increases from a to b.

The region under $\mathcal{C}$ and above the x-axis has area element given by $dA = y\, dx = g(t) f'(t)\, dt$, so its area (see Figure 8.29) is

$$A = \int_a^b g(t) f'(t)\, dt.$$

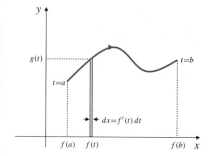

Figure 8.29

Similar arguments can be given for three other cases:

If $f'(t) \ge 0$ and $g(t) \le 0$ on $[a, b]$, then $A = -\int_a^b g(t) f'(t)\, dt$,

If $f'(t) \le 0$ and $g(t) \ge 0$ on $[a, b]$, then $A = -\int_a^b g(t) f'(t)\, dt$,

If $f'(t) \le 0$ and $g(t) \le 0$ on $[a, b]$, then $A = \int_a^b g(t) f'(t)\, dt$,

where A is the (positive) area bounded by $\mathcal{C}$, the x-axis, and the vertical lines $x = f(a)$ and $x = f(b)$. Combining these results we can see that

$$\int_a^b g(t) f'(t)\, dt = A_1 - A_2,$$

where A_1 is the area lying vertically between $\mathcal{C}$ and that part of the x-axis consisting of points $x = f(t)$ such that $g(t) f'(t) \ge 0$, and A_2 is a similar area corresponding to points where $g(t) f'(t) < 0$. This formula is valid for arbitrary continuous g and differentiable f. See Figure 8.30 for generic examples. In particular, if $\mathcal{C}$ is a non-self-intersecting closed curve, then the area of the region bounded by $\mathcal{C}$ is given by

$$A = \int_a^b g(t) f'(t) \, dt \qquad \text{if } C \text{ is traversed clockwise as } t \text{ increases,}$$

$$A = -\int_a^b g(t) f'(t) \, dt \qquad \text{if } C \text{ is traversed counterclockwise,}$$

both of which are illustrated in Figure 8.31.

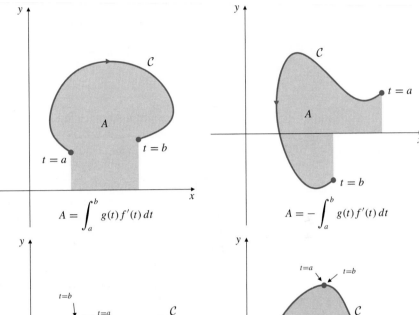

Figure 8.30   Areas defined by parametric curves

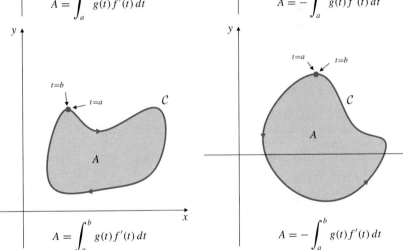

Figure 8.31   Areas bounded by closed parametric curves

Example 3   Find the area bounded by the ellipse $x = a \cos s$, $y = b \sin s$, $(0 \le s \le 2\pi)$.

Solution   This ellipse is traversed counterclockwise. (See Example 4 in Section 8.2.) The area enclosed is

$$A = -\int_0^{2\pi} b \sin s \, (-a \sin s) \, ds$$

$$= \frac{ab}{2} \int_0^{2\pi} (1 - \cos 2s) \, ds$$

$$= \frac{ab}{2} s \Big|_0^{2\pi} - \frac{ab}{4} \sin 2s \Big|_0^{2\pi} = \pi ab \text{ square units.}$$

Example 4   Find the area above the x-axis and under one arch of the cycloid $x = at - a \sin t$, $y = a - a \cos t$.

Solution   Part of the cycloid is shown in Figure 8.21 in Section 8.2. One arch corresponds to the parameter interval $0 \le t \le 2\pi$. Since $y = a(1 - \cos t) \ge 0$ and $dx/dt = a(1 - \cos t) \ge 0$, the area under one arch is

$$A = \int_0^{2\pi} a^2 (1 - \cos t)^2 \, dt = a^2 \int_0^{2\pi} \left(1 - 2\cos t + \frac{1 + \cos 2t}{2} \right) dt$$

$$= a^2 \left(t - 2 \sin t + \frac{t}{2} + \frac{\sin 2t}{4} \right) \Big|_0^{2\pi} = 3\pi a^2 \text{ square units.} \qquad \blacksquare$$

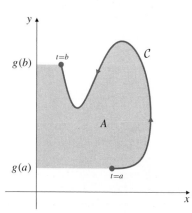

y

$g(b)$

$t=b$

$\mathcal{C}$

A

$g(a)$

$t=a$

x

Figure 8.32   The shaded area is

$$A = \int_a^b f(t)g'(t) \, dt$$

Similar arguments to those used above show that if f is continuous and g is differentiable, then we can also interpret

$$\int_a^b f(t)g'(t) \, dt = \int_{t=a}^{t=b} x \, dy = A_1 - A_2,$$

where A_1 is the area of the region lying *horizontally* between the parametric curve $x = f(t)$, $y = g(t)$, ($a \le t \le b$) and that part of the y-axis consisting of points $y = g(t)$ such that $f(t)g'(t) \ge 0$, and A_2 is the area of a similar region corresponding to $f(t)g'(t) < 0$. For example, the region shaded in Figure 8.32 has area $\int_a^b f(t)g'(t) \, dt$. Green's Theorem in Section 16.3 provides a more coherent approach to finding such areas.

Exercises 8.4

Find the lengths of the curves in Exercises 1–8.

1. $x = 3t^2$, $y = 2t^3$, $(0 \le t \le 1)$

2. $x = 1 + t^3$, $y = 1 - t^2$, $(-1 \le t \le 2)$

3. $x = a \cos^3 t$, $y = a \sin^3 t$, $(0 \le t \le 2\pi)$

4. $x = \ln(1 + t^2)$, $y = 2 \tan^{-1} t$, $(0 \le t \le 1)$

5. $x = t^2 \sin t$, $y = t^2 \cos t$, $(0 \le t \le 2\pi)$

6. $x = \cos t + t \sin t$, $y = \sin t - t \cos t$, $(0 \le t \le 2\pi)$

7. $x = t + \sin t$, $y = \cos t$, $(0 \le t \le \pi)$

8. $x = \sin^2 t$, $y = 2 \cos t$, $(0 \le t \le \pi/2)$

9. Find the length of one arch of the cycloid $x = at - a \sin t$, $y = a - a \cos t$. (One arch corresponds to $0 \le t \le 2\pi$.)

10. Find the area of the surfaces obtained by rotating one arch of the cycloid in Exercise 9 about (a) the x-axis, (b) the y-axis.

11. Find the area of the surface generated by rotating the curve $x = e^t \cos t$, $y = e^t \sin t$, $(0 \le t \le \pi/2)$ about the x-axis.

12. Find the area of the surface generated by rotating the curve of Exercise 11 about the y-axis.

13. Find the area of the surface generated by rotating the curve

$x = 3t^2$, $y = 2t^3$, $(0 \le t \le 1)$ about the y-axis.

14. Find the area of the surface generated by rotating the curve $x = 3t^2$, $y = 2t^3$, $(0 \le t \le 1)$ about the x-axis.

In Exercises 15–20, sketch and find the area of the region R described in terms of the given parametric curves.

15. R is the closed loop bounded by $x = t^3 - 4t$, $y = t^2$, $(-2 \le t \le 2)$.

16. R is bounded by the astroid $x = a \cos^3 t$, $y = a \sin^3 t$, $(0 \le t \le 2\pi)$.

17. R is bounded by the coordinate axes and the parabolic arc $x = \sin^4 t$, $y = \cos^4 t$.

18. R is bounded by $x = \cos s \sin s$, $y = \sin^2 s$, $(0 \le s \le \pi/2)$, and the y-axis.

19. R is bounded by the oval $x = (2 + \sin t) \cos t$, $y = (2 + \sin t) \sin t$.

∗ 20. R is bounded by the x-axis, the hyperbola $x = \sec t$, $y = \tan t$, and the ray joining the origin to the point $(\sec t_0, \tan t_0)$.

21. Show that the region bounded by the x-axis and the hyperbola $x = \cosh t$, $y = \sinh t$ (where $t > 0$), and the ray from the origin to the point $(\cosh t_0, \sinh t_0)$ has area $t_0/2$ square units. This proves a claim made at the beginning of Section 3.6.

22. Find the volume of the solid obtained by rotating about the x-axis the region bounded by that axis and one arch of the cycloid $x = at - a \sin t$, $y = a - a \cos t$. (See Example 8 in Section 8.2.)

23. Find the volume generated by rotating about the x-axis the region lying under the astroid $x = a \cos^3 t$, $y = a \sin^3 t$ and above the x-axis.

8.5 Polar Coordinates and Polar Curves

The **polar coordinate system** is an alternative to the rectangular (Cartesian) coordinate system for describing the location of points in a plane. Sometimes it is more important to know how far, and in what direction, a point is from the origin than it is to know its Cartesian coordinates. In the polar coordinate system there is an origin (or **pole**), O, and a **polar axis**, a ray (i.e., a half-line) extending from O horizontally to the right. The position of any point P in the plane is then determined by its polar coordinates $[r, \theta]$, where

(i) r is the distance from O to P, and

(ii) θ is the angle that the ray OP makes with the polar axis (counterclockwise angles being considered positive).

We will use square brackets $[\cdot, \cdot]$ for polar coordinates of a point to distinguish them from rectangular (Cartesian) coordinates. Figure 8.33 shows some points with their polar coordinates. The rectangular coordinate axes x and y are usually shown on a polar graph. The polar axis coincides with the positive x-axis.

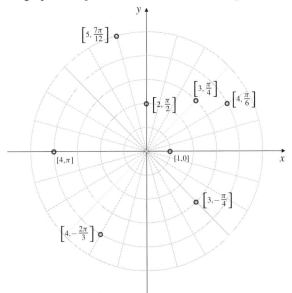

Figure 8.33   Polar coordinates of some points in the plane

Unlike rectangular coordinates, the polar coordinates of a point are not unique. The polar coordinates $[r, \theta_1]$ and $[r, \theta_2]$ represent the same point provided θ_1 and θ_2 differ by an integer multiple of 2π:

$$\theta_2 = \theta_1 + 2n\pi, \qquad \text{where } n = 0, \pm 1, \pm 2, \ldots.$$

For instance, the polar coordinates

$$\left[3, \frac{\pi}{4}\right], \qquad \left[3, \frac{9\pi}{4}\right], \qquad \text{and} \qquad \left[3, -\frac{7\pi}{4}\right]$$

all represent the same point with Cartesian coordinates $\left(\frac{3}{\sqrt{2}}, \frac{3}{\sqrt{2}}\right)$. Similarly, $[4, \pi]$ and $[4, -\pi]$ both represent the point with Cartesian coordinates $(-4, 0)$, and $[1, 0]$ and $[1, 2\pi]$ both represent the point with Cartesian coordinates $(1, 0)$. In addition, the origin O has polar coordinates $[0, \theta]$ for any value of θ. (If we go zero distance from O, it doesn't matter in what direction we go.)

Sometimes we need to interpret polar coordinates $[r, \theta]$, where $r < 0$. The appropriate interpretation for this "negative distance" r is that it represents a positive distance $-r$ measured in the *opposite direction* (i.e., in the direction $\theta + \pi$):

$$[r, \theta] = [-r, \theta + \pi].$$

For example, $[-1, \pi/4] = [1, 5\pi/4]$. Allowing $r < 0$ increases the number of different sets of polar coordinates that represent the same point.

If we want to consider both rectangular and polar coordinate systems in the same plane, and we choose the positive x-axis as the polar axis, then the relationships between the rectangular coordinates of a point and its polar coordinates are as shown in Figure 8.34.

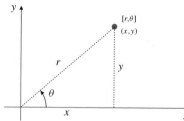

Figure 8.34   Relating Cartesian and polar coordinates of a point

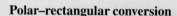

Polar–rectangular conversion

$$x = r \cos \theta \qquad\qquad x^2 + y^2 = r^2$$

$$y = r \sin \theta \qquad\qquad \tan \theta = \frac{y}{x}$$

A single equation in x and y generally represents a curve in the plane with respect to the Cartesian coordinate system. Similarly, a single equation in r and θ generally represents a curve with respect to the polar coordinate system. The conversion formulas above can be used to convert one representation of a curve into the other.

Example 1   The straight line $2x - 3y = 5$ has polar equation $r(2 \cos \theta - 3 \sin \theta) = 5$, or

$$r = \frac{5}{2 \cos \theta - 3 \sin \theta}.$$

Example 2   Find the Cartesian equation of the curve represented by the polar equation $r = 2a \cos \theta$; hence identify the curve.

Solution   The polar equation can be transformed to Cartesian coordinates if we first multiply it by r:

$$r^2 = 2ar \cos \theta$$
$$x^2 + y^2 = 2ax$$
$$(x - a)^2 + y^2 = a^2$$

The given polar equation $r = 2a \cos \theta$ thus represents a circle with centre $(a, 0)$ and radius a as shown in Figure 8.35. Observe from the equation that $r \to 0$ as $\theta \to \pm\pi/2$. In the figure, this corresponds to the fact that the circle approaches the origin in the vertical direction.

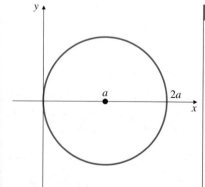

Figure 8.35   The circle $r = 2a \cos \theta$

Some Polar Curves

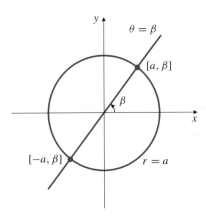

Figure 8.36   Coordinate curves for the polar coordinate system

Figure 8.36 shows the graphs of the polar equations $r = a$ and $\theta = \beta$, where a and β (Greek *beta*) are constants. These are, respectively, the circle with radius $|a|$ centred at the origin, and a line through the origin making angle β with the polar axis. Note that the line and the circle meet at two points, with polar coordinates $[a, \beta]$ and $[-a, \beta]$. The "coordinate curves" for polar coordinates, that is, the curves with equations $r = \text{constant}$ and $\theta = \text{constant}$, are circles centred at the origin and lines through the origin, respectively. The "coordinate curves" for Cartesian coordinates, $x = \text{constant}$ and $y = \text{constant}$, are vertical and horizontal straight lines. Cartesian graph paper is ruled with vertical and horizontal lines; polar graph paper is ruled with concentric circles and radial lines emanating from the origin, as shown in Figure 8.33.

The graph of an equation of the form $r = f(\theta)$ is called the **polar graph** of the function f. Some polar graphs can be recognized easily if the polar equation is transformed to rectangular form. For others, this transformation does not help; the rectangular equation may be too complicated to be recognizable. In these cases one must resort to constructing a table of values and plotting points.

Example 3   Sketch and identify the curve $r = 2a \cos(\theta - \theta_0)$.

Solution   We proceed as in Example 2.

$$r^2 = 2ar \cos(\theta - \theta_0) = 2ar \cos\theta_0 \cos\theta + 2ar \sin\theta_0 \sin\theta$$
$$x^2 + y^2 = 2a \cos\theta_0 x + 2a \sin\theta_0 y$$
$$x^2 - 2a \cos\theta_0 x + a^2 \cos^2\theta_0 + y^2 - 2a \sin\theta_0 y + a^2 \sin^2\theta_0 = a^2$$
$$(x - a \cos\theta_0)^2 + (y - a \sin\theta_0)^2 = a^2.$$

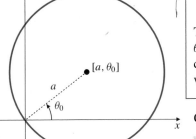

Figure 8.37   The circle $r = 2a \cos(\theta - \theta_0)$

This is a circle of radius a that passes through the origin in the directions $\theta = \theta_0 \pm \frac{\pi}{2}$, which make $r = 0$. (See Figure 8.37.) Its centre has Cartesian coordinates $(a \cos\theta_0, a \sin\theta_0)$ and hence polar coordinates $[a, \theta_0]$. For $\theta_0 = \pi/2$ we have $r = 2a \sin\theta$ as the equation of a circle of radius a centred on the y-axis. ∎

Comparing Examples 2 and 3, we are led to formulate the following principle.

> **Rotating a polar graph**
>
> The polar graph with equation $r = f(\theta - \theta_0)$ is the polar graph with equation $r = f(\theta)$ rotated through angle θ_0 about the origin.

Example 4   Sketch the polar curve $r = a(1 - \cos\theta)$, where $a > 0$.

Solution   Transformation to rectangular coordinates is not much help here; the resulting equation is $(x^2 + y^2 + ax)^2 = a^2(x^2 + y^2)$ (verify this), which we do not recognize. Therefore, we will make a table of values and plot some points.

Table 3.

θ	0	$\pm\dfrac{\pi}{6}$	$\pm\dfrac{\pi}{4}$	$\pm\dfrac{\pi}{3}$	$\pm\dfrac{\pi}{2}$	$\pm\dfrac{2\pi}{3}$	$\pm\dfrac{3\pi}{4}$	$\pm\dfrac{5\pi}{6}$	π
r	0	0.13a	0.29a	0.5a	a	1.5a	1.71a	1.87a	2a

Because it is shaped like a heart, this curve is called a **cardioid**. Observe the cusp at the origin in Figure 8.38. As in the previous example, the curve enters the origin in the directions θ that make $r = f(\theta) = 0$. In this case, the only such direction is $\theta = 0$. It is important, when sketching polar graphs, to show clearly any directions of approach to the origin.

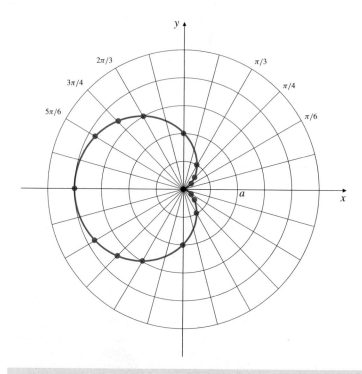

Figure 8.38 The cardioid
$r = a(1 - \cos\theta)$

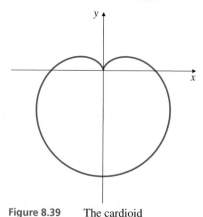

Figure 8.39 The cardioid
$r = a(1 - \sin\theta)$

Direction of a polar graph at the origin

A polar graph $r = f(\theta)$ approaches the origin from the direction θ for which $f(\theta) = 0$.

The equation $r = a(1 - \cos(\theta - \theta_0))$ represents a cardioid of the same size and shape as that in Figure 8.38 but rotated through an angle θ_0 counterclockwise about the origin. Its cusp is in the direction $\theta = \theta_0$. In particular, $r = a(1 - \sin\theta)$ has a vertical cusp, as shown in Figure 8.39.

It is not usually necessary to make a detailed table of values to sketch a polar curve with a simple equation of the form $r = f(\theta)$. It is essential to determine those values of θ for which $r = 0$ and indicate them on the graph with rays. It is also useful to determine points where the curve is farthest from the origin. (Where is $f(\theta)$ maximum or minimum?) Except possibly at the origin, polar curves will be smooth wherever $f(\theta)$ is a differentiable function of θ.

Example 5 Sketch the polar graphs (a) $r = \cos(2\theta)$, (b) $r = \sin(3\theta)$, and (c) $r^2 = \cos(2\theta)$.

Solution The graphs are shown in Figures 8.40–8.42. Observe how the curves (a) and (c) approach the origin in the directions $\theta = \pm\frac{\pi}{4}$ and $\theta = \pm\frac{3\pi}{4}$, and curve

(b) approaches in the directions $\theta = 0$, π, $\pm\frac{\pi}{3}$ and $\pm\frac{2\pi}{3}$. This curve is traced out twice as θ increases from $-\pi$ to π. So is curve (c) if we allow both square roots $r = \pm\sqrt{\cos(2\theta)}$. Note that there are no points on curve (c) between $\theta = \pm\frac{\pi}{4}$ and $\theta = \pm\frac{3\pi}{4}$ because r^2 cannot be negative.

Curve (c) is called a **lemniscate**. Lemniscates are curves consisting of points P such that the product of the distances from P to certain fixed points is constant. For the curve (c) these fixed points are $\left(\pm\frac{1}{\sqrt{2}}, 0\right)$.

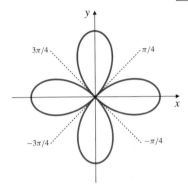

Figure 8.40   Curve (a): the polar curve $r = \cos(2\theta)$

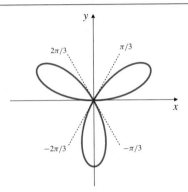

Figure 8.41   Curve (b): the polar curve $r = \sin(3\theta)$

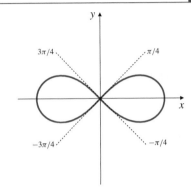

Figure 8.42   Curve (c): the lemniscate $r^2 = \cos(2\theta)$

In all of the examples above, the functions $f(\theta)$ are periodic and 2π is a period of each of them, so each line through the origin could meet the polar graph at most twice. (θ and $\theta + \pi$ determine the same line.) If $f(\theta)$ does not have period 2π, then the curve can wind around the origin many times. Two such *spirals* are shown in Figure 8.43, the **equiangular spiral** $r = \theta$ and the **exponential spiral** $r = e^{-\theta/3}$, each sketched for positive values of θ.

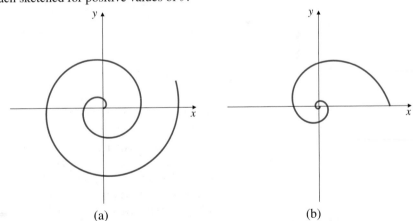

Figure 8.43

(a) The equiangular spiral $r = \theta$

(b) The exponential spiral $r = e^{-\theta/3}$

(a)   (b)

Remark   Maple has a `polarplot` routine as part of its "plots" package, which must be loaded prior to the use of polarplot. Here is how to get Maple to plot on the same graph the polar curves $r = 1$ and $r = 2\sin(3\theta)$, for $0 \le \theta \le 2\pi$:

```
>  with(plots):
>  polarplot([1,2*sin(3*t)],t=0..2*Pi,scaling=constrained
```

The option `scaling=constrained` is necessary with polar plots to force Maple to use the same distance unit on both axes (so a circle will appear circular).

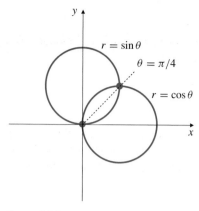

Figure 8.44

Intersections of Polar Curves

Because the polar coordinates of points are not unique, finding the intersection points of two polar curves can be more complicated than the similar problem for Cartesian graphs. Of course, the polar curves $r = f(\theta)$ and $r = g(\theta)$ will intersect at any points $[r_0, \theta_0]$ for which

$$f(\theta_0) = g(\theta_0) \qquad \text{and} \qquad r_0 = f(\theta_0),$$

but there may be other intersections as well. In particular, if both curves pass through the origin, then the origin will be an intersection point, even though it may not show up in solving $f(\theta) = g(\theta)$, because the curves may be at the origin for different values of θ. For example, the two circles $r = \cos\theta$ and $r = \sin\theta$ intersect at the origin and also at the point $[1/\sqrt{2}, \pi/4]$, even though only the latter point is obtained by solving the equation $\cos\theta = \sin\theta$. (See Figure 8.44.)

Example 6 Find the intersections of the curves $r = \sin\theta$ and $r = 1 - \sin\theta$.

Solution Since both functions of θ are periodic with period 2π, we need only look for solutions satisfying $0 \le \theta \le 2\pi$. Solving the equation

$$\sin\theta = 1 - \sin\theta,$$

we get $\sin\theta = 1/2$, so that $\theta = \pi/6$ or $\theta = 5\pi/6$. Both curves have $r = 1/2$ at these points, so the two curves intersect at $[1/2, \pi/6]$ and $[1/2, 5\pi/6]$. Also, the origin lies on the curve $r = \sin\theta$ (for $\theta = 0$ and $\theta = 2\pi$) and on the curve $r = 1 - \sin\theta$ (for $\theta = \pi/2$). Therefore, the origin is also an intersection point of the curves. (See Figure 8.45.) ∎

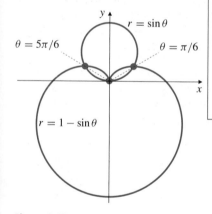

Figure 8.45

Finally, if negative values of r are allowed, then the curves $r = f(\theta)$, $y = g(\theta)$ will also intersect at $[r_1, \theta_1] = [r_2, \theta_2]$ if, for some integer k,

$$\theta_1 = \theta_2 + (2k + 1)\pi \qquad \text{and} \qquad r_1 = f(\theta_1) = -g(\theta_2) = -r_2.$$

See Exercise 28 for an example.

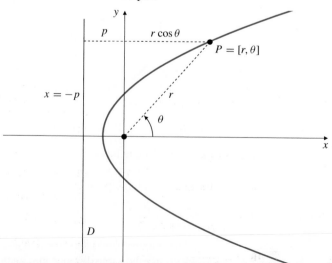

Figure 8.46 A conic curve with eccentricity ε, focus at the origin, and directrix $x = -p$

Polar Conics

Let D be the vertical straight line $x = -p$ and let ε be a positive real number. The set of points P in the plane that satisfy the condition

$$\frac{\text{distance of } P \text{ from the origin}}{\text{perpendicular distance from } P \text{ to } D} = \varepsilon$$

is a conic section with eccentricity ε, focus at the origin, and corresponding directrix D, as observed in Section 8.1. (It is an ellipse if $\varepsilon < 1$, a parabola if $\varepsilon = 1$, and a hyperbola if $\varepsilon > 1$.) If P has polar coordinates $[r, \theta]$, then the condition above becomes (see Figure 8.46)

$$\frac{r}{p + r\cos\theta} = \varepsilon,$$

or, solving for r,

$$r = \frac{\varepsilon p}{1 - \varepsilon\cos\theta}.$$

Examples of the three possibilities (ellipse, parabola, and hyperbola) are shown in Figures 8.47–8.49. Note that for the hyperbola, the directions of the asymptotes are the angles that make the denominator $1 - \varepsilon\cos\theta = 0$. We will have more to say about polar equations of conics, especially ellipses, in Section 11.6.

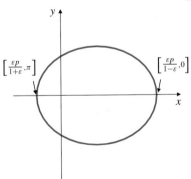

Figure 8.47   Ellipse: $\varepsilon < 1$

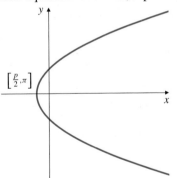

Figure 8.48   Parabola: $\varepsilon = 1$

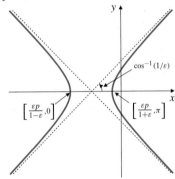

Figure 8.49   Hyperbola: $\varepsilon > 1$

Exercises 8.5

In Exercises 1–12, transform the given polar equation to rectangular coordinates and thus identify the curve represented.

1. $r = 3\sec\theta$

2. $r = -2\csc\theta$

3. $r = \dfrac{5}{3\sin\theta - 4\cos\theta}$

4. $r = \sin\theta + \cos\theta$

5. $r^2 = \csc 2\theta$

6. $r = \sec\theta\tan\theta$

7. $r = \sec\theta(1 + \tan\theta)$

8. $r = \dfrac{2}{\sqrt{\cos^2\theta + 4\sin^2\theta}}$

9. $r = \dfrac{1}{1 - \cos\theta}$

10. $r = \dfrac{2}{2 - \cos\theta}$

11. $r = \dfrac{2}{1 - 2\sin\theta}$

12. $r = \dfrac{2}{1 + \sin\theta}$

In Exercises 13–24, sketch the polar graphs of the given equations.

13. $r = 1 + \sin\theta$

14. $r = 1 - \cos(\theta + \frac{\pi}{4})$

15. $r = 1 + 2\cos\theta$

16. $r = 1 - 2\sin\theta$

17. $r = 2 + \cos\theta$

18. $r = 2\sin 2\theta$

19. $r = \cos 3\theta$

20. $r = 2\cos 4\theta$

21. $r^2 = 4\sin 2\theta$

22. $r^2 = 4\cos 3\theta$

23. $r^2 = \sin 3\theta$

24. $r = \ln\theta$

Find all intersections of the pairs of curves in Exercises 25–28.

25. $r = \sqrt{3}\cos\theta$, $r = \sin\theta$

26. $r^2 = 2\cos(2\theta)$, $r = 1$

27. $r = 1 + \cos\theta$, $r = 3\cos\theta$

* **28.** $r = \theta$, $r = \theta + \pi$

* **29.** Sketch the graph of the equation $r = 1/\theta$, $\theta > 0$. Show that this curve has a horizontal asymptote. Does $r = 1/(\theta - \alpha)$ have an asymptote?

30. How many leaves does the curve $r = \cos n\theta$ have? the curve $r^2 = \cos n\theta$? Distinguish the cases where n is odd and even.

31. Show that the polar graph $r = f(\theta)$ (where f is continuous) can be written as a parametric curve with parameter θ.

In Exercises 32–37, use computer graphing software or a graphing calculator to plot various members of the given families of polar curves, and try to observe patterns that would enable you to predict behaviour of other members of the families.

32. $r = \cos\theta\cos(m\theta)$, $m = 1, 2, 3, \ldots$

33. $r = 1 + \cos\theta\cos(m\theta)$, $m = 1, 2, 3, \ldots$

34. $r = \sin(2\theta)\sin(m\theta)$, $m = 2, 3, 4, 5, \ldots$

35. $r = 1 + \sin(2\theta)\sin(m\theta)$, $m = 2, 3, 4, 5, \ldots$

36. $r = C + \cos\theta\cos(2\theta)$ for $C = 0$, $C = 1$, values of C between 0 and 1, and values of C greater than 1

37. $r = C + \cos\theta\sin(3\theta)$ for $C = 0$, $C = 1$, values of C between 0 and 1, values of C less than 0, and values of C greater than 1

38. Plot the curve $r = \ln\theta$ for $0 < \theta \leq 2\pi$. It intersects itself at point P. Thus there are two values θ_1 and θ_2 between 0 and 2π for which $[f(\theta_1), \theta_1] = [f(\theta_2), \theta_2]$. What equations must be satisfied by θ_1 and θ_2? Find θ_1 and θ_2, and find the Cartesian coordinates of P correct to 6 decimal places.

39. Simultaneously plot the two curves $r = \ln\theta$ and $r = 1/\theta$, for $0 < \theta \leq 2\pi$. The two curves intersect at two points. What equations must be satisfied by the θ values of these points? What are their Cartesian coordinates to 6 decimal places?

8.6 Slopes, Areas, and Arc Lengths for Polar Curves

There is a simple formula that can be used to determine the direction of the tangent line to a polar curve $r = f(\theta)$ at a point $P = [r, \theta]$ other than the origin. Let Q be a point on the curve near P corresponding to polar angle $\theta + h$. Draw PS perpendicular to OQ. Observe that $PS = f(\theta)\sin h$ and $SQ = OQ - OS = f(\theta + h) - f(\theta)\cos h$. If the tangent line to $r = f(\theta)$ at P makes angle ψ (Greek *psi*) with the radial line OP as shown in Figure 8.50(a), then ψ is the limit of the angle SQP as $h \to 0$. Thus

$$\tan\psi = \lim_{h\to 0}\frac{PS}{SQ} = \lim_{h\to 0}\frac{f(\theta)\sin h}{f(\theta + h) - f(\theta)\cos h} \qquad \left[\frac{0}{0}\right]$$

$$= \lim_{h\to 0}\frac{f(\theta)\cos h}{f'(\theta + h) + f(\theta)\sin h} \qquad \text{(by l'Hôpital's Rule)}$$

$$= \frac{f(\theta)}{f'(\theta)} = \frac{r}{dr/d\theta}.$$

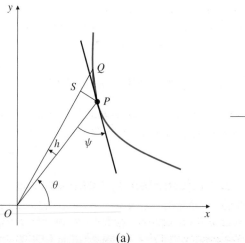

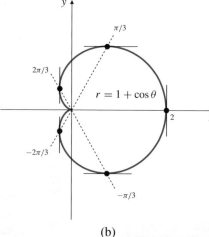

Figure 8.50

(a) The angle ψ is the limit of angle SQP as $h \to 0$

(b) Horizontal and vertical tangents to a cardioid

(a)

(b)

> ### Tangent direction for a polar curve
>
> At any point P other than the origin on the polar curve $r = f(\theta)$, the angle ψ between the radial line from the origin to P and the tangent to the curve is given by
>
> $$\tan \psi = \frac{f(\theta)}{f'(\theta)}.$$
>
> In particular, $\psi = \pi/2$ if $f'(\theta) = 0$.
> If $f(\theta_0) = 0$ and the curve has a tangent line at θ_0, then that tangent line has equation $\theta = \theta_0$.

The formula above can be used to find points where a polar graph has horizontal or vertical tangents:

$$\psi + \theta = \pi, \quad \text{so} \ \tan \psi = -\tan \theta \qquad \text{for a horizontal tangent,}$$

$$\psi + \theta = \frac{\pi}{2}, \quad \text{so} \ \tan \psi = \cot \theta \qquad \text{for a vertical tangent.}$$

Remark Since for parametric curves horizontal and vertical tangents correspond to $dy/dt = 0$ and $dx/dt = 0$, respectively, it is usually easier to find the critical points of $y = f(\theta) \sin \theta$ for horizontal tangents and of $x = f(\theta) \cos \theta$ for vertical tangents.

Example 1 Find the points on the cardioid $r = 1 + \cos \theta$, where the tangent lines are vertical or horizontal.

Solution We have $y = (1 + \cos \theta) \sin \theta$ and $x = (1 + \cos \theta) \cos \theta$. For horizontal tangents

$$\begin{aligned} 0 = \frac{dy}{d\theta} &= -\sin^2 \theta + \cos^2 \theta + \cos \theta \\ &= 2\cos^2 \theta + \cos \theta - 1 \\ &= (2\cos \theta - 1)(\cos \theta + 1). \end{aligned}$$

The solutions are $\cos \theta = \frac{1}{2}$ and $\cos \theta = -1$, that is, $\theta = \pm\pi/3$ and $\theta = \pi$. There are horizontal tangents at $\left[\frac{3}{2}, \pm\frac{\pi}{3}\right]$. At $\theta = \pi$, we have $r = 0$. The curve does not have a tangent line at the origin (it has a cusp). See Figure 8.50(b).

For vertical tangents

$$0 = \frac{dx}{d\theta} = -\sin \theta - 2\cos \theta \sin \theta = -\sin \theta (1 + 2\cos \theta).$$

The solutions are $\sin \theta = 0$ and $\cos \theta = -\frac{1}{2}$, that is, $\theta = 0, \pi, \pm 2\pi/3$. There are vertical tangent lines at $[2, 0]$ and $\left[\frac{1}{2}, \pm\frac{2\pi}{3}\right]$.

Areas Bounded by Polar Curves

The basic area problem in polar coordinates is that of finding the area A of the region R bounded by the polar graph $r = f(\theta)$ and the two rays $\theta = \alpha$ and $\theta = \beta$. We assume that $\beta > \alpha$ and that f is continuous for $\alpha \le \theta \le \beta$. See Figure 8.51.

A suitable area element in this case is a sector of angular width $d\theta$, as shown in Figure 8.51. For infinitesimal $d\theta$ this is just a sector of a circle of radius $r = f(\theta)$:

$$dA = \frac{d\theta}{2\pi} \pi r^2 = \frac{1}{2} r^2 \, d\theta = \frac{1}{2} \left(f(\theta)\right)^2 d\theta.$$

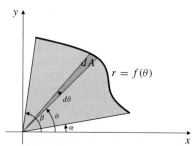

Figure 8.51   An area element in polar coordinates

Area in polar coordinates

The region bounded by $r = f(\theta)$ and the rays $\theta = \alpha$ and $\theta = \beta$, $(\alpha < \beta)$, has area

$$A = \frac{1}{2} \int_{\alpha}^{\beta} \left(f(\theta)\right)^2 d\theta.$$

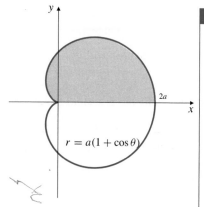

Figure 8.52

Example 2   Find the area bounded by the cardioid $r = a(1+\cos\theta)$, as illustrated in Figure 8.52.

Solution   By symmetry, the area is twice that of the top half:

$$A = 2 \times \frac{1}{2} \int_0^{\pi} a^2 (1 + \cos\theta)^2 \, d\theta$$

$$= a^2 \int_0^{\pi} (1 + 2\cos\theta + \cos^2\theta) \, d\theta$$

$$= a^2 \int_0^{\pi} \left(1 + 2\cos\theta + \frac{1 + \cos 2\theta}{2}\right) d\theta$$

$$= a^2 \left(\frac{3}{2}\theta + 2\sin\theta + \frac{1}{4}\sin 2\theta\right)\Big|_0^{\pi} = \frac{3}{2}\pi a^2 \text{ square units.}$$

Example 3   Find the area of the region that lies inside the circle $r = \sqrt{2}\sin\theta$ and inside the lemniscate $r^2 = \sin 2\theta$.

Solution   The region is shaded in Figure 8.53. Besides intersecting at the origin, the curves intersect at the first quadrant point satisfying

$$2\sin^2\theta = \sin 2\theta = 2\sin\theta\cos\theta.$$

Thus $\sin\theta = \cos\theta$ and $\theta = \pi/4$. The required area is

$$A = \frac{1}{2}\int_0^{\pi/4} 2\sin^2\theta \, d\theta + \frac{1}{2}\int_{\pi/4}^{\pi/2} \sin 2\theta \, d\theta$$

$$= \int_0^{\pi/4} \frac{1 - \cos 2\theta}{2} \, d\theta - \frac{1}{4}\cos 2\theta\Big|_{\pi/4}^{\pi/2}$$

$$= \frac{\pi}{8} - \frac{1}{4}\sin 2\theta\Big|_0^{\pi/4} + \frac{1}{4} = \frac{\pi}{8} - \frac{1}{4} + \frac{1}{4} = \frac{\pi}{8} \text{ square units.}$$

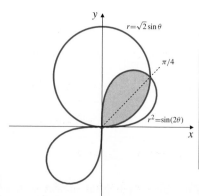

Figure 8.53

Arc Lengths for Polar Curves

The arc length element for the polar curve $r = f(\theta)$ can be determined from the differential triangle shown in Figure 8.54. The leg $r\,d\theta$ of the triangle is obtained as the arc length of a circular arc of radius r subtending angle $d\theta$ at the origin. We have

$$(ds)^2 = (dr)^2 + r^2(d\theta)^2 = \left[\left(\frac{dr}{d\theta}\right)^2 + r^2\right](d\theta)^2,$$

so we obtain the following formula:

> **Arc length element for a polar curve**
>
> The arc length element for the polar curve $r = f(\theta)$ is
>
> $$ds = \sqrt{\left(\frac{dr}{d\theta}\right)^2 + r^2}\,d\theta = \sqrt{\left(f'(\theta)\right)^2 + \left(f(\theta)\right)^2}\,d\theta.$$

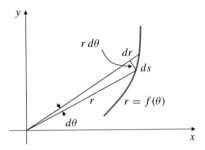

Figure 8.54   The arc length element for a polar curve

This arc length element can also be derived from that for a parametric curve. See Exercise 26 at the end of this section.

Example 4   Find the total length of the cardioid $r = a(1 + \cos\theta)$.

Solution   The total length is twice the length from $\theta = 0$ to $\theta = \pi$. (Review Figure 8.52.) Since $dr/d\theta = -a\sin\theta$ for the cardioid, the arc length is

$$s = 2\int_0^\pi \sqrt{a^2\sin^2\theta + a^2(1+\cos\theta)^2}\,d\theta$$

$$= 2\int_0^\pi \sqrt{2a^2 + 2a^2\cos\theta}\,d\theta \qquad \text{(but } 1+\cos\theta = 2\cos^2(\theta/2)\text{)}$$

$$= 2\sqrt{2}a\int_0^\pi \sqrt{2\cos^2\frac{\theta}{2}}\,d\theta$$

$$= 4a\int_0^\pi \cos\frac{\theta}{2}\,d\theta = 8a\sin\frac{\theta}{2}\Big|_0^\pi = 8a \text{ units.}$$

Exercises 8.6

In Exercises 1–11, sketch and find the areas of the given polar regions R.

1. R lies between the origin and the spiral $r = \sqrt{\theta}$, $0 \le \theta \le 2\pi$.

2. R lies between the origin and the spiral $r = \theta$, $0 \le \theta \le 2\pi$.

3. R is bounded by the curve $r^2 = a^2\cos 2\theta$.

4. R is one leaf of the curve $r = \sin 3\theta$.

5. R is bounded by the curve $r = \cos 4\theta$.

6. R lies inside both of the circles $r = a$ and $r = 2a\cos\theta$.

7. R lies inside the cardioid $r = 1 - \cos\theta$ and outside the circle $r = 1$.

8. R lies inside the cardioid $r = a(1 - \sin\theta)$ and inside the circle $r = a$.

9. R lies inside the cardioid $r = 1 + \cos\theta$ and outside the circle $r = 3\cos\theta$.

10. R is bounded by the lemniscate $r^2 = 2\cos 2\theta$ and is outside the circle $r = 1$.

11. R is bounded by the smaller loop of the curve $r = 1 + 2\cos\theta$.

Find the lengths of the polar curves in Exercises 12–14

12. $r = \theta^2$, $0 \le \theta \le \pi$ 13. $r = e^{a\theta}$, $-\pi \le \theta \le \pi$

14. $r = a\theta$, $0 \le \theta \le 2\pi$

15. Show that the total arc length of the lemniscate $r^2 = \cos 2\theta$
is $4 \displaystyle\int_0^{\pi/4} \sqrt{\sec 2\theta}\, d\theta$.

16. One leaf of the lemniscate $r^2 = \cos 2\theta$ is rotated (a) about the x-axis and (b) about the y-axis. Find the area of the surface generated in each case.

* 17. Determine the angles at which the straight line $\theta = \pi/4$ intersects the cardioid $r = 1 + \sin\theta$.

* 18. At what points do the curves $r^2 = 2\sin 2\theta$ and $r = 2\cos\theta$ intersect? At what angle do the curves intersect at each of these points?

* 19. At what points do the curves $r = 1 - \cos\theta$ and $r = 1 - \sin\theta$ intersect? At what angle do the curves intersect at each of these points?

In Exercises 20–25, find all points on the given curve where the tangent line is horizontal, vertical, or does not exist.

* 20. $r = \cos\theta + \sin\theta$ * 21. $r = 2\cos\theta$

* 22. $r^2 = \cos 2\theta$ * 23. $r = \sin 2\theta$

* 24. $r = e^\theta$ * 25. $r = 2(1 - \sin\theta)$

26. The polar curve $r = f(\theta)$, $(\alpha \le \theta \le \beta)$ can be parametrized:

$$x = r\cos\theta = f(\theta)\cos\theta, \qquad y = r\sin\theta = f(\theta)\sin\theta.$$

Derive the formula for the arc length element for the polar curve from that for a parametric curve.

Chapter Review

Key Ideas

• **What do the following terms and phrases mean?**

◇ a conic section   ◇ an ellipse

◇ a parabola   ◇ a hyperbola

◇ a parametric curve   ◇ a parametrization of a curve

◇ a smooth curve   ◇ a polar curve

• **What is the focus-directrix definition of a conic?**

• **How can you find the slope of a parametric curve?**

• **How can you find the length of a parametric curve?**

• **How can you find the length of a polar curve?**

• **How can you find the area bounded by a polar curve?**

Review Exercises

In Exercises 1–4, describe the conic having the given equation. Give its foci and principal axes and, if it is a hyperbola, its asymptotes.

1. $x^2 + 2y^2 = 2$ 2. $9x^2 - 4y^2 = 36$

3. $x + y^2 = 2y + 3$ 4. $2x^2 + 8y^2 = 4x - 48y$

Identify the parametric curves in Exercises 5–10.

5. $x = t$, $y = 2 - t$, $(0 \le t \le 2)$

6. $x = 2\sin 3t$, $y = 2\cos 3t$, $(0 \le t \le 1/2)$

7. $x = \cosh t$, $y = \sinh^2 t$,

8. $x = e^t$, $y = e^{-2t}$, $(-1 \le t \le 1)$

9. $x = \cos(t/2)$, $y = 4\sin(t/2)$, $(0 \le t \le \pi)$

10. $x = \cos t + \sin t$, $y = \cos t - \sin t$, $(0 \le t \le 2\pi)$

In Exercises 11–14, determine the points where the given parametric curves have horizontal and vertical tangents, and sketch the curves.

11. $x = \dfrac{4}{1 + t^2}$, $y = t^3 - 3t$

12. $x = t^3 - 3t$, $y = t^3 + 3t$

13. $x = t^3 - 3t$, $y = t^3$

14. $x = t^3 - 3t$, $y = t^3 - 12t$

15. Find the area bounded by the part of the curve $x = t^3 - t$, $y = |t^3|$ that forms a closed loop.

16. Find the volume of the solid of revolution generated by rotating the closed loop in the previous exercise about the y-axis.

17. Find the length of the curve $x = e^t - t$, $y = 4e^{t/2}$ from $t = 0$ to $t = 2$.

18. Find the area of the surface obtained by rotating the arc in the previous exercise about the x-axis.

Sketch the polar graphs of the equations in Exercises 19–24.

19. $r = \theta$, $\left(-\frac{3\pi}{2} \le \theta \le \frac{3\pi}{2}\right)$ 20. $r = |\theta|$, $(-2\pi \le \theta \le 2\pi)$

21. $r = 1 + \cos 2\theta$ 22. $r = 2 + \cos 2\theta$

23. $r = 1 + 2\cos 2\theta$ 24. $r = 1 - \sin 3\theta$

25. Find the area of one of the two larger loops of the curve in Exercise 23.

26. Find the area of one of the two smaller loops of the curve in Exercise 23.

27. Find the area of the smaller of the two loops enclosed by the curve $r = 1 + \sqrt{2}\sin\theta$.

28. Find the area of the region inside the cardioid $r = 1 + \cos\theta$ and to the left of the line $x = 1/4$.

Challenging Problems

1. A glass in the shape of a circular cylinder of radius 4 cm is more than half filled with water. If the glass is tilted by an angle θ from the vertical, where θ is small enough that no water spills out, find the surface area of the water.

2. Show that a plane that is not parallel to the axis of a circular cylinder intersects the cylinder in an ellipse.

Hint: you can do this by the same method used in Exercise 27 of Section 8.1.

3. Given two points F_1 and F_2 that are foci of an ellipse and a third point P on the ellipse, describe a geometric method (using straight edge and compass) for constructing the tangent line to the ellipse at P. *Hint:* think about the reflection property of ellipses.

4. Let C be a parabola with vertex V, and let P be any point on the parabola. Let R be the point where the tangent to the parabola at P intersects the axis of the parabola. (Thus the axis is the line RV.) Let Q be the point on RV such that PQ is perpendicular to RV. Show that V bisects the line segment RQ. How does this result suggest a geometric method for constructing a tangent to a parabola at a point on it, given the axis and vertex of the parabola?

5. A barrel has the shape of a solid of revolution obtained by rotating about its major axis the part of an ellipse lying between lines through its foci perpendicular to that axis. The barrel is 4 ft high and 2 ft in radius at its middle. What is its volume?

6. (a) Show that any straight line not passing through the origin can be written in polar form as

$$r = \frac{a}{\cos(\theta - \theta_0)},$$

where a and θ_0 are constants. What is the geometric significance of these constants?

(b) Let $r = g(\theta)$ be the polar equation of a straight line that does not pass through the origin. Show that

$$g^2 + 2(g')^2 - gg'' = 0.$$

(c) Let $r = f(\theta)$ be the polar equation of a curve, where f'' is continuous and $r \neq 0$ in some interval of values of θ. Let

$$F = f^2 + 2(f')^2 - ff''.$$

Show that the curve is turning toward the origin if $F > 0$ and away from the origin if $F < 0$. *Hint:* let $r = g(\theta)$ be the polar equation of a straight line tangent to the curve, and use part (b). How do f, f', and f'' relate to g, g', and g'' at the point of tangency?

7. (**Fast trip, but it might get hot**) If we assume that the density of the earth is uniform throughout, then it can be shown that the acceleration of gravity at a distance $r \leq R$ from the centre of the earth is directed toward the centre of the earth and has magnitude $a(r) = rg/R$, where g is the usual acceleration of gravity at the surface ($g \approx 32$ ft/s^2), and R is the radius of the earth ($R \approx 3960$ mi). Suppose that a straight tunnel AB is drilled through the earth between any two points A and B

on the surface, say Atlanta and Baghdad. (See Figure 8.55.) Suppose that a vehicle is constructed that can slide without friction or air resistance through this tunnel. Show that such a vehicle will, if released at one end of the tunnel, fall back and forth between A and B, executing simple harmonic motion with period $2\pi\sqrt{R/g}$. How many minutes will the round trip take? What is surprising here is that this period does not depend on where A and B are, or on the distance between them. *Hint:* Let the x-axis lie along the tunnel, with origin at the point closest to the centre of the earth. When the vehicle is at position with x-coordinate $x(t)$, its acceleration along the tunnel is the component of the gravitational acceleration along the tunnel, that is, $-a(r)\cos\theta$, where θ is the angle between the line of the tunnel and the line from the vehicle to the centre of the earth.

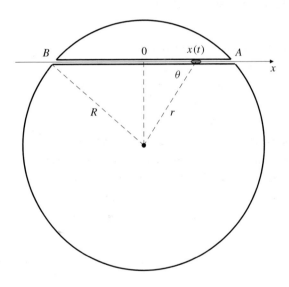

Figure 8.55

8. (**Search and Rescue**) Two coast guard stations pick up a distress signal from a ship and use radio direction finders to locate it. Station O observes that the distress signal is coming from the northeast (45° east of north), while station P, which is 100 miles north of station O, observes that the signal is coming from due east. Each station's direction finder is accurate to within ±3°.

(a) How large an area of the ocean must a rescue aircraft search to ensure that it finds the foundering ship?

(b) If the accuracy of the direction finders is within ±ε, how sensitive is the search area to changes in ε when $\varepsilon = 3°$? (Express your answer in square miles per degree.)

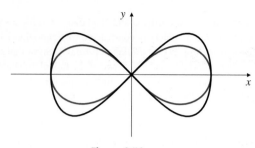

Figure 8.56

9. Figure 8.56 shows the graphs of the parametric curve $x = \sin t$, $y = \frac{1}{2} \sin(2t)$, $0 \le t \le 2\pi$, and the polar curve $r^2 = \cos(2\theta)$. Each has the shape of an "∞." Which curve is which? Find the area inside the outer curve and outside the inner curve.

CHAPTER 9

Sequences, Series, and Power Series

Introduction   An infinite series is a sum that involves infinitely many terms. Since addition is carried out on two numbers at a time, the evaluation of the sum of an infinite series necessarily involves finding a limit. Complicated functions $f(x)$ can frequently be expressed as series of simpler functions. For example, many of the transcendental functions we have encountered can be expressed as series of powers of x so that they resemble polynomials of infinite degree. Such series can be differentiated and integrated term by term, and they play a very important role in the study of calculus.

9.1 Sequences and Convergence

By a **sequence** (or an **infinite sequence**) we mean an ordered list having a first element but no last element. For our purposes, the elements (called **terms**) of a sequence will always be real numbers, although much of our discussion could be applied to complex numbers as well. Examples of sequences are:

$\{1,\ 2,\ 3,\ 4,\ 5,\ \ldots\}$ the sequence of positive integers,

$\left\{-\dfrac{1}{2},\ \dfrac{1}{4},\ -\dfrac{1}{8},\ \dfrac{1}{16},\ \ldots\right\}$ the sequence of positive integer powers of $-\dfrac{1}{2}$.

The terms of a sequence are usually listed in braces { } as shown. The ellipsis (. . .) should be read "and so on."

An infinite sequence is a special kind of function, one whose domain is a set of integers extending from some starting integer to infinity. The starting integer is usually 1, so the domain is the set of positive integers. The sequence $\{a_1,\ a_2,\ a_3,\ a_4,\ \ldots\}$ is the function f that takes the value $f(n) = a_n$ at each positive integer n. A sequence can be specified in three ways:

(i) We can list the first few terms followed by . . . *if the pattern is obvious.*

(ii) We can provide a formula for the **general term** a_n as a function of n.

(iii) We can provide a formula for calculating the term a_n as a function of earlier terms $a_1,\ a_2,\ \ldots,\ a_{n-1}$ and specify enough of the beginning terms so the process of computing higher terms can begin.

In each case it must be possible to determine any term of the sequence, although it may be necessary to calculate all the preceding terms first.

Example 1   (Some examples of sequences)

(a) $\{n\} = \{1,\ 2,\ 3,\ 4,\ 5,\ \ldots\}$

(b) $\left\{\left(-\dfrac{1}{2}\right)^n\right\} = \left\{-\dfrac{1}{2},\ \dfrac{1}{4},\ -\dfrac{1}{8},\ \dfrac{1}{16},\ \ldots\right\}$

(c) $\left\{\dfrac{n-1}{n}\right\} = \left\{0,\ \dfrac{1}{2},\ \dfrac{2}{3},\ \dfrac{3}{4},\ \dfrac{4}{5},\ \dots\right\}$

(d) $\{(-1)^{n-1}\} = \{\cos((n-1)\pi)\} = \{1,\ -1,\ 1,\ -1,\ 1,\ \dots\}$

(e) $\left\{\dfrac{n^2}{2^n}\right\} = \left\{\dfrac{1}{2},\ 1,\ \dfrac{9}{8},\ 1,\ \dfrac{25}{32},\ \dfrac{36}{64},\ \dfrac{49}{128},\ \dots\right\}$

(f) $\left\{\left(1+\dfrac{1}{n}\right)^n\right\} = \left\{2,\ \left(\dfrac{3}{2}\right)^2,\ \left(\dfrac{4}{3}\right)^3,\ \left(\dfrac{5}{4}\right)^4,\ \dots\right\}$

(g) $\left\{\dfrac{\cos(n\pi/2)}{n}\right\} = \left\{0,\ -\dfrac{1}{2},\ 0,\ \dfrac{1}{4},\ 0,\ -\dfrac{1}{6},\ 0,\ \dfrac{1}{8},\ 0,\ \dots\right\}$

(h) $a_1 = 1$, $a_{n+1} = \sqrt{6+a_n}$, $(n=1, 2, 3, \dots)$

In this case $\{a_n\} = \{1,\ \sqrt{7},\ \sqrt{6+\sqrt{7}},\ \dots\}$.

Note that there is no *obvious* formula for a_n as an explicit function of n here, but we can still calculate a_n for any desired value of n provided we first calculate all the earlier values $a_2, a_3, \dots, a_{n-1}$.

(i) $a_1 = 1$, $a_2 = 1$, $a_{n+2} = a_n + a_{n+1}$, $(n=1, 2, 3, \dots)$

Here $\{a_n\} = \{1,\ 1,\ 2,\ 3,\ 5,\ 8,\ 13,\ 21,\ \dots\}$.

This is called the **Fibonacci sequence**. Each term after the second is the sum of the previous two terms.

■

In parts (a)–(g) of Example 1, the formulas on the left sides define the general term of each sequence $\{a_n\}$ as an explicit function of n. In parts (h) and (i) we say the sequence $\{a_n\}$ is defined **recursively** or **inductively**; each term must be calculated from previous ones rather than directly as a function of n.

The following definition introduces terminology used to describe various properties of sequences.

DEFINITION   **1**

Terms for describing sequences

(a) The sequence $\{a_n\}$ is **bounded below** by L, and L is a **lower bound** for $\{a_n\}$, if $a_n \geq L$ for every $n = 1, 2, 3, \dots$. The sequence is **bounded above** by M, and M is an **upper bound**, if $a_n \leq M$ for every such n.

The sequence $\{a_n\}$ is **bounded** if it is both bounded above and bounded below. In this case there is a constant K such that $|a_n| \leq K$ for every $n = 1, 2, 3, \dots$. (We can take K to be the larger of $-L$ and M.)

(b) The sequence $\{a_n\}$ is **positive** if it is bounded below by zero, that is, if $a_n \geq 0$ for every $n = 1, 2, 3, \dots$; it is **negative** if $a_n \leq 0$ for every n.

(c) The sequence $\{a_n\}$ is **increasing** if $a_{n+1} \geq a_n$ for every $n = 1, 2, 3, \dots$; it is **decreasing** if $a_{n+1} \leq a_n$ for every such n. The sequence is said to be **monotonic** if it is either increasing or decreasing. (The terminology here is looser than that we have used for functions, where we would have used *nondecreasing* and *nonincreasing* to describe this behaviour.)

(d) The sequence $\{a_n\}$ is **alternating** if $a_n a_{n+1} < 0$ for every $n = 1, 2, \dots$, that is, if any two consecutive terms have opposite sign. Note that this definition requires $a_n \neq 0$ for each n.

Example 2   (Describing some sequences)

(a) The sequence $\{n\} = \{1, 2, 3, \ldots\}$ is positive, increasing, and bounded below. A lower bound for the sequence is 1 or any smaller number. The sequence is not bounded above.

(b) $\left\{\dfrac{n-1}{n}\right\} = \left\{0, \dfrac{1}{2}, \dfrac{2}{3}, \dfrac{3}{4}, \ldots\right\}$ is positive, bounded, and increasing. Here 0 is a lower bound and 1 is an upper bound.

(c) $\left\{\left(-\dfrac{1}{2}\right)^n\right\} = \left\{-\dfrac{1}{2}, \dfrac{1}{4}, -\dfrac{1}{8}, \dfrac{1}{16}, \ldots\right\}$ is bounded and alternating. Here $-1/2$ is a lower bound and $1/4$ is an upper bound.

(d) $\{(-1)^n n\} = \{-1, 2, -3, 4, -5, \ldots\}$ is alternating but not bounded either above or below.

When you want to show that a sequence is increasing, you can try to show that the inequality $a_{n+1} - a_n \geq 0$ holds for $n \geq 1$. Alternatively, if $a_n = f(n)$ for a differentiable function $f(x)$, you can show that f is a nondecreasing function on $[1, \infty[$ by showing that $f'(x) \geq 0$ there. Similar approaches are useful for showing that a sequence is decreasing.

Example 3   If $a_n = \dfrac{n}{n^2 + 1}$, show that the sequence $\{a_n\}$ is decreasing.

Solution   Since $a_n = f(n)$, where $f(x) = \dfrac{x}{x^2 + 1}$ and

$$f'(x) = \frac{(x^2 + 1)(1) - x(2x)}{(x^2 + 1)^2} = \frac{1 - x^2}{(x^2 + 1)^2} \leq 0 \quad \text{for } x \geq 1,$$

the function $f(x)$ is decreasing on $[1, \infty[$; therefore, $\{a_n\}$ is a decreasing sequence.

The sequence $\left\{\dfrac{n^2}{2^n}\right\} = \left\{\dfrac{1}{2}, 1, \dfrac{9}{8}, 1, \dfrac{25}{32}, \dfrac{36}{64}, \dfrac{49}{128}, \ldots\right\}$ is positive and therefore bounded below. It seems clear that from the fourth term on, all the terms are getting smaller. However, $a_2 > a_1$ and $a_3 > a_2$. Since $a_{n+1} \leq a_n$ only if $n \geq 3$, we say that this sequence is **ultimately decreasing**. The adverb *ultimately* is used to describe any termwise property of a sequence that the terms have from some point on, but not necessarily at the beginning of the sequence. Thus, the sequence

$$\{n - 100\} = \{-99, -98, \ldots, -2, -1, 0, 1, 2, 3, \ldots\}$$

is *ultimately positive* even though the first 99 terms are negative, and the sequence

$$\left\{(-1)^n + \frac{4}{n}\right\} = \left\{3, 3, \frac{1}{3}, 2, -\frac{1}{5}, \frac{5}{3}, -\frac{3}{7}, \frac{3}{2}, \ldots\right\}$$

is *ultimately alternating* even though the first few terms do not alternate.

Convergence of Sequences

Central to the study of sequences is the notion of convergence. The concept of the limit of a sequence is a special case of the concept of the limit of a function $f(x)$

as $x \to \infty$. We say that the sequence $\{a_n\}$ **converges to the limit** L, and we write $\lim_{n\to\infty} a_n = L$, provided the distance from a_n to L on the real line approaches 0 as n increases toward ∞. We state this definition more formally as follows:

DEFINITION 2

Limit of a sequence

We say that sequence $\{a_n\}$ converges to the limit L, and we write $\lim_{n\to\infty} a_n = L$, if for every positive real number ϵ there exists an integer N (which may depend on ϵ) such that if $n \geq N$, then $|a_n - L| < \epsilon$.

This definition is illustrated in Figure 9.1.

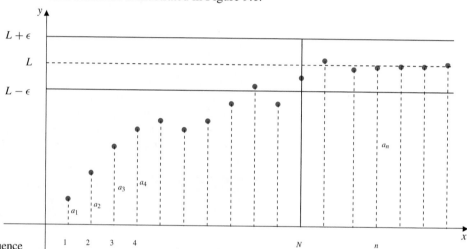

Figure 9.1   A convergent sequence

Example 4   Show that $\lim_{n\to\infty} \dfrac{c}{n^p} = 0$ for any real number c and any $p > 0$.

Solution   Let $\epsilon > 0$ be given. Then

$$\left| \frac{c}{n^p} \right| < \epsilon \quad \text{if} \quad n^p > \frac{|c|}{\epsilon},$$

that is, if $n \geq N$, the least integer greater than $(|c|/\epsilon)^{1/p}$. By Definition 2, $\lim_{n\to\infty} \dfrac{c}{n^p} = 0$.

Every sequence $\{a_n\}$ must either **converge** to a finite limit L or **diverge**. That is, either $\lim_{n\to\infty} a_n = L$ exists (is a real number) or $\lim_{n\to\infty} a_n$ does not exist. If $\lim_{n\to\infty} a_n = \infty$, we can say that the sequence diverges to ∞; if $\lim_{n\to\infty} a_n = -\infty$, we can say that it diverges to $-\infty$. If $\lim_{n\to\infty} a_n$ simply does not exist (but is not ∞ or $-\infty$), we can only say that the sequence diverges.

Example 5   (**Examples of convergent and divergent sequences**)

(a) $\{(n - 1)/n\}$ converges to 1; $\lim_{n\to\infty}(n - 1)/n = \lim_{n\to\infty}\bigl(1 - (1/n)\bigr) = 1$.

(b) $\{n\} = \{1, 2, 3, 4, \ldots\}$ diverges to ∞.

(c) $\{-n\} = \{-1, -2, -3, -4, \ldots\}$ diverges to $-\infty$.

(d) $\{(-1)^n\} = \{-1, 1, -1, 1, -1, \ldots\}$ simply diverges.

(e) $\{(-1)^n n\} = \{-1, 2, -3, 4, -5, \ldots\}$ diverges (but not to ∞ or $-\infty$ even though $\lim_{n\to\infty} |a_n| = \infty$).

The limit of a sequence is equivalent to the limit of a function as its argument approaches infinity:

If $\lim_{x\to\infty} f(x) = L$ and $a_n = f(n)$, then $\lim_{n\to\infty} a_n = L$.

Because of this, the standard rules for limits of functions (Theorems 2 and 4 of Section 1.2) also hold for limits of sequences, with the appropriate changes of notation. Thus, if $\{a_n\}$ and $\{b_n\}$ converge, then

$$\lim_{n\to\infty} (a_n \pm b_n) = \lim_{n\to\infty} a_n \pm \lim_{n\to\infty} b_n,$$

$$\lim_{n\to\infty} c a_n = c \lim_{n\to\infty} a_n,$$

$$\lim_{n\to\infty} a_n b_n = (\lim_{n\to\infty} a_n)(\lim_{n\to\infty} b_n),$$

$$\lim_{n\to\infty} \frac{a_n}{b_n} = \frac{\lim_{n\to\infty} a_n}{\lim_{n\to\infty} b_n} \qquad \text{assuming } \lim_{n\to\infty} b_n \neq 0.$$

If $a_n \leq b_n$ ultimately, then $\lim_{n\to\infty} a_n \leq \lim_{n\to\infty} b_n$.

If $a_n \leq b_n \leq c_n$ ultimately, and $\lim_{n\to\infty} a_n = L = \lim_{n\to\infty} c_n$, then $\lim_{n\to\infty} b_n = L$.

The limits of many explicitly defined sequences can be evaluated using these properties in a manner similar to the methods used for limits of the form $\lim_{x\to\infty} f(x)$ in Section 1.3.

Example 6 Calculate the limits of the sequences

(a) $\left\{\dfrac{2n^2 - n - 1}{5n^2 + n - 3}\right\}$,   (b) $\left\{\dfrac{\cos n}{n}\right\}$,   and (c) $\{\sqrt{n^2 + 2n} - n\}$.

Solution

(a) We divide the numerator and denominator of the expression for a_n by the highest power of n in the denominator, that is, by n^2:

$$\lim_{n\to\infty} \frac{2n^2 - n - 1}{5n^2 + n - 3} = \lim_{n\to\infty} \frac{2 - (1/n) - (1/n^2)}{5 + (1/n) - (3/n^2)} = \frac{2 - 0 - 0}{5 + 0 - 0} = \frac{2}{5},$$

since $\lim_{n\to\infty} 1/n = 0$ and $\lim_{n\to\infty} 1/n^2 = 0$. The sequence converges and its limit is 2/5.

(b) Since $|\cos n| \leq 1$ for every n, we have

$$-\frac{1}{n} \leq \frac{\cos n}{n} \leq \frac{1}{n} \quad \text{for} \quad n \geq 1.$$

Now, $\lim_{n\to\infty} -1/n = 0$ and $\lim_{n\to\infty} 1/n = 0$. Therefore, by the sequence version of the Squeeze Theorem, $\lim_{n\to\infty} (\cos n)/n = 0$. The given sequence converges to 0.

(c) For this sequence we multiply the numerator and the denominator (which is 1) by the conjugate of the expression in the numerator:

$$\lim_{n \to \infty} (\sqrt{n^2 + 2n} - n) = \lim_{n \to \infty} \frac{(\sqrt{n^2 + 2n} - n)(\sqrt{n^2 + 2n} + n)}{\sqrt{n^2 + 2n} + n}$$

$$= \lim_{n \to \infty} \frac{2n}{\sqrt{n^2 + 2n} + n} = \lim_{n \to \infty} \frac{2}{\sqrt{1 + (2/n)} + 1} = 1.$$

The sequence converges to 1.

Example 7   Evaluate $\lim_{n \to \infty} n \tan^{-1}\left(\dfrac{1}{n}\right)$.

Solution   For this example it is best to replace the nth term of the sequence by the corresponding function of a real variable x and take the limit as $x \to \infty$. We use l'Hôpital's Rule:

$$\lim_{n \to \infty} n \tan^{-1}\left(\frac{1}{n}\right) = \lim_{x \to \infty} x \tan^{-1}\left(\frac{1}{x}\right)$$

$$= \lim_{x \to \infty} \frac{\tan^{-1}\left(\dfrac{1}{x}\right)}{\dfrac{1}{x}} \qquad \begin{bmatrix} 0 \\ 0 \end{bmatrix}$$

$$= \lim_{x \to \infty} \frac{\dfrac{1}{1 + (1/x^2)}\left(-\dfrac{1}{x^2}\right)}{-\left(\dfrac{1}{x^2}\right)} = \lim_{x \to \infty} \frac{1}{1 + \dfrac{1}{x^2}} = 1.$$

THEOREM   1   If $\{a_n\}$ converges, then $\{a_n\}$ is bounded.

PROOF   Suppose $\lim_{n \to \infty} a_n = L$. According to Definition 2, for $\epsilon = 1$ there exists a number N such that if $n > N$, then $|a_n - L| < 1$; therefore $|a_n| < 1 + |L|$ for such n. (Why is this true?) If K denotes the largest of the numbers $|a_1|$, $|a_2|$, ..., $|a_N|$, and $1 + |L|$, then $|a_n| \le K$ for every $n = 1, 2, 3, \ldots$. Hence $\{a_n\}$ is bounded.

The converse of Theorem 1 is false; the sequence $\{(-1)^n\}$ is bounded but does not converge.

The *completeness property* of the real number system (see Section P.1) can be reformulated in terms of sequences to read as follows:

Bounded monotonic sequences converge

If the sequence $\{a_n\}$ is bounded above and is (ultimately) increasing, then it converges. The same conclusion holds if $\{a_n\}$ is bounded below and is (ultimately) decreasing.

Thus, a bounded, ultimately monotonic sequence is convergent. (See Figure 9.2.)

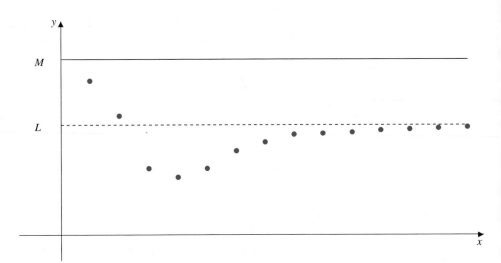

Figure 9.2   An ultimately increasing sequence that is bounded above

Example 8   Let a_n be defined recursively by

$$a_1 = 1, \qquad a_{n+1} = \sqrt{6 + a_n} \qquad (n = 1, 2, 3, \ldots).$$

Show that $\lim_{n \to \infty} a_n$ exists and find its value.

Solution   Observe that $a_2 = \sqrt{6 + 1} = \sqrt{7} > a_1$. If $a_{k+1} > a_k$, then $a_{k+2} = \sqrt{6 + a_{k+1}} > \sqrt{6 + a_k} = a_{k+1}$, so $\{a_n\}$ is increasing, by induction. Now observe that $a_1 = 1 < 3$. If $a_k < 3$, then $a_{k+1} = \sqrt{6 + a_k} < \sqrt{6 + 3} = 3$, so $a_n < 3$ for every n by induction. Since $\{a_n\}$ is increasing and bounded above, $\lim_{n \to \infty} a_n = a$ exists, by completeness. Since $\sqrt{6 + x}$ is a continuous function of x, we have

$$a = \lim_{n \to \infty} a_{n+1} = \lim_{n \to \infty} \sqrt{6 + a_n} = \sqrt{6 + \lim_{n \to \infty} a_n} = \sqrt{6 + a}.$$

Thus $a^2 = 6 + a$, or $a^2 - a - 6 = 0$, or $(a - 3)(a + 2) = 0$. This quadratic has roots $a = 3$ and $a = -2$. Since $a_n \geq 1$ for every n, we must have $a \geq 1$. Therefore, $a = 3$ and $\lim_{n \to \infty} a_n = 3$. ∎

There is a subtle point to note in this solution. Showing that $\{a_n\}$ is increasing is pretty obvious, but how did we know to try and show that 3 (rather than some other number) was an upper bound? The answer is that we actually did the last part first and showed that *if* $\lim a_n = a$ exists, *then* $a = 3$. It then makes sense to try and show that $a_n < 3$ for all n.

Example 9   Does $\left\{ \left(1 + \dfrac{1}{n} \right)^n \right\}$ converge or not?

Solution   We could make an effort to show that the given sequence is, in fact, increasing and bounded above. (See Exercise 32 at the end of this section.) However, we already know the answer. By Theorem 6 of Section 3.4,

$$\lim_{n \to \infty} \left(1 + \frac{1}{n} \right)^n = e^1 = e.$$

∎

THEOREM **2**   If $\{a_n\}$ is (ultimately) increasing, then either it is bounded above, and therefore convergent, or it is not bounded above and diverges to infinity.

The proof of this theorem is left as an exercise. A corresponding result holds for (ultimately) decreasing sequences.

The following theorem evaluates two important limits that find frequent application in the study of series.

THEOREM 3

(a) If $|x| < 1$, then $\lim_{n\to\infty} x^n = 0$.

(b) If x is any real number, then $\lim_{n\to\infty} \dfrac{x^n}{n!} = 0$.

PROOF   For part (a) observe that

$$\lim_{n\to\infty} \ln |x|^n = \lim_{n\to\infty} n \ln |x| = -\infty,$$

since $\ln |x| < 0$ when $|x| < 1$. Accordingly, since e^x is continuous,

$$\lim_{n\to\infty} |x|^n = \lim_{n\to\infty} e^{\ln |x|^n} = e^{\lim_{n\to\infty} \ln |x|^n} = 0.$$

Since $-|x|^n \le x^n \le |x|^n$, we have $\lim_{n\to\infty} x^n = 0$ by the Squeeze Theorem.

For part (b), pick any x and let N be an integer such that $N > |x|$. If $n > N$ we have

$$\left| \frac{x^n}{n!} \right| = \frac{|x|}{1} \frac{|x|}{2} \frac{|x|}{3} \cdots \frac{|x|}{N-1} \frac{|x|}{N} \frac{|x|}{N+1} \cdots \frac{|x|}{n}$$

$$< \frac{|x|^{N-1}}{(N-1)!} \frac{|x|}{N} \frac{|x|}{N} \frac{|x|}{N} \cdots \frac{|x|}{N}$$

$$= \frac{|x|^{N-1}}{(N-1)!} \left(\frac{|x|}{N} \right)^{n-N+1} = K \left(\frac{|x|}{N} \right)^n,$$

where $K = \dfrac{|x|^{N-1}}{(N-1)!} \left(\dfrac{|x|}{N} \right)^{1-N}$ is a constant that is independent of n. Since $|x|/N < 1$, we have $\lim_{n\to\infty} (|x|/N)^n = 0$ by part (a). Thus $\lim_{n\to\infty} |x^n/n!| = 0$, so $\lim_{n\to\infty} x^n/n! = 0$.

Example 10   Find $\lim_{n\to\infty} \dfrac{3^n + 4^n + 5^n}{5^n}$.

Solution $\lim_{n\to\infty} \dfrac{3^n + 4^n + 5^n}{5^n} = \lim_{n\to\infty} \left[\left(\dfrac{3}{5} \right)^n + \left(\dfrac{4}{5} \right)^n + 1 \right] = 0 + 0 + 1 = 1$, by Theorem 3(a).

Exercises 9.1

In Exercises 1–13, determine whether the given sequence is (a) bounded (above or below), (b) positive or negative (ultimately), (c) increasing, decreasing, or alternating, and (d) convergent, divergent, divergent to ∞ or $-\infty$.

1. $\left\{ \dfrac{2n^2}{n^2 + 1} \right\}$

2. $\left\{ \dfrac{2n}{n^2 + 1} \right\}$

3. $\left\{ 4 - \dfrac{(-1)^n}{n} \right\}$

4. $\left\{ \sin \dfrac{1}{n} \right\}$

5. $\left\{ \dfrac{n^2 - 1}{n} \right\}$

6. $\left\{ \dfrac{e^n}{\pi^n} \right\}$

7. $\left\{ \dfrac{e^n}{\pi^{n/2}} \right\}$

8. $\left\{ \dfrac{(-1)^n n}{e^n} \right\}$

9. $\left\{ \dfrac{2^n}{n^n} \right\}$

10. $\left\{ \dfrac{(n!)^2}{(2n)!} \right\}$

11. $\left\{ n \cos \left(\dfrac{n\pi}{2} \right) \right\}$

12. $\left\{ \dfrac{\sin n}{n} \right\}$

13. $\{1,\ 1,\ -2,\ 3,\ 3,\ -4,\ 5,\ 5,\ -6,\ \ldots\}$

In Exercises 14–29, evaluate, wherever possible, the limit of the sequence $\{a_n\}$.

14. $a_n = \dfrac{5 - 2n}{3n - 7}$

15. $a_n = \dfrac{n^2 - 4}{n + 5}$

16. $a_n = \dfrac{n^2}{n^3 + 1}$

17. $a_n = (-1)^n \dfrac{n}{n^3 + 1}$

18. $a_n = \dfrac{n^2 - 2\sqrt{n} + 1}{1 - n - 3n^2}$

19. $a_n = \dfrac{e^n - e^{-n}}{e^n + e^{-n}}$

20. $a_n = n \sin \dfrac{1}{n}$

21. $a_n = \left(\dfrac{n - 3}{n} \right)^n$

22. $a_n = \dfrac{n}{\ln(n + 1)}$

23. $a_n = \sqrt{n + 1} - \sqrt{n}$

24. $a_n = n - \sqrt{n^2 - 4n}$

25. $a_n = \sqrt{n^2 + n} - \sqrt{n^2 - 1}$

26. $a_n = \left(\dfrac{n - 1}{n + 1} \right)^n$

27. $a_n = \dfrac{(n!)^2}{(2n)!}$

28. $a_n = \dfrac{n^2 2^n}{n!}$

29. $a_n = \dfrac{\pi^n}{1 + 2^{2n}}$

30. Let $a_1 = 1$ and $a_{n+1} = \sqrt{1 + 2a_n}$ $(n = 1, 2, 3, \ldots)$. Show that $\{a_n\}$ is increasing and bounded above. (*Hint:* show that 3 is an upper bound.) Hence, conclude that the sequence converges, and find its limit.

∗ 31. Repeat Exercise 30 for the sequence defined by $a_1 = 3$, $a_{n+1} = \sqrt{15 + 2a_n}$, $n = 1, 2, 3, \ldots$. This time you will have to guess an upper bound.

∗ 32. Let $a_n = \left(1 + \dfrac{1}{n} \right)^n$ so that $\ln a_n = n \ln \left(1 + \dfrac{1}{n} \right)$. Use properties of the logarithm function to show that (a) $\{a_n\}$ is increasing and (b) e is an upper bound for $\{a_n\}$.

∗ 33. Prove Theorem 2. Also, state an analogous theorem pertaining to ultimately decreasing sequences.

34. If $\{|a_n|\}$ is bounded, prove that $\{a_n\}$ is bounded.

35. If $\lim_{n\to\infty} |a_n| = 0$, prove that $\lim_{n\to\infty} a_n = 0$.

∗ 36. Which of the following statements are TRUE and which are FALSE? Justify your answers.

(a) If $\lim_{n\to\infty} a_n = \infty$ and $\lim_{n\to\infty} b_n = L > 0$, then $\lim_{n\to\infty} a_n b_n = \infty$.

(b) If $\lim_{n\to\infty} a_n = \infty$ and $\lim_{n\to\infty} b_n = -\infty$, then $\lim_{n\to\infty} (a_n + b_n) = 0$.

(c) If $\lim_{n\to\infty} a_n = \infty$ and $\lim_{n\to\infty} b_n = -\infty$, then $\lim_{n\to\infty} a_n b_n = -\infty$.

(d) If neither $\{a_n\}$ nor $\{b_n\}$ converges, then $\{a_n b_n\}$ does not converge.

(e) If $\{|a_n|\}$ converges, then $\{a_n\}$ converges.

9.2 Infinite Series

An **infinite series**, usually just called a **series**, is a formal sum of infinitely many terms; for instance,

$$a_1 + a_2 + a_3 + a_4 + \cdots$$

is a series formed by adding the terms of the sequence $\{a_n\}$. This series is also denoted $\sum_{n=1}^{\infty} a_n$:

$$\sum_{n=1}^{\infty} a_n = a_1 + a_2 + a_3 + a_4 + \cdots$$

For example,

$$\sum_{n=1}^{\infty} \frac{1}{n} = 1 + \frac{1}{2} + \frac{1}{3} + \frac{1}{4} + \cdots$$

$$\sum_{n=1}^{\infty} \frac{(-1)^{n-1}}{2^{n-1}} = 1 - \frac{1}{2} + \frac{1}{4} - \frac{1}{8} + \frac{1}{16} - \cdots.$$

It is sometimes necessary or useful to start the sum from some index other than 1:

$$\sum_{n=0}^{\infty} a^n = 1 + a + a^2 + a^3 + \cdots$$

$$\sum_{n=2}^{\infty} \frac{1}{\ln n} = \frac{1}{\ln 2} + \frac{1}{\ln 3} + \frac{1}{\ln 4} + \cdots.$$

Note that the latter series would make no sense if we had started the sum from $n = 1$; the first term would have been undefined.

When necessary, we can change the index of summation to start at a different value. This is accomplished by a substitution as illustrated in Example 3 of Section 5.1. For instance, using the substitution $n = m - 2$, we can rewrite $\sum_{n=1}^{\infty} a_n$ in the form $\sum_{m=3}^{\infty} a_{m-2}$. Both sums give rise to the same expansion

$$\sum_{n=1}^{\infty} a_n = a_1 + a_2 + a_3 + \cdots = \sum_{m=3}^{\infty} a_{m-2}.$$

Addition is an operation that is carried out on two numbers at a time. If we want to calculate the finite sum

$$a_1 + a_2 + a_3,$$

we could proceed by adding $a_1 + a_2$ and then adding a_3 to this sum, or else we might first add $a_2 + a_3$ and then add a_1 to the sum. Of course, the associative law for addition assures us we will get the same answer both ways. This is the reason the symbol $a_1 + a_2 + a_3$ makes sense; we would otherwise have to write $(a_1 + a_2) + a_3$ or $a_1 + (a_2 + a_3)$. This reasoning extends to any sum $a_1 + a_2 + \cdots + a_n$ of finitely many terms, but it is not obvious what should be meant by a sum with infinitely many terms:

$$a_1 + a_2 + a_3 + a_4 + \cdots.$$

We no longer have any assurance that the terms can be added up in any order to yield the same sum. In fact, we will see in Section 9.4 that in certain circumstances, changing the order of terms in a series can actually change the sum of the series. The interpretation we place on the infinite sum is that of adding from left to right, as suggested by the grouping

$$\cdots ((((a_1 + a_2) + a_3) + a_4) + a_5) + \cdots.$$

We accomplish this by defining a new sequence $\{s_n\}$, called the **sequence of partial sums** of the series $\sum_{n=1}^{\infty} a_n$ so that s_n is the sum of the first n terms of the series:

$$s_1 = a_1$$
$$s_2 = s_1 + a_2 = a_1 + a_2$$
$$s_3 = s_2 + a_3 = a_1 + a_2 + a_3$$
$$\vdots$$
$$s_n = s_{n-1} + a_n = a_1 + a_2 + a_3 + \cdots + a_n = \sum_{j=1}^{n} a_j.$$
$$\vdots$$

We then define the sum of the infinite series to be the limit of this sequence of partial sums.

DEFINITION **3**

Convergence of a series

We say that the series $\sum_{n=1}^{\infty} a_n$ **converges to the sum** s, and we write

$$\sum_{n=1}^{\infty} a_n = s,$$

if $\lim_{n \to \infty} s_n = s$, where s_n is the nth partial sum of $\sum_{n=1}^{\infty} a_n$:

$$s_n = a_1 + a_2 + a_3 + \cdots + a_n = \sum_{j=1}^{n} a_j.$$

Thus, a *series* converges if and only if the *sequence* of its partial sums converges.

Similarly, a series is said to diverge to infinity, diverge to negative infinity, or simply diverge if its sequence of partial sums does so. It must be stressed that the convergence of the series $\sum_{n=1}^{\infty} a_n$ depends on the convergence of the sequence $\{s_n\} = \{\sum_{j=1}^{n} a_j\}$, *not* the sequence $\{a_n\}$.

Geometric Series

DEFINITION **4**

Geometric series

A series of the form $\sum_{n=1}^{\infty} a\, r^{n-1} = a + ar + ar^2 + ar^3 + \cdots$, whose nth term is $a_n = a\, r^{n-1}$, is called a **geometric series**. The number a is the first term. The number r is called the **common ratio** of the series, since it is the value of the ratio of the $(n+1)$st term to the nth term for any $n \geq 1$:

$$\frac{a_{n+1}}{a_n} = \frac{ar^n}{ar^{n-1}} = r, \qquad n = 1, 2, 3, \ldots.$$

The nth partial sum s_n of a geometric series is calculated as follows:

$$s_n = a + ar + ar^2 + ar^3 + \cdots + ar^{n-1}$$
$$r s_n = \quad ar + ar^2 + ar^3 + \cdots + ar^{n-1} + ar^n.$$

The second equation is obtained by multiplying the first by r. Subtracting these two equations (note the cancellations), we get $(1 - r)s_n = a - ar^n$. If $r \neq 1$, we can divide by $1 - r$ and get a formula for s_n.

Partial sums of geometric series

If $r = 1$, then the nth partial sum of a geometric series $\sum_{n=1}^{\infty} ar^{n-1}$ is $s_n = a + a + \cdots + a = na$. If $r \neq 1$, then

$$s_n = a + ar + ar^2 + \cdots + ar^{n-1} = \frac{a(1 - r^n)}{1 - r}.$$

If $a = 0$, then $s_n = 0$ for every n, and $\lim_{n \to \infty} s_n = 0$. Now suppose $a \neq 0$. If $|r| < 1$, then $\lim_{n \to \infty} r^n = 0$, so $\lim_{n \to \infty} s_n = a/(1 - r)$. If $r > 1$, then $\lim_{n \to \infty} r^n = \infty$, and $\lim_{n \to \infty} s_n = \infty$ if $a > 0$, or $\lim_{n \to \infty} s_n = -\infty$ if $a < 0$. The same conclusion holds if $r = 1$, since $s_n = na$ in this case. If $r \leq -1$, $\lim_{n \to \infty} r^n$ does not exist and neither does $\lim_{n \to \infty} s_n$. Hence we conclude that

$$\sum_{n=1}^{\infty} ar^{n-1} \begin{cases} \text{converges to } 0 & \text{if } a = 0 \\ \text{converges to } \dfrac{a}{1-r} & \text{if } |r| < 1 \\ \text{diverges to } \infty & \text{if } r \geq 1 \text{ and } a > 0 \\ \text{diverges to } -\infty & \text{if } r \geq 1 \text{ and } a < 0 \\ \text{diverges} & \text{if } r \leq -1 \text{ and } a \neq 0. \end{cases}$$

The representation of the function $1/(1 - x)$ as the sum of a geometric series,

$$\frac{1}{1-x} = \sum_{n=0}^{\infty} x^n = 1 + x + x^2 + x^3 + \cdots \quad \text{for } -1 < x < 1,$$

will be important in our discussion of power series later in this chapter.

Example 1 **(Examples of geometric series and their sums)**

(a) $1 + \dfrac{1}{2} + \dfrac{1}{4} + \dfrac{1}{8} + \cdots = \displaystyle\sum_{n=1}^{\infty} \left(\dfrac{1}{2}\right)^{n-1} = \dfrac{1}{1 - \dfrac{1}{2}} = 2$. Here $a = 1$ and $r = \dfrac{1}{2}$.

Since $|r| < 1$, the series converges.

(b) $\pi - e + \dfrac{e^2}{\pi} - \dfrac{e^3}{\pi^2} + \cdots = \displaystyle\sum_{n=1}^{\infty} \pi \left(-\dfrac{e}{\pi}\right)^{n-1}$ Here $a = \pi$ and $r = -\dfrac{e}{\pi}$.

$$= \dfrac{\pi}{1 - \left(-\dfrac{e}{\pi}\right)} = \dfrac{\pi^2}{\pi + e}.$$

The series converges since $\left|-\dfrac{e}{\pi}\right| < 1$.

(c) $1 + 2^{1/2} + 2 + 2^{3/2} + \cdots = \displaystyle\sum_{n=1}^{\infty} (\sqrt{2})^{n-1}$. This series diverges to ∞ since $a = 1 > 0$ and $r = \sqrt{2} > 1$.

(d) $1 - 1 + 1 - 1 + 1 - \cdots = \displaystyle\sum_{n=1}^{\infty} (-1)^{n-1}$. This series diverges since $r = -1$.

(e) Let $x = 0.323232 \cdots = 0.\overline{32}$; then

$$x = \dfrac{32}{100} + \dfrac{32}{100^2} + \dfrac{32}{100^3} + \cdots = \sum_{n=1}^{\infty} \dfrac{32}{100} \left(\dfrac{1}{100}\right)^{n-1} = \dfrac{32}{100} \dfrac{1}{1 - \dfrac{1}{100}} = \dfrac{32}{99}.$$

This is an alternative to the method of Example 1 of Section P.1 for representing repeating decimals as quotients of integers.

Example 2 If money earns interest at a constant effective rate of 8% per year, how much should you pay today for an annuity that will pay you (a) $1,000 at the end of each of the next 10 years and (b) $1,000 at the end of every year forever?

Solution A payment of $1,000 that is due to be received n years from now has present value $1,000 \times \left(\dfrac{1}{1.08}\right)^n$ (since $\$A$ would grow to $\$A(1.08)^n$ in n years). Thus $1,000 payments at the end of each of the next n years are worth $\$s_n$ at the present time, where

$$s_n = 1,000\left[\frac{1}{1.08} + \left(\frac{1}{1.08}\right)^2 + \cdots + \left(\frac{1}{1.08}\right)^n\right]$$

$$= \frac{1,000}{1.08}\left[1 + \frac{1}{1.08} + \left(\frac{1}{1.08}\right)^2 + \cdots + \left(\frac{1}{1.08}\right)^{n-1}\right].$$

$$= \frac{1,000}{1.08}\,\frac{1 - \left(\dfrac{1}{1.08}\right)^n}{1 - \dfrac{1}{1.08}} = \frac{1,000}{0.08}\left[1 - \left(\frac{1}{1.08}\right)^n\right]$$

(a) The present value of 10 future payments is $\$s_{10} = \$6,710.08$.

(b) The present value of future payments continuing forever is

$$\$\lim_{n\to\infty} s_n = \frac{\$1,000}{0.08} = \$12,500.$$

Telescoping Series and Harmonic Series

Example 3 Show that the series

$$\sum_{n=1}^{\infty} \frac{1}{n(n+1)} = \frac{1}{1 \times 2} + \frac{1}{2 \times 3} + \frac{1}{3 \times 4} + \frac{1}{4 \times 5} + \cdots$$

converges and find its sum.

Solution Since $\dfrac{1}{n(n+1)} = \dfrac{1}{n} - \dfrac{1}{n+1}$, we can write the partial sum s_n in the form

$$s_n = \frac{1}{1 \times 2} + \frac{1}{2 \times 3} + \frac{1}{3 \times 4} + \cdots + \frac{1}{(n-1)n} + \frac{1}{n(n+1)}$$

$$= \left(1 - \frac{1}{2}\right) + \left(\frac{1}{2} - \frac{1}{3}\right) + \left(\frac{1}{3} - \frac{1}{4}\right) + \cdots$$

$$\cdots + \left(\frac{1}{n-1} - \frac{1}{n}\right) + \left(\frac{1}{n} - \frac{1}{n+1}\right)$$

$$= 1 - \frac{1}{2} + \frac{1}{2} - \frac{1}{3} + \frac{1}{3} - \cdots - \frac{1}{n} + \frac{1}{n} - \frac{1}{n+1}$$

$$= 1 - \frac{1}{n+1}.$$

Therefore, $\lim_{n \to \infty} s_n = 1$ and the series converges to 1:

$$\sum_{n=1}^{\infty} \frac{1}{n(n+1)} = 1.$$

This is an example of a **telescoping series**, so called because the partial sums *fold up* into a simple form when the terms are expanded in partial fractions. Other examples can be found in the exercises at the end of this section. As these examples show, the method of partial fractions can be a useful tool for series as well as for integrals. ■

Example 4   Show that the **harmonic series**

$$\sum_{n=1}^{\infty} \frac{1}{n} = 1 + \frac{1}{2} + \frac{1}{3} + \frac{1}{4} + \cdots$$

diverges to infinity.

Solution   If s_n is the nth partial sum of the harmonic series, then

$$
\begin{aligned}
s_n &= 1 + \frac{1}{2} + \frac{1}{3} + \cdots + \frac{1}{n} \\
&= \text{sum of areas of rectangles shaded in Figure 9.3} \\
&> \text{area under } y = \frac{1}{x} \text{ from 1 to } n+1 \\
&= \int_1^{n+1} \frac{dx}{x} = \ln(n+1).
\end{aligned}
$$

Now $\lim_{n \to \infty} \ln(n+1) = \infty$. Therefore, $\lim_{n \to \infty} s_n = \infty$ and

$$\sum_{n=1}^{\infty} \frac{1}{n} = 1 + \frac{1}{2} + \frac{1}{3} + \cdots \qquad \text{diverges to infinity.}$$

■

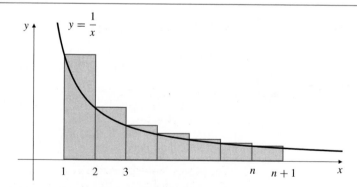

Figure 9.3   A partial sum of the harmonic series

Like geometric series, the harmonic series will often be encountered in subsequent sections.

Some Theorems About Series

THEOREM   4

If $\sum_{n=1}^{\infty} a_n$ converges, then $\lim_{n \to \infty} a_n = 0$.

PROOF   If $s_n = a_1 + a_2 + \cdots + a_n$, then $s_n - s_{n-1} = a_n$. If $\sum_{n=1}^{\infty} a_n$ converges, then $\lim_{n \to \infty} s_n = s$ exists, and $\lim_{n \to \infty} s_{n-1} = s$. Hence $\lim_{n \to \infty} a_n = s - s = 0$.

Remark   Theorem 4 is *very important* for the understanding of infinite series. Students often err either in forgetting that *a series cannot converge if its terms do not approach zero* or in confusing this result with its *converse*, which is false. The converse would say that if $\lim_{n \to \infty} a_n = 0$, then $\sum_{n=1}^{\infty} a_n$ must converge. The harmonic series is a counterexample showing the falsehood of this assertion:

$$\lim_{n \to \infty} \frac{1}{n} = 0 \qquad \text{but} \qquad \sum_{n=1}^{\infty} \frac{1}{n} \text{ diverges to infinity.}$$

When considering whether a given series converges, the first question you should ask yourself is: "Does the nth term approach 0 as n approaches ∞?" If the answer is *no*, then the series does *not* converge. If the answer is *yes*, then the series *may or may not* converge. If the sequence of terms $\{a_n\}$ tends to a nonzero limit L, then $\sum_{n=1}^{\infty} a_n$ diverges to infinity if $L > 0$ and diverges to negative infinity if $L < 0$.

Example 5

(a) $\sum_{n=1}^{\infty} \dfrac{n}{2n-1}$ diverges to infinity since $\lim_{n \to \infty} \dfrac{n}{2n-1} = 1/2 > 0$.

(b) $\sum_{n=1}^{\infty} (-1)^n n \sin(1/n)$ diverges since

$$\lim_{n \to \infty} \left| (-1)^n n \sin \frac{1}{n} \right| = \lim_{n \to \infty} \frac{\sin(1/n)}{1/n} = \lim_{x \to 0+} \frac{\sin x}{x} = 1 \neq 0.$$

The following theorem asserts that it is only the *ultimate* behaviour of $\{a_n\}$ that determines whether $\sum_{n=1}^{\infty} a_n$ converges. Any finite number of terms can be dropped from the beginning of a series without affecting the convergence; the convergence depends only on the *tail* of the series. Of course, the actual sum of the series depends on *all* the terms.

THEOREM 5

$\sum_{n=1}^{\infty} a_n$ converges if and only if $\sum_{n=N}^{\infty} a_n$ converges for any integer $N \geq 1$.

THEOREM 6

If $\{a_n\}$ is ultimately positive, then the series $\sum_{n=1}^{\infty} a_n$ must either converge (if its partial sums are bounded above) or diverge to infinity (if its partial sums are not bounded above).

The proofs of these two theorems are posed as exercises at the end of this section. The following theorem is just a reformulation of standard laws of limits.

THEOREM 7

If $\sum_{n=1}^{\infty} a_n$ and $\sum_{n=1}^{\infty} b_n$ converge to A and B, respectively, then

(a) $\sum_{n=1}^{\infty} c a_n$ converges to cA (where c is any constant);

(b) $\sum_{n=1}^{\infty} (a_n \pm b_n)$ converges to $A \pm B$;

(c) if $a_n \leq b_n$ for all $n = 1, 2, 3, \ldots$, then $A \leq B$.

Example 6 Find the sum of the series $\displaystyle\sum_{n=1}^{\infty} \frac{1+2^{n+1}}{3^n}$.

Solution The given series is the sum of two geometric series,

$$\sum_{n=1}^{\infty} \frac{1}{3^n} = \sum_{n=1}^{\infty} \frac{1}{3}\left(\frac{1}{3}\right)^{n-1} = \frac{1/3}{1-(1/3)} = \frac{1}{2} \quad \text{and}$$

$$\sum_{n=1}^{\infty} \frac{2^{n+1}}{3^n} = \sum_{n=1}^{\infty} \frac{4}{3}\left(\frac{2}{3}\right)^{n-1} = \frac{4/3}{1-(2/3)} = 4.$$

Thus its sum is $\dfrac{1}{2} + 4 = \dfrac{9}{2}$ by Theorem 7(b).

Exercises 9.2

In Exercises 1–18, find the sum of the given series, or show that the series diverges (possibly to infinity or negative infinity). Exercises 11–14 are telescoping series and should be done by partial fractions as suggested in Example 3 in this section.

1. $\dfrac{1}{3} + \dfrac{1}{9} + \dfrac{1}{27} + \cdots = \displaystyle\sum_{n=1}^{\infty} \frac{1}{3^n}$

2. $3 - \dfrac{3}{4} + \dfrac{3}{16} - \dfrac{3}{64} + \cdots = \displaystyle\sum_{n=1}^{\infty} 3\left(-\frac{1}{4}\right)^{n-1}$

3. $\displaystyle\sum_{n=5}^{\infty} \frac{1}{(2+\pi)^{2n}}$

4. $\displaystyle\sum_{n=0}^{\infty} \frac{5}{10^{3n}}$

5. $\displaystyle\sum_{n=2}^{\infty} \frac{(-5)^n}{8^{2n}}$

6. $\displaystyle\sum_{n=0}^{\infty} \frac{1}{e^n}$

7. $\displaystyle\sum_{k=0}^{\infty} \frac{2^{k+3}}{e^{k-3}}$

8. $\displaystyle\sum_{j=1}^{\infty} \pi^{j/2}\cos(j\pi)$

9. $\displaystyle\sum_{n=1}^{\infty} \frac{3+2^n}{2^{n+2}}$

10. $\displaystyle\sum_{n=0}^{\infty} \frac{3+2^n}{3^{n+2}}$

11. $\displaystyle\sum_{n=1}^{\infty} \frac{1}{n(n+2)} = \frac{1}{1\times 3} + \frac{1}{2\times 4} + \frac{1}{3\times 5} + \cdots$

12. $\displaystyle\sum_{n=1}^{\infty} \frac{1}{(2n-1)(2n+1)} = \frac{1}{1\times 3} + \frac{1}{3\times 5} + \frac{1}{5\times 7} + \cdots$

13. $\displaystyle\sum_{n=1}^{\infty} \frac{1}{(3n-2)(3n+1)} = \frac{1}{1\times 4} + \frac{1}{4\times 7} + \frac{1}{7\times 10} + \cdots$

* 14. $\displaystyle\sum_{n=1}^{\infty} \frac{1}{n(n+1)(n+2)}$

$= \dfrac{1}{1\times 2\times 3} + \dfrac{1}{2\times 3\times 4} + \dfrac{1}{3\times 4\times 5} + \cdots$

15. $\displaystyle\sum_{n=1}^{\infty} \frac{1}{2n-1}$

16. $\displaystyle\sum_{n=1}^{\infty} \frac{n}{n+2}$

17. $\displaystyle\sum_{n=1}^{\infty} n^{-1/2}$

18. $\displaystyle\sum_{n=1}^{\infty} \frac{2}{n+1}$

19. Obtain a simple expression for the partial sum s_n of the series $\sum_{n=1}^{\infty}(-1)^n$, and use it to show that the series diverges.

20. Find the sum of the series

$$\frac{1}{1} + \frac{1}{1+2} + \frac{1}{1+2+3} + \frac{1}{1+2+3+4} + \cdots.$$

21. When dropped, an elastic ball bounces back up to a height three-quarters of that from which it fell. If the ball is dropped from a height of 2 m and allowed to bounce up and down indefinitely, what is the total distance it travels before coming to rest?

22. If a bank account pays 10% simple interest into an account once a year, what is the balance in the account at the end of 8 years if $1,000 is deposited into the account at the beginning of each of the 8 years? (Assume there was no balance in the account initially.)

* 23. Prove Theorem 5. *** 24.** Prove Theorem 6.

* 25. State a theorem analogous to Theorem 6 but for a negative sequence.

In Exercises 26–31, decide whether the given statement is TRUE or FALSE. If it is true, prove it. If it is false, give a counterexample showing the falsehood.

* 26. If $a_n = 0$ for every n, then $\sum a_n$ converges.

* 27. If $\sum a_n$ converges, then $\sum(1/a_n)$ diverges to infinity.

* 28. If $\sum a_n$ and $\sum b_n$ both diverge, then so does $\sum(a_n + b_n)$.

* 29. If $a_n \geq c > 0$ for every n, then $\sum a_n$ diverges to infinity.

* **30.** If $\sum a_n$ diverges and $\{b_n\}$ is bounded, then $\sum a_n b_n$ diverges.

* **31.** If $a_n > 0$ and $\sum a_n$ converges, then $\sum (a_n)^2$ converges.

9.3 Convergence Tests for Positive Series

In the previous section we saw a few examples of convergent series (geometric and telescoping series) whose sums could be determined exactly because the partial sums s_n could be expressed in closed form as explicit functions of n whose limits as $n \to \infty$ could be evaluated. It is not usually possible to do this with a given series, and therefore it is not usually possible to determine the sum of the series exactly. However, there are many techniques for determining whether a given series converges and, if it does, for approximating the sum to any desired degree of accuracy.

In this section we deal exclusively with *positive series*, that is, series of the form

$$\sum_{n=1}^{\infty} a_n = a_1 + a_2 + a_3 + \cdots,$$

where $a_n \geq 0$ for all $n \geq 1$. As noted in Theorem 6, such a series will converge if its partial sums are bounded above and will diverge to infinity otherwise. All our results apply equally well to *ultimately* positive series since convergence or divergence depends only on the *tail* of a series.

The Integral Test

The integral test provides a means for determining whether an ultimately positive series converges or diverges by comparing it with an improper integral that behaves similarly. Example 4 in Section 9.2 is an example of the use of this technique. We formalize the method in the following theorem.

THEOREM 8

The integral test

Suppose that $a_n = f(n)$, where f is positive, continuous, and nonincreasing on an interval $[N, \infty[$ for some positive integer N. Then

$$\sum_{n=1}^{\infty} a_n \qquad \text{and} \qquad \int_N^{\infty} f(t)\, dt$$

either both converge or both diverge to infinity.

PROOF Let $s_n = a_1 + a_2 + \cdots + a_n$. If $n > N$, we have

$$s_n = s_N + a_{N+1} + a_{N+2} + \cdots + a_n$$
$$= s_N + f(N+1) + f(N+2) + \cdots + f(n)$$
$$= s_N + \text{sum of areas of rectangles shaded in Figure 9.4(a)}$$
$$\leq s_N + \int_N^{\infty} f(t)\, dt.$$

If the improper integral $\int_N^{\infty} f(t)\, dt$ converges, then the sequence $\{s_n\}$ is bounded above and $\sum_{n=1}^{\infty} a_n$ converges.

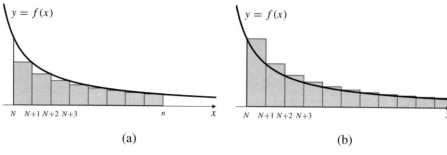

Figure 9.4

(a)   (b)

Conversely, suppose that $\sum_{n=1}^{\infty} a_n$ converges to the sum s. Then

$$\int_N^{\infty} f(t)\, dt = \text{area under } y = f(t) \text{ above } y = 0 \text{ from } t = N \text{ to } t = \infty$$

$$\leq \text{ sum of areas of shaded rectangles in Figure 9.4(b)}$$

$$= a_N + a_{N+1} + a_{N+2} + \cdots$$

$$= s - s_{N-1} < \infty,$$

so the improper integral represents a finite area and is thus convergent. (We omit the remaining details showing that $\lim_{R\to\infty} \int_N^R f(t)\, dt$ exists; like the series case, the argument depends on the completeness of the real numbers.)

Remark If $a_n = f(n)$, where f is positive, continuous, and nonincreasing on $[1, \infty[$, then Theorem 8 assures us that $\sum_{n=1}^{\infty} a_n$ and $\int_1^{\infty} f(x)\, dx$ both converge or both diverge to infinity. It does *not* tell us that the sum of the series is equal to the value of the integral. The two are not likely to be equal in the case of convergence. However, as we see below, integrals can help us approximate the sum of a series.

The principal use of the integral test is to establish the result of the following example concerning the series $\sum_{n=1}^{\infty} n^{-p}$, which is called a **$p$-series**. This result should be memorized; we will frequently compare the behaviour of other series with p-series later in this and subsequent sections.

Example 1 **(p-series)** Show that

$$\sum_{n=1}^{\infty} n^{-p} = \sum_{n=1}^{\infty} \frac{1}{n^p} \begin{cases} \text{converges if } p > 1 \\ \text{diverges to infinity if } p \leq 1. \end{cases}$$

Solution Observe that if $p > 0$, then $f(x) = x^{-p}$ is positive, continuous, and decreasing on $[1, \infty[$. By the integral test, the p-series converges for $p > 1$ and diverges for $0 < p \leq 1$ by comparison with $\int_1^{\infty} x^{-p}\, dx$. (See Theorem 2(a) of Section 6.5.) If $p \leq 0$, then $\lim_{n\to\infty}(1/n^p) \neq 0$, so the series cannot converge in this case. Being a positive series, it must diverge to infinity. ∎

Remark The harmonic series $\sum_{n=1}^{\infty} n^{-1}$ (the case $p = 1$ of the p-series) is on the borderline between convergence and divergence, although it diverges. While its terms decrease toward 0 as n increases, they do not decrease *fast enough* to allow the sum of the series to be finite. If $p > 1$, the terms of $\sum_{n=1}^{\infty} n^{-p}$ decrease toward zero fast enough that their sum is finite. We can refine the distinction between convergence and divergence at $p = 1$ by using terms that decrease faster than $1/n$, but not as fast as $1/n^q$ for any $q > 1$. If $p > 0$, terms $1/(n(\ln n)^p)$ have this property since $\ln n$ grows more slowly than any positive power of n as n increases. The question now arises whether $\sum_{n=2}^{\infty} 1/(n(\ln n)^p)$ converges or not. It does, provided again that $p > 1$; you can use the substitution $u = \ln x$ to check that

$$\int_2^{\infty} \frac{dx}{x(\ln x)^p} = \int_{\ln 2}^{\infty} \frac{du}{u^p},$$

which converges if $p > 1$ and diverges if $0 < p \leq 1$. This process of fine-tuning Example 1 can be extended even further. (See Exercise 36 below.)

Using Integral Bounds to Estimate the Sum of a Series

Suppose that $a_k = f(k)$ for $k = n+1, n+2, n+3, \ldots$, where f is a positive, continuous function, decreasing at least on the interval $[n, \infty[$. We have:

$$s - s_n = \sum_{k=n+1}^{\infty} f(k)$$

$$= \text{sum of areas of rectangles shaded in Figure 9.5(a)}$$

$$\leq \int_n^{\infty} f(x)\, dx.$$

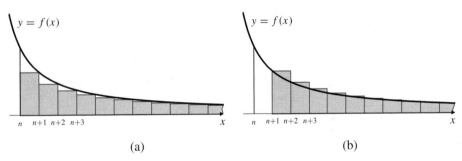

Figure 9.5

(a)   (b)

Similarly,

$$s - s_n = \text{sum of areas of rectangles in Figure 9.5(b)}$$

$$\geq \int_{n+1}^{\infty} f(x)\, dx.$$

If we define

$$A_n = \int_n^{\infty} f(x)\, dx,$$

then we can combine the above inequalities to obtain

$$A_{n+1} \leq s - s_n \leq A_n,$$

or, equivalently:

$$s_n + A_{n+1} \leq s \leq s_n + A_n.$$

The error in the approximation $s \approx s_n$ satisfies $0 \leq s - s_n \leq A_n$. However, since s must lie in the interval $[s_n + A_{n+1}, s_n + A_n]$, we can do better by using the midpoint s_n^* of this interval as an approximation for s. The error is then less than half the length $A_n - A_{n+1}$ of the interval:

A better integral approximation

The error $|s - s_n^*|$ in the approximation

$$s \approx s_n^* = s_n + \frac{A_{n+1} + A_n}{2}, \qquad \text{where} \quad A_n = \int_n^\infty f(x)\, dx,$$

satisfies $|s - s_n^*| \leq \dfrac{A_n - A_{n+1}}{2}.$

(Whenever a quantity is known to lie in a certain interval, the midpoint of that interval can be used to approximate the quantity, and the absolute value of the error in that approximation does not exceed half the length of the interval.)

Example 2   Find the best approximation s_n^* to the sum s of the series $\sum_{n=1}^\infty 1/n^2$, making use of the partial sum s_n of the first n terms. How large would n have to be to ensure that the approximation $s \approx s_n^*$ has error less than 0.001 in absolute value? How large would n have to be to ensure that the approximation $s \approx s_n$ has error less than 0.001 in absolute value?

Solution   Since $f(x) = 1/x^2$ is positive, continuous, and decreasing on $[1, \infty[$ for any $n = 1, 2, 3, \ldots$, we have

$$s_n + A_{n+1} \leq s \leq s_n + A_n,$$

where

$$A_n = \int_n^\infty \frac{dx}{x^2} = \lim_{R \to \infty} \left(-\frac{1}{x} \right) \bigg|_n^R = \frac{1}{n}.$$

The best approximation to s using s_n is

$$s_n^* = s_n + \frac{1}{2}\left(\frac{1}{n+1} + \frac{1}{n} \right) = s_n + \frac{2n+1}{2n(n+1)}$$

$$= 1 + \frac{1}{4} + \frac{1}{9} + \cdots + \frac{1}{n^2} + \frac{2n+1}{2n(n+1)}.$$

The error in this approximation satisfies

$$|s - s_n^*| \leq \frac{1}{2}\left(\frac{1}{n} - \frac{1}{n+1} \right) = \frac{1}{2n(n+1)} \leq 0.001,$$

provided $2n(n+1) \geq 1/0.001 = 1{,}000$. It is easily checked that this condition is satisfied if $n \geq 22$; the approximation

$$s \approx s_{22}^* = 1 + \frac{1}{4} + \frac{1}{9} + \cdots + \frac{1}{22^2} + \frac{45}{44 \times 23}$$

will have error with absolute value not exceeding 0.001. Had we used the approximation $s \approx s_n$ we could only have concluded that

$$0 \leq s - s_n \leq A_n = \frac{1}{n} < 0.001,$$

provided $n > 1,000$; we would need 1,000 terms of the series to get the desired accuracy.

■

Comparison Tests

The next test we consider for positive series is analogous to the comparison theorem for improper integrals. (See Theorem 3 of Section 6.5.) It enables us to determine the convergence or divergence of one series by comparing it with another series that is known to converge or diverge.

THEOREM 9

A comparison test

Let $\{a_n\}$ and $\{b_n\}$ be sequences for which there exists a positive constant K such that, ultimately, $0 \leq a_n \leq K b_n$.

(a) If the series $\sum_{n=1}^{\infty} b_n$ converges, then so does the series $\sum_{n=1}^{\infty} a_n$.

(b) If the series $\sum_{n=1}^{\infty} a_n$ diverges to infinity, then so does the series $\sum_{n=1}^{\infty} b_n$.

! BEWARE

Theorem 9 does *not* say that if $\sum a_n$ converges, then $\sum b_n$ converges. It is possible that the *smaller* sum may be finite while the *larger* one is infinite. (Do not confuse a theorem with its converse.)

PROOF Since a series converges if and only if its tail converges (Theorem 5), we can assume, without loss of generality, that the condition $0 \leq a_n \leq K b_n$ holds for all $n \geq 1$. Let $s_n = a_1 + a_2 + \cdots + a_n$ and $S_n = b_1 + b_2 + \cdots + b_n$. Then $s_n \leq K S_n$. If $\sum b_n$ converges, then $\{S_n\}$ is convergent and hence is bounded by Theorem 1. Hence $\{s_n\}$ is bounded above. By Theorem 6, $\sum a_n$ converges. Since the convergence of $\sum b_n$ guarantees that of $\sum a_n$, if the latter series diverges to infinity, then the former cannot converge either, so it must diverge to infinity too.

●

Example 3 Do the following series converge or not? Give reasons for your answer.

(a) $\displaystyle\sum_{n=1}^{\infty} \frac{1}{2^n + 1}$, (b) $\displaystyle\sum_{n=1}^{\infty} \frac{3n + 1}{n^3 + 1}$, (c) $\displaystyle\sum_{n=2}^{\infty} \frac{1}{\ln n}$.

Solution In each case we must find a suitable comparison series that we already know converges or diverges.

(a) Since $0 < \dfrac{1}{2^n + 1} < \dfrac{1}{2^n}$ for $n = 1, 2, 3, \ldots$, and since $\sum_{n=1}^{\infty} \dfrac{1}{2^n}$ is a convergent geometric series, the series $\sum_{n=1}^{\infty} \dfrac{1}{2^n + 1}$ also converges by comparison.

(b) Observe that $\dfrac{3n + 1}{n^3 + 1}$ behaves like $\dfrac{3}{n^2}$ for large n, so we would expect to compare the series with the convergent p-series $\sum_{n=1}^{\infty} n^{-2}$. We have, for $n \geq 1$,

$$\frac{3n + 1}{n^3 + 1} = \frac{3n}{n^3 + 1} + \frac{1}{n^3 + 1} < \frac{3n}{n^3} + \frac{1}{n^3} < \frac{3}{n^2} + \frac{1}{n^2} = \frac{4}{n^2}.$$

Thus, the given series converges by Theorem 9.

(c) For $n = 2, 3, 4, \ldots$, we have $0 < \ln n < n$. Thus $\dfrac{1}{\ln n} > \dfrac{1}{n}$. Since $\sum_{n=2}^{\infty} \dfrac{1}{n}$

diverges to infinity (it is a harmonic series), so does $\sum_{n=2}^{\infty} \dfrac{1}{\ln n}$ by comparison.

The following theorem provides a version of the comparison test that is not quite as general as Theorem 9 but is often easier to apply in specific cases.

THEOREM 10

A limit comparison test

Suppose that $\{a_n\}$ and $\{b_n\}$ are positive sequences and that

$$\lim_{n \to \infty} \frac{a_n}{b_n} = L,$$

where L is either a nonnegative finite number or $+\infty$.

(a) If $L < \infty$ and $\sum_{n=1}^{\infty} b_n$ converges, then $\sum_{n=1}^{\infty} a_n$ also converges.

(b) If $L > 0$ and $\sum_{n=1}^{\infty} b_n$ diverges to infinity, then so does $\sum_{n=1}^{\infty} a_n$.

PROOF If $L < \infty$, then for n sufficiently large, we have $b_n > 0$ and

$$0 \le \frac{a_n}{b_n} \le L + 1,$$

so $0 \le a_n \le (L+1)b_n$. Hence $\sum_{n=1}^{\infty} a_n$ converges if $\sum_{n=1}^{\infty} b_n$ converges, by Theorem 9(a).

If $L > 0$, then for n sufficiently large

$$\frac{a_n}{b_n} \ge \frac{L}{2}.$$

Therefore, $0 < b_n \le (2/L)a_n$, and $\sum_{n=1}^{\infty} a_n$ diverges to infinity if $\sum_{n=1}^{\infty} b_n$ does, by Theorem 9(b).

Example 4 Do the following series converge or not? Give reasons for your answers.

(a) $\displaystyle\sum_{n=1}^{\infty} \frac{1}{1 + \sqrt{n}}$, (b) $\displaystyle\sum_{n=1}^{\infty} \frac{n+5}{n^3 - 2n + 3}$.

Solution Again we must make appropriate choices for comparison series.

(a) The terms of this series decrease like $1/\sqrt{n}$. Observe that

$$L = \lim_{n \to \infty} \frac{\dfrac{1}{1 + \sqrt{n}}}{\dfrac{1}{\sqrt{n}}} = \lim_{n \to \infty} \frac{\sqrt{n}}{1 + \sqrt{n}} = \lim_{n \to \infty} \frac{1}{(1/\sqrt{n}) + 1} = 1.$$

Since the p-series $\sum_{n=1}^{\infty} \dfrac{1}{\sqrt{n}}$ diverges to infinity ($p = 1/2$), so does the series $\sum_{n=1}^{\infty} \dfrac{1}{1 + \sqrt{n}}$, by the limit comparison test.

(b) For large n, the terms behave like n/n^3, so let us compare the series with the p-series $\sum_{n=1}^{\infty} 1/n^2$, which we know converges.

$$L = \lim_{n \to \infty} \frac{\dfrac{n+5}{n^3 - 2n + 3}}{\dfrac{1}{n^2}} = \lim_{n \to \infty} \frac{n^3 + 5n^2}{n^3 - 2n + 3} = 1.$$

Since $L < \infty$, the series $\displaystyle\sum_{n=1}^{\infty} \frac{n+5}{n^3 - 2n + 3}$ also converges by the limit comparison test.

∎

In order to apply the original version of the comparison test (Theorem 9) successfully, it is important to have an intuitive feeling for whether the given series converges or diverges. The form of the comparison will depend on whether you are trying to prove convergence or divergence. For instance, if you did not know intuitively that

$$\sum_{n=1}^{\infty} \frac{1}{100n + 20{,}000}$$

would have to diverge to infinity, you might try to argue that

$$\frac{1}{100n + 20{,}000} < \frac{1}{n} \qquad \text{for } n = 1,\ 2,\ 3,\ \dots.$$

While true, this doesn't help at all. $\sum_{n=1}^{\infty} 1/n$ diverges to infinity; therefore Theorem 9 yields no information from this comparison. We could, of course, argue instead that

$$\frac{1}{100n + 20{,}000} \geq \frac{1}{20{,}100n} \qquad \text{if } n \geq 1,$$

and conclude by Theorem 9 that $\sum_{n=1}^{\infty} (1/(100n + 20{,}000))$ diverges to infinity by comparison with the divergent series $\sum_{n=1}^{\infty} 1/n$. An easier way is to use Theorem 10 and the fact that

$$L = \lim_{n \to \infty} \frac{\dfrac{1}{100n + 20{,}000}}{\dfrac{1}{n}} = \lim_{n \to \infty} \frac{n}{100n + 20{,}000} = \frac{1}{100} > 0.$$

However, the limit comparison test Theorem 10 has a disadvantage when compared to the ordinary comparison test Theorem 9. It can fail in certain cases because the limit L does not exist. In such cases it is possible that the ordinary comparison test may still work.

Example 5 Test the series $\displaystyle\sum_{n=1}^{\infty} \frac{1 + \sin n}{n^2}$ for convergence.

Solution   Since

$$\lim_{n\to\infty} \frac{\dfrac{1+\sin n}{n^2}}{\dfrac{1}{n^2}} = \lim_{n\to\infty} (1 + \sin n)$$

does not exist, the limit comparison test gives us no information. However, since $\sin n \le 1$, we have

$$0 \le \frac{1+\sin n}{n^2} \le \frac{2}{n^2} \qquad \text{for } n = 1, 2, 3, \dots.$$

The given series does, in fact, converge by comparison with $\sum_{n=1}^{\infty} 1/n^2$, using the ordinary comparison test.

■

The Ratio and Root Tests

THEOREM 11

The ratio test

Suppose that $a_n > 0$ (ultimately) and that $\rho = \lim_{n\to\infty} \dfrac{a_{n+1}}{a_n}$ exists or is $+\infty$.

(a) If $0 \le \rho < 1$, then $\sum_{n=1}^{\infty} a_n$ converges.

(b) If $1 < \rho \le \infty$, then $\lim_{n\to\infty} a_n = \infty$ and $\sum_{n=1}^{\infty} a_n$ diverges to infinity.

(c) If $\rho = 1$, this test gives no information; the series may either converge or diverge to infinity.

PROOF   Here ρ is the lowercase Greek letter *rho*, (pronounced "roe").

(a) Suppose $\rho < 1$. Pick a number r such that $\rho < r < 1$. Since we are given that $\lim_{n\to\infty} a_{n+1}/a_n = \rho$, we have $a_{n+1}/a_n \le r$ for n sufficiently large; that is, $a_{n+1} \le ra_n$ for $n \ge N$, say. In particular,

$$a_{N+1} \le ra_N$$
$$a_{N+2} \le ra_{N+1} \le r^2 a_N$$
$$a_{N+3} \le ra_{N+2} \le r^3 a_N$$
$$\vdots$$
$$a_{N+k} \le r^k a_N \qquad (k = 0, 1, 2, 3, \dots)$$

Hence, $\sum_{n=N}^{\infty} a_n$ converges by comparison with the convergent geometric series $\sum_{k=0}^{\infty} r^k$. It follows that $\sum_{n=1}^{\infty} a_n = \sum_{n=1}^{N-1} a_n + \sum_{n=N}^{\infty} a_n$ must also converge.

(b) Now suppose that $\rho > 1$. Pick a number r such that $1 < r < \rho$. Since $\lim_{n\to\infty} a_{n+1}/a_n = \rho$, we have $a_{n+1}/a_n \ge r$ for n sufficiently large, say for $n \ge N$. We assume N is chosen large enough that $a_N > 0$. It follows by an argument similar to that used in part (a) that $a_{N+k} \ge r^k a_N$ for $k = 0, 1, 2, \dots$, and since $r > 1$, $\lim_{n\to\infty} a_n = \infty$. Therefore $\sum_{n=1}^{\infty} a_n$ diverges to infinity.

(c) If ρ is computed for the series $\sum_{n=1}^{\infty} 1/n$ and $\sum_{n=1}^{\infty} 1/n^2$, we get $\rho = 1$ in each case. Since the first series diverges to infinity and the second converges, the ratio test cannot distinguish between convergence and divergence if $\rho = 1$.

All p-series fall into the indecisive category where $\rho = 1$, as does $\sum_{n=1}^{\infty} a_n$, where a_n is any rational function of n. The ratio test is most useful for series whose terms decrease at least exponentially fast. The presence of factorials in a term also suggests that the ratio test might be useful.

Example 6 Test the following series for convergence:

(a) $\displaystyle\sum_{n=1}^{\infty} \frac{99^n}{n!}$, (b) $\displaystyle\sum_{n=1}^{\infty} \frac{n^5}{2^n}$, (c) $\displaystyle\sum_{n=1}^{\infty} \frac{n!}{n^n}$, (d) $\displaystyle\sum_{n=1}^{\infty} \frac{(2n)!}{(n!)^2}$.

Solution We use the ratio test for each of these series.

(a) $\displaystyle\rho = \lim_{n\to\infty} \frac{99^{n+1}}{(n+1)!} \bigg/ \frac{99^n}{n!} = \lim_{n\to\infty} \frac{99}{n+1} = 0 < 1.$

Thus $\sum_{n=1}^{\infty} (99^n/n!)$ converges.

(b) $\displaystyle\rho = \lim_{n\to\infty} \frac{(n+1)^5}{2^{n+1}} \bigg/ \frac{n^5}{2^n} = \lim_{n\to\infty} \frac{1}{2} \left(\frac{n+1}{n}\right)^5 = \frac{1}{2} < 1.$

Hence $\sum_{n=1}^{\infty} (n^5/2^n)$ converges.

(c) $\displaystyle\rho = \lim_{n\to\infty} \frac{(n+1)!}{(n+1)^{n+1}} \bigg/ \frac{n!}{n^n} = \lim_{n\to\infty} \frac{(n+1)!\,n^n}{(n+1)^{n+1}n!} = \lim_{n\to\infty} \left(\frac{n}{n+1}\right)^n$

$\displaystyle = \lim_{n\to\infty} \frac{1}{\left(1+\dfrac{1}{n}\right)^n} = \frac{1}{e} < 1.$

Thus $\sum_{n=1}^{\infty} (n!/n^n)$ converges.

(d) $\displaystyle\rho = \lim_{n\to\infty} \frac{(2(n+1))!}{((n+1)!)^2} \bigg/ \frac{(2n)!}{(n!)^2} = \lim_{n\to\infty} \frac{(2n+2)(2n+1)}{(n+1)^2} = 4 > 1.$

Thus $\sum_{n=1}^{\infty} (2n)!/(n!)^2$ diverges to infinity.

The following theorem is very similar to the ratio test but is less frequently used. Its proof is left as an exercise. (See Exercise 37.) For examples of series to which it can be applied, see Exercises 38 and 39.

THEOREM 12

The root test

Suppose that $a_n > 0$ (ultimately) and that $\sigma = \lim_{n\to\infty} (a_n)^{1/n}$ exists or is $+\infty$.

(a) If $0 \le \sigma < 1$, then $\sum_{n=1}^{\infty} a_n$ converges.

(b) If $1 < \sigma \le \infty$, then $\lim_{n\to\infty} a_n = \infty$ and $\sum_{n=1}^{\infty} a_n$ diverges to infinity.

(c) If $\sigma = 1$, this test gives no information; the series may either converge or diverge to infinity.

Using Geometric Bounds to Estimate the Sum of a Series

Suppose that an inequality of the form

$$0 \le a_k \le Kr^k$$

holds for $k = n+1, n+2, n+3, \ldots$, where K and r are constants and $r < 1$. We can then use a geometric series to bound the tail of $\sum_{n=1}^{\infty} a_n$.

$$
\begin{aligned}
0 \le s - s_n = \sum_{k=n+1}^{\infty} a_k &\le \sum_{k=n+1}^{\infty} Kr^k \\
&= Kr^{n+1}(1 + r + r^2 + \cdots) \\
&= \frac{Kr^{n+1}}{1-r}.
\end{aligned}
$$

Since $r < 1$, the series converges and the error approaches 0 at an exponential rate as n increases.

Example 7 In Section 9.6 we will show that

$$
e = \frac{1}{0!} + \frac{1}{1!} + \frac{1}{2!} + \frac{1}{3!} + \cdots = \sum_{n=0}^{\infty} \frac{1}{n!}.
$$

(Recall that $0! = 1$.) Estimate the error if the sum s_n of the first n terms of the series is used to approximate e. Find e to 3-decimal-place accuracy using the series.

Solution We have

$$
\begin{aligned}
s_n &= \frac{1}{0!} + \frac{1}{1!} + \frac{1}{2!} + \frac{1}{3!} + \cdots + \frac{1}{(n-1)!} \\
&= 1 + 1 + \frac{1}{2} + \frac{1}{6} + \frac{1}{24} + \cdots + \frac{1}{(n-1)!}.
\end{aligned}
$$

(Since the series starts with the term for $n = 0$, the nth term is $1/(n-1)!$.) We can estimate the error in the approximation $s \approx s_n$ as follows:

$$
\begin{aligned}
0 < s - s_n &= \frac{1}{n!} + \frac{1}{(n+1)!} + \frac{1}{(n+2)!} + \frac{1}{(n+3)!} + \cdots \\
&= \frac{1}{n!}\left(1 + \frac{1}{n+1} + \frac{1}{(n+1)(n+2)} + \frac{1}{(n+1)(n+2)(n+3)} + \cdots\right) \\
&< \frac{1}{n!}\left(1 + \frac{1}{n+1} + \frac{1}{(n+1)^2} + \frac{1}{(n+1)^3} + \cdots\right)
\end{aligned}
$$

since $n+2 > n+1$, $n+3 > n+1$, and so on. The latter series is geometric, so

$$
0 < s - s_n < \frac{1}{n!}\,\frac{1}{1 - \dfrac{1}{n+1}} = \frac{n+1}{n!\,n}.
$$

If we want to evaluate e accurately to 3 decimal places, then we must ensure that the error is less than 5 in the fourth decimal place, that is, that the error is less than 0.0005. Hence we want

$$
\frac{n+1}{n}\frac{1}{n!} < 0.0005 = \frac{1}{2,000}.
$$

Since $7! = 5{,}040$, but $6! = 720$, we can use $n = 7$ but no smaller. We have

$$e \approx 1 + 1 + \frac{1}{2!} + \frac{1}{3!} + \frac{1}{4!} + \frac{1}{5!} + \frac{1}{6!}$$

$$= 2 + \frac{1}{2} + \frac{1}{6} + \frac{1}{24} + \frac{1}{120} + \frac{1}{720} \approx 2.718 \quad \text{to 3 decimal places.}$$

It is appropriate to use geometric series to bound the tails of positive series whose convergence would be demonstrated by the ratio test. Such series converge ultimately faster than any p-series $\sum_{n=1}^{\infty} n^{-p}$, for which the limit ratio is $\rho = 1$.

Exercises 9.3

In Exercises 1–26, determine whether the given series converges or diverges by using any appropriate test. The p-series can be used for comparison, as can geometric series. Be alert for series whose terms do not approach 0.

1. $\displaystyle\sum_{n=1}^{\infty} \frac{1}{n^2 + 1}$

2. $\displaystyle\sum_{n=1}^{\infty} \frac{n}{n^4 - 2}$

3. $\displaystyle\sum_{n=1}^{\infty} \frac{n^2 + 1}{n^3 + 1}$

4. $\displaystyle\sum_{n=1}^{\infty} \frac{\sqrt{n}}{n^2 + n + 1}$

5. $\displaystyle\sum_{n=1}^{\infty} \left| \sin \frac{1}{n^2} \right|$

6. $\displaystyle\sum_{n=8}^{\infty} \frac{1}{\pi^n + 5}$

7. $\displaystyle\sum_{n=2}^{\infty} \frac{1}{(\ln n)^3}$

8. $\displaystyle\sum_{n=1}^{\infty} \frac{1}{\ln(3n)}$

9. $\displaystyle\sum_{n=1}^{\infty} \frac{1}{\pi^n - n^\pi}$

10. $\displaystyle\sum_{n=0}^{\infty} \frac{1 + n}{2 + n}$

11. $\displaystyle\sum_{n=1}^{\infty} \frac{1 + n^{4/3}}{2 + n^{5/3}}$

12. $\displaystyle\sum_{n=1}^{\infty} \frac{n^2}{1 + n\sqrt{n}}$

13. $\displaystyle\sum_{n=3}^{\infty} \frac{1}{n \ln n \sqrt{\ln \ln n}}$

14. $\displaystyle\sum_{n=2}^{\infty} \frac{1}{n \ln n (\ln \ln n)^2}$

15. $\displaystyle\sum_{n=1}^{\infty} \frac{1 - (-1)^n}{n^4}$

16. $\displaystyle\sum_{n=1}^{\infty} \frac{1 + (-1)^n}{\sqrt{n}}$

17. $\displaystyle\sum_{n=1}^{\infty} \frac{1}{2^n (n + 1)}$

18. $\displaystyle\sum_{n=1}^{\infty} \frac{n^4}{n!}$

19. $\displaystyle\sum_{n=1}^{\infty} \frac{n!}{n^2 e^n}$

20. $\displaystyle\sum_{n=1}^{\infty} \frac{(2n)! 6^n}{(3n)!}$

21. $\displaystyle\sum_{n=2}^{\infty} \frac{\sqrt{n}}{3^n \ln n}$

22. $\displaystyle\sum_{n=0}^{\infty} \frac{n^{100} 2^n}{\sqrt{n!}}$

23. $\displaystyle\sum_{n=1}^{\infty} \frac{(2n)!}{(n!)^3}$

24. $\displaystyle\sum_{n=1}^{\infty} \frac{1 + n!}{(1 + n)!}$

25. $\displaystyle\sum_{n=4}^{\infty} \frac{2^n}{3^n - n^3}$

26. $\displaystyle\sum_{n=1}^{\infty} \frac{n^n}{\pi^n n!}$

In Exercises 27–30, use s_n and integral bounds to find the smallest interval that you can be sure contains the sum s of the series. If the midpoint s_n^* of this interval is used to approximate s, how large should n be chosen to ensure that the error is less than 0.001?

27. $\displaystyle\sum_{k=1}^{\infty} \frac{1}{k^4}$

28. $\displaystyle\sum_{k=1}^{\infty} \frac{1}{k^3}$

29. $\displaystyle\sum_{k=1}^{\infty} \frac{1}{k^{3/2}}$

30. $\displaystyle\sum_{k=1}^{\infty} \frac{1}{k^2 + 4}$

For each positive series in Exercises 31–34, find the best upper bound you can for the error $s - s_n$ encountered if the partial sum s_n is used to approximate the sum s of the series. How many terms of each series do you need to be sure that the approximation has error less than 0.001?

31. $\displaystyle\sum_{k=1}^{\infty} \frac{1}{2^k k!}$

32. $\displaystyle\sum_{n=1}^{\infty} \frac{1}{(2n - 1)!}$

33. $\displaystyle\sum_{n=0}^{\infty} \frac{2^n}{(2n)!}$

34. $\displaystyle\sum_{n=1}^{\infty} \frac{1}{n^n}$

35. Use the integral test to show that $\displaystyle\sum_{n=1}^{\infty} \frac{1}{1 + n^2}$ converges. Show that the sum s of the series is less than $\pi/2$.

* 36. Show that $\sum_{n=3}^{\infty} (1/(n \ln n (\ln \ln n)^p)$ converges if and only if $p > 1$. Generalize this result to series of the form

$$\sum_{n=N}^{\infty} \frac{1}{n (\ln n)(\ln \ln n) \cdots (\ln_j n)(\ln_{j+1} n)^p},$$

where $\ln_j n = \underbrace{\ln \ln \ln \ln \cdots \ln}_{j \ \ln's} n$.

∗ 37. Prove the root test. *Hint:* mimic the proof of the ratio test.

38. Use the root test to show that $\displaystyle\sum_{n=1}^{\infty} \frac{2^{n+1}}{n^n}$ converges.

∗ 39. Use the root test to test the following series for convergence:

$$\sum_{n=1}^{\infty} \left(\frac{n}{n+1}\right)^{n^2}.$$

40. Repeat Exercise 38, but use the ratio test instead of the root test.

∗ 41. Try to use the ratio test to determine whether $\displaystyle\sum_{n=1}^{\infty} \frac{2^{2n}(n!)^2}{(2n)!}$ converges. What happens? Now observe that

$$\frac{2^{2n}(n!)^2}{(2n)!} = \frac{[2n(2n-2)(2n-4)\cdots 6 \times 4 \times 2]^2}{2n(2n-1)(2n-2)\cdots 4 \times 3 \times 2 \times 1}$$

$$= \frac{2n}{2n-1} \times \frac{2n-2}{2n-3} \times \cdots \times \frac{4}{3} \times \frac{2}{1}.$$

Does the given series converge? Why or why not?

∗ 42. Determine whether the series $\displaystyle\sum_{n=1}^{\infty} \frac{(2n)!}{2^{2n}(n!)^2}$ converges. *Hint:* proceed as in Exercise 41. Show that $a_n \geq 1/(2n)$.

∗ 43. (a) Show that if $k > 0$ and n is a positive integer, then
$$n < \frac{1}{k}(1+k)^n.$$

(b) Use the estimate in (a) with $0 < k < 1$ to obtain an upper bound for the sum of the series $\sum_{n=0}^{\infty} n/2^n$. For

what value of k is this bound least?

(c) If we use the sum s_n of the first n terms to approximate the sum s of the series in (b), obtain an upper bound for the error $s - s_n$ using the inequality from (a). For given n, find k to minimize this upper bound.

∗ 44. (Improving the convergence of a series) We know that $\sum_{n=1}^{\infty} 1/(n(n+1)) = 1$. (See Example 3 of Section 9.2.) Since

$$\frac{1}{n^2} = \frac{1}{n(n+1)} + c_n, \quad \text{where} \quad c_n = \frac{1}{n^2(n+1)},$$

we have $\displaystyle\sum_{n=1}^{\infty} \frac{1}{n^2} = 1 + \sum_{n=1}^{\infty} c_n$.

The series $\sum_{n=1}^{\infty} c_n$ converges more rapidly than does $\sum_{n=1}^{\infty} 1/n^2$ because its terms decrease like $1/n^3$. Hence, fewer terms of that series will be needed to compute $\sum_{n=1}^{\infty} 1/n^2$ to any desired degree of accuracy than would be needed if we calculated with $\sum_{n=1}^{\infty} 1/n^2$ directly. Using integral upper and lower bounds, determine a value of n for which the modified partial sum s_n^* for the series $\sum_{n=1}^{\infty} c_n$ approximates the sum of that series with error less than 0.001 in absolute value. Hence, determine $\sum_{n=1}^{\infty} 1/n^2$ to within 0.001 of its true value.

(The technique exibited in this exercise is known as *improving the convergence* of a series. It can be applied to estimating the sum $\sum a_n$ if we know the sum $\sum b_n$ and if $a_n - b_n = c_n$, where $|c_n|$ decreases faster than $|a_n|$ as n tends to infinity.)

9.4 Absolute and Conditional Convergence

All of the series $\sum_{n=1}^{\infty} a_n$ considered in the previous section were ultimately positive; that is, $a_n \geq 0$ for n sufficiently large. We now drop this restriction and allow arbitrary real terms a_n. We can, however, always obtain a positive series from any given series by replacing all the terms with their absolute values.

DEFINITION 5

Absolute convergence

The series $\sum_{n=1}^{\infty} a_n$ is said to be **absolutely convergent** if $\sum_{n=1}^{\infty} |a_n|$ converges.

The series

$$s = \sum_{n=1}^{\infty} \frac{(-1)^n}{n^2} = -1 + \frac{1}{4} - \frac{1}{9} + \frac{1}{16} - \cdots$$

converges absolutely since

$$S = \sum_{n=1}^{\infty} \left| \frac{(-1)^n}{n^2} \right| = \sum_{n=1}^{\infty} \frac{1}{n^2} = 1 + \frac{1}{4} + \frac{1}{9} + \frac{1}{16} + \cdots$$

converges. It seems reasonable that the first series must converge, and its sum s should satisfy $-S \le s \le S$. In general, the cancellation that occurs because some terms are negative and others positive makes it *easier* for a series to converge than if all the terms are of one sign. We verify this insight in the following theorem.

THEOREM 13

If a series converges absolutely, then it converges.

PROOF Let $\sum_{n=1}^{\infty} a_n$ be absolutely convergent, and let $b_n = a_n + |a_n|$ for each n. Since $-|a_n| \le a_n \le |a_n|$, we have $0 \le b_n \le 2|a_n|$ for each n. Thus $\sum_{n=1}^{\infty} b_n$ converges by the comparison test. Therefore, $\sum_{n=1}^{\infty} a_n = \sum_{n=1}^{\infty} b_n - \sum_{n=1}^{\infty} |a_n|$ also converges.

Again you are cautioned not to confuse the statement of Theorem 13 with the converse statement, which is false. We will show later in this section that the **alternating harmonic series**

BEWARE

Although absolute convergence implies convergence, convergence does *not* imply absolute convergence.

$$\sum_{n=1}^{\infty} \frac{(-1)^{n-1}}{n} = 1 - \frac{1}{2} + \frac{1}{3} - \frac{1}{4} + \frac{1}{5} - \cdots$$

converges, although it does not converge absolutely. If we replace all the terms by their absolute values, we get the divergent harmonic series

$$\sum_{n=1}^{\infty} \frac{1}{n} = 1 + \frac{1}{2} + \frac{1}{3} + \frac{1}{4} + \cdots = \infty.$$

DEFINITION 6

Conditional convergence
If $\sum_{n=1}^{\infty} a_n$ is convergent, but not absolutely convergent, then we say that it is **conditionally convergent** or that it **converges conditionally**.

The alternating harmonic series is an example of a conditionally convergent series.

The comparison tests, the integral test, and the ratio test, can each be used to test for absolute convergence. They should be applied to the series $\sum_{n=1}^{\infty} |a_n|$. For the ratio test we calculate $\rho = \lim_{n \to \infty} |a_{n+1}/a_n|$. If $\rho < 1$, then $\sum_{n=1}^{\infty} a_n$ converges absolutely. If $\rho > 1$, then $\lim_{n \to \infty} |a_n| = \infty$, so both $\sum_{n=1}^{\infty} |a_n|$ and $\sum_{n=1}^{\infty} a_n$ must diverge. If $\rho = 1$, we get no information; the series $\sum_{n=1}^{\infty} a_n$ may converge absolutely, it may converge conditionally, or it may diverge.

Example 1   Test the following series for absolute convergence:

(a) $\displaystyle\sum_{n=1}^{\infty} \frac{(-1)^{n-1}}{2n-1}$,   (b) $\displaystyle\sum_{n=1}^{\infty} \frac{n\cos(n\pi)}{2^n}$.

Solution

(a) $\displaystyle\lim_{n\to\infty} \left| \frac{(-1)^{n-1}}{2n-1} \right| \Big/ \frac{1}{n} = \lim_{n\to\infty} \frac{n}{2n-1} = \frac{1}{2} > 0$. Since the harmonic series $\sum_{n=1}^{\infty}(1/n)$ diverges to infinity, the comparison test assures us that $\sum_{n=1}^{\infty}((-1)^{n-1}/(2n-1))$ does not converge absolutely.

(b) $\displaystyle\rho = \lim_{n\to\infty} \left| \frac{(n+1)\cos((n+1)\pi)}{2^{n+1}} \right| \Big/ \left| \frac{n\cos(n\pi)}{2^n} \right| = \lim_{n\to\infty} \frac{n+1}{2n} = \frac{1}{2} < 1$.

(Note that $\cos(n\pi)$ is just a fancy way of writing $(-1)^n$.) Therefore (ratio test) $\sum_{n=1}^{\infty}((n\cos(n\pi))/2^n)$ converges absolutely.   ∎

The Alternating Series Test

We cannot use any of the previously developed tests to show that the alternating harmonic series converges; all of those tests apply only to (ultimately) positive series, so they can test only for absolute convergence. Demonstrating convergence that is not absolute is generally harder to do. We present only one test that can establish such convergence; this test can only be used on a very special kind of series.

THEOREM 14

The alternating series test

Suppose that the sequence $\{a_n\}$ is positive, decreasing, and converges to 0, that is, suppose that

(i) $a_n \geq 0$ for $n = 1, 2, 3, \ldots$,

(ii) $a_{n+1} \leq a_n$ for $n = 1, 2, 3, \ldots$, and

(iii) $\lim_{n\to\infty} a_n = 0$.

Then the alternating series

$$\sum_{n=1}^{\infty} (-1)^{n-1} a_n = a_1 - a_2 + a_3 - a_4 + a_5 - \cdots$$

converges.

BEWARE

Read this proof slowly and think about why each statement is true.

PROOF   Since the sequence $\{a_n\}$ is decreasing, we have $a_{2n+1} \geq a_{2n+2}$. Therefore $s_{2n+2} = s_{2n} + a_{2n+1} - a_{2n+2} \geq s_{2n}$ for $n = 1, 2, 3, \ldots$; the even partial sums $\{s_n\}$ form an increasing sequence. Similarly $s_{2n+1} = s_{2n-1} - a_{2n} + a_{2n+1} \leq s_{2n-1}$, so the odd partial sums $\{s_{2n-1}\}$ form a decreasing sequence. Since $s_{2n} = s_{2n-1} - a_{2n} \leq s_{2n-1}$, we can say, for any $n \geq 1$, that

$$s_2 \leq s_4 \leq s_6 \leq \cdots \leq s_{2n} \leq s_{2n-1} \leq s_{2n-3} \leq \cdots \leq s_5 \leq s_3 \leq s_1.$$

Hence, s_2 is a lower bound for the decreasing sequence $\{s_{2n-1}\}$, and s_1 is an upper bound for the increasing sequence $\{s_{2n}\}$. Both of these sequences therefore converge by the completeness of the real numbers:

$$\lim_{n\to\infty} s_{2n-1} = s_{\text{odd}}, \qquad \lim_{n\to\infty} s_{2n} = s_{\text{even}}.$$

Now $a_{2n} = s_{2n-1} - s_{2n}$, so $0 = \lim_{n\to\infty} a_{2n} = \lim_{n\to\infty}(s_{2n-1} - s_{2n}) = s_{\text{odd}} - s_{\text{even}}$. Therefore $s_{\text{odd}} = s_{\text{even}} = s$, say. Every partial sum s_n is either of the form s_{2n-1} or of the form s_{2n}. Thus, $\lim_{n\to\infty} s_n = s$ exists and the series $\sum(-1)^{n-1}a_n$ converges to this sum s.

Remark Note that the series $\sum_{n=1}^{\infty}(-1)^{n-1}a_n$ begins with a positive term, a_1. The conclusion of Theorem 14 also holds for the series $\sum_{n=1}^{\infty}(-1)^n a_n$ which starts with a negative term, $-a_1$. (It is just the negative of the first series.) The theorem also remains valid if the conditions that $\{a_n\}$ is positive and decreasing are replaced by the corresponding *ultimate* versions:

$\quad$ (i) $\quad a_n \geq 0 \quad$ and $\quad$ (ii) $a_{n+1} \leq a_n \quad$ for $n = N, N+1, N+2, \ldots$.

Remark The proof of Theorem 14 shows that every even partial sum is less than or equal to s and every odd partial sum is greater than or equal to s. (The reverse is true if $(-1)^n$ is used instead of $(-1)^{n-1}$.) That is, s lies between s_{2n} and either s_{2n-1} or s_{2n+1}. It follows that, for odd or even n,

$$|s - s_n| \leq |s_{n+1} - s_n| = a_{n+1}.$$

This proves the following theorem.

THEOREM 15

Error estimate for alternating series

If the sequence $\{a_n\}$ satisfies the conditions of the alternating series test (Theorem 14), so that the series $\sum_{n=1}^{\infty}(-1)^{n-1}a_n$ (or $\sum_{n=1}^{\infty}(-1)^n a_n$) converges to the sum s, then for any $n \geq 1$, the nth partial sum s_n of the series satisfies

$$|s - s_n| \leq |a_{n+1}|.$$

That is, the size of the error involved in using s_n as an approximation to s is less than the size of the first omitted term.

Example 2 How many terms of the series $\displaystyle\sum_{n=1}^{\infty} \frac{(-1)^n}{1+2^n}$ are needed to compute the sum of the series with error less than 0.001?

Solution This series satisfies the hypotheses for Theorem 15. If we use the partial sum of the first n terms of the series to approximate the sum of the series, the error will satisfy

$$|\text{error}| \leq |\text{first omitted term}| = \frac{1}{1+2^{n+1}}.$$

This error is less than 0.001 if $1 + 2^{n+1} > 1{,}000$. Since $2^{10} = 1{,}024$, $n + 1 = 10$ will do; we need 9 terms of the series to compute the sum to within 0.001 of its actual value.

When determining the convergence of a given series, it is best to consider first whether the series converges absolutely. If it does not, then there remains the possibility of conditional convergence.

Example 3   Test the following series for absolute and conditional convergence:

(a) $\displaystyle\sum_{n=1}^{\infty} \frac{(-1)^{n-1}}{n}$,   (b) $\displaystyle\sum_{n=2}^{\infty} \frac{\cos(n\pi)}{\ln n}$,   (c) $\displaystyle\sum_{n=1}^{\infty} \frac{(-1)^{n-1}}{n^4}$.

Solution   The absolute values of the terms in series (a) and (b) are $1/n$ and $1/(\ln n)$, respectively. Since $1/(\ln n) > 1/n$, and $\sum_{n=1}^{\infty} 1/n$ diverges to infinity, neither series (a) nor (b) converges absolutely. However, both $\{1/n\}$ and $\{1/(\ln n)\}$ are positive, decreasing sequences that converge to 0. Therefore, both (a) and (b) converge by Theorem 14. Each of these series is conditionally convergent.

Series (c) is absolutely convergent because $|(-1)^{n-1}/n^4| = 1/n^4$, and $\sum_{n=1}^{\infty} 1/n^4$ is a convergent p-series ($p = 4 > 1$). We could establish its convergence using Theorem 14, but there is no need to do that since every absolutely convergent series is convergent (Theorem 13).

Example 4   For what values of x does the series $\displaystyle\sum_{n=1}^{\infty} \frac{(x-5)^n}{n\,2^n}$ converge absolutely? converge conditionally? diverge?

Solution   For such series whose terms involve functions of a variable x, it is usually wisest to begin testing for absolute convergence with the ratio test. We have

$$\rho = \lim_{n\to\infty} \left| \frac{(x-5)^{n+1}}{(n+1)2^{n+1}} \middle/ \frac{(x-5)^n}{n\,2^n} \right| = \lim_{n\to\infty} \frac{n}{n+1} \left| \frac{x-5}{2} \right| = \left| \frac{x-5}{2} \right|.$$

The series converges absolutely if $|(x-5)/2| < 1$. This inequality is equivalent to $|x-5| < 2$ (the distance from x to 5 is less than 2), that is, $3 < x < 7$. If $x < 3$ or $x > 7$, then $|(x-5)/2| > 1$. The series diverges; its terms do not approach zero.

If $x = 3$, the series is $\sum_{n=1}^{\infty}((-1)^n/n)$, which converges conditionally (it is an alternating harmonic series); if $x = 7$, the series is the harmonic series $\sum_{n=1}^{\infty} 1/n$, which diverges to infinity. Hence, the given series converges absolutely on the open interval $]3, 7[$, converges conditionally at $x = 3$, and diverges everywhere else.

Example 5   For what values of x does the series $\displaystyle\sum_{n=0}^{\infty}(n+1)^2 \left(\frac{x}{x+2} \right)^n$ converge absolutely? converge conditionally? diverge?

Solution   Again we begin with the ratio test.

$$\rho = \lim_{n\to\infty} \left| (n+2)^2 \left(\frac{x}{x+2} \right)^{n+1} \middle/ (n+1)^2 \left(\frac{x}{x+2} \right)^n \right|$$

$$= \lim_{n\to\infty} \left(\frac{n+2}{n+1} \right)^2 \left| \frac{x}{x+2} \right| = \left| \frac{x}{x+2} \right| = \frac{|x|}{|x+2|}.$$

The series converges absolutely if $|x|/|x+2| < 1$. This condition says that the distance from x to 0 is less than the distance from x to -2. Hence $x > -1$. The series diverges if $|x|/|x+2| > 1$, that is, if $x < -1$. If $x = -1$, the series is $\sum_{n=0}^{\infty}(-1)^n(n+1)^2$, which diverges. We conclude that the series converges absolutely for $x > -1$, converges conditionally nowhere, and diverges for $x \le -1$.

When using the alternating series test, it is important to verify (at least mentally) that *all three conditions* (i)–(iii) are satisfied. (As mentioned above, conditions (i) and (ii) need only be satisfied ultimately.)

Example 6   Test the following series for convergence:

(a) $\displaystyle\sum_{n=1}^{\infty}(-1)^{n-1}\frac{n+1}{n}$,

(b) $\displaystyle 1 - \frac{1}{4} + \frac{1}{3} - \frac{1}{16} + \frac{1}{5} - \cdots = \sum_{n=1}^{\infty}(-1)^{n-1}a_n$,   where

$$a_n = \begin{cases} 1/n & \text{if } n \text{ is odd}, \\ 1/n^2 & \text{if } n \text{ is even}. \end{cases}$$

Solution

(a) Here, $a_n = (n+1)/n$ is positive and decreases as n increases. However, $\lim_{n\to\infty} a_n = 1 \neq 0$. The alternating series test does not apply. In fact, the given series diverges because its terms do not approach 0.

(b) This series alternates, a_n is positive, and $\lim_{n\to\infty} a_n = 0$. However, $\{a_n\}$ is not decreasing (even ultimately). Once again, the alternating series test cannot be applied. In fact, since

$$-\frac{1}{4} - \frac{1}{16} - \cdots - \frac{1}{(2n)^2} - \cdots \qquad \text{converges, and}$$

$$1 + \frac{1}{3} + \frac{1}{5} + \cdots + \frac{1}{2n-1} + \cdots \qquad \text{diverges to infinity,}$$

it is readily seen that the given series diverges to infinity.   ∎

Rearranging the Terms in a Series

The basic difference between absolute and conditional convergence is that when a series $\sum_{n=1}^{\infty} a_n$ converges absolutely, it does so because its terms $\{a_n\}$ decrease in size fast enough that their sum can be finite even if no cancellation occurs due to terms of opposite sign. If cancellation is required to make the series converge (because the terms decrease slowly), then the series can only converge conditionally.

Consider the alternating harmonic series

$$1 - \frac{1}{2} + \frac{1}{3} - \frac{1}{4} + \frac{1}{5} - \frac{1}{6} + \cdots.$$

This series converges, but only conditionally. If we take the subseries containing only the positive terms, we get the series

$$1 + \frac{1}{3} + \frac{1}{5} + \frac{1}{7} + \cdots,$$

which diverges to infinity. Similarly, the subseries of negative terms

$$-\frac{1}{2} - \frac{1}{4} - \frac{1}{6} - \frac{1}{8} - \cdots$$

diverges to negative infinity.

If a series converges absolutely, the subseries consisting of positive terms and the subseries consisting of negative terms must each converge to a finite sum. If a series converges conditionally, the positive and negative subseries will both diverge, to ∞ and $-\infty$, respectively.

Using these facts we can answer a question raised at the beginning of Section 9.2. If we rearrange the terms of a convergent series so that they are added in a different order, must the rearranged series converge or not, and if it does will it converge to the same sum? The answer depends on whether the original series was absolutely convergent or merely conditionally convergent.

THEOREM **Convergence of rearrangements of a series**

(a) If the terms of an absolutely convergent series are rearranged so that addition occurs in a different order, the rearranged series still converges to the same sum as the original series.

(b) If a series is conditionally convergent, and L is any real number, then the terms of the series can be rearranged so as to make the series converge (conditionally) to the sum L. It can also be rearranged so as to diverge to ∞ or to $-\infty$, or just to diverge.

Part (b) shows that conditional convergence is a rather suspect kind of convergence, being dependent on the order in which the terms are added. We will not present a formal proof of the theorem but will give an example suggesting what is involved. (See also Exercise 30 below.)

Example 7   In Section 9.5 we will show that the alternating harmonic series

$$\sum_{n=1}^{\infty} \frac{(-1)^{n-1}}{n} = 1 - \frac{1}{2} + \frac{1}{3} - \frac{1}{4} + \frac{1}{5} - \frac{1}{6} + \frac{1}{7} - \cdots$$

converges (conditionally) to the sum $\ln 2$. Describe how to rearrange its terms so that it converges to 8 instead.

Solution   Start adding terms of the positive subseries

$$1 + \frac{1}{3} + \frac{1}{5} + \cdots,$$

and keep going until the partial sum exceeds 8. (It will, eventually, because the positive subseries diverges to infinity.) Then add the first term $-1/2$ of the negative subseries

$$-\frac{1}{2} - \frac{1}{4} - \frac{1}{6} - \cdots.$$

This will reduce the partial sum below 8 again. Now resume adding terms of the positive subseries until the partial sum climbs above 8 once more. Then add the second term of the negative subseries and the partial sum will drop below 8.

Keep repeating this procedure, alternately adding terms of the positive subseries to force the sum above 8 and then terms of the negative subseries to force it below 8. Since both subseries have infinitely many terms and diverge to ∞ and $-\infty$, respectively, eventually every term of the original series will be included, and the partial sums of the new series will oscillate back and forth around 8, converging to that number. Of course, any number other than 8 could also be used in place of 8. ∎

Exercises 9.4

Determine whether the series in Exercises 1–12 converge absolutely, converge conditionally, or diverge.

1. $\displaystyle\sum_{n=1}^{\infty} \frac{(-1)^{n-1}}{\sqrt{n}}$

2. $\displaystyle\sum_{n=1}^{\infty} \frac{(-1)^n}{n^2 + \ln n}$

3. $\displaystyle\sum_{n=1}^{\infty} \frac{\cos(n\pi)}{(n+1)\ln(n+1)}$

4. $\displaystyle\sum_{n=1}^{\infty} \frac{(-1)^{2n}}{2^n}$

5. $\displaystyle\sum_{n=0}^{\infty} \frac{(-1)^n (n^2 - 1)}{n^2 + 1}$

6. $\displaystyle\sum_{n=1}^{\infty} \frac{(-2)^n}{n!}$

7. $\displaystyle\sum_{n=1}^{\infty} \frac{(-1)^n}{n\pi^n}$

8. $\displaystyle\sum_{n=0}^{\infty} \frac{-n}{n^2 + 1}$

9. $\displaystyle\sum_{n=1}^{\infty} (-1)^n \frac{20n^2 - n - 1}{n^3 + n^2 + 33}$

10. $\displaystyle\sum_{n=1}^{\infty} \frac{100\cos(n\pi)}{2n + 3}$

11. $\displaystyle\sum_{n=1}^{\infty} \frac{n!}{(-100)^n}$

12. $\displaystyle\sum_{n=10}^{\infty} \frac{\sin(n + 1/2)\pi}{\ln \ln n}$

For the series in Exercises 13–16, find the smallest integer n that ensures that the partial sum s_n approximates the sum s of the series with error less than 0.001 in absolute value.

13. $\displaystyle\sum_{n=1}^{\infty} (-1)^{n-1} \frac{n}{n^2 + 1}$

14. $\displaystyle\sum_{n=0}^{\infty} \frac{(-1)^n}{(2n)!}$

15. $\displaystyle\sum_{n=1}^{\infty} (-1)^{n-1} \frac{n}{2^n}$

16. $\displaystyle\sum_{n=0}^{\infty} (-1)^n \frac{3^n}{n!}$

Determine the values of x for which the series in Exercises 17–24 converge absolutely, converge conditionally, or diverge.

17. $\displaystyle\sum_{n=0}^{\infty} \frac{x^n}{\sqrt{n+1}}$

18. $\displaystyle\sum_{n=1}^{\infty} \frac{(x-2)^n}{n^2 2^{2n}}$

19. $\displaystyle\sum_{n=0}^{\infty} (-1)^n \frac{(x-1)^n}{2n + 3}$

20. $\displaystyle\sum_{n=1}^{\infty} \frac{1}{2n - 1} \left(\frac{3x+2}{-5} \right)^n$

21. $\displaystyle\sum_{n=2}^{\infty} \frac{x^n}{2^n \ln n}$

22. $\displaystyle\sum_{n=1}^{\infty} \frac{(4x + 1)^n}{n^3}$

23. $\displaystyle\sum_{n=1}^{\infty} \frac{(2x + 3)^n}{n^{1/3} 4^n}$

24. $\displaystyle\sum_{n=1}^{\infty} \frac{1}{n} \left(1 + \frac{1}{x} \right)^n$

* **25.** Does the alternating series test apply directly to the series $\sum_{n=1}^{\infty} (1/n) \sin(n\pi/2)$? Determine whether the series converges.

* **26.** Show that the series $\sum_{n=1}^{\infty} a_n$ converges absolutely if $a_n = 10/n^2$ for even n and $a_n = -1/10n^3$ for odd n.

* **27.** Which of the following statements are TRUE and which are FALSE? Justify your assertion of truth, or give a counterexample to show falsehood.

  (a) If $\sum_{n=1}^{\infty} a_n$ converges, then $\sum_{n=1}^{\infty} (-1)^n a_n$ converges.

  (b) If $\sum_{n=1}^{\infty} a_n$ converges and $\sum_{n=1}^{\infty} (-1)^n a_n$ converges, then $\sum_{n=1}^{\infty} a_n$ converges absolutely.

  (c) If $\sum_{n=1}^{\infty} a_n$ converges absolutely, then

$$\sum_{n=1}^{\infty} (-1)^n a_n \text{ converges absolutely.}$$

* **28.** (a) Use a Riemann sum argument to show that

$$\ln n! \geq \int_1^n \ln t \, dt = n \ln n - n + 1.$$

  (b) For what values of x does the series $\sum_{n=1}^{\infty} \frac{n! x^n}{n^n}$ converge absolutely? converge conditionally? diverge? (*Hint:* First use the ratio test. To test the cases where $\rho = 1$, you may find the inequality in part (a) useful.)

* **29.** For what values of x does the series $\sum_{n=1}^{\infty} \frac{(2n)! x^n}{2^{2n} (n!)^2}$ converge absolutely? converge conditionally? diverge? *Hint:* see Exercise 42 of Section 9.3.

* **30.** Devise procedures for rearranging the terms of the alternating harmonic series so that the rearranged series (a) diverges to ∞,   (b) converges to -2.

9.5 Power Series

This section is concerned with a special kind of infinite series called a *power series*, which may be thought of as a polynomial of infinite degree.

DEFINITION 7

Power series

A series of the form

$$\sum_{n=0}^{\infty} a_n(x - c)^n = a_0 + a_1(x - c) + a_2(x - c)^2 + a_3(x - c)^3 + \cdots$$

is called a **power series in powers of** $x - c$ or a **power series about the point** $x = c$. The constants $a_0, a_1, a_2, \ldots$ are called the **coefficients** of the power series.

Since the terms of a power series are functions of a variable x, the series may or may not converge for each value of x. For those values of x for which the series does converge, the sum defines a function of x. For example, if $-1 < x < 1$, then

$$1 + x + x^2 + x^3 + \cdots = \frac{1}{1 - x}.$$

The geometric series on the left side is a power series *representation* of the function $1/(1 - x)$ in powers of x (or about the point $x = 0$). Note that the representation is valid only in the open interval $]-1, 1[$ even though $1/(1 - x)$ is defined for all real x except $x = 1$. For $x = -1$ and for $|x| > 1$ the series does not converge, so it cannot represent $1/(1 - x)$ at these points.

The point c is the **centre of convergence** of the power series $\sum_{n=0}^{\infty} a_n(x - c)^n$. The series certainly converges (to a_0) at $x = c$. (All the terms except possibly the first are 0.) Theorem 17 below shows that if the series converges anywhere else, then it converges on an interval (possibly infinite) centred at $x = c$, and it converges absolutely everywhere on that interval except possibly at one or both of the endpoints if the interval is finite. The geometric series

$$1 + x + x^2 + x^3 + \cdots$$

is an example of this behaviour. It has centre of convergence $c = 0$, and converges only on the interval $]-1, 1[$, centred at 0. The convergence is absolute at every point of the interval. Another example is the series

$$\sum_{n=1}^{\infty} \frac{1}{n \, 2^n} (x - 5)^n = \frac{x - 5}{2} + \frac{(x - 5)^2}{2 \times 2^2} + \frac{(x - 5)^3}{3 \times 2^3} + \cdots,$$

which we discussed in Example 4 of Section 9.4. We showed that this series converges on the interval $[3, 7[$, an interval with centre $x = 5$, and that the convergence is absolute on the open interval $]3, 7[$ but is only conditional at the endpoint $x = 3$.

THEOREM 17

For any power series $\sum_{n=0}^{\infty} a_n (x - c)^n$ one of the following alternatives must hold:

    (i) the series may converge only at $x = c$,

    (ii) the series may converge at every real number x, or

(iii) there may exist a positive real number R such that the series converges at every x satisfying $|x - c| < R$ and diverges at every x satisfying $|x - c| > R$. In this case the series may or may not converge at either of the two *endpoints* $x = c - R$ and $x = c + R$.

In each of these cases the convergence is absolute except possibly at the endpoints $x = c - R$ and $x = c + R$ in case (iii).

PROOF We observed above that every power series converges at its centre of convergence; only the first term can be nonzero so the convergence is absolute. To prove the rest of this theorem, it suffices to show that if the series converges at any number $x_0 \neq c$, then it converges absolutely at every number x closer to c than x_0 is, that is, at every x satisfying $|x - c| < |x_0 - c|$. This means that convergence at any $x_0 \neq c$ implies absolute convergence on $]c - x_0, c + x_0[$, so the set of points x where the series converges must be an interval centred at c.

Suppose, therefore, that $\sum_{n=0}^{\infty} a_n(x_0 - c)^n$ converges. Then $\lim a_n(x_0 - c)^n = 0$, so $|a_n(x_0 - c)^n| \leq K$ for all n, where K is some constant (Theorem 1 of Section 9.1). If $r = |x - c|/|x_0 - c| < 1$, then

$$\sum_{n=0}^{\infty} |a_n(x - c)^n| = \sum_{n=0}^{\infty} |a_n(x_0 - c)^n| \left| \frac{x - c}{x_0 - c} \right|^n \leq K \sum_{n=0}^{\infty} r^n = \frac{K}{1 - r} < \infty.$$

Thus $\sum_{n=0}^{\infty} a_n(x - c)^n$ converges absolutely.

By Theorem 17, the set of values x for which the power series $\sum_{n=0}^{\infty} a_n(x - c)^n$ converges is an interval centred at $x = c$. We call this interval the **interval of convergence** of the power series. It must be of one of the following forms:

(i) the isolated point $x = c$ (a degenerate closed interval $[c, c]$),

(ii) the entire line $]-\infty, \infty[$,

(iii) a finite interval centred at c:
$[c - R, c + R]$, or $[c - R, c + R[$, or $]c - R, c + R]$, or $]c - R, c + R[$.

The number R in (iii) is called the **radius of convergence** of the power series. In case (i) we say the radius of convergence is $R = 0$; in case (ii) it is $R = \infty$.

The radius of convergence, R, can often be found by using the ratio test on the power series: if

$$\rho = \lim_{n \to \infty} \left| \frac{a_{n+1}(x - c)^{n+1}}{a_n(x - c)^n} \right| = \left(\lim_{n \to \infty} \left| \frac{a_{n+1}}{a_n} \right| \right) |x - c|$$

exists, then the series $\sum_{n=0}^{\infty} a_n(x - c)^n$ converges absolutely where $\rho < 1$, that is, where

$$|x - c| < R = 1 \left/ \lim_{n \to \infty} \left| \frac{a_{n+1}}{a_n} \right| \right. .$$

The series diverges if $|x - c| > R$.

Radius of convergence

Suppose that $L = \lim_{n \to \infty} \left| \frac{a_{n+1}}{a_n} \right|$ exists or is ∞. Then the power series $\sum_{n=0}^{\infty} a_n(x - c)^n$ has radius of convergence $R = 1/L$. (If $L = 0$, then $R = \infty$; if $L = \infty$, then $R = 0$.)

Example 1   Determine the centre, radius, and interval of convergence of

$$\sum_{n=0}^{\infty} \frac{(2x+5)^n}{(n^2+1)3^n}.$$

$2x + 5 = 0$
$x = -\frac{5}{2}$

Solution   The series can be rewritten

$$\sum_{n=0}^{\infty} \left(\frac{2}{3}\right)^n \frac{1}{n^2+1} \left(x + \frac{5}{2}\right)^n.$$

The centre of convergence is $x = -5/2$. The radius of convergence, R, is given by

$$\frac{1}{R} = L = \lim \left| \frac{\left(\frac{2}{3}\right)^{n+1} \frac{1}{(n+1)^2+1}}{\left(\frac{2}{3}\right)^n \frac{1}{n^2+1}} \right| = \lim \frac{2}{3} \frac{n^2+1}{(n+1)^2+1} = \frac{2}{3}.$$

Thus $R = 3/2$. The series converges absolutely on $]-5/2 - 3/2, -5/2 + 3/2[=$ $]-4, -1[$, and it diverges on $]-\infty, -4[$ and on $]-1, \infty[$. At $x = -1$ the series is $\sum_{n=0}^{\infty} 1/(n^2+1)$; at $x = -4$ it is $\sum_{n=0}^{\infty}(-1)^n/(n^2+1)$. Both series converge (absolutely). The interval of convergence of the given power series is therefore $[-4, -1]$.

Example 2   Determine the radii of convergence of the series

(a) $\displaystyle\sum_{n=0}^{\infty} \frac{x^n}{n!}$ and   (b) $\displaystyle\sum_{n=0}^{\infty} n!x^n.$

Solution

(a) $L = \left| \lim \frac{1}{(n+1)!} \middle/ \frac{1}{n!} \right| = \lim \frac{n!}{(n+1)!} = \lim \frac{1}{n+1} = 0$. Thus $R = \infty$.

This series converges (absolutely) for all x. The sum is e^x, as will be shown in Example 1 in the next section.

(b) $L = \left| \lim \frac{(n+1)!}{n!} \right| = \lim(n+1) = \infty$. Thus $R = 0$

This series converges only at its centre of convergence, $x = 0$.

Algebraic Operations on Power Series

To simplify the following discussion, we will consider only power series with $x = 0$ as centre of convergence, that is, series of the form

$$\sum_{n=0}^{\infty} a_n x^n = a_0 + a_1 x + a_2 x^2 + a_3 x^3 + \cdots.$$

Any properties we demonstrate for such series extend automatically to power series of the form $\sum_{n=0}^{\infty} a_n(y - c)^n$ via the change of variable $x = y - c$.

First we observe that series having the same centre of convergence can be added or subtracted on whatever interval is common to their intervals of convergence. The following theorem is a simple consequence of Theorem 7 of Section 9.2 and does not require a proof.

THEOREM **18**

Let $\sum_{n=0}^{\infty} a_n x^n$ and $\sum_{n=0}^{\infty} b_n x^n$ be two power series with radii of convergence R_a and R_b, respectively, and let c be a constant. Then

(i) $\sum_{n=0}^{\infty}(ca_n) x^n$ has radius of convergence R_a, and

$$\sum_{n=0}^{\infty}(ca_n) x^n = c \sum_{n=0}^{\infty} a_n x^n$$

wherever the series on the right converges.

(ii) $\sum_{n=0}^{\infty}(a_n + b_n) x^n$ has radius of convergence R at least as large as the smaller of R_a and R_b ($R \geq \min\{R_a, R_b\}$), and

$$\sum_{n=0}^{\infty}(a_n + b_n) x^n = \sum_{n=0}^{\infty} a_n x^n + \sum_{n=0}^{\infty} b_n x^n$$

wherever both series on the right converge.

The situation regarding multiplication and division of power series is more complicated. We will mention only the results and will not attempt any proofs of our assertions. A textbook in mathematical analysis will provide more details.

Long multiplication of the form

$$(a_0 + a_1 x + a_2 x^2 + \cdots)(b_0 + b_1 x + b_2 x^2 + \cdots)$$
$$= a_0 b_0 + (a_0 b_1 + a_1 b_0)x + (a_0 b_2 + a_1 b_1 + a_2 b_0)x^2 + \cdots$$

leads us to conjecture the formula

$$\left(\sum_{n=0}^{\infty} a_n x^n \right) \left(\sum_{n=0}^{\infty} b_n x^n \right) = \sum_{n=0}^{\infty} c_n x^n,$$

where

$$c_n = a_0 b_n + a_1 b_{n-1} + \cdots + a_n b_0 = \sum_{j=0}^{n} a_j b_{n-j}.$$

The series $\sum_{n=0}^{\infty} c_n x^n$ is called the **Cauchy product** of the series $\sum_{n=0}^{\infty} a_n x^n$ and $\sum_{n=0}^{\infty} b_n x^n$. Like the sum, the Cauchy product also has radius of convergence at least equal to the lesser of those of the factor series.

Example 3 Since

$$\frac{1}{1 - x} = 1 + x + x^2 + x^3 + \cdots = \sum_{n=0}^{\infty} x^n$$

holds for $-1 < x < 1$, we can determine a power series representation for $1/(1-x)^2$

by taking the Cauchy product of this series with itself. Since $a_n = b_n = 1$ for $n = 0, 1, 2, \ldots$, we have

$$c_n = \sum_{j=0}^{n} 1 = n + 1$$

and

$$\frac{1}{(1-x)^2} = 1 + 2x + 3x^2 + 4x^3 + \cdots = \sum_{n=0}^{\infty} (n+1)x^n,$$

which must also hold for $-1 < x < 1$. The same series can be obtained by direct long multiplication of the series:

$$
\begin{array}{ccccccccc}
 & 1 & + & x & + & x^2 & + & x^3 & + & \cdots \\
\times & 1 & + & x & + & x^2 & + & x^3 & + & \cdots \\
\hline
 & 1 & + & x & + & x^2 & + & x^3 & + & \cdots \\
 & & & x & + & x^2 & + & x^3 & + & \cdots \\
 & & & & & x^2 & + & x^3 & + & \cdots \\
 & & & & & & & x^3 & + & \cdots \\
 & & & & & & & & & \cdots \\
\hline
 & 1 & + & 2x & + & 3x^2 & + & 4x^3 & + & \cdots
\end{array}
$$

Long division can also be performed on power series, but there is no simple rule for determining the coefficients of the quotient series. The radius of convergence of the quotient series is not less than the least of the three numbers R_1, R_2, and R_3, where R_1 and R_2 are the radii of convergence of the divisor and dividend series and R_3 is the distance from the centre of convergence to the nearest complex number where the divisor series has sum equal to 0. To illustrate this point, observe that 1 and $1 - x$ are both power series with infinite radii of convergence:

$$1 = 1 + 0x + 0x^2 + 0x^3 + \cdots \qquad \text{for all } x,$$
$$1 - x = 1 - x + 0x^2 + 0x^3 + \cdots \qquad \text{for all } x.$$

Their quotient, $1/(1-x)$, however, only has radius of convergence 1, the distance from the centre of convergence $x = 0$ to the point $x = 1$ where the denominator vanishes:

$$\frac{1}{1-x} = 1 + x + x^2 + x^3 + \cdots \qquad \text{for} \quad |x| < 1.$$

Differentiation and Integration of Power Series

If a power series has a positive radius of convergence, it can be differentiated or integrated term by term. The resulting series will converge to the appropriate derivative or integral of the sum of the original series everywhere except possibly at the endpoints of the interval of convergence of the original series. This very important fact ensures that, for purposes of calculation, power series behave just like polynomials, the easiest functions to differentiate and integrate. We formalize the differentiation and integration properties of power series in the following theorem.

THEOREM 19

Term by term differentiation and integration of power series

If the series $\sum_{n=0}^{\infty} a_n x^n$ converges to the sum $f(x)$ on an interval $]-R, R[$, where $R > 0$, that is,

$$f(x) = \sum_{n=0}^{\infty} a_n x^n = a_0 + a_1 x + a_2 x^2 + a_3 x^3 + \cdots, \qquad (-R < x < R),$$

then f is differentiable on $]-R, R[$ and

$$f'(x) = \sum_{n=1}^{\infty} n a_n x^{n-1} = a_1 + 2a_2 x + 3a_3 x^2 + \cdots, \qquad (-R < x < R).$$

Also, f is integrable over any closed subinterval of $]-R, R[$, and if $|x| < R$, then

$$\int_0^x f(t)\, dt = \sum_{n=0}^{\infty} \frac{a_n}{n+1} x^{n+1} = a_0 x + \frac{a_1}{2} x^2 + \frac{a_2}{3} x^3 + \cdots.$$

PROOF Let x satisfy $-R < x < R$ and choose $H > 0$ such that $|x| + H < R$. By Theorem 17 we then have[1]

$$\sum_{n=1}^{\infty} |a_n|(|x| + H)^n = K < \infty.$$

The Binomial Theorem (see Section 9.9) shows that if $n \geq 1$, then

$$(x + h)^n = x^n + n x^{n-1} h + \sum_{k=2}^{n} \binom{n}{k} x^{n-k} h^k.$$

Therefore, if $|h| \leq H$ we have

$$
\begin{aligned}
|(x + h)^n - x^n - n x^{n-1} h| &= \left| \sum_{k=2}^{n} \binom{n}{k} x^{n-k} h^k \right| \\
&\leq \sum_{k=2}^{n} \binom{n}{k} |x|^{n-k} \frac{|h|^k}{H^k} H^k \\
&\leq \frac{|h|^2}{H^2} \sum_{k=0}^{n} \binom{n}{k} |x|^{n-k} H^k \\
&= \frac{|h|^2}{H^2} (|x| + H)^n.
\end{aligned}
$$

Also

$$|n x^{n-1}| = \frac{n |x|^{n-1} H}{H} \leq \frac{1}{H} (|x| + H)^n.$$

Thus

$$\sum_{n=1}^{\infty} |n a_n x^{n-1}| \leq \frac{1}{H} \sum_{n=1}^{\infty} |a_n|(|x| + H)^n = \frac{K}{H} < \infty,$$

---

[1] This proof is due to R. Výborný, *American Mathematical Monthly*, April 1987.

so the series $\sum_{n=1}^{\infty} n a_n x^{n-1}$ converges (absolutely) to $g(x)$, say. Now

$$\left| \frac{f(x+h) - f(x)}{h} - g(x) \right| = \left| \sum_{n=1}^{\infty} \frac{a_n(x+h)^n - a_n x^n - n a_n x^{n-1} h}{h} \right|$$

$$\le \frac{1}{|h|} \sum_{n=1}^{\infty} |a_n| |(x+h)^n - x^n - n x^{n-1} h|$$

$$\le \frac{|h|}{H^2} \sum_{n=1}^{\infty} |a_n| (|x| + H)^n \le \frac{K|h|}{H^2}.$$

Letting h approach zero, we obtain $|f'(x) - g(x)| \le 0$, so $f'(x) = g(x)$, as required.

Now observe that since $|a_n/(n+1)| \le |a_n|$, the series

$$h(x) = \sum_{n=0}^{\infty} \frac{a_n}{n+1} x^{n+1}$$

converges (absolutely) at least on the interval $]-R, R[$. Using the differentiation result proved above, we obtain

$$h'(x) = \sum_{n=0}^{\infty} a_n x^n = f(x).$$

Since $h(0) = 0$, we have

$$\int_0^x f(t)\, dt = \int_0^x h'(t)\, dt = h(t) \Big|_0^x = h(x),$$

as required.

Together, these results imply that the termwise differentiated or integrated series have the same radius of convergence as the given series. In fact, as the following examples illustrate, the interval of convergence of the differentiated series is the same as that of the original series except for the *possible* loss of one or both endpoints if the original series converges at endpoints of its interval of convergence. Similarly, the integrated series will converge everywhere on the interval of convergence of the original series and possibly at one or both endpoints of that interval, even if the original series does not converge at the endpoints.

Being differentiable on $]-R, R[$, where R is the radius of convergence, the sum $f(x)$ of a power series is necessarily continuous on that open interval. If the series happens to converge at either or both of the endpoints $-R$ and R, then f is also continuous (on one side) up to these endpoints. This result is stated formally in the following theorem. We will not prove it here; the interested reader is referred to textbooks on mathematical analysis for a proof.

THEOREM 20

Abel's Theorem

The sum of a power series is a continuous function everywhere on the interval of convergence of the series. In particular, if $\sum_{n=0}^{\infty} a_n R^n$ converges for some $R > 0$, then

$$\lim_{x \to R-} \sum_{n=0}^{\infty} a_n x^n = \sum_{n=0}^{\infty} a_n R^n,$$

and if $\sum_{n=0}^{\infty} a_n(-R)^n$ converges, then

$$\lim_{x \to -R+} \sum_{n=0}^{\infty} a_n x^n = \sum_{n=0}^{\infty} a_n(-R)^n.$$

The following examples show how the above theorems are applied to obtain power series representations for functions.

Example 4 Find power series representations for the functions

(a) $\dfrac{1}{(1-x)^2}$, (b) $\dfrac{1}{(1-x)^3}$, and (c) $\ln(1+x)$

by starting with the geometric series

$$\frac{1}{1-x} = \sum_{n=0}^{\infty} x^n = 1 + x + x^2 + x^3 + \cdots \qquad (-1 < x < 1)$$

and using differentiation, integration, and substitution. Where is each series valid?

Solution

(a) Differentiate the geometric series term by term to obtain

$$\frac{1}{(1-x)^2} = \sum_{n=1}^{\infty} nx^{n-1} = 1 + 2x + 3x^2 + 4x^3 + \cdots \qquad (-1 < x < 1).$$

This is the same result obtained by multiplication of series in Example 3 above.

(b) Differentiate again to get, for $-1 < x < 1$,

$$\frac{2}{(1-x)^3} = \sum_{n=2}^{\infty} n(n-1) x^{n-2} = (1 \times 2) + (2 \times 3)x + (3 \times 4)x^2 + \cdots.$$

Now divide by 2:

$$\frac{1}{(1-x)^3} = \sum_{n=2}^{\infty} \frac{n(n-1)}{2} x^{n-2} = 1 + 3x + 6x^2 + 10x^3 + \cdots \qquad (-1 < x < 1).$$

(c) Substitute $-t$ in place of x in the original geometric series:

$$\frac{1}{1+t} = \sum_{n=0}^{\infty} (-1)^n t^n = 1 - t + t^2 - t^3 + t^4 - \cdots \qquad (-1 < t < 1).$$

Integrate from 0 to x, where $|x| < 1$, to get

$$\ln(1+x) = \int_0^x \frac{dt}{1+t} = \sum_{n=0}^{\infty} (-1)^n \int_0^x t^n \, dt$$

$$= \sum_{n=0}^{\infty} (-1)^n \frac{x^{n+1}}{n+1} = x - \frac{x^2}{2} + \frac{x^3}{3} - \frac{x^4}{4} + \cdots \qquad (-1 < x \le 1).$$

Note that the latter series converges (conditionally) at the endpoint $x = 1$ as well as on the interval $-1 < x < 1$. Since $\ln(1 + x)$ is continuous at $x = 1$, Theorem 20 assures us that the series must converge to that function at $x = 1$ also. In particular, therefore, the alternating harmonic series converges to $\ln 2$:

$$\ln 2 = 1 - \frac{1}{2} + \frac{1}{3} - \frac{1}{4} + \frac{1}{5} - \cdots = \sum_{n=0}^{\infty} \frac{(-1)^n}{n+1}.$$

This would not, however, be a very useful formula for calculating the value of $\ln 2$. (Why not?)

■

Example 5   Use the geometric series of the previous example to find a power series representation for $\tan^{-1} x$.

Solution   Substitute $-t^2$ for x in the geometric series. Since $0 \le t^2 < 1$ whenever $-1 < t < 1$, we obtain

$$\frac{1}{1+t^2} = 1 - t^2 + t^4 - t^6 + t^8 - \cdots \qquad (-1 < t < 1).$$

Now integrate from 0 to x, where $|x| < 1$:

$$\tan^{-1} x = \int_0^x \frac{dt}{1+t^2} = \int_0^x (1 - t^2 + t^4 - t^6 + t^8 - \cdots)\, dt$$

$$= x - \frac{x^3}{3} + \frac{x^5}{5} - \frac{x^7}{7} + \frac{x^9}{9} - \cdots$$

$$= \sum_{n=0}^{\infty} (-1)^n \frac{x^{2n+1}}{2n+1} \qquad (-1 < x < 1).$$

However, note that the series also converges (conditionally) at $x = -1$ and 1. Since $\tan^{-1}$ is continuous at ± 1, the above series representation for $\tan^{-1} x$ also holds for these values, by Theorem 20. Letting $x = 1$ we get another interesting result:

$$\frac{\pi}{4} = 1 - \frac{1}{3} + \frac{1}{5} - \frac{1}{7} + \frac{1}{9} - \cdots.$$

Again, however, this would not be a good formula with which to calculate a numerical value of π. (Why not?)

■

Example 6   Find the sum of the series $\sum_{n=1}^{\infty} \frac{n^2}{2^n}$ by first finding the sum of the power series

$$\sum_{n=1}^{\infty} n^2 x^n = x + 4x^2 + 9x^3 + 16x^4 + \cdots.$$

Solution   Observe (in Example 4(a)) how the process of differentiating the geometric series produces a series with coefficients 1, 2, 3, ... Start with the series obtained for $1/(1-x)^2$ and multiply it by x to obtain

$$\sum_{n=1}^{\infty} nx^n = x + 2x^2 + 3x^3 + 4x^4 + \cdots = \frac{x}{(1-x)^2}.$$

Now differentiate again to get a series with coefficients 1^2, 2^2, 3^2, ...:

$$\sum_{n=1}^{\infty} n^2 x^{n-1} = 1 + 4x + 9x^2 + 16x^3 + \cdots = \frac{d}{dx}\frac{x}{(x-1)^2} = \frac{1+x}{(1-x)^3}.$$

Multiplication by x again gives the desired power series:

$$\sum_{n=1}^{\infty} n^2 x^n = x + 4x^2 + 9x^3 + 16x^4 + \cdots = \frac{x(1+x)}{(1-x)^3}.$$

Differentiation and multiplication by x do not change the radius of convergence, so this series converges to the indicated function for $-1 < x < 1$. Putting $x = 1/2$, we get

$$\sum_{n=1}^{\infty} \frac{n^2}{2^n} = \frac{\frac{1}{2} \times \frac{3}{2}}{\frac{1}{8}} = 6.$$

The following example illustrates how substitution can be used to obtain power series representations of functions with centres of convergence different from 0.

Example 7   Find a series representation of $f(x) = 1/(2+x)$ in powers of $x - 1$. What is the interval of convergence of this series?

Solution   Let $t = x - 1$ so that $x = t + 1$. We have

$$\frac{1}{2+x} = \frac{1}{3+t} = \frac{1}{3}\frac{1}{1+\dfrac{t}{3}}$$

$$= \frac{1}{3}\left(1 - \frac{t}{3} + \frac{t^2}{3^2} - \frac{t^3}{3^3} + \cdots\right) \qquad (-1 < t/3 < 1)$$

$$= \sum_{n=0}^{\infty} (-1)^n \frac{t^n}{3^{n+1}} \qquad (-3 < t < 3)$$

$$= \sum_{n=0}^{\infty} (-1)^n \frac{(x-1)^n}{3^{n+1}} \qquad (-2 < x < 4).$$

Note that the radius of convergence of this series is 3, the distance from the centre of convergence, 1, to the point -2 where the denominator is 0. We could have predicted this in advance.

Maple Calculations

Maple can find the sums of many kinds of series, including absolutely and conditionally convergent numerical series and many power series. Even when Maple can't find the formal sum of a (convergent) series, it can provide a decimal approximation to the precision indicated by the current value of its variable `Digits`, which defaults to 10. Here are some examples.

```
>   sum(n^4/2^n, n=1..infinity);
```
$$150$$
```
>   sum(1/n^2, n=1..infinity);
```
$$\frac{1}{6}\pi^2$$
```
>   sum(exp(-n^2), n=0..infinity);
```
$$\sum_{n=0}^{\infty} e^{(-n^2)}$$
```
>   evalf(%);
```
$$1.386318602$$
```
>   f := x -> sum(x^(n-1)/n, n=1..infinity);
```
$$f := x \rightarrow \sum_{n=1}^{\infty} \frac{x^{(n-1)}}{n}$$
```
>   f(1); f(-1); f(1/2);
```
$$\infty$$
$$\ln(2)$$
$$2\ln(2)$$

Exercises 9.5

Determine the centre, radius, and interval of convergence of each of the power series in Exercises 1–8.

1. $\displaystyle\sum_{n=0}^{\infty} \frac{x^{2n}}{\sqrt{n+1}}$

2. $\displaystyle\sum_{n=0}^{\infty} 3n\,(x+1)^n$

3. $\displaystyle\sum_{n=1}^{\infty} \frac{1}{n}\left(\frac{x+2}{2}\right)^n$

4. $\displaystyle\sum_{n=1}^{\infty} \frac{(-1)^n}{n^4 2^{2n}}\,x^n$

5. $\displaystyle\sum_{n=0}^{\infty} n^3(2x-3)^n$

6. $\displaystyle\sum_{n=1}^{\infty} \frac{e^n}{n^3}(4-x)^n$

7. $\displaystyle\sum_{n=0}^{\infty} \frac{(1+5^n)}{n!}\,x^n$

8. $\displaystyle\sum_{n=1}^{\infty} \frac{(4x-1)^n}{n^n}$

9. Use multiplication of series to find a power series representation of $1/(1-x)^3$ valid in the interval $]-1, 1[$.

10. Determine the Cauchy product of the series $1 + x + x^2 + x^3 + \cdots$ and $1 - x + x^2 - x^3 + \cdots$. On what interval and to what function does the product series converge?

11. Determine the power series expansion of $1/(1-x)^2$ by formally dividing $1 - 2x + x^2$ into 1.

Starting with the power series representation

$$\frac{1}{1-x} = 1 + x + x^2 + x^3 + \cdots, \qquad (-1 < x < 1),$$

determine power series representations for the functions indicated in Exercises 12–20. On what interval is each representation valid?

12. $\dfrac{1}{2-x}$ in powers of x

13. $\dfrac{1}{(2-x)^2}$ in powers of x

14. $\dfrac{1}{1+2x}$ in powers of x

15. $\ln(2-x)$ in powers of x

16. $\dfrac{1}{x}$ in powers of $x-1$

17. $\dfrac{1}{x^2}$ in powers of $x+2$

18. $\dfrac{1-x}{1+x}$ in powers of x

19. $\dfrac{x^3}{1-2x^2}$ in powers of x

20. $\ln x$ in powers of $x-4$

Determine the interval of convergence and the sum of each of the series in Exercises 21–26.

21. $1 - 4x + 16x^2 - 64x^3 + \cdots = \displaystyle\sum_{n=0}^{\infty}(-1)^n(4x)^n$

* **22.** $3 + 4x + 5x^2 + 6x^3 + \cdots = \sum_{n=0}^{\infty} (n+3)x^n$

* **23.** $\dfrac{1}{3} + \dfrac{x}{4} + \dfrac{x^2}{5} + \dfrac{x^3}{6} + \cdots = \sum_{n=0}^{\infty} \dfrac{x^n}{n+3}$

* **24.** $1 \times 3 - 2 \times 4x + 3 \times 5x^2 - 4 \times 6x^3 + \cdots$
$$= \sum_{n=0}^{\infty} (-1)^n (n+1)(n+3)\, x^n$$

* **25.** $2 + 4x^2 + 6x^4 + 8x^6 + 10x^8 + \cdots = \sum_{n=0}^{\infty} 2(n+1)\, x^{2n}$

* **26.** $1 - \dfrac{x^2}{2} + \dfrac{x^4}{3} - \dfrac{x^6}{4} + \dfrac{x^8}{5} - \cdots = \sum_{n=0}^{\infty} \dfrac{(-1)^n x^{2n}}{n+1}$

Use the technique (or the result) of Example 6 to find the sums of the numerical series in Exercises 27–32.

27. $\displaystyle\sum_{n=1}^{\infty} \dfrac{n}{3^n}$

28. $\displaystyle\sum_{n=0}^{\infty} \dfrac{n+1}{2^n}$

* **29.** $\displaystyle\sum_{n=0}^{\infty} \dfrac{(n+1)^2}{\pi^n}$

* **30.** $\displaystyle\sum_{n=1}^{\infty} \dfrac{(-1)^n n(n+1)}{2^n}$

31. $\displaystyle\sum_{n=1}^{\infty} \dfrac{(-1)^{n-1}}{n2^n}$

32. $\displaystyle\sum_{n=3}^{\infty} \dfrac{1}{n2^n}$

9.6 Taylor and Maclaurin Series

If a power series $\sum_{n=0}^{\infty} a_n(x-c)^n$ has a positive radius of convergence R, then the sum of the series defines a function $f(x)$ on the interval $]c-R, c+R[$. We say that the power series is a **representation** of $f(x)$ on that interval. What relationship exists between the function $f(x)$ and the coefficients $a_0,\ a_1,\ a_2,\ \ldots$ of the power series? The following theorem answers this question.

THEOREM 21

Suppose the series

$$f(x) = \sum_{n=0}^{\infty} a_n(x-c)^n = a_0 + a_1(x-c) + a_2(x-c)^2 + a_3(x-c)^3 + \cdots$$

converges to $f(x)$ for $c - R < x < c + R$, where $R > 0$. Then

$$a_k = \frac{f^{(k)}(c)}{k!} \qquad \text{for } k = 0, 1, 2, 3, \ldots.$$

PROOF This proof requires that we differentiate the series for $f(x)$ term by term several times, a process justified by Theorem 19 (suitably reformulated for powers of $x - c$):

$$f'(x) = \sum_{n=1}^{\infty} na_n(x-c)^{n-1} = a_1 + 2a_2(x-c) + 3a_3(x-c)^2 + \cdots$$

$$f''(x) = \sum_{n=2}^{\infty} n(n-1)a_n(x-c)^{n-2} = 2a_2 + 6a_3(x-c) + 12a_4(x-c)^2 + \cdots$$

$$\vdots$$

$$f^{(k)}(x) = \sum_{n=k}^{\infty} n(n-1)(n-2)\cdots(n-k+1)a_n(x-c)^{n-k}$$

$$= k!a_k + \frac{(k+1)!}{1!}a_{k+1}(x-c) + \frac{(k+2)!}{2!}a_{k+2}(x-c)^2 + \cdots.$$

Each series converges for $c - R < x < c + R$. Setting $x = c$, we obtain $f^{(k)}(c) = k!a_k$, which proves the theorem.

Theorem 21 shows that a function $f(x)$ that has a power series representation with centre at c and positive radius of convergence must have derivatives of all orders in an interval around $x = c$, and it can have only one representation as a power series in powers of $x - c$, namely

$$f(x) = \sum_{n=0}^{\infty} \frac{f^{(n)}(c)}{n!} (x - c)^n = f(c) + f'(c)(x - c) + \frac{f''(c)}{2!}(x - c)^2 + \cdots .$$

Such a series is called a Taylor series or, if $c = 0$, a Maclaurin series.

DEFINITION   8

Taylor and Maclaurin series

If $f(x)$ has derivatives of all orders at $x = c$ (i.e., if $f^{(k)}(c)$ exists for $k = 0, 1, 2, 3, \ldots$), then the series

$$\sum_{k=0}^{\infty} \frac{f^{(k)}(c)}{k!} (x - c)^k$$

$$= f(c) + f'(c)(x - c) + \frac{f''(c)}{2!}(x - c)^2 + \frac{f^{(3)}(c)}{3!}(x - c)^3 + \cdots$$

is called the **Taylor series of f about $x = c$** (or the **Taylor series of f in powers of $x - c$**). If $c = 0$, the term **Maclaurin series** is usually used in place of Taylor series.

Note that the partial sums of such Taylor (or Maclaurin) series are just the Taylor (or Maclaurin) polynomials studied in Section 4.8.

The Taylor series is a power series as defined in the previous section. Theorem 17 implies that c must be the centre of any interval on which such a series converges, but the definition of Taylor series makes no requirement that the series should converge anywhere except at the point $x = c$ where the series is just $f(0) + 0 + 0 + \cdots$. The series exists provided all the derivatives of f exist at $x = c$; in practice this means that each derivative must exist in an open interval containing $x = c$. (Why?) However, the series may converge nowhere except at $x = c$, and if it does converge elsewhere, it may converge to something other than $f(x)$. (See Exercise 40 at the end of this section for an example where this happens.) If the Taylor series does converge to $f(x)$ in an open interval containing c, then we will say that f is analytic at $x = c$.

DEFINITION   9

Analytic functions

A function $f(x)$ is **analytic** at $x = c$ if $f(x)$ is the sum of a power series in powers of $x - c$ having positive radius of convergence. (The series is its Taylor series.) If f is analytic at each point of an open interval I, then we say it is analytic on the interval I.

Most, but not all, of the elementary functions encountered in calculus are analytic wherever they have derivatives of all orders. On the other hand, whenever a power series converges on an open interval containing c, then its sum $f(x)$ is analytic at c, and the given series is the Taylor series of $f(x)$ about $x = c$.

Maclaurin Series for Some Elementary Functions

Calculating Taylor and Maclaurin series for a function f directly from Definition 8 is practical only when we can find a formula for the nth derivative of f. Examples of such functions include $(ax + b)^r$, e^{ax+b}, $\ln(ax + b)$, $\sin(ax + b)$, $\cos(ax + b)$, and sums of such functions.

Example 1   Find the Taylor series for e^x about $x = c$. Where does the series converge to e^x? Where is e^x analytic? What is the Maclaurin series for e^x?

Solution   Since all the derivatives of $f(x) = e^x$ are e^x, we have $f^{(n)}(c) = e^c$ for every integer $n \geq 0$. Thus the Taylor series for e^x about $x = c$ is

$$\sum_{n=0}^{\infty} \frac{e^c}{n!}(x - c)^n = e^c + e^c(x - c) + \frac{e^c}{2!}(x - c)^2 + \frac{e^c}{3!}(x - c)^3 + \cdots.$$

The radius of convergence R of this series is given by

$$\frac{1}{R} = \lim_{n \to \infty} \left| \frac{e^c/(n+1)!}{e^c/n!} \right| = \lim_{n \to \infty} \frac{n!}{(n+1)!} = \lim_{n \to \infty} \frac{1}{n+1} = 0.$$

Thus the radius of convergence is $R = \infty$ and the series converges for all x. Suppose the sum is $g(x)$:

$$g(x) = e^c + e^c(x - c) + \frac{e^c}{2!}(x - c)^2 + \frac{e^c}{3!}(x - c)^3 + \cdots.$$

By Theorem 19, we have

$$g'(x) = 0 + e^c + \frac{e^c}{2!}2(x - c) + \frac{e^c}{3!}3(x - c)^2 + \cdots$$

$$= e^c + e^c(x - c) + \frac{e^c}{2!}(x - c)^2 + \cdots = g(x).$$

Also, $g(c) = e^c + 0 + 0 + \cdots = e^c$. Since $g(x)$ satisfies the differential equation $g'(x) = g(x)$ of exponential growth, we have $g(x) = Ce^x$. Substituting $x = c$ gives $e^c = g(c) = Ce^c$, so $C = 1$. Thus the Taylor series for e^x in powers of $x - c$ converges to e^x for every real number x:

$$e^x = \sum_{n=0}^{\infty} \frac{e^c}{n!}(x - c)^n$$

$$= e^c + e^c(x - c) + \frac{e^c}{2!}(x - c)^2 + \frac{e^c}{3!}(x - c)^3 + \cdots \qquad \text{(for all } x\text{)}.$$

In particular, setting $c = 0$ we obtain the Maclaurin series for e^x:

$$e^x = \sum_{n=0}^{\infty} \frac{x^n}{n!} = 1 + x + \frac{x^2}{2!} + \frac{x^3}{3!} + \cdots \qquad \text{(for all } x\text{)}.$$

Example 2 Find the Maclaurin series for (a) $\sin x$ and (b) $\cos x$. Where does each series converge?

Solution Let $f(x) = \sin x$. Then we have $f(0) = 0$ and

$$
\begin{aligned}
f'(x) &= \cos x & f'(0) &= 1 \\
f''(x) &= -\sin x & f''(0) &= 0 \\
f^{(3)}(x) &= -\cos x & f^{(3)}(0) &= -1 \\
f^{(4)}(x) &= \sin x & f^{(4)}(0) &= 0 \\
f^{(5)}(x) &= \cos x & f^{(5)}(0) &= 1
\end{aligned}
$$

$$\vdots \qquad\qquad\qquad \vdots$$

Thus, the Maclaurin series for $\sin x$ is

$$g(x) = 0 + x + 0 - \frac{x^3}{3!} + 0 + \frac{x^5}{5!} + 0 - \cdots$$

$$= x - \frac{x^3}{3!} + \frac{x^5}{5!} - \frac{x^7}{7!} + \cdots = \sum_{n=0}^{\infty} \frac{(-1)^n}{(2n+1)!} x^{2n+1}.$$

We have denoted the sum by $g(x)$ since we don't yet know whether the series converges to $\sin x$. The series does converge for all x by the ratio test:

$$\lim_{n\to\infty} \left| \frac{\dfrac{(-1)^{n+1}}{(2(n+1)+1)!} x^{2(n+1)+1}}{\dfrac{(-1)^n}{(2n+1)!} x^{2n+1}} \right| = \lim_{n\to\infty} \frac{(2n+1)!}{(2n+3)!} |x|^2$$

$$= \lim_{n\to\infty} \frac{|x|^2}{(2n+3)(2n+2)} = 0.$$

Now we can differentiate the function $g(x)$ twice to get

$$g'(x) = 1 - \frac{x^2}{2!} + \frac{x^4}{4!} - \frac{x^6}{6!} + \cdots$$

$$g''(x) = -x + \frac{x^3}{3!} - \frac{x^5}{5!} + \frac{x^7}{7!} - \cdots = -g(x).$$

Thus, $g(x)$ satisfies the differential equation $g''(x) + g(x) = 0$ of simple harmonic motion. The general solution of this equation, as observed in Section 3.7, is

$$g(x) = A \cos x + B \sin x.$$

Observe, from the series, that $g(0) = 0$ and $g'(0) = 1$. These values determine that $A = 0$ and $B = 1$. Thus, $g(x) = \sin x$ and $g'(x) = \cos x$ for all x.

We have therefore demonstrated that

$$\sin x = \sum_{n=0}^{\infty} \frac{(-1)^n}{(2n+1)!} x^{2n+1} = x - \frac{x^3}{3!} + \frac{x^5}{5!} - \frac{x^7}{7!} + \cdots \qquad \text{(for all } x\text{)},$$

$$\cos x = \sum_{n=0}^{\infty} \frac{(-1)^n}{(2n)!} x^{2n} = 1 - \frac{x^2}{2!} + \frac{x^4}{4!} - \frac{x^6}{6!} + \cdots \qquad \text{(for all } x\text{)}.$$

Theorem 21 shows that we can use any available means to find a power series converging to a given function on an interval, and the series obtained will turn out to be the Taylor series. In Section 9.5 several series were constructed by manipulating a geometric series. These include:

Some Maclaurin series

$$\frac{1}{1-x} = \sum_{n=0}^{\infty} x^n = 1 + x + x^2 + x^3 + \cdots \qquad (-1 < x < 1)$$

$$\frac{1}{(1-x)^2} = \sum_{n=1}^{\infty} nx^{n-1} = 1 + 2x + 3x^2 + 4x^3 + \cdots \qquad (-1 < x < 1)$$

$$\ln(1+x) = \sum_{n=1}^{\infty} \frac{(-1)^{n-1}}{n} x^n = x - \frac{x^2}{2} + \frac{x^3}{3} - \frac{x^4}{4} + \cdots \qquad (-1 < x \le 1)$$

$$\tan^{-1} x = \sum_{n=0}^{\infty} \frac{(-1)^n}{2n+1} x^{2n+1} = x - \frac{x^3}{3} + \frac{x^5}{5} - \frac{x^7}{7} + \cdots \quad (-1 \le x \le 1)$$

These series, together with the intervals on which they converge, are frequently used hereafter and should be memorized.

Other Maclaurin and Taylor Series

Series can be combined in various ways to generate new series. For example, we can find the Maclaurin series for e^{-x} by replacing x with $-x$ in the Maclaurin series for e^x:

$$e^{-x} = \sum_{n=0}^{\infty} \frac{(-1)^n}{n!} x^n = 1 - x + \frac{x^2}{2!} - \frac{x^3}{3!} + \cdots \quad \text{(for all } x\text{)}.$$

The series for e^x and e^{-x} can then be subtracted or added and the results divided by 2 to obtain Maclaurin series for the hyperbolic functions $\sinh x$ and $\cosh x$:

$$\sinh x = \frac{e^x - e^{-x}}{2} = \sum_{n=0}^{\infty} \frac{x^{2n+1}}{(2n+1)!} = x + \frac{x^3}{3!} + \frac{x^5}{5!} + \cdots \quad \text{(for all } x\text{)}$$

$$\cosh x = \frac{e^x + e^{-x}}{2} = \sum_{n=0}^{\infty} \frac{x^{2n}}{(2n)!} = 1 + \frac{x^2}{2!} + \frac{x^4}{4!} + \cdots \quad \text{(for all } x\text{)}.$$

Remark   Observe the similarity between the series for $\sin x$ and $\sinh x$ and between those for $\cos x$ and $\cosh x$. If we were to allow complex numbers (numbers of the form $z = x + iy$, where $i^2 = -1$ and x and y are real; see Appendix I) as arguments for our functions, and if we were to demonstrate that our operations on series could be extended to series of complex numbers, we would see that $\cos x = \cosh(ix)$ and $\sin x = -i \sinh(ix)$. In fact,

$$e^{ix} = \cos x + i \sin x \qquad \text{and} \qquad e^{-ix} = \cos x - i \sin x,$$

so

$$\cos x = \frac{e^{ix} + e^{-ix}}{2} \qquad \text{and} \qquad \sin x = \frac{e^{ix} - e^{-ix}}{2i}.$$

Such formulas are encountered in the study of functions of a complex variable; from the complex point of view the trigonometric and exponential functions are just different manifestations of the same basic function, a complex exponential $e^z = e^{x+iy}$. We content ourselves here with having mentioned the interesting relationships above and invite the reader to verify them formally by calculating with series. (Such formal calculations do not, of course, constitute a proof, since we have not established the various rules covering series of complex numbers.)

Example 3 Obtain Maclaurin series for the following functions:

(a) $e^{-x^2/3}$, (b) $\dfrac{\sin(x^2)}{x}$, (c) $\sin^2 x$.

Solution

(a) We substitute $-x^2/3$ for x in the Maclaurin series for e^x:

$$e^{-x^2/3} = 1 - \frac{x^2}{3} + \frac{1}{2!}\left(\frac{x^2}{3}\right)^2 - \frac{1}{3!}\left(\frac{x^2}{3}\right)^3 + \cdots$$

$$= \sum_{n=0}^{\infty} (-1)^n \frac{1}{3^n n!} x^{2n} \qquad \text{(for all real } x\text{)}.$$

(b) For all $x \neq 0$ we have

$$\frac{\sin x^2}{x} = \frac{1}{x}\left(x^2 - \frac{(x^2)^3}{3!} + \frac{(x^2)^5}{5!} - \cdots\right)$$

$$= x - \frac{x^5}{3!} + \frac{x^9}{5!} - \cdots = \sum_{n=0}^{\infty} (-1)^n \frac{x^{4n+1}}{(2n+1)!}.$$

Note that $f(x) = (\sin(x^2))/x$ is not defined at $x = 0$ but does have a limit (namely 0) as x approaches 0. If we define $f(0) = 0$ (the continuous extension of $f(x)$ to $x = 0$), then the series converges to $f(x)$ for all x.

(c) We use a trigonometric identity to express $\sin^2 x$ in terms of $\cos 2x$ and then use the Maclaurin series for $\cos x$ with x replaced by $2x$.

$$\sin^2 x = \frac{1 - \cos 2x}{2} = \frac{1}{2} - \frac{1}{2}\left(1 - \frac{(2x)^2}{2!} + \frac{(2x)^4}{4!} - \cdots\right)$$

$$= \frac{1}{2}\left(\frac{(2x)^2}{2!} - \frac{(2x)^4}{4!} + \frac{(2x)^6}{6!} - \cdots\right)$$

$$= \sum_{n=0}^{\infty} (-1)^n \frac{2^{2n+1}}{(2n+2)!} x^{2n+2} \qquad \text{(for all real } x\text{)}.$$

Taylor series about points other than 0 can often be obtained from known Maclaurin series by a change of variable.

Example 4 Find the Taylor series for $\ln x$ in powers of $x - 2$. Where does the series converge to $\ln x$?

Solution Note that if $t = (x - 2)/2$, then

$$\ln x = \ln(2 + (x - 2)) = \ln\left[2\left(1 + \frac{x-2}{2}\right)\right] = \ln 2 + \ln(1 + t).$$

We use the known Maclaurin series for $\ln(1+t)$:

$$\ln x = \ln 2 + \ln(1+t)$$

$$= \ln 2 + t - \frac{t^2}{2} + \frac{t^3}{3} - \frac{t^4}{4} - \cdots$$

$$= \ln 2 + \frac{x-2}{2} - \frac{(x-2)^2}{2 \times 2^2} + \frac{(x-2)^3}{3 \times 2^3} - \frac{(x-2)^4}{4 \times 2^4} + \cdots$$

$$= \ln 2 + \sum_{n=1}^{\infty} \frac{(-1)^{n-1}}{n\,2^n} (x-2)^n.$$

Since the series for $\ln(1+t)$ is valid for $-1 < t \le 1$, this series for $\ln x$ is valid for $-1 < (x-2)/2 \le 1$, that is, for $0 < x \le 4$.

■

Example 5   Find the Taylor series for $\cos x$ about the point $x = \pi/3$. Where is the series valid?

Solution   We use the addition formula for cosine:

$$\cos x = \cos\left(x - \frac{\pi}{3} + \frac{\pi}{3}\right) = \cos\left(x - \frac{\pi}{3}\right)\cos\frac{\pi}{3} - \sin\left(x - \frac{\pi}{3}\right)\sin\frac{\pi}{3}$$

$$= \frac{1}{2}\left[1 - \frac{1}{2!}\left(x - \frac{\pi}{3}\right)^2 + \frac{1}{4!}\left(x - \frac{\pi}{3}\right)^4 - \cdots\right]$$

$$- \frac{\sqrt{3}}{2}\left[\left(x - \frac{\pi}{3}\right) - \frac{1}{3!}\left(x - \frac{\pi}{3}\right)^3 + \cdots\right]$$

$$= \frac{1}{2} - \frac{\sqrt{3}}{2}\left(x - \frac{\pi}{3}\right) - \frac{1}{2}\frac{1}{2!}\left(x - \frac{\pi}{3}\right)^2 + \frac{\sqrt{3}}{2}\frac{1}{3!}\left(x - \frac{\pi}{3}\right)^3$$

$$+ \frac{1}{2}\frac{1}{4!}\left(x - \frac{\pi}{3}\right)^4 - \cdots.$$

This series representation is valid for all x. A similar calculation would enable us to expand $\cos x$ or $\sin x$ in powers of $x - c$ for any real c; both functions are analytic at every point of the real line.

■

Sometimes it is quite difficult, if not impossible, to find a formula for the general term of a Maclaurin or Taylor series. In such cases it is usually possible to obtain the first few terms before the calculations get too cumbersome. Had we attempted to solve Example 3(c) by multiplying the series for $\sin x$ by itself we might have found ourselves in this bind. Other examples occur when it is necessary to substitute one series into another or to divide one by another.

Example 6   Obtain the first three nonzero terms of the Maclaurin series for (a) $\ln\cos x$, and (b) $\tan x$.

Solution

(a) $\ln\cos x = \ln\left(1 + \left(-\frac{x^2}{2!} + \frac{x^4}{4!} - \frac{x^6}{6!} + \cdots\right)\right)$

$$= \left(-\frac{x^2}{2!} + \frac{x^4}{4!} - \frac{x^6}{6!} + \cdots\right) - \frac{1}{2}\left(-\frac{x^2}{2!} + \frac{x^4}{4!} - \frac{x^6}{6!} + \cdots\right)^2$$

$$+ \frac{1}{3}\left(-\frac{x^2}{2!} + \frac{x^4}{4!} - \frac{x^6}{6!} + \cdots\right)^3 - \cdots$$

$$= -\frac{x^2}{2} + \frac{x^4}{24} - \frac{x^6}{720} + \cdots - \frac{1}{2}\left(\frac{x^4}{4} - \frac{x^6}{24} + \cdots\right)$$
$$+ \frac{1}{3}\left(-\frac{x^6}{8} + \cdots\right) - \cdots$$
$$= -\frac{x^2}{2} - \frac{x^4}{12} - \frac{x^6}{45} - \cdots.$$

Note that at each stage of the calculation we kept only enough terms to ensure that we could get all the terms with powers up to x^6. Being an even function, $\ln \cos x$ has only even powers in its Maclaurin series. We cannot find the general term of this series, and only with considerable computational effort can we find many more terms than we have already found. We could also try to calculate terms by using the formula $a_k = f^{(k)}(0)/k!$ but even this becomes difficult after the first few values of k.

(b) $\tan x = (\sin x)/(\cos x)$. We can obtain the first three terms of the Maclaurin series for $\tan x$ by long division of the series for $\cos x$ into that for $\sin x$:

$$
\begin{array}{r}
x + \dfrac{x^3}{3} + \dfrac{2}{15}x^5 + \cdots \\[4pt]
\hline
1 - \dfrac{x^2}{2} + \dfrac{x^4}{24} \,\big)\, x - \dfrac{x^3}{6} + \dfrac{x^5}{120} - \cdots \\[4pt]
x - \dfrac{x^3}{2} + \dfrac{x^5}{24} - \cdots \\[4pt]
\hline
\dfrac{x^3}{3} - \dfrac{x^5}{30} + \cdots \\[4pt]
\dfrac{x^3}{3} - \dfrac{x^5}{6} + \cdots \\[4pt]
\hline
\dfrac{2x^5}{15} - \cdots \\[4pt]
\dfrac{2x^5}{15} - \cdots \\[4pt]
\hline
\end{array}
$$

Thus $\tan x = x + \dfrac{1}{3}x^3 + \dfrac{2}{15}x^5 + \cdots$.

Again, we cannot easily find all the terms of the series. This Maclaurin series for $\tan x$ converges for $|x| < \pi/2$, but we cannot demonstrate this fact by the techniques we have at our disposal now. Note that the series for $\tan x$ could also have been derived from that of $\ln \cos x$ obtained in part (a) because we have

$$\tan x = -\frac{d}{dx}\ln \cos x.$$

∎

Exercises 9.6

Find Maclaurin series representations for the functions in Exercises 1–14. For what values of x is each representation valid?

1. e^{3x+1}

2. $\cos(2x^3)$

3. $\sin(x - \pi/4)$

4. $\cos(2x - \pi)$

5. $x^2 \sin(x/3)$

6. $\cos^2(x/2)$

7. $\sin x \cos x$

8. $\tan^{-1}(5x^2)$

9. $\dfrac{1+x^3}{1+x^2}$

10. $\ln(2+x^2)$

11. $\ln \dfrac{1-x}{1+x}$

12. $(e^{2x^2}-1)/x^2$

13. $\cosh x - \cos x$

14. $\sinh x - \sin x$

Find the required Taylor series representations of the functions in Exercises 15–26. Where is each series representation valid?

15. $f(x) = e^{-2x}$ about the point $x = -1$

16. $f(x) = \sin x$ about the point $x = \pi/2$

17. $f(x) = \cos x$ in powers of $x - \pi$

18. $f(x) = \ln x$ in powers of $x - 3$

19. $f(x) = \ln(2+x)$ in powers of $x - 2$

20. $f(x) = e^{2x+3}$ in powers of $x + 1$

21. $f(x) = \sin x - \cos x$ about $x = \dfrac{\pi}{4}$

22. $f(x) = \cos^2 x$ about $x = \dfrac{\pi}{8}$

23. $f(x) = 1/x^2$ in powers of $x + 2$

24. $f(x) = \dfrac{x}{1+x}$ in powers of $x - 1$

25. $f(x) = x \ln x$ in powers of $x - 1$

26. $f(x) = xe^x$ in powers of $x + 2$

Find the first three nonzero terms in the Maclaurin series for the functions in Exercises 27–30.

27. $\sec x$

28. $\sec x \tan x$

29. $\tan^{-1}(e^x - 1)$

30. $e^{\tan^{-1} x} - 1$

* 31. Use the fact that $(\sqrt{1+x})^2 = 1 + x$ to find the first three nonzero terms of the Maclaurin series for $\sqrt{1+x}$.

32. Does $\csc x$ have a Maclaurin series? Why? Find the first three nonzero terms of the Taylor series for $\csc x$ about the point $x = \pi/2$.

Find the sums of the series in Exercises 33–36.

33. $1 + x^2 + \dfrac{x^4}{2!} + \dfrac{x^6}{3!} + \dfrac{x^8}{4!} + \cdots$

* 34. $x^3 - \dfrac{x^9}{3! \times 4} + \dfrac{x^{15}}{5! \times 16} - \dfrac{x^{21}}{7! \times 64} + \dfrac{x^{27}}{9! \times 256} - \cdots$

35. $1 + \dfrac{x^2}{3!} + \dfrac{x^4}{5!} + \dfrac{x^6}{7!} + \dfrac{x^8}{9!} + \cdots$

* 36. $1 + \dfrac{1}{2 \times 2!} + \dfrac{1}{4 \times 3!} + \dfrac{1}{8 \times 4!} + \cdots$

37. Let $P(x) = 1 + x + x^2$. Find (a) the Maclaurin series for $P(x)$ and (b) the Taylor series for $P(x)$ about $x = 1$.

* 38. Verify by direct calculation that $f(x) = 1/x$ is analytic at $x = a$ for every $a \neq 0$.

* 39. Verify by direct calculation that $\ln x$ is analytic at $x = a$ for every $a > 0$.

* 40. Review Exercise 41 of Section 4.3. It shows that the function

$$f(x) = \begin{cases} e^{-1/x^2} & \text{if } x \neq 0 \\ 0 & \text{if } x = 0 \end{cases}$$

has derivatives of all orders at every point of the real line, and $f^{(k)}(0) = 0$ for every positive integer k. What is the Maclaurin series for $f(x)$? What is the interval of convergence of this Maclaurin series? On what interval does the series converge to $f(x)$? Is f analytic at $x = 0$?

* 41. By direct multiplication of the Maclaurin series for e^x and e^y show that $e^x e^y = e^{x+y}$.

9.7 Applications of Taylor and Maclaurin Series

Approximating the Values of Functions

We saw in Section 4.8 how Taylor and Maclaurin polynomials (the partial sums of Taylor and Maclaurin series) can be used as polynomial approximations to more complicated functions. In Example 4 of that section we used the Lagrange remainder in Taylor's Formula to determine how many terms of the Maclaurin series for e^x are needed to calculate $e^1 = e$ correct to t3 decimal places. (We will reconsider Taylor's Formula in the next section.) For comparison, we obtained the same result in Example 7 in Section 9.3 by using a geometric series to bound the tail of the series for e.

The following example shows how the error bound associated with the alternating series test (see Theorem 15 in Section 9.4) can also be used for such approximations: when the terms a_n of a series (i) alternate in sign, (ii) decrease steadily in size, and (iii) approach zero as $n \to \infty$, then the error involved in using a partial sum of the series as an approximation to the sum of the series has the same sign as, and is smaller in absolute value than, the first omitted term.

Example 1 Find $\cos 43°$ with error less than $1/10,000$.

Solution We give two alternative solutions:

Method I. We can use the Maclaurin series:

$$\cos 43° = \cos \frac{43\pi}{180} = 1 - \frac{1}{2!}\left(\frac{43\pi}{180}\right)^2 + \frac{1}{4!}\left(\frac{43\pi}{180}\right)^4 - \cdots.$$

Now $43\pi/180 \approx 0.75049 \cdots < 1$, so the series above must satisfy the conditions (i)–(iii) mentioned above. If we truncate the series after the nth term

$$(-1)^{n-1}\frac{1}{(2n-2)!}\left(\frac{43\pi}{180}\right)^{2n-2},$$

then the error E will satisfy

$$|E| \le \frac{1}{(2n)!}\left(\frac{43\pi}{180}\right)^{2n} < \frac{1}{(2n)!}.$$

The error will not exceed $1/10,000$ if $(2n)! > 10,000$, so $n = 4$ will do ($8! = 40,320$).

$$\cos 43° \approx 1 - \frac{1}{2!}\left(\frac{43\pi}{180}\right)^2 + \frac{1}{4!}\left(\frac{43\pi}{180}\right)^4 - \frac{1}{6!}\left(\frac{43\pi}{180}\right)^6 \approx 0.73135\cdots$$

Method II. Since $43°$ is close to $45°$, we can do a bit better by using the Taylor series about $x = \pi/4$ instead of the Maclaurin series:

$$\cos 43° = \cos\left(\frac{\pi}{4} - \frac{\pi}{90}\right)$$
$$= \cos\frac{\pi}{4}\cos\frac{\pi}{90} + \sin\frac{\pi}{4}\sin\frac{\pi}{90}$$
$$= \frac{1}{\sqrt{2}}\left[\left(1 - \frac{1}{2!}\left(\frac{\pi}{90}\right)^2 + \frac{1}{4!}\left(\frac{\pi}{90}\right)^4 - \cdots\right)\right.$$
$$\left. + \left(\frac{\pi}{90} - \frac{1}{3!}\left(\frac{\pi}{90}\right)^3 + \cdots\right)\right].$$

Since

$$\frac{1}{4!}\left(\frac{\pi}{90}\right)^4 < \frac{1}{3!}\left(\frac{\pi}{90}\right)^3 < \frac{1}{20,000},$$

we need only the first two terms of the first series and the first term of the second series:

$$\cos 43° \approx \frac{1}{\sqrt{2}}\left(1 + \frac{\pi}{90} - \frac{1}{2}\left(\frac{\pi}{90}\right)^2\right) \approx 0.731358\cdots.$$

(In fact, $\cos 43° = 0.7313537\cdots$.)

When finding approximate values of functions, it is best, whenever possible, to use a power series about a point as close as possible to the point where the approximation is desired.

Functions Defined by Integrals

Many functions that can be expressed as simple combinations of elementary functions cannot be antidifferentiated by elementary techniques; their antiderivatives are not simple combinations of elementary functions. We can, however, often find the Taylor series for the antiderivatives of such functions and hence approximate their definite integrals.

Example 2   Find the Maclaurin series for

$$E(x) = \int_0^x e^{-t^2}\, dt,$$

and use it to evaluate $E(1)$ correct to 3 decimal places.

Solution   The Maclaurin series for $E(x)$ is given by

$$E(x) = \int_0^x \left(1 - t^2 + \frac{t^4}{2!} - \frac{t^6}{3!} + \frac{t^8}{4!} - \cdots\right) dt$$

$$= \left(t - \frac{t^3}{3} + \frac{t^5}{5 \times 2!} - \frac{t^7}{7 \times 3!} + \frac{t^9}{9 \times 4!} - \cdots\right)\Bigg|_0^x$$

$$= x - \frac{x^3}{3} + \frac{x^5}{5 \times 2!} - \frac{x^7}{7 \times 3!} + \frac{x^9}{9 \times 4!} - \cdots = \sum_{n=0}^{\infty} (-1)^n \frac{x^{2n+1}}{(2n+1)n!},$$

and is valid for all x because the series for e^{-t^2} is valid for all t. Therefore,

$$E(1) = 1 - \frac{1}{3} + \frac{1}{5 \times 2!} - \frac{1}{7 \times 3!} + \cdots$$

$$\approx 1 - \frac{1}{3} + \frac{1}{5 \times 2!} - \frac{1}{7 \times 3!} + \cdots + \frac{(-1)^{n-1}}{(2n-1)(n-1)!}.$$

We stopped with the nth term. The error in this approximation does not exceed the first omitted term, so it will be less than 0.0005, provided $(2n+1)n! > 2,000$. Since $13 \times 6! = 9,360$, $n = 6$ will do. Thus,

$$E(1) \approx 1 - \frac{1}{3} + \frac{1}{10} - \frac{1}{42} + \frac{1}{216} - \frac{1}{1,320} \approx 0.747,$$

rounded to three decimal places.

Indeterminate Forms

Examples 1 and 2 of Section 4.9 showed how Maclaurin polynomials could be used for evaluating the limits of indeterminate forms. Here are two more examples, this time using the series directly and keeping enough terms to allow cancellation of the [0/0] factors.

Example 3 Evaluate (a) $\lim_{x \to 0} \dfrac{x - \sin x}{x^3}$ and (b) $\lim_{x \to 0} \dfrac{(e^{2x} - 1)\ln(1 + x^3)}{(1 - \cos 3x)^2}$.

Solution

(a) $\displaystyle\lim_{x \to 0} \frac{x - \sin x}{x^3} \qquad \left[\frac{0}{0}\right]$

$$= \lim_{x \to 0} \frac{x - \left(x - \dfrac{x^3}{3!} + \dfrac{x^5}{5!} - \cdots\right)}{x^3}$$

$$= \lim_{x \to 0} \frac{\dfrac{x^3}{3!} - \dfrac{x^5}{5!} + \cdots}{x^3}$$

$$= \lim_{x \to 0} \left(\frac{1}{3!} - \frac{x^2}{5!} + \cdots\right) = \frac{1}{3!} = \frac{1}{6}.$$

(b) $\displaystyle\lim_{x \to 0} \frac{(e^{2x} - 1)\ln(1 + x^3)}{(1 - \cos 3x)^2} \qquad \left[\frac{0}{0}\right]$

$$= \lim_{x \to 0} \frac{\left(1 + (2x) + \dfrac{(2x)^2}{2!} + \dfrac{(2x)^3}{3!} + \cdots - 1\right)\left(x^3 - \dfrac{x^6}{2} + \cdots\right)}{\left(1 - \left(1 - \dfrac{(3x)^2}{2!} + \dfrac{(3x)^4}{4!} - \cdots\right)\right)^2}$$

$$= \lim_{x \to 0} \frac{2x^4 + 2x^5 + \cdots}{\left(\dfrac{9}{2}x^2 - \dfrac{3^4}{4!}x^4 + \cdots\right)^2}$$

$$= \lim_{x \to 0} \frac{2 + 2x + \cdots}{\left(\dfrac{9}{2} - \dfrac{3^4}{4!}x^2 + \cdots\right)^2} = \frac{2}{\left(\dfrac{9}{2}\right)^2} = \frac{8}{81}.$$

You can check that the second of these examples is much more difficult if attempted using l'Hôpital's Rule. ∎

Exercises 9.7

Use Maclaurin or Taylor series to calculate the function values indicated in Exercises 1–12, with error less than 5×10^{-5} in absolute value.

1. $e^{0.2}$

2. $1/e$

3. $e^{1.2}$

4. $\sin(0.1)$

5. $\cos 5°$

6. $\ln(6/5)$

7. $\ln(0.9)$

8. $\sin 80°$

9. $\cos 65°$

10. $\tan^{-1} 0.2$

11. $\cosh(1)$

12. $\ln(3/2)$

Find Maclaurin series for the functions in Exercises 13–17.

13. $I(x) = \displaystyle\int_0^x \frac{\sin t}{t}\, dt$

14. $J(x) = \displaystyle\int_0^x \frac{e^t - 1}{t}\, dt$

15. $K(x) = \displaystyle\int_1^{1+x} \frac{\ln t}{t - 1}\, dt$

16. $L(x) = \displaystyle\int_0^x \cos(t^2)\, dt$

17. $M(x) = \displaystyle\int_0^x \frac{\tan^{-1} t^2}{t^2}\, dt$

18. Find $L(0.5)$ correct to 3 decimal places, with L defined as in Exercise 16.

19. Find $I(1)$ correct to 3 decimal places, with I defined as in Exercise 13.

Evaluate the limits in Exercises 20–25.

20. $\displaystyle\lim_{x \to 0} \frac{\sin(x^2)}{\sinh x}$

21. $\displaystyle\lim_{x \to 0} \frac{1 - \cos(x^2)}{(1 - \cos x)^2}$

22. $\lim\limits_{x\to 0} \dfrac{(e^x - 1 - x)^2}{x^2 - \ln(1 + x^2)}$ **23.** $\lim\limits_{x\to 0} \dfrac{2\sin 3x - 3\sin 2x}{5x - \tan^{-1} 5x}$ **24.** $\lim\limits_{x\to 0} \dfrac{\sin(\sin x) - x}{x(\cos(\sin x) - 1)}$ **25.** $\lim\limits_{x\to 0} \dfrac{\sinh x - \sin x}{\cosh x - \cos x}$

9.8 Taylor's Formula Revisited

Theorem 10 of Section 4.8 (Taylor's Theorem with Lagrange remainder) provides a formula for the error involved when the Taylor polynomial

$$P_n(x) = \sum_{k=0}^{n} \frac{f^{(k)}(c)}{k!}(x - c)^k$$

of a function $f(x)$ about $x = c$ is used to approximate $f(x)$ for values of $x \neq c$. Specifically, it states the following:

THEOREM 22

Taylor's Theorem with Lagrange remainder

If the $(n + 1)$st derivative of f exists on an interval containing c and x, and if $P_n(x)$ is the Taylor polynomial of degree n for f about the point $x = c$, then Taylor's Formula

$$f(x) = P_n(x) + E_n(x)$$

holds, where the error term $E_n(x)$ is given by

$$E_n(x) = \frac{f^{(n+1)}(X)}{(n + 1)!}(x - c)^{n+1}$$

for some X between c and x. ($E_n(x)$ is called the Lagrange remainder in Taylor's Formula.)

Observe that the Lagrange form of the remainder, $E_n(x)$, looks just like the $(n+1)$st degree term in $P_{n+1}(x)$, except that c in $f^{(n+1)}(c)$ has been replaced by an unknown number X between c and x. The cases $n = 0$ and $n = 1$ of Taylor's Formula with Lagrange remainder are just the Mean-Value Theorem (Theorem 11 of Section 2.6) and the error formula for linear approximation (Theorem 9 of Section 4.7), respectively.

Example 1 Use Taylor's Theorem to determine how many terms of the Maclaurin series for $\cos x$ are needed to calculate $\cos 10°$ correctly to 5 decimal places.

Solution Being an even function, $f(x) = \cos x$ has only even degree terms in its Maclaurin series. The Maclaurin polynomials P_{2n} and P_{2n+1} for $f(x)$ are therefore equal:

$$P_{2n}(x) = P_{2n+1}(x) = \sum_{j=0}^{n} \frac{(-1)^j}{(2j)!} x^{2j}$$

$$1 - \frac{x^2}{2!} + \frac{x^4}{4!} - \cdots + \frac{(-1)^n}{(2n)!} x^{2n}.$$

It makes good sense to use the remainder E_{2n+1} rather than the remainder E_{2n}; it is likely to be smaller and therefore assure us of more accuracy for any given value of n. Since $f^{(2n+2)}(x) = (-1)^{n+1} \cos x$, we have, for some X between 0 and x,

$$|E_{2n+1}(x)| = \left| \frac{(-1)^{n+1} \cos X}{(2n+2)!} x^{2n+2} \right| \le \frac{|x|^{2n+2}}{(2n+2)!}.$$

For $x = 10° = \pi/18 \approx 0.174533 < 0.2$ radians, we will have 5 decimal place accuracy if

$$\frac{0.2^{2n+2}}{(2n+2)!} < 0.000005.$$

This is satisfied if $n = 2$ ($0.2^6/6! < 9 \times 10^{-8}$), but not $n = 1$. Thus,

$$\cos 10° = \cos \frac{\pi}{18} \approx 1 - \frac{1}{2} \left(\frac{\pi}{18} \right)^2 + \frac{1}{24} \left(\frac{\pi}{18} \right)^4 \approx 0.98481$$

to 5 decimal places.

Using Taylor's Theorem to Find Taylor and Maclaurin Series

If a function f has derivatives of all orders, then we can write Taylor's Formula for any n:

$$f(x) = P_n(x) + E_n(x).$$

If we can show that $\lim_{n \to \infty} E_n(x) = 0$ for all x in some interval I, then we are entitled to conclude, for x in I, that

$$f(x) = \lim_{n \to \infty} P_n(x) = \sum_{k=0}^{\infty} \frac{f^{(k)}(c)}{k!} (x - c)^k,$$

that is, we will have expressed $f(x)$ as the sum of an infinite series of terms which are multiples of positive integer powers of $x - c$, and the series converges for all x in I. This series is the **Taylor series** representation of f in powers of $x - c$ (or the **Maclaurin series** if $c = 0$).

Example 2 Use Taylor's Theorem to find the Maclaurin series for $f(x) = e^x$. Where does the series converge to $f(x)$?

Solution Since e^x is positive and increasing, $e^X \le e^{|x|}$ for any $X \le |x|$. Since $f^{(k)}(x) = e^x$ for any k we have, taking $c = 0$ in the Lagrange remainder in Taylor's Formula,

$$|E_n(x)| = \left| \frac{f^{(n+1)}(X)}{(n+1)!} x^{n+1} \right|$$

$$\leq \frac{e^X}{(n+1)!} |x|^{n+1} \leq e^{|x|} \frac{|x|^{n+1}}{(n+1)!} \to 0 \text{ as } n \to \infty$$

for any real x, as shown in Theorem 3(b) of Section 9.1. Therefore,

$$e^x = \sum_{k=0}^{\infty} \frac{x^k}{k!} = 1 + x + \frac{x^2}{2!} + \frac{x^3}{3!} + \cdots,$$

and the series converges to e^x for all real numbers x. ∎

Taylor's Theorem with Integral Remainder

The following theorem is another version of Taylor's Theorem, where the remainder in Taylor's Formula is expressed as an integral.

THEOREM 23

Taylor's Theorem with integral remainder

If the $(n+1)$st derivative of f exists on an interval containing c and x, and if $P_n(x)$ is the Taylor polynomial of degree n for f about the point $x = c$, then the remainder $E_n(x) = f(x) - P_n(x)$ in Taylor's Formula is given by

$$E_n(x) = \frac{1}{n!} \int_c^x (x - t)^n f^{(n+1)}(t) \, dt.$$

PROOF   We start with the Fundamental Theorem of Calculus written in the form

$$f(x) = f(c) + \int_c^x f'(t) \, dt = P_0(x) + E_0(x).$$

(Note that the Fundamental Theorem is just the special case $n = 0$ of Taylor's Formula with integral remainder.) We now apply integration by parts to the integral, setting

$$U = f'(t), \qquad dV = dt,$$
$$dU = f''(t) \, dt, \qquad V = -(x - t).$$

(We have broken our usual rule about not including a constant of integration with V. In this case we have included the constant $-x$ in V in order to have V vanish when $t = x$.) We have

$$f(x) = f(c) - f'(t)(x - t) \Big|_{t=c}^{t=x} + \int_c^x (x - t) f''(t) \, dt$$

$$= f(c) + f'(c)(x - c) + \int_c^x (x - t) f''(t) \, dt$$

$$= P_1(x) + E_1(x).$$

We have thus proved the case $n = 1$ of Taylor's Formula with integral remainder.

Let us complete the proof for general n by mathematical induction. Suppose that Taylor's Formula holds with integral remainder for some $n = k$:

$$f(x) = P_k(x) + E_k(x) = P_k(x) + \frac{1}{k!} \int_c^x (x - t)^k f^{(k+1)}(t) \, dt.$$

Again we integrate by parts. Let

$$U = f^{(k+1)}(t), \qquad dV = (x-t)^k \, dt,$$
$$dU = f^{(k+2)}(t) \, dt, \qquad V = \frac{-1}{k+1}(x-t)^{k+1}.$$

We have

$$f(x) = P_k(x) + \frac{1}{k!}\left(-\frac{f^{(k+1)}(t)(x-t)^{k+1}}{k+1}\Big|_{t=c}^{t=x} + \int_c^x \frac{(x-t)^{k+1} f^{(k+2)}(t)}{k+1} \, dt \right)$$

$$= P_k(x) + \frac{f^{(k+1)}(c)}{(k+1)!}(x-c)^{k+1} + \frac{1}{(k+1)!}\int_c^x (x-t)^{k+1} f^{(k+2)}(t) \, dt$$

$$= P_{k+1}(x) + E_{k+1}(x).$$

Thus Taylor's Formula with integral remainder is valid for $n = k+1$ if it is valid for $n = k$. Having been shown to be valid for $n = 0$ (and $n = 1$), it must therefore be valid for every positive integer n for which $E_n(x)$ exists.

Remark   Using one or the other of the versions of Taylor's Theorem given in this section, all the basic Maclaurin and Taylor series given in Section 9.6 can be verified without having to use the theory of power series.

Exercises 9.8

1. Estimate the error if the Maclaurin polynomial of degree 5 for $\sin x$ is used to approximate $\sin(0.2)$.

2. Estimate the error if the Maclaurin polynomial of degree 6 for $\cos x$ is used to approximate $\cos(1)$.

3. Estimate the error if the Maclaurin polynomial of degree 4 for e^{-x} is used to approximate $e^{-0.5}$.

4. Estimate the error if the Maclaurin polynomial of degree 2 for $\sec x$ is used to approximate $\sec(0.2)$.

5. Estimate the error if the Maclaurin polynomial of degree 3 for $\ln(\cos x)$ is used to approximate $\ln(\cos 0.1)$.

6. Estimate the error if the Taylor polynomial of degree 3 for $\tan^{-1} x$ in powers of $x - 1$ is used to approximate $\tan^{-1} 0.99$.

7. Estimate the error if the Taylor polynomial of degree 4 for $\ln x$ in powers of $x - 2$ is used to approximate $\ln(1.95)$.

Use Taylor's Formula to establish the Maclaurin series for the functions in Exercises 8–15.

8. e^{-x}

9. 2^x

10. $\cos x$

11. $\sin x$

12. $\sin^2 x$

13. $\dfrac{1}{1-x}$

** **14.** $\ln(1+x)$ (Use the integral remainder.)

* **15.** $\dfrac{x}{2+3x}$ (Use Exercise 13.)

Use Taylor's Formula to obtain the Taylor series indicated in Exercises 16–21.

16. for e^x in powers of $x - a$

17. for $\sin x$ in powers of $x - (\pi/6)$

18. for $\cos x$ in powers of $x - (\pi/4)$

* **19.** for $\ln x$ in powers of $x - 1$ (Use the integral remainder.)

* **20.** for $\ln x$ in powers of $x - 2$

21. for $1/x$ in powers of $x + 2$ (Use Exercise 13.)

9.9 The Binomial Theorem and Binomial Series

Example 1 Use Taylor's Formula to prove the Binomial Theorem: if n is a positive integer, then

$$(a + x)^n = a^n + n\,a^{n-1}x + \frac{n(n-1)}{2!}a^{n-2}x^2 + \cdots + n\,ax^{n-1} + x^n$$

$$= \sum_{k=0}^{n} \binom{n}{k} a^{n-k}x^k,$$

where $\binom{n}{k} = \dfrac{n!}{(n-k)!\,k!}$.

Solution Let $f(x) = (a+x)^n$. Then

$$f'(x) = n(a+x)^{n-1} = \frac{n!}{(n-1)!}(a+x)^{n-1}$$

$$f''(x) = \frac{n!}{(n-1)!}(n-1)(a+x)^{n-2} = \frac{n!}{(n-2)!}(a+x)^{n-2}$$

$$\vdots$$

$$f^{(k)}(x) = \frac{n!}{(n-k)!}(a+x)^{n-k} \qquad (0 \le k \le n).$$

In particular, $f^{(n)}(x) = \dfrac{n!}{0!}(a+x)^{n-n} = n!$, a constant, and

$$f^{(k)}(x) = 0 \qquad \text{for all } x, \text{ if } k > n.$$

For $0 \le k \le n$ we have $f^{(k)}(0) = \dfrac{n!}{(n-k)!}a^{n-k}$. Thus, by Taylor's Theorem with Lagrange remainder,

$$(a+x)^n = f(x) = \sum_{k=0}^{n} \frac{f^{(k)}(0)}{k!}x^k + \frac{f^{(n+1)}(X)}{(n+1)!}x^{n+1}$$

$$= \sum_{k=0}^{n} \frac{n!}{(n-k)!\,k!}a^{n-k}x^k + 0 = \sum_{k=0}^{n} \binom{n}{k} a^{n-k}x^k.$$

This is, in fact, the Maclaurin *series* for $(a+x)^n$, not just the Maclaurin polynomial of degree n. Since all higher-degree terms are zero, the series has only finitely many nonzero terms and so converges for all x. ∎

Remark If $f(x) = (a+x)^r$, where $a > 0$ and r is any real number, then calculations similar to those above show that the Maclaurin polynomial of degree n for f is

$$P_n(x) = a^r + \sum_{k=1}^{n} \frac{r(r-1)(r-2)\cdots(r-k+1)}{k!}a^{r-k}x^k.$$

However, if r is not a positive integer, then there will be no positive integer n for which the remainder $E_n(x) = f(x) - P_n(x)$ vanishes identically, and the corresponding Maclaurin series will not be a polynomial.

The Binomial Series

To simplify the discussion of the function $(a + x)^r$ when r is not a positive integer, we take $a = 1$ and consider the function $(1 + x)^r$. Results for the general case follow via the identity

$$(a + x)^r = a^r \left(1 + \frac{x}{a}\right)^r,$$

valid for any $a > 0$.

If r is any real number and $x > -1$, then the kth derivative of $(1 + x)^r$ is

$$r(r - 1)(r - 2) \cdots (r - k + 1)(1 + x)^{r-k}, \qquad (k = 1, 2, \ldots).$$

Thus, the Maclaurin series for $(1 + x)^r$ is

$$1 + \sum_{k=1}^{\infty} \frac{r(r - 1)(r - 2) \cdots (r - k + 1)}{k!} x^k,$$

which is called the **binomial series**. The following theorem shows that the binomial series does, in fact, converge to $(1 + x)^r$ if $|x| < 1$. We could accomplish this by writing Taylor's Formula for $(1 + x)^r$ with $c = 0$ and showing that the remainder $E_n(x) \to 0$ as $n \to \infty$. (We would need to use the integral form of the remainder to prove this for all $|x| < 1$.) However, we will use an easier method, similar to the one used for the exponential and trigonometric functions in Section 9.6.

THEOREM 24

The binomial series

If $|x| < 1$, then

$$(1 + x)^r = 1 + rx + \frac{r(r - 1)}{2!} x^2 + \frac{r(r - 1)(r - 2)}{3!} x^3 + \cdots$$

$$= 1 + \sum_{n=1}^{\infty} \frac{r(r - 1)(r - 2) \cdots (r - n + 1)}{n!} x^n \quad (-1 < x < 1).$$

PROOF   If $|x| < 1$, then the series

$$f(x) = 1 + \sum_{n=1}^{\infty} \frac{r(r - 1)(r - 2) \cdots (r - n + 1)}{n!} x^n$$

converges by the ratio test, since

$$\rho = \lim_{n \to \infty} \left| \frac{\dfrac{r(r - 1)(r - 2) \cdots (r - n + 1)(r - n)}{(n + 1)!} x^{n+1}}{\dfrac{r(r - 1)(r - 2) \cdots (r - n + 1)}{n!} x^n} \right|$$

$$= \lim_{n \to \infty} \left| \frac{r - n}{n + 1} \right| |x| = |x| < 1.$$

Note that $f(0) = 1$. We need to show that $f(x) = (1 + x)^r$ for $|x| < 1$.

By Theorem 19, we can differentiate the series for $f(x)$ termwise on $|x| < 1$ to obtain

$$f'(x) = \sum_{n=1}^{\infty} \frac{r(r-1)(r-2)\cdots(r-n+1)}{(n-1)!} x^{n-1}$$

$$= \sum_{n=0}^{\infty} \frac{r(r-1)(r-2)\cdots(r-n)}{n!} x^{n}.$$

We have replaced n with $n+1$ to get the second version of the sum from the first version. Adding the second version to x times the first version, we get

$$(1+x)f'(x) = \sum_{n=0}^{\infty} \frac{r(r-1)(r-2)\cdots(r-n)}{n!} x^{n}$$

$$+ \sum_{n=1}^{\infty} \frac{r(r-1)(r-2)\cdots(r-n+1)}{(n-1)!} x^{n}$$

$$= r + \sum_{n=1}^{\infty} \frac{r(r-1)(r-2)\cdots(r-n+1)}{n!} x^{n} \left[(r-n)+n\right]$$

$$= r\, f(x).$$

The differential equation $(1+x)f'(x) = rf(x)$ implies that

$$\frac{d}{dx} \frac{f(x)}{(1+x)^r} = \frac{(1+x)^r f'(x) - r(1+x)^{r-1} f(x)}{(1+x)^{2r}} = 0$$

for all x satisfying $|x| < 1$. Thus, $f(x)/(1+x)^r$ is constant on that interval, and since $f(0) = 1$, the constant must be 1. Thus $f(x) = (1+x)^r$.

Remark For some values of r the binomial series may converge at the endpoints $x = 1$ or $x = -1$. As observed above, if r is a positive integer, the series has only finitely many nonzero terms, and so converges for all x.

Example 2 Find the Maclaurin series for $\dfrac{1}{\sqrt{1+x}}$.

Solution Here $r = -(1/2)$:

$$\frac{1}{\sqrt{1+x}} = (1+x)^{-1/2}$$

$$= 1 - \frac{1}{2}x + \frac{1}{2!}\left(-\frac{1}{2}\right)\left(-\frac{3}{2}\right)x^2 + \frac{1}{3!}\left(-\frac{1}{2}\right)\left(-\frac{3}{2}\right)\left(-\frac{5}{2}\right)x^3 + \cdots$$

$$= 1 - \frac{1}{2}x + \frac{1 \times 3}{2^2 2!}x^2 - \frac{1 \times 3 \times 5}{2^3 3!}x^3 + \cdots$$

$$= 1 + \sum_{n=1}^{\infty}(-1)^n \frac{1 \times 3 \times 5 \times \cdots \times (2n-1)}{2^n n!} x^n.$$

This series converges for $-1 < x \le 1$. (Use the alternating series test to get the endpoint $x = 1$.)

Example 3   Find the Maclaurin series for $\sin^{-1} x$.

Solution   Replace x with $-t^2$ in the series obtained in the previous example to get

$$\frac{1}{\sqrt{1-t^2}} = 1 + \sum_{n=1}^{\infty} \frac{1 \times 3 \times 5 \times \cdots \times (2n-1)}{2^n n!} t^{2n} \qquad (-1 < t < 1).$$

Now integrate t from 0 to x:

$$\sin^{-1} x = \int_0^x \frac{dt}{\sqrt{1-t^2}} = \int_0^x \left(1 + \sum_{n=1}^{\infty} \frac{1 \times 3 \times 5 \times \cdots \times (2n-1)}{2^n n!} t^{2n} \right) dt$$

$$= x + \sum_{n=1}^{\infty} \frac{1 \times 3 \times 5 \times \cdots \times (2n-1)}{2^n n! (2n+1)} x^{2n+1}$$

$$= x + \frac{x^3}{6} + \frac{3}{40} x^5 + \cdots \qquad (-1 < x < 1).$$

■

Exercises 9.9

Find Maclaurin series representations for the functions in Exercises 1–6. Use the binomial series to calculate the answers.

1. $\sqrt{1+x}$

2. $x\sqrt{1-x}$

3. $\sqrt{4+x}$

4. $\dfrac{1}{\sqrt{4+x^2}}$

5. $(1-x)^{-2}$

6. $(1+x)^{-3}$

* **7. (Binomial coefficients)** Show that the binomial coefficients

$$\binom{n}{k} = \frac{n!}{k!\,(n-k)!}$$

satisfy

(i) $\dbinom{n}{0} = \dbinom{n}{n} = 1$ for every n, and

(ii) if $0 \le k \le n$, then $\dbinom{n}{k-1} + \dbinom{n}{k} = \dbinom{n+1}{k}$.

It follows that, for fixed $n \ge 1$, the binomial coefficients

$\dbinom{n}{0}, \dbinom{n}{1}, \dbinom{n}{2}, \ldots, \dbinom{n}{n}$ are the elements of the nth row of **Pascal's triangle**:

```
            1   1
          1   2   1
        1   3   3   1
      1   4   6   4   1
    1   5   10  10  5   1
```

where each element with value > 1 is the sum of the two diagonally above it.

* **8. (An inductive proof of the Binomial Theorem)** Use mathematical induction and the results of Exercise 7 to prove the Binomial Theorem:

$$(a+b)^n = \sum_{k=0}^{n} \binom{n}{k} a^{n-k} b^k$$

$$= a^n + na^{n-1}b + \binom{n}{2} a^{n-2}b^2 + \binom{n}{3} a^{n-3}b^3 + \cdots + b^n.$$

* **9. (The Leibniz Rule)** Use mathematical induction, the Product Rule, and Exercise 7 to verify the Leibniz Rule for the nth derivative of a product of two functions:

$$(fg)^{(n)} = \sum_{k=0}^{n} \binom{n}{k} f^{(n-k)} g^{(k)}$$

$$= f^{(n)} g + n f^{(n-1)} g' + \binom{n}{2} f^{(n-2)} g''$$

$$+ \binom{n}{3} f^{(n-3)} g^{(3)} + \cdots + f g^{(n)}.$$

9.10 Series Solutions of Differential Equations

In Section 3.7 we developed a recipe for solving second-order, linear, homogeneous differential equations with constant coefficients:

$$ay'' + by' + cy = 0.$$

Many of the second-order, linear, homogeneous differential equations that arise in applications do not have constant coefficients. If the coefficient functions of such an equation are sufficiently well behaved, we can often find solutions in the form of power series (Taylor series). Such series solutions are frequently used to define new functions, whose properties are deduced partly from the fact that they solve particular differential equations. For example, Bessel functions of order ν are defined to be certain series solutions of Bessel's differential equation

$$x^2 y'' + xy' + (x^2 - \nu^2)y = 0.$$

Series solutions for second-order homogeneous linear differential equations are most easily found near an **ordinary point** of the equation. This is a point $x = a$ such that the equation can be expressed in the form

$$y'' + p(x)y' + q(x)y = 0,$$

where the functions $p(x)$ and $q(x)$ are **analytic** at $x = a$. (Recall that a function f is analytic at $x = a$ if $f(x)$ can be expressed as the sum of its Taylor series in powers of $x - a$ in an interval of positive radius centred at $x = a$.) Thus we assume

$$p(x) = \sum_{n=0}^{\infty} p_n (x - a)^n,$$

$$q(x) = \sum_{n=0}^{\infty} q_n (x - a)^n,$$

with both series converging in some interval of the form $a - R < x < a + R$. Frequently $p(x)$ and $q(x)$ are polynomials, so are analytic everywhere. A change of independent variable $\xi = x - a$ will put the point $x = a$ at the origin $\xi = 0$, so we can assume that $a = 0$.

The following example illustrates the technique of series solution around an ordinary point.

Example 1 Find two independent solutions in powers of x for the Hermite equation

$$y'' - 2xy' + \nu y = 0.$$

For what values of ν does the equation have a polynomial solution?

Solution We try for a power series solution of the form

$$y = \sum_{n=0}^{\infty} a_n x^n = a_0 + a_1 x + a_2 x^2 + a_3 x^3 + \cdots, \qquad \text{so that}$$

$$y' = \sum_{n=1}^{\infty} n a_n x^{n-1}$$

$$y'' = \sum_{n=2}^{\infty} n(n-1) a_n x^{n-2} = \sum_{n=0}^{\infty} (n+2)(n+1) a_{n+2} x^n.$$

(We have replaced n by $n + 2$ in order to get x^n in the sum for y''.) We substitute these expressions into the differential equation to get

$$\sum_{n=0}^{\infty} (n+2)(n+1) a_{n+2} x^n - 2 \sum_{n=1}^{\infty} n a_n x^n + \nu \sum_{n=0}^{\infty} a_n x^n = 0$$

or $\quad 2a_2 + \nu a_0 + \displaystyle\sum_{n=1}^{\infty} \Big[(n+2)(n+1) a_{n+2} - (2n - \nu) a_n \Big] x^n = 0.$

This identity holds for all x provided that the coefficient of every power of x vanishes; that is,

$$a_2 = -\frac{\nu a_0}{2}, \qquad a_{n+2} = \frac{(2n - \nu) a_n}{(n+2)(n+1)}, \qquad (n = 1, 2, \cdots).$$

The latter of these formulas is called a **recurrence relation**.

We can choose a_0 and a_1 to have any values; then the above conditions determine all the remaining coefficients a_n, $(n \geq 2)$. We can get one solution by choosing, for instance, $a_0 = 1$ and $a_1 = 0$. Then, by the recurrence relation,

$$a_3 = 0, \quad a_5 = 0, \quad a_7 = 0, \quad \cdots, \qquad \text{and}$$

$$a_2 = -\frac{\nu}{2}$$

$$a_4 = \frac{(4 - \nu) a_2}{4 \times 3} = -\frac{\nu(4 - \nu)}{2 \times 3 \times 4} = -\frac{\nu(4 - \nu)}{4!}$$

$$a_6 = \frac{(8 - \nu) a_4}{6 \times 5} = -\frac{\nu(4 - \nu)(8 - \nu)}{6!}$$

$$\cdots$$

The pattern is obvious here:

$$a_{2n} = -\frac{\nu(4 - \nu)(8 - \nu) \cdots (4n - 4 - \nu)}{(2n)!}, \qquad (n = 1, 2, \cdots).$$

One solution to the Hermite equation is

$$y_1 = 1 + \sum_{n=1}^{\infty} -\frac{\nu(4-\nu)(8-\nu)\cdots(4n-4-\nu)}{(2n)!} x^{2n}.$$

We observe that if $\nu = 4n$ for some nonnegative integer n, then y_1 is an even polynomial of degree $2n$, because $a_{2n+2} = 0$ and all subsequent even coefficients therefore also vanish.

The second solution, y_2, can be found in the same way, by choosing $a_0 = 0$ and $a_1 = 1$. It is

$$y_2 = x + \sum_{n=1}^{\infty} \frac{(2-\nu)(6-\nu)\cdots(4n-2-\nu)}{(2n+1)!} x^{2n+1},$$

and it is an odd polynomial of degree $2n + 1$ if $\nu = 4n + 2$.

Both of these series solutions converge for all x. The ratio test can be applied directly to the recurrence relation. Since consecutive nonzero terms of each series are of the form $a_n x^n$ and $a_{n+2} x^{n+2}$, we calculate

$$\rho = \lim_{n \to \infty} \left| \frac{a_{n+2} x^{n+2}}{a_n x^n} \right| = |x|^2 \lim_{n \to \infty} \left| \frac{a_{n+2}}{a_n} \right| = |x|^2 \lim_{n \to \infty} \left| \frac{2n - \nu}{(n+2)(n+1)} \right| = 0$$

for every x, so the series converges by the ratio test. ∎

If $x = a$ is not an ordinary point of the equation

$$y'' + p(x)y' + q(x)y = 0,$$

then it is called a **singular point** of that equation. This means that at least one of the functions $p(x)$ and $q(x)$ is not analytic at $x = a$. If, however, $(x - a)p(x)$ and $(x - a)^2 q(x)$ are analytic at $x = a$, then the singular point is said to be a **regular singular point**. For example, the origin $x = 0$ is a regular singular point of Bessel's equation,

$$x^2 y'' + xy' + (x^2 - \nu^2)y = 0,$$

since $p(x) = 1/x$ and $q(x) = (x^2 - \nu^2)/x^2$ satisfy $xp(x) = 1$ and $x^2 q(x) = x^2 - \nu^2$, which are both polynomials and therefore analytic.

The solutions of differential equations are usually not analytic at singular points. However, it is still possible to find at least one series solution about such a point. The method involves searching for a series solution of the form x^μ times a power series, that is,

$$y = (x - a)^\mu \sum_{n=0}^{\infty} a_n (x - a)^n = \sum_{n=0}^{\infty} a_n (x - a)^{n+\mu}, \qquad \text{where } a_0 \neq 0.$$

Substitution into the differential equation produces a quadratic **indicial equation**, which determines one or two values of μ for which such solutions can be found, and a **recurrence relation** enabling the coefficients a_n to be calculated for $n \geq 1$. If the indicial roots are not equal and do not differ by an integer, two independent solutions can be calculated. If the indicial roots are equal or differ by an integer, one such solution can be calculated (corresponding to the larger indicial root), but finding a second independent solution (and so the general solution) requires techniques beyond the scope of this book. The reader is referred to standard texts on differential equations for more discussion and examples. We will content ourselves here with one final example.

Example 2   Find one solution, in powers of x, of Bessel's equation of order $v = 1$, namely,

$$x^2 y'' + xy' + (x^2 - 1)y = 0.$$

Solution   We try

$$y = \sum_{n=0}^{\infty} a_n x^{\mu+n}$$

$$y' = \sum_{n=0}^{\infty} (\mu + n)a_n x^{\mu+n-1}$$

$$y'' = \sum_{n=0}^{\infty} (\mu + n)(\mu + n - 1)a_n x^{\mu+n-2}.$$

Substituting these expressions into the Bessel equation, we get

$$\sum_{n=0}^{\infty} \left[\left((\mu + n)(\mu + n - 1) + (\mu + n) - 1 \right)a_n x^n + a_n x^{n+2} \right] = 0$$

$$\sum_{n=0}^{\infty} \left[(\mu + n)^2 - 1 \right]a_n x^n + \sum_{n=2}^{\infty} a_{n-2} x^n = 0$$

$$(\mu^2 - 1)a_0 + \left((\mu + 1)^2 - 1 \right)a_1 x + \sum_{n=2}^{\infty} \left[\left((\mu + n)^2 - 1 \right)a_n + a_{n-2} \right]x^n = 0.$$

All of the terms must vanish. Since $a_0 \neq 0$ (we may take $a_0 = 1$) we obtain

$$\mu^2 - 1 = 0, \qquad\qquad\qquad \text{the indicial equation}$$

$$[(\mu + 1)^2 - 1]a_1 = 0,$$

$$a_n = -\frac{a_{n-2}}{(\mu + n)^2 - 1}, \quad (n \geq 2). \qquad \text{the recurrence relation}$$

Evidently $\mu = \pm 1$; therefore $a_1 = 0$. If we take $\mu = 1$, then the recurrence relation is $a_n = -a_{n-2}/(n)(n + 2)$. Thus,

$$a_3 = 0, \quad a_5 = 0, \quad a_7 = 0, \quad \cdots$$

$$a_2 = \frac{-1}{2 \times 4}, \quad a_4 = \frac{1}{2 \times 4 \times 4 \times 6}, \quad a_6 = \frac{-1}{2 \times 4 \times 4 \times 6 \times 6 \times 8}, \quad \cdots.$$

Again the pattern is obvious:

$$a_{2n} = \frac{(-1)^n}{2^{2n} n!(n + 1)!},$$

and one solution of the Bessel equation of order 1 is

$$y = \sum_{n=0}^{\infty} \frac{(-1)^n}{2^{2n} n!(n + 1)!} x^{2n+1}.$$

By the ratio test, this series converges for all x.

Remark   Observe that if we tried to calculate a second solution using $\mu = -1$ we would get the recurrence relation

$$a_n = -\frac{a_{n-2}}{n(n-2)},$$

and we would be unable to calculate a_2. This shows what can happen if the indicial roots differ by an integer.

Exercises 9.10

1. Find the general solution of $y'' = (x-1)^2 y$ in the form of a power series $y = \sum_{n=0}^{\infty} a_n (x-1)^n$.

2. Find the general solution of $y'' = xy$ in the form of a power series $y = \sum_{n=0}^{\infty} a_n x^n$ with a_0 and a_1 arbitrary.

3. Solve the initial-value problem

$$\begin{cases} y'' + xy' + 2y = 0 \\ y(0) = 1 \\ y'(0) = 2. \end{cases}$$

4. Find the solution of $y'' + xy' + y = 0$ that satisfies $y(0) = 1$ and $y'(0) = 0$.

5. Find the first three nonzero terms in a power series solution in powers of x for the initial-value problem $y'' + (\sin x)y = 0$, $y(0) = 1$, $y'(0) = 0$.

6. Find the solution, in powers of x, for the initial-value problem

$$(1-x^2)y'' - xy' + 9y = 0, \quad y(0) = 0, \quad y'(0) = 1.$$

7. Find two power series solutions in powers of x for $3xy'' + 2y' + y = 0$.

8. Find one power series solution for the Bessel equation of order $\nu = 0$, that is, the equation $xy'' + y' + xy = 0$.

Chapter Review

Key Ideas

• **What does it mean to say that the sequence $\{a_n\}$**

  ◇ is bounded above?   ◇ is ultimately positive?
  ◇ is alternating?   ◇ is increasing?
  ◇ converges?   ◇ diverges to infinity?

• **What does it mean to say that the series $\sum_{n=1}^{\infty} a_n$**

  ◇ converges?   ◇ diverges?
  ◇ is geometric?   ◇ is telescoping?
  ◇ is a p-series?   ◇ is positive?
  ◇ converges absolutely?   ◇ converges conditionally?

• **State the following convergence tests for series.**

  ◇ the integral test   ◇ the comparison test
  ◇ the limit comparison test ◇ the ratio test
  ◇ the alternating series test

• **How can you find bounds for the tail of a series?**
• **What is a bound for the tail of an alternating series?**
• **What do the following terms and phrases mean?**

  ◇ a power series   ◇ interval of convergence

  ◇ radius of convergence   ◇ centre of convergence
  ◇ a Taylor series   ◇ a Maclaurin series
  ◇ a Taylor polynomial   ◇ a binomial series
  ◇ an analytic function

• **Where is the sum of a power series differentiable?**
• **Where does the integral of a power series converge?**
• **Where is the sum of a power series continuous?**
• **State Taylor's Theorem with Lagrange remainder.**
• **State Taylor's Theorem with integral remainder.**
• **What is the binomial theorem?**

Review Exercises

In Exercises 1–4, determine whether the given sequence does or does not converge, and find its limit if it does converge.

1. $\left\{ \dfrac{(-1)^n e^n}{n!} \right\}$

2. $\left\{ \dfrac{n^{100} + 2^n \pi}{2^n} \right\}$

3. $\left\{ \dfrac{\ln n}{\tan^{-1} n} \right\}$

4. $\left\{ \dfrac{(-1)^n n^2}{\pi n(n-\pi)} \right\}$

5. Let $a_1 > \sqrt{2}$, and let

$$a_{n+1} = \frac{a_n}{2} + \frac{1}{a_n} \quad \text{for} \quad n = 1, 2, 3, \ldots$$

Show that $\{a_n\}$ is decreasing and that $a_n > \sqrt{2}$ for $n \geq 1$. Why must $\{a_n\}$ converge? Find $\lim_{n\to\infty} a_n$.

6. Find the limit of the sequence $\{\ln \ln(n + 1) - \ln \ln n\}$.

Evaluate the sums of the series in Exercises 7–10.

7. $\displaystyle\sum_{n=1}^{\infty} 2^{-(n-5)/2}$

8. $\displaystyle\sum_{n=0}^{\infty} \frac{4^{n-1}}{(\pi - 1)^{2n}}$

9. $\displaystyle\sum_{n=1}^{\infty} \frac{1}{n^2 - \frac{1}{4}}$

10. $\displaystyle\sum_{n=1}^{\infty} \frac{1}{n^2 - \frac{9}{4}}$

Determine whether the series in Exercises 11–16 converge or diverge. Give reasons for your answers.

11. $\displaystyle\sum_{n=1}^{\infty} \frac{n - 1}{n^3}$

12. $\displaystyle\sum_{n=1}^{\infty} \frac{n + 2^n}{1 + 3^n}$

13. $\displaystyle\sum_{n=1}^{\infty} \frac{n}{(1 + n)(1 + n\sqrt{n})}$

14. $\displaystyle\sum_{n=1}^{\infty} \frac{n^2}{(1 + 2^n)(1 + n\sqrt{n})}$

15. $\displaystyle\sum_{n=1}^{\infty} \frac{3^{2n+1}}{n!}$

16. $\displaystyle\sum_{n=1}^{\infty} \frac{n!}{(n + 2)! + 1}$

Do the series in Exercises 17–20 converge absolutely, converge conditionally, or diverge?

17. $\displaystyle\sum_{n=1}^{\infty} \frac{(-1)^{n-1}}{1 + n^3}$

18. $\displaystyle\sum_{n=1}^{\infty} \frac{(-1)^n}{2^n - n}$

19. $\displaystyle\sum_{n=10}^{\infty} \frac{(-1)^{n-1}}{\ln \ln n}$

20. $\displaystyle\sum_{n=1}^{\infty} \frac{n^2 \cos(n\pi)}{1 + n^3}$

For what values of x do the series in Exercises 21–22 converge absolutely? converge conditionally? diverge?

21. $\displaystyle\sum_{n=1}^{\infty} \frac{(x - 2)^n}{3^n \sqrt{n}}$

22. $\displaystyle\sum_{n=1}^{\infty} \frac{(5 - 2x)^n}{n}$

Determine the sums of the series in Exercises 23–24 to within 0.001.

23. $\displaystyle\sum_{n=1}^{\infty} \frac{1}{n^3}$

24. $\displaystyle\sum_{n=1}^{\infty} \frac{1}{4 + n^2}$

In Exercises 25–32, find Maclaurin series for the given functions. State where each series converges to the function.

25. $\dfrac{1}{3 - x}$

26. $\dfrac{x}{3 - x^2}$

27. $\ln(e + x^2)$

28. $\dfrac{1 - e^{-2x}}{x}$

29. $x \cos^2 x$

30. $\sin(x + (\pi/3))$

31. $(8 + x)^{-1/3}$

32. $(1 + x)^{1/3}$

Find Taylor series for the functions in Exercises 33–34 about the indicated points $x = c$.

33. $1/x$, $\quad c = \pi$

34. $\sin x + \cos x$, $\quad c = \pi/4$

Find the Maclaurin polynomial of the indicated degree for the functions in Exercises 35–38.

35. e^{x^2+2x}, degree 3

36. $\sin(1 + x)$, degree 3

37. $\cos(\sin x)$, degree 4

38. $\sqrt{1 + \sin x}$, degree 4

39. What function has Maclaurin series

$$1 - \frac{x}{2!} + \frac{x^2}{4!} - \cdots = \sum_{n=0}^{\infty} \frac{(-1)^n x^n}{(2n)!}?$$

40. A function $f(x)$ has Maclaurin series

$$1 + x^2 + \frac{x^4}{2^2} + \frac{x^6}{3^2} + \cdots = 1 + \sum_{n=1}^{\infty} \frac{x^{2n}}{n^2}.$$

Find $f^{(k)}(0)$ for all positive integers k.

Find the sums of the series in 41–44.

41. $\displaystyle\sum_{n=0}^{\infty} \frac{n + 1}{\pi^n}$

* 42. $\displaystyle\sum_{n=0}^{\infty} \frac{n^2}{\pi^n}$

43. $\displaystyle\sum_{n=1}^{\infty} \frac{1}{ne^n}$

* 44. $\displaystyle\sum_{n=2}^{\infty} \frac{(-1)^n \pi^{2n-4}}{(2n - 1)!}$

45. If $S(x) = \displaystyle\int_0^x \sin(t^2)\, dt$, find $\displaystyle\lim_{x\to 0} \frac{x^3 - 3S(x)}{x^7}$.

46. Use series to evaluate $\displaystyle\lim_{x\to 0} \frac{(x - \tan^{-1}x)(e^{2x} - 1)}{2x^2 - 1 + \cos(2x)}$.

47. How many nonzero terms in the Maclaurin series for e^{-x^4} are needed to evaluate $\int_0^{1/2} e^{-x^4}\, dx$ correct to 5 decimal places? Evaluate the integral to that accuracy.

48. Estimate the size of the error if the Taylor polynomial of degree 4 about $x = \pi/2$ for $f(x) = \ln \sin x$ is used to approximate $\ln \sin(1.5)$.

Challenging Problems

1. (**A refinement of the ratio test**) Suppose $a_n > 0$ and $a_{n+1}/a_n \geq n/n + 1$ for all n. Show that $\sum_{n=1}^{\infty} a_n$ diverges. *Hint:* $a_n \geq K/n$ for some constant K.

* 2. (**Summation by parts**) Let $\{u_n\}$ and $\{v_n\}$ be two sequences, and let $s_n = \sum_{k=1}^{n} v_k$.

(a) Show that $\sum_{k=1}^{n} u_k v_k = u_{n+1}s_n + \sum_{k=1}^{n} (u_k - u_{k+1})s_n$.
(*Hint:* write $v_n = s_n - s_{n-1}$, with $s_0 = 0$, and rearrange the sum.)

(b) If $\{u_n\}$ is positive, decreasing, and convergent to 0, and if $\{v_n\}$ has bounded partial sums, $|s_n| \leq K$ for all n, where K is a constant, show that $\sum_{n=1}^{\infty} u_n v_n$ converges. (*Hint:* show that the series $\sum_{n=1}^{\infty} (u_n - u_{n+1})s_n$ converges by comparing it to the telescoping series $\sum_{n=1}^{\infty} (u_n - u_{n+1})$.)

* 3. Show that $\sum_{n=1}^{\infty}(1/n)\sin(nx)$ converges for every x. *Hint:* if x is an integer multiple of π, all the terms in the series are 0 so there is nothing to prove. Otherwise, $\sin(x/2) \neq 0$. In this case show that

$$\sum_{n=1}^{N}\sin(nx) = \frac{\cos(x/2) - \cos((N+1/2)x)}{2\sin(x/2)}$$

using the identity

$$\sin a \sin b = \frac{\cos(a-b) - \cos(a+b)}{2}$$

to make the sum telescope. Then apply the result of Exercise 2(b) with $u_n = 1/n$ and $v_n = \sin(nx)$.

4. Let $a_1, a_2, a_3, \ldots$ be those positive integers that do not contain the digit 0 in their decimal representations. Thus $a_1 = 1$, $a_2 = 2, \ldots a_9 = 9$, $a_{10} = 11, \ldots a_{18} = 19$, $a_{19} = 21$, $\ldots a_{90} = 99$, $a_{91} = 111$, etc. Show that the series $\sum_{n=1}^{\infty}\frac{1}{a_n}$ converges and that the sum is less than 90. (*Hint:* How many of these integers have m digits? Each term $1/a_n$, where a_n has m digits, is less than 10^{-m+1}.)

* 5. (Using an integral to improve convergence) Recall the error formula for the Midpoint Rule, according to which

$$\int_{k-1/2}^{k+1/2} f(x)\,dx - f(k) = \frac{f''(c)}{24},$$

where $k - (1/2) \leq c \leq k + (1/2)$.

(a) If $f''(x)$ is a decreasing function of x, show that

$$f'(k+\tfrac{3}{2}) - f'(k+\tfrac{1}{2}) \leq f''(c) \leq f'(k-\tfrac{1}{2}) - f'(k-\tfrac{3}{2}).$$

(b) If (i) $f''(x)$ is a decreasing function of x,
  (ii) $\int_{N+1/2}^{\infty} f(x)\,dx$ converges, and (iii) $f'(x) \to 0$ as $x \to \infty$, show that

$$\frac{f'(N-\tfrac{1}{2})}{24} \leq \sum_{n=N+1}^{\infty} f(n) - \int_{N+1/2}^{\infty} f(x)\,dx \leq \frac{f'(N+\tfrac{3}{2})}{24}.$$

(c) Use the result of part (b) to approximate $\sum_{n=1}^{\infty} 1/n^2$ to within 0.001.

* 6. (The number e is irrational.) Start with $e = \sum_{n=0}^{\infty} 1/n!$.
(a) Use the technique of Example 7 in Section 9.3 to show that for any $n > 0$,

$$0 < e - \sum_{j=0}^{n}\frac{1}{j!} < \frac{1}{n!n}.$$

(Note that the sum here has $n + 1$ terms rather than n terms.)

(b) Suppose that e is a rational number, say $e = M/N$ for certain positive integers M and N. Show that

$$N!\left(e - \sum_{j=0}^{N}(1/j!)\right) \text{ is an integer.}$$

(c) Combine parts (a) and (b) to show that there is an integer between 0 and $1/N$. Why is this not possible? Conclude that e cannot be a rational number.

7. Let

$$f(x) = \sum_{k=0}^{\infty}\frac{2^{2k}k!}{(2k+1)!}x^{2k+1}$$

$$= x + \frac{2}{3}x^3 + \frac{4}{3\times 5}x^5 + \frac{8}{3\times 5\times 7}x^7 + \cdots.$$

(a) Find the radius of convergence of this power series.
(b) Show that $f'(x) = 1 + 2xf(x)$.
(c) What is $\dfrac{d}{dx}\left(e^{-x^2}f(x)\right)$?
(d) Express $f(x)$ in terms of an integral.

* 8. (The number π is irrational) Problem 6 above shows how to prove that e is irrational by assuming the contrary and deducing a contradiction. In this problem you will show that π is also irrational. The proof for π is also by contradiction but is rather more complicated, so it will be broken down into several parts.

(a) Let $f(x)$ be a polynomial, and let

$$g(x) = f(x) - f''(x) + f^{(4)}(x) - f^{(6)}(x) + \cdots$$

$$= \sum_{j=0}^{\infty}(-1)^j f^{(2j)}(x).$$

(Since f is a polynomial, all but a finite number of terms in the above sum are identically zero, so there are no convergence problems.) Verify that

$$\frac{d}{dx}\left(g'(x)\sin x - g(x)\cos x\right) = f(x)\sin x,$$

and hence that

$$\int_0^{\pi} f(x)\sin x\,dx = g(\pi) + g(0).$$

(b) Suppose that π is rational, say $\pi = m/n$, where m and n are positive integers. You will show that this leads to a contradiction and thus cannot be true. Choose a positive integer k such that $(\pi m)^k/k! < 1/2$. (Why is this possible?) Consider the polynomial

$$f(x) = \frac{x^k(m - nx)^k}{k!} = \frac{1}{k!}\sum_{j=0}^{k}\binom{k}{j}m^{k-j}(-n)^j x^{j+k}.$$

Show that $0 < f(x) < 1/2$ for $0 < x < \pi$, and hence that

$$0 < \int_0^\pi f(x) \sin x \, dx < 1.$$

Thus, $0 < g(\pi) + g(0) < 1$, where $g(x)$ is defined as in part (a).

(c) Show that the ith derivative of $f(x)$ is given by

$$f^{(i)}(x) = \frac{1}{k!} \sum_{j=0}^k \binom{k}{j} m^{k-j} (-n)^j \frac{(j+k)!}{(j+k-i)!} x^{j+k-i}.$$

(d) Show that $f^{(i)}(0)$ is an integer for $i = 0, 1, 2, \ldots$. (*Hint:* Observe for $i < k$ that $f^{(i)}(0) = 0$, and for $i > 2k$ that $f^{(i)}(x) = 0$ for all x. For $k \le i \le 2k$, show that only one term in the sum for $f^{(i)}(0)$ is not 0, and that this term is an integer. You will need the fact that the binomial coefficients $\binom{k}{j}$ are integers.)

(e) Show that $f(\pi - x) = f(x)$ for all x, and hence that $f^{(i)}(\pi)$ is also an integer for each $i = 0, 1, 2, \ldots$. Therefore, if $g(x)$ is defined as in (a), then $g(\pi) + g(0)$ is an integer. This contradicts the conclusion of part (b) and so shows that π cannot be rational.

* **9. (An asymptotic series)** Use integration by parts to show that

$$\int_0^x e^{-1/t} \, dt = e^{-1/x} \sum_{n=2}^N (-1)^n (n-1)! x^n$$

$$+ (-1)^{N+1} N! \int_0^x t^{N-1} e^{-1/t} \, dt.$$

Why can't you just use a Maclaurin series to approximate this integral? Using $N = 5$, find an approximate value for $\int_0^{0.1} e^{-1/t} \, dt$, and estimate the error. Estimate the error for $N = 10$ and $N = 20$.

Note that the series $\sum_{n=2}^\infty (-1)^n (n-1)! x^n$ **diverges** for any $x \ne 0$. This is an example of what is called an **asymptotic series**. Even though it diverges, a properly chosen partial sum gives a good approximation to our function when x is small.

Appendix I

Complex Numbers

Many of the problems to which mathematics is applied involve the solution of equations. Over the centuries the number system had to be expanded many times to provide solutions for more and more kinds of equations. The natural numbers

$$\mathbb{N} = \{1,\ 2,\ 3,\ 4,\ \ldots\}$$

are inadequate for the solutions of equations of the form

$$x + n = m, \qquad (m,\ n \in \mathbb{N}).$$

Zero and negative numbers can be added to create the integers

$$\mathbb{Z} = \{\ldots,\ -3,\ -2,\ -1,\ 0,\ 1,\ 2,\ 3,\ \ldots\}$$

in which that equation has the solution $x = m - n$ even if $m < n$. (Historically, this extension of the number system came much later than some of those mentioned below.) Some equations of the form

$$nx = m, \qquad (m,\ n \in \mathbb{Z}, \quad n \neq 0)$$

cannot be solved in the integers. Another extension is made to to include numbers of the form m/n, thus producing the set of rational numbers

$$\mathbb{Q} = \left\{ \frac{m}{n} \ :\ m,\ n \in \mathbb{Z}, \quad n \neq 0 \right\}.$$

Every linear equation

$$ax = b, \qquad (a,\ b \in \mathbb{Q}, \quad a \neq 0)$$

has a solution $x = b/a$ in $\mathbb{Q}$, but the quadratic equation

$$x^2 = 2$$

has no solution in $\mathbb{Q}$, as was shown in Section P.1. Another extension enriches the rational numbers to the real numbers $\mathbb{R}$ in which some equations like $x^2 = 2$ have solutions. However, other quadratic equations, for instance,

$$x^2 = -1$$

do not have solutions, even in the real numbers, so the extension process is not complete. In order to be able to solve any quadratic equation, we need to extend the real number system to a larger set, which we call **the complex number system**. In this appendix we will define complex numbers and develop some of their basic properties.

Definition of Complex Numbers

We begin by defining the symbol i, called **the imaginary unit**[1], to have the property

$$i^2 = -1.$$

Thus, we could also call i the **square root of** -1 and denote it $\sqrt{-1}$. Of course, i is not a real number; no real number has a negative square.

<table>
<tr><td>**DEFINITION**   **1**</td><td>

A **complex number** is an expression of the form

$$a + bi \qquad \text{or} \qquad a + ib$$

where a and b are *real numbers*, and i is the imaginary unit.

</td></tr>
</table>

For example, $3 + 2i$, $\frac{7}{2} - \frac{2}{3}i$, $i\pi = 0 + i\pi$, and $-3 = -3 + 0i$ are all complex numbers. The last of these examples shows that every real number can be regarded as a complex number. (We will normally use $a + bi$ unless b is a complicated expression, in which case we will write $a + ib$ instead. Either form is acceptable.)

It is often convenient to represent a complex number by a single letter; w and z are frequently used for this purpose. If a, b, x, and y are real numbers, and

$$w = a + bi \qquad \text{and} \qquad z = x + yi,$$

then we can refer to the complex numbers w and z. Note that $w = z$ if and only if $a = x$ and $b = y$. Of special importance are the complex numbers

$$0 = 0 + 0i, \qquad 1 = 1 + 0i, \qquad \text{and} \qquad i = 0 + 1i.$$

<table>
<tr><td>**DEFINITION**   **2**</td><td>

If $z = x + yi$ is a complex number (where x and y are real), we call x the **real part** of z and denote it Re (z). We call y the **imaginary part** of z and denote it Im (z):

$$\text{Re}\,(z) = \text{Re}\,(x + yi) = x, \qquad \text{Im}\,(z) = \text{Im}\,(x + yi) = y.$$

</td></tr>
</table>

Note that both the real and imaginary parts of a complex number are real numbers.

$$\text{Re}\,(3 - 5i) = 3 \qquad\qquad\qquad \text{Im}\,(3 - 5i) = -5$$
$$\text{Re}\,(2i) = \text{Re}\,(0 + 2i) = 0 \qquad \text{Im}\,(2i) = \text{Im}\,(0 + 2i) = 2$$
$$\text{Re}\,(-7) = \text{Re}\,(-7 + 0i) = -7 \qquad \text{Im}\,(-7) = \text{Im}\,(-7 + 0i) = 0.$$

Graphical Representation of Complex Numbers

Since complex numbers are constructed from pairs of real numbers (their real and imaginary parts), it is natural to represent complex numbers graphically as points in a Cartesian plane. We use the point with coordinates (a, b) to represent the complex number $w = a + ib$. In particular, the origin $(0, 0)$ represents the complex number 0, the point $(1, 0)$ represents the complex number $1 = 1 + 0i$, and the point $(0, 1)$ represents the point $i = 0 + 1i$. (See Figure I.1.)

---

[1]   In some fields, for example, electrical engineering, the imaginary unit is denoted j instead of i. Like "negative," "surd," and "irrational," the term "imaginary" suggests the distrust that greeted the new kinds of numbers when they were first introduced.

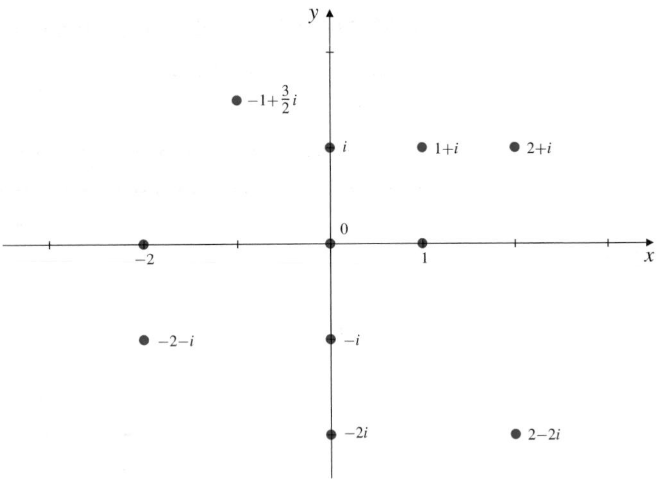

Figure I.1   An Argand diagram
representing the complex plane

Such a representation of complex numbers as points in a plane is called an **Argand diagram**. Since each complex number is represented by a unique point in the plane, the set of all complex numbers is often referred to as **the complex plane**. The symbol $\mathbb{C}$ is used to represent the set of all complex numbers and, equivalently, the complex plane:

$$\mathbb{C} = \{x + yi \; : \; x, \; y, \; \in \mathbb{R}\} \,.$$

The points on the x-axis of the complex plane correspond to real numbers ($x = x + 0i$), so the x-axis is called the **real axis**. The points on the y-axis correspond to **pure imaginary** numbers ($yi = 0 + yi$), so the y-axis is called the **imaginary axis**.

It can be helpful to use the *polar coordinates* of a point in the complex plane.

DEFINITION 3

The distance from the origin to the point (a, b) corresponding to the complex number $w = a + bi$ is called the **modulus** of w and is denoted by $|w|$ or $|a + bi|$:

$$|w| = |a + bi| = \sqrt{a^2 + b^2}.$$

If the line from the origin to (a, b) makes angle θ with the positive direction of the real axis (with positive angles measured counterclockwise), then we call θ an **argument** of the complex number $w = a + bi$ and denote it by $\arg(w)$ or $\arg(a + bi)$. (See Figure I.2.)

The modulus of a complex number is always real and nonnegative. It is positive unless the complex number is 0. Modulus plays a similar role for complex numbers that absolute value does for real numbers. Indeed, sometimes modulus is called absolute value.

Arguments of complex numbers are not unique. If $w = a + bi \neq 0$, then any two possible values for $\arg(w)$ differ by an integer multiple of 2π. The symbol $\arg(w)$ actually represents not a single number, but a set of numbers. When we write $\arg(w) = \theta$, we are saying that the set $\arg(w)$ contains all numbers of the form $\theta + 2k\pi$, where k is an integer. Similarly, the statement $\arg(z) = \arg(w)$ says that two sets are identical.

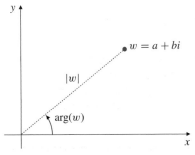

Figure I.2   The modulus and
argument of a complex number

If $w = a + bi$, where $a = \text{Re}\,(w) \neq 0$, then

$$\tan \arg\,(w) = \tan \arg\,(a + bi) = \frac{b}{a}.$$

This means that $\tan \theta = b/a$ for every θ in the set $\arg\,(w)$.

It is sometimes convenient to restrict $\theta = \arg\,(w)$ to an interval of length 2π, say, the interval $0 \leq \theta < 2\pi$, or $-\pi < \theta \leq \pi$, so that nonzero complex numbers will have unique arguments. We will call the value of $\arg\,(w)$ in the interval $-\pi < \theta \leq \pi$ the **principal argument** of w and denote it $\text{Arg}\,(w)$. Every complex number w except 0 has a unique principal argument $\text{Arg}\,(w)$.

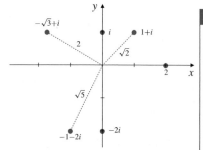

Figure I.3   Some complex numbers with their moduli

Example 1   **(Some moduli and principal arguments)** See Figure I.3.

$$|2| = 2 \qquad\qquad \text{Arg}\,(2) = 0$$
$$|1 + i| = \sqrt{2} \qquad\qquad \text{Arg}\,(1 + i) = \pi/4$$
$$|i| = 1 \qquad\qquad \text{Arg}\,(i) = \pi/2$$
$$|-2i| = 2 \qquad\qquad \text{Arg}\,(-2i) = -\pi/2$$
$$|-\sqrt{3} + i| = 2 \qquad\qquad \text{Arg}\,(-\sqrt{3} + i) = 5\pi/6$$
$$|-1 - 2i| = \sqrt{5} \qquad\qquad \text{Arg}\,(-1 - 2i) = -\pi + \tan^{-1}(2).$$

Remark   If $z = x + yi$ and $\text{Re}\,(z) = x > 0$, then $\text{Arg}\,(z) = \tan^{-1}(y/x)$. Many computer spreadsheets implement a two-variable arctan function denoted atan2(x, y) which gives the polar angle of (x, y) in the interval $]-\pi, \pi]$. Thus

$$\text{Arg}\,(x + yi) = \text{atan2}(x, y).$$

Given the modulus $r = |w|$ and any value of the argument $\theta = \arg\,(w)$ of a complex number $w = a + bi$, we have $a = r \cos \theta$ and $b = r \sin \theta$, so w can be expressed in terms of its modulus and argument as

$$w = r \cos \theta + i\, r \sin \theta.$$

The expression on the right side is called the **polar representation** of w.

DEFINITION   **4**

The **conjugate** or **complex conjugate** of a complex number $w = a + bi$ is another complex number, denoted $\overline{w}$, given by

$$\overline{w} = a - bi.$$

Example 2 $\overline{2 - 3i} = 2 + 3i, \qquad \overline{3} = 3, \qquad \overline{2i} = -2i.$

Observe that

$$\text{Re}\,(\overline{w}) = \text{Re}\,(w) \qquad\qquad |\overline{w}| = |w|$$
$$\text{Im}\,(\overline{w}) = -\text{Im}\,(w) \qquad\qquad \arg\,(\overline{w}) = -\arg\,(w).$$

In an Argand diagram the point $\overline{w}$ is the reflection of the point w in the real axis. (See Figure I.4.)

Note that w is real ($\text{Im}(w) = 0$) if and only if $\overline{w} = w$. Also, w is pure imaginary ($\text{Re}(w) = 0$) if and only if $\overline{w} = -w$. (Here, $-w = -a - bi$ if $w = a + bi$.)

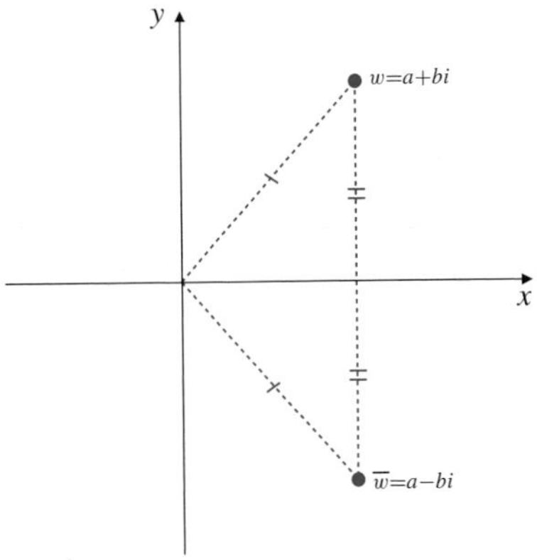

Complex Arithmetic

Like real numbers, complex numbers can be added, subtracted, multiplied, and divided. Two complex numbers are added or subtracted as though they are two-dimensional vectors whose components are their real and imaginary parts.

> **The sum and difference of complex numbers**
>
> If $w = a + bi$ and $z = x + yi$, where a, b, x, and y are real numbers, then
>
> $$w + z = (a + x) + (b + y)i$$
> $$w - z = (a - x) + (b - y)i.$$

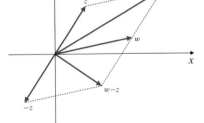

Figure I.4   A complex number and its conjugate are mirror images of each other in the real axis

In an Argand diagram the points $w + z$ and $w - z$ are the points whose position vectors are, respectively, the sum and difference of the position vectors of the points w and z. (See Figure I.5.) In particular, the complex number $a + bi$ is the sum of the real number $a = a + 0i$ and the pure imaginary number $bi = 0 + bi$.

Complex addition obeys the same rules as real addition: if w_1, w_2, and w_3 are three complex numbers, the following are easily verified:

$w_1 + w_2 = w_2 + w_1$	Addition is commutative.
$(w_1 + w_2) + w_3 = w_1 + (w_2 + w_3)$	Addition is associative.
$|w_1 \pm w_2| \le |w_1| + |w_2|$	the triangle inequality

Figure I.5   Complex numbers are added and subtracted vectorially. Observe the parallelograms

Note that $|w_1 - w_2|$ is the distance between the two points w_1 and w_2 in the complex plane. Thus, the triangle inequality says that in the triangle with vertices w_1, $\mp w_2$ and 0, the length of one side is less than the sum of the other two.

It is also easily verified that the conjugate of a sum (or difference) is the sum (or difference) of the conjugates:

$$\overline{w + z} = \overline{w} + \overline{z}.$$

Example 3

(a) If $w = 2 + 3i$ and $z = 4 - 5i$, then

$$w + z = (2 + 4) + (3 - 5)i = 6 - 2i$$
$$w - z = (2 - 4) + (3 - (-5))i = -2 + 8i.$$

(b) $3i + (1 - 2i) - (2 + 3i) + 5 = 4 - 2i$.

Multiplication of the complex numbers $w = a + bi$ and $z = x + yi$ is carried out by formally multiplying the binomial expressions and replacing i^2 by -1:

$$wz = (a + bi)(x + yi) = ax + ayi + bxi + byi^2$$
$$= (ax - by) + (ay + bx)i.$$

> **The product of complex numbers**
>
> If $w = a + bi$ and $z = x + yi$, where $a, b, x,$ and y are real numbers, then
>
> $$wz = (ax - by) + (ay + bx)i.$$

Example 4

(a) $(2 + 3i)(1 - 2i) = 2 - 4i + 3i - 6i^2 = 8 - i.$

(b) $i(5 - 4i) = 5i - 4i^2 = 4 + 5i.$

(c) $(a + bi)(a - bi) = a^2 - abi + abi - b^2i^2 = a^2 + b^2.$

Part (c) of the example above shows that the square of the modulus of a complex number is the product of that number with its complex conjugate:

$$w\,\overline{w} = |w|^2.$$

Complex multiplication has many properties in common with real multiplication. In particular, if $w_1, w_2,$ and w_3 are complex numbers, then

$$w_1 w_2 = w_2 w_1 \qquad \text{Multiplication is commutative.}$$
$$(w_1 w_2)w_3 = w_1(w_2 w_3) \qquad \text{Multiplication is associative.}$$
$$w_1(w_2 + w_3) = w_1 w_2 + w_1 w_3 \qquad \text{Multiplication distributes over addition.}$$

The conjugate of a product is the product of the conjugates:

$$\overline{wz} = \overline{w}\,\overline{z}.$$

To see this, let $w = a + bi$ and $z = x + yi$. Then

$$\overline{wz} = \overline{(ax - by) + (ay + bx)i}$$
$$= (ax - by) - (ay + bx)i$$
$$= (a - bi)(x - yi) = \overline{w}\,\overline{z}.$$

It is particularly easy to determine the product of complex numbers expressed in polar form. If

$$w = r(\cos\theta + i\,\sin\theta) \qquad \text{and} \qquad z = s(\cos\phi + i\,\sin\phi),$$

where $r = |w|$, $\theta = \arg(w)$, $s = |z|$, and $\phi = \arg(z)$, then

$$wz = rs(\cos\theta + i\,\sin\theta)(\cos\phi + i\,\sin\phi)$$
$$= rs\big((\cos\theta\cos\phi - \sin\theta\sin\phi) + i(\sin\theta\cos\phi + \cos\theta\sin\phi)\big)$$
$$= rs\big(\cos(\theta + \phi) + i\,\sin(\theta + \phi)\big).$$

(See Figure I.6.) Since arguments are only determined up to integer multiples of 2π, we have proved that

> **The modulus and argument of a product**
>
> $$|wz| = |w||z| \qquad \text{and} \qquad \arg(wz) = \arg(w) + \arg(z).$$

The second of these equations says that the set $\arg(wz)$ consists of all numbers $\theta + \phi$, where θ belongs to the set $\arg(w)$ and ϕ to the set $\arg(z)$.

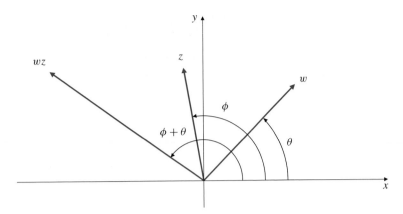

Figure I.6   The argument of a product is the sum of the arguments of the factors

More generally, if w_1, w_2, ... w_n are complex numbers, then

$$|w_1 w_2 \cdots w_n| = |w_1||w_2| \cdots |w_n|$$
$$\arg(w_1 w_2 \cdots w_n) = \arg(w_1) + \arg(w_2) + \cdots + \arg(w_n).$$

Multiplication of a complex number by i has a particularly simple geometric interpretation in an Argand diagram. Since $|i| = 1$ and $\arg(i) = \pi/2$, multiplication of $w = a + bi$ by i leaves the modulus of w, unchanged but increases its argument by $\pi/2$. (See Figure I.7.) Thus, multiplication by i rotates the position vector of w counterclockwise by $90°$ about the origin.

Let $z = \cos\theta + i\sin\theta$. Then $|z| = 1$ and $\arg(z) = \theta$. Since the modulus of a product is the product of the moduli of the factors and the argument of a product is the sum of the arguments of the factors, we have $|z^n| = |z|^n = 1$ and $\arg(z^n) = n\arg(z) = n\theta$. Thus,

$$z^n = \cos n\theta + i\sin n\theta,$$

and we have proved

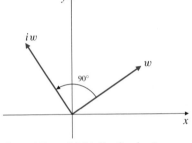

Figure I.7   Multiplication by i corresponds to counterclockwise rotation by $90°$

THEOREM **1**   **de Moivre's Theorem**

$$\left(\cos\theta + i\sin\theta\right)^n = \cos n\theta + i\sin n\theta.$$

Remark   The study of complex-valued functions of a complex variable is beyond the scope of this book. However, we point out that there is a complex version of the exponential function having the following property: if $z = x + iy$ (where x and y are real), then

$$e^z = e^{x+iy} = e^x e^{iy} = e^x(\cos y + i\sin y).$$

Thus the modulus of e^z is $e^{\text{Re}(z)}$ and $\text{Im}(z)$ is a value of $\arg(e^z)$. In this context, de Moivre's Theorem just says

$$(e^{i\theta})^n = e^{in\theta}.$$

Example 5   Express $(1 + i)^5$ in the form $a + bi$.

Solution   Since $|(1 + i)^5| = |1 + i|^5 = (\sqrt{2})^5 = 4\sqrt{2}$, and
$\arg\left((1 + i)^5\right) = 5 \arg(1 + i) = \dfrac{5\pi}{4}$, we have

$$(1 + i)^5 = 4\sqrt{2}\left(\cos\frac{5\pi}{4} + i\sin\frac{5\pi}{4}\right) = 4\sqrt{2}\left(-\frac{1}{\sqrt{2}} - \frac{1}{\sqrt{2}}i\right) = -4 - 4i.$$

de Moivre's Theorem can be used to generate trigonometric identities for multiples of an angle. For example, for $n = 2$ we have

$$\cos 2\theta + i\sin 2\theta = \left(\cos\theta + i\sin\theta\right)^2 = \cos^2\theta - \sin^2\theta + 2i\cos\theta\sin\theta.$$

Thus, $\cos 2\theta = \cos^2\theta - \sin^2\theta$, and $\sin 2\theta = 2\sin\theta\cos\theta$.

The **reciprocal** of the nonzero complex number $w = a + bi$ can be calculated by multiplying the numerator and denominator of the reciprocal expression by the conjugate of w:

$$w^{-1} = \frac{1}{w} = \frac{1}{a + bi} = \frac{a - bi}{(a + bi)(a - bi)} = \frac{a - bi}{a^2 + b^2} = \frac{\overline{w}}{|w|^2}.$$

Since $|\overline{w}| = |w|$, and $\arg(\overline{w}) = -\arg(w)$, we have

$$\left|\frac{1}{w}\right| = \frac{|\overline{w}|}{|w|^2} = \frac{1}{|w|} \qquad \text{and} \qquad \arg\left(\frac{1}{w}\right) = -\arg(w).$$

The **quotient** z/w of two complex numbers $z = x + yi$ and $w = a + bi$ is the product of z and $1/w$, so

$$\frac{z}{w} = \frac{z\overline{w}}{|w|^2} = \frac{(x + yi)(a - bi)}{a^2 + b^2} = \frac{xa + yb + i(ya - xb)}{a^2 + b^2}.$$

We have

> **The modulus and argument of a quotient**
>
> $$\left|\frac{z}{w}\right| = \frac{|z|}{|w|} \qquad \text{and} \qquad \arg\left(\frac{z}{w}\right) = \arg(z) - \arg(w).$$

The set $\arg(z/w)$ consists of all numbers $\theta - \phi$ where θ belongs to the set $\arg(z)$ and ϕ to the set $\arg(w)$.

Example 6   Simplify   (a) $\dfrac{2 + 3i}{4 - i}$ and   (b) $\dfrac{i}{1 + i\sqrt{3}}$.

Solution

(a) $\dfrac{2 + 3i}{4 - i} = \dfrac{(2 + 3i)(4 + i)}{(4 - i)(4 + i)} = \dfrac{8 - 3 + (2 + 12)i}{4^2 + 1^2} = \dfrac{5}{17} + \dfrac{14}{17}i.$

(b) $\dfrac{i}{1+i\sqrt{3}} = \dfrac{i(1-i\sqrt{3})}{(1+i\sqrt{3})(1-i\sqrt{3})} = \dfrac{\sqrt{3}+i}{1^2+3} = \dfrac{\sqrt{3}}{4} + \dfrac{1}{4}i.$

Alternatively, since $|1+i\sqrt{3}| = 2$ and $\arg(1+i\sqrt{3}) = \tan^{-1}\sqrt{3} = \dfrac{\pi}{3}$, the quotient in (b) has modulus $\dfrac{1}{2}$ and argument $\dfrac{\pi}{2} - \dfrac{\pi}{3} = \dfrac{\pi}{6}$. Thus

$$\dfrac{i}{1+i\sqrt{3}} = \dfrac{1}{2}\left(\cos\dfrac{\pi}{6} + i\sin\dfrac{\pi}{6}\right) = \dfrac{\sqrt{3}}{4} + \dfrac{1}{4}i.$$

Roots of Complex Numbers

If a is a positive real number, there are two distinct real numbers whose square is a. These are usually denoted

$\sqrt{a}$ (the positive square root of a and)

$-\sqrt{a}$ (the negative square root of a).

Every nonzero complex number $z = x + yi$ (where $x^2 + y^2 > 0$) also has two square roots; if w_1 is a complex number such that $w_1^2 = z$, then $w_2 = -w_1$ also satisfies $w_2^2 = z$. Again, we would like to single out one of these roots and call it $\sqrt{z}$.

Let $r = |z|$, so that $r > 0$. Let $\theta = \text{Arg}(z)$. Thus $-\pi < \theta \le \pi$. Since

$$z = r\big(\cos\theta + i\sin\theta\big),$$

the complex number

$$w = \sqrt{r}\left(\cos\dfrac{\theta}{2} + i\sin\dfrac{\theta}{2}\right)$$

clearly satisfies $w^2 = z$. We call this w the **principal square root** of z and denote it $\sqrt{z}$. The two solutions of the equation $w^2 = z$ are, thus, $w = \sqrt{z}$ and $w = -\sqrt{z}$. Observe that the real part of $\sqrt{z}$ is always nonnegative since $\cos(\theta/2) \ge 0$ for $-\pi/2 < \theta \le \pi/2$. In this interval $\sin(\theta/2) = 0$ only if $\theta = 0$ in which case $\sqrt{z}$ is real and positive.

Example 7

(a) $\sqrt{4} = \sqrt{4(\cos 0 + i\sin 0)} = 2.$

(b) $\sqrt{i} = \sqrt{1\left(\cos\dfrac{\pi}{2} + i\sin\dfrac{\pi}{2}\right)} = \cos\dfrac{\pi}{4} + i\sin\dfrac{\pi}{4} = \dfrac{1}{\sqrt{2}} + \dfrac{1}{\sqrt{2}}i.$

(c) $\sqrt{-4i} = \sqrt{4\left[\cos\left(-\dfrac{\pi}{2}\right) + i\sin\left(-\dfrac{\pi}{2}\right)\right]} = 2\left[\cos\left(-\dfrac{\pi}{4}\right) + i\sin\left(-\dfrac{\pi}{4}\right)\right]$
$= \sqrt{2} - i\sqrt{2}.$

(d) $\sqrt{-\dfrac{1}{2} + i\dfrac{\sqrt{3}}{2}} = \sqrt{\cos\dfrac{2\pi}{3} + i\sin\dfrac{2\pi}{3}} = \cos\dfrac{\pi}{3} + i\sin\dfrac{\pi}{3} = \dfrac{1}{2} + \dfrac{\sqrt{3}}{2}i.$

Given a nonzero complex number z we can find n distinct complex numbers w that satisfy $w^n = z$. These n numbers are called nth roots of z. For example, if $z = 1 = \cos 0 + i \sin 0$, then each of the numbers

$$w_1 = 1$$

$$w_2 = \cos \frac{2\pi}{n} + i \sin \frac{2\pi}{n}$$

$$w_3 = \cos \frac{4\pi}{n} + i \sin \frac{4\pi}{n}$$

$$w_4 = \cos \frac{6\pi}{n} + i \sin \frac{6\pi}{n}$$

$$\vdots$$

$$w_n = \cos \frac{2(n-1)\pi}{n} + i \sin \frac{2(n-1)\pi}{n}$$

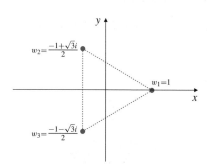

Figure I.8   The cube roots of unity

satisfies $w^n = 1$ so is an nth root of 1. (These numbers are usually called the nth roots of unity.) Figure I.8 shows the three cube roots of 1. Observe that they are at the three vertices of an equilateral triangle with centre at the origin and one vertex at 1. In general, the n nth roots of unity lie on a circle of radius 1 centred at the origin, at the vertices of a regular n-sided polygon with one vertex at 1.

If z is any nonzero complex number, and θ is the principal argument of z $(-\pi < \theta \le \pi)$, then the number

$$w_1 = |z|^{1/n} \left(\cos \frac{\theta}{n} + i \sin \frac{\theta}{n} \right)$$

is called the **principal** nth root of z. All the nth roots of z are on the circle of radius $|z|^{1/n}$ centred at the origin and are at the vertices of a regular n-sided polygon with one vertex at w_1. (See Figure I.9.) The other nth roots are

$$w_2 = |z|^{1/n} \left(\cos \frac{\theta + 2\pi}{n} + i \sin \frac{\theta + 2\pi}{n} \right)$$

$$w_3 = |z|^{1/n} \left(\cos \frac{\theta + 4\pi}{n} + i \sin \frac{\theta + 4\pi}{n} \right)$$

$$\vdots$$

$$w_n = |z|^{1/n} \left(\cos \frac{\theta + 2(n-1)\pi}{n} + i \sin \frac{\theta + 2(n-1)\pi}{n} \right).$$

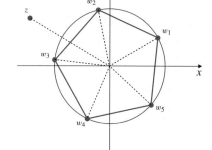

Figure I.9   The five 5th roots of z

We can obtain all n of the nth roots of z by multiplying the principal nth root by the nth roots of unity.

Example 8   Find the 4th roots of -4. Sketch them in an Argand diagram.

Solution   Since $|-4|^{1/4} = \sqrt{2}$ and $\arg(-4) = \pi$, the principal 4th root of -4 is

$$w_1 = \sqrt{2} \left(\cos \frac{\pi}{4} + i \sin \frac{\pi}{4} \right) = 1 + i.$$

The other three 4th roots are at the vertices of a square with centre at the origin and one vertex at $1 + i$. (See Figure I.10.) Thus the other roots are

$$w_2 = -1 + i, \qquad w_3 = -1 - i, \qquad w_4 = 1 - i.$$

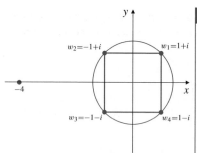

Figure I.10   The four 4th roots of -4

Exercises

In Exercises 1–4, find the real and imaginary parts (Re (z) and Im (z)) of the given complex numbers z, and sketch the position of each number in the complex plane (i.e., in an Argand diagram).

1. $z = -5 + 2i$ **2.** $z = 4 - i$

3. $z = -\pi i$ **4.** $z = -6$

In Exercises 5–15, find the modulus $r = |z|$ and the principal argument $\theta = \text{Arg}\,(z)$ of each given complex number z, and express z in terms of r and θ.

5. $z = -1 + i$ **6.** $z = -2$

7. $z = 3i$ **8.** $z = -5i$

9. $z = 1 + 2i$ **10.** $z = -2 + i$

11. $z = -3 - 4i$ **12.** $z = 3 - 4i$

13. $z = \sqrt{3} - i$ **14.** $z = -\sqrt{3} - 3i$

15. $z = 3\cos\dfrac{4\pi}{5} + 3i\sin\dfrac{4\pi}{5}$

16. If $\text{Arg}\,(z) = 3\pi/4$ and $\text{Arg}\,(w) = \pi/2$, find $\text{Arg}\,(zw)$.

17. If $\text{Arg}\,(z) = -5\pi/6$ and $\text{Arg}\,(w) = \pi/4$, find $\text{Arg}\,(z/w)$.

In Exercises 18–23, express in the form $z = x + yi$ the complex number z whose modulus and argument are given.

18. $|z| = 2$, $\text{arg}\,(z) = \pi$ **19.** $|z| = 5$, $\text{arg}\,(z) = \tan^{-1}\dfrac{3}{4}$

20. $|z| = 1$, $\text{arg}\,(z) = \dfrac{3\pi}{4}$ **21.** $|z| = \pi$, $\text{arg}\,(z) = \dfrac{\pi}{6}$

22. $|z| = 0$, $\text{arg}\,(z) = 1$ **23.** $|z| = \dfrac{1}{2}$, $\text{arg}\,(z) = -\dfrac{\pi}{3}$

In Exercises 24–27, find the complex conjugates of the given complex numbers.

24. $5 + 3i$ **25.** $-3 - 5i$

26. $4i$ **27.** $2 - i$

Describe geometrically (or make a sketch of) the set of points z in the complex plane satisfying the given equations or inequalities in Exercises 28–33.

28. $|z| = 2$ **29.** $|z| \le 2$

30. $|z - 2i| \le 3$ **31.** $|z - 3 + 4i| \le 5$

32. $\text{arg}\,z = \dfrac{\pi}{3}$ **33.** $\pi \le \text{arg}\,(z) \le \dfrac{7\pi}{4}$

Simplify the expressions in Exercises 34–43.

34. $(2 + 5i) + (3 - i)$ **35.** $i - (3 - 2i) + (7 - 3i)$

36. $(4 + i)(4 - i)$ **37.** $(1 + i)(2 - 3i)$

38. $(a + bi)(\overline{2a - bi})$ **39.** $(2 + i)^3$

40. $\dfrac{2 - i}{2 + i}$ **41.** $\dfrac{1 + 3i}{2 - i}$

42. $\dfrac{1 + i}{i(2 + 3i)}$ **43.** $\dfrac{(1 + 2i)(2 - 3i)}{(2 - i)(3 + 2i)}$

44. Prove that $\overline{z + w} = \overline{z} + \overline{w}$.

45. Prove that $\overline{\left(\dfrac{z}{w}\right)} = \dfrac{\overline{z}}{\overline{w}}$.

46. Express each of the complex numbers $z = 3 + i\sqrt{3}$ and $w = -1 + i\sqrt{3}$ in polar form (i.e., in terms of its modulus and argument). Use these expressions to calculate zw and z/w.

47. Repeat the previous exercise for $z = -1 + i$ and $w = 3i$.

48. Use de Moivre's Theorem to find a trigonometric identity for $\cos 3\theta$ in terms of $\cos\theta$ and one for $\sin 3\theta$ in terms of $\sin\theta$.

49. Describe the solutions, if any, of the equations (a) $\overline{z} = 2/z$ and (b) $\overline{z} = -2/z$.

50. For positive real numbers a and b it is always true that $\sqrt{ab} = \sqrt{a}\sqrt{b}$. Does a similar identity hold for $\sqrt{zw}$, where z and w are complex numbers? *Hint:* consider $z = w = -1$.

51. Find the three cube roots of -1.

52. Find the three cube roots of $-8i$.

53. Find the three cube roots of $-1 + i$.

54. Find all the fourth roots of 4.

55. Find all complex solutions of the equation $z^4 + 1 - i\sqrt{3} = 0$.

56. Find all solutions of $z^5 + a^5 = 0$, where a is a positive real number.

* 57. Show that the sum of the n nth roots of unity is zero. *Hint:* show that these roots are all powers of the principal root.

Appendix II

Continuous Functions

The development of calculus depends in an essential way on the concept of limit of a function and thereby on properties of the real number system. In Chapter 1 we presented these notions in an intuitive way and did not attempt to prove them except in Section 1.5, where the *formal* definition of limit was given and used to verify some elementary limits and prove some simple properties of limits.

Many of the results on limits and continuity of functions stated in Chapter 1 may seem quite obvious; most students and users of calculus are not bothered by applying them without proof. Nevertheless, mathematics is a highly logical and rigorous discipline, and any statement, however obvious, that cannot be proved by strictly logical arguments from acceptable assumptions must be considered suspect. In this appendix we build upon the formal definition of limit given in Section 1.5, and combine it with the notion of *completeness* of the real number system first encountered in Section P.1 to give formal proofs of the very important results about continuous functions stated in Theorems 8 and 9 of Section 1.4, the Max-Min Theorem and the Intermediate-Value Theorem. Most of our development of calculus in this book depends essentially on these two theorems.

The branch of mathematics that deals with proofs such as these is called mathematical analysis. This subject is usually not pursued by students in introductory calculus courses but is postponed to higher years and studied by students in majors or honours programs in mathematics. It is hoped that some of this material will be of value to honours-level calculus courses and individual students with a deeper interest in understanding calculus.

Limits of Functions

At the heart of mathematical analysis is the formal definition of limit, Definition 9 in Section 1.5, which we restate as follows:

The formal definition of limit

We say that $\lim_{x \to a} f(x) = L$ if for every positive number ϵ there exists a positive number δ, depending on ϵ (i.e., $\delta = \delta(\epsilon)$), such that

$$0 < |x - a| < \delta \implies |f(x) - L| < \epsilon.$$

Section 1.5 was marked "optional" because understanding the material presented there was not essential for learning calculus. However, that material is an *essential* prerequisite for this appendix. It is highly recommended that you go back to Section 1.5 and read it carefully, paying special attention to Examples 2 and 4, and attempt at least Exercises 31–36. These exercises provide proofs for the standard laws of limits stated in Section 1.2.

Continuous Functions

Consider the following definitions of continuity, which are equivalent to those given in Section 1.4.

DEFINITION 1

Continuity of a function at a point

A function f, defined on an open interval containing the point a, is said to be continuous at the point a if

$$\lim_{x \to a} f(x) = f(a),$$

that is, if for every $\epsilon > 0$ there exists $\delta > 0$ such that if $|x - a| < \delta$, then $|f(x) - f(a)| < \epsilon$.

DEFINITION 2

Continuity of a function on an interval

A function f is continuous on an interval if it is continuous at every point of that interval. In the case of an endpoint of a closed interval, f need only be continuous on one side. Thus, f is continuous on the interval $[a, b]$ if

$$\lim_{t \to x} f(t) = f(x)$$

for each x satisfying $a < x < b$, and

$$\lim_{t \to a+} f(t) = f(a) \qquad \text{and} \qquad \lim_{t \to b-} f(t) = f(b).$$

These concepts are illustrated in Figure II.1.

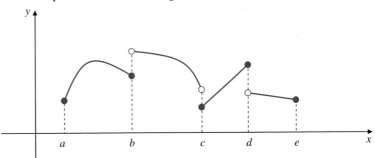

Figure II.1 f is continuous on the intervals $[a, b]$, $]b, c[$, $[c, d]$, and $]d, e]$

Some important results about continuous functions are collected in Theorems 6 and 7 of Section 1.4, which we restate here:

THEOREM 1

Combining continuous functions

(a) If f and g are continuous at the point a, then so are $f + g$, $f - g$, fg, and, if $g(a) \neq 0$, f/g.

(b) If f is continuous at the point L and if $\lim_{x \to a} g(x) = L$, then we have

$$\lim_{x \to a} f\big(g(x)\big) = f(L) = f\Big(\lim_{x \to a} g(x)\Big).$$

In particular, if g is continuous at the point a (so that $L = g(a)$), then $\lim_{x \to a} f(g(x)) = f(g(a))$, that is, $f \circ g(x) = f(g(x))$ is continuous at $x = a$.

(c) The functions $f(x) = C$ (constant) and $g(x) = x$ are continuous on the whole real line.

(d) For any rational number r the function $f(x) = x^r$ is continuous at every real number where it is defined.

PROOF   Part (a) is just a restatement of various rules for combining limits; for example,

$$\lim_{x \to a} f(x)g(x) = (\lim_{x \to a} f(x))(\lim_{x \to a} g(x)) = f(a)g(a).$$

Part (b) can be proved as follows. Let $\epsilon > 0$ be given. Since f is continuous at L, there exists $k > 0$ such that $|f(g(x)) - f(L)| < \epsilon$ whenever $|g(x) - L| < k$. Since $\lim_{x \to a} g(x) = L$, there exists $\delta > 0$ such that if $0 < |x - a| < \delta$, then $|g(x) - L| < k$. Hence, if $0 < |x - a| < \delta$, then $|f(g(x)) - f(L)| < \epsilon$, and $\lim_{x \to a} f(g(x)) = f(L)$.

The proofs of (c) and (d) are left to the student in Exercises 3–9 at the end of this appendix.

Completeness and Sequential Limits

DEFINITION 3

A real number u is said to be an **upper bound** for a nonempty set S of real numbers if $x \le u$ for every x in S.
The number u^* is called the **least upper bound** of S if u^* is an upper bound for S and $u^* \le u$ for every upper bound u of S.
Similarly, ℓ is a **lower bound** for S if $\ell \le x$ for every x in S. The number ℓ^* is the **greatest lower bound** of S if ℓ^* is a lower bound and $\ell \le \ell^*$ for every lower bound ℓ of S.

Example 1   Set $S_1 = [2, 3]$ and $S_2 =]2, \infty[$. Any number $u \ge 3$ is an upper bound for S_1. S_2 has no upper bound; we say that it is not bounded above. The least upper bound of S_1 is 3. Any real number $\ell \le 2$ is a lower bound for both S_1 and S_2. $\ell^* = 2$ is the greatest lower bound for each set. Note that the least upper bound and greatest lower bound of a set may or may not belong to that set.

We now recall the completeness axiom for the real number system, which we discussed briefly in Section P.1.

The completeness axiom for the real numbers

A nonempty set of real numbers that has an upper bound must have a least upper bound.
Equivalently, a nonempty set of real numbers having a lower bound must have a greatest lower bound.

We stress that this is an *axiom* to be assumed without proof. It cannot be deduced from the more elementary algebraic and order properties of the real numbers. These other properties are shared by the rational numbers, a set that is not complete. The completeness axiom is essential for the proof of the most important results about continuous functions, in particular, for the Max-Min Theorem and the Intermediate-Value Theorem. Before attempting these proofs, however, we must develop a little more machinery.

In Section 9.1 we stated a version of the completeness axiom that pertains to *sequences* of real numbers; specifically, that an increasing sequence that is bounded above converges to a limit. We begin by verifying that this follows from the version stated above. (Both statements are, in fact, equivalent.) As noted in Section 9.1, the sequence

$$\{x_n\} = \{x_1, \ x_2, \ x_3, \ \ldots\}$$

is a function on the positive integers, that is, $x_n = x(n)$. We say that the sequence converges to the limit L, and we write $\lim x_n = L$, if the corresponding function $x(t)$ satisfies $\lim_{t \to \infty} x(t) = L$ as defined above. More formally,

DEFINITION 4

> **Limit of a sequence**
>
> We say that $\lim x_n = L$ if for every positive number ϵ there exists a positive number $N = N(\epsilon)$ such that $|x_n - L| < \epsilon$ holds whenever $n \geq N$.

THEOREM 2

If $\{x_n\}$ is an increasing sequence that is bounded above, that is,

$$x_{n+1} \geq x_n \qquad \text{and} \qquad x_n \leq K \qquad \text{for } n = 1, \ 2, \ 3, \ \ldots,$$

then $\lim x_n = L$ exists. (Equivalently, if $\{x_n\}$ is decreasing and bounded below, then $\lim x_n$ exists.)

PROOF   Let $\{x_n\}$ be increasing and bounded above. The set S of real numbers x_n has an upper bound, K, and so has a least upper bound, say L. Thus $x_n \leq L$ for every n, and if $\epsilon > 0$, then there exists a positive integer N such that $x_N > L - \epsilon$. (Otherwise, $L - \epsilon$ would be an upper bound for S that is lower than the least upper bound.) If $n \geq N$, then we have $L - \epsilon < x_N \leq x_n \leq L$, so $|x_n - L| < \epsilon$. Thus $\lim x_n = L$. The proof for a decreasing sequence that is bounded below is similar.

THEOREM 3

If $a \leq x_n \leq b$ for each n, and if $\lim x_n = L$, then $a \leq L \leq b$.

PROOF   Suppose that $L > b$. Let $\epsilon = L - b$. Since $\lim x_n = L$, there exists n such that $|x_n - L| < \epsilon$. Thus $x_n > L - \epsilon = L - (L - b) = b$, which is a contradiction since we are given that $x_n \leq b$. Thus $L \leq b$. A similar argument shows that $L \geq a$.

THEOREM 4

If f is continuous on $[a, b]$, if $a \leq x_n \leq b$ for each n, and if $\lim x_n = L$, then $\lim f(x_n) = f(L)$.

The proof is similar to that of Theorem 1(b), and is left as Exercise 15 at the end of this appendix.

Continuous Functions on a Closed, Finite Interval

We are now in a position to prove the main results about continuous functions on closed, finite intervals.

THEOREM 5

The Boundedness Theorem

If f is continuous on $[a, b]$, then f is bounded there; that is, there exists a constant K such that $|f(x)| \leq K$ if $a \leq x \leq b$.

PROOF   We show that f is bounded above; a similar proof shows that f is bounded below. For each positive integer n let S_n be the set of points x in $[a, b]$ such that $f(x) > n$:

$$S_n = \{x : a \leq x \leq b \quad \text{and} \quad f(x) > n\}.$$

We would like to show that S_n is empty for some n. It would then follow that $f(x) \leq n$ for all x in $[a, b]$; that is, n would be an upper bound for f on $[a, b]$.

Suppose, to the contrary, that S_n is nonempty for every n. We will show that this leads to a contradiction. Since S_n is bounded below (a is a lower bound), by completeness S_n has a greatest lower bound; call it x_n. (See Figure II.2.) Evidently $a \leq x_n$. Since $f(x) > n$ at some point of $[a, b]$ and f is continuous at that point, $f(x) > n$ on some interval contained in $[a, b]$. Hence $x_n < b$. It follows that $f(x_n) \geq n$. (If $f(x_n) < n$, then by continuity $f(x) < n$ for some distance to the right of x_n, and x_n could not be the greatest lower bound of S_n.)

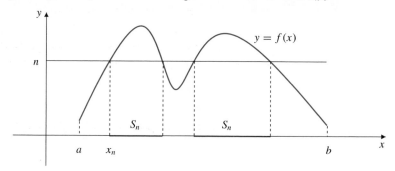

Figure II.2   The set S_n

For each n we have $S_{n+1} \subset S_n$. Therefore, $x_{n+1} \geq x_n$ and $\{x_n\}$ is an increasing sequence. Being bounded above (b is an upper bound) this sequence converges, by Theorem 2. Let $\lim x_n = L$. By Theorem 3, $a \leq L \leq b$. Since f is continuous at L, $\lim f(x_n) = f(L)$ exists by Theorem 4. But since $f(x_n) \geq n$, $\lim f(x_n)$ cannot exist. This contradiction completes the proof.

THEOREM 6

The Max-Min Theorem

If f is continuous on $[a, b]$, then there are points v and u in $[a, b]$ such that for any x in $[a, b]$ we have

$$f(v) \leq f(x) \leq f(u);$$

that is, f assumes maximum and minimum values on $[a, b]$.

PROOF   By Theorem 5 we know that the set $S = \{f(x) : a \le x \le b\}$ has an upper bound and, therefore, by the completeness axiom, a least upper bound. Call this least upper bound M. Suppose that there exists no point u in $[a, b]$ such that $f(u) = M$. Then by Theorem 1(a), $1/(M - f(x))$ is continuous on $[a, b]$. By Theorem 5, there exists a constant K such that $1/(M - f(x)) \le K$ for all x in $[a, b]$. Thus $f(x) \le M - 1/K$, which contradicts the fact that M is the *least* upper bound for the values of f. Hence, there must exist some point u in $[a, b]$ such that $f(u) = M$. Since M is an upper bound for the values of f on $[a, b]$, we have $f(x) \le f(u) = M$ for all x in $[a, b]$.

The proof that there must exist a point v in $[a, b]$ such that $f(x) \ge f(v)$ for all x in $[a, b]$ is similar.

THEOREM 7

The Intermediate-Value Theorem

If f is continuous on $[a, b]$ and s is a real number lying between the numbers $f(a)$ and $f(b)$, then there exists a point c in $[a, b]$ such that $f(c) = s$.

PROOF   To be specific, we assume that $f(a) < s < f(b)$. (The proof for the case $f(a) > s > f(b)$ is similar.) Let $S = \{x : a \le x \le b \text{ and } f(x) \le s\}$. S is nonempty (a belongs to S) and bounded above (b is an upper bound), so by completeness S has a least upper bound; call it c.

Suppose that $f(c) > s$. Then $c \ne a$ and, by continuity, $f(x) > s$ on some interval $]c - \delta, c]$ where $\delta > 0$. But this says $c - \delta$ is an upper bound for S lower than the least upper bound, which is impossible. Thus $f(c) \le s$.

Suppose $f(c) < s$. Then $c \ne b$ and, by continuity, $f(x) < s$ on some interval of the form $[c, c + \delta[$ for some $\delta > 0$. But this says that $[c, c + \delta[\subset S$, which contradicts the fact that c is an upper bound for S. Hence we cannot have $f(c) < s$. Therefore, $f(c) = s$.

For more discussion of these theorems, and some applications, see Section 1.4.

Exercises

1. Let $a < b < c$ and suppose that $f(x) \le g(x)$ for $a \le x \le c$. If $\lim_{x \to b} f(x) = L$ and $\lim_{x \to b} g(x) = M$, prove that $L \le M$. *Hint:* assume that $L > M$ and deduce that $f(x) > g(x)$ for all x sufficiently near b. This contradicts the condition that $f(x) \le g(x)$ for $a \le x \le b$.

2. If $f(x) \le K$ on the intervals $[a, b)$ and $(b, c]$, and if $\lim_{x \to b} f(x) = L$, prove that $L \le K$.

3. Use the formal definition of limit to prove that $\lim_{x \to 0+} x^r = 0$ for any positive, rational number r.

Prove the assertions in Exercises 4–9.

4. $f(x) = C$ (constant) and $g(x) = x$ are both continuous on the whole real line.

5. Every polynomial is continuous on the whole real line.

6. A rational function (quotient of polynomials) is continuous everywhere except where the denominator is 0.

7. If n is a positive integer and $a > 0$, then $f(x) = x^{1/n}$ is continuous at $x = a$.

8. If $r = m/n$ is a rational number, then $g(x) = x^r$ is continuous at every point $a > 0$.

9. If $r = m/n$, where m and n are integers and n is odd, show that $g(x) = x^r$ is continuous at every point $a < 0$. If $r \ge 0$, show that g is continuous at 0 also.

10. Prove that $f(x) = |x|$ is continuous on the real line.

Use the definitions from Chapter 3 for the functions in Exercises 11–14 to show that these functions are continuous on their respective domains.

11. $\sin x$

12. $\cos x$

13. $\ln x$

14. e^x

15. Prove Theorem 4.

16. Suppose that every function that is continuous and bounded on $[a, b]$ must assume a maximum value and a minimum value on that interval. Without using Theorem 5, prove that every function f that is continuous on $[a, b]$ must be bounded on that interval. *Hint:* show that $g(t) = t/(1 + |t|)$ is continuous and increasing on the real line. Then consider $g(f(x))$.

Appendix III

The Riemann Integral

In Section 5.3 we defined the definite integral $\int_a^b f(x)\, dx$ of a function f that is continuous on the finite, closed interval $[a, b]$. The integral was defined as a kind of "limit" of Riemann sums formed by partitioning the interval $[a, b]$ into small subintervals. In this appendix we will reformulate the definition of the integral so that it can be used for functions that are not necessarily continuous; in the following discussion we assume only that f is **bounded** on $[a, b]$. Later we will prove Theorem 2 of Section 5.3, which asserts that any continuous function is integrable.

Recall that a **partition** P of $[a, b]$ is a finite, ordered set of points $P = \{x_0, x_1, x_2, \ldots, x_n\}$, where $a = x_0 < x_1 < x_2 < \cdots < x_{n-1} < x_n = b$. Such a partition subdivides $[a, b]$ into n subintervals $[x_0, x_1], [x_1, x_2], \ldots, [x_{n-1}, x_n]$, where $n = n(P)$ depends on the partition. The length of the jth subinterval $[x_{j-1}, x_j]$ is $\Delta x_j = x_j - x_{j-1}$.

Suppose that the function f is bounded on $[a, b]$. Given any partition P, the n sets $S_j = \{f(x) : x_{j-1} \le x \le x_j\}$ have least upper bounds M_j and greatest lower bounds m_j, $(1 \le j \le n)$, so that

$$m_j \le f(x) \le M_j \qquad \text{on} \qquad [x_{j-1}, x_j].$$

We define upper and lower Riemann sums for f corresponding to the partition P to be

$$U(f, P) = \sum_{j=1}^{n(P)} M_j \Delta x_j \qquad \text{and}$$

$$L(f, P) = \sum_{j=1}^{n(P)} m_j \Delta x_j.$$

(See Figure III.1.) Note that if f is continuous on $[a, b]$, then m_j and M_j are, in fact, the minimum and maximum values of f over $[x_{j-1}, x_j]$ (by Theorem 6 of Appendix II); that is, $m_j = f(l_j)$ and $M_j = f(u_j)$, where $f(l_j) \le f(x) \le f(u_j)$ for $x_{j-1} \le x \le x_j$.

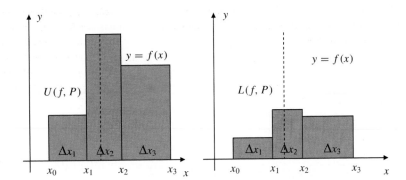

Figure III.1 Upper and lower sums corresponding to the partition $P = \{x_0, x_1, x_2, x_3\}$

If P is any partition of $[a, b]$ and we create a new partition P^* by adding new subdivision points to those of P, thus subdividing the subintervals of P into smaller ones, then we call P^* a **refinement** of P.

THEOREM 1

If P^* is a refinement of P, then $L(f, P^*) \geq L(f, P)$ and $U(f, P^*) \leq U(f, P)$.

PROOF If S and T are sets of real numbers, and $S \subset T$, then any lower bound (or upper bound) of T is also a lower bound (or upper bound) of S. Hence, the greatest lower bound of S is at least as large as that of T, and the least upper bound of S is no greater than that of T.

Let P be a given partition of $[a, b]$ and form a new partition P' by adding one subdivision point to those of P, say the point k dividing the jth subinterval $[x_{j-1}, x_j]$ of P into two subintervals $[x_{j-1}, k]$ and $[k, x_j]$. (See Figure III.2.) Let m_j, m'_j, and m''_j be the greatest lower bounds of the sets of values of $f(x)$ on the intervals $[x_{j-1}, x_j]$, $[x_{j-1}, k]$, and $[k, x_j]$, respectively. Then $m_j \leq m'_j$ and $m_j \leq m''_j$. Thus $m_j(x_j - x_{j-1}) \leq m'_j(k - x_{j-1}) + m''_j(x_j - k)$, so $L(f, P) \leq L(f, P')$.

If P^* is a refinement of P, it can be obtained by adding one point at a time to those of P and thus $L(f, P) \leq L(f, P^*)$. We can prove that $U(f, P) \geq U(f, P^*)$ in a similar manner.

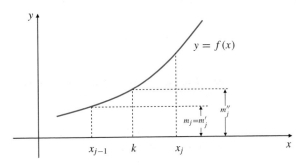

Figure III.2 Adding one point to a partition

THEOREM 2

If P and P' are any two partitions of $[a, b]$, then $L(f, P) \leq U(f, P')$.

PROOF Combine the subdivision points of P and P' to form a new partition P^*, which is a refinement of both P and P'. Then by Theorem 1,

$$L(f, P) \leq L(f, P^*) \leq U(f, P^*) \leq U(f, P').$$

No lower sum can exceed any upper sum.

Theorem 2 shows that the set of values of $L(f, P)$ for fixed f and various partitions P of $[a, b]$ is a bounded set; any upper sum is an upper bound for this set. By completeness, the set has a least upper bound, which we shall denote I_*. Thus, $L(f, P) \leq I_*$ for any partition P. Similarly, there exists a greatest lower bound I^* for the set of values of $U(f, P)$ corresponding to different partitions P. It follows that $I_* \leq I^*$. (See Exercise 4 at the end of this appendix.)

DEFINITION 1

> **The Riemann integral**
>
> If f is bounded on $[a, b]$ and $I_* = I^*$, then we say that f is **Riemann integrable**, or simply **integrable** on $[a, b]$, and denote by
>
> $$\int_a^b f(x)\, dx = I_* = I^*$$
>
> the **(Riemann) integral** of f on $[a, b]$.

The following theorem provides a convenient test for determining whether a given bounded function is integrable:

THEOREM 3

The bounded function f is integrable on $[a, b]$ if and only if for every positive number ϵ there exists a partition P of $[a, b]$ such that $U(f, P) - L(f, P) < \epsilon$.

PROOF   Suppose that for every $\epsilon > 0$ there exists a partition P of $[a, b]$ such that $U(f, P) - L(f, P) < \epsilon$, then

$$I^* \leq U(f, P) < L(f, P) + \epsilon \leq I_* + \epsilon.$$

Since $I^* < I_* + \epsilon$ must hold for every $\epsilon > 0$, it follows that $I^* \leq I_*$. Since we already know that $I^* \geq I_*$, we have $I^* = I_*$ and f is integrable on $[a, b]$.

Conversely, if $I^* = I_*$ and $\epsilon > 0$ are given, we can find a partition P' such that $L(f, P') > I_* - \epsilon/2$, and another partition P'' such that $U(f, P'') < I^* + \epsilon/2$. If P is a common refinement of P' and P'', then by Theorem 1 we have that $U(f, P) - L(f, P) \leq U(f, P'') - L(f, P') < (\epsilon/2) + (\epsilon/2) = \epsilon$, as required.

Example 1   Let $f(x) = \begin{cases} 0 & \text{if } 0 \leq x < 1 \text{ or } 1 < x \leq 2 \\ 1 & \text{if } x = 1. \end{cases}$

Show that f is integrable on $[0, 2]$ and find $\int_0^2 f(x)\, dx$.

Solution   Let $\epsilon > 0$ be given. Let $P = \{0,\ 1-\epsilon/3,\ 1+\epsilon/3,\ 2\}$. Then $L(f, P) = 0$ since $f(x) = 0$ at points of each of these subintervals into which P subdivides $[0, 2]$. (See Figure III.3.) Since $f(1) = 1$, we have

$$U(f, P) = 0\left(1 - \frac{\epsilon}{3}\right) + 1\left(\frac{2\epsilon}{3}\right) + 0\left(2 - \left(1 + \frac{\epsilon}{3}\right)\right) = \frac{2\epsilon}{3}.$$

Hence, $U(f, P) - L(f, P) < \epsilon$ and f is integrable on $[0, 2]$. Since $L(f, P) = 0$ for every partition, $\int_0^2 f(x)\, dx = I_* = 0$.

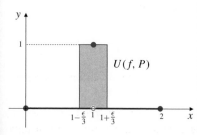

Figure III.3

Example 2   Let $f(x)$ be defined on $[0, 1]$ by

$$f(x) = \begin{cases} 1 & \text{if } x \text{ is rational} \\ 0 & \text{if } x \text{ is irrational.} \end{cases}$$

Show that f is not integrable on $[0, 1]$.

PROOF   Every subinterval of $[0, 1]$ having positive length contains both rational and irrational numbers. Hence, for any partition P of $[0, 1]$ we have $L(f, P) = 0$ and $U(f, P) = 1$. Thus $I_* = 0$ and $I^* = 1$, so f is not integrable on $[0, 1]$. ∎

Uniform Continuity

When we assert that a function f is continuous on the interval $[a, b]$, we imply that for every x in that interval and every $\epsilon > 0$, we can find a positive number δ (depending on *both* x and ϵ) such that $|f(y) - f(x)| < \epsilon$ whenever $|y - x| < \delta$ and y lies in $[a, b]$. In fact, however, it is possible to find a number δ *depending only on* ϵ such that $|f(y) - f(x)| < \epsilon$ holds whenever x and y belong to $[a, b]$ and satisfy $|y - x| < \delta$. We describe this phenomenon by saying that f is **uniformly continuous** on the interval $[a, b]$.

THEOREM 4

If f is continuous on the closed, finite interval $[a, b]$, then f is uniformly continuous on that interval.

PROOF   Let $\epsilon > 0$ be given. Define numbers x_n in $[a, b]$ and subsets S_n of $[a, b]$ as follows:

$$x_1 = a$$
$$S_1 = \left\{ x : x_1 < x \leq b \text{ and } |f(x) - f(x_1)| \geq \frac{\epsilon}{3} \right\}.$$

If S_1 is empty, stop; otherwise, let

$$x_2 = \text{ the greatest lower bound of } S_1$$
$$S_2 = \left\{ x : x_2 < x \leq b \text{ and } |f(x) - f(x_2)| \geq \frac{\epsilon}{3} \right\}.$$

If S_2 is empty, stop; otherwise, proceed to define x_3 and S_3 analogously. We proceed in this way as long as we can; if x_n and S_n have been defined and S_n is not empty, we define

$$x_{n+1} = \text{ the greatest lower bound of } S_n$$
$$S_{n+1} = \left\{ x : x_{n+1} < x \leq b \text{ and } |f(x) - f(x_{n+1})| \geq \frac{\epsilon}{3} \right\}.$$

At any stage where S_n is not empty, the continuity of f at x_n assures us that $x_{n+1} > x_n$ and $|f(x_{n+1}) - f(x_n)| = \epsilon/3$.

We must consider two possibilities for the above procedure: either S_n is empty for some n, or S_n is nonempty for every n.

Suppose S_n is nonempty for every n. Then we have constructed an infinite, increasing sequence $\{x_n\}$ in $[a, b]$ that, being bounded above (by b), must have a limit by completeness (Theorem 2 of Appendix II). Let $\lim x_n = x^*$. We have $a \leq x^* \leq b$. Since f is continuous at x^*, there exists $\delta > 0$ such that

$|f(x) - f(x^*)| < \epsilon/8$ whenever $|x - x^*| < \delta$ and x lies in $[a, b]$. Since $\lim x_n = x^*$, there exists a positive integer N such that $|x_n - x^*| < \delta$ whenever $n \geq N$. For such n we have

$$\frac{\epsilon}{3} = |f(x_{n+1}) - f(x_n)| = |f(x_{n+1}) - f(x^*) + f(x^*) - f(x_n)|$$
$$\leq |f(x_{n+1}) - f(x^*)| + |f(x_n) - f(x^*)|$$
$$< \frac{\epsilon}{8} + \frac{\epsilon}{8} = \frac{\epsilon}{4},$$

which is clearly impossible. Thus S_n must, in fact, be empty for some n.

Suppose that S_N is empty. Thus, S_n is nonempty for $n < N$, and the procedure for defining x_n stops with x_N. Since S_{N-1} is not empty, $x_N < b$. In this case define $x_{N+1} = b$ and let

$$\delta = \min\{x_2 - x_1, \ x_3 - x_2, \ \ldots, \ x_{N+1} - x_N\}.$$

The minimum of a finite set of positive numbers is a positive number, so $\delta > 0$. If x lies in $[a, b]$, then x lies in one of the intervals $[x_1, x_2], [x_2, x_3], \ldots, [x_N, x_{N+1}]$. Suppose x lies in $[x_k, x_{k+1}]$. If y is in $[a, b]$ and $|y - x| < \delta$, then y lies in either the same subinterval as x or in an adjacent one; that is, y lies in $[x_j, x_{j+1}]$, where $j = k - 1, k,$ or $k + 1$. Thus,

$$|f(y) - f(x)| = |f(y) - f(x_j) + f(x_j) - f(x_k) + f(x_k) - f(x)|$$
$$\leq |f(y) - f(x_j)| + |f(x_j) - f(x_k)| + |f(x_k) - f(x)|$$
$$< \frac{\epsilon}{3} + \frac{\epsilon}{3} + \frac{\epsilon}{3} = \epsilon,$$

which was to be proved.

We are now in a position to prove that a continuous function is integrable.

THEOREM 5   If f is continuous on $[a, b]$, then f is integrable on $[a, b]$.

PROOF   By Theorem 4, f is uniformly continuous on $[a, b]$. Let $\epsilon > 0$ be given. Let $\delta > 0$ be such that $|f(x) - f(y)| < \epsilon/(b - a)$ whenever $|x - y| < \delta$ and x and y belong to $[a, b]$. Choose a partition $P = \{x_0, x_1, \ldots, x_n\}$ of $[a, b]$ for which each subinterval $[x_{j-1}, x_j]$ has length $\Delta x_j < \delta$. Then the greatest lower bound, m_j, and the least upper bound, M_j, of the set of values of $f(x)$ on $[x_{j-1}, x_j]$ satisfy $M_j - m_j < \epsilon/(b - a)$. Accordingly,

$$U(f, P) - L(f, P) < \frac{\epsilon}{b - a} \sum_{j=1}^{n(P)} \Delta x_j = \frac{\epsilon}{b - a} (b - a) = \epsilon.$$

Thus f is integrable on $[a, b]$, as asserted.

Exercises

1. Let $f(x) = \begin{cases} 1 & \text{if } 0 \le x \le 1 \\ 0 & \text{if } 1 < x \le 2 \end{cases}$. Prove that f is integrable on $[0, 2]$ and find the value of $\int_0^2 f(x)\,dx$.

2. Let $f(x) = \begin{cases} 1 & \text{if } x = 1/n, \quad n = 1, 2, 3, \ldots \\ 0 & \text{for all other values of } x \end{cases}$.

Show that f is integrable over $[0, 1]$ and find the value of the integral $\int_0^1 f(x)\,dx$.

* **3.** Let $f(x) = 1/n$ if $x = m/n$, where m, n are integers having no common factors, and let $f(x) = 0$ if x is an irrational number. Thus, $f(1/2) = 1/2$, $f(1/3) = f(2/3) = 1/3$, $f(1/4) = f(3/4) = 1/4$, etc. Show that f is integrable on $[0, 1]$ and find $\int_0^1 f(x)\,dx$. *Hint:* show that for any $\epsilon > 0$, only finitely many points of the graph of f over $[0, 1]$ lie above the line $y = \epsilon$.

4. Prove that I_* and I^* defined in the paragraph following Theorem 2 satisfy $I_* \le I^*$ as claimed there.

5. Prove parts (c), (d), (e), (f), (g), and (h) of Theorem 3 in Section 5.4 for the Riemann integral.

6. Use the definition of uniform continuity given in the paragraph preceding Theorem 4 to prove that $f(x) = \sqrt{x}$ is uniformly continuous on $[0, 1]$. Do not use Theorem Theorem 4 itself.

7. Show directly from the definition of uniform continuity (without using Theorem 5 of Appendix II) that a function f uniformly continuous on a closed, finite interval is necessarily bounded there.

8. If f is bounded and integrable on $[a, b]$, prove that $F(x) = \int_a^x f(t)\,dt$ is uniformly continuous on $[a, b]$. (If f were continuous, we would have a stronger result; F would be differentiable on (a, b) and $F'(x) = f(x)$ (which is the Fundamental Theorem of Calculus).)

Appendix IV

Differential Equations

Introduction A **differential equation** (or **DE**) is an equation that involves one or more derivatives of an unknown function. Solving the differential equation means finding a function (or every such function) that satisfies the differential equation.

Many physical laws and relationships between quantities studied in various scientific disciplines are expressed mathematically as differential equations. For example, Newton's Second Law of Motion ($F = ma$) states that the position $x(t)$ at time t of an object of constant mass m subjected to a force $F(t)$ must satisfy the differential equation (equation of motion):

$$m\frac{d^2x}{dt^2} = F(t).$$

Similarly, the biomass $m(t)$ at time t of a bacterial culture growing in a uniformly supporting medium changes at a rate proportional to the biomass:

$$\frac{dm}{dt} = km(t),$$

which is the differential equation of exponential growth (or, if $k < 0$, exponential decay). Because differential equations arise so extensively in the abstract modelling of concrete phenomena, such equations and techniques for solving them are at the heart of applied mathematics. Indeed, most of the existing mathematical literature is either directly involved with differential equations or is motivated by problems arising in the study of such equations. Because of this, various differential equations, terms for their description, and techniques for their solution are introduced throughout *Calculus: A Complete Course*. This appendix provides some introductory background not covered elsewhere in the book. However, students of mathematics and its applications usually take one or more full courses on differential equations, and even then hardly scratch the surface of the subject.

Classifying Differential Equations

Differential equations are classified in several ways. The most significant classification is based on the number of variables with respect to which derivatives appear in the equation. An **ordinary differential equation (ODE)** is one that involves derivatives with respect to only one variable. Both of the examples given above are ordinary differential equations. A **partial differential equation (PDE)** is one that involves partial derivatives of the unknown function with respect to more than one variable. For example, the **one-dimensional wave equation** of Section 12.4

$$\frac{\partial^2 u}{\partial t^2} = c^2 \frac{\partial^2 u}{\partial x^2}$$

models the lateral displacement $u(x, t)$ at position x at time t of a stretched vibrating string. We will not discuss partial differential equations in this appendix.

Differential equations are also classified with respect to **order**. The order of a differential equation is the order of the highest-order derivative present in the equation. The one-dimensional wave equation is a second-order PDE. The following example records the order of two ODEs.

Example 1	$\dfrac{d^2 y}{dx^2} + x^3 y = \sin x$	has order 2,
	$\dfrac{d^3 y}{dx^3} + 4x \left(\dfrac{dy}{dx} \right)^2 = y \dfrac{d^2 y}{dx^2} + e^y$	has order 3.

Like any equation, a differential equation can be written in the form $F = 0$, where F is a function. For an ODE, the function F can depend on the independent variable (usually called x or t), the unknown function (usually y), and any derivatives of the unknown function up to the order of the equation. For instance, an nth-order ODE can be written in the form

$$F(x, y, y', y'', \dots, y^{(n)}) = 0.$$

Linear ODEs

An important special class of differential equations consists of those that are **linear**. An nth-order linear ODE has the form

$$a_n(x)y^{(n)}(x) + a_{n-1}(x)y^{(n-1)}(x) + \cdots$$
$$+ a_2(x)y''(x) + a_1(x)y'(x) + a_0(x)y(x) = f(x),$$

or, more simply,

$$P_n(D)y(x) = f(x),$$

where $P_n(D)$ is the nth-order differential operator

$$P_n(D) = a_n(x)D^n + a_{n-1}(x)D^{n-1} + \cdots + a_2(x)D^2 + a_1(x)D + a_0(x)$$

obtained by substituting the differential operator $D = d/dx$ for the variable r in the nth-degree polynomial

$$P_n(r) = a_n(x)r^n + a_{n-1}(x)r^{n-1} + \cdots + a_2(x)r^2 + a_1(x)r + a_0(x),$$

having coefficients depending on the variable x. It is often useful to write linear DEs in terms of differential operators in this way.

Each term in the expression on the left side of the linear DE is the product of a *coefficient* that is a function of x and a second factor that is either y or one of the derivatives of y. The term $f(x)$ on the right does not depend on y; it is called the **nonhomogeneous term**.

A linear ODE is said to be **homogeneous** if all of its terms involve the unknown function y, that is, if $f(x)$ is identically zero. If $f(x)$ is not identically zero, the equation is **nonhomogeneous**.

Example 2   The first DE in Example 1,

$$\frac{d^2y}{dx^2} + x^3 y = \sin x,$$

is linear and nonhomogeneous. Here, the coefficients are $a_2(x) = 1$, $a_1(x) = 0$, and $a_0(x) = x^3$, and the nonhomogeneous term is $f(x) = \sin x$. Although it can be written in the form

$$\frac{d^3y}{dx^3} + 4x\left(\frac{dy}{dx}\right)^2 - y\frac{d^2y}{dx^2} - e^y = 0,$$

the second equation in Example 1 is *not linear* (we say it is **nonlinear**) because the second term involves the square of a derivative of y, the third term involves the product of y and one of its derivatives, and the fourth term is not y times a function of x. The equation

$$(1 + x^2)\frac{d^3y}{dx^3} + \sin x \frac{d^2y}{dx^2} - 4\frac{dy}{dx} + y = 0$$

is a linear equation of order 3. The coefficients are $a_3(x) = 1 + x^2$, $a_2(x) = \sin x$, $a_1(x) = -4$, and $a_0(x) = 1$. Since $f(x) = 0$, this equation is *homogeneous*.

The following theorem states that any *linear combination* of solutions of a linear, homogeneous DE is also a solution. This is an extremely important fact about linear, homogeneous DEs.

THEOREM **1**   If $y = y_1(x)$ and $y = y_2(x)$ are two solutions of the linear, homogeneous DE

$$a_n y^{(n)} + a_{n-1} y^{(n-1)} + \cdots + a_2 y'' + a_1 y' + a_0 y = 0,$$

then so is the linear combination

$$y = A y_1(x) + B y_2(x)$$

for any values of the constants A and B.

PROOF   We are given that

$$a_n y_1^{(n)} + a_{n-1} y_1^{(n-1)} + \cdots + a_2 y_1'' + a_1 y_1' + a_0 y_1 = 0 \qquad \text{and}$$

$$a_n y_2^{(n)} + a_{n-1} y_2^{(n-1)} + \cdots + a_2 y_2'' + a_1 y_2' + a_0 y_2 = 0.$$

Multiplying the first equation by A and the second by B and adding the two gives

$$a_n (A y_1^{(n)} + B y_2^{(n)}) + a_{n-1} (A y_1^{(n-1)} + B y_2^{(n-1)})$$
$$+ \cdots + a_2 (A y_1'' + B y_2'') + a_1 (A y_1' + B y_2') + a_0 (A y_1 + B y_2) = 0.$$

Thus, $y = A y_1(x) + B y_2(x)$ is also a solution of the equation.

The same kind of proof can be used to verify the following theorem.

THEOREM 2

If $y = y_1(x)$ is a solution of the linear, homogeneous equation

$$a_n y^{(n)} + a_{n-1} y^{(n-1)} + \cdots + a_2 y'' + a_1 y' + a_0 y = 0$$

and $y = y_2(x)$ is a solution of the linear, nonhomogeneous equation

$$a_n y^{(n)} + a_{n-1} y^{(n-1)} + \cdots + a_2 y'' + a_1 y' + a_0 y = f(x),$$

then $y = y_1(x) + y_2(x)$ is also a solution of the same linear, nonhomogeneous equation.

We made extensive use of these two facts when we discussed second-order linear equations with constant coefficients in Section 3.7.

First-Order ODEs

We have discussed techniques for solving several kinds of first-order DEs in various sections of this book:

- Equations of the form $\dfrac{dy}{dx} = f(x)$ were discussed in Section 2.10.

- Equations of the form $\dfrac{dy}{dx} = f(x)g(y)$ (called **separable equations**) were discussed in Section 7.9

- Equations of the form $\dfrac{dy}{dx} + p(x)y = q(x)$ (which are **linear** and **nonhomogeneous**) were also treated in Section 7.9.

Unfortunately, the term *homogeneous* is used in more than one way in the study of differential equations. Certain first-order ODEs that are not necessarily linear are called homogeneous for a different reason than the one applying for linear equations above. A first-order DE of the form

$$\frac{dy}{dx} = f\left(\frac{y}{x}\right)$$

is said to be **homogeneous** because y/x and, therefore, $g(x, y) = f(y/x)$ are *homogeneous of degree 0* in the sense described in Section 12.5. Such a homogeneous equation can be transformed into a separable equation (and therefore solved) by means of a change of dependent variable. If we set

$$v = \frac{y}{x}, \qquad \text{or, equivalently,} \qquad y = xv(x),$$

then we have

$$\frac{dy}{dx} = v + x\frac{dv}{dx},$$

and the original differential equation transforms into

$$\frac{dv}{dx} = \frac{f(v) - v}{x},$$

which is separable.

Example 3   Solve the equation

$$\frac{dy}{dx} = \frac{x^2 + xy}{xy + y^2}.$$

Solution   The equation is homogeneous. (Divide the numerator and denominator of the right-hand side by x^2 to see this.) If $y = vx$, the equation becomes

$$v + x\frac{dv}{dx} = \frac{1 + v}{v + v^2} = \frac{1}{v},$$

or

$$x\frac{dv}{dx} = \frac{1 - v^2}{v}.$$

Separating variables and integrating, we calculate

$$\int \frac{v\,dv}{1 - v^2} = \int \frac{dx}{x} \qquad \text{Let } u = 1 - v^2.$$

$$-\frac{1}{2}\int \frac{du}{u} = \int \frac{dx}{x}$$

$$-\ln|u| = 2\ln|x| + C_1 = \ln C_2 x^2 \qquad\qquad (C_1 = \ln C_2).$$

$$\frac{1}{|u|} = C_2 x^2$$

$$|1 - v^2| = \frac{C_3}{x^2} \qquad (C_3 = 1/C_2).$$

$$\left|1 - \frac{y^2}{x^2}\right| = \frac{C_3}{x^2}.$$

The solution is best expressed in the form $x^2 - y^2 = C_4$. However, near points where $y \neq 0$, the equation can be solved for y as a function of x.

Exact Equations

A first-order differential equation expressed in differential form as

$$M(x, y)\,dx + N(x, y)\,dy = 0,$$

which is equivalent to $\dfrac{dy}{dx} = -\dfrac{M(x, y)}{N(x, y)}$, is said to be **exact** if the left-hand side is the differential of a function $\phi(x, y)$:

$$d\phi(x, y) = M(x, y)\, dx + N(x, y)\, dy.$$

The function ϕ is called an **integral function** of the differential equation. The level curves $\phi(x, y) = C$ of ϕ are the **solution curves** of the differential equation. For example, the differential equation

$$x\, dx + y\, dy = 0$$

has solution curves given by

$$x^2 + y^2 = C$$

since $d(x^2 + y^2) = 2(x\, dx + y\, dy) = 0$.

Remark   The condition that the differential equation $M\, dx + N\, dy = 0$ should be exact is just the condition that the vector field

$$\mathbf{F} = M(x, y)\,\mathbf{i} + N(x, y)\,\mathbf{j}$$

should be *conservative*; the integral function of the differential equation is then the potential function of the vector field. (See Section 15.2.)

A **necessary condition** for the exactness of the DE $M\, dx + N\, dy = 0$ is that

$$\frac{\partial M}{\partial y} = \frac{\partial N}{\partial x};$$

this just says that the mixed partial derivatives $\dfrac{\partial^2 \phi}{\partial x \partial y}$ and $\dfrac{\partial^2 \phi}{\partial y \partial x}$ of the integral function ϕ must be equal.

Once you know that an equation is exact, you can often guess the integral function. In any event, ϕ can always be found by the same method used to find the potential of a conservative vector field in Section 15.2.

Example 4   Verify that the DE

$$(2x + \sin y - ye^{-x})\, dx + (x \cos y + \cos y + e^{-x})\, dy = 0$$

is exact and find its solution curves.

Solution   Here, $M = 2x + \sin y - ye^{-x}$ and $N = x \cos y + \cos y + e^{-x}$. Since

$$\frac{\partial M}{\partial y} = \cos y - e^{-x} = \frac{\partial N}{\partial x},$$

the DE is exact. We want to find ϕ so that

$$\frac{\partial \phi}{\partial x} = M = 2x + \sin y - ye^{-x} \quad \text{and} \quad \frac{\partial \phi}{\partial y} = N = x \cos y + \cos y + e^{-x}.$$

Integrate the first equation with respect to x, being careful to allow the constant of integration to depend on y:

$$\phi(x, y) = \int (2x + \sin y - ye^{-x})\, dx = x^2 + x \sin y + ye^{-x} + C_1(y).$$

Now substitute this expression into the second equation:

$$x \cos y + \cos y + e^{-x} = \frac{\partial \phi}{\partial y} = x \cos y + e^{-x} + C_1'(y).$$

Thus $C_1'(y) = \cos y$, and $C_1(y) = \sin y + C_2$. (It is because the original DE was exact that the equation for $C_1'(y)$ turned out to be independent of x; this had to happen or we could not have found C_1 as a function of y only.) Choosing $C_2 = 0$, we find that $\phi(x, y) = x^2 + x \sin y + ye^{-x} + \sin y$ is an integral function for the given DE. The solution curves for the DE are the level curves

$$x^2 + x \sin y + ye^{-x} + \sin y = C.$$

$\blacksquare$

Integrating Factors

Any ordinary differential equation of order 1 and degree 1 can be expressed in differential form: $M\, dx + N\, dy = 0$. However, this latter equation will usually not be exact. It *may* be possible to multiply the equation by an **integrating factor** $\mu(x, y)$ so that the resulting equation

$$\mu(x, y)\, M(x, y)\, dx + \mu(x, y)\, N(x, y)\, dy = 0$$

is exact. In general, such integrating factors are difficult to find; they must satisfy the partial differential equation

$$M(x, y)\, \frac{\partial \mu}{\partial y} - N(x, y)\, \frac{\partial \mu}{\partial x} = \mu(x, y) \left(\frac{\partial N}{\partial x} - \frac{\partial M}{\partial y} \right),$$

which follows from the necessary condition for exactness stated above. We will not try to solve this equation here.

Sometimes it happens that a differential equation has an integrating factor depending on only one of the two variables. Suppose, for instance, that $\mu(x)$ is an integrating factor for $M\, dx + N\, dy = 0$. Then $\mu(x)$ must satisfy the ordinary differential equation

$$N(x, y)\, \frac{d\mu}{dx} = \mu(x) \left(\frac{\partial M}{\partial y} - \frac{\partial N}{\partial x} \right),$$

or

$$\frac{1}{\mu(x)} \frac{d\mu}{dx} = \frac{\dfrac{\partial M}{\partial y} - \dfrac{\partial N}{\partial x}}{N(x, y)}.$$

This equation can be solved (by integration) for μ as a function of x alone *provided that the right-hand side is independent of y.*

Example 5   Show that $(x + y^2)\,dx + xy\,dy = 0$ has an integrating factor depending only on x, find it, and solve the equation.

Solution   Here $M = x + y^2$ and $N = xy$. Since

$$\frac{\dfrac{\partial M}{\partial y} - \dfrac{\partial N}{\partial x}}{N(x, y)} = \frac{2y - y}{xy} = \frac{1}{x}$$

does not depend on y, the equation has an integrating factor depending only on x. This factor is given by $d\mu/\mu = dx/x$. Evidently $\mu = x$ is a suitable integrating factor; if we multiply the given differential equation by x, we obtain

$$0 = (x^2 + xy^2)\,dx + x^2 y\,dy = d\left(\frac{x^3}{3} + \frac{x^2 y^2}{2}\right).$$

The solution is therefore $2x^3 + 3x^2 y^2 = C$. ∎

Remark   Of course, it may be possible to find an integrating factor depending on y instead of x. See Exercises 34–36 below. It is also possible to look for integrating factors that depend on specific combinations of x and y, for instance, xy. See Exercise 37.

Slope Fields and Solution Curves

A general first-order differential equation of the form

$$\frac{dy}{dx} = f(x, y)$$

specifies a slope $f(x, y)$ at every point (x, y) in the domain of f and therefore represents a **slope field**. Such a slope field can be represented graphically by drawing short line segments of the indicated slope at many points in the xy-plane. Slope fields resemble vector fields, but the segments are usually drawn having the same length and without arrowheads. Figure IV.1 portrays the slope field for the differential equation

$$\frac{dy}{dx} = x - y.$$

Solving a typical initial-value problem

$$\begin{cases} \dfrac{dy}{dx} = f(x, y) \\ y(x_0) = y_0 \end{cases}$$

involves finding a function $y = \phi(x)$ such that

$$\phi'(x) = f\left(x, \phi(x)\right) \qquad \text{and} \qquad \phi(x_0) = y_0.$$

The graph of the equation $y = \phi(x)$ is a curve passing through (x_0, y_0) that is tangent to the slope-field at each point. Such curves are called **solution curves** of the differential equation. Figure IV.1 shows four solution curves for $y' = x - y$ corresponding to the initial conditions $y(0) = C$, where $C = -2, -1, 0$, and 1.

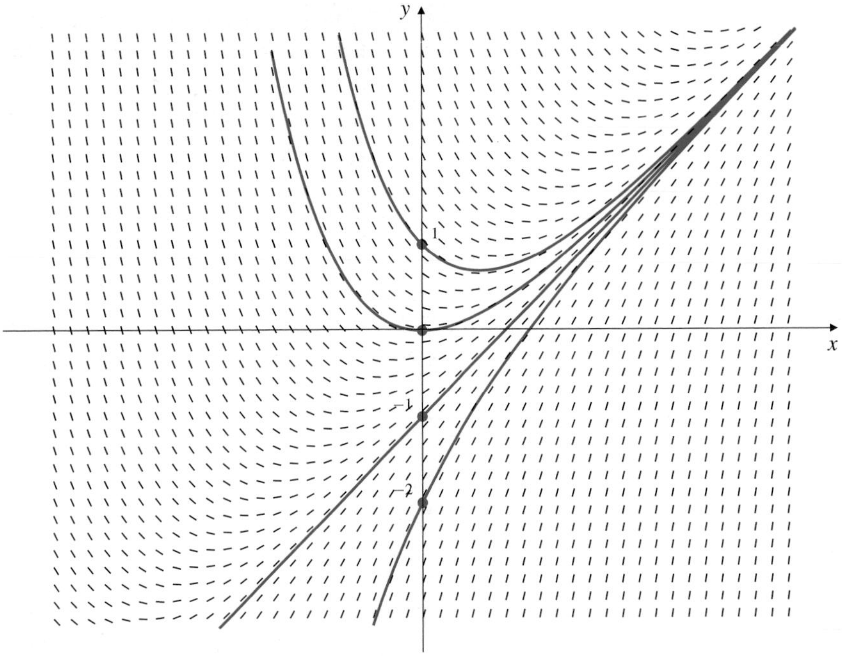

Figure IV.1 The slope field for the DE $y' = x - y$ and four solution curves for this DE

The DE $y' = x - y$ is linear and can be solved explicitly by the method of Section 7.9. Indeed, the solution satisfying $y(0) = C$ is $y = x - 1 + (C + 1)e^{-x}$. Most differential equations of the form $y' = f(x, y)$ cannot be solved for y as an explicit function of x, so we must use numerical approximation methods to find the value of a solution function $\phi(x)$ at particular points.

Existence and Uniqueness of Solutions

Even if we cannot calculate an explicit solution of an initial-value problem, it is important to know when the problem has a solution and whether that solution is unique.

THEOREM 3

An existence and uniqueness theorem for first-order initial-value problems

Suppose that $f(x, y)$ and $f_2(x, y) = (\partial/\partial y)f(x, y)$ are continuous on a rectangle R of the form $a \le x \le b$, $c \le y \le d$, containing the point (x_0, y_0) in its interior. Then there exists a number $\delta > 0$ and a *unique* function $\phi(x)$ defined and having a continuous derivative on the interval $(x_0 - \delta, x_0 + \delta)$ such that $\phi(x_0) = y_0$ and $\phi'(x) = f(x, \phi(x))$ for $x_0 - \delta < x < x_0 + \delta$. In other words, the initial-value problem

$$\begin{cases} \dfrac{dy}{dx} = f(x, y) \\ y(x_0) = y_0 \end{cases} \tag{$*$}$$

has a unique solution on $(x_0 - \delta, x_0 + \delta)$.

We give only an outline of the proof here. Any solution $y = \phi(x)$ of the initial-value problem ($*$) must also satisfy the **integral equation**

$$\phi(x) = y_0 + \int_{x_0}^{x} f(t, \phi(t)) \, dt, \tag{$**$}$$

and, conversely, any solution of the integral equation (✳✳) must also satisfy the initial-value problem (✳). A sequence of approximations $\phi_n(x)$ to a solution of (✳✳) can be constructed as follows:

$$\phi_0(x) = y_0$$

$$\phi_{n+1}(x) = y_0 + \int_{x_0}^{x} f\big(t, \phi_n(t)\big)\, dt \qquad \text{for} \quad n = 0, 1, 2, \ldots .$$

(These are called **Picard iterations**.) The proof of Theorem 3 involves showing that

$$\lim_{n \to \infty} \phi_n(x) = \phi(x)$$

exists on an interval $(x_0 - \delta, x_0 + \delta)$ and that the resulting limit $\phi(x)$ satisfies the integral equation (✳✳). The details can be found in more advanced texts on differential equations and analysis.

Remark Some initial-value problems can have nonunique solutions. For example, the functions $y_1(x) = x^3$ and $y_2(x) = 0$ both satisfy the initial-value problem

$$\begin{cases} \dfrac{dy}{dx} = 3y^{2/3} \\ y(0) = 0. \end{cases}$$

In this case $f(x, y) = 3y^{2/3}$ is continuous on the whole xy-plane. However, $\partial f / \partial y = 2y^{-1/3}$ is not continuous on the x-axis and is therefore not continuous on any rectangle containing $(0, 0)$ in its interior. The conditions of Theorem 3 are not satisfied and the initial-value problem has a solution but not a unique one.

Remark The unique solution $y = \phi(x)$ to the initial-value problem (✳) guaranteed by Theorem 3 may not be defined on the whole interval $[a, b]$, because it can "escape" from the rectangle R through the top or bottom edges. Even if $f(x, y)$ and $(\partial/\partial y)f(x, y)$ are continuous on the whole xy-plane, the solution may not be defined on the whole real line. For example

$$y = \frac{1}{1 - x} \qquad \text{satisfies the initial-value problem} \qquad \begin{cases} \dfrac{dy}{dx} = y^2 \\ y(0) = 1 \end{cases}$$

but only for $x < 1$. Starting from $(0, 1)$, we can follow the solution curve as far as we want to the left of $x = 0$, but to the right of $x = 0$ the curve recedes to ∞ as $x \to 1-$. It makes no sense to regard the part of the curve to the right of $x = 1$ as part of the solution curve to the initial-value problem. (See Figure IV.2.)

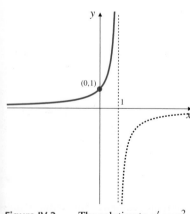

Figure IV.2 The solution to $y' = y^2$, $y(0) = 1$ is the part of the curve $y = 1/(1 - x)$ to the left of the vertical asymptote at $x = 1$

Numerical Methods

Suppose that the conditions of Theorem 3 are satisfied, so we know that the initial-value problem

$$\begin{cases} \dfrac{dy}{dx} = f(x, y) \\ y(x_0) = y_0 \end{cases}$$

has a unique solution $y = \phi(x)$ on some interval containing x_0. Even if we cannot solve the differential equation and find $\phi(x)$ explicitly, we can still try to find approximate values y_n for $\phi(x_n)$ at a sequence of points

$$x_0, \quad x_1 = x_0 + h, \quad x_2 = x_0 + 2h, \quad x_3 = x_0 + 3h, \quad \ldots$$

starting at x_0. Here $h > 0$ (or $h < 0$) is called the **step size** of the approximation scheme. In the remainder of this section we will describe three methods for constructing the approximations $\{y_n\}$, namely

1. The Euler method,
2. The improved Euler method, and
3. The fourth-order Runge–Kutta method.

Each of these methods starts with the given value of y_0 and provides a formula for constructing y_{n+1} when you know y_n. The three methods are listed above in increasing order of the complexity of their formulas, but the more complicated formulas produce much better approximations for any given step size h.

The Euler method involves approximating the solution curve $y = \phi(x)$ by a polygonal line (a sequence of straight line segments joined end to end), where each segment has horizontal length h and has slope determined by the value of $f(x, y)$ at the end of the previous segment. Thus, if $x_n = x_0 + nh$, then

$$y_1 = y_0 + f(x_0, y_0)h$$
$$y_2 = y_1 + f(x_1, y_1)h$$
$$y_3 = y_2 + f(x_2, y_2)h$$

and, in general,

Iteration formulas for Euler's method

$$x_{n+1} = x_n + h, \qquad y_{n+1} = y_n + hf(x_n, y_n).$$

Example 6   Use Euler's method to find approximate values for the solution of the initial-value problem

$$\begin{cases} \dfrac{dy}{dx} = x - y \\ y(0) = 1 \end{cases}$$

on the interval $[0, 1]$ using

(a) 5 steps of size $h = 0.2$ and

(b) 10 steps of size $h = 0.1$.

Calculate the error at each step, given that the problem (which involves a linear equation, so can be solved explicitly) has solution $y = \phi(x) = x - 1 + 2e^{-x}$.

Solution

(a) Here, we have $f(x, y) = x - y$, $x_0 = 0$, $y_0 = 1$, and $h = 0.2$, so that

$$x_n = \frac{n}{5}, \qquad y_{n+1} = y_n + 0.2(x_n - y_n),$$

and the error is $e_n = \phi(x_n) - y_n$ for $n = 0, 1, 2, 3, 4,$ and 5. The results of the calculation, which was done easily using a computer spreadsheet program, are presented in Table 1.

Table 1.   Euler approximations with $h = 0.2$

n	x_n	y_n	$f(x_n, y_n)$	y_{n+1}	$e_n = \phi(x_n) - y_n$
0	0.0	1.000000	−1.000000	0.800000	0.000000
1	0.2	0.800000	−0.600000	0.680000	0.037462
2	0.4	0.680000	−0.280000	0.624000	0.060640
3	0.6	0.624000	−0.024000	0.619200	0.073623
4	0.8	0.619200	0.180800	0.655360	0.079458
5	1.0	0.655360	0.344640		0.080399

The exact solution $y = \phi(x)$ and the polygonal line representing the Euler approximation are shown in Figure IV.3. The approximation lies below the solution curve, as is reflected in the positive values in the last column of Table 1, representing the error at each step.

(b) Here, we have $h = 0.1$, so that

$$x_n = \frac{n}{10}, \qquad y_{n+1} = y_n + 0.1(x_n - y_n)$$

for $n = 0, 1, \ldots, 10$. Again we present the results in tabular form:

Table 2.   Euler approximations with $h = 0.1$

n	x_n	y_n	$f(x_n, y_n)$	y_{n+1}	$e_n = \phi(x_n) - y_n$
0	0.0	1.000000	−1.000000	0.900000	0.000000
1	0.1	0.900000	−0.800000	0.820000	0.009675
2	0.2	0.820000	−0.620000	0.758000	0.017462
3	0.3	0.758000	−0.458000	0.712200	0.023636
4	0.4	0.712200	−0.312200	0.680980	0.028440
5	0.5	0.680980	−0.180980	0.662882	0.032081
6	0.6	0.662882	−0.062882	0.656594	0.034741
7	0.7	0.656594	0.043406	0.660934	0.036577
8	0.8	0.660934	0.139066	0.674841	0.037724
9	0.9	0.674841	0.225159	0.697357	0.038298
10	1.0	0.697357	0.302643		0.038402

Observe that the error at the end of the first step is about one-quarter of the error at the end of the first step in part (a), but the final error at $x = 1$ is only about half as large as in part (a). This behaviour is characteristic of Euler's method.

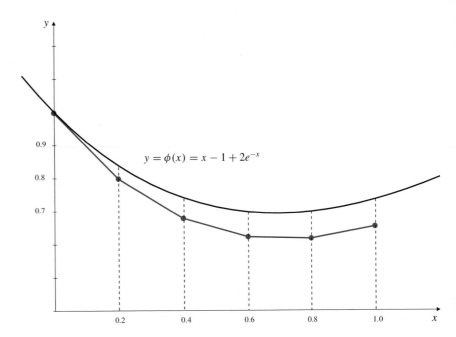

Figure IV.3 The solution $y = \phi(x)$ to $y' = x - y$, $y(0) = 1$, and an Euler approximation to it on $[0, 1]$ with step size $h = 0.2$

$y = \phi(x) = x - 1 + 2e^{-x}$

If we decrease the step size h, it takes more steps ($n = |x - x_0|/h$) to get from the starting point x_0 to a particular value x where we want to know the value of the solution. For Euler's method it can be shown that the error at each step decreases, on average, proportionally to h^2, but the errors can accumulate from step to step, so the error at x can be expected to decrease proportionally to $nh^2 = |x - x_0|h$. This is consistent with the results of Example 6. Decreasing h and so increasing n is costly in terms of computing resources, so we would like to find ways of reducing the error without decreasing the step size. This is similar to developing better techniques than the Trapezoid Rule for evaluating definite integrals numerically.

The improved Euler method is a step in this direction. The accuracy of the Euler method is hampered by the fact that the slope of each segment in the approximating polygonal line is determined by the value of $f(x, y)$ at one endpoint of the segment. Since f varies along the segment, we would expect to do better by using, say, the average value of $f(x, y)$ at the two ends of the segment, that is, by calculating y_{n+1} from the formula

$$y_{n+1} = y_n + h \frac{f(x_n, y_n) + f(x_{n+1}, y_{n+1})}{2}.$$

Unfortunately, y_{n+1} appears on both sides of this equation, and we can't usually solve the equation for y_{n+1}. We can get around this difficulty by replacing y_{n+1} on the right side by its Euler approximation $y_n + hf(x_n, y_n)$. The resulting formula is the basis for the improved Euler method.

Iteration formulas for the improved Euler method

$$x_{n+1} = x_n + h$$
$$u_{n+1} = y_n + h\, f(x_n, y_n)$$
$$y_{n+1} = y_n + h \frac{f(x_n, y_n) + f(x_{n+1}, u_{n+1})}{2}.$$

Example 7   Use the improved Euler method with $h = 0.2$ to find approximate values for the solution to the initial-value problem of Example 6 on $[0, 1]$. Compare the errors with those obtained by the Euler method.

Solution   Table 3 summarizes the calculation of five steps of the improved Euler method for $f(x, y) = x - y$, $x_0 = 0$, and $y_0 = 1$.

Table 3.   Improved Euler approximations with $h = 0.2$

n	x_n	y_n	u_{n+1}	y_{n+1}	$e_n = \phi(x_n) - y_n$
0	0.0	1.000000	0.800000	0.840000	0.000000
1	0.2	0.840000	0.712000	0.744800	−0.002538
2	0.4	0.744800	0.675840	0.702736	−0.004160
3	0.6	0.702736	0.682189	0.704244	−0.005113
4	0.8	0.704244	0.723395	0.741480	−0.005586
5	1.0	0.741480	0.793184		−0.005721

Observe that the errors are considerably less than 1/10 those obtained in Example 6(a). Of course, more calculations are necessary at each step, but the number of evaluations of $f(x, y)$ required is only twice the number required for Example 6(a). As for numerical integration, if f is complicated, it is these function evaluations that constitute most of the computational "cost" of computing numerical solutions. ∎

Remark   It can be shown for well-behaved functions f that the error at each step in the improved Euler method is bounded by a multiple of h^3, rather than h^2 as for the (unimproved) Euler method. Thus, the cumulative error at x can be bounded by a constant times $|x - x_0|h^2$. If Example 7 is repeated with 10 steps of size $h = 0.1$, the error at $n = 10$ (i.e., at $x = 1$) is -0.001323, which is about 1/4 the size of the error at $x = 1$ with $h = 0.2$.

The **fourth-order Runge–Kutta method** further improves upon the improved Euler method, but at the expense of requiring more complicated calculations at each step. It requires four evaluations of $f(x, y)$ at each step, but the error at each step is less than a constant times h^5, so the cumulative error decreases like h^4 as h decreases. Like the improved Euler method, this method involves calculating a certain kind of average slope for each segment in the polygonal approximation to the solution to the initial-value problem. We present the appropriate formulas below but cannot derive them here.

Iteration formulas for the Runge–Kutta method

$$x_{n+1} = x_n + h$$

$$p_n = f(x_n, y_n)$$

$$q_n = f\left(x_n + \frac{h}{2}, y_n + \frac{h}{2}p_n\right)$$

$$r_n = f\left(x_n + \frac{h}{2}, y_n + \frac{h}{2}q_n\right)$$

$$s_n = f(x_n + h, y_n + hr_n)$$

$$y_{n+1} = y_n + h\frac{p_n + 2q_n + 2r_n + s_n}{6}.$$

Example 8   Use the fourth-order Runge–Kutta method with $h = 0.2$ to find approximate values for the solution to the initial-value problem of Example 6 on [0, 1]. Compare the errors with those obtained by the Euler and improved Euler methods.

Solution   Table 4 summarizes the calculation of five steps of the Runge–Kutta method for $f(x, y) = x - y$, $x_0 = 0$, and $y_0 = 1$ according to the formulas above. The table does not show the values of the intermediate quantities p_n, q_n, r_n, and s_n, but columns for these quantities were included in the spreadsheet in which the calculations were made.

Table 4.   Fourth-order Runge–Kutta approximations with $h = 0.2$

n	x_n	y_n	$e_n = \phi(x_n) - y_n$
0	0.0	1.000000	0.0000000
1	0.2	0.837467	−0.0000052
2	0.4	0.740649	−0.0000085
3	0.6	0.697634	−0.0000104
4	0.8	0.698669	−0.0000113
5	1.0	0.735770	−0.0000116

The errors here are about 1/500 of the size of the errors obtained with the improved Euler method and about 1/7,000 of the size of the errors obtained with the Euler method. This great improvement was achieved at the expense of doubling the number of function evaluations required in the improved Euler method and quadrupling the number required in the Euler method. If we use 10 steps of size $h = 0.1$ in the Runge–Kutta method, the error at $x = 1$ is reduced to -6.66482×10^{-7}, which is less than 1/16 of its value when $h = 0.2$.   ■

Our final example shows what can happen with numerical approximations to a solution that is unbounded.

Example 9   Obtain solutions at $x = 0.4$, $x = 0.8$, and $x = 1.0$ for solutions to the initial-value problem

$$\begin{cases} y' = y^2 \\ y(0) = 1 \end{cases}$$

using all three methods described above, and using step sizes $h = 0.2$, $h = 0.1$, and $h = 0.05$ for each method. What do the results suggest about the values of the solution at these points? Compare the results with the actual solution $y = 1/(1-x)$.

Solution   The various approximations are calculated using the various formulas described above for $f(x, y) = y^2$, $x_0 = 0$, and $y_0 = 1$. The results are presented in Table 5.

Table 5.   Comparing methods and step sizes for $y' = y^2$, $y(0) = 1$

	$h = 0.2$	$h = 0.1$	$h = 0.05$
Euler			
$x = 0.4$	1.488000	1.557797	1.605224
$x = 0.8$	2.676449	3.239652	3.793197
$x = 1.0$	4.109124	6.128898	9.552668
Improved Euler			
$x = 0.4$	1.640092	1.658736	1.664515
$x = 0.8$	4.190396	4.677726	4.897519
$x = 1.0$	11.878846	22.290765	43.114668
Runge–Kutta			
$x = 0.4$	1.666473	1.666653	1.666666
$x = 0.8$	4.965008	4.996628	4.999751
$x = 1.0$	41.016258	81.996399	163.983395

Little useful information can be read from the Euler results. The improved Euler results suggest that the solution exists at $x = 0.4$ and $x = 0.8$, but likely not at $x = 1$. The Runge–Kutta results confirm this and suggest that $y(0.4) = 5/3$ and $y(0.8) = 5$, which are the correct values provided by the actual solution $y = 1/(1 - x)$. They also suggest very strongly that the solution "blows up" at (or near) $x = 1$.

Exercises

In Exercises 1–10, state the order of the given DE and whether it is linear or nonlinear. If it is linear, is it homogeneous or nonhomogeneous?

1. $\dfrac{dy}{dx} = 5y$

2. $\dfrac{d^2y}{dx^2} + x = y$

3. $y\dfrac{dy}{dx} = x$

4. $y''' + xy' = x \sin x$

5. $y'' + x \sin x \, y' = y$

6. $y'' + 4y' - 3y = 2y^2$

7. $\dfrac{d^3y}{dt^3} + t\dfrac{dy}{dt} + t^2y = t^3$

8. $\cos x \dfrac{dx}{dt} + x \sin t = 0$

9. $y^{(4)} + e^x y'' = x^3 y'$

10. $x^2 y'' + e^x y' = \dfrac{1}{y}$

11. Verify that $y = \cos x$ and $y = \sin x$ are solutions of the DE $y'' + y = 0$. Are any of the following functions solutions:
(a) $\sin x - \cos x$, (b) $\sin(x + 3)$, and (c) $\sin 2x$? Justify your answers.

12. Verify that $y = e^x$ and $y = e^{-x}$ are solutions of the DE $y'' - y = 0$. Are any of the following functions solutions:
(a) $\cosh x = \frac{1}{2}(e^x + e^{-x})$, (b) $\cos x$, and (c) x^e? Justify your answers.

13. $y_1 = \cos(kx)$ is a solution of $y'' + k^2y = 0$. Guess and verify another solution y_2 that is not a multiple of y_1. Then find a solution that satisfies $y(\pi/k) = 3$ and $y'(\pi/k) = 3$.

14. $y_1 = e^{kx}$ is a solution of $y'' - k^2y = 0$. Guess and verify another solution y_2 that is not a multiple of y_1. Then find a solution that satisfies $y(1) = 0$ and $y'(1) = 2$.

15. Find a solution of $y'' + y = 0$ that satisfies $y(\pi/2) = 2y(0)$ and $y(\pi/4) = 3$. *Hint:* see Exercise 11.

16. Find two values of r such that $y = e^{rx}$ is a solution of $y'' - y' - 2y = 0$. Then find a solution of the equation that satisfies $y(0) = 1$, $y'(0) = 2$.

17. Verify that $y = x$ is a solution of $y'' + y = x$, and find a solution y of this DE that satisfies $y(\pi) = 1$ and $y'(\pi) = 0$. *Hint:* use Exercise 11 and Theorem 2.

18. Verify that $y = -e$ is a solution of $y'' - y = e$, and find a solution y of this DE that satisfies $y(1) = 0$ and $y'(1) = 1$. *Hint:* use Exercise 12 and Theorem 2.

Solve the differential equations in Exercises 19–24.

19. $\dfrac{dy}{dx} = \dfrac{x + y}{x - y}$

20. $\dfrac{dy}{dx} = \dfrac{xy}{x^2 + 2y^2}$

21. $\dfrac{dy}{dx} = \dfrac{x^2 + xy + y^2}{x^2}$

22. $\dfrac{dy}{dx} = \dfrac{x^3 + 3xy^2}{3x^2y + y^3}$

23. $x\dfrac{dy}{dx} = y + x \cos^2\left(\dfrac{y}{x}\right)$

24. $\dfrac{dy}{dx} = \dfrac{y}{x} - e^{-y/x}$

25. Find an equation of the curve in the xy-plane that passes through the point $(1, 3)$ and has, at every point (x, y) on it, slope equal to $1 + (2y/x)$.

26. Show that the change of variables $\xi = x - x_0$, $\eta = y - y_0$ transforms the equation

$$\frac{dy}{dx} = \frac{ax + by + c}{ex + fy + g}$$

into the homogeneous equation

$$\frac{d\eta}{d\xi} = \frac{a\xi + b\eta}{e\xi + f\eta},$$

provided (x_0, y_0) is the solution of the system

$$ax + by + c = 0$$
$$ex + fy + g = 0.$$

27. Use the technique of the previous exercise to solve the equation $\dfrac{dy}{dx} = \dfrac{x + 2y - 4}{2x - y - 3}$.

Show that the DEs in Exercises 28–31 are exact, and solve them.

28. $(xy^2 + y) \, dx + (x^2 y + x) \, dy = 0$

29. $(e^x \sin y + 2x) \, dx + (e^x \cos y + 2y) \, dy = 0$

30. $e^{xy}(1 + xy) \, dx + x^2 e^{xy} \, dy = 0$

31. $\left(2x + 1 - \dfrac{y^2}{x^2}\right) dx + \dfrac{2y}{x} \, dy = 0$

Show that the DEs in Exercises 32–33 admit integrating factors that are functions of x alone. Then solve the equations.

32. $(x^2 + 2y) \, dx - x \, dy = 0$

33. $(xe^x + x \ln y + y) \, dx + \left(\dfrac{x^2}{y} + x \ln x + x \sin y\right) dy = 0$

34. What condition must the coefficients $M(x, y)$ and $N(x, y)$ satisfy if the equation $M \, dx + N \, dy = 0$ is to have an integrating factor of the form $\mu(y)$, and what DE must the integrating factor satisfy?

35. Find an integrating factor of the form $\mu(y)$ for the equation

$$2y^2(x + y^2) \, dx + xy(x + 6y^2) \, dy = 0,$$

and hence solve the equation. *Hint:* see Exercise 34.

36. Find an integrating factor of the form $\mu(y)$ for the equation $y \, dx - (2x + y^3 e^y) \, dy = 0$, and hence solve the equation. *Hint:* see Exercise 34.

37. What condition must the coefficients $M(x, y)$ and $N(x, y)$ satisfy if the equation $M \, dx + N \, dy = 0$ is to have an integrating factor of the form $\mu(xy)$, and what DE must the integrating factor satisfy?

38. Find an integrating factor of the form $\mu(xy)$ for the equation

$$\left(x \cos x + \dfrac{y^2}{x}\right) dx - \left(\dfrac{x \sin x}{y} + y\right) dy = 0,$$

and hence solve the equation. *Hint:* see Exercise 37.

A computer is almost essential for doing Exercises 39–44. The calculations are easily done with a spreadsheet program in which formulas for calculating the various quantities involved can be replicated down columns to automate the iteration process.

39. Use the Euler method with step sizes (a) $h = 0.2$, (b) $h = 0.1$, and (c) $h = 0.05$ to approximate $y(2)$ given that $y' = x + y$ and $y(1) = 0$.

40. Repeat Exercise 39 using the improved Euler method.

41. Repeat Exercise 39 using the Runge–Kutta method.

42. Use the Euler method with step sizes (a) $h = 0.2$, and (b) $h = 0.1$ to approximate $y(2)$ given that $y' = xe^{-y}$ and $y(0) = 0$.

43. Repeat Exercise 42 using the improved Euler method.

44. Repeat Exercise 42 using the Runge–Kutta method.

Solve the integral equations in Exercises 45–46 by rephrasing them as initial-value problems.

45. $y(x) = 2 + \displaystyle\int_1^x \left(y(t)\right)^2 dt$. *Hint:* find $\dfrac{dy}{dx}$ and $y(1)$.

46. $u(x) = 1 + 3 \displaystyle\int_2^x t^2 u(t) \, dt$. *Hint:* find $\dfrac{du}{dx}$ and $u(2)$.

47. The methods of this section can be used to approximate definite integrals numerically. For example,

$$I = \int_a^b f(x) \, dx$$

is given by $I = y(b)$, where

$$y' = f(x) \qquad \text{and} \qquad y(a) = 0.$$

Show that one step of the Runge–Kutta method with $h = b - a$ gives the same result for I as does Simpson's Rule with two subintervals of length $h/2$.

48. If $\phi(0) = A \geq 0$ and $\phi'(x) \geq k\phi(x)$ on $[0, X]$, where $k > 0$ and $X > 0$ are constants, show that $\phi(x) \geq Ae^{kx}$ on $[0, X]$. *Hint:* calculate $(d/dx)(\phi(x)/e^{kx})$.

* 49. Consider the three initial-value problems

(A)	$u' = u^2$	$u(0) = 1$
(B)	$y' = x + y^2$	$y(0) = 1$
(C)	$v' = 1 + v^2$	$v(0) = 1$

(a) Show that the solution of (B) remains between the solutions of (A) and (C) on any interval $[0, X]$ where solutions of all three problems exist. *Hint:* we must have $u(x) \geq 1$, $y(x) \geq 1$, and $v(x) \geq 1$ on $[0, X]$. (Why?) Apply the result of Exercise 48 to $\phi = y - u$ and to $\phi = v - y$.

(b) Find explicit solutions for problems (A) and (C). What can you conclude about the solution to problem (B)?

(c) Use the Runge–Kutta method with $h = 0.05$, $h = 0.02$, and $h = 0.01$ to approximate the solution to (B) on $[0, 1]$. What can you conclude now?

Doing Calculus with Maple

Computer algebra systems like Maple and Mathematica are capable of doing most of the tedious calculations involved in doing calculus, especially the very intensive calculations required by many applied problems. (They cannot, of course, do the thinking for you; you must still fully understand what you are doing and what are the limitations of such programs.) Throughout this text we have inserted material illustrating how to use **Maple** to do common calculus-oriented calculations. These insertions range in length from single paragraphs and remarks to entire sections. To help you locate the Maple material appropriate for specific topics, we include below a list pointing to the text sections containing Maple examples and the pages on which they start.

Note, however, that this material assumes you are familiar with the basics of starting a Maple session, preferably with a graphical user interface which typically displays the prompt ">" when it is waiting for your input. In this book the input is shown in colour. It normally concludes with a semicolon ";" followed by pressing the <enter> key, which we omit from our examples. The output is typically printed by Maple centred in the window; we show it in black. For instance,

```
>   factor(x^2-x-2);
```

$$(x + 1)(x - 2)$$

The author used Maple V, Release 5, and Maple 6 for preparing these examples. Some of the examples involve procedure definition and worksheet files available from the website for this text:

```
http://www.pearsoned.ca/text/adams_calc
```

Two of the Maple procedures used in Section 13.7 for finding roots of systems of nonlinear equations and for finding and classifying critical points of functions of several variables are quite lengthy, and rather than list them there, we have included their definitions later on in this Appendix.

The Maple examples in this book are by no means complete or exhaustive. For a more complete treatment of Maple as a tool for doing calculus, the author highly recommends the excellent Maple lab manual *Calculus: The Maple Way* written by his colleague, Professor Robert Israel of the University of British Columbia. Like this book, it is published by Pearson Canada under the Addison-Wesley logo.

List of Maple Examples and Discussion

Topic	Section	Page
Defining and Graphing Functions	P.4	34
Calculating with Trigonometric Functions	P.6	52
Calculating Limits	1.3	76
Solving Equations with `fsolve`	1.4	87
Finding Derivatives	2.4	123
Higher-Order Derivatives	2.8	150
Derivatives of Implicit Functions	2.9	155
Solving DEs with `dsolve`	3.7	228
More Graph Plotting	4.4	261
Calculating Sums	5.1	307
Integrating Functions	6.4	373
Numerical Integration	6.4	374
Plotting Parametric Curves	8.2	491
Plotting Polar Curves	8.5	509
Infinite Series	9.5	564
Vector and Matrix Calculations	10.7	642
Velocity, Acceleration, Curvature, Torsion	11.5	688
Three-Dimensional Graphing	12.1	710
Partial Derivatives	12.4	728
The Jacobian Matrix	12.6	750
Gradients	12.7	760
Taylor Polynomials	12.9	777
Multivariable Newton's Method	13.7	828
Double and Multiple Integrals	14.2	848
Gradient, Divergence, Curl, Laplacian	16.2	959

Several of the topics in the above list are covered over several pages. Only the first page is listed.

The "newtroot" Procedure of Section 13.7

Here is a listing of the Maple procedures `newtroot` discussed in Section 13.7. You can learn much about Maple by reading this listing and trying to understand what it is doing.

```
newtroot:=proc(F::procedure,v,m::integer,tol::float)
local i,j,k,v0,v1,w,FV,JF,A,b,error,n;
error := tol + 1.0; # tol = desired accuracy
i := 1; v0 := v;
if type(v,list) then
    convert(v0,vector);
```

```
        n := vectdim(v0);
    else n := 1 fi;
 w := vector(n);
 if n = 1 then
     while tol < error and i < m +1 do
       v1:= evalf(v0-F(v0)/(D(F)(v0))); #Newton iteration
       error := abs(v1-v0);
       v0 := v1; #v0 becomes the new approximation
       print(i, v0, F(v0), error);
       i:= i+1;
     od
   else FV := v -> F(seq(v[j],j=1..n));
     JF := proc(FF::procedure,vv::vector)
       jacobian(FF(vv),vv);
     end;
     while tol < error and i < m +1 do
       A:= subs(seq(w[k]=v0[k],k=1..n),JF(FV,w));
            # A = JF(v0).
       b:= FV(v0);
       v1 := evalf(evalm(v0 - linsolve(A,b)));
            #Newton iteration
       error := norm(v0-v1);
       v0 := evalm(v1);
            #v0 becomes the new approximation
       print(i, v0,FV(v0),error);
       i := i + 1;
     od
  fi;
 if error <= tol then
     RETURN(evalm(v0))
   else print('FAILED_TO_FIND_ROOT',error);
     RETURN(evalm(v0))
   fi;
end;
```

The scalar case $n = 1$ evidently calculates the Newton's Method approximation v_1 from v_0 and then renames v_0 to be this new approximation before doing another iteration. So does the vector case $n > 1$, handled by the "else" clause. However, it does not use determinants (i.e., Cramer's Rule) to calculate the next approximation as we did in Section 13.6. Instead, it uses the **Jacobian matrix** $\mathcal{A} = JF(\mathbf{v}_0)$ of **F** at $\mathbf{v}_0$. The next approximation $\mathbf{v}_1$ satisfies the system of equations

$$\mathcal{A}(\mathbf{v}_1 - \mathbf{v}_0) + \mathbf{b} = \mathbf{0}, \qquad \text{where} \quad \mathbf{b} = \mathbf{F}(\mathbf{v}_0).$$

The Maple function linsolve(A,b) determines the solution $\mathbf{x} = \mathcal{A}^{-1}\mathbf{b}$ of the system $\mathcal{A}\mathbf{x} = \mathbf{b}$, so the procedure calculates $\mathbf{v}_1 = \mathbf{v}_0 - \mathcal{A}^{-1}\mathbf{b}$. The evalm and evalf operators then convert the resulting list of solutions to a vector with real components.

The "newtcp" Procedure of Section 13.7

This procedure, also discussed in Section 13.7, is a variant of newtroot used to find and classify the critical points of a function of several variables.

```
newtcp:=proc(F::procedure,v,m::integer,tol::float)
local i,j,k,v0,v1,w,FV,GRADF,HF,A,b,error,n;
error := tol + 1.0; # tol = desired accuracy
i := 1; v0 := v;
if type(v,list) then
    convert(v0,vector);
    n := vectdim(v0);
  else n := 1 fi;
w := vector(n);
if n = 1 then
    while tol < error and i < m +1 do
      v1:= evalf(v0-D(F)(v0)/(D@@2)(F)(v0));
      error := abs(v1-v0);
      v0 := v1;
      print(i,v0,F(v0),error);
      i:= i+1;
    od;
  else
    FV := v -> F(seq(v[j],j=1..n));
    GRADF := proc(FF::procedure,vv::vector)
      grad(FF(vv),vv);
    end;
    HF := proc(FF::procedure,vv::vector)
      hessian(FF(vv),vv);
    end;
    while tol < error and i < m +1 do
      A:= subs(seq(w[k]=v0[k],k=1..n),HF(FV,w));
              #A=HF(v0).
      b:= subs(seq(w[k]=v0[k],k=1..n),GRADF(FV,w));
      v1 := evalf(evalm(v0 - linsolve(A,b)));
              #Newton iteration
      error := norm(v0-v1);
      v0 := evalm(v1);
              #v0 becomes the new approximation
      print(i,v0,FV(v0),error);
      i := i + 1;
    od
  fi;
if (error <= tol) then
    if (n=1) then
      print('Second_deriv',evalf((D@@2)(F)(v0)));
      RETURN(evalm(v0),F(v0))
    else print('Eigenvalues',evalf(eigenvals(A))) fi;
    RETURN(evalm(v0),FV(v0))
  else print('FAILED',error);
    RETURN(evalm(v0))
fi;
```

```
end;
```

This procedure functions in much the same way as newtroot. Note that the Jacobian matrix of the gradient of the scalar field F is just the Hessian matrix of F.

Answers to Odd-Numbered Exercises

Chapter P

Preliminaries

1. $0.\overline{2}$ **3.** $4/33$

5. $1/7 = 0.\overline{142857}, 2/7 = 0.\overline{285714},$
   $3/7 = 0.\overline{428571}, 4/7 = 0.\overline{571428},$
   $5/7 = 0.\overline{714285}, 6/7 = 0.\overline{857142}$

7. $[0, 5]$ **9.** $]-\infty, -6[\cup]-5, \infty[$

11. $]-2, \infty[$ **13.** $]-\infty, -2[$

15. $(-\infty, 5/4]$ **17.** $]0, \infty[$

19. $]-\infty, 5/3[\cup]2, \infty[$ **21.** $[0, 2]$

23. $]-2, 0[\cup]2, \infty[$ **25.** $[-2, 0[\cup [4, \infty[$

27. $x = -3, 3$ **29.** $t = -1/2, -9/2$

31. $s = -1/3, 17/3$ **33.** $(-2, 2)$

35. $[-1, 3]$ **37.** $\left]\dfrac{5}{3}, 3\right[$

39. $[0, 4]$ **41.** $x > 1$

43. true if $a \geq 0$, false if $a < 0$

27. 4, 3,

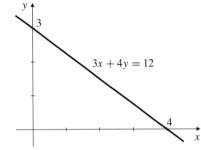

29. $\sqrt{2}, -2/\sqrt{3}$

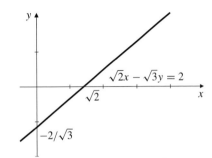

31. (a) $y = x - 1$, (b) $y = -x + 3$

33. $(2, -3)$ **37.** 5

39. $\$23,000$ **43.** $(-2, -2)$

45. $\left(\frac{1}{3}(x_1 + 2x_2), \frac{1}{3}(y_1 + 2y_2)\right)$

47. circle, centre $(2, 0)$, radius 4

49. perp. if $k = -8$, parallel if $k = 1/2$

1. $\Delta x = 4, \Delta y = -3$, dist $= 5$

3. $\Delta x = -4, \Delta y = -4$, dist $= 4\sqrt{2}$

5. $(2, -4)$

7. circle, centre $(0, 0)$, radius 1

9. points inside and on circle, centre $(0, 0)$, radius 1

11. points on and above the parabola $y = x^2$

13. (a) $x = -2$, (b) $y = 5/3$

15. $y = x + 2$ **17.** $y = 2x + b$

19. above   **21.** $y = 3x/2$

23. $y = (7 - x)/3$ **25.** $y = \sqrt{2} - 2x$

1. $x^2 + y^2 = 16$ **3.** $x^2 + y^2 + 4x = 5$

5. $(1, 0), 2$ **7.** $(1, -2), 3$

9. exterior of circle, centre $(0, 0)$, radius 1

11. closed disk, centre $(-1, 0)$, radius 2

13. washer shaped region between the circles of radius 1
   and 2 centred at $(0, 0)$

15. first octant region lying inside the two circles of radius
   1 having centres at $(1, 0)$ and $(0, 1)$

17. $x^2 + y^2 + 2x - 4y < 1$

19. $x^2 + y^2 < 2, x \geq 1$ **21.** $x^2 = 16y$

23. $y^2 = 8x$

25. $(0, 1/2)$, $y = -1/2$

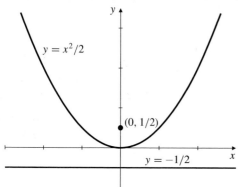

27. $(-1, 0)$, $x = 1$

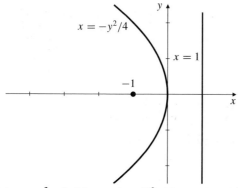

29. (a) $y = x^2 - 3$, (b) $y = (x-4)^2$, (c) $y = (x-3)^2 + 3$,
      (d) $y = (x-4)^2 - 2$

31. $y = \sqrt{(x/3) + 1}$ **33.** $y = \sqrt{(3x/2) + 1}$

35. $y = -(x+1)^2$ **37.** $y = (x-2)^2 - 2$

39. $(2, 7)$, $(1, 4)$ **41.** $(4, -3)$, $(-4, 3)$

43. ellipse, centre $(0, 0)$, semiaxes $2, 1$

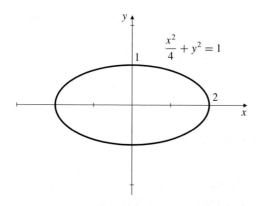

45. ellipse, centre $(3, -2)$, semiaxes $3, 2$

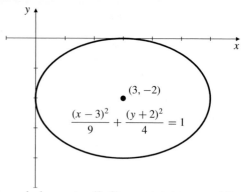

47. hyperbola, centre $(0, 0)$, asymptotes $x = \pm 2y$, vertices $(\pm 2, 0)$

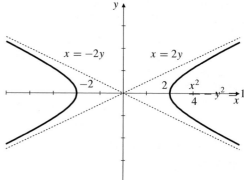

49. rectangular hyperbola, asymptotes $x = 0$ and $y = 0$,
      vertices $(2, -2)$ and $(-2, 2)$

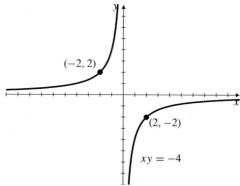

51. (a) reflecting the graph in the y-axis, (b) reflecting the graph in the x-axis.

53.

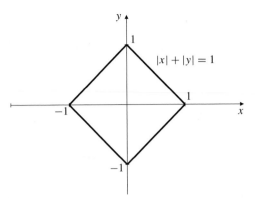

$|x| + |y| = 1$

Section P.4   (page 35)

1. $\mathcal{D}(f) = \mathbb{R}, \mathcal{R}(f) = [1, \infty[$

3. $\mathcal{D}(G) =]-\infty, 4], \mathcal{R}(g) = [0, \infty[$

5. $\mathcal{D}(h) =]-\infty, 2[, \mathcal{R}(h) =]-\infty, \infty[$

7. Only (ii) is the graph of a function. Vertical lines can meet the others more than once.

11. even, sym. about y-axis

13. odd, sym. about $(0, 0)$ **15.** sym. about $(2, 0)$

17. sym. about $x = 3$

19. even, sym. about y-axis

21. no symmetry

23.

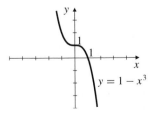

$y = -x^2$

25.

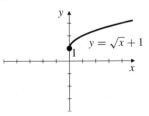

$y = (x - 1)^2$

27.

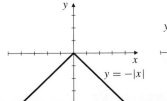

$y = 1 - x^3$

29.

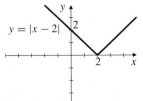

$y = \sqrt{x} + 1$

31.

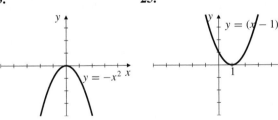

$y = -|x|$

33.

$y = |x - 2|$

35.

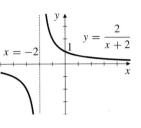

$x = -2$ $y = \dfrac{2}{x+2}$

37.

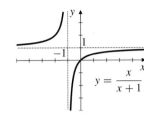

$y = \dfrac{x}{x+1}$

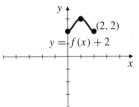

$y = f(x) + 2$ $(2, 2)$

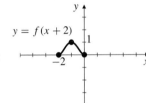

$y = f(x + 2)$

39. $D = [0, 2], R = [2, 3]$

41. $D = [-2, 0], R = [0, 1]$

43. $D = [0, 2], R = [-1, 0]$

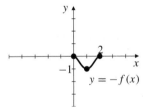

$y = -f(x)$

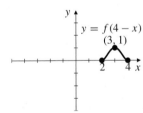

$y = f(4 - x)$ $(3, 1)$

45. $D = [2, 4], R = [0, 1]$ **47.** $[-0.18, 0.68]$

49. $y = 3/2$

51. $(2, 1), \ y = x - 1, \ y = 3 - x$

53. $f(x) = 0$

Section P.5   (page 41)

1. The domains of $f + g, f - g, fg,$ and g/f are $[1, \infty[$.
The domain of f/g is $]1, \infty[$.
$(f + g)(x) = x + \sqrt{x - 1}$
$(f - g)(x) = x - \sqrt{x - 1}$
$(fg)(x) = x\sqrt{x - 1}$
$(f/g)(x) = x/\sqrt{x - 1}$
$(g/f)(x) = \sqrt{x - 1}/x$

3.

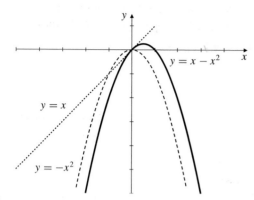

5.

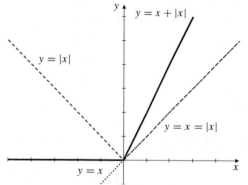

7. (a) 2, (b) 22, (c) $x^2 + 2$, (d) $x^2 + 10x + 22$, (e) 5, (f) -2, (g) $x + 10$, (h) $x^4 - 6x^2 + 6$

9. (a) $(x - 1)/x$, $x \neq 0$, 1,
(b) $1/(1 - \sqrt{x - 1})$ on $[1, 2[\cup]2, \infty[$,
(c) $\sqrt{x/(1 - x)}$, on $[0, 1[$
(d) $\sqrt{\sqrt{x - 1} - 1}$, on $[2, \infty[$

11. $(x + 1)^2$ **13.** x^2

15. $1/(x - 1)$ **19.** $D = [0, 2]$, $R = [0, 2]$

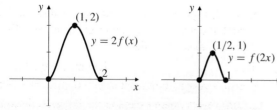

21. $D = [0, 1]$, $R = [0, 1]$

23. $D = [-4, 0]$, $R = [1, 2]$

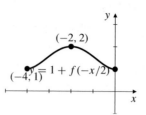

25.

27. (a) $A = 0$, B arbitrary, or $A = 1$, $B = 0$
(b) $A = -1$, B arbitrary, or $A = 1$, $B = 0$

29. all integers

31.

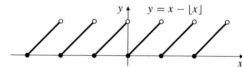

33. f^2, g^2, $f \circ f$, $f \circ g$, $g \circ f$ are even
fg, f/g, g/f, $g \circ g$ are odd
$f + g$ is neither, unless either $f(x) = 0$ or $g(x) = 0$.

Section P.6   (page 55)

1. $-1/\sqrt{2}$ **3.** $\sqrt{3}/2$

5. $(\sqrt{3} - 1)/(2\sqrt{2})$ **7.** $-\cos x$

9. $-\cos x$ **11.** $1/(\sin x \cos x)$

17. $3 \sin x - 4 \sin^3 x$

19. period π

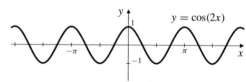

21. period 2

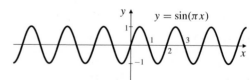

23.

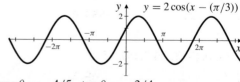

25. $\cos \theta = -4/5$, $\tan \theta = -3/4$

27. $\sin \theta = -2\sqrt{2}/3$, $\tan \theta = -2\sqrt{2}$

29. $\cos \theta = -\sqrt{3}/2$, $\tan \theta = 1/\sqrt{3}$

31. $a = 1$, $b = \sqrt{3}$

33. $b = 5/\sqrt{3}$, $c = 10/\sqrt{3}$

35. $a = b \tan A$ **37.** $a = b \cot B$

39. $c = b \sec A$ **41.** $\sin A = \sqrt{c^2 - b^2}/c$

43. $\sin B = 3/(4\sqrt{2})$ **45.** $\sin B = \sqrt{135}/16$

47. $6/(1 + \sqrt{3})$

49. $b = 4 \sin 40°/\sin 70° \approx 2.736$

51. approx. 16.98 m

Chapter 1
Limits and Continuity

Section 1.1   (page 61)

1. $((t + h)^2 - t^2)/h$ m/s   **3.** 4 m/s

5. -3 m/s, 3 m/s, 0 m/s

7. to the left, stopped, to the right

9. height 2, moving down

11. -1 ft/s, weight moving downward

13. day 45

Section 1.2   (page 70)

1. (a) 1, (b) 0, (c) 1   **3.** 1

5. 0   **7.** 1

9. $2/3$ **11.** 0

13. 0   **15.** does not exist

17. $1/6$ **19.** 0

21. -1 **23.** does not exist

25. 2   **27.** $3/8$

29. $-1/2$ **31.** $8/3$

33. $1/4$ **35.** $1/\sqrt{2}$

37. $2x$ **39.** $-1/x^2$

41. $1/(2\sqrt{x})$ **43.** 1

45. $1/2$ **47.** 1

49. 0   **51.** 2

53. does not exist   **55.** does not exist

57. $-1/(2a)$ **59.** 0

61. -2 **63.** π^2

65. (a) 0, (b) 8, (c) 9, (d) -3

67. 5   **69.** 1

71. 0.7071   **73.** $\lim_{x \to 0} f(x) = 0$

75. 2

77. $x^{1/3} < x^3$ on $]-1, 0[$ and $]1, \infty[$,
   $x^{1/3} > x^3$ on $]-\infty, -1[$ and $]0, 1[$,
   $\lim_{x \to a} h(x) = a$ for $a = -1, 0,$ and 1

Section 1.3   (page 77)

1. $1/2$ **3.** $-3/5$

5. 0   **7.** -3

9. $-2/\sqrt{3}$ **11.** does not exist

13. $+\infty$ **15.** 0

17. $-\infty$ **19.** $-\infty$

21. ∞ **23.** $-\infty$

25. ∞ **27.** $-\sqrt{2}/4$

29. -2 **31.** -1

33. horiz: $y = 0$, $y = -1$, vert: $x = 0$

35. 1   **37.** 1

39. $-\infty$ **41.** 2

43. -1 **45.** 1

47. 3   **49.** does not exist

51. 1

53. $C(t)$ has a limit at every real t except at the integers. $\lim_{t \to t_0-} C(t) = C(t_0)$ everywhere, but

$$\lim_{t \to t_0+} C(t) = \begin{cases} C(t_0) & \text{if } t_0 \text{ not integral} \\ C(t_0) + 1.5 & \text{if } t_0 \text{ an integer} \end{cases}$$

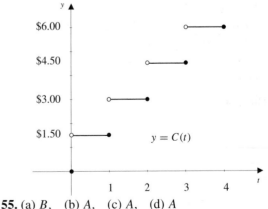

55. (a) B,   (b) A,   (c) A,   (d) A

Section 1.4   (page 87)

1. at -2, right cont. and cont., at -1 disc., at 0 disc. but left cont., at 1 disc. and right cont., at 2 disc.

3. no abs. max, abs. min 0

5. no   **7.** cont. everywhere

9. cont. everywhere except at $x = 0$, disc. at $x = 0$

11. cont. everywhere except at the integers, discontinuous but left-continuous at the integers

13. 4, $x + 2$

15. $1/5$, $(t - 2)/(t + 2)$

17. $k = 8$

19. no max, min $= 0$

21. 16

23. 5

25. f positive on $]-1, 0[$ and $]1, \infty[$; f negative on $]-\infty, -1[$ and $]0, 1[$

27. f positive on $]-\infty, -2[$, $]-1, 1[$ and $]2, \infty[$; f negative on $]-2, -1[$ and $]1, 2[$

35. max 1.593 at -0.831, min -0.756 at 0.629

37. max $31/3 \approx 10.333$ at $x = 3$, min 4.762 at $x = 1.260$

39. 0.682

41. -0.6367326508, $\quad 1.409624004$

Section 1.5   (page 94)

1. between $12°C$ and $20°C$

3. $(1.99, 2.01)$

5. $(0.81, 1.21)$

7. $\delta = 0.01$

9. $\delta \approx 0.0165$

Review Exercises   (page 95)

1. 13

3. 12

5. 4

7. does not exist

9. does not exist

11. $-\infty$

13. $12\sqrt{3}$

15. 0

17. does not exist

19. $-1/3$

21. $-\infty$

23. ∞

25. does not exist

27. 0

29. 2

31. no disc.

33. disc. and left cont. at 2

35. disc. and right cont. at $x = 1$

37. no disc.

Challenging Problems   (page 96)

1. to the right

3. $-1/4$

5. 3

7. T, F, T, F, F

Chapter 2
Differentiation

Section 2.1   (page 102)

1. $y = 3x - 1$

3. $y = 8x - 13$

5. $y = 12x + 24$

7. $x - 4y = -5$

9. $x - 4y = -2$

11. $y = 2x_0 x - x_0^2$

13. no

15. yes, $x = -2$

17. yes, $x = 0$

19. (a) $3a^2$;   (b) $y = 3x - 2$ and $y = 3x + 2$

21. $(1, 1)$, $(-1, 1)$

23. $k = 3/4$

25. horiz. tangent at $(0, 0)$, $(3, 108)$, $(5, 0)$

27. horiz. tangent at $(-0.5, 1.25)$, no tangents at $(-1, 1)$ and $(1, -1)$

29. horiz. tangent at $(0, -1)$

31. no, consider $y = x^{2/3}$ at $(0, 0)$

Section 2.2   (page 110)

1.

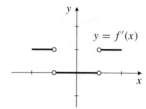

3.

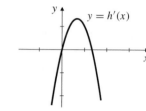

5. on $[-2, 2]$ except at $x = -1$ and $x = 1$

7. slope positive for $x < 1.5$, negative for $x > 1.5$; horizontal tangent at $x = 1.5$

9. singular points at $x = -1$, 0, 1, horizontal tangents at about $x = \pm 0.57$

11. $2x - 3$

13. $3x^2$

15. $\dfrac{1}{\sqrt{2t + 1}}$

17. $1 - \dfrac{1}{x^2}$

19. $-\dfrac{x}{(1 + x^2)^{3/2}}$

21. $-\dfrac{1}{2(1 + x)^{3/2}}$

23. Define $f(0) = 0$, f is not differentiable at 0

25. at $x = -1$ and $x = -2$

27.

x	$\dfrac{f(x) - f(2)}{x - 2}$	x	$\dfrac{f(x) - f(2)}{x - 2}$
1.9	-0.26316	2.1	-0.23810
1.99	-0.25126	2.01	-0.24876
1.999	-0.25013	2.001	-0.24988
1.9999	-0.25001	2.0001	-0.24999

$$\frac{d}{dx}\left(\frac{1}{x}\right)\bigg|_{x=2} = -\frac{1}{4}$$

29. $x - 6y = -15$

31. $y = \dfrac{2}{a^2 + a} - \dfrac{2(2a + 1)}{(a^2 + a)^2}(t - a)$

33. $22t^{21}$, all t

35. $-(1/3)x^{-4/3}$, $x \neq 0$

37. $(119/4)s^{115/4}$, $s \geq 0$

39. -16

41. $1/(8\sqrt{2})$

43. $y = a^2 x - a^3 + \dfrac{1}{a}$

45. $y = 6x - 9$ and $y = -2x - 1$

47. $\dfrac{1}{2\sqrt{2}}$

51. $f'(x) = \frac{1}{3} x^{-2/3}$

Section 2.3   (page 119)

1. $6x - 5$ **3.** $2Ax + B$

5. $\frac{1}{3}s^4 - \frac{1}{5}s^2$

7. $\frac{1}{3}t^{-2/3} + \frac{1}{2}t^{-3/4} + \frac{3}{5}t^{-4/5}$

9. $x^{2/3} + x^{-8/5}$ **11.** $\dfrac{5}{2\sqrt{x}} - \dfrac{3}{2}\sqrt{x} - \dfrac{5}{6}x^{3/2}$

13. $-\dfrac{2x+5}{(x^2+5x)^2}$ **15.** $\dfrac{\pi^2}{(2-\pi t)^2}$

17. $(4x^2 - 3)/x^4$

19. $-t^{-3/2} + (1/2)t^{-1/2} + (3/2)\sqrt{t}$

21. $-\dfrac{24}{(3+4x)^2}$ **23.** $\dfrac{1}{\sqrt{t}(1-\sqrt{t})^2}$

25. $\dfrac{ad-bc}{(cx+d)^2}$

27. $10 + 70x + 150x^2 + 96x^3$

29. $2x(\sqrt{x}+1)(5x^{2/3}-2) + \dfrac{1}{2\sqrt{x}}(x^2+4)(5x^{2/3}-2)$

$\quad + \dfrac{10}{3}x^{-1/3}(x^2+4)(\sqrt{x}+1)$

31. $\dfrac{6x+1}{(6x^2+2x+1)^2}$ **33.** -1

35. 20 **37.** $-\dfrac{1}{2}$

39. $-\dfrac{1}{18\sqrt{2}}$ **41.** $y = 4x - 6$

43. $(1, 2)$ and $(-1, -2)$ **45.** $\left(-\frac{1}{2}, \frac{4}{3}\right)$

47. $y = b - \dfrac{b^2 x}{4}$

49. $y = 12x - 16,\ y = 3x + 2$

51. $x/\sqrt{x^2+1}$

Section 2.4   (page 125)

1. $12(2x+3)^5$ **3.** $-20x(4-x^2)^9$

5. $\dfrac{30}{t^2}\left(2+\dfrac{3}{t}\right)^{-11}$ **7.** $\dfrac{12}{(5-4x)^2}$

9. $-2x\,\mathrm{sgn}\,(1-x^2)$ **11.** $\begin{cases} 8 & \text{if } x > 1/4 \\ 0 & \text{if } x < 1/4 \end{cases}$

13. $\dfrac{-3}{2\sqrt{3x+4}(2+\sqrt{3x+4})^2}$

15. $-\dfrac{5}{3}\left(1-\dfrac{1}{(u-1)^2}\right)\left(u+\dfrac{1}{u-1}\right)^{-8/3}$

17.

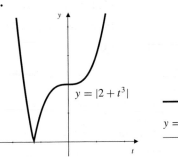

23. $(5-2x)f'(5x-x^2)$ **25.** $\dfrac{f'(x)}{\sqrt{3+2f(x)}}$

27. $\dfrac{1}{\sqrt{x}}f'(3+2\sqrt{x})$

29. $15f'(4-5t)f'(2-3f(4-5t))$

31. $\dfrac{3}{2\sqrt{2}}$ **33.** 102

35. $-6\left(1 - \frac{15}{2}(3x)^4\left((3x)^5-2\right)^{-3/2}\right)$

$\quad \times \left(x + \left((3x)^5-2\right)^{-1/2}\right)^{-7}$

37. $y = 2^{3/2} - \sqrt{2}(x+1)$ **39.** $y = \frac{1}{27} + \frac{5}{162}(x+2)$

41. $\dfrac{x(x^4+2x^2-2)}{(x^2+1)^{5/2}}$ **43.** $857{,}592$

45. no; yes; both functions are equal to x^2.

Section 2.5   (page 131)

3. $-3\sin 3x$ **5.** $\pi\sec^2\pi x$

7. $3\csc^2(4-3x)$ **9.** $r\sin(s-rx)$

11. $2\pi x\cos(\pi x^2)$ **13.** $\dfrac{-\sin x}{2\sqrt{1+\cos x}}$

15. $-(1+\cos x)\sin(x+\sin x)$

17. $(3\pi/2)\sin^2(\pi x/2)\cos(\pi x/2)$

19. $a\cos 2at$ **21.** $2\cos(2x) + 2\sin(2x)$

23. $\sec^2 x - \csc^2 x$ **25.** $\tan^2 x$

27. $-t\sin t$ **29.** $1/(1+\cos x)$

31. $2x\cos(3x) - 3x^2\sin(3x)$

33. $2x[\sec(x^2)\tan^2(x^2) + \sec^3(x^2)]$

35. $-\sec^2 t\,\sin(\tan t)\cos(\cos(\tan t))$

39. $y = \pi - x,\ y = x - \pi$

41. $y = 1 - (x-\pi)/4,\ y = 1 + 4(x-\pi)$

43. $y = \dfrac{1}{\sqrt{2}} + \dfrac{\pi}{180\sqrt{2}}(x-45)$

45. $\pm(\pi/4, 1)$ **49.** yes, (π, π)

51. yes, $(2\pi/3, (2\pi/3)+\sqrt{3}),\ (4\pi/3, (4\pi/3)-\sqrt{3})$

53. 2 **55.** 1

57. $1/2$

59. infinitely many, $0.336508, 0.161228$

Section 2.6 (page 139)

1. $c = \dfrac{a+b}{2}$

3. $c = \pm\dfrac{2}{\sqrt{3}}$

9. inc. on $\left]-\infty, -\dfrac{2}{\sqrt{3}}\right[$ and $\left]\dfrac{2}{\sqrt{3}}, \infty\right[$, dec. on $\left]-\dfrac{2}{\sqrt{3}}, \dfrac{2}{\sqrt{3}}\right[$

11. inc. on $]-2, 0[$ and $]2, \infty[$; dec. on $]-\infty, -2[$ and $]0, 2[$

13. inc. on $]-\infty, 3[$ and $]5, \infty[$; dec. on $]3, 5[$

15. inc. on $]-\infty, \infty[$

17. The two separate applications of MVT cannot be expected to give the same value of c.

Section 2.7 (page 145)

1. 4%

3. -4%

5. 1%

7. 6%

9. $8 \text{ ft}^2/\text{ft}$

11. $1/\sqrt{\pi A}$ units/square unit

13. $16\pi \text{ m}^3/\text{m}$

15. $\dfrac{dC}{dA} = \sqrt{\dfrac{\pi}{A}}$ length units/area unit

17. CP. $x = 0$, incr. $x > 0$, decr. $x < 0$

19. CP. $x = 0$, $x = -4$, incr. on $]-\infty, -4[$ and $]0, \infty[$, decr. on $]-4, 0[$

23. 0.535898, 7.464102 **25.** $0, -0.518784$

27. (a) $10,500$ L/min, $3,500$ L/min, (b) $7,000$ L/min

29. decreases at $1/8$ lb/mi

31. (a) $\$300$, (b) $C(101) - C(100) = \$299.50$

33. (a) $-\$2.00$, (b) $\$9.11$

Section 2.8 (page 150)

1. $\begin{cases} y' = -14(3-2x)^6, \\ y'' = 168(3-2x)^5, \\ y''' = -1680(3-2x)^4 \end{cases}$

3. $\begin{cases} y' = -12(x-1)^{-3}, \\ y'' = 36(x-1)^{-4}, \\ y''' = -144(x-1)^{-5} \end{cases}$

5. $\begin{cases} y' = \frac{1}{3}x^{-2/3} + \frac{1}{3}x^{-4/3}, \\ y'' = -\frac{2}{9}x^{-5/3} - \frac{4}{9}x^{-7/3} \\ y''' = \frac{10}{27}x^{-8/3} + \frac{28}{27}x^{-10/3} \end{cases}$

7. $\begin{cases} y' = \frac{5}{2}x^{3/2} + \frac{3}{2}x^{-1/2} \\ y'' = \frac{15}{4}x^{1/2} - \frac{3}{4}x^{-3/2} \\ y''' = \frac{15}{8}x^{-1/2} + \frac{9}{8}x^{-5/2} \end{cases}$

9. $y' = \sec^2 x$, $y'' = 2\sec^2 x \tan x$, $y''' = 4\sec^2 x \tan^2 x + 2\sec^4 x$

11. $y' = -2x\sin(x^2)$, $y'' = -2\sin(x^2) - 4x^2\cos(x^2)$, $y''' = -12x\cos(x^2) + 8x^3\sin(x^2)$

13. $(-1)^n n! x^{-(n+1)}$ **15.** $n!(2-x)^{-(n+1)}$

17. $(-1)^n n! b^n (a+bx)^{-(n+1)}$

19. $f^{(n)} = \begin{cases} (-1)^k a^n \cos(ax) & \text{if } n = 2k \\ (-1)^{k+1} a^n \sin(ax) & \text{if } n = 2k+1 \end{cases}$ where $k = 0, 1, 2, \ldots$

21. $f^{(n)} = (-1)^k[a^n x \sin(ax) - na^{n-1}\cos(ax)]$ if $n = 2k$, or $(-1)^k[a^n x \cos(ax) + na^{n-1}\sin(ax)]$ if $n = 2k+1$, where $k = 0, 1, 2, \ldots$

23. $-\dfrac{1 \times 3 \times 5 \times \cdots \times (2n-3)}{2^n} 3^n (1-3x)^{-(2n-1)/2}$, $(n = 2, 3, \ldots)$

31. If $f^{(n)}$ exists on an interval I and f vanishes at $n+1$ distinct points of I then $f^{(n)}$ vanishes at at least one point of I.

Section 2.9 (page 156)

1. $\dfrac{1-y}{2+x}$

3. $\dfrac{2x+y}{3y^2-x}$

5. $\dfrac{2-2xy^3}{3x^2y^2+1}$

7. $-\dfrac{3x^2+2xy}{x^2+4y}$

9. $2x + 3y = 5$

11. $y = x$

13. $y = 1 - \dfrac{4}{4-\pi}\left(x - \dfrac{\pi}{4}\right)$

15. $y = 2 - x$

17. $\dfrac{2(y-1)}{(1-x)^2}$

19. $\dfrac{(2-6y)(1-3x^2)^2}{(3y^2-2y)^3} - \dfrac{6x}{3y^2-2y}$

21. $-a^2/y^3$

23. 0

25. -26

Section 2.10 (page 162)

1. $5x + C$

3. $\frac{2}{3}x^{3/2} + C$

5. $\frac{1}{4}x^4 + C$

7. $-\cos x + C$

9. $a^2 x - \frac{1}{3}x^3 + C$

11. $\frac{4}{3}x^{3/2} + \frac{9}{4}x^{4/3} + C$

13. $\frac{1}{12}x^4 - \frac{1}{6}x^3 + \frac{1}{2}x^2 - x + C$

15. $\frac{1}{2}\sin(2x) + C$

17. $\dfrac{-1}{1+x} + C$

19. $\frac{1}{3}(2x+3)^{3/2} + C$

21. $-\cos(x^2) + C$

23. $\tan x - x + C$

25. $(x + \sin x \cos x)/2 + C$

27. $y = \frac{1}{2}x^2 - 2x + 3$, all x

29. $y = 2x^{3/2} - 15$, $(x > 0)$

31. $y = \dfrac{A}{3}(x^3 - 1) + \dfrac{B}{2}(x^2 - 1) + C(x - 1) + 1$, (all x)

33. $y = \sin x + (3/2)$, (all x)

35. $y = 1 + \tan x$, $-\pi/2 < x < \pi/2$

37. $y = x^2 + 5x - 3$, (all x)

39. $y = \dfrac{x^5}{20} - \dfrac{x^2}{2} + 8$, (all x)

41. $y = 1 + x - \cos x$, (all x)

43. $y = 3x - \dfrac{1}{x}$, $(x > 0)$

45. $y = -\dfrac{7\sqrt{x}}{2} + \dfrac{18}{\sqrt{x}}$, $(x > 0)$

Section 2.11   (page 169)

1. (a) $t > 2$,   (b) $t < 2$,   (c) all t,   (d) no t,
(e) $t > 2$,   (f) $t < 2$,   (g) 2,   (h) 0

3. (a) $t < -2/\sqrt{3}$ or $t > 2/\sqrt{3}$,
(b) $-2/\sqrt{3} < t < 2/\sqrt{3}$,   (c) $t > 0$,   (d) $t < 0$,
(e) $t > 2/\sqrt{3}$ or $-2/\sqrt{3} < t < 0$,
(f) $t < -2/\sqrt{3}$ or $0 < t < 2/\sqrt{3}$,
(g) $\pm 12/\sqrt{3}$ at $t = \pm 2/\sqrt{3}$,   (h) 12

5. acc $= 9.8$ m/s^2 downward at all times;
max height $= 4.9$ m; ball strikes ground at 9.8 m/s

7. time 27.8 s; distance 771.6 m

9. $4h$ m, $\sqrt{2}v_0$ m/s   **11.** 400 ft

13. 0.833 km

15. $v = \begin{cases} 2t & \text{if } 0 < t \le 2 \\ 4 & \text{if } 2 < t < 8 \\ 20 - 2t & \text{if } 8 \le t < 10 \end{cases}$
v is continuous for $0 < t < 10$.
$a = \begin{cases} 2 & \text{if } 0 < t < 2 \\ 0 & \text{if } 2 < t < 8 \\ -2 & \text{if } 8 < t < 10 \end{cases}$
a is continuous except at $t = 2$ and $t = 8$.
Maximum velocity 4 is attained for $2 \le t \le 8$.

17. 7 s   **19.** 448 ft

Review Exercises   (page 171)

1. $18x + 6$ **3.** -1

5. $6\pi x + 12y = 6\sqrt{3} + \pi$

7. $\dfrac{\cos x - 1}{(x - \sin x)^2}$ **9.** $x^{-3/5}(4 - x^{2/5})^{-7/2}$

11. $-2\theta \sec^2 \theta \tan \theta$ **13.** $20x^{19}$

15. $-\sqrt{3}$ **17.** $-2xf'(3 - x^2)$

19. $2f'(2x)\sqrt{g(x/2)} + \dfrac{f(2x)\,g'(x/2)}{4\sqrt{g(x/2)}}$

21. $f'(x + (g(x))^2)(1 + 2g(x)g'(x))$

23. $\cos x\, f'(\sin x)\, g(\cos x) - \sin x\, f(\sin x)\, g'(\cos x)$

25. $7x + 10y = 24$ **27.** $\dfrac{x^3}{3} - \dfrac{1}{x} + C$

29. $2\tan x + 3\sec x + C$ **31.** $4x^3 + 3x^4 - 7$

33. $I_1 = x\sin x + \cos x + C$, $I_2 = \sin x - x\cos x + C$

35. $y = 3x$

37. points $k\pi$ and $k\pi/(n + 1)$, where k is any integer

39. $(0, 0)$, $(\pm 1/\sqrt{2}, 1/2)$, dist. $= \sqrt{3}/2$ units

41. (a) $k = g/R$ **43.** 15.3 m

45. 80 ft/s, or about 55 mph

Challenging Problems   (page 172)

3. (a) 0, (b) 3/8, (c) 12, (d) -48, (e) 3/7, (f) 21

13. $f(m) = C - (m - B)^2/(4A)$

17. (a) $3b^2 > 8ac$

19. (a) 3 s, (b) $t = 7$ s, (c) $t = 12$ s, (d) about 13.07 m/s^2,
(e) 197.5 m, (f) 60.3 m.

Chapter 3
Transcendental Functions

Section 3.1   (page 181)

1. $f^{-1}(x) = x + 1$
$\mathcal{D}(f^{-1}) = \mathcal{R}(f) = \mathcal{R}(f^{-1}) = \mathcal{D}(f) = \mathbb{R}$

3. $f^{-1}(x) = x^2 + 1$, $\mathcal{D}(f^{-1}) = \mathcal{R}(f) = [0, \infty[$,
$\mathcal{R}(f^{-1}) = \mathcal{D}(f) = [1, \infty[$

5. $f^{-1}(x) = x^{1/3}$
$\mathcal{D}(f^{-1}) = \mathcal{R}(f) = \mathcal{R}(f^{-1}) = \mathcal{D}(f) = \mathbb{R}$

7. $f^{-1}(x) = -\sqrt{x}$, $\mathcal{D}(f^{-1}) = \mathcal{R}(f) = [0, \infty[$,
$\mathcal{R}(f^{-1}) = \mathcal{D}(f) =]-\infty, 0]$

9. $f^{-1}(x) = \dfrac{1}{x} - 1$, $\mathcal{D}(f^{-1}) = \mathcal{R}(f) = \{x : x \ne 0\}$,
$\mathcal{R}(f^{-1}) = \mathcal{D}(f) = \{x : x \ne -1\}$

11. $f^{-1}(x) = \dfrac{1 - x}{2 + x}$,
$\mathcal{D}(f^{-1}) = \mathcal{R}(f) = \{x : x \ne -2\}$,
$\mathcal{R}(f^{-1}) = \mathcal{D}(f) = \{x : x \ne -1\}$

13. $g^{-1}(x) = f^{-1}(x + 2)$ **15.** $k^{-1}(x) = f^{-1}\left(-\dfrac{x}{3}\right)$

17. $p^{-1}(x) = f^{-1}\left(\dfrac{1}{x} - 1\right)$

19. $r^{-1}(x) = \dfrac{1}{4}\left(3 - f^{-1}\left(\dfrac{1 - x}{2}\right)\right)$

21. $f^{-1}(x) = \begin{cases} \sqrt{x - 1} & \text{if } x >= 1 \\ x - 1 & \text{if } x < 1 \end{cases}$

23. $h^{-1}(x) = \begin{cases} \sqrt{x-1} & \text{if } x \geq 1 \\ \sqrt{1-x} & \text{if } x < 1 \end{cases}$

25. $g^{-1}(1) = 2$ **29.** $1/[6(f^{-1}(x))^2]$

31. 2.23362   **33.** $\mathbb{R}$, 1

35. $c = 1$, a, b arbitrary, or $a = b = 0$, $c = -1$.

37. no

Section 3.2   (page 185)

1. $\sqrt{3}$ **3.** x^6

5. 3   **7.** $-2x$

9. x **11.** 1

13. 1   **15.** 2

17. $\log_a(x^4 + 4x^2 + 3)$ **19.** $4.728804\ldots$

21. $x = (\log_{10} 5)/(\log_{10}(4/5)) \approx -7.212567$

23. $x = 3^{1/5} = 10^{(\log_{10} 3)/5} \approx 1.24573$

29. $1/2$ **31.** 0

33. ∞

Section 3.3   (page 195)

1. $\sqrt{e}$ **3.** x^5

5. $-3x$ **7.** $\ln \dfrac{64}{81}$

9. $\ln(x^2(x-2)^5)$ **11.** $x = \dfrac{\ln 2}{\ln(3/2)}$

13. $x = \dfrac{\ln 5 - 9\ln 2}{2\ln 2}$ **15.** $0 < x < 2$

17. $3 < x < 7/2$ **19.** $5e^{5x}$

21. $(1 - 2x)e^{-2x}$ **23.** $\dfrac{3}{3x - 2}$

25. $\dfrac{e^x}{1 + e^x}$ **27.** $\dfrac{e^x - e^{-x}}{2}$

29. e^{x+e^x} **31.** $\dfrac{e^x}{(1 + e^x)^2}$

33. $e^x(\sin x - \cos x)$ **35.** $\dfrac{1}{x \ln x}$

37. $2x \ln x$ **39.** $(2 \ln 5)5^{2x+1}$

41. $t^x x^t \ln t + t^{x+1} x^{t-1}$ **43.** $\dfrac{b}{(bs + c)\ln a}$

45. $x^{\sqrt{x}} \left(\dfrac{1}{\sqrt{x}} \left(\tfrac{1}{2} \ln x + 1 \right) \right)$

47. $\sec x$ **49.** $-\dfrac{1}{\sqrt{x^2 + a^2}}$

51. $f^{(n)}(x) = e^{ax}(na^{n-1} + a^n x)$, $n = 1, 2, 3, \ldots$

53. $y' = 2xe^{x^2}$, $y'' = 2(1 + 2x^2)e^{x^2}$, $y''' = 4(3x + 2x^3)e^{x^2}$, $y^{(4)} = 4(3 + 12x^2 + 4x^4)e^{x^2}$

55. $f'(x) = x^{x^2+1}(2\ln x + 1)$, $g'(x) = x^{x^x} x^x \left(\ln x + (\ln x)^2 + \dfrac{1}{x} \right)$; g grows more rapidly than does f.

57. $f'(x) = f(x) \left(\dfrac{1}{x-1} + \dfrac{1}{x-2} + \dfrac{1}{x-3} + \dfrac{1}{x-4} \right)$

59. $f'(2) = \dfrac{556}{3675}$, $f'(1) = \dfrac{1}{6}$

61. f inc. for $x < 1$, dec. for $x > 1$

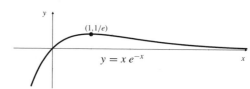

63. $y = ex$ **65.** $y = 2e \ln 2(x - 1)$

67. $-1/e^2$

69. $f'(x) = (A + B)\cos \ln x + (B - A)\sin \ln x$, $\int \cos \ln x \, dx = \dfrac{x}{2}(\cos \ln x + \sin \ln x)$, $\int \sin \ln x \, dx = \dfrac{x}{2}(\sin \ln x - \cos \ln x)$

71. (a) $F_{2B,-2A}(x)$;   (b) $-2e^x(\cos x + \sin x)$

Section 3.4   (page 203)

1. 0   **3.** 2

5. 0   **7.** 0

9. 566   **11.** 29.15 years

13. 160.85 years   **15.** 4,139 g

17. $7,557.84   **19.** about 14.7 years

21. (a) $f(x) = Ce^{bx} - (a/b)$,   (b) $y = (y_0 + (a/b))e^{bx} - (a/b)$

23. $22.35°C$ **25.** 6.84 min

29. $(0, -(1/k)\ln(y_0/(y_0 - L)))$, solution $\to -\infty$

31. about 7,671 cases, growing at about 3,028 cases/week

Section 3.5   (page 212)

1. $\pi/3$ **3.** $-\pi/4$

5. 0.7   **7.** $-\pi/3$

9. $\dfrac{\pi}{2} + 0.2$ **11.** $2/\sqrt{5}$

13. $\sqrt{1 - x^2}$ **15.** $\dfrac{1}{\sqrt{1 + x^2}}$

17. $\dfrac{\sqrt{1 - x^2}}{x}$ **19.** $\dfrac{1}{\sqrt{2 + x - x^2}}$

21. $\dfrac{-\operatorname{sgn} a}{\sqrt{a^2 - (x-b)^2}}$

23. $\tan^{-1} t + \dfrac{t}{1+t^2}$

25. $2x \tan^{-1} x + 1$

27. $\dfrac{\sqrt{1-4x^2}\,\sin^{-1} 2x - 2\sqrt{1-x^2}\,\sin^{-1} x}{\sqrt{1-x^2}\sqrt{1-4x^2}\,\left(\sin^{-1} 2x\right)^2}$

29. $\dfrac{x}{\sqrt{(1-x^4)\,\sin^{-1} x^2}}$

31. $\sqrt{\dfrac{a-x}{a+x}}$

33. $\dfrac{\pi - 2}{\pi - 1}$

37. $\dfrac{d}{dx}\csc^{-1} x = -\dfrac{1}{|x|\sqrt{x^2-1}}$

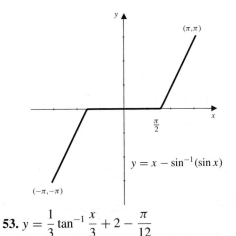

39. $\tan^{-1} x + \cot^{-1} x = -\dfrac{\pi}{2}$ for $x < 0$

41. cont. everywhere, differentiable except at $n\pi$ for integers n

43. continuous and differentiable everywhere except at odd multiples of $\pi/2$

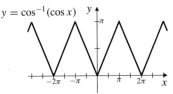

$y = \cos^{-1}(\cos x)$

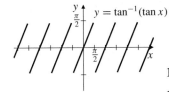

$y = \tan^{-1}(\tan x)$

49. $\tan^{-1}\left(\dfrac{x-1}{x+1}\right) - \tan^{-1} x = \dfrac{3\pi}{4}$ on $(-\infty, -1)$

51. $f'(x) = 1 - \operatorname{sgn}(\cos x)$

$y = x - \sin^{-1}(\sin x)$

53. $y = \dfrac{1}{3}\tan^{-1}\dfrac{x}{3} + 2 - \dfrac{\pi}{12}$

55. $y = 4\sin^{-1}\dfrac{x}{5}$

Section 3.6   (page 218)

3. $\tanh(x+y) = \dfrac{\tanh x + \tanh y}{1 + \tanh x \tanh y}$

$\tanh(x-y) = \dfrac{\tanh x - \tanh y}{1 - \tanh x \tanh y}$

5. $\dfrac{d}{dx}\sinh^{-1}(x) = \dfrac{1}{\sqrt{x^2+1}}$,

$\dfrac{d}{dx}\cosh^{-1}(x) = \dfrac{1}{\sqrt{x^2-1}}$,

$\dfrac{d}{dx}\tanh^{-1}(x) = \dfrac{1}{1-x^2}$,

$\displaystyle\int \dfrac{dx}{\sqrt{x^2+1}} = \sinh^{-1}(x) + C$,

$\displaystyle\int \dfrac{dx}{\sqrt{x^2-1}} = \cosh^{-1}(x) + C \quad (x > 1)$,

$\displaystyle\int \dfrac{dx}{1-x^2} = \tanh^{-1}(x) + C \quad (-1 < x < 1)$

7. (a) $\dfrac{x^2-1}{2x}$; (b) $\dfrac{x^2+1}{2x}$; (c) $\dfrac{x^2-1}{x^2+1}$; (d) x^2

9. $\coth^{-1} x = \tanh^{-1}\dfrac{1}{x} = \dfrac{1}{2}\ln\left(\dfrac{x+1}{x-1}\right)$, domain: all x such that $|x| > 1$, range: all $y \neq 0$, derivative: $-1/(x^2-1)$

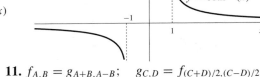

$y = \coth^{-1}(x)$

11. $f_{A,B} = g_{A+B, A-B}$; $g_{C,D} = f_{(C+D)/2, (C-D)/2}$

13. $y = y_0 \cosh k(x-a) + \dfrac{v_0}{k}\sinh k(x-a)$

Section 3.7   (page 228)

1. $y = Ae^{-5t} + Be^{-2t}$ **3.** $y = A + Be^{-2t}$

5. $y = (A + Bt)e^{-4t}$

7. $y = (A\cos t + B\sin t)e^{3t}$

9. $y = (A\cos 2t + B\sin 2t)e^{-t}$

11. $y = (A\cos\sqrt{2}t + B\sin\sqrt{2}t)e^{-t}$

13. $y = \dfrac{6}{7}e^{t/2} + \dfrac{1}{7}e^{-3t}$

15. $y = e^{-2t}(2\cos t + 6\sin t)$

25. $y = \dfrac{3}{10}\sin(10t)$, circ freq 10, freq $\dfrac{10}{2\pi}$, per $\dfrac{2\pi}{10}$, amp $\dfrac{3}{10}$

33. $y = e^{3-t}[2\cos(2(t-3)) + \sin(2(t-3))]$

35. $y = -\dfrac{1}{2} + C_1 e^t + C_2 e^{-2t}$

37. $y = -\dfrac{1}{2}e^{-t} + C_1 e^t + C_2 e^{-2t}$

39. $y = -e^x \sin x + 3e^x \cos x + C_1 e^x + C_2 e^{-2x}$

41. $y = C_1 t^{r_1} + C_2 t^{r_2}$ **45.** $y = C_1 t^2 + C_2 t^2 \ln t$

Review Exercises   (page 230)

1. $1/3$ **3.** both limits are 0

5. max $1/\sqrt{2e}$, min $-1/\sqrt{2e}$

7. $f(x) = 3e^{(x^2/2)-2}$

9. (a) about 13.863%, (b) about 68 days

11. e^{2x} **13.** $y = x$

15. 13.8165% approx.

17. $\cos^{-1} x = \frac{\pi}{2} - \sin^{-1} x, \cot^{-1} x = \operatorname{sgn} x \sin^{-1}(1/\sqrt{x^2+1})$,
$\csc^{-1} x = \sin^{-1}(1/x)$

19. 15°C

Chapter 4
Some Applications of Derivatives

Section 4.1   (page 237)

1. 32 cm²/min

3. increasing at 160π cm²/s

5. (a) $1/(6\pi r)$ km/hr, (b) $1/(6\sqrt{\pi A})$ km/hr

7. $1/(180\pi)$ cm/s   **9.** 2 cm²/s

11. increasing at 2 cm³/s   **13.** increasing at rate 12

15. increasing at rate $2/\sqrt{5}$

17. $45\sqrt{3}$ km/h   **19.** 1/3 m/s, 5/6 m/s

21. 100 tons/day   **23.** $16\frac{4}{11}$ min after 3:00

25. $1/(18\pi)$ m/min

27. $9/(6250\pi)$ m/min, 4.64 m

29. 8 m/min   **31.** dec. at 126.9 km/h

33. 1/8 units/s   **35.** $\sqrt{3}/16$ m/min

37. (a) down at 24/125 m/s, (b) right at 7/125 m/s

39. dec. at 0.0197 rad/s   **41.** 0.047 rad/s

Section 4.2   (page 246)

1. abs min 1 at $x = -1$; abs max 3 at $x = 1$

3. abs min 1 at $x = -1$; no max

5. abs min -1 at $x = 0$; abs max 8 at $x = 3$; loc max 3
at $x = -2$

7. abs min $a^3 + a - 4$ at $x = a$; abs max $b^3 + b - 4$ at
$x = b$

9. abs max $b^5 + b^3 + 2b$ at $x = b$; no min value

11. no max or min values

13. max 3 at $x = -2$, min 0 at $x = 1$

15. abs max 1 at $x = 0$; no min value

17. no max or min value

19. loc max at $x = -1$; loc min at $x = 1$

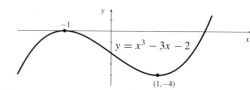

21. loc max at $x = \frac{1}{3}$; loc min at $x = 1$

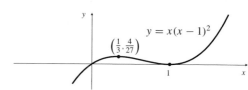

23. loc max at $x = \frac{3}{5}$; loc min at $x = 1$;
critical point $x = 0$ is neither max nor min

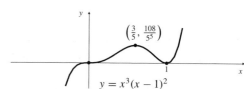

25. loc max at $x = -1$ and $x = 1/\sqrt{5}$; loc min at $x = 1$
and $x = -1/\sqrt{5}$

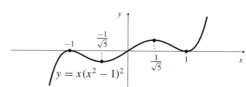

27. abs min at $x = 0$

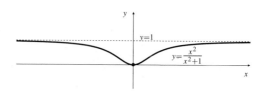

29. loc min at CP $x = -1$ and endpoint SP $x = \sqrt{2}$;
loc max at CP $x = 1$ and endpoint SP $x = -\sqrt{2}$

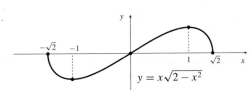

$y = x\sqrt{2 - x^2}$

31. loc max at $x = 2n\pi - \dfrac{\pi}{3}$; loc min at $x = 2n\pi + \dfrac{\pi}{3}$
$(n = 0, \pm 1, \pm 2, \ldots)$

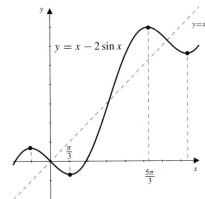

$y = x - 2\sin x$

$y = x$

33. loc max at CP $x = \sqrt{3}/2$ and endpoint SP $x = -1$;
loc min at CP $x = -\sqrt{3}/2$ and endpoint SP $x = 1$

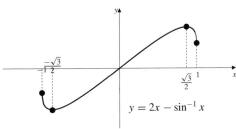

$y = 2x - \sin^{-1} x$

35. abs max at $x = 1/\ln 2$

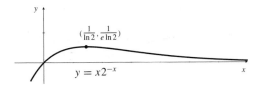

$\left(\dfrac{1}{\ln 2}, \dfrac{1}{e\ln 2}\right)$

$y = x2^{-x}$

37. abs max at $x = e$

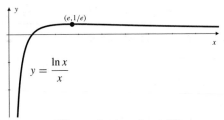

$(e, 1/e)$

$y = \dfrac{\ln x}{x}$

39. loc max at CP $x = 0$; abs min at SPs
$x = \pm 1$

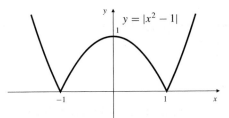

$y = |x^2 - 1|$

41. abs max at CPs $x = (2n + 1)\pi/2$; abs min at SPs
$x = n\pi$ $(n = 0, \pm 1, \pm 2, \ldots)$

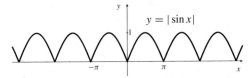

$y = |\sin x|$

43. no max or min   **45.** max 2, min -2

47. has min, no max   **49.** yes, no

Section 4.3   (page 251)

1. conc down on $]0, \infty[$ **3.** conc up on $\mathbb{R}$

5. conc down on $]-1, 0[$ and $]1, \infty[$; conc up on
$]-\infty, -1[$ and $]0, 1[$; infl $x = -1, 0, 1$

7. conc down on $]-1, 1[$; conc up on $]-\infty, -1[$ and
$]1, \infty[$; infl $x = \pm 1$

9. conc down on $]-2, -2/\sqrt{5}[$ and $]2/\sqrt{5}, 2[$; conc
up on $]-\infty, -2[$, $]-2/\sqrt{5}, 2/\sqrt{5}[$ and $]2, \infty[$; infl
$x = \pm 2, \ \pm 2/\sqrt{5}$

11. conc down on $]2n\pi, (2n + 1)\pi[$; conc up on
$](2n - 1)\pi, 2n\pi[$, $(n = 0, \pm 1, \pm 2, \ldots)$; infl $x = n\pi$

13. conc down on $]n\pi, (n + \frac{1}{2})\pi[$;
conc up on $](n - \frac{1}{2})\pi, n\pi[$; infl $x = n\pi/2$,
$(n = 0, \pm 1, \pm 2, \ldots)$

15. conc down on $]0, \infty[$, up on $]-\infty, 0[$; infl $x = 0$

17. conc down on $]-1/\sqrt{2}, 1/\sqrt{2}[$, up on $]-\infty, -1/\sqrt{2}[$
and $]1/\sqrt{2}, \infty[$; infl $x = \pm 1/\sqrt{2}$

19. conc down on $]-\infty, -1[$ and $]1, \infty[$; conc up on
$]-1, 1[$; infl $x = \pm 1$

21. conc down on $]-\infty, 4[$, up on $]4, \infty[$; infl $x = 4$

23. no concavity, no inflections

25. loc min at $x = 2$; loc max at $x = \frac{2}{3}$

27. loc min at $x = 1/\sqrt[4]{3}$; loc max at $-1/\sqrt[4]{3}$

29. loc max at $x = 1$; loc min at $x = -1$ (both abs)

31. loc (and abs) min at $x = 1/e$

33. loc min at $x = 0$; inflections at $x = \pm 2$ (not discernible by Second Derivative Test)

35. abs min at $x = 0$; abs max at $x = \pm 1/\sqrt{2}$

39. If n is even, f_n has a min and g_n has a max at $x = 0$. If n is odd both have inflections at $x = 0$.

Section 4.4 (page 261)

1. (a) g, (b) f'', (c) f, (d) f'

3. (a) $k(x)$, (b) $g(x)$, (c) $f(x)$, (d) $h(x)$

5.

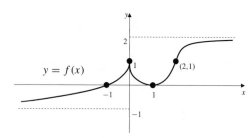

7.

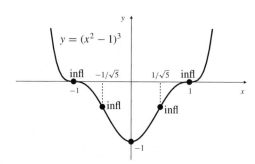

9.

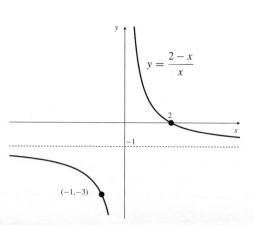

11.

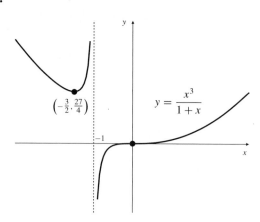

13.

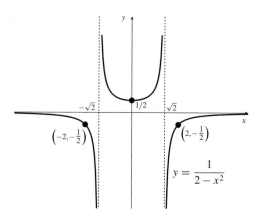

15.

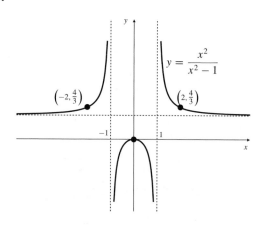

17.

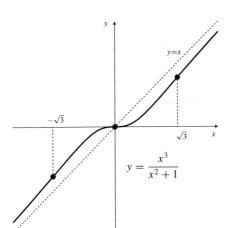

$$y = \frac{x^3}{x^2 + 1}$$

19.

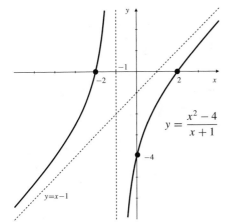

$$y = \frac{x^2 - 4}{x + 1}$$

21.

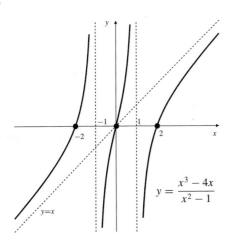

$$y = \frac{x^3 - 4x}{x^2 - 1}$$

23.

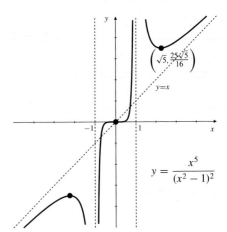

$$y = \frac{x^5}{(x^2 - 1)^2}$$

25.

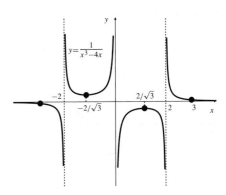

27.

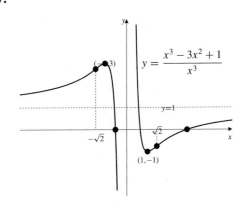

29.

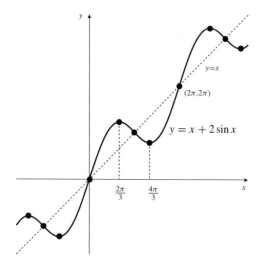

31.

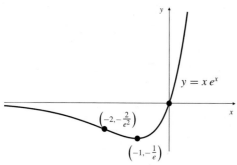

33.

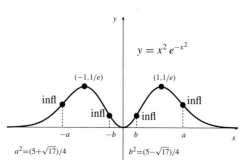

35.

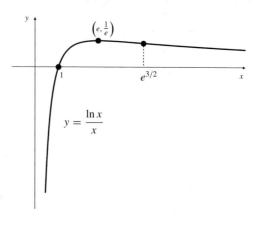

37.

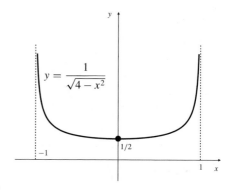

39.

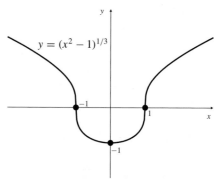

41. $y = 0$. curve crosses asymptote at $x = n\pi$ for every integer n.

Section 4.5   (page 269)

1. $49/4$ **3.** 20 and 40

5. 71.45 **11.** R^2 sq. units

13. $2ab$ un^2

15. width $8 + 10\sqrt{2}$ m, height $4 + 5\sqrt{2}$ m

17. rebate $250 **19.** point 5 km east of A

21. (a) 0 m, (b) $\pi/(4 + \pi)$ m

23. $8\sqrt{3}$ units

25. $\left[(a^{2/3} + b^{2/3})^3 + c^2\right]^{1/2}$ units

27. $3^{1/2}/2^{1/3}$ units

29. height $\dfrac{2R}{\sqrt{3}}$, radius $\sqrt{\dfrac{2}{3}}\,R$ units

31. base 2m×2m, height 1m

33. width $\dfrac{20}{4+\pi}$ m, height $\dfrac{10}{4+\pi}$ m

37. width R, depth $\sqrt{3}R$ **39.** $Q = 3L/8$

41. $2\sqrt{6}$ ft   **43.** $\dfrac{2\pi}{9\sqrt{3}}R^3$ cubic units

Section 4.6 (page 278)

1. 1.41421356237   **3.** 0.453397651516

5. 1.64809536561, 2.352392647658

7. 0.510973429389

9. infinitely many, 4.49340945791

13. max 1, min $-0.11063967219\ldots$

15. $x_1 = -a$, $x_2 = a = x_0$. Look for a root half way between x_0 and x_1

17. $x_n = (-1/2)^n \to 0$ (root) as $n \to \infty$.

19. 0.95025   **21.** 0.45340

23. $N(x_n)$ is the Newton's Method approximation x_{n+1}

Section 4.7 (page 284)

1. $6x - 9$ **3.** $2 - (x/4)$

5. $(7 - 2x)/27$ **7.** $\pi - x$

9. $(1/4) + (\sqrt{3}/2)(x - (\pi/6))$

11. about 8 cm^2 **13.** about 62.8 mi

15. $\sqrt{50} \approx \frac{99}{14} \approx 7.071429$, error < 0,
$|\text{error}| < \frac{1}{2744} \approx 0.0003644$,]7.07106, 7.071429[

17. $\sqrt[4]{85} \approx \frac{82}{27}$, error < 0, $|\text{error}| < \frac{1}{2\times3^6}$,]3.03635, 3.03704[

19. $\cos 46° \approx \dfrac{1}{\sqrt{2}}\left(1 - \dfrac{\pi}{180}\right) \approx 0.694765$, error < 0,
$|\text{error}| < \dfrac{1}{2\sqrt{2}}\left(\dfrac{\pi}{180}\right)^2$,]0.694658, 0.694765[

21. $\sin(3.14) \approx \pi - 3.14$, error < 0,
$|\text{error}| < (\pi - 3.14)^3/2 < 2.02 \times 10^{-9}$,
$(\pi - 3.14 - (\pi - 3.14)^3/2, \pi - 3.14)$

23.]7.07106, 7.07108[, $\sqrt{50} \approx 7.07107$

25.]0.80891, 0.80921[, $\sqrt[4]{85} \approx 0.80906$

27. $3 \le f(3) \le 13/4$

29. $g(1.8) \approx 0.6$, $|\text{error}| < 0.0208$

31. about 1005 cm^3

Section 4.8 (page 292)

1. $1 - x + \frac{1}{2}x^2 - \frac{1}{6}x^3 + \frac{1}{24}x^4$

3. $1 + \dfrac{x - e}{e} - \dfrac{(x - e)^2}{2e^2} + \dfrac{(x - e)^3}{3e^3} - \dfrac{(x - e)^4}{4e^4}$

5. $2 + \dfrac{x - 4}{4} - \dfrac{(x - 4)^2}{64} + \dfrac{3(x - 4)^3}{1536}$

7. $x^{1/3} \approx 2 + \frac{1}{12}(x - 8) - \frac{1}{288}(x - 8)^2$, $9^{1/3} \approx 2.07986$,
$0 < \text{error} \le 5/(81 \times 256)$,
$2.07986 < 9^{1/3} < 2.08010$

9. $\dfrac{1}{x} \approx 1 - (x - 1) + (x - 1)^2$, $\dfrac{1}{1.02} \approx 0.9804$,
$-(0.02)^3 \le \text{error} < 0$, $0.980392 \le \dfrac{1}{1.02} < 0.9804$

11. $e^x \approx 1 + x + \frac{1}{2}x^2$, $e^{-0.5} \approx 0.625$,
$-\frac{1}{6}(0.5)^3 \le \text{error} < 0$, $0.604 \le e^{-0.5} < 0.625$

13. $\sin x = x - \dfrac{x^3}{3!} + \dfrac{x^5}{5!} - \dfrac{x^7}{7!} + R_7$;
$R_7 = \dfrac{\sin X}{8!}x^8$ for some X between 0 and x

15. $\sin x = \dfrac{1}{\sqrt{2}}\left[1 + \left(x - \dfrac{\pi}{4}\right) - \dfrac{1}{2!}\left(x - \dfrac{\pi}{4}\right)^2\right.$
$\left. - \dfrac{1}{3!}\left(x - \dfrac{\pi}{4}\right)^3 + \dfrac{1}{4!}\left(x - \dfrac{\pi}{4}\right)^4\right] + R_4$;
where $R_4 = \dfrac{\cos X}{5!}\left(x - \dfrac{\pi}{4}\right)^5$ for some X between x and $\pi/4$

17. $\ln x = (x - 1) - \dfrac{(x - 1)^2}{2} + \dfrac{(x - 1)^3}{3} - \dfrac{(x - 11)^4}{4}$
$+ \dfrac{(x - 1)^5}{5} - \dfrac{(x - 1)^6}{6} + R_6$;
where $R_6 = \dfrac{(x - 1)^7}{7X^7}$ for some X between 1 and x

19. $\dfrac{1}{e^3} + \dfrac{3}{e^3}(x + 1) + \dfrac{9}{2e^3}(x + 1)^2 + \dfrac{9}{2e^3}(x + 1)^3$

21. $x^2 - \frac{1}{3}x^4$ **23.** $1 - 2x^2 + 4x^4 - 8x^6$

25. $x + \dfrac{x^3}{3!} + \dfrac{x^5}{5!} + \cdots + \dfrac{x^{2n+1}}{(2n + 1)!}$

27. $e^{-x} = 1 - x + \dfrac{x^2}{2!} - \dfrac{x^3}{3!} + \cdots + (-1)^n\dfrac{x^n}{n!} + R_n$;
where $R_n = (-1)^{n+1}\dfrac{e^{-X}x^{n+1}}{(n + 1)!}$ for some X between 0 and x;
$\dfrac{1}{e} \approx 1 - \dfrac{1}{2!} + \dfrac{1}{3!} + \cdots + \dfrac{1}{8!} \approx 0.36788$

29. $1 - 2x + x^2$ (f is its own best quadratic approximation); (error $= 0$). $g(x) \approx 4 + 3x + 2x^2$; error $= x^3$; since $g'''(x) = 6 = 3!$, therefore error $= \dfrac{g'''(X)}{3!}x^3$; no improvement possible.

Section 4.9 (page 298)

1. 3/4   **3.** a/b

5. 1

9. 0

13. 1

17. ∞

21. -2

25. 1

29. e^{-2}

33. $f''(x)$

7. 1

11. $-3/2$

15. $-1/2$

19. $2/\pi$

23. a

27. $-1/2$

31. 0

Review Exercises   (page 299)

1. 6%/min

3. (a) $-1,600$ ohms/min, (b) $-1,350$ ohms/min

5. 2,000

7. $32\pi R^3/81$ un^3

9. 9000 cm^3

11. approx 0.057 rad/s

13. about 9.69465 cm

15. 2.06%

17. $\frac{\pi}{4} + 0.0475 \approx 0.83290$, |error| < 0.00011

19. 0, 1.4055636328

21. approx. $(-1.1462, 0.3178)$

Challenging Problems   (page 301)

1. (a) $\dfrac{dx}{dt} = \dfrac{k}{3}(x_0^3 - x^3)$, (b) $V_0/2$

3. (b) 11

5. (c) $y_0(1 - (t/T))^2$, (d) $(1 - (1/\sqrt{2}))T$

7. $P^2(3 - 2\sqrt{2})/4$

9. (a) $\cos^{-1}(r_2/r_1)^2$, (b) $\cos^{-1}(r_2/r_1)^4$.

11. approx 921 cm^3

Chapter 5
Integration

Section 5.1   (page 307)

1. $1^3 + 2^3 + 3^3 + 4^3$

3. $3 + 3^2 + 3^3 + \cdots + 3^n$

5. $\dfrac{(-2)^3}{1^2} + \dfrac{(-2)^4}{2^2} + \dfrac{(-2)^5}{3^2} + \cdots + \dfrac{(-2)^n}{(n-2)^2}$

7. $\sin\dfrac{\pi}{3k} + \sin\dfrac{2\pi}{3k} + \sin\dfrac{3\pi}{3k} + \cdots + + \sin\dfrac{k\pi}{3k}$

9. $\sum_{i=5}^{9} i$

11. $\sum_{i=2}^{99} (-1)^i i^2$

13. $\sum_{i=0}^{n} x^i$

15. $\sum_{i=1}^{n} (-1)^{i-1}/i^2$

17. $\sum_{i=1}^{100} \sin(i - 1)$

19. 15

21. $n(n+1)(2n+7)/6$

23. $\dfrac{\pi(\pi^n - 1)}{\pi - 1} - 3n$

25. $\ln(n!)$

27. 400

29. $(x^{2n+1} + 1)/(x + 1)$

31. $-4,949$

35. $2^m - 1$

37. $n/(n + 1)$

Section 5.2   (page 314)

1. 3/2 sq. un.

3. 6 sq. un.

5. 26/3 sq. un.

7. 15 sq. un.

9. 4 sq. un.

11. 32/3 sq. un.

13. $3/(2\ln 2)$ sq. un.

15. $\ln(b/a)$, follows from definition of ln

17. 0

19. $\pi/4$

Section 5.3   (page 320)

1. $L(f, P_8) = 7/4$, $U(f, P_8) = 9/4$

3. $L(f, P_4) = \dfrac{e^4 - 1}{e^2(e - 1)} \approx 4.22$,

   $U(f, P_4) = \dfrac{e^4 - 1}{e(e - 1)} \approx 11.48$

5. $L(f, P_6) = \dfrac{\pi}{6}(1 + \sqrt{3}) \approx 1.43$,

   $U(f, P_6) = \dfrac{\pi}{6}(3 + \sqrt{3}) \approx 2.48$

7. $L(f, P_n) = \dfrac{n-1}{2n}$, $U(f, P_n) = \dfrac{n+1}{2n}$,

   $\int_0^1 x\, dx = \dfrac{1}{2}$

9. $L(f, P_n) = \dfrac{(n-1)^2}{4n^2}$, $U(f, P_n) = \dfrac{(n+1)^2}{4n^2}$,

   $\int_0^1 x^3\, dx = \dfrac{1}{4}$

11. $\int_0^1 \sqrt{x}\, dx$

13. $\int_0^\pi \sin x\, dx$

15. $\int_0^1 \tan^{-1}x\, dx$

Section 5.4   (page 327)

1. 0

3. 8

5. $(b^2 - a^2)/2$

7. π

9. 0

11. 2π

13. 0

15. $(2\pi + 3\sqrt{3})/6$

17. 16

19. 32/3

21. $(4 + 3\pi)/12$

23. $\ln 2$

25. $\ln 3$

27. 4

29. 1

31. $\pi/2$

33. 1

35. 11/6

37. $\dfrac{\pi}{3} - \sqrt{3}$

39. 41/2

41. 3/4

43. $k = \overline{f}$

Section 5.5   (page 333)

1. 4

3. 1

5. 9

7. $80\frac{4}{5}$

9. $\dfrac{2-\sqrt{2}}{2\sqrt{2}}$

11. $(1/\sqrt{2})-(1/2)$

13. $e^{\pi}-e^{-\pi}$

15. $(a^e-1)/\ln a$

17. $\pi/2$

19. $\dfrac{\pi}{3}$

21. $\frac{1}{5}$ sq. un.

23. $\frac{32}{3}$ sq. un.

25. $\frac{1}{6}$ sq. un.

27. $\frac{1}{3}$ sq. un.

29. $\frac{1}{12}$ sq. un.

31. 2π sq. un.

33. 3

35. $\frac{16}{3}$

37. $e-1$

39. $\dfrac{\sin x}{x}$

41. $-2\,\dfrac{\sin x^2}{x}$

43. $\dfrac{\cos t}{1+t^2}$

45. $(\cos x)/(2\sqrt{x})$

47. $f(x)=\pi e^{\pi(x-1)}$

49. $1/x^2$ is not continuous (or even defined) at $x=0$ so the Fundamental Theorem cannot be applied over $[-1,1]$; since $1/x^2>0$ on its domain, we would expect the integral to be positive if it exists at all. (It doesn't.)

51. $F(x)$ has a maximum value at $x=1$ but no minimum value.

53. 2

Section 5.6   (page 341)

1. $-\frac{1}{2}e^{5-2x}+C$

3. $\frac{2}{9}(3x+4)^{3/2}+C$

5. $-\frac{1}{32}(4x^2+1)^{-4}+C$

7. $\frac{1}{2}e^{x^2}+C$

9. $\frac{1}{2}\tan^{-1}\left(\frac{1}{2}\sin x\right)+C$

11. $2\ln\left|e^{x/2}-e^{-x/2}\right|+C=\ln\left|e^x-2+e^{-x}\right|+C$

13. $-\frac{2}{5}\sqrt{4-5s}+C$

15. $\frac{1}{2}\sin^{-1}\left(\dfrac{t^2}{2}\right)+C$

17. $-\ln\left(1+e^{-x}\right)+C$

19. $-\frac{1}{2}(\ln\cos x)^2+C$

21. $\frac{1}{2}\tan^{-1}\dfrac{x+3}{2}+C$

23. $\frac{1}{8}\cos^8 x-\frac{1}{6}\cos^6 x+C$

25. $-\dfrac{1}{3a}\cos^3 ax+C$

27. $\frac{5}{16}x-\frac{1}{4}\sin 2x+\frac{3}{64}\sin 4x+\frac{1}{48}\sin^3 2x+C$

29. $\frac{1}{5}\sec^5 x+C$

31. $\frac{2}{3}(\tan x)^{3/2}+\frac{2}{7}(\tan x)^{7/2}+C$

33. $\frac{3}{8}\sin x-\frac{1}{4}\sin(2\sin x)+\frac{1}{32}\sin(4\sin x)+C$

35. $\frac{1}{3}\tan^3 x+C$

37. $-\frac{1}{9}\csc^9 x+\frac{2}{7}\csc^7 x-\frac{1}{5}\csc^5 x+C$

39. $\frac{14}{3}\sqrt{17}+\frac{2}{3}$

41. $3\pi/16$

43. $\ln 2$

45. $2,\ 2(\sqrt{2}-1)$

47. $\pi/32$ sq. un.

Section 5.7   (page 346)

1. $\dfrac{1}{6}$ sq. units

3. $\dfrac{64}{3}$ sq. units

5. $\dfrac{125}{12}$ sq. units

7. $\dfrac{1}{2}$ sq. units

9. $\dfrac{5}{12}$ sq. units

11. $\dfrac{15}{8}-2\ln 2$ sq. units

13. $\dfrac{\pi}{2}-\dfrac{1}{3}$ sq. units

15. $\dfrac{4}{3}$ sq. units

17. $2\sqrt{2}$ sq. units

19. $1-\pi/4$ sq. units

21. $(\pi/8)-\ln\sqrt{2}$ sq. units

23. $(4\pi/3)-2\ln(2+\sqrt{3})$ sq. units

25. $(4/\pi)-1$ sq. units

27. $\dfrac{4}{3}$ sq. units

29. $\dfrac{e}{2}-1$ sq. units

Review Exercises   (page 347)

1. sum is $n(n+2)/(n+1)^2$

3. $20/3$

5. 4π

7. 0

9. 2

11. $\sin(t^2)$

13. $-4e^{\sin(4s)}$

15. $f(x)=-\frac{1}{2}e^{(3/2)(1-x)}$

17. $9/2$ sq. units

19. $3/10$ sq. units

21. $(3\sqrt{3}/4)-1$ sq. units

23. $(\frac{1}{6}\sin(2x^3+1)+C$

25. $98/3$

27. $(\pi/8)-(1/2)\tan^{-1}(1/2)$

29. $-\cos\sqrt{2s+1}+C$

31. min $-\pi/4$, no max

Chapter 6
Techniques of Integration

Section 6.1   (page 355)

1. $x\sin x+\cos x+C$

3. $\dfrac{1}{\pi}x^2\sin\pi x+\dfrac{2}{\pi^2}x\cos\pi x-\dfrac{2}{\pi^3}\sin\pi x+C$

5. $\frac{1}{4}x^4\ln x-\frac{1}{16}x^4+C$

7. $x\tan^{-1}x-\frac{1}{2}\ln(1+x^2)+C$

9. $\left(\frac{1}{2}x^2-\frac{1}{4}\right)\sin^{-1}x+\frac{1}{4}x\sqrt{1-x^2}+C$

11. $\frac{7}{8}\sqrt{2}+\frac{3}{8}\ln(1+\sqrt{2})$

13. $\frac{1}{13}e^{2x}(2\sin 3x - 3\cos 3x) + C$

15. $\ln(2 + \sqrt{3}) - \frac{\pi}{6}$

17. $x\tan x - \ln|\sec x| + C$

19. $\frac{x}{2}[\cos(\ln x) + \sin(\ln x)] + C$

21. $\ln x(\ln(\ln x) - 1) + C$

23. $x\cos^{-1}x - \sqrt{1 - x^2} + C$

25. $\frac{2\pi}{3} - \ln(2 + \sqrt{3})$

27. $\frac{1}{2}(x^2 + 1)(\tan^{-1}x)^2 - x\tan^{-1}x + \frac{1}{2}\ln(1 + x^2) + C$

29. $\dfrac{1 + e^{-\pi}}{2}$ square units

31. $I_n = x(\ln x)^n - nI_{n-1}$,
$I_4 = x[(\ln x)^4 - 4(\ln x)^3 + 12(\ln x)^2 - 24(\ln x) + 24] + C$

33. $I_n = -\frac{1}{n}\sin^{n-1}x\cos x + \frac{n-1}{n}I_{n-2}$,
$I_6 = \frac{5x}{16} - \cos x[\frac{1}{6}\sin^5 x + \frac{5}{24}\sin^3 x + \frac{5}{16}\sin x] + C$,
$I_7 = -\cos x[\frac{1}{7}\sin^6 x + \frac{6}{35}\sin^4 x + \frac{8}{35}\sin^2 x + \frac{16}{35}] + C$

35. $I_n = \dfrac{x}{2a^2(n-1)(x^2 + a^2)^{n-1}} + \dfrac{2n-3}{2a^2(n-1)}I_{n-1}$,
$I_3 = \dfrac{x}{4a^2(x^2 + a^2)^2} + \dfrac{3x}{8a^4(x^2 + a^2)} + \dfrac{3}{8a^5}\tan^{-1}\frac{x}{a} + C$

37. Any conditions that guarantee that
$f(b)g'(b) - f'(b)g(b) = f(a)g'(a) - f'(a)g(a)$
will suffice.

Section 6.2   (page 363)

1. $\frac{1}{2}\sin^{-1}(2x) + C$

3. $\frac{9}{2}\sin^{-1}\frac{x}{3} - \frac{1}{2}x\sqrt{9 - x^2} + C$

5. $-\dfrac{\sqrt{9 - x^2}}{9x} + C$

7. $-\sqrt{9 - x^2} + \sin^{-1}\frac{x}{3} + C$

9. $\frac{1}{3}(9 + x^2)^{3/2} - 9\sqrt{9 + x^2} + C$

11. $\dfrac{1}{a^2}\dfrac{x}{\sqrt{a^2 - x^2}} + C$

13. $\dfrac{x}{\sqrt{a^2 - x^2}} - \sin^{-1}\frac{x}{a} + C$

15. $\frac{1}{2}\sec^{-1}\frac{x}{2} + C$ **17.** $\frac{1}{3}\tan^{-1}\frac{x+1}{3} + C$

19. $\frac{1}{32}\tan^{-1}\frac{2x+1}{2} + \frac{1}{16}\frac{2x+1}{4x^2 + 4x + 5} + C$

21. $a\sin^{-1}\dfrac{x - a}{a} - \sqrt{2ax - x^2} + C$

23. $\dfrac{3 - x}{4\sqrt{3 - 2x - x^2}} + C$

25. $\frac{3}{8}\tan^{-1}x + \dfrac{3x^3 + 5x}{8(1 + x^2)^2} + C$

27. $\frac{1}{2}\ln\left(1 + \sqrt{1 - x^2}\right) - \frac{1}{2}\ln|x| - \dfrac{\sqrt{1 - x^2}}{2x^2} + C$

29. $2\sqrt{x} - 4\ln(2 + \sqrt{x}) + C$

31. $\frac{6}{7}x^{7/6} - \frac{6}{5}x^{5/6} + \frac{3}{2}x^{2/3} + 2x^{1/2} - 3x^{1/3} - 6x^{1/6} + 3\ln(1 + x^{1/3}) + 6\tan^{-1}x^{1/6} + C$

33. $\dfrac{\pi}{6} - \dfrac{\sqrt{3}}{8}$ **35.** $\pi/3$

37. $\dfrac{2}{\sqrt{3}}\tan^{-1}\left(\dfrac{2\tan(\theta/2) + 1}{\sqrt{3}}\right) + C$

39. $\dfrac{2}{\sqrt{5}}\tan^{-1}\left(\dfrac{\tan(\theta/2)}{\sqrt{5}}\right) + C$

41. $\dfrac{9}{2\sqrt{2}}\tan^{-1}\dfrac{1}{\sqrt{2}} - \dfrac{1}{2}$ square units

43. $a^2\cos^{-1}\left(\dfrac{b}{a}\right) - b\sqrt{a^2 - b^2}$ square units

45. $\dfrac{25}{2}\left(\sin^{-1}\frac{4}{5} - \sin^{-1}\frac{3}{5}\right) - 12\ln\frac{4}{3}$ square units

47. $\dfrac{\ln(Y + \sqrt{1 + Y^2})}{2}$ sq. units

Section 6.3   (page 372)

1. $\ln|2x - 3| + C$

3. $\dfrac{x}{\pi} - \dfrac{2}{\pi^2}\ln|\pi x + 2| + C$

5. $\frac{1}{6}\ln\left|\dfrac{x - 3}{x + 3}\right| + C$ **7.** $\frac{1}{2a}\ln\left|\dfrac{a + x}{a - x}\right| + C$

9. $x - \frac{4}{3}\ln|x + 2| + \frac{1}{3}\ln|x - 1| + C$

11. $3\ln|x + 1| - 2\ln|x| + C$

13. $\dfrac{1}{3(1 - 3x)} + C$

15. $-\frac{1}{9}x - \dfrac{13}{54}\ln|2 - 3x| + \frac{1}{6}\ln|x| + C$

17. $\dfrac{1}{2a^2}\ln\dfrac{|x^2 - a^2|}{x^2} + C$

19. $x + \dfrac{a}{3}\ln|x - a| - \dfrac{a}{6}\ln(x^2 + ax + a^2) - \dfrac{a}{\sqrt{3}}\tan^{-1}\dfrac{2x + a}{\sqrt{3}a} + C$

21. $\frac{1}{3}\ln|x| - \frac{1}{2}\ln|x - 1| + \frac{1}{6}\ln|x - 3| + C$

23. $\dfrac{1}{4}\ln\left|\dfrac{x + 1}{x - 1}\right| - \dfrac{x}{2(x^2 - 1)} + C$

25. $\dfrac{1}{27}\ln\left|\dfrac{x - 3}{x}\right| + \dfrac{1}{9x} + \dfrac{1}{6x^2} + C$

27. $\dfrac{t - 1}{4(t^2 + 1)} - \frac{1}{4}\ln|t + 1| + \frac{1}{8}\ln(t^2 + 1) + C$

29. $\dfrac{1}{3}\ln\left|\dfrac{1-\sqrt{1-x^2}}{x}\right|+\dfrac{1}{12}\ln\left(\dfrac{(2+\sqrt{1-x^2})^2}{3+x^2}\right)+C$

31. $\dfrac{1}{\sqrt{1+x^2}}+\dfrac{1}{2}\ln\left|\dfrac{1-\sqrt{1+x^2}}{1+\sqrt{1+x^2}}\right|+C$

33. $\dfrac{1-2x^2}{x\sqrt{x^2-1}}+C$

Section 6.4   (page 376)

5. $\dfrac{x\sqrt{x^2-2}}{2}+\ln|x+\sqrt{x^2-2}|+C$

7. $-\sqrt{3t^2+5}/(5t)+C$

9. $(x^5/3125)(625(\ln x)^4-500(\ln x)^3+300(\ln x)^2$
$-120\ln x+24)+C$

11. $(1/6)(2x^2-x-3)\sqrt{2x-x^2}-(1/2)\sin^{-1}(1-x)+C$

13. $(x-2)/(4\sqrt{4x-x^2})+C$

Section 6.5   (page 384)

1. $1/2$

3. $1/2$

5. $3\times 2^{1/3}$

7. $3/2$

9. 3

11. π

13. $1/2$

15. diverges to ∞

17. 2

19. diverges

21. 0

23. 1 sq. units

25. $2\ln 2$ square units

29. 2

31. diverges to ∞

33. converges

35. diverges to ∞

37. diverges to ∞

39. diverges

41. diverges to ∞

Section 6.6   (page 392)

1. $T_4=4.75,$
$M_4=4.625,$
$T_8=4.6875,$
$M_8=4.65625,$
$T_{16}=4.671875,$
Actual errors:
$I-T_4\approx-0.0833333,$
$I-M_4\approx0.0416667,$
$I-T_8\approx-0.0208333,$
$I-M_8\approx0.0104167,$
$I-T_{16}\approx-0.0052083$
Error estimates:
$|I-T_4|\le 0.0833334,$
$|I-M_4|\le 0.0416667,$
$|I-T_8|\le 0.0208334,$
$|I-M_8|\le 0.0104167,$
$|I-T_{16}|\le 0.0052084$

3. $T_4=0.9871158,$
$M_4=1.0064545,$
$T_8=0.9967852,$
$M_8=1.0016082,$
$T_{16}=0.9991967,$
Actual errors:
$I-T_4\approx0.0128842,$
$I-M_4\approx-0.0064545,$
$I-T_8\approx0.0032148,$
$I-M_8\approx-0.0016082,$
$I-T_{16}\approx0.0008033$
Error estimates:
$|I-T_4|\le 0.020186,$
$|I-M_4|\le 0.010093,$
$|I-T_8|\le 0.005047,$
$|I-M_8|\le 0.002523,$
$|I-T_{16}|\le 0.001262$

5. $T_4=46,\ T_8=46.7$

7. $T_4=3,000\ \text{km}^2,\ T_8=3,400\ \text{km}^2$

9. $T_4\approx 2.02622,\quad M_4\approx 2.03236,$
$T_8\approx 2.02929,\quad M_8\approx 2.02982,$
$T_{16}\approx 2.029555$

11. $M_8\approx 1.3714136,\quad T_{16}\approx 1.3704366,\ I\approx 1.371$

Section 6.7   (page 397)

1. $S_4=S_8=I,\quad$ Errors $=0$

3. $S_4\approx 1.0001346,\quad S_8\approx 1.0000083,$
$I-S_4\approx-0.0001346,\quad I-S_8\approx-0.0000083$

5. 46.93

7. For $f(x)=e^{-x}$:
$|I-S_4|\le 0.000022,\ |I-S_8|\le 0.0000014;$
for $f(x)=\sin x,$
$|I-S_4|\le 0.00021,$
$|I-S_8|\le 0.000013$

9. $S_4\approx 2.0343333,\quad S_8\approx 2.0303133,$
$S_{16}\approx 2.0296433$

Section 6.8   (page 403)

1. $3\displaystyle\int_0^1\dfrac{u\,du}{1+u^3}$

3. $\displaystyle\int_{-\pi/2}^{\pi/2}e^{\sin\theta}\,d\theta,\quad$ or $\quad 2\displaystyle\int_0^1\dfrac{e^{1-u^2}+e^{u^2-1}}{\sqrt{2-u^2}}\,du$

5. $4\displaystyle\int_0^1\dfrac{dv}{\sqrt{(2-v^2)(2-2v^2+v^4)}}$

7. $T_2\approx 0.603553\quad T_4\approx 0.643283,$
$T_8\approx 0.658130,\quad T_{16}\approx 0.663581;$
Errors: $I-T_2\approx 0.0631,\quad I-T_4\approx 0.0234,$
$I-T_8\approx 0.0085,\quad I-T_{16}\approx 0.0031;$
Errors do not decrease like $1/n^2$ because the second
derivative of $f(x)=\sqrt{x}$ is not bounded on $[0,1]$.

9. $I \approx 0.74684$ with error less than 10^{-4}; seven terms of the series are needed.

11. $A = 1, u = 1/\sqrt{3}$

13. $A = 5/9, B = 8/9, u = \sqrt{3/5}$

15. $R_1 \approx 0.7471805$, $R_2 \approx 0.7468337$,
    $R_3 \approx 0.7468241$, $I \approx 0.746824$

17. $R_2 = \dfrac{2h}{45}(7y_0 + 32y_1 + 12y_2 + 32y_3 + 7y_4)$

Review Exercises (Techniques of Integration)   (page 404)

1. $\frac{2}{3}\ln|x+2| - \frac{1}{6}\ln|2x+1| + C$

3. $\frac{1}{4}\sin^4 x - \frac{1}{6}\sin^6 x + C$ **5.** $\frac{3}{4}\ln\left|\dfrac{2x-1}{2x+1}\right| + C$

7. $-\dfrac{1}{3}\left(\dfrac{\sqrt{1-x^2}}{x}\right)^3 + C$ **9.** $\frac{1}{5}(5x^3 - 2)^{1/3} + C$

11. $\frac{1}{16}\tan^{-1}\dfrac{x}{2} + \dfrac{x}{8(4+x^2)} + C$

13. $\dfrac{1}{2\ln 2}\left(2^x\sqrt{1+4^x} + \ln(2^x + \sqrt{1+4^x})\right) + C$

15. $\frac{1}{4}\tan^4 x + \frac{1}{6}\tan^6 x + C$

17. $-e^{-x}\left(\frac{2}{5}\cos 2x + \frac{1}{5}\sin 2x\right) + C$

19. $\dfrac{x}{10}\left(\cos(3\ln x) + 3\sin(3\ln x)\right) + C$

21. $\frac{1}{4}\left(\ln(1+x^2)\right)^2 + C$

23. $\sin^{-1}\dfrac{x}{\sqrt{2}} - \dfrac{x\sqrt{2-x^2}}{2} + C$

25. $\dfrac{1}{64}\left(\dfrac{-1}{7(4x+1)^7} + \dfrac{1}{4(4x+1)^8} - \dfrac{1}{9(4x+1)^9}\right) + C$

27. $-\frac{1}{4}\cos 4x + \frac{1}{6}\cos^3 4x - \frac{1}{20}\cos^5 4x + C$

29. $-\frac{1}{2}\ln(2e^{-x} + 1) + C$

31. $-\frac{1}{2}\sin^2 x - 2\sin x - 4\ln(2 - \sin x) + C$

33. $-\dfrac{\sqrt{1-x^2}}{x} + C$

35. $\frac{1}{48}(1 - 4x^2)^{3/2} - \frac{1}{16}\sqrt{1-4x^2} + C$

37. $\sqrt{x^2+1} + \ln(x + \sqrt{x^2+1}) + C$

39. $x + \frac{1}{3}\ln|x| + \frac{4}{3}\ln|x-3| - \frac{5}{3}\ln|x+3| + C$

41. $-\frac{1}{10}\cos^{10}x + \frac{1}{6}\cos^{12}x - \frac{1}{14}\cos^{14}x + C$

43. $\dfrac{1}{2}\ln|x^2 + 2x - 1| - \dfrac{1}{2\sqrt{2}}\ln\left|\dfrac{x+1-\sqrt{2}}{x+1+\sqrt{2}}\right| + C$

45. $\frac{1}{3}x^3\sin^{-1}2x + \frac{1}{24}\sqrt{1-4x^2} - \frac{1}{72}(1-4x^2)^{3/2} + C$

47. $\frac{1}{128}\left(3x - \sin(4x) + \frac{1}{8}\sin(8x)\right)$

49. $\tan^{-1}\dfrac{\sqrt{x}}{2} + C$

51. $\dfrac{x^2}{2} - 2x + \dfrac{1}{4}\ln|x| + \dfrac{1}{2x} + \dfrac{15}{4}\ln|x+2| + C$

53. $-\frac{1}{2}\cos(2\ln x) + C$ **55.** $\frac{1}{2}\exp(2\tan^{-1}x) + C$

57. $\frac{1}{4}\left(\ln(3+x^2)\right)^2 + C$ **59.** $\frac{1}{2}\left(\sin^{-1}(x/2)\right)^2 + C$

61. $\sqrt{x^2 + 6x + 10} - 2\ln(x + 3 + \sqrt{x^2 + 6x + 10}) + C$

63. $\dfrac{2}{5(2+x^2)^{5/2}} - \dfrac{1}{3(2+x^2)^{3/2}} + C$

65. $\frac{6}{7}x^{7/6} - \frac{6}{5}x^{5/6} + 2\sqrt{x} - 6x^{1/6} + 6\tan^{-1}x^{1/6} + C$

67. $\frac{2}{3}x^{3/2} - x + 4\sqrt{x} - 4\ln(1 + \sqrt{x}) + C$

69. $\dfrac{1}{2(4-x^2)} + C$

71. $\frac{1}{3}x^3\tan^{-1}x - \frac{1}{6}x^2 + \frac{1}{6}\ln(1+x^2) + C$

73. $\dfrac{1}{5}\ln\left|\dfrac{3\tan(x/2)-1}{\tan(x/2)+3}\right| + C$

75. $\frac{1}{2}\ln|\tan(x/2)| - \frac{1}{4}\left(\tan^{-1}(x/2)\right)^2 + C$
$= \dfrac{1}{4}\left(\ln\left|\dfrac{1-\cos x}{1+\cos x}\right| - \dfrac{1-\cos x}{1+\cos x}\right) + C$

77. $2\sqrt{x} - 2\tan^{-1}\sqrt{x} + C$

79. $\dfrac{1}{2}x^2 + \dfrac{4}{3}\ln|x-2| - \frac{2}{3}\ln(x^2 + 2x + 4)$
$+ \dfrac{4}{\sqrt{3}}\tan^{-1}\dfrac{x+1}{\sqrt{3}} + C$

Review Exercises (Other)   (page 405)

1. $I = \frac{1}{2}\left(xe^x\cos x + (x-1)e^x\sin x\right)$,
    $J = \frac{1}{2}\left((1-x)e^x\cos x + xe^x\sin x\right)$

3. diverges to ∞ **5.** $-4/9$

9. 367,000 m^3

11. $T_8 = 1.61800,\ S_8 = 1.62092,\ I \approx 1.62$

13. (a) $T_4 = 5.526,\ S_4 = 5.504$; (b) $S_8 = 5.504$; (c) yes, because $S_4 = S_8$, and Simpson's Rule is exact for cubics.

Challenging Problems   (page 406)

1. (c) $I = \dfrac{1}{630},\ \dfrac{22}{7} - \dfrac{1}{630} < \pi < \dfrac{22}{7} - \dfrac{1}{1260}$.

3. (a) $\dfrac{1}{\sqrt{3}}\tan^{-1}\left(\dfrac{2x+1}{\sqrt{3}}\right) + \dfrac{1}{\sqrt{3}}\tan^{-1}\left(\dfrac{2x-1}{\sqrt{3}}\right)$,

(b) $\dfrac{1}{\sqrt{2}}\tan^{-1}(\sqrt{2}x + 1) + \dfrac{1}{\sqrt{2}}\tan^{-1}(\sqrt{2}x - 1)$

7. (a) $a = 7/90, b = 16/45, c = 2/15$.
(b) one interval: approx 0.6321208750, two intervals: approx 0.6321205638, true val: 0.6321205588

Chapter 7
Applications of Integration

Section 7.1   (page 416)

1. $\dfrac{\pi}{5}$ cu. units

3. $\dfrac{3\pi}{10}$ cu. units

5. (a) $\dfrac{16\pi}{15}$ cu. units,   (b) $\dfrac{8\pi}{3}$ cu. units

7. (a) $\dfrac{27\pi}{2}$ cu. units,   (b) $\dfrac{108\pi}{5}$ cu. units

9. (a) $\dfrac{15\pi}{4} - \dfrac{\pi^2}{8}$ cu. units,   (b) $\pi(2 - \ln 2)$ cu. units

11. $\dfrac{10\pi}{3}$ cu. units

13. about 35%

15. $\dfrac{\pi h}{3}\left(b^2 - 3a^2 + \dfrac{2a^3}{b}\right)$ cu. units

17. $\dfrac{\pi}{3}(a - b)^2(2a + b)$ cu. units

19. $\dfrac{4\pi ab^2}{3}$ cu. units

21. (a) $\pi/2$ cu. units,   (b) 2π cu. units

23. $k > 2$

25. about $1,537$ cu. units

27. $8192\pi/105$ cu. units

29. $R = \dfrac{h \sin \alpha}{\sin \alpha + \cos 2\alpha}$

Section 7.2   (page 420)

1. 6 m^3

3. $\pi/3$ units^3

5. 132 ft^3

7. $\pi a^2 h/2$ cm^3

9. $3z^2$ sq. units

11. $\dfrac{16r^3}{3}$ cu. units

13. 72π cm^3

15. $\pi r^2(a + b)/2$ cu. units

17. $\dfrac{16{,}000}{3}$ cu. units

19. $12\pi\sqrt{2}$ in^3

21. approx 97.28 cm^3

Section 7.3   (page 428)

1. $2\sqrt{5}$ units

3. $52/3$ units

5. $(2/27)(13^{3/2} - 8)$ units

7. 6 units

9. $(e^2 + 1)/4$ units

11. $\sinh a$ units.

13. $\sqrt{17} + \frac{1}{4}\ln(4 + \sqrt{17})$ units

15. $6a$ units

17. 1.0338 units

19. 1.0581

21. $(10^{3/2} - 1)\pi/27$ sq. units

23. $\dfrac{64\pi}{81}\left[\dfrac{(13/4)^{5/2} - 1}{5} - \dfrac{(13/4)^{3/2} - 1}{3}\right]$ sq. units

25. $2\pi\left(\sqrt{2} + \ln(1 + \sqrt{2})\right)$ sq. units

27. $2\pi\left(\dfrac{255}{16} + \ln 4\right)$ sq. units

29. $4\pi^2 ab$ sq. units

31. $8\pi\left(1 + \dfrac{\ln(2 + \sqrt{3})}{2\sqrt{3}}\right)$ sq. units

33. $s = \dfrac{5}{\pi}\sqrt{4 + \pi^2}\,E\left(\dfrac{\pi}{\sqrt{4 + \pi^2}}\right)$

35. $k > -1$

37. (a) π cu. units;   (c) "Covering" a surface with paint requires putting on a layer of constant thickness. Far enough to the right, the horn is thinner than any prescribed constant, so it can contain less paint than would be necessary to paint its surface.

Section 7.4   (page 436)

1. mass $\dfrac{2L}{\pi}$; centre of mass at $\bar{s} = \dfrac{L}{2}$

3. $m = \frac{1}{4}\pi\delta_0 a^2$; $\bar{x} = \bar{y} = \dfrac{4a}{3\pi}$

5. $m = \dfrac{256k}{15}$; $\bar{x} = 0$, $\bar{y} = \dfrac{16}{7}$

7. $m = \dfrac{ka^3}{2}$; $\bar{x} = \dfrac{2a}{3}$, $\bar{y} = \dfrac{a}{2}$

9. $m = \int_a^b \delta(x)\big(g(x) - f(x)\big)\,dx$;
$M_{x=0} = \int_a^b x\delta(x)\big(g(x) - f(x)\big)\,dx$, $\bar{x} = M_{x=0}/m$,
$M_{y=0} = \frac{1}{2}\int_a^b \delta(x)\big((g(x))^2 - (f(x))^2\big)\,dx$,
$\bar{y} = M_{y=0}/m$

11. Mass is $\frac{8}{3}\pi R^4$ kg. The centre of mass is along the line through the centre of the ball perpendicular to the plane, at a distance $R/10$ m from the centre of the ball on the side opposite the plane.

13. $m = \frac{1}{8}\pi\delta_0 a^4$; $\bar{x} = 16a/(15\pi)$, $\bar{y} = 0$, $\bar{z} = 8a/15$

15. $m = \frac{1}{3}k\pi a^3$; $\bar{x} = 0$, $y = \dfrac{3a}{2\pi}$

17. about $5.57 C/k^{3/2}$

Section 7.5   (page 442)

1. $\left(\dfrac{4r}{3\pi}, \dfrac{4r}{3\pi}\right)$

3. $\left(\dfrac{\sqrt{2} - 1}{\ln(1 + \sqrt{2})}, \dfrac{\pi}{8\ln(1 + \sqrt{2})}\right)$

5. $\left(0, \dfrac{9\sqrt{3} - 4\pi}{4\pi - 3\sqrt{3}}\right)$

7. $\left(\dfrac{19}{9}, -\dfrac{1}{3}\right)$

9. The centroid is on the axis of symmetry of the hemisphere half way between the base plane and the vertex.

11. The centroid is on the axis of the cone, one-quarter of the cone's height above the base plane.

13. $(8/9, 11/9)$

15. $(0, 2/(3(\pi + 2)))$

17. $\left(\dfrac{\pi}{2}, \dfrac{\pi}{8}\right)$

19. $\left(\dfrac{2r}{\pi}, \dfrac{2r}{\pi}\right)$

21. $(1, -2)$

23. $\dfrac{5\pi}{3}$ cu. units

25. $(0.71377, 0.26053)$

27. $\left(1, \frac{1}{5}\right)$

29. $\bar{x} = \dfrac{M_{x=0}}{A}, \bar{y} = \dfrac{M_{y=0}}{A}$,

where $A = \displaystyle\int_c^d \big(g(y) - f(y)\big)\, dy$,

$M_{x=0} = \frac{1}{2}\displaystyle\int_c^d \big((g(y))^2 - (f(y))^2\big)\, dy$,

$M_{y=0} = \displaystyle\int_c^d y\big(g(y) - f(y)\big)\, dy$

31. diamond orientation, edge upward

Section 7.6 (page 449)

1. (a) 235,200 N, (b) 352,800 N

3. 6.12×10^8 N

5. 8.92×10^6 N

7. 7.056×10^5 N·m

9. $2450\pi a^3\left(a + \dfrac{8h}{3}\right)$ N·m

Section 7.7 (page 453)

1. $11,000

3. $8(\sqrt{x} - \ln(1 + \sqrt{x}))$

5. $9,063.46

7. $5,865.64

9. $50,000

11. $11,477.55

13. $64,872.10

15. $\int_0^T e^{-\lambda(t)} P(t)\, dt$

17. about 23,300, $11,890

Section 7.8 (page 464)

1. (a) $\dfrac{2}{9}$, (b) $\mu = 2, \sigma^2 = \dfrac{1}{2}, \sigma = \dfrac{1}{\sqrt{2}}$,

(c) $\dfrac{8}{9\sqrt{2}} \approx 0.63$

3. (a) 3, (b) $\mu = \dfrac{3}{4}, \sigma^2 = \dfrac{3}{80}, \sigma = \sqrt{\dfrac{3}{80}}$,

(c) $\dfrac{69}{20}\sqrt{\dfrac{3}{80}} \approx 0.668$

5. (a) 6, (b) $\mu = \dfrac{1}{2}, \sigma^2 = \dfrac{1}{20}, \sigma = \sqrt{\dfrac{1}{20}}$,

(c) $\dfrac{7}{5\sqrt{5}} \approx 0.626$

7. (a) $\dfrac{2}{\sqrt{\pi}}$, (b) $\mu = \dfrac{1}{\sqrt{\pi}} \approx 0.0.564, \sigma^2 = \dfrac{\pi - 2}{2\pi}$,

$\sigma = \sqrt{\dfrac{\pi - 2}{2\pi}} \approx 0.426$, (c) Pr$\approx 0.68$

11. (a) 0, (b) $e^{-3} \approx 0.05$, (c) ≈ 0.046

13. approximately 0.006

Section 7.9 (page 472)

1. $y^2 = Cx$

3. $x^3 - y^3 = C$

5. $Y = Ce^{t^2/2}$

7. $y = \dfrac{Ce^{2x} - 1}{Ce^{2x} + 1}$

9. $y = -\ln\left(Ce^{-2t} - \frac{1}{2}\right)$

11. $y = x^3 + Cx^2$

13. $y = \frac{3}{2} + Ce^{-2x}$

15. $y = x - 1 + Ce^{-x}$

17. $y = \sqrt{4 + x^2}$

19. $y = \dfrac{2x}{1 + x}$, $(x > 0)$

21. If $a = b$, the given solution is indeterminate 0/0; in this case the solution is $x = a^2kt/(1 + akt)$.

23. $v = \sqrt{\dfrac{mg}{k}} \dfrac{e^{2\sqrt{kg/mt}} - 1}{e^{2\sqrt{kg/mt}} + 1}$, $v \to \sqrt{\dfrac{mg}{k}}$

25. the hyperbolas $x^2 - y^2 = C$

Review Exercises (page 473)

1. about 833

3. $a \approx 1.1904, b \approx 0.0476$

5. $a = 2.1773$

7. $\left(\dfrac{8}{3\pi}, \dfrac{4}{3\pi}\right)$

9. about 27,726 N·cm

11. $y = 4(x - 1)^3$

13. $8,798.85

Challenging Problems (page 473)

1. (b) $\ln 2/(2\pi)$, (c) $\pi/(4k(k^2 + 1))$

3. $y = (r/h^3)x^3 - 3(r/h^2)x^2 + 3(r/h)x$

5. $b = -a = 27/2$

7. $1/\pi$

9. (a) $S(a, a, c) = 2\pi a^2 + \dfrac{2\pi ac^2}{\sqrt{a^2 - c^2}} \ln\left(\dfrac{a + \sqrt{a^2 - c^2}}{c}\right)$.

(b) $S(a, c, c) = 2\pi c^2 + \dfrac{2\pi a^2 c}{\sqrt{a^2 - c^2}} \cos^{-1}\left(\dfrac{c}{a}\right)$.

(c) $S(a, b, c) \approx \dfrac{b - c}{a - c} S(a, a, c) + \dfrac{a - b}{a - c} S(a, c, c)$.

(d) $S(3, 2, 1) \approx 49.595$.

Chapter 8
Conics, Parametric Curves, and Polar Curves

Section 8.1 (page 487)

1. $(x^2/5) + (y^2/9) = 1$

3. $(x - 2)^2 = 16 - 4y$

5. $3y^2 - x^2 = 3$

7. single point $(-1, 0)$

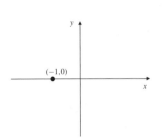

9. ellipse, centre $(0,2)$
11. parabola, vertex $(-1, -4)$

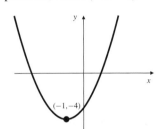

13. hyperbola, centre $\left(-\frac{3}{2}, 1\right)$
 asymptotes
 $2x+3 = \pm 2^{3/2}(y-1)$

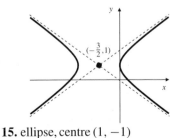

15. ellipse, centre $(1, -1)$

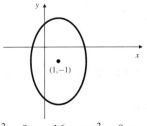

17. $y^2 - 8y = 16x$ or $y^2 - 8y = -4x$

19. rectangular hyperbola, centre $(1, -1)$,
 semiaxes $a = b = \sqrt{2}$,
 eccentricity $\sqrt{2}$,
 foci $(\sqrt{2}+1, \sqrt{2}-1)$,
 $(-\sqrt{2}+1, -\sqrt{2}-1)$,
 asymptotes $x = 1, y = -1$

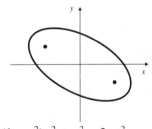

21. ellipse, centre $(0,0)$,
 semi-axes $a = 2, b = 1$,
 foci $\pm \left(2\sqrt{\frac{3}{5}}, -\sqrt{\frac{3}{5}}\right)$

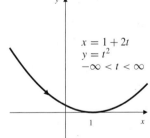

23. $(1 - \varepsilon^2)x^2 + y^2 - 2p\varepsilon^2 x = \varepsilon^2 p^2$

Section 8.2 (page 494)

1. $y = (x - 1)^2/4$

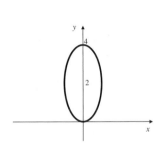

$x = 1 + 2t$
$y = t^2$
$-\infty < t < \infty$

3. $y = (1/x) - 1$

5. $x^2 + y^2 = 9$

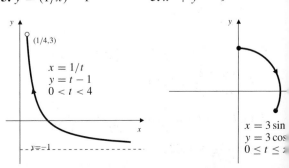

$x = 1/t$
$y = t - 1$
$0 < t < 4$

$x = 3\sin$
$y = 3\cos$
$0 \le t \le$

7. $\dfrac{x^2}{9} + \dfrac{y^2}{16} = 1$

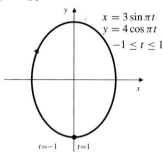

$x = 3 \sin \pi t$
$y = 4 \cos \pi t$
$-1 \le t \le 1$

9. $x^{2/3} + y^{2/3} = 1$

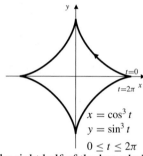

$x = \cos^3 t$
$y = \sin^3 t$
$0 \le t \le 2\pi$

11. the right half of the hyperbola $x^2 - y^2 = 1$

13. the curve starts at the origin and spirals twice counterclockwise around the origin to end at $(4\pi, 0)$

15. $x = m/2, \quad y = m^2/4, \quad (-\infty < m < \infty)$

17. $x = a \sec t, \quad y = a \sin t$;
$y^2 = a^2(x^2 - a^2)/x^2$

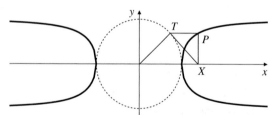

19. $x^3 + y^3 = 3xy$

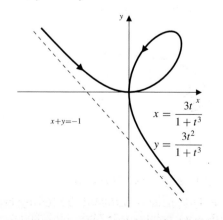

$x + y = -1$

$x = \dfrac{3t}{1 + t^3}$

$y = \dfrac{3t^2}{1 + t^3}$

Section 8.3 (page 499)

1. vertical at $(1, -4)$

3. horizontal at $(0, -16)$ and $(8, 16)$; vertical at $(-1, -11)$

5. horizontal at $(0, 1)$, vertical at $(\pm 1/\sqrt{e}, 1/e)$

7. horiz. at $(0, \pm 1)$, vert. at $(\pm 1, 1/\sqrt{2})$ and $(\pm 1, -1/\sqrt{2})$

9. $-3/4$ **11.** $-1/2$

13. $x = t - 2, \; y = 4t - 2$ **15.** slopes ± 1

17. not smooth at $t = 0$

19. not smooth at $t = 0$

21.   **23.**

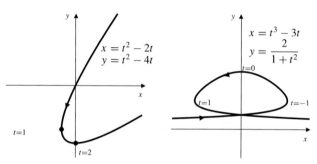

$x = t^2 - 2t$
$y = t^2 - 4t$

$x = t^3 - 3t$
$y = \dfrac{2}{1 + t^2}$

25.

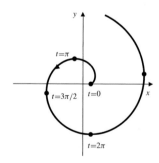

Section 8.4 (page 504)

1. $4\sqrt{2} - 2$ units   **3.** $6a$ units

5. $\frac{8}{3}\left((1 + \pi^2)^{3/2} - 1\right)$ units

7. 4 units   **9.** $8a$ units

11. $2\sqrt{2}\pi(1 + 2e^\pi)/5$ sq. units

13. $72\pi(1 + \sqrt{2})/15$ sq. units

15. $256/15$ sq. units   **17.** $1/6$ sq. units   **17.**   **19.**

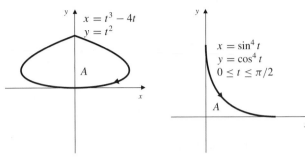

$x = t^3 - 4t$
$y = t^2$

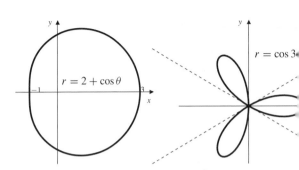

$x = \sin^4 t$
$y = \cos^4 t$
$0 \le t \le \pi/2$

$r = 2 + \cos\theta$

$r = \cos 3\theta$

19. $9\pi/2$ sq. units

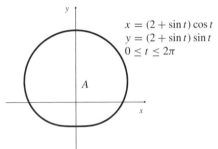

$x = (2 + \sin t)\cos t$
$y = (2 + \sin t)\sin t$
$0 \le t \le 2\pi$

21.   **23.** $r = \pm\sqrt{\sin 3\theta}$

23. $32\pi a^3/105$ cu. units

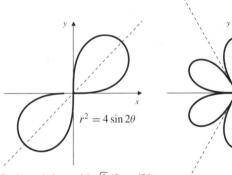

$r^2 = 4\sin 2\theta$

$r^2 = \sin$

Section 8.5   (page 511)

1. $x = 3$, vertical straight line
3. $3y - 4x = 5$, straight line
5. $2xy = 1$, rectangular hyperbola
7. $y = x^2 - x$, a parabola
9. $y^2 = 1 + 2x$, a parabola
11. $x^2 - 3y^2 - 8y = 4$, a hyperbola
13.   **15.**

25. the origin and $[\sqrt{3}/2, \pi/3]$
27. the origin and $[3/2, \pm\pi/3]$
29. asymptote $y = 1$,
    $r = 1/(\theta - \alpha)$ has
    asymptote $(\cos\alpha)y - (\sin\alpha)x = 1$

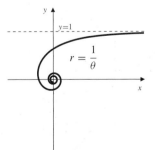

$y=1$

$r = \dfrac{1}{\theta}$

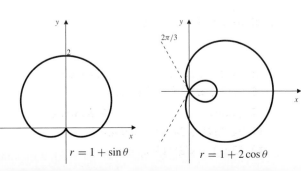

$r = 1 + \sin\theta$ $r = 1 + 2\cos\theta$

31. $x = f(\theta)\cos\theta, \quad y = f(\theta)\sin\theta$

39. $\ln\theta_1 = 1/\theta_1$, point $(-0.108461, 0.556676)$; $\ln\theta_2 = -1/(\theta_2 + \pi)$, point $(-0.182488, -0.178606)$

Section 8.6   (page 515)

1. π^2 sq. units

3. a^2 sq. units

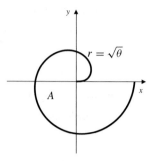

$r = \sqrt{\theta}$

A

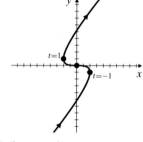

$\pi/4$

$r^2 = a^2 \cos 2\theta$

5. $\pi/2$ sq. units

7. $2 + (\pi/4)$ sq. units

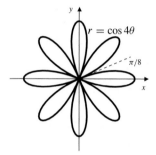

$r = \cos 4\theta$

$\pi/8$

$r = 1 - \cos\theta$

$r = 1$

9. $\pi/4$ sq. units

11. $\pi - \frac{3}{2}\sqrt{3}$ sq. units

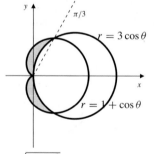

$\pi/3$

$r = 3\cos\theta$

$r = 1 + \cos\theta$

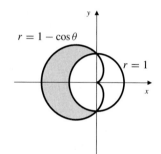

$2\pi/3$

$r = 1 + 2\cos\theta$

13. $\dfrac{\sqrt{1+a^2}}{a}\left(e^{a\pi} - e^{-a\pi}\right)$ units

17. $67.5°, -22.5°$

19. $90°$ at $(0,0)$,

$\pm 45°$ at $\left(1 - \dfrac{1}{\sqrt{2}}, \dfrac{\pi}{4}\right)$,

$\pm 135°$ at $\left(1 + \dfrac{1}{\sqrt{2}}, \dfrac{5\pi}{4}\right)$

21. horizontal at $\left(\pm\frac{\pi}{4}, \sqrt{2}\right)$, vertical at $(2,0)$ and the origin

23. horizontal at $(0,0)$, $\left(\frac{2}{3}\sqrt{2}, \pm\tan^{-1}\sqrt{2}\right)$, $\left(\frac{2}{3}\sqrt{2}, \pi \pm \tan^{-1}\sqrt{2}\right)$, vertical at $\left(0, \frac{\pi}{2}\right)$, $\left(\frac{2}{3}\sqrt{2}, \pm\tan^{-1}(1/\sqrt{2})\right)$, $\left(\frac{2}{3}\sqrt{2}, \pi \pm \tan^{-1}(1/\sqrt{2})\right)$

25. horizontal at $\left(4, -\frac{\pi}{2}\right)$, $\left(1, \frac{\pi}{6}\right)$, $\left(1, \frac{5\pi}{6}\right)$, vertical at $\left(3, -\frac{\pi}{6}\right)$, $\left(3, -\frac{5\pi}{6}\right)$, no tangent at $\left(0, \frac{\pi}{2}\right)$

Review Exercises   (page 516)

1. ellipse, foci $(\pm 1, 0)$, semi-major axis $\sqrt{2}$, semi-minor axis 1

3. parabola, vertex $(4, 1)$, focus $(15/4, 1)$

5. straight line from $(0, 2)$ to $(2, 0)$

7. the parabola $y = x^2 - 1$ left to right

9. first quadrant part of ellipse $16x^2 + y^2 = 16$ from $(1, 0)$ to $(0, 4)$

11. horizontal tangents at $(2, \pm 2)$ (i.e. $t = \pm 1$) vertical tangent at $(4, 0)$ (i.e. $t = 0$)

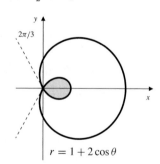

$t=-1$

$t=\pm\sqrt{3}$ $t=0$

$t=1$

$t=1$

$t=-1$

13. horizontal tangent at $(0, 0)$ (i.e. $t = 0$) vertical tangents at $(2, -1)$ and $(-2, 1)$ (i.e. $t = \pm 1$)

15. $1/2$ sq. units

17. $1 + e^2$ units

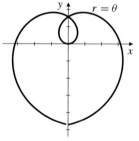

$r = \theta$

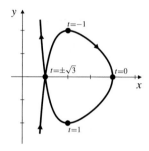

$r = 1 + \cos 2\theta$

19. $r = \theta$

21. $r = 1 + \cos 2\theta$

23. $r = 1 + 2\cos 2\theta$

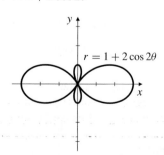

$r = 1 + 2\cos 2\theta$

25. $\pi + (3\sqrt{3}/4)$ sq. units   **27.** $(\pi - 3)/2$ sq. units

Challenging Problems   (page 516)

1. $16\pi \sec\theta$ cm^2 **5.** $40\pi/3$ ft^3

7. about 84.65 min

9. $r^2 = \cos(2\theta)$ is the inner curve; area between curves is 1/3 sq. units

Chapter 9
Sequences, Series, and Power Series

Section 9.1   (page 526)

1. bounded, positive, increasing, convergent to 2

3. bounded, positive, convergent to 4

5. bounded below, positive, increasing, divergent to infinity

7. bounded below, positive, increasing, divergent to infinity

9. bounded, positive, decreasing, convergent to 0

11. divergent   **13.** divergent

15. ∞ **17.** 0

19. 1   **21.** e^{-3}

23. 0   **25.** 1/2

27. 0   **29.** 0

31. $\lim_{n\to\infty} a_n = 5$

33. If $\{a_n\}$ is (ultimately) decreasing, then either it is bounded below and therefore convergent, or it is unbounded below and therefore divergent to negative infinity.

Section 9.2   (page 534)

1. $\dfrac{1}{2}$

3. $\dfrac{1}{(2+\pi)^8\big((2+\pi)^2 - 1\big)}$

5. $\dfrac{25}{4,416}$ **7.** $\dfrac{8e^4}{e-2}$

9. diverges to ∞ **11.** $\dfrac{3}{4}$

13. $\dfrac{1}{3}$ **15.** div. to ∞

17. div. to ∞ **19.** diverges

21. 14 m

25. If $\{a_n\}$ is ultimately negative, then the series $\sum a_n$ must either converge (if its partial sums are bounded below), or diverge to $-\infty$ (if its partial sums are not bounded below).

27. false, e.g., $\sum \dfrac{(-1)^n}{2^n}$ **29.** true

31. true

Section 9.3   (page 545)

1. converges   **3.** diverges to ∞

5. converges   **7.** diverges to ∞

9. converges   **11.** diverges to ∞

13. diverges to ∞ **15.** converges

17. converges   **19.** diverges to ∞

21. converges   **23.** converges

25. converges

27. $s_n + \dfrac{1}{3(n+1)^3} \le s \le s_n + \dfrac{1}{3n^3}; \quad n = 6$

29. $s_n + \dfrac{2}{\sqrt{n+1}} \le s \le s_n + \dfrac{2}{\sqrt{n}}; \quad n = 63$

31. $0 < s - s_n \le \dfrac{n+2}{2^n(n+1)!(2n+3)}; \quad n = 4$

33. $0 < s - s_n \le \dfrac{2^n(4n^2 + 6n + 2)}{(2n)!(4n^2 + 6n)}; \quad n = 4$

39. converges, $a_n^{1/n} \to (1/e) < 1$

41. no info from ratio test, but series diverges to infinity since all terms exceed 1.

43. (b) $s \le \dfrac{2}{k(1-k)}, k = \tfrac{1}{2}$,

    (c) $0 < s - s_n < \dfrac{(1+k)^{n+1}}{2^n k(1-k)}, k = \dfrac{n+2-\sqrt{n^2+8}}{2(n-1)}$ for $n \ge 2$

Section 9.4   (page 553)

1. conv. conditionally   **3.** conv. conditionally

5. diverges   **7.** conv. absolutely

9. conv. conditionally   **11.** diverges

13. 999   **15.** 13

17. converges absolutely if $-1 < x < 1$, conditionally if $x = -1$, diverges elsewhere

19. converges absolutely if $0 < x < 2$, conditionally if $x = 2$, diverges elsewhere

21. converges absolutely if $-2 < x < 2$, conditionally if $x = -2$, diverges elsewhere

23. converges absolutely if $-\tfrac{7}{2} < x < \tfrac{1}{2}$, conditionally if $x = -\tfrac{7}{2}$, diverges elsewhere

25. AST does not apply directly, but does if we remove all the 0 terms; series converges conditionally

27. (a) false, e.g., $a_n = \dfrac{(-1)^n}{n}$,

    (b) false, e.g., $a_n = \dfrac{\sin(n\pi/2)}{n}$ (see Exercise 25),

    (c) true

29. converges absolutely for $-1 < x < 1$, conditionally if $x = -1$, diverges elsewhere

Section 9.5   (page 564)

1. centre 0, radius 1, interval $]-1, 1[$

3. centre -2, radius 2, interval $[-4, 0[$

5. centre $\frac{3}{2}$, radius $\frac{1}{2}$, interval $]1, 2[$

7. centre 0, radius ∞, interval $]-\infty, \infty[$

9. $\dfrac{1}{(1-x)^3} = \displaystyle\sum_{n=0}^{\infty} \dfrac{(n+1)(n+2)}{2} x^n, \ (-1 < x < 1)$

11. $\dfrac{1}{(1-x)^2} = \displaystyle\sum_{n=0}^{\infty} (n+1)x^n, \quad (-1 < x < 1)$

13. $\dfrac{1}{(2-x)^2} = \displaystyle\sum_{n=0}^{\infty} \dfrac{n+1}{2^{n+2}} x^n, \quad (-2 < x < 2)$

15. $\ln(2-x) = \ln 2 - \displaystyle\sum_{n=1}^{\infty} \dfrac{x^n}{2^n n}, \quad (-2 \le x < 2)$

17. $\dfrac{1}{x^2} = \displaystyle\sum_{n=0}^{\infty} \dfrac{n+1}{2^{n+2}} (x+2)^n, \quad (-4 < x < 0)$

19. $\dfrac{x^3}{1-2x^2} = \displaystyle\sum_{n=0}^{\infty} 2^n x^{2n+3}, \quad \left(-\dfrac{1}{\sqrt{2}} < x < \dfrac{1}{\sqrt{2}}\right)$

21. $\left(-\frac{1}{4}, \frac{1}{4}\right); \quad \dfrac{1}{1+4x}$

23. $[-1, 1); \quad \frac{1}{3}$ if $x = 0$,

$-\dfrac{1}{x^3} \ln(1-x) - \dfrac{1}{x^2} - \dfrac{1}{2x}$ otherwise

25. $(-1, 1); \quad \dfrac{2}{(1-x^2)^2}$ **27.** 3/4

29. $\pi^2(\pi + 1)/(\pi - 1)^3$ **31.** $\ln(3/2)$

Section 9.6   (page 572)

1. $e^{3x+1} = \displaystyle\sum_{n=0}^{\infty} \dfrac{3^n e}{n!} x^n, \ (\text{all } x)$

3. $\sin\left(x - \dfrac{\pi}{4}\right)$

$= \dfrac{1}{\sqrt{2}} \displaystyle\sum_{n=0}^{\infty} (-1)^n \left[-\dfrac{x^{2n}}{(2n)!} + \dfrac{x^{2n+1}}{(2n+1)!} \right], \ (\text{all } x)$

5. $x^2 \sin\left(\dfrac{x}{3}\right) = \displaystyle\sum_{n=0}^{\infty} \dfrac{(-1)^n}{3^{2n+1}(2n+1)!} x^{2n+3}, \ (\text{all } x)$

7. $\sin x \cos x = \displaystyle\sum_{n=0}^{\infty} \dfrac{(-1)^n 2^{2n}}{(2n+1)!} x^{2n+1}, \ (\text{all } x)$

9. $\dfrac{1+x^3}{1+x^2} = 1 - x^2 + \displaystyle\sum_{n=2}^{\infty} (-1)^n \left(x^{2n-1} + x^{2n} \right),$

$(-1 < x < 1)$

11. $\ln \dfrac{1-x}{1+x} = -2 \displaystyle\sum_{n=1}^{\infty} \dfrac{x^{2n-1}}{2n-1}, \quad (-1 < x < 1)$

13. $\cosh x - \cos x = 2 \displaystyle\sum_{n=0}^{\infty} \dfrac{x^{4n+2}}{(4n+2)!}, \ (\text{all } x)$

15. $e^{-2x} = e^2 \displaystyle\sum_{n=0}^{\infty} \dfrac{(-1)^n 2^n}{n!} (x+1)^n, \ (\text{all } x)$

17. $\cos x = \displaystyle\sum_{n=0}^{\infty} \dfrac{(-1)^{n+1}}{(2n)!} (x-\pi)^{2n}, \ (\text{all } x)$

19. $\ln 4 + \displaystyle\sum_{n=1}^{\infty} \dfrac{(-1)^{n-1}}{4^n n} (x-2)^n, \quad (-2 < x \le 6)$

21. $\sin x - \cos x =$

$\sqrt{2} \displaystyle\sum_{n=0}^{\infty} \dfrac{(-1)^n}{(2n+1)!} \left(x - \dfrac{\pi}{4}\right)^{2n+1}, \quad (\text{all } x)$

23. $\dfrac{1}{x^2} = \dfrac{1}{4} \displaystyle\sum_{n=0}^{\infty} \dfrac{n+1}{2^n} (x+2)^n, \quad (-4 < x < 0)$

25. $(x-1) + \displaystyle\sum_{n=2}^{\infty} \dfrac{(-1)^n}{n(n-1)} (x-1)^n, \quad (0 \le x \le 2)$

27. $1 + \dfrac{x^2}{2} + \dfrac{5x^4}{24}$ **29.** $x + \dfrac{x^2}{2} - \dfrac{x^3}{6}$

31. $1 + \dfrac{x}{2} - \dfrac{x^2}{8}$ **33.** e^{x^2} (all x)

35. $\dfrac{e^x - e^{-x}}{2x} = \dfrac{\sinh x}{x}$ if $x \ne 0$, 1 if $x = 0$

37. (a) $1 + x + x^2$,   (b) $3 + 3(x-1) + (x-1)^2$

Section 9.7   (page 576)

1. 1.22140   **3.** 3.32011

5. 0.99619   **7.** -0.10533

9. 0.42262   **11.** 1.54306

13. $I(x) = \displaystyle\sum_{n=0}^{\infty} \dfrac{(-1)^n}{(2n+1)(2n+1)!} x^{2n+1}, \ (\text{all } x)$

15. $K(x) = \displaystyle\sum_{n=0}^{\infty} \dfrac{(-1)^n}{(n+1)^2} x^{n+1}, \quad (-1 \le x \le 1)$

17. $M(x) = \displaystyle\sum_{n=0}^{\infty} \dfrac{(-1)^n}{(2n+1)(4n+1)} x^{4n+1},$

$(-1 \le x \le 1)$

19. 0.946   **21.** 2

23. $-3/25$ **25.** 0

Section 9.8   (page 580)

1. $\dfrac{1}{720}(0.2)^7$ **3.** $\dfrac{1}{120}(0.5)^5$

5. $\dfrac{4\sec^2(0.1)\tan^2(0.1) + 2\sec^4(0.1)}{4! \, 10^4}$

7. $\dfrac{24}{120(1.95)^5(20)^5}$

9. $2^x = \displaystyle\sum_{n=0}^{\infty} \dfrac{(x \ln 2)^n}{n!}, \quad$ all x

11. $\sin x = \displaystyle\sum_{n=0}^{\infty} (-1)^n \dfrac{x^{2n+1}}{(2n+1)!}, \quad$ all x

13. $\dfrac{1}{1-x} = \displaystyle\sum_{n=0}^{\infty} x^n, \ -1 < x < 1$

15. $\dfrac{x}{2+3x} = \displaystyle\sum_{n=1}^{\infty} (-1)^{n-1} 3^{n-1} \left(\dfrac{x}{2}\right)^n, \ -\dfrac{2}{3} < x < \dfrac{2}{3}$

17. $\sin x = \dfrac{1}{2} \displaystyle\sum_{n=0}^{\infty} \dfrac{c_n}{n!} \left(x - \dfrac{\pi}{6}\right)^n$, (for all x), where
$c_n = (-1)^{n/2}$ if n is even, and $c_n = (-1)^{(n-1)/2}\sqrt{3}$ if n is odd

19. $\ln x = \displaystyle\sum_{n=1}^{\infty} (-1)^{n-1} \dfrac{(x-1)^n}{n}, \ 0 < x \le 2$

21. $\dfrac{1}{x} = -\dfrac{1}{2} \displaystyle\sum_{n=0}^{\infty} \left(\dfrac{x+2}{2}\right)^n, \ -4 < x < 0$

Section 9.9   (page 584)

1. $\sqrt{1+x}$
$= 1 + \displaystyle\sum_{n=1}^{\infty} \dfrac{(-1)^{n-1} 1 \times 3 \times 5 \times \cdots \times (2n-3)}{2^n n!} x^n$
$|x| < 1$

3. $\sqrt{4+x}$
$= 2 + \dfrac{x}{4} + 2 \displaystyle\sum_{n=2}^{\infty} (-1)^{n-1} \dfrac{1 \times 3 \times 5 \times \cdots \times (2n-3)}{2^{3n} n!} x^n$,
$(-4 < x \le 4)$

5. $\displaystyle\sum_{n=0}^{\infty} (n+1)x^n, \quad |x| < 1$

Section 9.10   (page 589)

1. $y = a_0 \left(1 + \displaystyle\sum_{k=1}^{\infty} \dfrac{(x-1)^{4k}}{4(k!)(3)(7) \cdots (4k-1)}\right)$
$+ a_1 \left(x - 1 + \displaystyle\sum_{k=1}^{\infty} \dfrac{(x-1)^{4k+1}}{4(k!)(5)(9) \cdots (4k+1)}\right)$

3. $y = \displaystyle\sum_{n=0}^{\infty} (-1)^n \left[\dfrac{2^n n!}{(2n)!} x^{2n} + \dfrac{1}{2^{n-1} n!} x^{2n+1}\right]$

5. $y = 1 - \dfrac{1}{6}x^3 + \dfrac{1}{120}x^5 + \cdots$

7. $y_1 = 1 + \displaystyle\sum_{k=1}^{\infty} \dfrac{(-1)^k x^k}{(k!)(2)(5)(8) \cdots (3k-1)}$,
$y_2 = x^{1/3} \left(1 + \displaystyle\sum_{k=0}^{\infty} \dfrac{(-1)^k x^k}{(k!)(4)(7) \cdots (3k+1)}\right)$

Review Exercises   (page 589)

1. conv. to 0
3. div. to ∞
5. $\lim_{n \to \infty} a_n = \sqrt{2}$
7. $4\sqrt{2}/(\sqrt{2} - 1)$
9. 2
11. converges
13. converges
15. converges
17. conv. abs.
19. conv. cond.
21. conv. abs.   for x in $(-1, 5)$, cond. for $x = -1$, div. elsewhere

23. 1.202
25. $\sum_{n=0}^{\infty} x^n/3^{n+1}, \ |x| < 3$
27. $1 + \sum_{n=1}^{\infty} (-1)^{n-1} x^{2n}/(ne^n), \ -\sqrt{e} < x \le \sqrt{e}$
29. $x + \sum_{n=1}^{\infty} (-1)^n 2^{2n-1} x^{2n+1}/(2n)!, \ $ all x
31. $(1/2) + \sum_{n=1}^{\infty} \dfrac{(-1)^n 1 \times 4 \times 7 \times \cdots \times (3n-2)x^n}{2 \times 24^n n!}$,
$-8 < x \le 8$
33. $\sum_{n=0}^{\infty} (-1)^n (x-\pi)^n/\pi^{n+1}, \ 0 < x < 2\pi$
35. $1 + 2x + 3x^2 + \dfrac{10}{3}x^3$
37. $1 - \dfrac{1}{2}x^2 + \dfrac{5}{24}x^4$
39. $\begin{cases} \cos\sqrt{x} & \text{if } x \ge 0 \\ \cosh\sqrt{|x|} & \text{if } x < 0 \end{cases}$
41. $\pi^2/(\pi - 1)^2$
43. $\ln(e/(e-1))$
45. $1/14$
47. 3, 0.49386

Challenging Problems   (page 590)

5. (c) 1.645
7. (a) ∞, (c) e^{-x^2}, (d) $f(x) = e^{x^2} \int_0^x e^{-t^2}\,dt$

Appendix I   Complex Numbers
(page A-11)

1. $\Re(z) = -5, \ \Im(z) = 2$
3. $\Re(z) = 0, \ \Im(z) = -\pi$
5. $|z| = \sqrt{2}, \ \theta = 3\pi/4$
7. $|z| = 3, \ \theta = \pi/2$
9. $|z| = \sqrt{5}, \ \theta = \tan^{-1} 2$
11. $|z| = 5, \ \theta = \pi + \tan^{-1}(4/3)$
13. $|z| = 2, \ \theta = 11\pi/6$
15. $|z| = 3, \ \theta = 4\pi/5$
17. $23\pi/12$
19. $4 + 3i$
21. $\dfrac{\pi\sqrt{3}}{2} + \dfrac{\pi}{2}i$
23. $\dfrac{1}{4} - \dfrac{\sqrt{3}}{4}i$
25. $-3 + 5i$
27. $2 + i$
29. closed disk, radius 2, centre 0
31. closed disk, radius 5, centre $3 - 4i$
33. closed plane sector lying under $y = 0$ and to the left of $y = -x$
35. 4
37. $5 - i$
39. $2 + 11i$
41. $-\dfrac{1}{5} + \dfrac{7}{5}i$
43. 1
47. $zw = -3 - 3i, \ \dfrac{z}{w} = \dfrac{1+i}{3}$
49. (a) circle $|z| = \sqrt{2}$, (b) no solutions
51. $-1, \ \dfrac{1}{2} \pm \dfrac{\sqrt{3}}{2}i$
53. $2^{1/6}(\cos\theta + i\sin\theta)$ where $\theta = \pi/4, \ 11\pi/12, \ 19\pi/12$
55. $\pm 2^{1/4}\left(\dfrac{\sqrt{3}}{2} + \dfrac{1}{2}i\right), \ \pm 2^{1/4}\left(\dfrac{1}{2} - \dfrac{\sqrt{3}}{2}i\right)$

Appendix IV   Differential Equations
(page A-38)

1. 1, linear, homogeneous   **3.** 1, nonlinear

5. 2, linear, homogeneous

7. 3, linear, nonhomogeneous

9. 4, linear, homogeneous

11. (a) and (b) are solutions, (c) is not

13. $y_2 = \sin(kx)$, $y = -3(\cos(kx) + (3/k)\sin(kx))$

15. $y = \sqrt{2}(\cos x + 2\sin x)$

17. $y = x + \sin x + (\pi - 1)\cos x$

19. $2\tan^{-1}(y/x) = \ln(x^2 + y^2) + C$

21. $\tan^{-1}(y/x) = \ln|x| + C$

23. $y = x\tan^{-1}(\ln|Cx|)$ **25.** $y^3 + 3y - 3x^2 = 24$

25. $x + y = 4x^2$

27. $4\tan^{-1}\frac{y-1}{x-2} = \ln\left((y-1)^2 + (x-2)^2\right) + C$

29. $e^x \sin y + x^2 + y^2 = C$

31. $x^2 + x + \dfrac{y^2}{x} = C$

33. $e^x + x\ln y + y\ln x - \cos y = C$

35. $\mu(y) = \dfrac{1}{y}$, $x^2 y + 2xy^3 = C$

37. $x\mu'(xy)M + \mu(xy)\dfrac{\partial M}{\partial y} = y\mu'(xy)N + \mu(xy)\dfrac{\partial N}{\partial x}$

39. (a) 1.97664, (b) 2.187485, (c) 2.306595

41. (a) 2.436502, (b) 2.436559, (c) 2.436563

43. (a) 1.097897, (b) 1.098401

45. $y = 2/(3 - 2x)$

49. (b) $u = 1/(1-x)$, $v = \tan(x + \frac{\pi}{4})$. $y(x)$ is defined at least on $[0, \pi/4)$ and satisfies $1/(1-x) \le y(x) \le \tan(x + \frac{\pi}{4})$ there.

Index

INTEGRATION RULES

$$\int (Af(x) + Bg(x))\,dx = A \int f(x)\,dx + B \int g(x)\,dx$$

$$\int f'(g(x))\,g'(x)\,dx = f(g(x)) + C$$

$$\int U(x)\,dV(x) = U(x)\,V(x) - \int V(x)\,dU(x)$$

$$\int_a^b f'(x)\,dx = f(b) - f(a)$$

$$\frac{d}{dx} \int_a^x f(t)\,dt = f(x)$$

ELEMENTARY INTEGRALS

$$\int x^r\,dx = \frac{1}{r+1} x^{r+1} + C \text{ if } r \neq -1$$

$$\int \frac{dx}{x} = \ln|x| + C$$

$$\int e^x\,dx = e^x + C$$

$$\int a^x\,dx = \frac{a^x}{\ln a} + C$$

$$\int \sin x\,dx = -\cos x + C$$

$$\int \cos x\,dx = \sin x + C$$

$$\int \sec^2 x\,dx = \tan x + C$$

$$\int \csc^2 x\,dx = -\cot x + C$$

$$\int \sec x \tan x\,dx = \sec x + C$$

$$\int \csc x \cot x\,dx = -\csc x + C$$

$$\int \tan x\,dx = \ln|\sec x| + C$$

$$\int \cot x\,dx = \ln|\sin x| + C$$

$$\int \sec x\,dx = \ln|\sec x + \tan x| + C$$

$$\int \csc x\,dx = \ln|\csc x - \cot x| + C$$

$$\int \frac{dx}{\sqrt{a^2 - x^2}} = \sin^{-1}\frac{x}{a} + C \quad (a > 0,\ |x| < a)$$

$$\int \frac{dx}{a^2 + x^2} = \frac{1}{a} \tan^{-1}\frac{x}{a} + C \quad (a > 0)$$

$$\int \frac{dx}{a^2 - x^2} = \frac{1}{2a} \ln\left|\frac{x+a}{x-a}\right| + C \quad (a > 0)$$

$$\int \frac{dx}{x\sqrt{x^2 - a^2}} = \frac{1}{a} \sec^{-1}\left|\frac{x}{a}\right| + C \quad (a > 0,\ |x| > a)$$

TRIGONOMETRIC INTEGRALS

$$\int \sin^2 x\,dx = \frac{x}{2} - \frac{1}{4}\sin 2x + C$$

$$\int \cos^2 x\,dx = \frac{x}{2} + \frac{1}{4}\sin 2x + C$$

$$\int \tan^2 x\,dx = \tan x - x + C$$

$$\int \cot^2 x\,dx = -\cot x - x + C$$

$$\int \sec^3 x\,dx = \frac{1}{2}\sec x \tan x + \frac{1}{2}\ln|\sec x + \tan x| + C$$

$$\int \csc^3 x\,dx = -\frac{1}{2}\csc x \cot x + \frac{1}{2}\ln|\csc x - \cot x| + C$$

$$\int \sin ax \sin bx\,dx = \frac{\sin(a-b)x}{2(a-b)} - \frac{\sin(a+b)x}{2(a+b)} + C \text{ if } a^2 \neq b^2$$

$$\int \cos ax \cos bx\,dx = \frac{\sin(a-b)x}{2(a-b)} + \frac{\sin(a+b)x}{2(a+b)} + C \text{ if } a^2 \neq b^2$$

$$\int \sin ax \cos bx\,dx = -\frac{\cos(a-b)x}{2(a-b)} - \frac{\cos(a+b)x}{2(a+b)} + C \text{ if } a^2 \neq b^2$$

$$\int \sin^n x\,dx = -\frac{1}{n}\sin^{n-1} x \cos x + \frac{n-1}{n}\int \sin^{n-2} x\,dx$$

$$\int \cos^n x\,dx = \frac{1}{n}\cos^{n-1} x \sin x + \frac{n-1}{n}\int \cos^{n-2} x\,dx$$

$$\int \tan^n x\,dx = \frac{1}{n-1}\tan^{n-1} x - \int \tan^{n-2} x\,dx \text{ if } n \neq 1$$

$$\int \cot^n x\,dx = \frac{-1}{n-1}\cot^{n-1} x - \int \cot^{n-2} x\,dx \text{ if } n \neq 1$$

$$\int \sec^n x\,dx = \frac{1}{n-1}\sec^{n-2} x \tan x + \frac{n-2}{n-1}\int \sec^{n-2} x\,dx \text{ if } n \neq 1$$

$$\int \csc^n x\,dx = \frac{-1}{n-1}\csc^{n-2} x \cot x + \frac{n-2}{n-1}\int \csc^{n-2} x\,dx \text{ if } n \neq 1$$

$$\int \sin^n x \cos^m x\,dx = -\frac{\sin^{n-1} x \cos^{m+1} x}{n+m} + \frac{n-1}{n+m}\int \sin^{n-2} x \cos^m x\,dx \text{ if } n \neq -m$$

$$\int \sin^n x \cos^m x\,dx = \frac{\sin^{n+1} x \cos^{m-1} x}{n+m} + \frac{m-1}{n+m}\int \sin^n x \cos^{m-2} x\,dx \text{ if } m \neq -n$$

$$\int x \sin x\,dx = \sin x - x \cos x + C$$

$$\int x \cos x\,dx = \cos x + x \sin x + C$$

$$\int x^n \sin x\,dx = -x^n \cos x + n \int x^{n-1} \cos x\,dx$$

$$\int x^n \cos x\,dx = x^n \sin x - n \int x^{n-1} \sin x\,dx$$

INTEGRALS INVOLVING $\sqrt{x^2 \pm a^2}$ $(a > 0)$

(If $\sqrt{x^2 - a^2}$, assume $x > a > 0$.)

$$\int \sqrt{x^2 \pm a^2}\, dx = \frac{x}{2}\sqrt{x^2 \pm a^2} \pm \frac{a^2}{2}\ln|x + \sqrt{x^2 \pm a^2}| + C$$

$$\int \frac{dx}{\sqrt{x^2 \pm a^2}} = \ln|x + \sqrt{x^2 \pm a^2}| + C$$

$$\int \frac{\sqrt{x^2 + a^2}}{x}\, dx = \sqrt{x^2 + a^2} - a\ln\left|\frac{a + \sqrt{x^2 + a^2}}{x}\right| + C$$

$$\int \frac{\sqrt{x^2 - a^2}}{x}\, dx = \sqrt{x^2 - a^2} - a\tan^{-1}\frac{\sqrt{x^2 - a^2}}{a} + C$$

$$\int x^2\sqrt{x^2 \pm a^2}\, dx = \frac{x}{8}(2x^2 \pm a^2)\sqrt{x^2 \pm a^2} - \frac{a^4}{8}\ln|x + \sqrt{x^2 \pm a^2}| + C$$

$$\int \frac{x^2}{\sqrt{x^2 \pm a^2}}\, dx = \frac{x}{2}\sqrt{x^2 \pm a^2} \mp \frac{a^2}{2}\ln|x + \sqrt{x^2 \pm a^2}| + C$$

$$\int \frac{\sqrt{x^2 \pm a^2}}{x^2}\, dx = -\frac{\sqrt{x^2 \pm a^2}}{x} + \ln|x + \sqrt{x^2 \pm a^2}| + C$$

$$\int \frac{dx}{x^2\sqrt{x^2 \pm a^2}} = \mp\frac{\sqrt{x^2 \pm a^2}}{a^2 x} + C$$

$$\int \frac{dx}{(x^2 \pm a^2)^{3/2}} = \frac{\pm x}{a^2\sqrt{x^2 \pm a^2}} + C$$

$$\int (x^2 \pm a^2)^{3/2}\, dx = \frac{x}{8}(2x^2 \pm 5a^2)\sqrt{x^2 \pm a^2} + \frac{3a^4}{8}\ln|x + \sqrt{x^2 \pm a^2}| + C$$

INTEGRALS INVOLVING $\sqrt{a^2 - x^2}$ $(a > 0, |x| < a)$

$$\int \sqrt{a^2 - x^2}\, dx = \frac{x}{2}\sqrt{a^2 - x^2} + \frac{a^2}{2}\sin^{-1}\frac{x}{a} + C$$

$$\int \frac{\sqrt{a^2 - x^2}}{x}\, dx = \sqrt{a^2 - x^2} - a\ln\left|\frac{a + \sqrt{a^2 - x^2}}{x}\right| + C$$

$$\int \frac{x^2}{\sqrt{a^2 - x^2}}\, dx = -\frac{x}{2}\sqrt{a^2 - x^2} + \frac{a^2}{2}\sin^{-1}\frac{x}{a} + C$$

$$\int x^2\sqrt{a^2 - x^2}\, dx = \frac{x}{8}(2x^2 - a^2)\sqrt{a^2 - x^2} + \frac{a^4}{8}\sin^{-1}\frac{x}{a} + C$$

$$\int \frac{dx}{x^2\sqrt{a^2 - x^2}} = -\frac{\sqrt{a^2 - x^2}}{a^2 x} + C$$

$$\int \frac{\sqrt{a^2 - x^2}}{x^2}\, dx = -\frac{\sqrt{a^2 - x^2}}{x} - \sin^{-1}\frac{x}{a} + C$$

$$\int \frac{dx}{x\sqrt{a^2 - x^2}} = -\frac{1}{a}\ln\left|\frac{a + \sqrt{a^2 - x^2}}{x}\right| + C$$

$$\int \frac{dx}{(a^2 - x^2)^{3/2}} = \frac{x}{a^2\sqrt{a^2 - x^2}} + C$$

$$\int (a^2 - x^2)^{3/2}\, dx = \frac{x}{8}(5a^2 - 2x^2)\sqrt{a^2 - x^2} + \frac{3a^4}{8}\sin^{-1}\frac{x}{a} + C$$

INTEGRALS OF INVERSE TRIGONOMETRIC FUNCTIONS

$$\int \sin^{-1}x\, dx = x\sin^{-1}x + \sqrt{1 - x^2} + C$$

$$\int \tan^{-1}x\, dx = x\tan^{-1}x - \frac{1}{2}\ln(1 + x^2) + C$$

$$\int \sec^{-1}x\, dx = x\sec^{-1}x - \ln|x + \sqrt{x^2 - 1}| + C \quad (x > 1)$$

$$\int x\sin^{-1}x\, dx = \frac{1}{4}(2x^2 - 1)\sin^{-1}x + \frac{x}{4}\sqrt{1 - x^2} + C$$

$$\int x\tan^{-1}x\, dx = \frac{1}{2}(x^2 + 1)\tan^{-1}x - \frac{x}{2} + C$$

$$\int x\sec^{-1}x\, dx = \frac{x^2}{2}\sec^{-1}x - \frac{1}{2}\sqrt{x^2 - 1} + C \quad (x > 1)$$

$$\int x^n\sin^{-1}x\, dx = \frac{x^{n+1}}{n+1}\sin^{-1}x - \frac{1}{n+1}\int \frac{x^{n+1}}{\sqrt{1 - x^2}}\, dx + C \text{ if } n \neq -1$$

$$\int x^n\tan^{-1}x\, dx = \frac{x^{n+1}}{n+1}\tan^{-1}x - \frac{1}{n+1}\int \frac{x^{n+1}}{1 + x^2}\, dx + C \text{ if } n \neq -1$$

$$\int x^n\sec^{-1}x\, dx = \frac{x^{n+1}}{n+1}\sec^{-1}x - \frac{1}{n+1}\int \frac{x^n}{\sqrt{x^2 - 1}}\, dx + C \quad (n \neq -1, x > 1)$$

EXPONENTIAL AND LOGARITHMIC INTEGRALS

$$\int xe^x\, dx = (x - 1)e^x + C$$

$$\int x^n e^x\, dx = x^n e^x - n\int x^{n-1}e^x\, dx$$

$$\int \ln x\, dx = x\ln x - x + C$$

$$\int x^n\ln x\, dx = \frac{x^{n+1}}{n+1}\ln x - \frac{x^{n+1}}{(n+1)^2} + C, \quad (n \neq -1)$$

$$\int x^n(\ln x)^m\, dx = \frac{x^{n+1}}{n+1}(\ln x)^m - \frac{m}{n+1}\int x^n(\ln x)^{m-1}\, dx \quad (n \neq -1)$$

$$\int e^{ax}\sin bx\, dx = \frac{e^{ax}}{a^2 + b^2}(a\sin bx - b\cos bx) + C$$

$$\int e^{ax}\cos bx\, dx = \frac{e^{ax}}{a^2 + b^2}(a\cos bx + b\sin bx) + C$$

INTEGRALS OF HYPERBOLIC FUNCTIONS

$$\int \sinh x\, dx = \cosh x + C$$

$$\int \cosh x\, dx = \sinh x + C$$

$$\int \tanh x\, dx = \ln(\cosh x) + C$$

$$\int \coth x\, dx = \ln|\sinh x| + C$$

$$\int \text{sech} x\, dx = 2\tan^{-1}(e^x) + C$$

$$\int \text{csch} x\, dx = \ln\left|\tanh\frac{x}{2}\right| + C$$

$$\int \sinh^2 x\, dx = \frac{1}{4}\sinh 2x - \frac{x}{2} + C$$

$$\int \cosh^2 x\, dx = \frac{1}{4}\sinh 2x + \frac{x}{2} + C$$

$$\int \tanh^2 x\, dx = x - \tanh x + C$$

$$\int \coth^2 x\, dx = x - \coth x + C$$

$$\int \text{sech}^2 x\, dx = \tanh x + C$$

$$\int \text{csch}^2 x\, dx = -\coth x + C$$

$$\int \text{sech} x\, \tanh x\, dx = -\text{sech} x + C$$

$$\int \text{csch} x\, \coth x\, dx = -\text{csch} x + C$$